Advances in High-Pressure Technology for Geophysical Applications

The illustrations on the cover show (clockwise from top left):

Photograph of the GHz-ultrasonic diamond anvil cell attached to the interferometer and schematic of the high-pressure ultrasonic experiment in shear geometry. A purely-polarized converted S-wave enters the [100]-oriented diamond anvil, and interferometry is carried out by overlapping echoes from the diamond culet (diamond echo) and the far-end of the sample (sample echo). See details in the article by Jacobsen *et al.*

Schematic of D-DIA. The D-DIA high pressure device is driven by one main ram. This ram drives the top and bottom anvils together and the side anvils (two of the four are illustrated) are driven into the sample chamber by a wedge type effect. The upper and lower anvils have each an additional ram that can drive them independently of the main ram and hence independently of the side anvils. During deformation, the main ram may need to be backed off of the sample in order to maintain a constant pressure (which can be monitored by the diffraction observations). See details in the articles by Weidner *et al.* and Uchida *et al.*

Advances in High-Pressure Technology for Geophysical Applications

Editors

Jiuhua Chen
Mineral Physics Institute, Department of Geosciences
SUNY at Stony Brook, NY, USA

Yanbin Wang
CARS, University of Chicago
Chicago, IL, USA

Thomas S. Duffy
Department of Geosciences, Princeton University
Princeton, NJ, USA

Guoyin Shen
CARS, University of Chicago
Chicago, IL, USA

Larissa F. Dobrzhinetskaya
Department of Earth Sciences, UCA at Riverside
Riverside, CA, USA

2005

ELSEVIER

Amsterdam – Boston – Heidelberg – London – New York – Oxford
Paris – San Diego – San Francisco – Singapore – Sydney – Tokyo

ELSEVIER B.V.
Radarweg 29
P.O. Box 211, 1000 AE, Amsterdam
The Netherlands

ELSEVIER Inc.
525 B Street, Suite 1900
San Diego, CA 92101-4495
USA

ELSEVIER Ltd
The Boulevard, Langford Lane
Kidlington, Oxford OX5 1GB
UK

ELSEVIER Ltd
84 Theobalds Road
London WC1X 8RR
UK

First edition 2005

Library of Congress Cataloging in Publication Data
A catalog record is available from the Library of Congress.

British Library Cataloguing in Publication Data
A catalogue record is available from the British Library.

ISBN: 0-444-51979-3

♾ The paper used in this publication meets the requirements of ANSI/NISO Z39.48-1992 (Permanence of Paper).

Transferred to Digital Print 2011
Printed and bound in the United Kingdom

Contents

Contributors

Yoshitaka Aizawa Institute for Study of the Earth's Interior, Okayama University, Misasa, Tottori-ken 682-0193 Japan

Jeffrey R. Allwardt Department of Geological and Environmental Sciences, Stanford University; Bayerisches Geoinstitut, Universität Bayreuth, Germany; Department of Geological and Environmental Sciences, Bldg 320, Stanford University, Stanford, CA, 94305-2115, USA
E-mail: allwardt@pangea.stanford.edu

R. Ando Institute of Mineralogy, Petrology, and Economic Geology, Tohoku University, Sendai 980-8578, Japan

Ross J. Angel Virginia Tech Crystallography Laboratory, Department of GeoSciences, Virginia Tech, Blacksburg, VA 24060, USA
E-mail: rangel@vt.edu, Web: www.crystal.vt.edu

Krassimir N. Bozhilov Institute of Geophysics and Planetary Physics, University of California, Riverside, CA 92521, USA

Robert C. Burruss U.S. Geological Survey, 956 National Center, Reston, VA 20192, USA

Jiuhua Chen Mineral Physics Institute, Department of Geosciences, State University of New York at Stony Brook, Stony Brook, NY 11794-2100, USA
E-mail: Jiuhua.Chen@sunysb.edu

I-Ming Chou U.S. Geological Survey, 954 National Center, Reston, VA 20192, USA
E-mail: imchou@usgs.gov

George D. Cody Geophysical Laboratory, Carnegie Institution of Washington, 5251 Broad Branch Rd., Washington DC 20015, USA

Wilson A. Crichton European Synchrotron Radiation Facility, B.P. 220, 38043 Grenoble cedex, France
E-mail: crichton@esrf.fr

Jonathan C. Crowhurst Lawrence Livermore National Laboratory, 7000 East Avenue, Livermore, CA 94551, USA
E-mail: crowhurst1@llnl.gov

Luke. L. Daemen Los Alamos Neutron Science Center, Los Alamos National Laboratory, Los Alamos, NM 87545, USA

Larissa F. Dobrzhinetskaya Department of Earth Sciences, University of California, Riverside CA 92521, USA
E-mail: larissa@ucr.edu

Natalia Dubrovinskaia Bayerisches Geoinstitut, Universität Bayreuth, 95440 Bayreuth, Germany
E-mail: Natalia.Dubrovinakaia@uni-bayreuth.de

Leonid Dubrovinsky Bayerisches Geoinstitut, Universität Bayreuth, 95440 Bayreuth, Germany
E-mail: Leonid.Dubrovinsky@uni-bayreuth.de

William Durham Lawrence Livermore National Laboratory, Livermore, CA 94550, USA

Yingwei Fei Geophysical Laboratory, Carnegie Institution of Washington, 5251 Broad Branch Rd., Washington DC 20015, USA

Daniel J. Frost Bayerisches Geoinstitut, Universität Bayreuth, Germany

Ken-ichi Funakoshi Japan Synchrotron Radiation Research Institute, Kouto1-1-1, Mikazuki, Hyogo-ken, 678-5198 Japan

Stuart A. Gilder Laboratoire de Paléomagnétisme, Institut de Physique du Globe de Paris, 75252 Paris Cedex 05 France
E-mail: gilder@ipgp.jussieu.fr

Alexander F. Goncharov Lawrence Livermore National Laboratory, 7000 East Avenue, Livermore, CA 94551, USA

Harry W. Green Department of Earth Sciences, University of California, Riverside CA 92521, USA; Institute of Geophysics and Planetary Physics, University of California, Riverside, CA 92521, USA

Duanwei He Los Alamos Neutron Science Center, Los Alamos National Laboratory, Los Alamos, NM 87545, USA

Russell J. Hemley Geophysical Laboratory, Carnegie Institution of Washington, 5251 Broad Branch Road, NW, Washington, DC, 20015, USA

Eiji Ito Institute for Study of the Earth's Interior, Okayama University, Misasa, Tottori-ken 682-0193 Japan
E-mail: eiito@misasa.okayma-u.ac.jp

Ian Jackson Research School of Earth Sciences, Australian National University, Canberra, ACT 0200, Australia
E-mail: Ian.Jackson@anu.edu.au

Steven D. Jacobsen Geophysical Laboratory, Carnegie Institution of Washington, 5251 Broad Branch Rd. NW, Washington DC 20015, USA
E-mail: s.jacobsen@gl.ciw.edu

Hiroshi Kaneko Japan Atomic Energy Research Institute, Japan

Anastasia Kantor Bayerisches Geoinstitut, Universität Bayreuth, D-95440 Bayreuth, Germany

Innokenty Kantor Bayerisches Geoinstitut, Universität Bayreuth, 95440 Bayreuth, Germany

Shun-ichiro Karato Yale University, Department of Geology and Geophysics, New Haven, CT 06520, USA
E-mail: shun-ichiro.karato@yale.edu

Y. Katayama Synchrotron Radiation Research Center, Japan Atomic Energy Research Institute, Mikazuki 679-5148, Japan

Tomoo Katsura Institute for Study of the Earth's Interior, Okayama University, Misasa, Tottori-ken 682-0193 Japan

Kazuyuki Kawabe Institute for Study of the Earth's Interior, Okayama University, Misasa, Tottori-ken 682-0193 Japan

Atsushi Kubo Institute for Study of the Earth's Interior, Okayama University, Misasa, Tottori-ken 682-0193 Japan; Now at Department of Geosciences, Princeton University, Princeton, New Jersey 08544 USA

Jennifer Kung Mineral Physics Institute and Department of Geosciences, Stony Brook University, Stony Brook, NY 11794-2100, USA

Alexei Kuznetsov Bayerisches Geoinstitut, Universität Bayreuth, 95440 Bayreuth, Germany

Christian Lathe GeoForschungs Zentrum Potsdam, Dept. 4, Telegraphenberg, D-14473 Potsdam, Germany

Joern Lauterjung GeoForschungs Zentrum Potsdam, Dept. 4, Telegraphenberg, D-14473 Potsdam, Germany

Maxime LeGoff Laboratoire de Paléomagnétisme, Institut de Physique du Globe de Paris, 75252 Paris Cedex 05 France

Baosheng Li Mineral Physics Institute and Department of Geosciences, Stony Brook University, Stony Brook, NY 11794-2100, USA
E-mail: Baosheng.Li@sunysb.edu

Li Li Laboratoire de Structure et Propriétés de l'État Solide (associated to CNRS), Université des Sciences et Technologies de Lille, F-59655, Villeneuve d'Ascq Cedex, France; Mineral Physics Institute, Department of Geosciences, State University of New York at Stony Brook, Stony Brook, NY, 11794-2100, USA

Jung-Fu Lin Geophysical Laboratory, Carnegie Institution of Washington, 5251 Broad Branch Road, NW, Washington, DC 20015, USA
E-mail: j.lin@gl.ciw.edu

Konstantin A. Lokshin Los Alamos Neutron Science Center, Los Alamos National Laboratory, Los Alamos, NM 87545, USA

Wanjun Lu Guangzhou Institute of Geochemistry, Chinese Academy of Sciences, Guangzhou 510640, China

Ho-kwang Mao Geophysical Laboratory, Carnegie Institution of Washington, 5251 Broad Branch Road, NW, Washington, DC 20015, USA

Mohamed Mezouar European Synchrotron Radiation Facility, B.P. 220, 38043 Grenoble cedex, France

Hans J. Mueller GeoForschungs Zentrum Potsdam, Dept. 4, Telegraphenberg, D-14473 Potsdam, Germany
E-mail: Hans-Joachim.Mueller@gfz-potsdam.de

Bjorn O. Mysen Geophysical Laboratory, Carnegie Institution of Washington, 5251 Broad Branch Rd., Washington DC 20015, USA

Yu Nishihara Yale University, Department of Geology and Geophysics, New Haven, CT 06520, USA

Norimasa Nishiyama Center for Advanced Radiation Sources, The University of Chicago, USA

Akifumi Nozawa Japan Synchrotron Radiation Research Institute, Kouto1-1-1, Mikazuki, Hyogo-ken, 678-5198 Japan

E. Ohtani Institute of Mineralogy, Petrology, and Economic Geology, Tohoku University, Sendai 980-8578, Japan
E-mail: ohtani@mail.tains.tohoku.ac.jp

Takuo Okuchi Geophysical Laboratory, Carnegie Institution of Washington, 5251 Broad Branch Road, NW, Washington, DC, 20015 USA; Graduate School of Environmental Studies, Nagoya University Furo-cho, Chikusa, Nagoya 464-8601, Japan
E-mail: t.okuchi@gl.ciw.edu

Cristian Pantea Los Alamos Neutron Science Center, Los Alamos National Laboratory, Los Alamos, NM 87545, USA

Jiang Qian Los Alamos Neutron Science Center, Los Alamos National Laboratory, Los Alamos, NM 87545, USA

Hans J. Reichmann Geoforschungszentrum Potsdam, Telegrafenberg, Division 4, 14473 Potsdam, Germany

Alex P. Renfro Department of Earth Sciences, University of California, Riverside CA 92521, USA

Mark L. Rivers Center for Advanced Radiation Sources, The University of Chicago, USA

Mario Santoro LENS, European Laboratory for Non Linear Spectroscopy and INFM, Via N. Carrara 1, I-50019 Sesto Fiorentino, Firenze, Italy; Geophysical Laboratory, Carnegie Institution of Washington, 5251 Broad Branch Road, NW, Washington, D.C. 20015, USA
E-mail: santoro@lens.unifi.it, m.santoro@gl.ciw.edu

Frank R. Schilling GeoForschungs Zentrum Potsdam, Dept. 4, Telegraphenberg, D-14473 Potsdam, Germany

Burkhard C. Schmidt Bayerisches Geoinstitut, Universität Bayreuth, Germany

Guoyin Shen Consortium for Advanced Radiation Sources, The University of Chicago, Chicago, IL 60637, USA

Hartmut A. Spetzler Cooperative Institute for Research in Environmental Sciences and Department of Geological Sciences, University of Colorado, Boulder, CO 80309-0216, USA

Jonathan F. Stebbins Department of Geological and Environmental Sciences, Stanford University, Stanford, CA 94305-2115, USA

Viktor V. Struzhkin Geophysical Laboratory, Carnegie Institution of Washington, 5251 Broad Branch Road, NW, Washington, DC 20015, USA

Wolfgang Sturhahn Advanced Photon Source, Argonne National Laboratory, 9700 South Cass Avenue, Argonne, IL 60439, USA

Sung Keun Lee Geophysical Laboratory, Carnegie Institution of Washington, 5251 Broad Branch Rd., Washington DC 20015, USA; School of Earth and Environmental Sciences, Seoul National University, Seoul, Korea, 151-742
E-mail: sungklee@snu.ac.kr

Steve R. Sutton Center for Advanced Radiation Sources, The University of Chicago, USA

A. Suzuki Institute of Mineralogy, Petrology, and Economic Geology, Tohoku University, Sendai 980-8578, Japan

Takeyuki Uchida Center for Advanced Radiation Sources, The University of Chicago, USA
E-mail: uchida@cars.uchicago.edu

S. Urakawa Department of Earth Sciences, Okayama University, Okayama 700-8530, Japan

Michael T. Vaughan Mineral Physics Institute, State University of New York at Stony Brook, Stony Brook, NY 11794-2100, USA

Robert Von Dreele Advanced Photon Source and Intense Pulse Neutron Source, Argonne National Laboratory, USA

Liping Wang Mineral Physics Institute, State University of New York at Stony Brook, Stony Brook, NY 11794-2100, USA

Yanbin Wang Center for Advanced Radiation Sources, The University of Chicago, USA
E-mail: wang@cars.uchicago.edu

Donald J. Weidner Laboratoire de Structure et Propriétés de l'État Solide (associated to CNRS), Université des Sciences et Technologies de Lille, F-59655, Villeneuve d'Ascq Cedex, France; Mineral Physics Institute, Department of Geosciences, State University of New York at Stony Brook, Stony Brook, NY, 11794-2100, USA
E-mail: dweidner@sunysb.edu

Yousheng Xu Yale University, Department of Geology and Geophysics, New Haven, CT 06520, USA

Sho Yokoshi Institute for Study of the Earth's Interior, Okayama University, Misasa, Tottori-ken 682-0193, Japan

Christopher E. Young Mineral Physics Institute, State University of New York at Stony Brook, Stony Brook, NY 11794-2100, USA

Joseph M. Zaug Lawrence Livermore National Laboratory, 7000 East Avenue, Livermore, CA 94551, USA

Jianzhong Zhang Los Alamos Neutron Science Center, Los Alamos National Laboratory, Los Alamos, NM 87545, USA

Jiyong Zhao Advanced Photon Source, Argonne National Laboratory, 9700 South Cass Avenue, Argonne, IL 60439, USA

Yusheng Zhao LANSCE-12, MS-H805, Los Alamos Neutron Science Center, Los Alamos National Laboratory, Los Alamos, NM 87545, USA
E-mail: yzhao@lanl.gov

Preface

High-pressure mineral physics is a field that is strongly driven by the development of new technology. Fifty years ago, when experimentally achievable pressures were limited to just 25 GPa, little was known about the mineralogy of the Earth's lower mantle. Silicate perovskite, the likely dominant mineral of the deep Earth, was identified only when the high-pressure techniques broke the pressure barrier of 25 GPa in the 1970s. However, as the maximum achievable pressure reached beyond 1 Megabar (100 GPa) and even to the pressure of Earth's core on minute samples, new discoveries were increasingly fostered by the development of new analytical techniques and improvements in sensitivity and precision of existing techniques. Today, as this volume highlights, we can measure a wide range of physical and chemical properties at ever higher pressure and temperature conditions on increasingly complex samples.

Over the past decade, some of the main developments in high-pressure technology have focused on integration of analytical techniques with high-pressure apparatus, and pressure cell modifications for enlarging sample volumes and generating controlled stress fields. In particular, the technological frontiers are moving rapidly as the high-pressure community has expanded its access to large-scale radiation (synchrotron and neutron) sources worldwide. The exciting new developments in the high-pressure field have drawn growing attention from scientists in geophysics and other fields. We are therefore publishing this volume which aims to collect details of selected recent developments and to serve as a technical reference and handbook for both graduate or advanced undergraduate students as well as more senior researchers interested in keeping abreast of these new developments.

The book consists of six sections which group the chapters according to their main topics: (a) Elastic and Anelastic Properties; (b) Rheology; (c) Melt and Glass Properties; (d) Structural and Magnetic Properties; (e) Diffraction and Spectroscopy; (f) Pressure Calibration and Generation. As many chapters actually cover multiple topics, readers may find chapters of interest in different sections. All chapters are prepared with an emphasis on technical details suitable for a technical reference. Many on-line software resources are also listed in as detailed a manner as possible. However, the URL of the software sites may be subject to change without notice.

The editors would like to thank all the authors for their excellent contributions and timely cooperation during the editorial process. The book would not be possible without the efforts of numerous referees, including many anonymous reviewers, who provided critical, constructive reviews of the contents to ensure the quality of the book.

Jiuhua Chen
Yanbin Wang
Thomas S. Duffy
Guoyin Shen
Larissa Dobrzhinetskaya

Reviewers

Ross Angel
James Badro
Nonna Bakun-Czubarow
William Bassett
Andrew Campbell
Marc Conradi
Przemyslaw Dera
Jason Diefenbacher
Leonid Dubrovinsky
Thomas Duffy
William Durham
Gary Ernst
Ivan Getting
Douglas Green
Donald Isaak
Yoshinori Katayama
Gunter Kletetscka
David Kohlstedt
Martin Kunz
Li Li
J. G. (Louie) Liou
John Loveday
Malcolm McMahon
Sébastien Merkel
Ozden Odmir
Eiji Ohtani
Vitali Prakapenka
Chrystele Sanloup
Frank Schilling
Richard Secco
Anurag Sharma
Sang-Heon (Dan) Shim
Isaak Silvera
Stas Sinogeikin
Marc Smith
Hartmut Spetzler
Gerd Steinle-Neumann
Victor Struzhkin
Marc Toplis
Stanely Tozer
Takeyuki Uchida
Wataru Utsumi
Robert von Dreele
Liping Wang
Donald Weidner
Gerhard Wortmann
Xianyu Xue
Akira Yoneda
Choong-shik Yoo
Chang-Sheng Zha

* Some anonymous reviewers are not listed.

I
Elastic and anelastic properties

Advances in High-Pressure Technology for Geophysical Applications
Jiuhua Chen, Yanbin Wang, T.S. Duffy, Guoyin Shen, L.F. Dobrzhinetskaya, editors

Chapter 1
Direct measurements of the elastic properties of iron and cobalt to 120 GPa – implications for the composition of Earth's core

Jonathan C. Crowhurst, Alexander F. Goncharov and Joseph M. Zaug

Abstract

We discuss the use of impulsive stimulated light scattering (ISLS) to determine acoustic velocities of metals under high pressure in the diamond anvil cell. We discuss the experimental details of the method, and analyze its advantages and disadvantages compared with other methods. As illustrations we present data we have obtained on iron and cobalt to a static pressure of nearly 120 GPa. We extrapolate the velocities of iron to inner core pressures and correct them to the corresponding temperatures. The results are found to match well the velocities of the preliminary reference Earth model (PREM). This observation suggests an inner core composition that is almost pure iron. A similar comparison with outer core velocities suggests a composition consistent with the presence of light impurities. These results hold over the wide temperature range that corresponds to the uncertainty in the geotherm. The elastic properties of Co reveal a softening of the shear and compressional elastic constants, which we relate to a gradual loss of magnetic moments under pressure.

1. Introduction

Knowledge of the precise chemical composition of the Earth's core is required in order to fully understand the origin of various phenomena that are of direct importance to the planet's inhabitants. These phenomena include tectonic motion, earthquakes, and volcanism, and also the Earth's magnetic field, which is responsible for shielding our atmosphere from the solar wind. Since iron is believed to be the major constituent of the core, such a requirement implies the need for detailed knowledge of its elastic, thermodynamic, transport, and vibrational properties under extreme conditions of pressure and temperature. It is only when armed with this information that we can interpret the results of seismological and geomagnetic investigations of the Earth's deep interior, and thus begin to place quantitative limits on core composition.

The density distribution within the Earth had been determined some 30 years ago (see Anderson, 1989, for references), and has been recently reanalyzed by Masters and Gubbins (2003). This determination is essentially model dependent and may be improved in future. Seismic sound velocities are known very accurately and can be used to make comparisons with laboratory data where the latter is available. The inner core has been found to exhibit acoustic anisotropy (see for example, Stixrude and Brown, 1998), which is presumably related to the crystal structure of the major constituent.

Several striking advances in experimental techniques have recently been made. They are mainly related to the advent of a new generation of synchrotron sources that are capable of generating highly monochromatic and brilliant radiation, and also of offering very high spatial resolution. These qualities are ideally matched to the requirements of diamond anvil cell (DAC) research at ultrahigh (Mbar) pressure. X-ray diffraction studies, for example, have become feasible at enormous pressures (~300 GPa – see Mao et al., 1990). Recently, these measurements have also been extended to high temperature (to 1400 K) by Dubrovinsky et al. (2000) (The pressure at the center of the Earth is approximately 360 GPa, while the temperature is between 4500 and 7000 K). Moreover, analysis of the Debye–Waller factor by Dubrovinsky et al. (2001) has also made it possible to estimate the sound velocities of Fe under these extreme conditions. Radial X-ray diffraction reported by Mao et al. (1998) and Singh et al. (1998) has been used to estimate elastic properties by analyzing non-hydrostatic stress conditions in the DAC. Two additional synchrotron-based techniques have recently been employed to measure sound velocities: nuclear resonant inelastic X-ray scattering (NRIXS) (Lübbers et al., 2000; Mao et al., 2001), and inelastic X-ray scattering (IXS) (Fiquet et al., 2001). The former has been used to determine the density of phonon states. Combined with the equation of state (EOS), this method can be used to estimate aggregate sound velocities. IXS has been used to measure phonon dispersions from which zone-center acoustic frequencies can be inferred. NRIXS experiments in the laser-heated DAC have also recently been reported. We also mention the Raman spectroscopic measurements reported by Merkel et al. (2000) and Olijnyk et al. (2001). In those experiments the measured frequency of the zone-boundary phonon (in the extended-zone model) was used to infer single-crystal elastic tensor elements.

Data obtained from experiments in which high pressure is achieved dynamically (i.e. via shock wave compression) have provided a wealth of information on the EOS, sound velocities and other thermodynamic parameters of iron. To compare the data with the results of measurements made on statically compressed samples (i.e. in the DAC or multi-anvil press), and also with theoretical calculations, direct measurement of temperature is desirable, but this poses a formidable experimental challenge. We are only aware of a few reports in the literature (Bass et al., 1987; Ahrens et al., 1990; Bass et al., 1990; Yoo et al., 1993) that describe the experimental determination of temperature in the shock compression of iron. The knowledge of the dependence of sound speeds on temperature is also required to extrapolate and compare to seismic measurements. Furthermore, pressure and temperature are difficult to control independently in shock wave compression experiments.

Theoretical investigations have also been vigorously pursued in recent years. Nevertheless there still exists a significant discrepancy between, for example, the experimental and theoretically derived EOS. The agreement is better if the magnetic properties of iron in the hcp phase up to approximately 60 GPa are considered, as has been shown by Cohen and Mukherjee (2004). However, even at high pressure (where magnetic properties are believed to be unimportant), theory predicts phonon frequencies and elastic moduli that are higher than measured values.

The results of measurements performed under conditions of static pressure are also in many cases significantly inconsistent with each other. This may in part be due to the fact that techniques such as IXS, NRIXS and Raman spectroscopy do not measure acoustic velocities or elastic moduli directly, as described above. *Direct* measurements of sound velocities (using the conceptually familiar ultrasonics technique) of iron have been limited to low pressures (16.5 GPa, see Mao et al., 1998), due to the requirement of relatively large samples volumes. Assumptions (Merkel et al., 2000; Olijnyk et al., 2001) or extrapolations (Fiquet et al., 2001) are required in order to infer elastic moduli from measured phonon frequencies, or to make comparisons with the measured velocities of long-wavelength seismic waves.

Here we present the results of impulsive stimulated light scattering (ISLS) measurements of the velocity of an acoustic excitation that is stimulated at the interface between diamond and iron, up to a static pressure of 115 GPa. Since the corresponding acoustic wavelength was more than 2 μm, we are confident that our results represent an accurate determination of the acoustic velocities of the bulk sample (please see Section 2 below). Our data also directly reflect zone-center acoustic behavior. The chapter is organized as following. We first describe the technique of ISLS and its application to opaque materials under high pressure. The remainder of the article will be devoted to a description of our high-pressure investigations of polycrystalline iron and cobalt.

2. Impulsive stimulated light scattering

In the present context ISLS can be considered to be a light-induced ultrasonics technique. It is one of several similar techniques in which laser light is used to coherently excite and then probe material excitations of various types (see for example Brown et al., 1988; Abramson et al., 1999, and references therein). In our system (Fig. 1) we employ the Nd:YAG fundamental (1064 nm) as the excitation source and the second harmonic (532 nm) as the probe (our laser is a Time–Bandwidth Jaguar). These wavelengths are chosen for convenience, but many other combinations can be used. In general, the excitation radiation is pulsed, but the probe may be either pulsed or continuous. In the continuous case the excitations may be probed in real time, which is preferable if their frequencies and

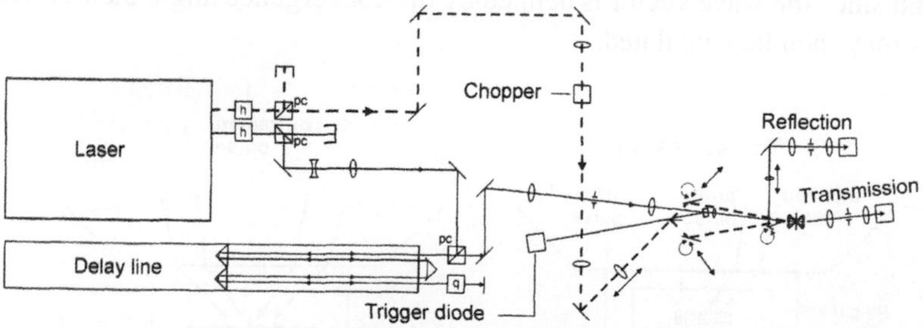

Figure 1. The ISLS apparatus. The solid line represents the path of the probe pulse, the dashed line(s) the path(s) of the excitation pulse(s). Attenuation of the excitation and probe beams are achieved with a combination of half-wave plates (h) and polarizing beam splitters (pc). The excitation beam is diverted to make up for the delay in the emission from the laser of the second harmonic pulse (some ns). The probe beam is introduced and recovered from the delay line via a combination of a polarizing beam splitter and a quarter-wave plate (q). The delay line travel is 1.5 m, but a triple retro-reflector arrangement is used to increase the time-of-flight delay to approximately 40 ns. Final steering optics are on rails and/or rotation stages to allow adjustment of convergence angle and hence wave vector. The maximum dimension of the apparatus is approximately 3.5 m. Other components are discussed in the text.

intensities are appropriate. However, high frequencies and low intensities may make a pulsed probe preferable. In this case the frequency resolution of the system is ultimately limited by the pulse width (which in our case is approximately 100 ps). Our system, to which the present discussion will be limited, is always operated in the pulsed configuration.

The excitations themselves are created by combining in the sample, two excitation pulses at some known angle of convergence. This produces an interference pattern, which is followed by optical absorption and the consequent creation of a regular array of heated volumes (Fig. 2). Rapid thermal expansion in these volumes leads in turn to a periodic variation in the local dielectric function, i.e. a diffraction grating. Since the material has some finite elasticity, the sudden expansion almost simultaneously launches counter-propagating acoustic waves, which also behave as gratings. In general many types of grating may be excited, but it is only the thermal and acoustic that are of present interest.

Some fraction of an incident probe pulse will be diffracted by the gratings, and the corresponding measured intensity will depend on the delay between the excitation pulses and the arrival of the probe pulse. In the experiment this delay is achieved by smoothly varying the time of flight of the probe for each excitation/diffraction process. A series of snapshots of grating evolution, forming a time series, is thus acquired (Fig. 3). In the case of materials like cobalt and iron in air the excitations typically disappear within some tens of ns (a function of the diameter of the excited spot amongst other things), and since the separation between pulses is not less than 250 μs, corresponding to a laser repetition rate of 4 kHz, each subsequent pair of excitation pulses encounter a quiescent system. The time series does not in general exhibit a monotonically decreasing intensity as might be expected. The various diffracted components interfere with one another to produce modulations of different frequencies. The component due to the static thermal grating and that due to the acoustic wave will interfere to produce a modulation at exactly the acoustic frequency, for example, while two counter-propagating acoustic waves of the same frequency will produce a modulation at twice the acoustic frequency. A fast Fourier transform (FFT) of the measured time series then yields the frequencies of the acoustic waves, and since the wave vector is defined by the convergence angle their corresponding velocities may then be calculated.

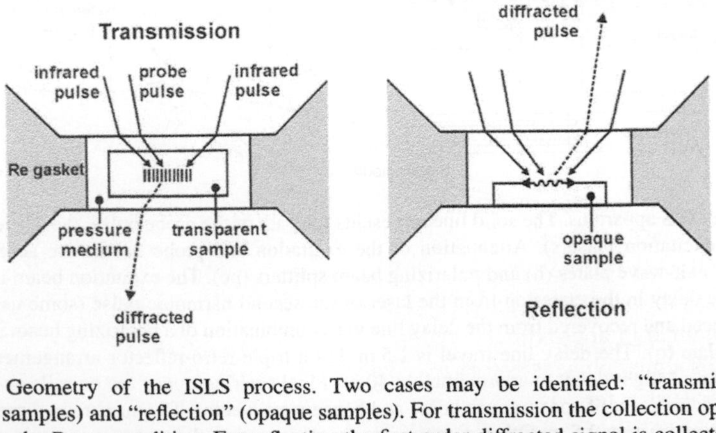

Figure 2. Geometry of the ISLS process. Two cases may be identified: "transmission" (largely transparent samples) and "reflection" (opaque samples). For transmission the collection optics are aligned according to the Bragg condition. For reflection the first order diffracted signal is collected.

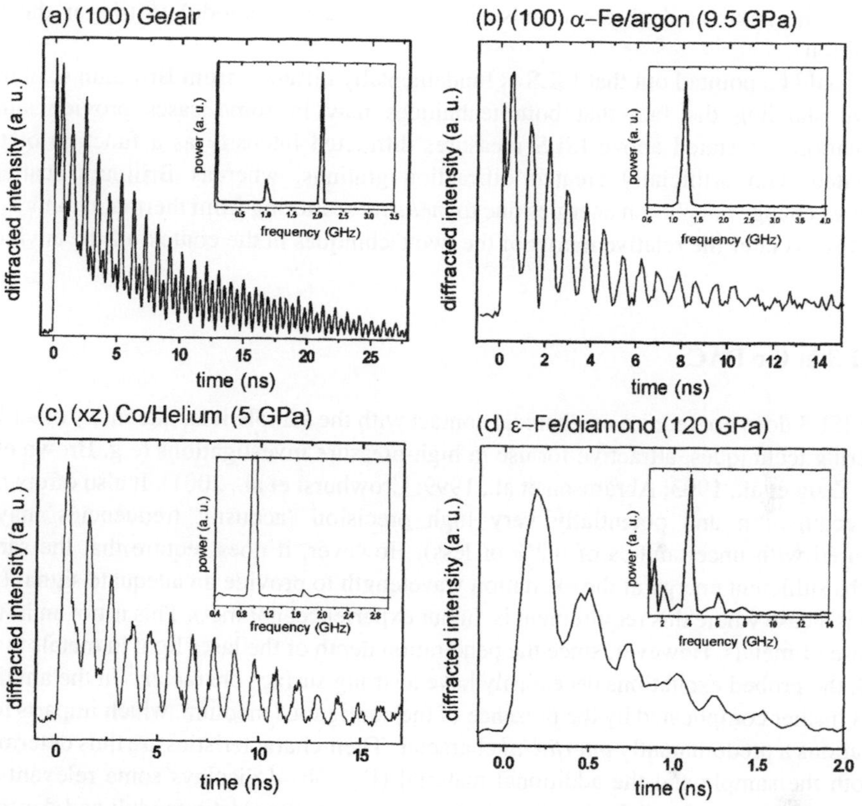

Figure 3. Selected time series and power spectra. (a) Single crystal germanium. Propagation in the (100) plane at 1 bar. (b) Single crystal iron. Propagation in the (100) plane at 9.5 GPa in solid Argon. (c) Single crystal cobalt. Propagation in the *xz* plane in liquid helium. (d) Polycrystalline iron on diamond at 120 GPa. 1 GHz corresponds to approximately 3000 m/s for (a)–(c) and 2000 m/s for (d), respectively. The small feature in (c) at 1.7 GHz is due to propagation of a bulk-like excitation in helium. The additional small feature at just under 2 GHz is the frequency doubled analogue of the interfacial wave. The different time scale in (d) should be noted.

In practice, in order to improve the signal to noise ratio hundreds or thousands of excitation/diffraction processes are performed and the corresponding intensities are averaged to produce each point of the time series. The rate at which the delay is increased must then be slow enough to ensure an adequate sampling rate for the expected frequencies. If the time scale of other sources of noise is appropriate, it may be useful to average the entire time series. Our system also employs an optical chopper wheel that is synched to the laser repetition rate. It is used to cut out every second excitation pulse, which then allows parasitic scattering (i.e. diffuse scattered light from sample asperities or defects) to be subtracted from the raw signal. In our experiment the complete time series is typically obtained in approximately 1 min.

The 100 ps signal pulse is converted to a voltage by the PMT detector. This voltage in turn lasts for just a few ns. Fast acquire-and-hold electronics are needed to store this voltage for the minimum period of 250 μs, which is sufficiently long for the computer to sample. The next incident probe pulse is used (via a photodiode) to trigger the next acquisition.

It should be pointed out that ISLS is fundamentally different from Brillouin scattering, not withstanding the fact that both techniques may in some cases provide similar information. As stated above ISLS measures diffracted intensity as a function of time associated with artificially created diffraction gratings, whereas Brillouin scattering measures changes in photon energies due to *inelastic* scattering from thermal phonons. We shall discuss later the relative merits of the two techniques in the context of the anvil cell.

3. ISLS in the DAC

Since ISLS does not require mechanical contact with the sample it is, like many other light scattering techniques, attractive for use in high-pressure investigations (e.g. Brown et al., 1988; Zaug et al., 1993; Abramson et al., 1999; Crowhurst et al., 2001). It also offers rapid data acquisition and potentially very high precision (acoustic frequencies may be measured with uncertainties of 0.2% or less). However, it does require that the sample absorbs sufficient energy at the excitation wavelength to provide an adequate signal (pure MgO is a case where this requirement is, in our experience, not met). This is not an issue in the case of metals. However, since the penetration depth of the laser light in metals is very small, the probed excitations necessarily have a strong surface character.[1] In the anvil cell this is further complicated by the presence of the contiguous medium, which imparts to the excitations a predominantly *interfacial* character. Their characteristics are thus determined by both the sample and the additional material (Fig. 3b–d displays some relevant time series and power spectra). In the case of acoustic waves, the elastic moduli and density of the latter must be known beforehand in order to extract the desired properties of the sample from the measured velocities. An additional drawback is that in most cases the individual elements of the elastic tensor cannot all be obtained separately from surface waves velocities (in some case this can be done – see for example Crowhurst and Zaug, 2004). However, in the case of our high-pressure polycrystalline cobalt and iron measurements, we have combined the known EOS with the interfacial velocities to yield separately the compressional and shear moduli (obviously under the assumption of elastic isotropy).

It is important to point out here that although the waves in question are *surface-like*, they are not necessarily sensitive to *surface* properties. It depends on to what extent and to what depth the properties of the sample are assumed different from those of its bulk.

---

[1] The character of the waves which are stimulated in an ISLS experiment depend in large part on the degree to which the sample may be considered opaque to the excitation and probing radiation. In the case of water for example, longitudinal bulk waves will be excited. In a highly opaque and elastically isotropic material in air or vacuum the major feature is the surface Rayleigh wave. The associated particle motion is elliptical and the energy flow is strictly parallel to the sample surface. It has both longitudinal and transverse components and depends on both the bulk compressional and shear moduli. In the case of opaque crystals the excitation again has a strong surface character, although depending on crystal symmetry, plane of propagation, and elastic anisotropy, it may leak energy into the bulk material.

Surface acoustic waves have penetration depths that are comparable to their wavelengths, which in our case is typically between 2 and 3 μm. This is much larger than what might be expected for the depth to which polishing damage and other effects might extend. Crowhurst and Zaug (2004) have for example demonstrated perfect agreement (within the measured precision) between the elastic stiffnesses obtained from ISLS surface wave measurements on a germanium single crystal (Fig. 3a), and the known bulk values.

In order to model the surface or interfacial acoustic response a prescription is needed that relates the power spectrum of the excitations to the acoustic parameters (i.e. the elastic tensor, mass density, crystal symmetry, and plane and direction of propagation). To fully compare the calculated response with experiment, one must also know the associated optical and thermal parameters (e.g. the dielectric function, Pockels coefficients, scattering geometry, coefficient of thermal expansion etc.) Since a complete expression is not available, and the various parameters are not in any case all known under high pressure, approximation must suffice.

For metals in air an approximation to the scattering process known is ripple scattering is a particularly good one. In this limit, light is diffracted only from surface ripples and does not interact at all with the sample bulk. Of course in the DAC, the interfacial acoustic motion extends into both the opaque metal and the nearly transparent contiguous medium, so one should also expect an elasto-optic contribution to the scattering, which may be expected to alter the relative intensities of the excitations. However, since we are only interested in the acoustic frequencies, the ripple-scattering approximation is sufficient.

As described in Crowhurst and Zaug (2004), we begin by stating the connection between the surface displacement field $\mathbf{u}(\mathbf{x},t)$, and an applied impulsive line force $\mathbf{F}$. The connection may be expressed as

$$U_i(\mathbf{x}, t) = G_{ij}(\mathbf{x}, t)F_j$$

where $G_{ij}(\mathbf{x}, t)$ is the elastodynamic Green's function. It may then be shown that the power spectrum is proportional to $(1/\varpi)^2|\hat{G}_{ij}(\mathbf{q}, \varpi)|^2$, where $\hat{G}_{ij}(\mathbf{q}, \varpi)$ is the Fourier transform of the Green's function, $\mathbf{q}$ is the surface wave vector (defined by the experimental scattering geometry), and ϖ is the frequency (Maznev et al., 1999). For pure ripple scattering, only the displacement component perpendicular to the surface is relevant. Also, since the applied force is parallel to the surface, the relevant Green's function component is $\hat{G}_{31}(\mathbf{q}, \varpi)$. The Green's function may be evaluated using a method well described in the literature (e.g. Every et al., 1997; Every and Briggs, 1998). The method is applicable to any combination of crystalline symmetry and elastic properties, and may also be extended to multicomponents systems e.g. supported thin films, two half spaces in welded contact, etc. (Every and Briggs, 1998; Maznev et al., 1999; Crowhurst et al., 2001). Calculated results are compared with experiment in Figure 4 for the case of propagation in the (100) plane of germanium.

4. High pressure examples

Figure 5 (Crowhurst et al., 2001) shows measured velocities as a function of direction of the acoustic waves (essentially Scholte waves) that propagate along the interface formed

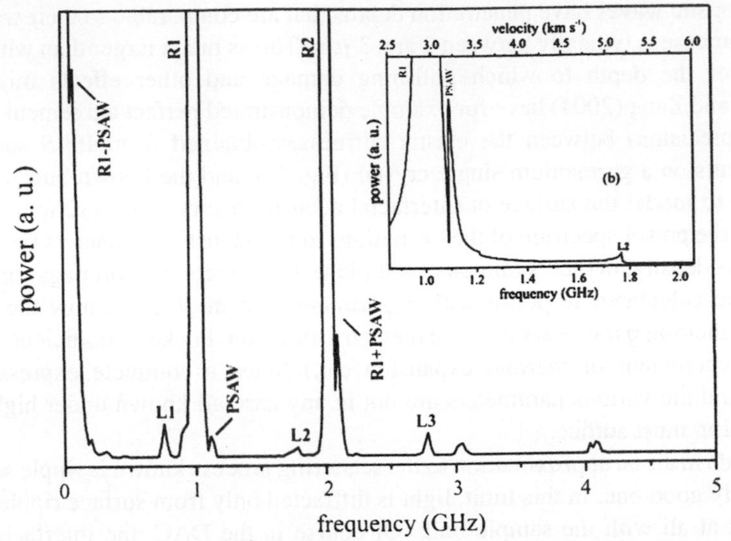

Figure 4. Measured and calculated (inset) power spectra for propagation along a non-symmetry direction in the plane of (100) Ge in air. The most intense features are due to propagation of the Rayleigh wave. Also apparent is the pseudo-surface acoustic wave (pSAW) and a longitudinal resonance, *L*. Sum and difference modes are apparent (see Crowhurst and Zaug, 2004).

by a single crystal of germanium in contact with a methanol pressure medium. It also shows the corresponding case for germanium in air. Since the acoustic parameters of both Ge and methanol are known at these low pressures, it was straightforward to calculate the expected dispersion (lines). The agreement with the experimental data testifies to the reliability of the calculation. Clearly visible in Figure 5a are two distinct acoustic branches. One is the propagation of the familiar Rayleigh wave (solid line) and the other is

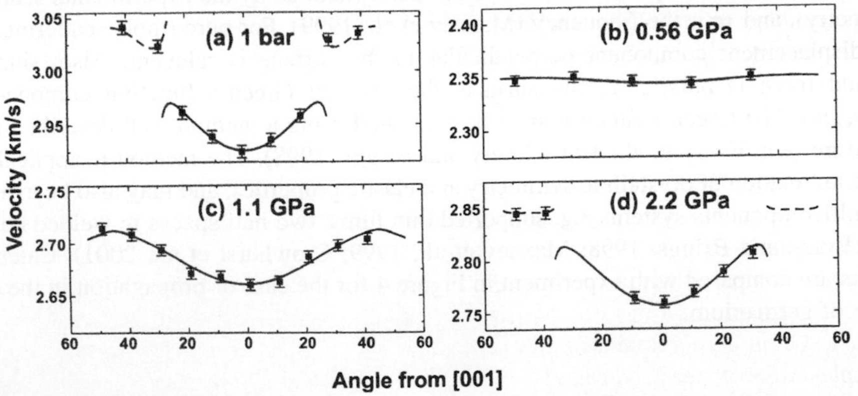

Figure 5. Dependence of surface/interfacial wave velocity with pressure in a (100)-oriented Ge crystal in (a) in air, (b–d) in methanol at the indicated applied pressures. Lines are the results of calculations using known values of the acoustic parameters. Solid lines: Rayleigh or Rayleigh-type branch, dashed lines: pseudo-surface acoustic wave.

that of the pseudo-surface acoustic wave (pSAW). This behavior is typical for propagation in this plane (100) of a cubic crystal with an anisotropy ratio (i.e. $2c_{44}/(c_{11} - c_{12})$) greater than one. It is instructive to examine the behavior of the interfacial wave (Fig. 5b–d) with increasing pressure. At low pressure it exhibits almost no dispersion and has a velocity very close to that of the bulk longitudinal wave of methanol. As the pressure is increased the speed of sound in methanol increases rapidly, and the interfacial wave gradually takes on the characteristics of the ambient pressure Ge dispersion. In Figure 5d the two acoustic branches are clearly separated. Except for a mass loading (note the lower values) the velocities are now almost exclusively dependent on the crystalline solid. (A detailed description of this type of behavior may be found in the theoretical treatment of Every and Briggs.)

Thus far our work on single crystals has been limited to pressures of around 10 GPa. We have obtained complete dispersion curves in the case of tantalum and α-iron in a crystalline argon pressure medium. Naturally, the presence of a non-zero shear modulus in the pressure medium makes the analysis more complicated. Fortunately, the elastic tensor of argon is known to at least 30 GPa (Grimsditch et al., 1986). We will report our findings for iron up to the bcc–hcp transition elsewhere.

Perhaps the most important practical requirement for the successful application of ISLS to opaque materials is the optical quality of the sample surface. In our experience, if the surface is highly specular, an adequate signal is assured. While we have been able to fabricate suitable single crystals for low-pressure work the surfaces of the samples have not been sufficiently well preserved above approximately 15 GPa to permit measurements. All our experiments were performed in argon and it is presumably the non-uniformity of pressure in this medium that has been responsible for the deterioration of the surfaces. The deterioration may also be exacerbated in the case of iron due to the α–ε transition. The use of helium as a pressure medium will hopefully reduce the severity of this problem.

Faced with this difficulty we have been forced to adopt a less than satisfactory approach for performing ISLS measurements in the Mbar regime. This approach has been to fill the gasket cavity with the metal in the form of a powder or foil and then squeeze it directly between the diamonds. This naturally guarantees that the sample surface is smooth at high pressures. Sample fabrication and loading is thus trivial and the entire experiment (as in the case of cobalt and iron) could be performed in one day. Using this method we have made direct measurements of acoustic velocities to the highest static pressures to date (Fig. 3d). There are several disadvantages to this approach, however, apart from the lack of hydrostaticity. Even if the sample does not undergo phase transitions it is not a single crystal and only aggregate elastic quantities can be measured. One must then make assumptions or attempt to measure (we did not) the texture and grain size *in situ* because of their potential effect on the acoustic velocities. As we mention in this work, however, the effect of texturing on the velocities is not expected to be severe in the case of iron and cobalt. Grain size remains a problem, although repeated measurements on different samples were at least consistent within the measurement precision. Perhaps the most serious difficulty is a result of the physics of interfacial wave propagation. A metal in contact with a material as fast as diamond will only support an interfacial wave that is very short lived (as predicted by the calculation described above. The acoustic energy in fact leaks rapidly into the surrounding bulk material. This is in contrast to the case of the Rayleigh wave of a free surface, which has in principle an unlimited lifetime and an energy flow that is strictly parallel to the surface). This means that our measured time series yield only a relatively poorly defined frequency (we estimate more than an order of magnitude increase in the uncertainty). For the experiment to be useful large pressure changes must be achieved. Since diamond is so hard the excitation is also very weak. To be sure we correctly identified the mode we carried out several tests. First we changed the convergence angle of the excitation pulses and confirmed that the frequency increased linearly, hence yielding the same velocity, as indeed it should. We also noted that the measured frequencies increased nearly monotonically with pressure. Finally we conducted experiments on iron and tantalum at the same pressure in contact with the same diamond and in the same orientation (Fig. 6). Tantalum is much slower than iron and the mode frequency should correspondingly be lower. This indeed was the case. Also by extrapolating the pressure derivatives determined by Katahara et al. (1976) and calculating the Hill average we found that our measured interfacial velocity was in close agreement with calculation.

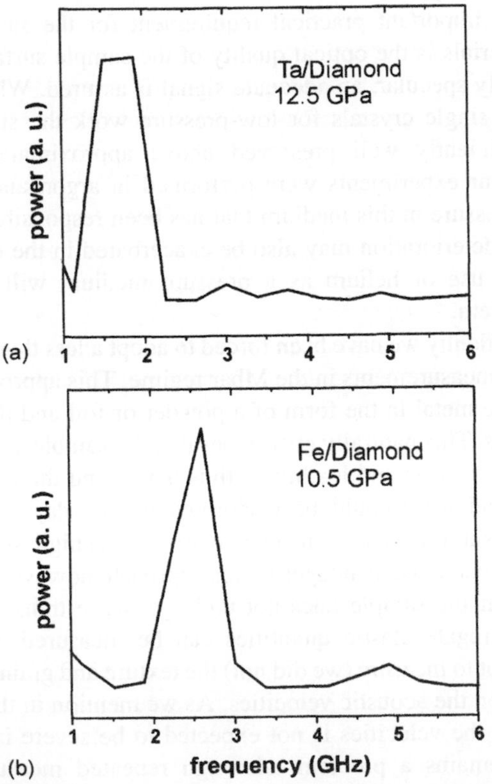

Figure 6. Interfacial power spectra for two different polycrystalline metals in contact with diamond: (a) tantalum, (b) iron. 1 GHz corresponds to about 2500 m/s. The spectra were obtained at the same wave vector at nearly the same pressures using the same diamond in the same orientation. The interfacial excitation supported by tantalum is clearly slower than that supported by iron. This is in accordance with calculation.

5. Brillouin scattering compared with ISLS

We conclude this section with a few brief remarks on what we perceive to be the relative merits of high-pressure Brillouin scattering and ISLS. Brillouin light scattering is an elegant and versatile technique that has historically been the method of choice for measuring acoustic velocities of transparent materials above static pressures of 10 GPa or so. Unlike ISLS, coupling to the sample is not an issue, and signals are typically strong. The experimental apparatus is modest and its central component, the Fabry–Perot interferometer, is an "off-the-shelf" item. Previously, this has been an advantage over ISLS systems, which must be built entirely by the users and which have been significantly more complicated. However, thanks largely to improvements in laser technology, our ISLS apparatus is comparatively simple and has required only basic knowledge of optics and electronics to assemble.

Brillouin scattering has been performed at very high pressures (~70 GPa, see for example Shimizu et al., 2001), but the requirement of a platelet scattering geometry in which incident and scattered beam angles deviate significantly from the normal (i.e. the cell axis) presumably presents a difficulty for reaching still higher pressures (this requirement is necessary to ensure a sufficiently large change in frequency and thus adequate separation from the quasi-elastic peak). ISLS measurements, on the other hand, can be performed using angles smaller than 10° with respect to the cell axis. We are currently attempting to make measurements on various relevant minerals in the Mbar regime.

As far as surface wave studies are concerned there is only one report in the literature of which we are aware dealing with the application of Brillouin scattering under high pressure (Crowhurst et al., 1999). The complicated nature of the system studied in that work – the sample was effectively a triple-component arrangement consisting of glass, a gold film, and the pressure medium – meant it was, in any case, largely a proof-of-principles investigation. Generally speaking, the Brillouin scattering efficiency of surface waves is weak and much smaller than that of the bulk waves traditionally studied in the DAC (e.g. four orders of magnitude). The scattering from the latter via multiple geometries may often be expected to swamp the weak surface wave signals. Furthermore, an even larger deviation from normal incidence of the laser light must be arranged (assuming the traditional back-scattering geometry used in surface wave studies). This is obviously a difficulty when working with DACs designed for very high pressure and also introduces optical aberrations when attempting to tightly focus on an opaque surface. None of these are issues for ISLS measurements.

6. High-pressure elastic moduli of iron and cobalt. Physical implications of measured quantities

Fe and Co samples (of 99.99% purity) in the form of foil or powder (Fe) were loaded into a rhenium or inconel gasket and compressed in the DAC. Several experimental configurations were adopted, depending on the desired pressure range. Within the experimental uncertainty, measured velocities to 40 GPa did not depend on the form of the sample, or whether it was contained in a cavity or compressed between one of the diamond anvils and the gasket. For the highest pressure measurements (up to 115 GPa), one of the two identical flat diamonds (with a culet diameter of 500 μm) was replaced by a beveled anvil (having a bevel angle of 9°) with a 100 μm central flat. The sample in the form of powder was loaded into a 100 μm-diameter cavity in a rhenium gasket. Several ruby chips (with dimensions of microns) embedded into the sample served as pressure sensors. The pressure was determined by averaging the results of several sensors. The pressure gradient across the probed sample area did not exceed 8% of the nominal pressure. The corresponding gradients for the low-pressure experiments were substantially smaller. An additional systematic contribution to the ISLS experimental error is due to the curvature of the anvil culets at high pressure. This effect was estimated to translate into an uncertainty in the measured wave velocity of less than 1%.

Figure 7 shows the compressional and shear elastic moduli of Fe and Co as a function of pressure. To compare with available literature data for single crystals, corresponding aggregate quantities were calculated using the Voigt–Reuss–Hill averaging scheme. Data we have obtained for the low-pressure body-centered-cubic (bcc) phase of iron are in good agreement with the results of inelastic neutron scattering (Klotz and Braden, 2000). The elastic constants of Fe and Co in hcp phases show monotonic and sub-linear pressure dependencies within the experimental precision. The elastic properties of Co are smaller than those of Fe, which is partly related to the presence of large magnetic moments in ferromagnetic Co (Steinle-Neumann et al., 1999), while only small magnetic moments (if any) may be present in hcp Fe (see below)). Magnetization in 3*d* metals is related

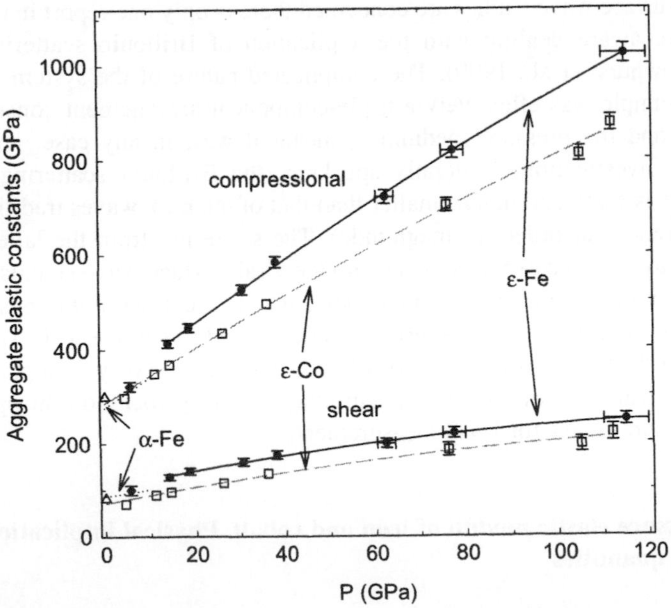

Figure 7. ISLS data for the shear and compressional aggregate elastic constants of polycrystalline iron (solid circles and solid line) and cobalt (open squares and gray dashed line). The triangles represent isotropic averages for cobalt under ambient pressure (Schober and Dederichs, 1979) as a function of pressure. The dotted line is a guide to the eye for low-pressure iron (i.e. in the α-phase). The other lines are second order polynomials.

to a nearly complete filling of the majority 3*d* band, while the minority 3*d* band is only partially filled. It is the filling factor of the latter band, which determines the structure and elastic properties of these metals (see, e.g. Söderlind et al., 1994). The effect of pressure is to widen the electronic bands, which in turn leads to a decrease of the density of states at the Fermi level, N_F, such that the stability condition, given by the Stoner equation ($N_F S > 1$, where S is the Stoner factor) for the magnetic state is no longer met. (See also Goncharov et al., 2004.) That work also makes a comparison between ISLS data and the Raman data we obtained for cobalt. See also Antonangeli et al. (2004a) which reports IXS measurements of the entire elastic tensor to 39 GPa.)

The ε-Co phase is stable over a wide pressure range at room temperature, but it transforms to the fcc phase in the pressure range of 105–150 GPa (Yoo et al., 2000). There is no measurable change of volume at the transition, but thereafter the slope of the compression curve is reduced, indicating a net increase of the elastic stiffness (Fig. 8). This fact was interpreted by Yoo et al. (2000) as an indication of the non-magnetic nature of the high-pressure phase (β) in contrast to the ambient pressure magnetic phase. The compression curve of Fe in the hcp phase (Jephcoat et al., 1986; Mao et al., 1990) (Fig. 8) is smooth, but an increased K' of 5.33 over the typical value ($K' = 4$) is necessary to match the available experimental data. This fact may be related to a gradual disappearance of magnetic moments within the hcp phase of Fe. It is the difference in compression curves of Fe and Co, which is primarily responsible for the qualitatively dissimilar behavior of sound velocities as a function of density (Fig. 9). Fe data appears largely linear, while a significant amount of curvature is present in the case of Co. The shear velocity of Co is almost independent of density when approaching the hcp–fcc transition and in the transition region. We used the EOS of hcp Co when determining the sound velocities at the

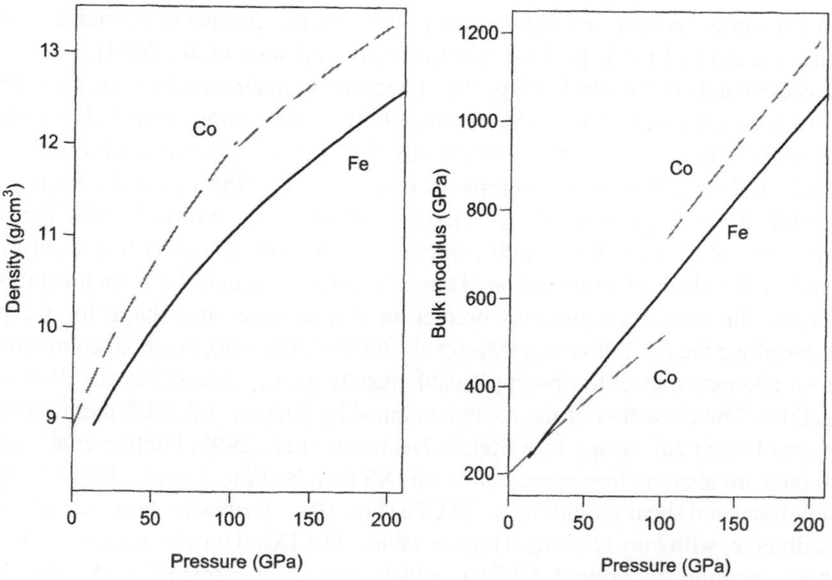

Figure 8. Equation of state of Fe after Mao et al. (1990) and Co after Yoo et al. (2000). Solid line – Fe, dashed gray line – Co.

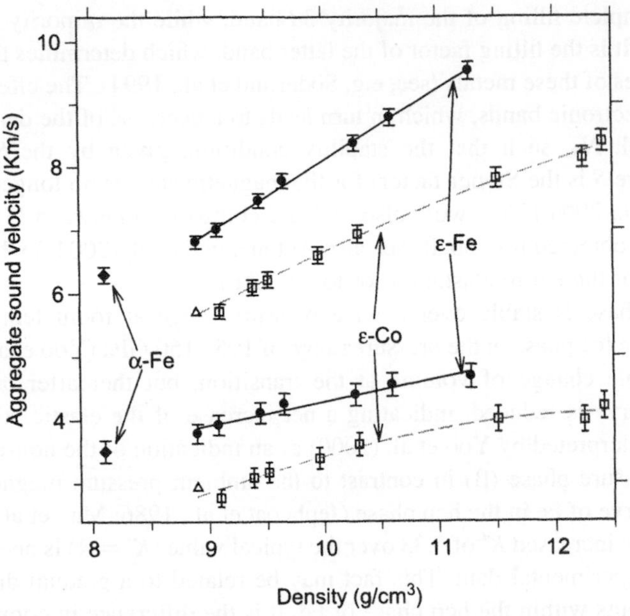

Figure 9. Shear and compressional aggregate sound velocities of ε-iron and cobalt as a function of density. The full circles and solid lines are our data for iron, with the density calculated according to Jephcoat et al. (1986) and Mao et al. (1990). The open squares and dashed lines are our data for cobalt, with the density calculated according to Fujihisa and Takemura (1996) and Yoo et al. (2000). Open triangle – literature data for cobalt at ambient pressure. The dashed lines are guides to the eye.

highest pressures because our Raman data indicated the persistent presence of the E_{2g} phonon, indicative of the hcp phase (see below) (Goncharov et al., 2004).

The present data in the hcp-Fe (Fig. 10a) agree well with direct ultrasonic measurements made in a large press at 16.5 GPa (Mao et al., 1998) for the compressional elastic modulus, but less well in the case of the shear modulus. Our measured shear modulus also agrees reasonably well with Raman measurements (Merkel et al., 2000) over the entire pressure range studied. A single-crystal c_{44} elastic modulus was inferred from the Raman measurements of Merkel et al. (2000), whereas the present data refer to a shear modulus averaged in the plane of propagation. Thus, the agreement may be coincidental. On the other hand, theoretical calculations predict an almost pure sine shape for the phonon dispersion along the Γ–A direction (Mao et al., 2001; L. Vocadlo, personal communication, 2001), so one may expect to obtain plausible results for c_{44} (Merkel et al., 2000; Olijnyk et al., 2001). The closeness of the results obtained by Raman and ISLS methods suggests only a small shear anisotropy (see Steinle-Neumann et al., 1999; Giefers et al., 2002).

Our data are also in close agreement with IXS data by Fiquet et al. (2001) for both the compressional and shear moduli up to 30 GPa (Fig. 10a). Thereafter however, the two sets of data diverge, with ours tending to higher values. The IXS data also imply a maximum in the shear modulus at around 60 GPa, which may be contrasted with our observed monotonic increase. However, the more recent IXS data of Antonangeli et al. (2004b) have yielded moduli that are in very close agreement with ours (see Fig. 10).

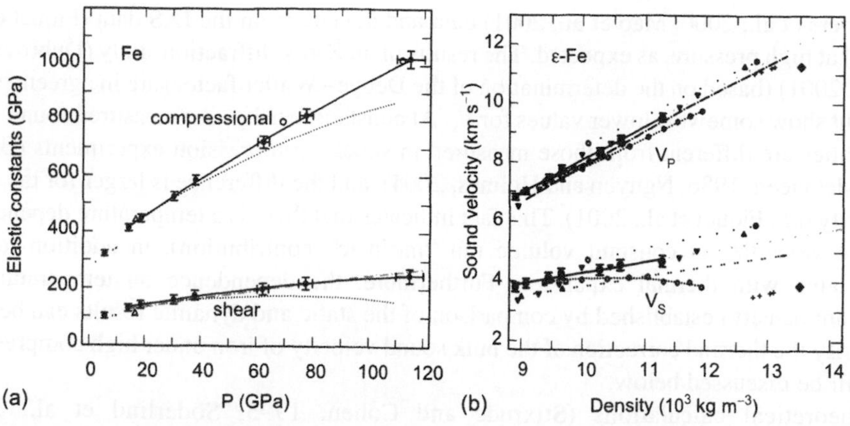

Figure 10. (a) Shear and compressional aggregate elastic constants of polycrystalline iron (solid circles and solid line) as a function of pressure. The dotted line is a polynomial fit to the IXS data (Fiquet et al., 2001) (We obtained a shear modulus from the compressional velocities reported by Fiquet et al. (2001) using the equation of state reported by Mao et al. (1990). Open hexagon – recent improved IXS data on Fe – Antonangeli et al. (2004b). The thick gray dashed-dot line is the result of Raman measurements from Merkel et al. (2000). The values at approximately 5 GPa (solid diamonds) were obtained in the bcc phase and are compared with the results of a neutron scattering study (Klotz and Braden, 2000) (dashed lines). The agreement (within 3%) attests to the reliability of the ISLS technique. Open up triangles – ultrasonic data from Mao et al. (1998). (b) Shear and compressional aggregate sound velocities of ε-iron as a function of density. The full circles and bold lines are our data for iron, with the density calculated according to Mao et al. (1990). Solid diamonds: our extrapolated to the Earth's inner core and temperature corrected results (see text). Open hexagons – shock wave data by Brown and McQueen (1986). Open up triangles – radial diffraction and ultrasonic data by Mao et al. (1998). Open down triangles and open squares – NRIXS data by Mao et al. (2001) and Lübbers et al. (2000), respectively. Open diamonds – IXS data (Fiquet et al., 2001). Solid hexagons latest IXS data by Antonangeli et al. (2004b). Dash-dotted lines represent the results of X-ray diffraction study by Dubrovinsky et al. (2001). Open circles and long dashed line – first principles calculations by Stixrude and Cohen (1995). The results of other calculations (see text) are congruent with first principle calculations. Crosses – seismic data from Dziewonski and Anderson (1981).

The above comparisons have been made with the implicit assumption of a perfectly isotropic randomly orientated aggregate. While this supposition is valid at low pressures, at higher pressures a preferred orientation is expected to develop (see e.g. Wenk et al., 2000) with the *c*-axis of the crystallites predominantly parallel to the load direction. Fortunately, the elastic anisotropy of iron is relatively small. The results of recent NRIXS measurements by Giefers et al. (2002) and theoretical calculations by Stixrude and Cohen (1995) yield the following results: up to 3% in Vd (Debye velocity measured from the phonon density of states) at 40 GPa (Giefers et al., 2002), 6% in V_p and 13% in V_s (Stixrude and Cohen, 1995) at 360 GPa. The differences between the random and textured averages are much smaller (see the discussion in the paper by Fiquet et al., 2001), and are believed to be within our experimental precision.

Our experimental sound velocities in Fe at room temperature (Fig. 10b) exhibit a linear dependence on density (Birch's law) within the experimental precision. This observation suggests it is reasonable to extrapolate our velocities to higher pressures. Our results for V_p and V_s are generally consistent with the radial XRD (Mao et al., 1998) and NRIXS

(Lübbers et al., 2000; Mao et al., 2001) data and deviate from the IXS data (Fiquet et al., 2001) at high pressure, as expected. The results of an X-ray diffraction study (Dubrovinsky et al., 2001) (based on the determination of the Debye–Waller factor) are in agreement for V_p, but show somewhat lower values for V_s. At constant density, our measured sound wave velocities are different from those measured in shock-compression experiments (Brown and McQueen, 1986; Nguyen and Holmes, 2004), and the difference is larger for the shear velocity (*cf.*, Fiquet et al., 2001). This fact indicates that there is a temperature dependence of the velocities at constant volume (an "intrinsic" contribution), in addition to that associated with thermal expansion. Furthermore, the dependence on temperature (at constant density) established by comparison of the static and dynamic results can be used to verify the thermal correction of the bulk sound velocity of iron under high compression, as will be discussed below.

Theoretical calculations (Stixrude and Cohen, 1995; Söderlind et al., 1996; Steinle-Neumann et al., 1999; Laio et al., 2000) produce systematically higher velocities than those determined by us. Reliable measurements of phonon frequencies show (Merkel et al., 2000; Olijnyk et al., 2001; Mao et al., 2001; Giefers et al., 2002) this discrepancy is most pronounced at low pressure, a fact that may be related to the currently incomplete theoretical description of ε-iron (Steinle-Neumann et al., 1999) in the vicinity of the transition from the bcc to the hcp structure. Under these conditions iron is thought to possess anomalous magnetic properties (Cohen and Mukherjee, 2004). It has been demonstrated that including these properties in the theoretical model greatly improves the agreement between the experimental and theoretical EOS.

In order to compare static high-pressure, room-temperature results with seismic observations and shock wave data, we performed a temperature correction in a semi-empirical manner following the formalism suggested by Brown and McQueen (1986) and McQueen et al. (1970). To facilitate comparison, we performed a temperature correction to the bulk sound velocity, C, (defined by the relation $C^2 = K_s/\rho$ where K_s is the adiabatic bulk modulus; Jeanloz, 1979; Brown and McQueen, 1986), a property which is intimately related to other thermodynamic properties of the material, and then made corrections for temperature to V_p and V_s on the basis of physically reasonable assumptions. The bulk velocity is given in terms of the experimental compressional and shear sound velocities as $C^2 = V_p^2 - (4/3)V_s^2$. This dependence was then extrapolated to higher pressure and corrected for elevated temperature T using the expression given by Brown and McQueen (1986), i.e. $C^2 = C_{300K}^2 + P_{th}/\rho^2$. Here P_{th} is the thermal pressure, and ρ is the density at temperature T. We estimated the thermal pressure term at high pressure by constructing a thermal EOS based on empirical data.[2] We used a Burch–Murnagan EOS, which we fit by generating the temperature dependence of the unit cell volume $V_{T,0}$ at each pressure according to the expression $V_{T,P} = V_{0P}\exp(\int \alpha_{T,P}dT)$, where V_{0P} and $\alpha_{T,P}$ are, respectively, the unit cell volume and the thermal expansivity at pressure P (assumed to follow the Debye model). To determine the pressure dependence of $\alpha_{T,P}$, we adopted the empirical approximation that the product of the coefficient of

[2] There is also an electronic contribution (about 20 GPa at 7500 K), which we did not take into account (Anderson and Isaak, 2002; Isaak and Anderson, 2003). We do not believe this affects our conclusions.

thermal expansion and the bulk modulus is constant. The necessary inputs for the calculations of the pressure dependences of the Debye temperature has been taken from the work of Mao et al. (2001). We used the values of the linear compressibilities measured by Goncharov et al. (2002) in an X-ray diffraction study at a pressure of 23 GPa. We further assumed that the c/a ratio does not depend on P and T (Jeaphcoat et al., 1986; Mao et al., 1990; Steinle-Neumann et al., 2001; Goncharov et al., 2002; Belonoshko et al., 2003). The corresponding values of $\alpha_c = 2.25 \times 10^{-5}$, $\alpha_a = 1.5 \times 10^{-5}\ K^{-1}$ are very close to those reported by Manghnani et al. (1987). The results of the calculations were found to be in agreement with the available experimental (Brown and McQueen, 1986; Dubrovinsky et al., 2000) and theoretical data (Belonoshko et al., 2000; Belonoshko et al., 2003). To determine the temperature corrected values of V_p and V_s we assumed a linear dependence of the shear elastic modulus on the isobaric dilatation (see paper of Laio et al., 2000, and references therein) and we used the theoretically calculated volume change due to melting (e.g. papers of Belonoshko et al., 2000; Laio et al., 2000). The bulk sound velocity was assumed to experience no discontinuity at the melting point.

As can be seen from Figure 11, our calculations allow us to make suggestions concerning core composition. Under core conditions, V_p and V_s values match those obtained from seismic measurements of the inner core, implying that the latter consists of pure iron. Such a conclusion may appear to contradict PREM data (Dziewonski and Anderson, 1981) for the inner core density, which, according to PREM, is lower than that of iron at high temperature. As has been pointed out by Jeanloz (1979), Anderson (1989) and Masters and Gubbins (2003), however, the determination of core density from seismic measurements is somewhat ambiguous. Under the assumption that the inner core consists of pure iron (the presence of a small amount of heavy and light impurities do not change the results), the gradient of sound velocities for the inner core can also be essentially reproduced (Jeanloz, 1979), although some discrepancy in V_p can be seen at the inner–outer core boundary. In the outer core the situation is different. The discrepancies between pure iron and PREM velocities indicate the presence of light impurities and thus a compositional change at the inner–outer core boundary. The PREM velocities are also almost parallel to our values for C (Fig. 11), suggesting an outer core composition that is nearly uniform (Jeanloz, 1979).

These observations are in general agreement with conclusions made on the basis of shock-wave studies (Jeanloz, 1979; Brown and McQueen, 1986), but disagree with those of recent static pressure NRIXS (Mao et al., 2001) and IXS (Fiquet et al., 2001) studies regarding inner core composition. This disagreement may be imputed to differences between measured sound velocities (Fig. 10), as well as to the different procedures adopted to perform temperature corrections. An important finding of our work is that our temperature-corrected velocities match the PREM data for the wide range of temperatures that correspond to the uncertainty of the inner core geotherm (*cf.*, paper of Mao et al., 2001). Thus, the results presented here are not expected to be sensitive to future refinements of the estimate of core temperature.

It is also interesting to mention that our results, temperature-corrected to shock-wave conditions (Fig. 11), show almost perfect agreement with the shock-wave data, further helping to establish the validity of our correction procedure and the chosen thermodynamic parameters.

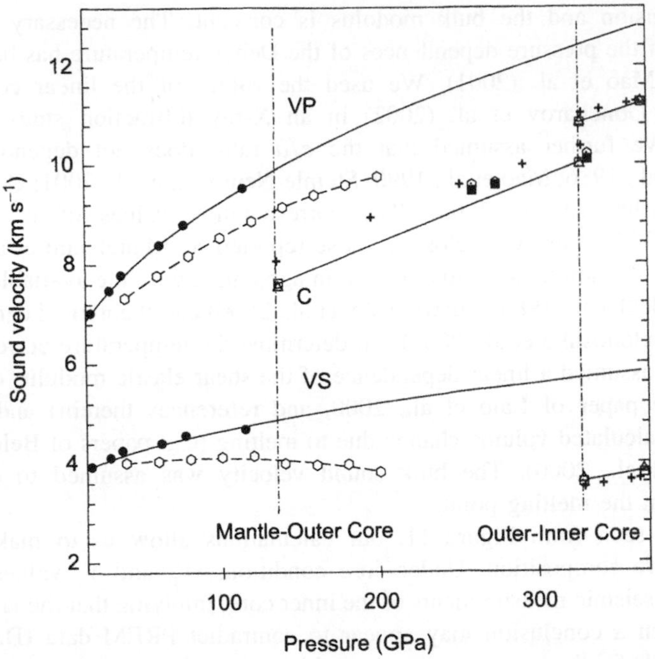

Figure 11. Sound velocities of ε-iron as a function of pressure compared to seismic data (Dziewonski and Anderson, 1981), and the results of shock-wave measurements (Brown and McQueen, 1986). All filled symbols refer to our data: circles – at room temperature, squares – extrapolated and temperature corrected to shock wave conditions; solid lines – extrapolations of V_p and V_s to core pressure. Open hexagons – shock-wave data for V_p and V_s, dotted hexagons – shock wave data for the bulk velocity C measurements (Brown and McQueen, 1986), crosses – Earth's seismic model (Dziewonski and Anderson, 1981). Dashed lines through shock-wave data are guides to the eye. Open triangles (melting curve from Belonoshko et al., 2000, 2003) and squares (melting curve from Alfè et al., 2002) – represent our extrapolated and temperature-corrected data for V_p (C), and V_s, respectively, at the Earth's center, outer–inner core boundary, and core–mantle boundary; solid gray lines through these data – guides to the eye.

Acknowledgements

We thank the following for many valuable discussions: R. Jeanloz, J.M. Brown, R.J. Hemley, V. Stuzhkin, E.H. Abramson, H.-K. Mao, J. Badro, M. Krisch, L. Dubrovinsky and R. L. Simpson. This work was performed under the auspices of the US Department of Energy by the University of California, Lawrence Livermore National Laboratory under Contract No. W-7405-Eng-48.

References

Abramson, E.H., Brown, J.M., Slutsky, L.J., 1999. Application of impulsive stimulated scattering in the Earth and planetary sciences. Annu. Rev. Phys. Chem. 50, 279–313.

Ahrens, J.R., Tan, H., Bass, J.D., 1990. Analysis of shock temperature data for iron. High Pressure Res. 2, 145–156.

Alfè, D., Price, G.D., Gillan, M.J., 2002. Iron under Earth's core conditions: liquid-state thermodynamics and high-pressure melting curve from *ab initio* calculations. Phys. Rev. B 65, 165118.

Anderson, D.L., 1989. Theory of the Earth. Blackwell Scientific Publications, Cambridge.

Anderson, O.L., Isaak, D.G., 2002. Another look at the core density deficit of the Earth's outer core. Phys. Earth Planet. Interiors 131, 19–27.

Antonangeli, D., Krisch, M., Fiquet, G., Farber, D.L., Aracne, C.M., Badro, J., Occelli, F., Requardt, H., 2004a. Elasticity of cobalt at high pressure studied by inelastic X-ray scattering. Phys. Rev. Lett. 93, 215505.

Antonangeli, D., Occelli, F., Requardta, H., Badro, J., Fiquet, G., Krisch, M., 2004b. Elastic anisotropy in textured hcp-iron to 112 GPa from sound wave propagation measurements. Earth Planet. Sci. Lett. 225, 243–251.

Bass, J.D., Svendsen, B., Ahrens, T.J., 1987. The temperature of shock-compressed iron. In: Manghnani, M.H., Syono, Y. (Eds), High Pressure Research in Mineral Physics, Terra Scientific, Tokyo, pp. 393–402.

Bass, J.D., Ahrens, T.J., Abelson, J.R., 1990. Shock temperature measurements in metals: new results for an Fe alloy. J. Geophys. Res. 95, 21767–21776.

Belonoshko, A.B., Ahuja, R., Johansson, B., 2000. Quasi-ab initio molecular dynamic study of Fe melting. Phys. Rev. Lett. 84, 3638–3641.

Belonoshko, A.B., Ahuja, R., Johansson, B., 2003. Stability of the body-centred-cubic phase of iron in the Earth's inner core. Nature 424, 1032–1034.

Brown, J.M., McQueen, R.G., 1986. Phase transitions, Gruneisen parameters, and elasticity of shocked iron between 77 GPa and 400 GPa. J. Geophys. Res. 91, 7485–7494.

Brown, J.M., Slutsky, L.J., Nelson, K.A., Cheng, L.-T., 1988. Sound velocities in olivine at Earth mantle pressures. Science 241, 65–68.

Cohen, R.E., Mukherjee, S., 2004. Non-collinear magnetism in iron at high pressures. Phys. Earth Planet. Interiors 143–144, 445–453.

Crowhurst, J.C., Zaug, J.M., 2004. Surface acoustic waves in germanium single crystals. Phys. Rev. B 69, 052301-1, 4 pages.

Crowhurst, J.C., Hearne, G.R., Comins, J.D., Every, A.G., Stoddart, P.R., 1999. Surface Brillouin scattering at high pressure: application to a thin supported gold film. Phys. Rev. B 60, R14990–R14993.

Crowhurst, J.C., Abramson, E.H., Slutsky, L.J., Brown, J.M., Zaug, J.M., Harrell, M.D., 2001. Surface acoustic waves in the diamond anvil cell: an application of impulsive stimulated light scattering. Phys. Rev. B 64, 100103 (R) 4 pages.

Dubrovinsky, L.S., Saxena, S.K., Tutti, F., Rekhi, S., LeBehan, T., 2000. In situ X-ray study of thermal expansion and phase transition of iron at multimegabar pressure. Phys. Rev. Lett. 84, 1720–1723.

Dubrovinsky, L.S., Dubrovinskaia, N.A., Le Behan, T., 2001. Aggregate sound velocities and acoustic Grüneisen parameter of iron up to 300 GPa and 1,200 K. PNAS 98, 9484–9489.

Dziewonski, A.M., Anderson, D.L., 1981. Preliminary reference Earth model. Phys. Earth Planet. Interiors 25, 297–356.

Every, A.G., Briggs, G.A.D., 1998. Surface response of a fluid-loaded solid to impulsive line and point forces: application to scanning acoustic microscopy. Phys. Rev. B 58, 1601–1612.

Every, A.G., Kim, K.Y., Mazenev, A.A., 1997. The elastodynamic response of a semi-infinite anisotropic solid to sudden surface loading. J. Acoust. Soc. Am. 102, 1346–1355.

Fiquet, G., Badro, J., Guyot, F., Requardt, H., Krisch, M., 2001. Sound velocities in iron to 110 gigapascals. Science 291, 468–471.

Fujihisa, H., Takemura, K., 1996. Equation of state of cobalt up to 79 GPa. Phys. Rev. B 54, 5–7.

Giefers, H., Lübbers, R., Rupprecht, K., Wortmann, G., Alfè, D., Chumakov, A.I., 2002. Phonon spectroscopy of oriented hcp iron. High Pressure Res. 22, 501–506.

Goncharov, A.F., Struzhkin, V.V., Gregoryanz, E., Maddury, S., Huang, E., Hemley, R.J., Mao, H.K., 2002. Raman and X-ray diffraction study of iron and iron–nickel alloys at varying $P-T$ conditions. EOS Trans. AGU 83 (19), Spring Meet. Suppl., Abstract, M32A-08.

Goncharov, A.F., Crowhurst, J.C., Zaug, J.M., 2004. Elastic and vibrational properties of cobalt to 120 GPa. Phys. Rev. Lett. 92, 115502.

Grimsditch, M., Loubeyre, P., Polian, A., 1986. Brillouin scattering and three-body forces in argon at high pressures. Phys. Rev. B 33, 7192–7200.

Isaak, D.G., Anderson, O.L., 2003. Thermal expansivity of hcp iron at very high pressure and temperature. Physica B 328, 345–354.

Jeanloz, R., 1979. Properties of iron at high pressure and the state of the core. J. Geophys. Res. 84, 6059–6069.

Jephcoat, A.P., Mao, H.K., Bell, P.M., 1986. Static compression of iron to 78 GPa with rare gas solids as pressure-transmitting media. J. Geophys. Res. 91, 4677–4784.

Katahara, K.W., Manghnani, M.H., Fisher, E.S., 1976. Pressure derivatives of the elastic moduli of niobium and tantalum. J. Appl. Phys. 47, 434–439.

Klotz, S., Braden, M., 2000. Phonon dispersion of bcc iron to 10 GPa. Phys. Rev. Lett. 85, 3209–3212.

Laio, A., Bernard, S., Chiarotti, G.L., Scandolo, S., Tosatti, E., 2000. Physics of iron at Earth's core conditions. Science 287, 1027–1030.

Lübbers, R., Grünsteudel, H.F., Chumakov, A.I., Wortmann, G., 2000. Density of phonon states in iron at high pressure. Science 287, 1250–1253.

Manghnani, M.H., Ming, L.C., Nakagiri, N., 1987. High Pressure Research in Mineral Physics. TERRAPUB, Tokyo, pp. 155–163.

Mao, H-K., Wu, Y., Chen, Y., Shu, J., Jephcoat, A.P., 1990. Static compression of iron to 300 GPa and $Fe_{0.8}Ni_{0.2}$ alloy to 260 GPa: implications for composition of the core. J. Geophys. Res. 95, 21737–21742.

Mao, H.K., Shu, J., Shen, G., Hemley, R.J., Li, B., Singh, A.K., 1998. Elasticity and rheology of iron above 220 GPa and the nature of the Earth's inner core. Nature 396, 741–745; Nature, 399, 280.

Mao, H.K., Xu, J., Struzhkin, V.V., Shu, J., Hemley, R.J., Sturhahn, W., Hu, M.Y., Alp, E.E., Vocadlo, L., Alfè, D., Price, G.D., Gillan, M.J., Schwoerer-Böhning, M., Häusermann, D., Eng, P., Shen, G., Giefers, H., Lübbers, R., Wortmann, G., 2001. Phonon density of states of iron up to 153 gigapascals. Science 292, 914–916.

Masters, G., Gubbins, D., 2003. On the resolution of density within the Earth. Phys. Earth Planet. Interiors 140, 159–167.

Maznev, A.A., Akthakul, A., Nelson, K.A., 1999. Surface acoustic modes in thin films on anisotropic substrates. J. Appl. Phys. 86, 2818–2824.

McQueen, R.G., Marsh, S.P., Taylor, J.W., Fritz, J.N., Carter, W.J., 1970. The equation of state of solids from shock wave studies. In: Kingslow, R. (Ed.), High-Pressure Impact Phenomena. Academic, Orlando, FL, pp. 244–419.

Merkel, S., Goncharov, A.F., Mao, H-K., Gillet, P., Hemley, R.J., 2000. Raman spectroscopy of iron to 152 gigapascals: implications for Earth's inner core. Science 288, 1626–1629.

Nguyen, J.N., Holmes, N.J., 2004. Melting of iron at the physical conditions of the Earth's core. Nature 427, 339–342.

Olijnyk, H., Jephcoat, A.P., Refson, K., 2001. On optical phonons and elasticity in the hcp transition metals Fe, Ru and Re at high pressure. Europhys. Lett. 53, 504–510.

Schober, H.R., Dederichs, H., 1979. In: Hellwege, K.-H., Hellwege, A.W., Börnstedt, L. (Eds), Elastic Piezoelectric Pyroelectric Piezooptic Electrooptic Constants and Non-linear Dielectric Susceptibilities of Crystals, New Series III, Vol. 11a. Springer, Berlin.

Shimizu, H., Tashiro, H., Kume, T., Sasaki, S., 2001. High-pressure elastic properties of solid argon to 70 GPa. Phys. Rev. Lett. 86, 4568–4571.

Singh, A.K., Mao, H-K., Shu, J.F., Hemley, R.J., 1998. Estimation of single-crystal elastic moduli from polycrystalline X-ray diffraction at high pressure: application to FeO and iron. Phys. Rev. Lett. 80, 2157–2160.

Söderlind, P., Ahuja, R., Eriksson, O., Wills, J.M., Johansson, B., 1994. Crystal structure and elastic-constant anomalies in the magnetic 3d transition metals. Phys. Rev. B 50, 5918–5927.

Söderlind, P., Moriarty, J.A., Wills, J.M., 1996. First-principles theory of iron up to Earth-core pressures: structural, vibrational, and elastic properties. Phys. Rev. B 53, 14063–14072.

Steinle-Neumann, G., Stixrude, L., Cohen, R.E., 1999. First-principles elastic constants for the hcp transition metals Fe, Co, and Re at high pressure. Phys. Rev. B 60, 791–799.

Steinle-Neumann, G., Stixrude, L., Cohen, R.E., Gulseren, O., 2001. Elasticity of iron at the temperature of the Earth's inner core. Nature 413, 57–60.

Stixrude, L., Brown, J.M., 1998. The Earth's core. In: Hemley, R.J. (Ed.), Chapter 8 of Ultrahigh-Pressure Mineralogy: Physics and Chemistry of the Earth's Deep Interior, Reviews in Mineralogy, Vol. 37.

Stixrude, L., Cohen, R.E., 1995. High-pressure elasticity of iron and anisotropy of the Earth's inner core. Science 267, 1972–1975.

Vocadlo, L., 2001. Private communication.

Wenk, H.-R., Matthies, S., Hemley, R.J., Mao, H-K., Li, B., Singh, A.K., 2000. The plastic deformation of iron at pressures of the Earth's inner core. Nature 405, 1044–1047.

Yoo, C.S., Holmes, N.C., Ross, M., Webb, D.J., Pike, C., 1993. Shock temperatures and melting of iron at Earth core conditions. Phys. Rev. Lett. 70, 3931–3934.

Yoo, C.S., Cynn, H., Söderlind, P., Iota, V., 2000. New beta (fcc)-cobalt to 210 GPa. Phys. Rev. Lett. 84, 4132–4135.

Zaug, J.M., Abramson, E., Brown, J.M., Slutsky, L.J., 1993. Velocity of sound and equations of state for methanol and ethanol in diamond-anvil cell. Science 260, 1487–1490.

Advances in High-Pressure Technology for Geophysical Applications
Jiuhua Chen, Yanbin Wang, T.S. Duffy, Guoyin Shen, L.F. Dobrzhinetskaya, editors

Chapter 2

A gigahertz ultrasonic interferometer for the diamond anvil cell and high-pressure elasticity of some iron-oxide minerals

Steven D. Jacobsen, Hans J. Reichmann, Anastasia Kantor and Hartmut A. Spetzler

Abstract

A second-generation high-frequency acoustic interferometer has been developed for high-pressure and high-temperature elasticity measurements in the diamond anvil cell. The instrument measures single-crystal compressional and shear-wave travel times, which are converted to sound velocities and elastic moduli for direct application to problems in geophysics. The second-order elastic constants (c_{ij}) of several iron-bearing oxide minerals has been measured under hydrostatic pressures to ~10 GPa. Pressure-induced c_{44} mode softening is observed in magnetite (Fe_3O_4), wüstite ($Fe_{0.95}O$) and in iron-rich magnesiowüstite-(Mg,Fe)O, indicating that strong magnetoelastic coupling is common among these iron-rich oxides well ahead of known structural phase transitions. In (Mg,Fe)O, the pressure derivative of c_{44} is highly sensitive to composition and switches sign between 1.2 ± 0.2 at 25 mol% FeO to −0.96 ± 0.3 at 75 mol% FeO, and is about zero for (Mg,Fe)O containing 50 mol% FeO. In wüstite, a discontinuity in the pressure derivatives of c_{11} and c_{12} at ~5 GPa is interpreted to result from the onset of magnetic ordering, implying that a partially ordered but still cubic phase of FeO exists between ~5 GPa and where the rhombohedral distortion occurs at ~17 GPa.

1. Introduction

Over 50 years ago, Birch (1952) recognized that the constitution of Earth's inaccessible interior could be interpreted from seismological observation, provided that the elasticity of its constituent materials could be measured or calculated at relevant conditions of pressure and temperature in the laboratory. A new gigahertz ultrasonic interferometer with shear-wave capabilities has been developed for high-pressure and high-temperature elasticity measurements in the diamond anvil cell (DAC). The acoustic instrumentation can be interfaced to nearly any type of DAC, and measures single-crystal compressional and shear-wave travel times with high precision. Measured travel times and their $P-T$ derivatives are converted to sound velocities and elastic moduli for direct application to problems in geophysics.

The emerging picture of Earth's deep interior from seismic tomography indicates more complexity than previously thought. The presence of seismic anisotropy in Earth's upper mantle (Ekström and Dziewonski, 1998), transition zone (Chen and Brudzinski, 2003), and lower mantle (van der Hilst and Kárason, 1999) highlights the importance of fully anisotropic elasticity data from mineral physics. The elastic tensor (c_{ijkl}) relating stress

(σ_{ij}) to strain (ε_{kl}) in Hooke's law of elasticity provides not only the necessary information for interpreting the bulk mineralogy of the interior from seismological observation (e.g. Duffy et al., 1995), but also for understanding textures and fabrics resulting from Earth's dynamic mantle flow (e.g. Christensen et al., 2001).

Several methods have been newly developed or applied to measuring the elastic constants of materials at elevated pressures and temperatures. Elastic constants can be estimated from lattice strain in polycrystalline samples under non-hydrostatic compression with radial X-ray diffraction (Mao et al., 1998; Singh et al., 1998; Dubrovinsky et al., 2000). The lattice strain technique offers the distinct advantage that experiments can be carried out to megabar pressures in the DAC, but uncertainties in the absolute value of c_{ij} can be on the order of 10–20%. Individual c_{ij} have also been determined from single-crystal inelastic neutron scattering (INS). INS typically requires large (≥10 mm^3) samples, thus precluding the study of tiny high-pressure single-crystals, although recent INS experiments on FeO have been carried out to 12 GPa in the Paris–Edinburgh cell (Klotz, 2001). Inelastic X-ray scattering (IXS) and nuclear resonant inelastic X-ray scattering (NRIXS) is now possible at third-generation synchrotron sources and have been used with the DAC to megabar pressures at room temperature (Mao et al., 2001a,b; Fiquet et al., 2004), and recently NRIXS has been interfaced with resistively heated (Shen et al., 2004) and laser-heated diamond cells (Lin et al., this volume). The c_{ij} of a single-crystal of argon was measured to 20 GPa using IXS (Occelli et al., 2001). Mao et al. (2001b) determined the aggregate sound velocities (V_P and V_S) in hcp-iron powder to 153 GPa from the Debye sound velocity measured with NRIXS, and Fiquet et al. (2001) reported the longitudinal sound velocity in iron to 110 GPa with IXS.

Efforts to obtain accurate pressure derivatives of single-crystal elastic constants ultrasonically and under hydrostatic conditions in the gigapascal pressure range were pioneered by Jackson and Neisler (1982) in a piston–cylinder apparatus to 3 GPa, and by Yoneda (1990) to 8 GPa using a hybrid fluid-filled assembly for the multi-anvil press. Using solid pressure-transmitting media, MHz-frequency ultrasonic interferometry has been carried out on single-crystal samples to 13 GPa in the multi-anvil press (Chen et al., 1996a) and to 6 GPa in the Paris–Edinburgh cell (Gauthier et al., 2003) with pulse superposition. Single-crystal ultrasonic measurements at simultaneous high P–T conditions follow those of Spetzler (1970) on MgO to 0.8 GPa and 800 K; in the DAC to 4.4 GPa and 500 K (Shen et al., 1998) and in the multi-anvil press to 8 GPa and 1600 K (Chen et al., 1998). *In situ* X-radiation techniques (X-ray diffraction and X-radiography) through the multi-anvil press assembly now permit direct observation of the sample length and volume during ultrasonic experiments (Li et al., 2001; Kung et al., 2002) and hold exciting potential for future single-crystal equation of state studies.

Brillouin scattering measurements in the DAC have been carried out on single-crystal mineral samples to 55 GPa at room temperature (Zha et al., 2000) and to 2500 K at room pressure by laser heating in a novel high-temperature cell for Brillouin scattering (Sinogeikin et al., 2004). Impulsive stimulated light scattering (ISLS) has been used to measure the elastic constants of minerals to 20 GPa (Chai et al., 1997; Abramson et al., 1999). The ISLS technique can also be used to measure surface-wave velocities in opaque materials (Crowhurst et al., 2001), from which bulk and shear moduli have been determined to 120 GPa in polycrystalline cobalt (Goncharov et al., 2004) and in hcp-iron to 115 GPa (Crowhurst et al., 2004), but prior knowledge of the bulk modulus and its pressure derivative is required to extract all the moduli.

With resonant ultrasound spectroscopy (RUS), it is possible to determine the elastic constants of minerals with high precision and to temperatures in excess of 1800 K (Goto et al., 1989; Isaak et al., 1989). Using right-rectangular parallelepiped resonance (RPR), the complete elastic tensor of crystals with low symmetry (e.g. monoclinic) have been measured (Isaak and Ohno, 2003). Another important development for RUS with application to geophysics is the resonant sphere technique, which has been employed in a gas–pressure apparatus to 0.2 GPa (Ohno et al., 2000), and was recently used to measure the temperature dependence of the adiabatic bulk and shear moduli of wadsleyite (Mayama et al., 2004).

Gigahertz-ultrasonic interferometry was initially developed by Spetzler et al. (1993), although the idea of using high-frequency ultrasound (at >100 MHz) in order to study micro-crystals and to eliminate unwanted diffraction effects can be traced back to McSkimin (1950). By extending ultrasonic measurements to thousands of MHz (i.e. GHz), the acoustic wavelengths in minerals are reduced to a few micrometers (μm), permitting single-crystal ultrasonic travel-time studies on samples as thin as a few 10's of μm in length. Elastic-wave travel times through thin films only ~1 μm or less in thickness have been measured using the bonded/non-bonded buffer rod amplitude-ratio technique described by Chen et al. (1997). Since its initial application in the gas-loading piston–cylinder at Colorado (Chen et al., 1996b), the GHz technique, at various stages of development, has been used successfully in the DAC with P-waves (Spetzler et al., 1996; Shen et al., 1998; Reichmann et al., 1998; Bassett et al., 2000).

Our main objective here is to describe the major advances in GHz-ultrasonic interferometry having taken place over the past 2 years. A new method of generating GHz shear-waves has been developed for ultrasonic interferometry (Jacobsen et al., 2002a), and has already been used to determine the complete elastic tensor of a high-pressure phase (Fo_{90} ringwoodite) recovered from the multi-anvil press (Jacobsen et al., 2004a). Shear-waves have since been transmitted into the DAC for acoustic interferometry (Jacobsen et al., 2004b), and the elastic tensor of some opaque iron-oxide minerals including magnetite (Reichmann and Jacobsen, 2004), magnesiowüstite, and wüstite (Kantor et al., 2004a) have been determined to ~10 GPa. Pressure-induced mode softening in the iron-rich oxides indicates that strong magnetoelastic coupling occurs well ahead of known structural phase transitions.

2. Transducers

Ultrasonic interferometry is a single-transducer delay-line experiment. The basic principles are unchanged from MHz-ultrasonic interferometry, with the major exceptions being within the electronics and transducer technology, discussed in this section. GHz-frequency electronics require special attention because high-frequency electromagnetic fields suffer higher attenuation in cables and are extremely sensitive to impedance mismatch. There are additional complexities for high-pressure ultrasonics because the travel-time of the carrier signal through the cables and circuitry approaches that of sound velocities through thin mineral plates, causing additional interference effects.

For example, the carrier signal traveling at about 3×10^8 m/s (30 cm/ns through) a 1-meter cable has roughly the same travel-time as a P-wave (round trip) through a 15 μm thick mineral sample in the DAC. The design challenges for ultrasonics at GHz frequencies have therefore mainly to do with impedance matching to avoid unwanted reflections in the circuitry.

Contact to the delicate transducers, which are only about 1 μm thick, is made through transmission-line connectors, designed to match the nominal 50-ohm impedance of the system as closely as possible. Transmission lines consist of a conductor–insulator–conductor sandwich, like the board of a printed circuit. Our 50-ohm transmission lines measure about 1 mm thick and 3 mm wide, and are soldered to standard SMA connectors. The lead contact consists of a polished pin mounted onto the end of a leaf spring, which protrudes through the insulating and ground layers. Contact to the transducer is made by gently pushing the transmission line against the transducer. With these connectors, no micro-soldering is necessary and different buffer rods can be switched in-and-out of the same connector.

The GHz-transducers are sputtered thin-films of ZnO, developed in collaboration with K. Müller and H. Ohlmeyer at the University of Bayreuth. The anatomy of a P-wave transducer is illustrated schematically in Figure 1. Thin films of chrome (5–10 nm thick) and then gold (~500 nm thick) are vacuum-coated onto a single-crystal buffer rod substrates, usually made of *c*-axis sapphire (Al_2O_3) for P-waves or [100]-oriented cubic yttrium aluminum garnet ($Y_3Al_5O_{12}$ or YAG) in the case for shear-waves. Deposition of the zinc oxide (ZnO) transducer is a sputtering process, requiring about 2 h for each micron of ZnO growth. To achieve transducers that can be driven at ~0.5–2.5 GHz, the ideal ZnO transducer thickness is 1.5 μm. Finally, a second Cr–Au layer is precision-coated onto the transducer for the lead contact. The diameter of the ZnO is normally about 1 mm, and the diameter of the top electrode (i.e. the effective transducer diameter) is normally ~0.3 mm (Fig. 1).

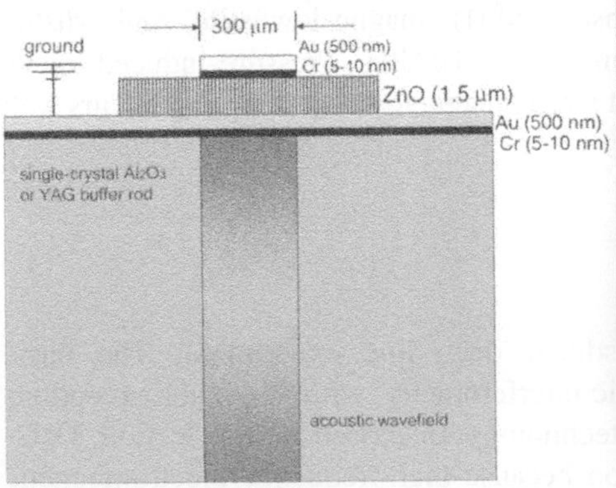

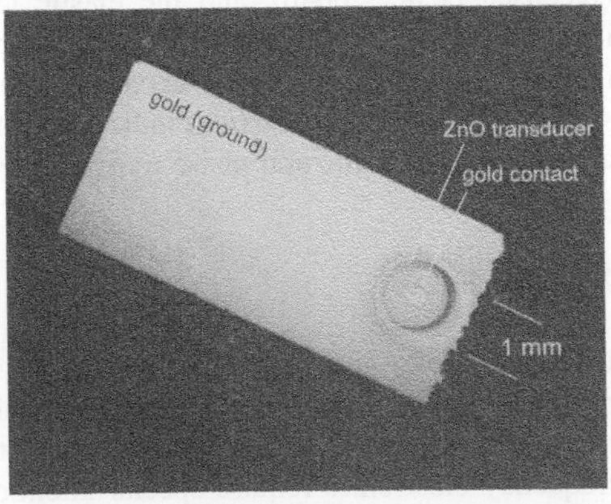

Figure 1. Gigahertz-ultrasonic interferometry has been made possible through the development of thin-film piezoelectric transducers, shown in schematic at left. The ZnO P-wave transducers function over a broad bandwidth, typically from ~0.5 to ~2.0 GHz, and are sputtered directly onto single-crystal buffer rods for ultrasonic interferometry. A reflected-light photograph of a transducer is shown at right.

The quality of compressional wave energy from the sputtered thin-film transducers is exceptionally high, but shear-wave transducers for GHz-frequencies are not possible with this method, nor are they commercially available. Earlier attempts to thin down a commercial MHz-frequency ultrasonic shear transducer produced near-GHz shear-waves, but of relatively low quality, and only over a narrow frequency range of 590–605 MHz (Chen et al., 1997). To get around this problem, a GHz shear-wave buffer rod was developed that produces purely polarized shear-waves by P-to-S conversion (Jacobsen et al., 2002a). The original prototype was made of MgO, which is too fragile for use against diamond anvils, so a similar P-to-S conversion buffer rod has been developed from YAG for ultrasonic interferometry in the DAC (Fig. 2).

The GHz shear-wave buffer rod works on the principle of Snell's Law. An incident P-wave strikes a polished conversion facet on a YAG single crystal at an angle of incidence (i) such that the converted S-wave is orthogonal to the incident P-wave (Fig. 2). If the incident P-wave is traveling in the [100] direction of the cubic YAG crystal, from Snell's law, the velocities are:

$$\frac{V_{\mathrm{P}}^{[100]}}{\sin(i)} = \frac{V_{\mathrm{S}}^{[010]}}{\sin(j)} \tag{1}$$

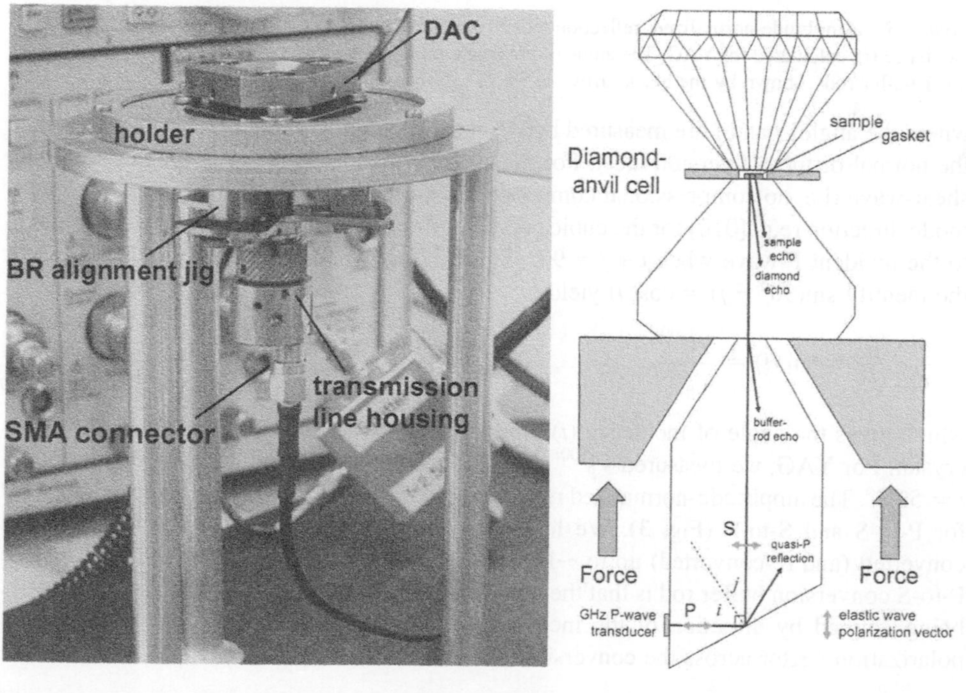

Figure 2. (Left) Photograph of the GHz-ultrasonic diamond anvil cell attached to the interferometer. (Right) Schematic of the high-pressure ultrasonic experiment in shear geometry (not to scale). A purely polarized converted S-wave enters the [100]-oriented diamond anvil, and interferometry is carried out by overlapping echoes from the diamond culet (diamond echo) and the far-end of the sample (sample echo). The P-to-S conversion and other experimental details are described in the text.

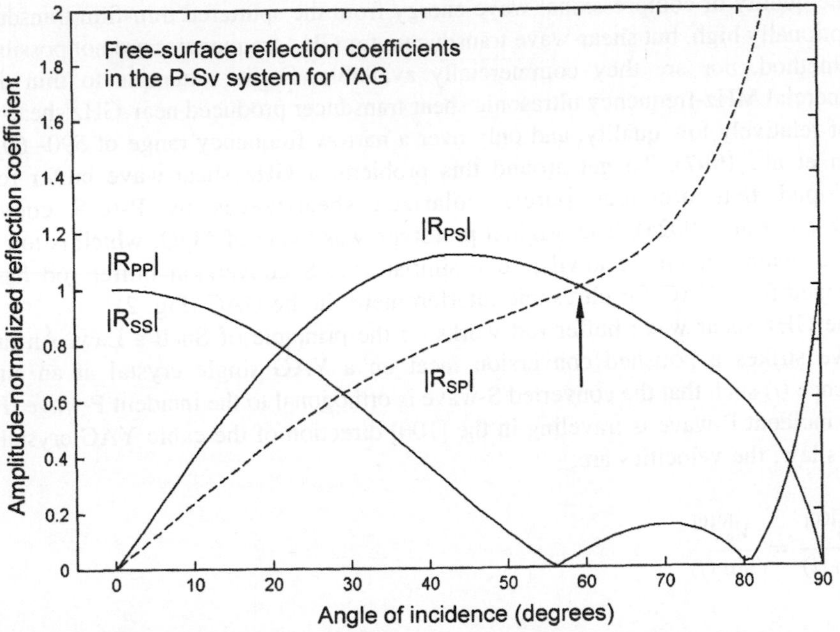

Figure 3. Amplitude-normalized reflection coefficients in the *P–S*v system for various angles of incidence (*i*), calculated for YAG. The angle of incidence producing a symmetric P-to-S conversion in the shear buffer rods, shown by the black arrow, is 59.5°.

where the angles *i* and *j* are measured between the incident P and converted S-wave from the normal of the conversion facet. For the converted shear-wave to propagate as a pure shear-wave (i.e. no compressional component or splitting), it must travel in another pure-mode direction (e.g. [010] for the cubic crystal buffer rod). The S-wave will be orthogonal to the incident P-wave when $i + j = 90°$. Substituting $i = 90° - j$ into Eq. (1) and using the identity $\sin(90° - j) = \cos(j)$ yields:

$$\frac{\sin(i)}{\cos(i)} = \tan(i) = \frac{V_P^{[100]}}{V_S^{[010]}}, \tag{2}$$

which gives the angle of incidence (*i*) producing an orthogonal conversion in any cubic crystal. For YAG, we measured $V_P^{[100]}$ and $V_S^{[010]}$ velocities on the bench and determined $i = 59.5°$. The amplitude-normalized reflection coefficients for this geometry are about 1.0 for P-to-S and S-to-P (Fig. 3). We have observed a single shot within the buffer rod converted (and re-converted) up to ~14 times before dissipation. Another useful feature P-to-S conversion buffer rod is that the polarization of the shear-wave is known precisely; being defined by direction of the incident P-wave because there is no rotation of the polarization vector across the conversion.

3. Instrumentation

The high-frequency ultrasonic interferometer consists of four main components that can be purchased separately off the shelf: (1) a radio-frequency (RF) generator or microwave synthesizer, (2) a pulse generator – either internal to the microwave synthesizer or external to the RF signal generator, (3) a high-frequency digitizing main-frame oscilloscope with trigger and channel set, and (4) a PC with IEEE 488.2 interface to drive the system and save the raw amplitude data.

The signal generator provides a highly stable RF source to the interferometer. This is akin to the use of a laser in an optical interferometry, and allows the measure ultrasonic travel times to a small part of an interference fringe, or about 1 part in 10^4 if the round-trip travel-time through a sample is about 100 ns (samples 0.3–0.5 mm thick) and to about 1 part in 10^3 if the sample thickness is an order of magnitude shorter (in the DAC). The GHz-system uses the *Gigatronics 6062A* (formerly FLUKE) signal generator, with 10 MHz to 2.1 GHz bandwidth, 0.1 kHz resolution, output power up to 12 dB, and high-stability 10-MHz quartz-oscillator reference.

The continuous output from the RF generator is gated by a pulse generator, providing phase-coherent pulses of programmable duration and delay. The GHz-system uses the Stanford Research Systems *DG535* Digital Delay/Pulse Generator, but in practice the instrument could be internal to a microwave synthesizer provided that it can produce gated signals as short as ~100 ns in duration and with appropriately fast rise times of about 3–5 ns.

The digitizing oscilloscope has GHz-bandwidth and picosecond resolution. The interferometer now at the Geophysical Lab uses the HP54122T system (i.e. HP54120B scope with the HP54122A 18 GHz trigger module). In Bayreuth and Potsdam, the HP54750 system is used (i.e. HP54750A scope with the 20 GHz HP54751A trigger module). The systems provide a maximum of 500 ps/div resolution (5 ns full window) when the trigger rate is 10 kHz.

4. Data collection

Ultrasonic P- and S-waves are delivered to the sample through the buffer rod and one of the [100]-oriented diamond anvils. Acoustic coupling between the buffer rod and the diamond is achieved by optically aligning the buffer rod (discussed in Section 5) and applying a small mechanical force. The sample must in-turn have good acoustic coupling to the acoustic diamond. A gated sinusoidal signal typically ~100 ns in duration and containing ~100 cycles is introduced into the system at a rate of about 10 kHz. Impedance contrasts at the buffer rod–diamond interface, the diamond–sample interface, and at the sample–pressure medium interface produce reflections which are detected by the source transducer. The entire round trip through the system is typically about 2.5 μs. An echo train at 1 GHz from the shear-wave buffer rod attached to the diamond cell is shown in Figure 4a. The time difference between the diamond and the first sample echo is usually very short (8–20 ns) compared to the duration of the input signal (100 ns), resulting in a complex echo containing both signals, as well as higher order sample signals (Fig. 4b). An interference pattern is produced by measuring the amplitude of the combined signal at a position where there is first-order interference between the diamond and sample echoes and scanning the frequency (Fig. 4c). Travel times are determined from each fitted frequency maxima and minima of interference by experimentally fitting the integer

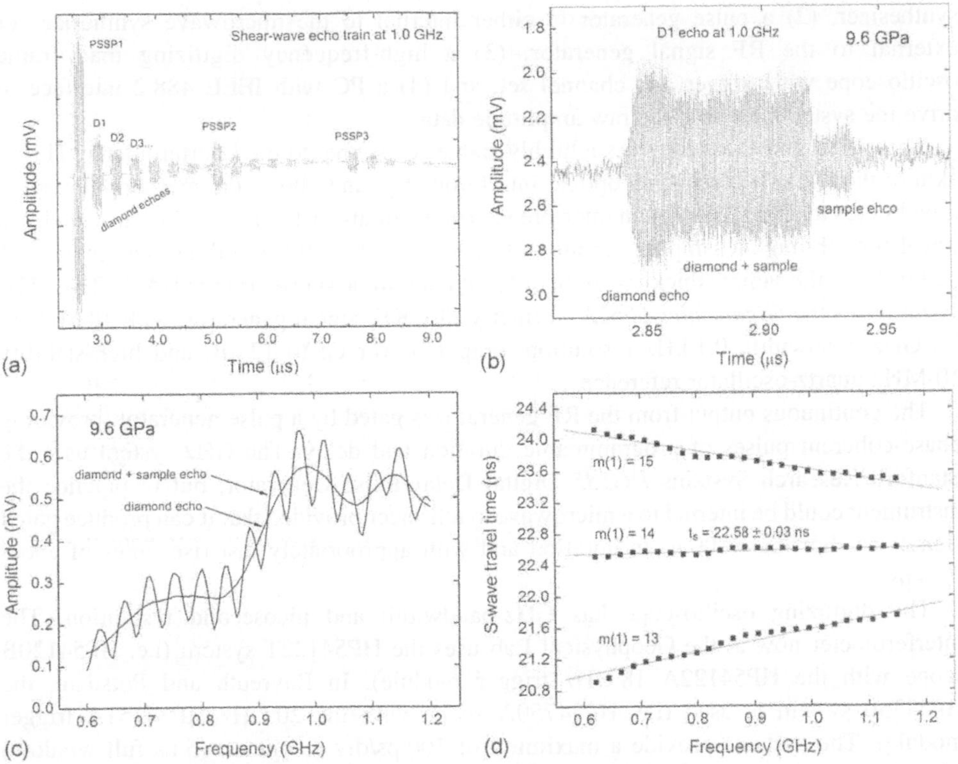

Figure 4. Shear-wave interferometry data from a crystal of $(Mg_{0.22}Fe_{0.78})O$ at ~9.6 GPa in the diamond anvil cell. (a) Echo train at 1 GHz showing reflections internal to the buffer rod (labeled PSSP) and multiple S-wave echoes in the diamond (labeled D1, D2, etc.). (b) Expansion of the D1 echo, about 100 ns in duration, reveals the interference between the diamond and sample echoes. The amplitude is measured as a function of frequency at two positions, first before the sample echo arrives (diamond echo) and secondly where there is first-order interference between the diamond and sample. These two curves are shown and labeled in (c). The diamond echo curve is used to demodulate the raw interference pattern from system responses. (d) S-wave travel times are calculated from each fitted frequency maximum and minimum for several values of $m(1)$, the fitted integer number of wavelengths in the round trip for the first extremum.

number of wavelengths in the round trip through sample (Fig. 4d). Additional details on data reduction are given by Spetzler et al. (1993).

5. Ultrasonic diamond cells

GHz-ultrasonic interferometry can be interfaced to nearly any type of diamond cell, provided that the buffer rod fits into the conical access of the seat for direct contact to the back of the anvil. In the hydrothermal DAC (Bassett et al., 2000), ultrasonic interferometry has been carried out to ~9 GPa at room temperature (Reichmann and Jacobsen, 2004), and to 500 K at ~3 GPa (Jacobsen et al., 2002c). A modified three-pin Merill-Bassett diamond cell has also been used for ultrasonic measurements to ~10 GPa (Jacobsen et al., 2004b).

We have also fitted the acoustical instruments to a miniature six-pin DAC with resistive whole-cell heater (Dubrovinskaia and Dubrovinsky, 2003) for future measurements at high $P-T$. All of the GHz-ultrasonic diamond cells are highly versatile and can be used also for X-ray diffraction or spectroscopy between ultrasonic measurements.

Alignment of the buffer rods is achieved by visually inspecting the polished flat of the buffer rod against the acoustic anvil using a zoom stereo microscope with on-axis illumination. While the diamond cell is still open, the polished flat of the buffer rod, typically about 250–300 μm in diameter is carefully brought close to the diamond table. Newton interference fringes between the flat and the diamond are used to guide the buffer rod parallel to the diamond table. Parallelism is achieved using a spherical rocker on the buffer rod attachment, while translation is made with set screws. Finally, once the buffer rod flat appears gray or black against the diamond, the buffer rod is pushed against the diamond with an advancing screw. The buffer rod alignment jig is attached to the DAC with a separate tight-fitting ring so the alignment procedure need only be carried out once before the experiment while the cell is still open. In this way, the P-wave and S-wave buffer rods can be switched out at each pressure without the dangerous re-alignment procedure when the cell is closed. If the end of a buffer rod is broken, both ends must be re-polished to achieve parallelism and the transducer must be re-sputtered.

The samples are prepared as oriented single-crystal plates; polished with parallel faces and flat to about $\sim 1/10\lambda$ with thicknesses ranging from about 30 to 50 μm. The plates have a finishing polish of optical quality, and are placed directly on the culet of the acoustic transmitting anvil. Samples are oriented with single-crystal X-ray diffraction, resulting in a typical uncertainty of about $\pm 1°$, due mainly to transferring of the crystal from the diffractometer to a glass slide for polishing. The number of orientations required to obtain the complete tensor depends on the crystal system of the mineral. In the cubic system, just one crystal oriented on [110] could be used to determine all three c_{ij} because there is one compressional and two shear modes, but it is preferable to have two or three pure-mode directions for redundancy (see Eqs. (5)–(11) below). The number of measurements required to obtain all the c_{ij} increases with decreasing symmetry such that six different orientations would be required to obtain the nine unique c_{ij} in the orthorhombic system. Although we have not used GHz-ultrasonic interferometry to study lower symmetry minerals, it is only a matter of effort and time in sample preparation.

Samples are kept in place on the acoustic anvil by adding a small amount of silica aerogel to the usual 16:3:1 methanol:ethanol:water pressure-transmitting medium. Aerogel is a highly compressible porous material with the lowest density of any known solid ($\rho \sim 0.1$ g/cm^3). Once wetted, the gel acts like a transparent sponge to gently press the sample against the anvil, while the overall pressure-transmitting medium remains alcohol.

With liquid pressure mediums, a fairly standard setup is used for the 10 GPa range: 500 μm-diameter flat culets, 250 μm-thick steel or rhenium gasket pre-indented to 80–100 μm thickness and a 250 μm-diameter hole for samples that are typically 30–50 μm thick and 150–200 μm wide. Pressures are determined using the ruby fluorescence scale (Piermarini et al., 1975; Mao et al., 1986). Pressures greater than 10 GPa are routinely reached, but the aerogel–alcohol mixture is fluid to only about 9 GPa, where the glass transition has been detected by the onset of shear-wave transmission completely through the sample chamber (Fig. 5).

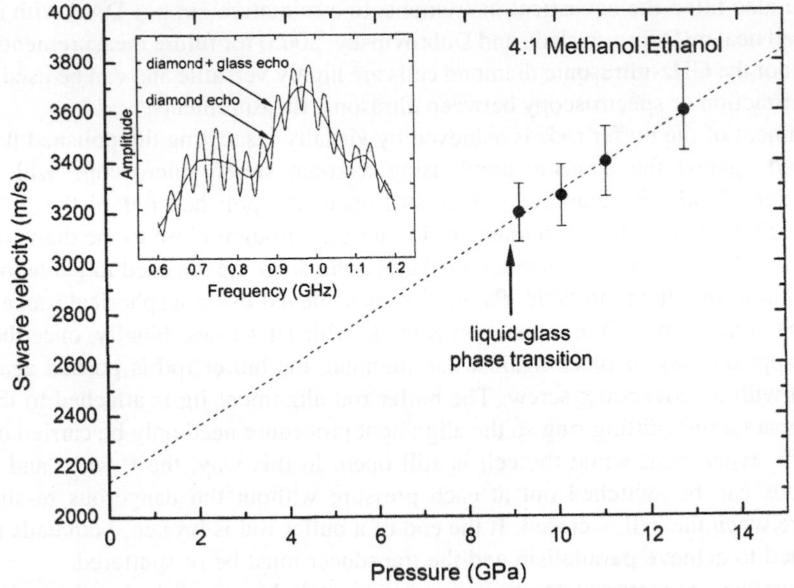

Figure 5. GHz-ultrasonic interferometry has also been used to study glasses at high pressure. A DAC was loaded with a 4:1 methanol–ethanol pressure medium and compressed without a sample. The onset of shear-wave propagation through the pressure medium (at the glass transition) was detected by the appearance of a diamond–diamond acoustic interference spectra (shown inset) at ~9.2 GPa. Variation of the shear-wave velocity in the 4:1 methanol–ethanol glass was estimated from measured distance between the diamond culets.

6. Calculating elastic constants and moduli from single-crystal ultrasonic data

This section contains an informal cookbook for calculating elastic constants and aggregate moduli from single-crystal travel-time data in the cubic system because they will be applied to the oxides in Section 7. For a thorough treatment of tensor properties of single crystals, the reader is referred to the classic texts on the subject (e.g. Musgrave, 1970; Nye, 1972). Compressional (V_P) and shear (V_S) velocities are related to the sample length (L) and the measured round-trip travel times (t) by:

$$V = (2L/t) \tag{3}$$

At ambient pressure, the length can be measured with a micrometer if the sample is sufficiently thick (i.e. >~0.1 mm). In the DAC, samples are typically only 30–50 μm in length, so often the initial length is calculated from the measured zero-pressure travel-time using a known velocity measured on a thicker sample. In the case that a thick sample is not available, the initial thickness can be measured with a laser interferometer. The elastic constants at ambient or high pressures are determined from the acoustic travel times from similar identities:

$$c_{ij} = c_{ij}^0(\rho/\rho_0)(L/L_0)^2(t/t_0)^2 \tag{4}$$

where the zero subscript or superscript indicates an initial (ambient pressure) value, ρ is the density, and the subscripts *ij* indicate the element of the elastic tensor corresponding to the various pure-mode directions in the single crystal. In the cubic system:

$$c_{11} = \rho(V_{\mathrm{P}}^{[100]})^2 \tag{5}$$

$$c_{44} = \rho(V_{\mathrm{S}}^{[100]})^2 \tag{6}$$

$$c_{11} + 4c_{44} + 2c_{12} = 3\rho(V_{\mathrm{P}}^{[111]})^2 \tag{7}$$

$$c_{11} + c_{44} - c_{12} = 3\rho(V_{\mathrm{S}}^{[111]})^2 \tag{8}$$

$$c_{11} + 2c_{44} + c_{12} = 2\rho(V_{\mathrm{P}}^{[110]})^2 \tag{9}$$

$$c_{11} - c_{12} = 2\rho(V_{\mathrm{S}}^{[110]}\mathrm{pol}^{[-110]})^2 \tag{10}$$

$$c_{44} = \rho(V_{\mathrm{S}}^{[110]}\mathrm{pol}^{[001]})^2 \tag{11}$$

The superscripts to V indicate the crystallographic direction of wave propagation and the superscripts to *pol* indicate the direction of shear-wave polarization if it is not degenerate. Readers looking for equations for pure-modes in one of the other six crystal systems are directed to Brugger, (1965). The change in length with pressure in a cubic mineral is given by:

$$L(P) = L_0[\rho(P)/\rho_0]^{-1/3} \tag{12}$$

where ρ(P) can either be measured *in situ* with X-ray diffraction, or, calculated from a known $P-V$ equation of state such as the third-order Birch-Murnaghan:

$$P = \frac{3}{2}\left[\left(\frac{\rho}{\rho_0}\right)^{7/3} - 1\right]\left(\frac{\rho}{\rho_0}\right)^{5/3} K_{\mathrm{T0}}\left\{1 + \frac{3}{4}(K_0' - 4)\left[\left(\frac{\rho}{\rho_0}\right)^{2/3} - 1\right]\right\} \tag{13}$$

where K_{T0} is the zero-pressure isothermal bulk modulus and $K_0{}'$ is the first pressure derivative. While the $P-V$ equation of state is a practical way to calculate the length-change of well-characterized materials (with known K_{T0} and K'), ideally the cell parameters should be measured at each pressure for the complete travel-time equation of state (Spetzler and Yoneda, 1993; Li et al., 2001).

Once the full elastic tensor is known (c_{11}, c_{44}, c_{12}), the isotropic equivalent or polycrystalline bulk (K_{S}) and shear (G) moduli can be calculated. The methods of Voigt and Reuss assume uniform strain, or uniform stress, respectively. Hill (1952) showed that the Voigt and Reuss limits represent upper and lower bounds of the isotropic moduli, so the widely used Voigt–Reuss–Hill (VRH) method simply refers to the average, although there is no physical basis for it. The Voigt and Reuss bounds are equivalent for the bulk modulus:

$$K_{\mathrm{S}} = 1/3(c_{11} + 2c_{12}) \tag{14}$$

The Voigt (G_{V}) and Reuss (G_{R}) bounds on the shear modulus are:

$$G_{\mathrm{V}} = 1/5(2c_{\mathrm{S}} + 3c_{44}) \tag{15}$$

$$G_R = \frac{5c_S c_{44}}{2c_{44} + 3c_S} \tag{16}$$

where

$$c_S = 1/2(c_{11} - c_{12}) \tag{17}$$

Hashin and Shtrikman (1962) formalized a variational approach to polycrystalline averaging of elastic constants that were shown to place much tighter bounds on the elastic moduli. In the cubic system, the Hashin–Shtrikman (HS) bounds in the cubic system are given by:

$$G_{HS}\text{lower} = c_S + 3\left(\frac{5}{c_{44} - c_S} - 4\beta_1\right)^{-1} \tag{18}$$

$$G_{HS}\text{upper} = c_{44} + 2\left(\frac{5}{c_S - c_{44}} - 6\beta_2\right)^{-1} \tag{19}$$

where

$$\beta_1 = -\left(\frac{3(K + 2c_S)}{5c_S(3K + 4c_S)}\right) \tag{20}$$

$$\beta_2 = -\left(\frac{3(K + 2c_{44})}{5c_{44}(3K + 4c_{44})}\right) \tag{21}$$

HS bounds for aggregates of crystals with lower symmetry have been derived for the orthorhombic (Watt, 1979), hexagonal, trigonal, tetragonal (Watt and Peselnick, 1980) and even monoclinic systems (Watt, 1980). Once K_S and G have been calculated, the aggregate sound velocities can be calculated from the well-known equations:

$$V_P = \sqrt{\left(\frac{K_S + 4/3G}{\rho}\right)} \tag{22}$$

$$V_S = \sqrt{\left(\frac{G}{\rho}\right)} \tag{23}$$

as well as other useful forms of the polycrystalline moduli, such as Young's modulus (E) and Poisson's ratio (υ)

$$E = \frac{9KG}{3K + G} \tag{24}$$

$$\upsilon = \frac{3K - 2G}{2(3K + G)} \tag{25}$$

7. High-pressure elasticity of some iron-oxide minerals

Oxygen constitutes roughly 43% of the crust and mantle by weight, making it the most abundant element in the planet. The Earth-forming minerals are basically oxides of Si, Fe and Mg with minor amounts of Al, Ca, and Na, but those with Si are classified as silicates. After silicates, oxides are the next most abundant minerals of the crust and mantle. Due to their structural simplicity, single and binary oxides have played an important role in deriving empirical concepts of crystal chemistry such ionic radii (Shannon and Prewitt, 1970) and bond compressibility (Hazen et al., 2000). In this section, we present recent results from GHz-ultrasonic interferometry on the single-crystal elasticity of several iron-rich oxides to ~10 GPa, including magnetite (Reichmann and Jacobsen, 2004), magnesiowüstite-(Mg,Fe)O (Jacobsen et al., 2004b), and wüstite-FeO (Kantor et al., 2004a).

7.1. Magnetite

Magnetite (Fe_3O_4) occurs naturally in both primary and secondary mineralization throughout Earth's uppermost mantle and crust, forming spinel solid solutions at high temperature with hercynite ($FeAl_2O_4$) and ulvöspinel (Fe_2TiO_4), among others. It is also one of the most important ferrimagnetic materials in industry. Magnetite has the inverse spinel structure (*Fd*3*m*) and is stable to at least 10 GPa at 1000 K (Haavik et al., 2000). Magnetite undergoes a sluggish first-order phase transformation starting at ~21 GPa, but there remains some uncertainty on the exact structure of the high-pressure modification. Recent *in situ* diffraction experiments suggest the high-pressure form of magnetite (h-Fe_3O_4) is orthorhombic (*Bbmm*) with the $CaTi_2O_4$-type structure (Haavik et al., 2000; Dubrovinsky et al., 2003).

There is an unusually large range in reported values for the bulk modulus (K_{0T}) of magnetite from isothermal compression studies of powders. Throughout the literature reviewed by Haavik et al. (2000), values for K_{0T} range from about 140 to 220 GPa, with K' ranging from 4 to over 7.5. There is no clear explanation for the broad range in observed values of K_{0T} for magnetite, so further studies on the high-pressure behavior of magnetite, and especially those determining the elastic constants (c_{ij}) might elucidate this unusual behavior. Isida et al. (1996) reported P-wave velocities along the [100], [111] and [110] directions to 1.2 GPa, and in the temperature range from 123 to 280 K, but without any S-wave velocities. We have determined the pressure dependence of all three single-crystal elastic constants (c_{11}, c_{12} and c_{44}) and the isotropic equivalent bulk (K_S) and shear (G) moduli for magnetite to ~9 GPa in the DAC using GHz-ultrasonic interferometry (Reichmann and Jacobsen, 2004).

The magnetite used in this study was a natural untwined and euhedral single-crystal from the Ural Mountains at Yekaterinburg, Russia, with an approximate composition of $(Fe_{2.99}Al_{0.007}Ti_{0.003})O_4$ from microprobe analysis. The measured cell parameter is $a = 8.39639(14)$ Å, for a calculated density of $\rho = 5.196(1)$ g/cm^3 (Table 1).

Prior to the high-pressure experiments, the room-pressure sound velocities of magnetite were determined using a 398 μm-thick (100) plate and a 325 μm-thick (111) plate. The sample lengths were measured with a micrometer, and have an estimated

Table 1. Elastic properties of several iron-oxide minerals from gigahertz ultrasonic interferometry.

	ρ_0 (kg/m^3)	c_{11} (GPa)	c_{44} (GPa)	c_{12} (GPa)	K_S (GPa)	G (GPa)	V_P (m/s)	V_S (m/s)
Magnetite[a]								
Fe_3O_4	5196(5)	261(1)	63(2)	148(3)	186(3)	60(3)	7157(30)	3407(20)
dM/dP		5.1(1)	−0.13(4)	5.4(1)	5.1(1)	−0.1(1)	47.4(4)	−12.5(3)
Ferropericlase								
$(Mg_{0.76}Fe_{0.24})O$	4157(6)	266(2)	123(1)	114(4)	165(2)	102(3)	8500(25)	4942(23)
dM/dP		9.3(2)	1.2(1)	1.3(6)	4.17[b]	2.7(1)	80(1)	48(2)
Magnesiowüstite								
$(Mg_{0.44}Fe_{0.56})O$	4847(19)	244(2)	83(1)	117(3)	159(2)	75(1)	7312(22)	3937(15)
dM/dP		9.6(4)	−0.16(9)	1.5(4)	4.17[b]	1.5(1)	61(1)	26(1)
Wüstite[c]								
$Fe_{0.95}O$	5721(6)	218(1)	46.1(2)	122(1)	153(1)	46.8(4)	6138(25)	2853(12)
dM/dP ($P < 4.7$ GPa)		10.5(3)	−1.13(3)	3.5(7)	6.1(4)	0.6(1)	76(1)	9(2)
dM/dP ($4.7 < P < 10$ GPa)		6.6(2)	−1.13(3)	0.27(7)	3.7(3)	−0.22(3)	26(1)	−14.0(1)

[a] Reichmann and Jacobsen (2004).
[b] Fixed from Kung et al. (2002); used to calculate c_{12} and G from the data of Jacobsen et al. (2002b, 2004b).
[c] Jacobsen et al. (2002b); Kantor et al. (2004a).

accuracy of ± 1 μm. The resulting velocities are $V_P^{[100]} = 7081 \pm 19$ m/s and $V_P^{[111]} = 7214 \pm 24$ m/s. The S-wave velocity in the [100] direction is $V_S^{[100]} = 3496 \pm 22$ m/s. The elastic constants c_{ij} were calculated from Eqs. (5)–(8). We obtain: $c_{11} = 260.5 \pm 1.0$, $c_{44} = 63.3 \pm 1.5$ and $c_{12} = 148.3 \pm 3.0$ GPa, resulting in $K_S = 185.7 \pm 3.0$ GPa and $G_0 = 60.3 \pm 3.0$ GPa using the VRH average.

The longitudinal sound velocity in the [111] direction was obtained only to a pressure of 6.3 GPa, but $V_P^{[100]}$ and $V_S^{[100]}$ were determined to 8 GPa or higher, so the $V_P^{[111]}$ data were extrapolated to 10 GPa using a linear fit in order to fit the complete elastic tensor to the highest pressure. Pressure derivatives of the elastic constants of magnetite are: $dc_{11}/dP = 5.14 \pm 0.13$, $dc_{12}/dP = 5.39 \pm 0.10$ and $dc_{44}/dP = -0.13 \pm 0.4$. Pressure derivatives of the VRH-averaged adiabatic bulk and shear moduli are $K_S{}' = 5.1 \pm 0.1$ and $G' = -0.1 \pm 0.1$. Variation of density and sample length with pressure were determined from its isothermal $P-V$ equation of state, measured on the same sample (during a separate experiment) using single-crystal X-ray diffraction, resulting in EoS parameters $K_{0T} = 180 \pm 1$ GPa and $K' = 5.2 \pm 0.4$. The elastic constants and aggregate moduli of magnetite are plotted as a function of pressure in Figure 6.

The reported bulk modulus of magnetite from high-pressure powder diffraction studies, often going up to the transition pressure at ~20 GPa, give values for K_T that range from about 141 to 222 GPa, with $K_T{}'$ varying from 4 to 7.5 (Haavik et al., 2000). Single-crystal compression studies listed by Haavik et al. (2000) and including this one (Reichmann and Jacobsen, 2004) result in a much smaller range in K_T between ~180 and 190 GPa, which fall perhaps coincidently near the mean value of that from powder studies. It is possible that a non-uniform strain field acting on the grains of a compressed powder might introduce errors in the compression data of magnetite. Dubrovinsky (personal communication) reported seeing wavy Debye–Scherrer lines in the unrolled image plate projections of diffraction patterns from powdered magnetite compressed in a DAC. Also, the single-crystal data are typically measured at lower pressures and under quasi-hydrostatic conditions (up to 10 GPa), so it is possible that some gradual change is occurring closer to the structural transition to cause large variations in the $P-V$ data.

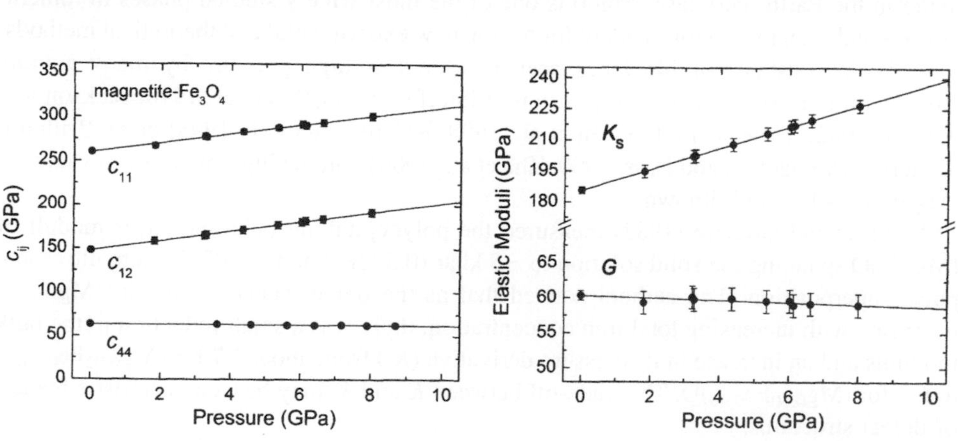

Figure 6. Variation of the elastic constants (left) and aggregate moduli (right) of magnetite with pressure from Reichmann and Jacobsen (2004). A linear pressure-induced c_{44} soft mode is observed in magnetite with $dc_{44}/dP = -0.13 \pm 0.04$ over this pressure range.

Another potential reason for the discrepancies in K_T and K' amongst experimental studies may be due to varying chemical compositions between synthetic and natural samples.

Shear-mode softening in magnetite reported here at high pressure is consistent with that observed on decreasing temperature (below 300 K), well ahead of the first-order, cubic-to-monoclinic (insulator-to-metal) Verwey transition at 120 K (Verwey, 1939; Isida et al., 1996; Schwenk et al., 2000; Gasparov et al., 2000). The mode softening in magnetite at low temperature appears to result from a bilinear coupling of the elastic strain to a charge ordering process between Fe^{2+} and Fe^{3+} on the octahedral site, as is observed in analog compounds (Schwenk et al., 2000; Paolone et al., 2003), but it is unlikely that increasing pressure would drive such charge coupling processes. Instead, we speculate that the negative slope of c_{44} throughout the experimental pressure range results from magnetoelastic interactions potentially driving the first-order phase transition above 21 GPa. It is suggested for future studies to measure magnetic ordering in magnetite as a function of pressure, perhaps with neutron scattering.

Yoneda (1990) and Chopelas (1996) reported the pressure dependence of the shear modulus for $MgAl_2O_4$ spinel of 0.36 and 0.072, respectively. The relatively low values of dG/dP for spinel, and the negative values of dc_{44}/dP and dG/dP for magnetite obtained in this study highlight the relatively low kinetic barrier for structural phase transformations in spinels. A linear extrapolation of our magnetite c_{44} data to higher pressures puts the high-pressure phase transformation to over 50 GPa, being much higher than the observed onset around 21 GPa. This indicates that at higher pressures c_{44} should decrease non-linearly with pressure.

7.2. Magnesiowüstite

Magnesiowüstite-(Mg,Fe)O is the dense magnesium–iron monoxide coexisting with silicate perovskite-(Mg,Fe,Al)SiO_3 at the P–T conditions of Earth's lower mantle (e.g. Shim et al., 2001). (Mg,Fe)O is therefore believed to be the most abundant non-silicate oxide in the Earth. Periclase (MgO) is one of the most widely studied phases in mineral physics and is an important standard for testing new experimental and theoretical methods. The elastic constants of MgO has been measured at high pressure by many authors either with ultrasonic pulse superposition or interferometry (Spetzler, 1970; Jackson and Niesler, 1982; Yoneda, 1990; Chen et al., 1998; Reichmann et al., 1998) or by Brillouin scattering (Sinogeikin and Bass, 2000; Zha et al., 2000), but the high-pressure elasticity of (Mg,Fe)O is less well known.

Bonczar and Graham (1982) measured the polycrystalline bulk and shear moduli of (Mg,Fe)O spanning the solid solution to ~5 kbar (0.5 GPa) and to 100°C using ultrasonic pulse superposition. The authors noticed that as the defect concentration in (Mg,Fe)O increased with increasing total iron concentration, there was a small reduction in the bulk modulus and an increase in its pressure derivative (K') from about 3.7 for $(Mg_{0.92}Fe_{0.08})O$ to 6.2 for $(Mg_{0.60}Fe_{0.40})O$. The trade-off between K and K' may be a characteristic feature of defect structures.

Kung et al. (2002) measured the elasticity of polycrystalline $(Mg_{0.83}Fe_{0.17})O$ to 9 GPa with ultrasonic interferometry in the multi-anvil press and obtain $K_S = 165.5 \pm 1.2$ GPa and $G = 112.4 \pm 0.4$ GPa, with $K' = 4.17(20)$ and $G' = 1.89(6)$. These values for

K and K' do not differ significantly from MgO, but G and G' are reduced by 14 and 24%, respectively, from MgO (Kung et al., 2002). The elastic tensor (and thus anisotropy) of (Mg,Fe)O has not previously been measured at high pressure, although measurements on (Mg,Fe)O containing about 6% iron, single-crystal Brillouin scattering are underway (Jackson et al., 2003).

The addition of iron into MgO has a profound affect on the elastic constants at ambient conditions (Fig. 7a). Jacobsen et al. (2002a) measured c_{ij} across the (Mg,Fe)O solid solution and found a non-linear reduction in c_{11} and c_{44} of 27 and 70%, respectively, while c_{12} increases by 32% between MgO and FeO. The effect of iron on the elasticity of (Mg,Fe)O is most pronounced up to about 30 mol% FeO, the composition where the c_{44} and c_{12} terms cross. In going from MgO to FeO, the anisotropy factor, $A = [(2c_{44} + c_{12})/c_{11}] - 1$ is reduced from 0.36 ± 0.02 to about zero (isotropic) for FeO. Jackson et al. (1990) measured the elastic constants of FeO to ~3 GPa using the ultrasonic phase comparison technique and showed that c_{44} in FeO is a soft mode with $dc_{44}/dP = -1.03 \pm 0.02$. The reduction in c_{44} by about 7%

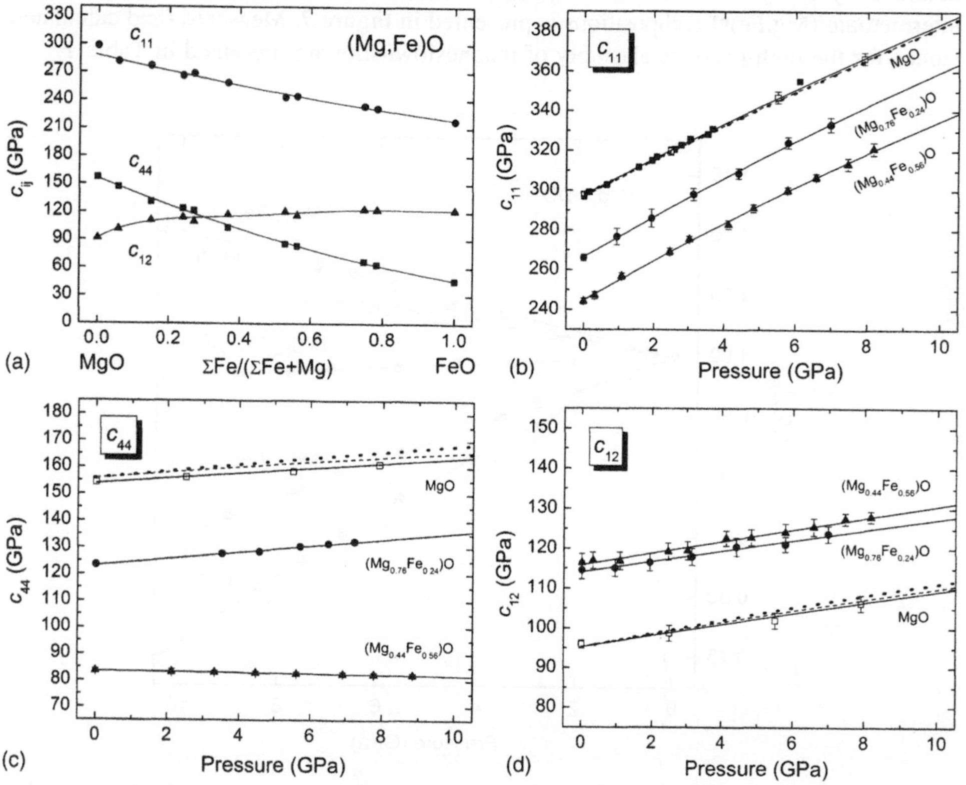

Figure 7. (a) Elastic tensor of (Mg,Fe)O as a function of composition from Jacobsen et al. (2002a). (b–d) Various terms of the elastic tensor as a function of pressure for several compositions of (Mg,Fe)O. Filled circles and triangles with solid-line fits are data from this study. Filled squares in (b) for MgO are also data from GHz-ultrasonic interferometry (Reichmann et al., 1998). Open squares with solid-line fits for MgO are data from Brillouin scattering (Sinogeikin and Bass, 2000). Additional ultrasonic data for MgO are shown by the dashed curves (Jackson and Niesler, 1982) and dotted curves (Yoneda, 1990).

at 3 GPa is due to strong magnetoelastic coupling driving the *B*1 to rhombohedral at around 17 GPa (Jackson et al., 1990; Mao et al., 1996; Struzhkin et al., 2001). This poses the question; if c_{44} decreases with pressure for FeO, and increases for MgO, what is the evolution of dc_{44}/dP with composition and thus potential phase transitions in (Mg,Fe)O?

We measured the c_{44} elastic constant of FeO and (Mg,Fe)O containing about 24, 56, and 78 mol% FeO to a maximum pressure of 9.6 GPa (Jacobsen et al., 2004a; Fig. 8). The pressure derivative of c_{44} changes from 1.2 ± 0.1 for $(Mg_{0.76}Fe_{0.24})O$ to -1.12 ± 0.02 for $Fe_{0.95}O$, switching sign at about 50 mol% FeO. The sign reversal in dc_{44}/dP points to a change in the topology of the (Mg,Fe)O phase diagram at around 50–60 mol% FeO, consistent with recent studies of high-pressure phase transitions in (Mg,Fe)O (Lin et al., 2002). We also measured the pressure dependence of the c_{11} elastic constant for $(Mg_{0.76}Fe_{0.24})O$ and $(Mg_{0.44}Fe_{0.56})O$ (Jacobsen et al., 2002b). Using our high-pressure data for c_{11} and c_{44}, we calculated the c_{12} elastic constant to 10 GPa from our measured values of K_S on the same samples (Jacobsen et al., 2002a) and the pressure derivative K' obtained by Kung et al. (2002) for $(Mg_{0.17}Fe_{0.83})O$. The complete tensor for the two intermediate (Mg,Fe)O compositions is presented in Figure 7. Measured (and calculated) values for the high-pressure elasticity of magnesiowüstite are presented in Table 1.

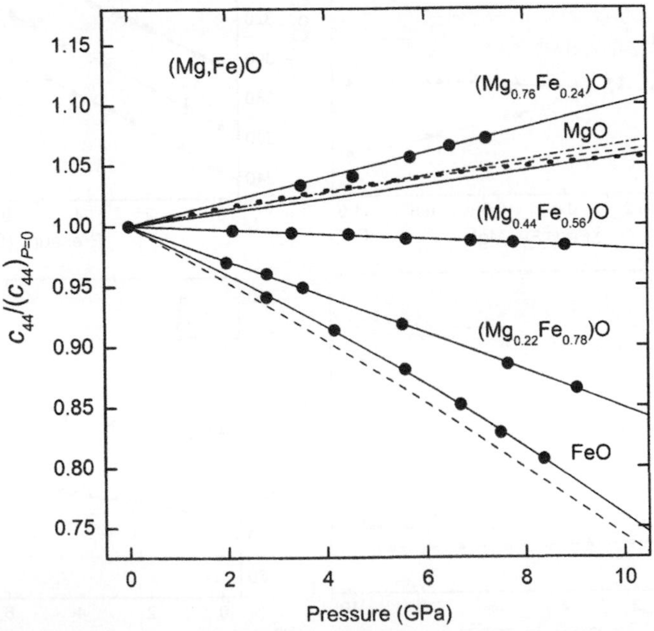

Figure 8. Variation of $c_{44}/(c_{44})_{P=0}$ in (Mg,Fe)O as a function of composition and pressure after Jacobsen et al. (2004b). Data from this study are shown by filled circles with solid-line fits. Data for MgO from Brillouin scattering is shown by the solid curve (Sinogeikin and Bass, 2000) and dash-dotted curve (Zha et al., 2000). Additional data for MgO from ultrasonics is shown by the dashed curve (Jackson and Niesler, 1982) and dotted curve (Yoneda, 1990). The dashed curve for FeO comes from (Jackson et al., 1990). This term of the elastic tensor is most sensitive to the rhombohedral distortion observed in FeO at ~17 GPa (Mao et al., 1996) and in $(Mg_{0.25}Fe_{0.75})O$ above ~60 GPa (Lin et al., 2002).

7.3. *FeO-wüstite*

The iron monoxide FeO is an important member of the highly correlated transition metal oxide group including NiO, CoO, and MnO. $Fe_{1-x}O$ is the classic non-stoichiometric oxide, where x typically ranges from 0.90 to 0.95 at room pressure. Stoichiometries as high as 0.98 or 0.99 have been achieved at pressures above 10 GPa (Zhang, 2000). The properties and defect structure of $Fe_{1-x}O$ have been reviewed elsewhere (Hazen and Jeanloz, 1984).

A single crystal of FeO was synthesized by the floating zone technique (Berthon et al., 1979). The cell parameter was measured by single-crystal X-ray diffraction is $a = 4.3068(1)$ Å, corresponding to $Fe_{0.946}O$ (McCammon and Liu, 1984). After initial bench-top experiments (Jacobsen et al., 2002b), polished pieces oriented on [100] and [111] measuring about 40 μm in thickness were prepared for high-pressure work.

Measured single-crystal travel-time data $t_P^{[100]}$, $t_S^{[100]}$, and $t_S^{[111]}$ were converted to sound wave velocities using the calculated sample thickness bench-top velocities measurements using Eq. (1). An isothermal equation of state for FeO (Hazen, 1981) was used to calculate the change in sample length with pressure. The cubic elastic constants (c_{11}, c_{12}, c_{44}) were calculated using Eqs. (5)–(8) and are plotted in Figure 9. The isotropic equivalent bulk and shear moduli were calculated using the variational approach of Hashin and Strikman and are plotted in Figure 9.

A discontinuity was detected in the pressure derivatives of c_{11} and c_{12} at 4.7 ± 0.2 GPa (Fig. 9). We tested and confirmed the reversibility of the anomaly by measuring $t_P^{[100]}$ on both compression and decompression. This unusual behavior is also clear in the isotropic moduli (Fig. 9). The reason for this discontinuity may be due to the onset magnetic

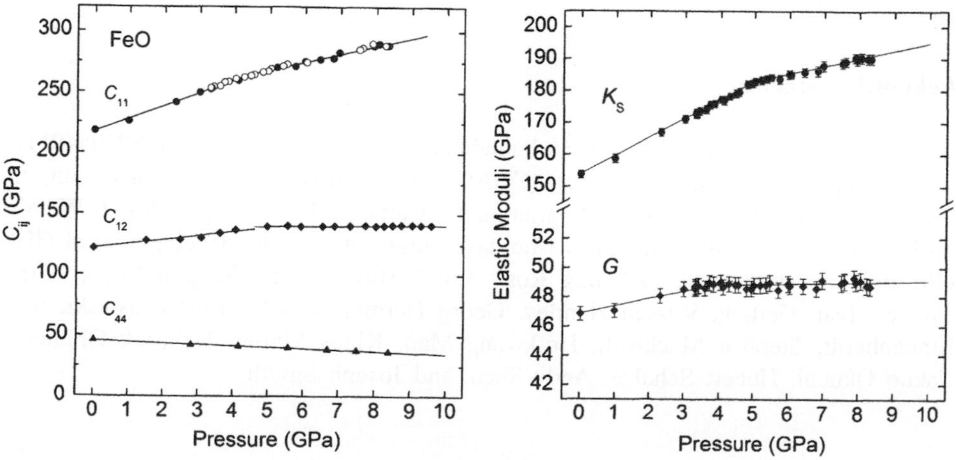

Figure 9. Elastic constants and aggregate moduli of wüstite to ~10 GPa after Kantor et al., (2004a). A discontinuity in the c_{11} and c_{12} pressure derivatives at around 4.7 GPa is interpreted to result from the onset of magnetic ordering in $Fe_{0.95}O$, which observed at ~5 GPa by high-pressure ^{57}Fe Mössbauer spectroscopy (Kantor et al., 2004b). The reversibility of the transition was tested by taking data on compression (filled circles) and decompression (open circles) during the c_{11} measurement. The results indicate there may be a partially ordered although still cubic phase of FeO between ~5 GPa and where the cubic–rhombohedral phase transition occurs at ~17 GPa (Mao et al., 1996).

ordering, which as been observed in high-pressure Mössbauer studies to begin at 5.0 ± 0.5 GPa (Kantor et al., 2004a,b). The results indicate that at room temperature and between ~5 and ~17 GPa there is an intermediate partially ordered but still cubic phase of FeO.

8. Summary

In this chapter, we have reported several major advances in gigahertz-ultrasonic interferometry having occurred during the past 2 years. A pilot study on San Carlos olivine to 500 K at ~3 GPa demonstrates that the acoustic signals can be maintained at high $P-T$ in a fluid pressure medium (Jacobsen et al., 2002c). The generation of shear-waves with near-optical wavelengths has been achieved using a P-to-S conversion (Jacobsen et al., 2002a), and S-waves have been transmitted into a DAC using a new high-pressure shear-wave buffer rod based on that principle (Jacobsen et al., 2004b). The pressure range over which ultrasonic interferometry has been carried out in the diamond cell has been effectively doubled to ~10 GPa, and the complete tensor of several important iron-oxide minerals has been measured over this pressure range (Reichmann and Jacobsen, 2004; Kantor et al., 2004a). The sound velocities and elastic constants of high-pressure crystals synthesized in the multi-anvil press can now be measured with GHz-ultrasonic interferometry (Jacobsen et al., 2004a,b). The versatility of GHz-ultrasonic interferometry to work with all kinds of diamond cells and samples of any optical character holds exciting potential for future studies aimed at measuring elastic anisotropy in high-pressure silicates, oxides and metals in the 30–50 GPa range or higher with application to problems in geophysics and materials science.

Acknowledgements

Supported by the National Science Foundation (EAR-0135540 and EAR-0440112), the Carnegie/DOE Alliance Center (CDAC), the German Science Foundation, the Bayerisches Geoinstitut Visitors Program and a Carnegie Fellowship to S.D.J. We also wish to thank the many individuals who have contributed to the development of GHz-ultrasonic interferometry, including Ross Angel, Bill Bassett, Ganglin Chen, Jeffrey Cooper, Ivan Getting, Russell Hemley, Georg Hermannsdörfer, Kurt Klasinski, Sven Lindenhardt, Stephen Mackwell, Ho-kwang Mao, Klaus Müller, Heinrick Ohlmeyer, Takuo Okuchi, Hubert Schulze, Andy Shen, and Joseph Smyth.

References

Abramson, E.H., Brown, J.M., Slutsky, L.J., 1999. Application of impulsive stimulated scattering in the Earth and planetary sciences. Annu. Rev. Phys. Chem. 50, 279–313.

Bassett, W.A., Reichmann, H.J., Angel, R.J., Spetzler, H., Smyth, J.R., 2000. New diamond anvil cells for gigahertz ultrasonic interferometry and X-ray diffraction. Am. Mineral. 85, 283–287.

Berthon, J., Revcolevschi, A., Morikawa, H., Touzelin, B., 1979. Growth of wüstite $Fe_{1-x}O$ crystals of various stoichiometries. J. Cryst. Growth 47, 736–738.

Birch, F., 1952. Elasticity and constitution of the Earth's interior. J. Geophys. Res. 57, 227–285.

Bonczar, L.J., Graham, E.K., 1982. The pressure and temperature dependence of the elastic properties of polycrystal magnesiowüstite. J. Geophys. Res. 87, 1061–1078.

Brugger, K., 1965. Pure modes for elastic waves in crystals. J. Appl. Phys. 36, 759–768.

Chai, M., Brown, J.M., Slutsky, L.J., 1997. The elastic constants of a pyrope–grossular–almandine garnet to 20 GPa. Geophys. Res. Lett. 24, 523–526.

Chen, W.P., Brudzinski, M.R., 2003. Seismic anisotropy in the mantle transition zone beneath Fiji-Tonga. Geophys. Res. Lett. 30, 1682, doi: 10.1029/2002GL016330.

Chen, G., Li, B., Liebermann, R.C., 1996a. Selected elastic moduli of single-crystal olivines from ultrasonic experiments to mantle pressures. Science 272, 979–980.

Chen, G., Yoneda, A., Getting, I.C., Spetzler, H.A., 1996b. Cross pressure and temperature derivatives of selected elastic moduli for olivine from gigahertz ultrasonic interferometry. J. Geophys. Res. 101, 161–171.

Chen, G., Miletich, R., Mueller, K., Spetzler, H.A., 1997. Shear and compressional mode measurements with GHz ultrasonic interferometry and velocity-composition systematics for the pyrope-almandine solid solution series. Phys. Earth Planet. Interiors 99, 273–287.

Chen, G., Liebermann, R.C., Weidner, D.J., 1998. Elasticity of single-crystal MgO to 8 gigapascals and 1600 Kelvin. Science 280, 1913–1916.

Chopelas, A., 1996. The fluorescence sideband method for obtaining acoustic velocities at high pressure: application to MgO and $MgAl_2O_4$. Phys. Chem. Miner. 23, 25–37.

Christensen, N.I., Medaris, L.G. Jr., Wang, H.F., Jelínek, E., 2001. Depth variation of seismic anisotropy and petrology in central European lithosphere: a tectonothermal synthesis from spinel lherzolite. J. Geophys. Res. 106, 645–664.

Crowhurst, J.C., Abramson, E.H., Slutsky, L.J., Brown, M.J., Zaug, J.M., Harrell, M.D., 2001. Surface acoustic waves in the diamond anvil cell: an application of impulsive stimulated light scattering. Phys. Rev. B 64, doi: 10.1103/PhysRevB.64.100103.

Crowhurst, J.C., Goncharov, A.F., Zaug, J.M., 2004. Impulsive stimulated light scattering from opaque materials at high pressure. J. Phys.: Condens. Matter. 16, doi: 10.1088/0953-8984/16/14/023.

Dubrovinskaia, N., Dubrovinsky, L., 2003. Whole-cell heater for the diamond anvil cell. Rev. Sci. Instrum. 74, 3433–3437.

Dubrovinsky, L., Dubrovinskaia, N., Saxena, S., LiBehan, T., 2000. X-ray diffraction under non-hydrostatic conditions in experiments with diamond anvil cell: wüstite (FeO) as an example. Mater. Sci. Eng. 288, 187–190.

Dubrovinsky, L.S., Dubrovinskaia, N.A., McCammon, C., Rozenberg, G.Kh., Ahuja, R., Osorio-Guillen, J.M., Dmitriev, V., Weber, H.P., Le Bihan, T., Johansson, B., 2003. The structure of the metallic high-pressure Fe_3O_4 polymorph: experimental and theoretical study. J. Phys.: Condens. Matter 15, 7696–7706.

Duffy, T.S., Zha, C.S., Downs, R.T., Mao, H.K., Hemley, R.J., 1995. Elasticity of forsterite to 16 GPa and the composition of the upper mantle. Nature 378, 170–173.

Ekström, G., Dziewonski, A.M., 1998. The unique anisotropy of the Pacific upper mantle. Nature 394, 168–172.

Fiquet, G., Badro, J., Guyot, F., Requardt, H., Krisch, M., 2001. Sound velocities in iron to 110 gigapascals. Science 291, 468–471.

Fiquet, G., Badro, J., Guyot, F., Bellen, C., Krisch, M., Antonangeli, D., Requardt, H., Mermet, A., Farber, D., Aracne-Ruddle, C., Zhang, J., 2004. Application of inelastic X-ray scattering to measurements of acoustic wave velocities in geophysical materials at very high pressure. Phys. Earth Planet. Interiors 143–144, 5–18.

Gasparov, L.V., Tanner, D.B., Romero, D.B., Berger, H., Margaritondo, G., Forro, L., 2000. Infrared and Raman studies of the Verwey transition in magnetite. Phys. Rev. 62, 7939–7944.

Gauthier, M., Lheureux, D., Decremps, F., Fischer, M., Itié, J.P., Syfosse, G., Polian, A., 2003. High-pressure ultrasonic setup using the Paris–Edinburgh press: elastic properties of single crystalline germanium up to 6 GPa. Rev. Sci. Instrum. 74, 3712–3716.

Goncharov, A.F., Crowhurst, J., Zaug, J.M., 2004. Elastic and vibrational properties of cobalt to 120 GPa. Phys. Rev. Lett. 92, doi: 10.1103/Phys. Rev. Lett. 92.115502.

Goto, T., Anderson, O.L., Ohno, I., Yamamoto, S., 1989. Elastic constants of corundum up to 1825 K. J. Geophys. Res. 94, 7588–7602.

Haavik, C., Stolen, S., Fjellvag, H., Hanfland, M., Häusermann, D., 2000. Equation of state of magnetite and its high-pressure modification: thermodynamics of the Fe–O system at high pressure. Am. Mineral. 85, 514–523.

Hashin, Z., Shtrikman, A., 1962. A variational approach to the theory of the elastic behavior of polycrystals. J. Mech. Phys. Solids 29, 81–95.

Hazen, R.M., 1981. Systematic variation of bulk modulus of wüstite with stoichiometry, Year Book, Vol. 80. Carnegie Institute Washington, Washington, pp. 277–280.

Hazen, R.M., Jeanloz, R., 1984. Wüstite ($Fe_{1-x}O$): a review of its defect structure and physical properties. Rev. Geophys. Space Phys. 22, 37–46.

Hazen, R.M., Downs, R.T., Prewitt, C.T., 2000. Principles of comparative crystal chemistry. In: Hazen, R.M., Downs, R.T. (Eds), High-temperature and High-pressure Crystal Chemistry, Rev. Miner. Geochem., 41, pp. 1–33.

Hill, R., 1952. The elastic behavior of a crystalline aggregate. Proc. Phys. Soc. London 64, 349–354.

Isaak, D.G., Ohno, I., 2003. Elastic constants of chrome-diopside: application of resonant ultrasound spectroscopy to monoclinic single-crystals. Phys. Chem. Miner. 30, 430–439.

Isaak, D.G., Anderson, O.L., Goto, T., 1989. Measured elastic moduli of single-crystal MgO up to 1800 K. Phys. Chem. Miner. 16, 704–713.

Isida, S., Suzuki, M., Todo, S., Mori, N., Siratori, K., 1996. Pressure effect on the elastic constants of magnetite. Physica B 219–220, 638–640.

Jackson, I., Niesler, H., 1982. The elasticity of periclase to 3 GPa and some geophysical implications. In: Akimoto, S., Manghnani, M.H. (Eds), High-Pressure Research in Geophysics. Center for Academic Pub., Japan, pp. 93–113.

Jackson, I., Khanna, S.K., Revcolevschi, A., Berthon, J., 1990. Elasticity, shear-mode softening and high-pressure polymorphism of wüstite ($Fe_{1-x}O$). J. Geophys. Res. 95, 21671–21685.

Jackson, J.M., Bass, J.D., Sinogeikin, S.V., Jacobsen, S.D., Reichmann, H.J., Mackwell, S.J., 2003. High-pressure Brillouin measurements on single-crystal ferropericlase, $(Mg_{0.94}Fe_{0.06})O$: implications for Earth's lower mantle. Eos Trans. Am. Geophys. Union 84 (46), Fall Meeting Suppl. Abstract T21A-07.

Jacobsen, S.D., Reichmann, H.J., Spetzler, H.A., Mackwell, S.J., Smyth, J.R., Angel, R.J., McCammon, C.A., 2002a. Structure and elasticity of single-crystal (Mg,Fe)O and a new method of generating shear waves for gigahertz ultrasonic interferometry. J. Geophys. Res. 107, 2037, doi: 10.1029/2001JB000490.

Jacobsen, S.D., Spetzler, H.A., Reichmann, H.J., Mackwell, S.J., Smyth, J.R., 2002b. Effects of iron and pressure on the c_{11} elastic constant of (Mg,Fe)O using a new GHz-ultrasonic diamond cell with in-situ X-ray diffraction to 10 GPa. Eos Trans. Am. Geophys. Union 83 (47), Fall Meeting Suppl. Abstract MR62B-1074.

Jacobsen, S.D., Spetzler, H.A., Reichmann, H.J., Smyth, J.R., Mackwell, S.J., Angel, R.J., Bassett, W.A., 2002c. Gigahertz ultrasonic interferometry at high P and T: new tools for obtaining a thermodynamic equation of state. J. Phys.: Condens. Matter 14, 11525–11530.

Jacobsen, S.D., Smyth, J.R., Spetzler, H.A., Holl, C.A., Frost, D.J., 2004a. Sound velocities and elastic constants of iron-bearing hydrous ringwoodite. Phys. Earth Planet. Interiors 143–144, 47–56.

Jacobsen, S.D., Spetzler, H.A., Reichmann, H.J., Smyth, J.R., 2004b. Shear waves in the diamond-anvil cell reveal pressure-induced instability in (Mg,Fe)O. Proc. Natl Acad. Sci. USA 101, 5867–5871.

Kantor, A., Jacobsen, S.D., Kantor, I.Y., Dubrovinsky, L.S., McCammon, C.A., Reichmann, H.J., Goncharenko, I.N., 2004a. Pressure-induced magnetization in FeO: evidence from elasticity and Mössbauer spectroscopy. Phys. Rev. Lett. 93, 215502, doi: 10.1103/PhysRevLett. 93.215502.

Kantor, I.Y., McCammon, C.A., Dubrovinsky, L.S., 2004b. Mössbauer spectroscopic study of pressure-induced magnetization in wüstite (FeO). J. Alloys Compd. 376, 5–8.

Klotz, S., 2001. Phonon dispersion curves by inelastic neutron scattering to 12GPa. Z. Kristallogr. 216, 420–429.

Kung, J., Li, B., Weidner, D.J., Zhang, J., Liebermann, R.C., 2002. Elasticity of $(Mg_{0.82}Fe_{0.17})O$ ferropericlase at high pressure: ultrasonic measurements in conjunction with X-radiation techniques. Earth Planet. Sci. Lett. 203, 557–566.

Li, B., Liebermann, R.C., Weidner, D.J., 2001. $P-V-V_p-V_s$–t measurements on wadsleyite to 7 GPa and 873 K: implications for the 410-km seismic discontinuity. J. Geophys. Res. 106, 30575–30591.

Lin, J.F., Heinz, D.L., Mao, H.K., Hemley, R.J., Devine, J.M., Li, J., Shen, G., 2002. Stability of magnesiowustite in Earth's lower mantle. Proc. Natl Acad. Sci. USA 100, 4405–4408.

Lin, J.F., Sturhahn, W., Zhao, J., Shen, G., Mao, H.K., Hemley, R.J., 2005. Nuclear resonant inelastic X-ray scattering and synchrotron Mössbauer spectroscopy with laser-heated diamond anvil cells. In: Chen, J., Wang, Y., Duffy, T.S., Shen, G., Dobrzhinetskaya, L.F. (Eds), Advances in High-Pressure Technology for Geophysical Applications. Elsevier, Amsterdam, 397–411.

Mayama, N., Suzuki, I., Saito, T., Ohno, I., Katsura, T., Yoneda, A., 2004. Temperature dependence of elastic moduli of β-$(Mg,Fe)_2SiO_4$. Geophys. Res. Lett. 31, L04612, doi: 10.1029/2003GL019247.

Mao, H.K., Xu, J.A., Bell, P.M., 1986. Calibration of the ruby pressure gauge to 800 kbar under quasi-hydrostatic conditions. J. Geophys. Res. 91, 4673–4676.

Mao, H.K., Shu, J., Fei, Y., Hu, J., Hemley, R.J., 1996. The wüstite enigma. Phys. Earth Planet. Interiors 96, 135–145.

Mao, H.K., Shu, J., Shen, G., Hemley, R.J., Li, B., Singh, A.K., 1998. Elasticity and rheology of iron above 220 GPa and the nature of the Earth's inner core. Nature 396, 741–743.

Mao, H.K., Kao, C., Hemley, R.J., 2001a. Inelastic X-ray scattering at ultrahigh pressures. J. Phys.: Condens. Matter 13, 7847–7858.

Mao, H.K., Xu, J., Struzhkin, V.V., Shu, J., Hemley, R.J., Sturhahn, W., Hu, M.Y., Alp, E.E., Vocadlo, L., Alfè, D., Price, G.D., Gillan, M.J., Schwoerer-Böhning, M., Häusermann, D., Eng, P., Shen, G., Giefers, H., Lübbers, R., Wortmann, G., 2001b. Phonon density of states of iron up to 153 gigapascals. Science 292, 914–916.

McCammon, C.A., Liu, L.G., 1984. The effects of pressure and temperature on nonstoichiometric wustite, Fe_xO; the iron-rich phase-boundary. Phys. Chem. Miner. 10, 106–113.

McSkimin, H.J., 1950. Ultrasonic measurement techniques applicable to small solid specimens. J. Acoust. Soc. Am. 22, 413–418.

Musgrave, M.J.P., 1970. Crystal Acoustics. Holden-Day, Inc., San Francisco, CA, p. 288.

Nye, J.F., 1972. Physical Properties of Crystals. Oxford University Press, Oxford, UK, p. 322.

Occelli, F., Krisch, M., Loubeyre, P., Sette, F., Le Toullec, R., Masciovecchio, C., Rueff, J.P., 2001. Phonon dispersion curves in argon single crystal at high pressure by inelastic X-ray scattering. Phys. Rev. B 63, doi: 10.1103/PhysRevB.63.224306.

Ohno, I., Abe, M., Kimura, M., Hanayama, Y., Oda, H., Suzuki, I., 2000. Elasticity measurements of silica glass under gas pressure. Am. Mineral. 85, 288–291.

Paolone, A., Cantelli, R., Rousse, G., Masquelier, C., 2003. The charge order transition and elastic/anelastic properties of $LiMn_2O_4$. J. Phys.: Condens. Matter 15, 457–465.

Piermarini, G.J., Block, S., Barnet, J.D., Forman, R.A., 1975. Calibration of the pressure dependence of the R_1 ruby fluorescence line to 195 kbar. J. Appl. Phys. 46, 2774–2780.

Reichmann, H.J., Jacobsen, S.D., 2004. High-pressure elasticity of a natural magnetite crystal. Am. Mineral. 89, 1061–1066.

Reichmann, H.J., Angel, R.J., Spetzler, H., Bassett, W.A., 1998. Ultrasonic interferometry and X-ray measurements on MgO in a new diamond anvil cell. Am. Mineral. 83, 1357–1360.

Schwenk, H., Bareiter, S., Hinkel, C., Luethi, B., Kakol, Z., Koslowski, A., Honig, J.M., 2000. Charge ordering and elastic constants in $Fe_{3-x}Zn_xO_4$. Eur. Phys. J. B13, 491–494.

Shannon, R.D., Prewitt, C.T., 1970. Revised values of effective ionic radii. Acta Crystallogr. B26, 1046–1048.

Shen, A.H., Reichmann, H.J., Chen, G., Angel, R.J., Bassett, W.A., Spetzler, H.A., 1998. GHz ultrasonic interferometry in a diamond anvil cell: P-wave velocities in periclase to 4.4 GPa and 207°C. In: Manghnani, M.H., Yagi, T. (Eds), Properties of Earth and Planetary Materials at High Pressure and Temperature, Geophysics Monograph 101. Am. Geophys. Union, Washington, DC, pp. 71–77.

Shen, G., Sturhahn, W., Alp, E.E., Zhao, J., Toellner, T.S., Prakapenka, V.B., Meng, Y., Mao, H.K., 2004. Phonon density of states in iron at high pressures and temperatures. Phys. Chem. Miner. 31, 353–359.

Shim, S.H., Duffy, T.S., Shen, G., 2001. The post-spinel transformation in Mg_2SiO_4 and its relation to the 660-km seismic discontinuity. Nature 411, 571–574.

Singh, A.K., Mao, H.K., Shu, J., Hemley, R.J., 1998. Estimation of single-crystal elastic moduli from polycrystalline X-ray diffraction at high pressure: application to FeO and iron. Phys. Rev. Lett. 80, 2157–2160.

Sinogeikin, S.V., Bass, J.D., 2000. Single-crystal elasticity of pyrope and MgO to 20 GPa by Brillouin scattering in the diamond cell. Phys. Earth Planet. Interiors 120, 43–62.

Sinogeikin, S.V., Lakshtanov, D.L., Nicholas, J.D., Bass, J.D., 2004. Sound velocity measurements on laser-heated MgO and Al_2O_3. Phys. Earth Planet. Interiors 143–144, 575–586.

Spetzler, H., 1970. Equation of state of polycrystalline and single-crystal MgO to 8 kilobars and 800 K. J. Geophys. Res. 75, 2073–2087.

Spetzler, H., Yoneda, A., 1993. Performance of the complete travel-time equation of state at simultaneous high pressure and temperature. Pure Appl. Geophys. 141, 379–392.

Spetzler, H.A., Chen, G., Whitehead, S., Getting, I.C., 1993. A new ultrasonic interferometer for determination of equation of state parameters of sub-millimeter single crystals. Pure Appl. Geophys. 141, 341–377.

Spetzler, H.A., Shen, A., Chen, G., Herrmannsdoerfer, G., Schulze, H., Weigel, R., 1996. Ultrasonic measurements in a diamond anvil cell. Phys. Earth Planet Interiors 98, 93–99.

Struzhkin, V.V., Mao, H.K., Hu, J., Schwoerer-Bohning, M., Shu, J., Hemley, R.J., Sturhahn, W., Hu, M.Y., Alp, E.E., Eng, P., Shen, G., 2001. Nuclear inelastic X-ray scattering of FeO to 48 GPa. Phys. Rev. Lett. 87, 255501, doi: 10.1103/PhysRevLett.87.255501.

van der Hilst, R.D., Kárason, H., 1999. Compositional heterogeneity in the bottom 1000 kilometers of Earth's mantle: toward a hybrid convection model. Science 283, 1885–1891.

Verwey, E.J.W., 1939. Electronic conduction of magnetite (Fe_3O_4) and its transition point at low temperatures. Nature 144, 327–328.

Watt, J.P., 1979. Hashin–Shtrikman bounds on the effective elastic moduli of polycrystals with orthorhombic symmetry. J. Appl. Phys. 50, 6290–6295.

Watt, J.P., 1980. Hashin–Shtrikman bounds on the effective elastic moduli of polycrystals with monoclinic symmetry. J. Appl. Phys. 51, 1520–1524.

Watt, J.P., Peselnick, L., 1980. Clarification of the Hashin–Shtrikman bounds on the effective elastic moduli of polycrystals with hexagonal, trigonal, and tetragonal symmetries. J. Appl. Phys. 51, 1525–1531.

Yoneda, A., 1990. Pressure derivatives of elastic constants of single crystal MgO and $MgAl_2O_4$. J. Phys. Earth 38, 19–55.

Zha, C.S., Mao, H.K., Hemley, R.J., 2000. Elasticity of MgO and a primary pressure scale to 55 GPa. Proc. Natl Acad. Sci. USA 97, 13494–13499.

Zhang, J., 2000. Effect of defects on the elastic properties of wüstite. Phys Rev. Lett. 84, 507–510.

Advances in High-Pressure Technology for Geophysical Applications
Jiuhua Chen, Yanbin Wang, T.S. Duffy, Guoyin Shen, L.F. Dobrzhinetskaya, editors

Chapter 3

Simultaneous equation of state, pressure calibration and sound velocity measurements to lower mantle pressures using multi-anvil apparatus

Baosheng Li, Jennifer Kung, T. Uchida and Yanbin Wang

Abstract

Multiple techniques for measuring acoustic travel times using ultrasonic interferometry, sample unit-cell volume using X-ray diffraction, and specimen length using X-radiography were adapted simultaneously to a double-stage, large-volume high-pressure apparatus, allowing measurements of acoustic velocities under high pressure and temperature. Combined analysis of the ultrasonic velocities and density data enables us to determine elastic properties and their pressure derivatives independent of pressure up to lower mantle conditions. Furthermore, sample pressure can be directly calculated using the measured velocity and density data and finite-strain equations of state and compared with those obtained from a pressure standard adjacent to the sample, providing a means to calibrate the pressure scales currently in use. Complete experimental procedures and data analysis are demonstrated using data on a polycrystalline wadsleyite sample to 20 GPa. Finite-strain fitting of wadsleyite data gives $K_{0S} = 174(2)$ *GPa,* $K'_0 = 4.2(1)$*,* $G_0 = 111(1)$ *GPa and* $G'_0 = 1.5(1)$ *for the elastic bulk and shear moduli , while the pressure calculated using these data differs from the NaCl scale by as much as* $\sim 12.0(1.2)\%$ *at 20 GPa. Using these techniques, it is possible to directly measure sound velocities of mantle minerals at lower mantle pressures to study the composition of the Earth's interior.*

1. Introduction

Seismic studies provide information about density, compressional (P) and shear (S) wave velocity profiles as a function of depth as well as their lateral variations as a function of depth. Successful interpretation of these seismic observations in terms of chemical composition, mineralogy, and/or temperature requires experimental and theoretical data on the elastic properties of Earth materials under elevated conditions of pressure and temperature. So far, elastic bulk moduli and their pressure and temperature dependence for many mantle and core materials have been obtained by static compression studies using X-ray diffraction. Shear properties, however, are still poorly understood, especially at high pressure and high temperature. Without the knowledge of shear properties, the sensitivity in distinguishing various Earth composition models is severely limited (Jackson and Rigden, 1998) using the seismic bulk sound speed data to constrain mantle composition.

Ultrasonic interferometry (e.g. McSkimin, 1950) has been used to obtain elastic bulk and shear properties of Earth materials at high pressure and/or high temperature (e.g. Gieske and Barsch, 1966; Schreiber and Anderson, 1966; Kumazawa and Anderson, 1969;

Jackson and Niesler, 1982; Fujisawa and Ito, 1985; Webb, 1989; Gwanmesia et al., 1990; Yoneda, 1990; Yoneda and Morioka, 1992; Fujisawa, 1998; Knoche et al., 1998; Li et al., 1998a; Mueller et al., 2002). In the last few years, the advancement of ultrasonic techniques in multi-anvil apparatus has enabled acoustic velocity measurements for many phases to mantle pressures using both polycrystalline and single crystal samples up to 14 GPa (Chen et al., 1996; Flesch et al., 1998; Gwanmesia et al., 1998; Liebermann and Li, 1998; Li and Liebermann, 2000; Liu et al., 2000). Recently, state-of-art techniques using combined acoustic, X-ray diffraction and X-ray imaging techniques in a DIA-type, cubic-anvil apparatus (SAM 85) have been described (e.g. Li et al., 1998b, 2004). Using these techniques, simultaneous P and S wave velocities and X-radiation measurements can be performed to 10 GPa and 1600 K on either single crystal or polycrystalline specimens (e.g. Liebermann and Li, 1998).

The feasibility of performing measurements to pressures above 10 GPa and temperatures above 1300 K using the Kawai type (MA-8), double-stage high-pressure apparatus has been demonstrated by earlier ultrasonic experiments (e.g. Knoche et al., 1998; Li et al., 1998a). Two major sources of uncertainty in these experiments are (1) the conventional method for determining pressure, (i.e. a calibration curve fitted through a few fixed points calibrated using known phase transformations) is not precise enough for the determination of pressure and temperature derivatives of the elastic moduli and cannot readily be applied at high temperatures, and (2) the estimation of the sample length at high pressure and high temperature using previous equation-of-state data is not reliable, especially when the sample environment significantly deviates from hydrostatic conditions and/or plastic deformation or other mechanisms (e.g. phase transition, melting) occur at high temperatures in which sample length change might become unpredictable.

In this chapter, we describe the implementation of the combined ultrasonic interferometry, X-ray diffraction and X-radiography techniques in the double-stage multi-anvil apparatus (T-25) installed at 13-ID-D, GSECARS, Advanced Photon Source (Argonne, IL, USA). Experimental results for a polycrystalline Mg_2SiO_4 wadsleyite are used to demonstrate a simultaneous study of equation of state, pressure calibration, and sound velocities using these combined techniques to extend our experimental studies on the elasticity of deep Earth materials towards the pressure and temperature conditions of the Earth's lower mantle.

2. Experimental set-up

2.1. High-pressure apparatus

Figure 1a shows the layout of the experimental facility installed at 13-ID-D/GSECARS for conducting the experiments described in this study. The key elements of this experimental

Figure 1. (a) Schematic diagram showing the layout of the experimental set-up for simultaneous acoustic, X-ray diffraction and X-ray imaging measurements in T-25 apparatus installed at beamline 13-ID-D, GSECARS, Advanced Photon Source. (b) Left: Pictures of the T-25 Modules outside the 1000-ton press; Right: Assembled WC cubic assembly sits on the bottom guide block and the cable for ultrasonic measurements. (c) Schematic diagram of the details of the cell assembly for the 10/4 cell assembly. Tc: thermocouple.

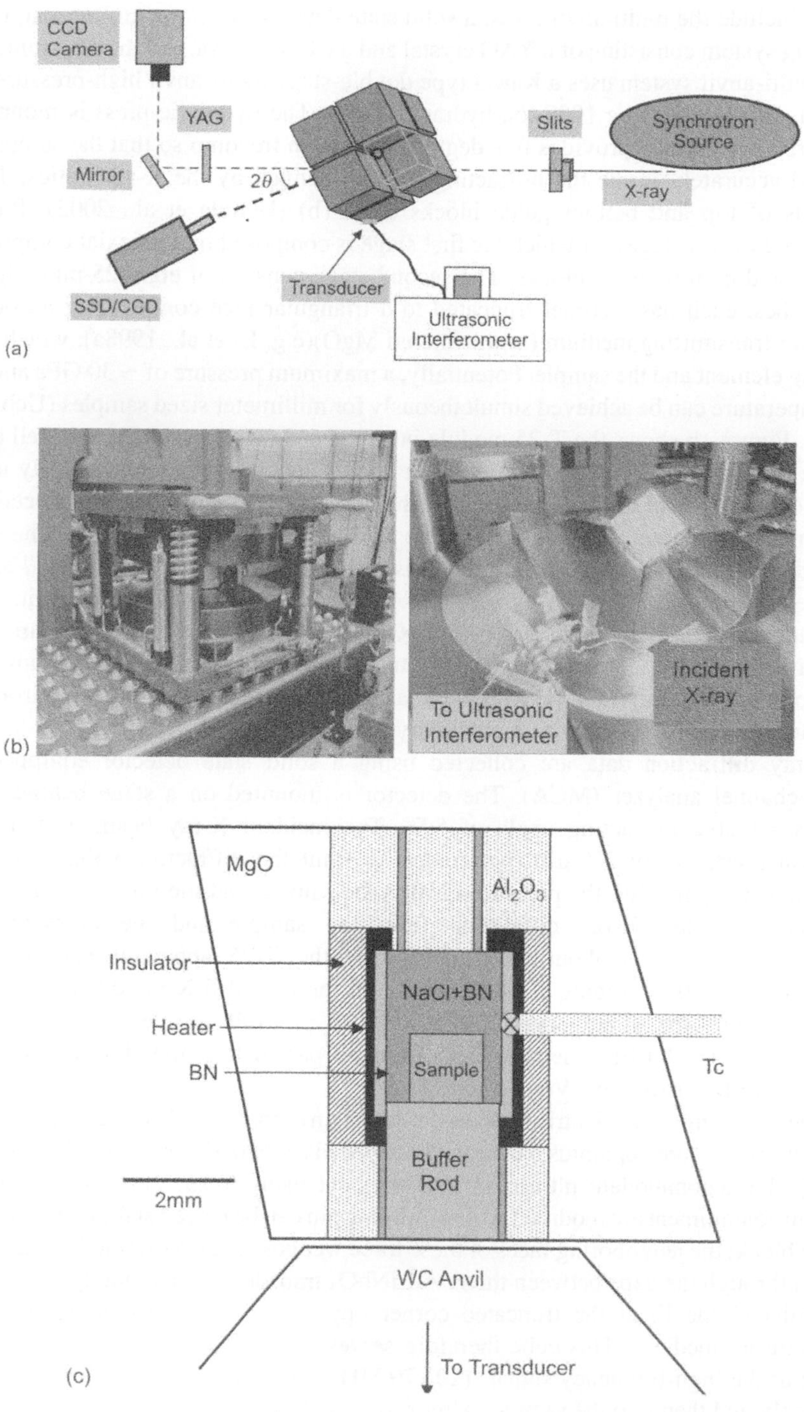
CCD Camera
YAG
Slits
Synchrotron Source
Mirror
2θ
X-ray
Transducer
SSD/CCD
Ultrasonic Interferometer
(a)
(b)
To Ultrasonic Interferometer
Incident X-ray
MgO
Al_2O_3
Insulator
NaCl+BN
Heater
BN
Sample
Tc
Buffer Rod
2mm
WC Anvil
To Transducer
(c)

set-up include the multi-anvil press, a solid state detector for X-ray diffraction, the X-ray imaging system consisting of a YAG crystal and a CCD camera, and the ultrasonic system. The multi-anvil system uses a Kawai type double-stage multi-anvil high-pressure module (T-25), compressed in a 1000-ton hydraulic press. The hydraulic press is mounted by a support frame, which provides five degrees of motion freedom so that the sample can be located accurately inside the diffracting volume defined by the X-ray optics. The T-25 consists of top and bottom guide blocks (Fig. 1b) (Uchida et al., 2002). Pressure is generated in two stages, of which the first stage is composed of a uniaxial compression of the top and bottom guide blocks. The second stage consists of eight 25-mm edge length WC cubes, each has a corner truncated to a triangular face compressing an octahedral pressure transmitting medium (semi sintered MgO)(e.g. Li et al., 1998a), which contains heating element and the sample. Potentially, a maximum pressure of ~30 GPa and 2500 K in temperature can be achieved simultaneously for millimeter sized samples (Uchida et al., 2002). Figure 1b shows the T-25 module installed and the assembled cubic cell assembly sitting on the lower guide block. Figure 1c shows details of the cell assembly inside the MgO octahedron for the 10/4 system, in which the edge length of the MgO octahedron is 10 mm and the corner of the WC cubes are truncated to 4 mm edge length. The sample is placed next to an acoustic buffer rod and is surrounded by a mixture of NaCl and boron nitride (BN) (NaCl:BN = 4:1 by weight). A cylindrical heater made of rhenium (25 μm thickness) and a thermal insulator ($LaCrO_3$ or ZrO_2) outside the heater are used for experiments at high temperatures. The mixture of NaCl plus BN around the sample serves as two important purposes: (1) to provide a pseudo-hydrostatic pressure environment for the sample, and (2) to serve as a secondary pressure standard.

X-ray diffraction data are collected using a solid state detector equipped with a multi-channel analyzer (MCA). The detector is mounted on a stage behind the press and is set at a diffracting angle of 5.5°. The incident X-ray beam is collimated to 100 μm (vertical) by 200 μm (horizontal) to limit the diffracting volume. During the experiment, by moving the press to position the sample and the pressure standard at the diffracting center, X-ray diffraction from the sample and the pressure standard are recorded. Details about the operation of the T-25 apparatus can be found in Uchida et al. (2002). Figure 2 is an example of the recorded X-ray diffraction pattern at high pressure for a polycrystalline wadsleyite. A full profile fitting using GSAS package and the La Bail method yields unit-cell parameters (therefore unit-cell volume and density) (Larson and Von Dreele, 1988).

The ultrasonic piezoelectric transducer–tungsten carbide anvil arrangement in a double-stage high pressure apparatus has been described elsewhere (Li et al., 1996; Knoche et al., 1998). To accommodate ultrasonic measurement using T-25, three wedges inside the bottom containment are modified so that a sliding pin can be placed at the center of the lower guide block; the neighboring faces of these three wedges are also trimmed so that wires can be fed through the gaps between them. A $LiNbO_3$ transducer is mounted on one WC cube (Toshiba Grade F) at the truncated corner opposed to the one compressing pressure transmitting medium. This cube therefore serves as an additional acoustic buffer rod to transmit the high-frequency signals (20–70 MHz) into the buffer rod enclosed in the cell assembly and then into the sample. After reaching the rear end of the sample, the acoustic waves will reverse the above propagation path and are received by the same transducer. The signals are recorded using an oscilloscope (Tektronix 5014, 5GS/s), consisting of a series of

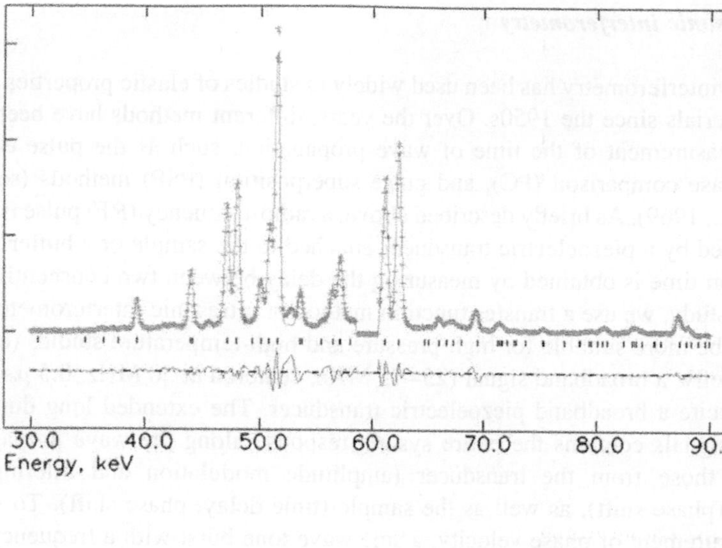
30.0
40.0
50.0
60.0
70.0
80.0
90.0
Energy, keV

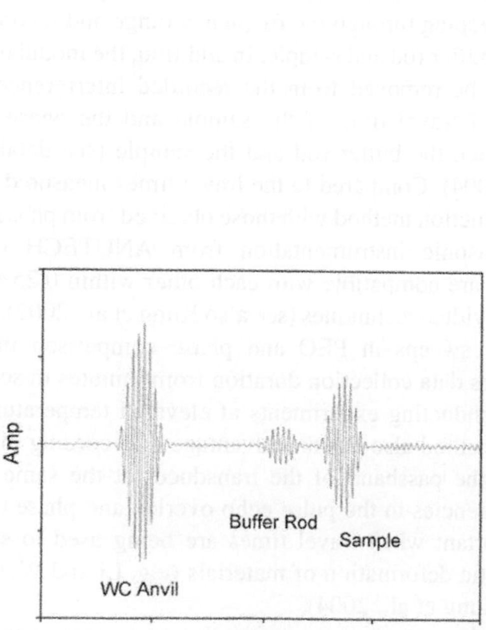
Amp
WC Anvil
Buffer Rod
Sample
27.0
27.5
28.0
28.5
29.0
Time Delay (µsec)

2.2. *Ultrasonic interferometry*

Ultrasonic interferometry has been used widely in studies of elastic properties of solid and liquid materials since the 1950s. Over the years, different methods have been developed for the measurement of the time of wave propagation, such as the pulse echo overlap (PEO), phase comparison (PC), and pulse superposition (PSP) methods (see details in Truell et al., 1969). As briefly described above, a radio-frequency (RF) pulse is transmitted and received by a piezoelectric transducer attached to the sample or a buffer rod, and the propagation time is obtained by measuring the delay between two consecutive echoes.

In this study, we use a transfer function method of ultrasonic interferometry which has proved to be more suitable for high-pressure and high-temperature studies (e.g. Li et al., 2002). Briefly, a broadband signal (25–75 MHz, centered at 50 MHz, 0.5 μs duration) is used to excite a broadband piezoelectric transducer. The extended long duration of the recorded signals contains the entire system response along the wave propagation path, including those from the transducer (amplitude modulation and filtering), bonding materials (phase shift), as well as the sample (time delay, phase shift). To simulate the PEO measurement of phase velocity, a sine wave tone burst with a frequency within the bandwidth of the input signal is convoluted with the transfer function, thus, the system response to such a single frequency input signal is reproduced (Fig. 3). The time delay caused by traveling inside the sample is obtained by measuring the time shift of the second copy of the echo pattern relative to the original one until the buffer rod and sample echoes overlap.

The phase comparison measurements can also be reproduced using the recorded transfer function by sweeping through the frequency range and recording the interference amplitude between the buffer rod and sample. In addition, the modulation of the transducer response envelope can be removed from the recorded interference spectrum, yielding precise determination of travel time of the sample and the phase shift caused by the bonding material between the buffer rod and the sample (see details in Spetzler et al., 1993; Li et al., 2002, 2004). Compared to the travel times measured for the same sample using current transfer function method with those obtained from phase comparison method using a different ultrasonic instrumentation from ANUTECH (Australian National University), the results are compatible with each other within 0.25%, comparable to the uncertainties of the individual techniques (see also Kung et al., 2002). However, compared to real-time frequency sweeps in PEO and phase comparison methods, the transfer function method reduces data collection duration from minutes to seconds. This is a very valuable attribute in conducting experiments at elevated temperatures and/or pressures. The transfer function method also has the advantage of recording the system response to all frequencies inside the passband of the transducer at the same time, in contrast to sweeping through frequencies in the pulse echo overlap and phase comparison methods; this is especially important when travel times are being used to study time-dependent processes, such as plastic deformation of materials (e.g. Li and Weidner, 1999) or phase transformations (e.g. Kung et al., 2004).

Another important feature of current experiment is that both P and S wave are measured at the same time in a single experiment. Normally, in experiments on polycrystalline samples, P and S wave travel times are measured using pure mode P and S wave transducers separately in different runs; one difficulty is that it is not simple, in practice, to ensure that the P and S travel times are measured at the same cell pressures in separate experimental runs. Alternatively, we used a dual-mode $LiNbO_3$ transducers (10° Y-cut) to simultaneously generate longitudinal and transverse acoustic signals despite the fact that the transverse signals have unspecified polarization direction. The WC cube (~35 mm acoustic path length along its diagonal direction) acts as a long delay line, resulting in well separated P and S wave echo trains in time axis. As demonstrated by Li et al. (2004), the travel times measured using dual mode transducers completely agree with those measured using pure mode transducers within 0.1% (see Figs 11 and 12 in Li et al., 2004).

2.3. *X-ray imaging*

The X-ray imaging system consists of a YAG crystal, a mirror, and a CCD camera as shown in Figure 1a. After recording the X-ray diffraction data, the incident slits are driven out of the way of the X-ray path to maximize the size of the X-ray beam. After the X-ray passes through the gap between WC cubes and the sample cell assembly, it illuminates the fluorescent YAG crystal behind the press. The visible light generated by the YAG crystal is captured by the CCD camera where an image of the cell assembly is obtained (Fig. 4). To enhance the visual effect, Figure 4 shows the negative of the original image.

In these X-radiographic images, the brightness contrast from region to region results from the difference in the X-ray absorption coefficients of the various materials in the high-pressure cell assembly. If the absorption coefficient for the sample is sufficiently different from those of the adjacent materials (buffer rod and NaCl), the sample region can be clearly distinguished from the rest of the cell assembly. In cases where the sample and the surrounding parts of the cell assembly have similar absorption coefficients, metal foils with high X-ray absorption (e.g. gold) are inserted at both the top and bottom surfaces of

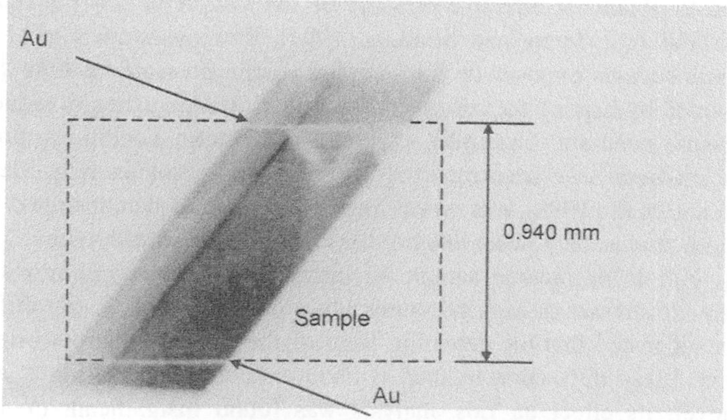

Figure 4. X-ray image of the wadsleyite specimen at 14 GPa (negative). The dashed line represents the approximate entire sample and the dark region is the sample region visible by X-rays through gaps between WC anvils. The white lines are gold foils placed at both surfaces of the sample.

the sample. In our ultrasonic measurements, such gold foils (0.002 mm thick) are routinely inserted at the sample/buffer rod interface to enhance the mechanical coupling between the buffer rod and sample; these also serve well in delineating the sample boundaries, as seen by the white lines (dark lines in original images) in Figure 4.

To calibrate the sample length with respect to pixels in an image, one can compare the pixel numbers from the last image (after the experiment has reached zero pressure) with the length of the recovered sample measured using a precise micrometer. This last image can then be used as the starting image to retrieve the change in pixels between neighboring $P-T$ conditions by cross-correlation of the center of the intensity profiles at sample boundaries. Typically, a change of 0.25 pixel can be resolved. For a sample with a dimension of about 1000 pixels, a precision to resolve length change at 2.5×10^{-4} (or 0.025%) can be achieved. In contrast, when the sample dimension in pixels is directly measured from an image, it has a total uncertainty about 2–4 pixels, which gives 0.2–0.4% in precision, regardless of the sample state at high pressure and high temperature.

3. Data analysis

3.1. X-ray diffraction data and equation of state

As illustrated in Figure 2, the unit-cell volumes at high pressures can be obtained by fitting the recorded X-ray diffraction spectrum. Typically, an uncertainty of 0.1% or less in the derived unit-cell volumes can be achieved. As in static compression studies, we can analyze these data using equations of state (e.g. Birch–Murnaghan) to constrain isothermal bulk modulus and its pressure derivative. In this approach, hydrostatic pressure and accurate measurements of sample pressures and volumes are crucial in deriving reliable equation of state results. Peak broadening, commonly observed in powder X-ray diffraction, can be used as a stress indicator (Warren, 1990). By analyzing the full width and half maximum (FWHM) of the diffractions lines, the measured pressures maybe corrected to obtain the effective pressure on the sample to perform reliable equation of state analysis (e.g. Meng and Weidner, 1993). Previous studies have shown that non-hydrostatic stresses imposed on the sample by solid pressure medium can be minimized and removed by heating the cell assembly to high temperatures to induce plastic flow of the pressure medium. Examples of quasi-hydrostatic conditions provided by NaCl pressure medium have been reported in previous Eos studies (e.g. Meng and Weidner, 1993; Wang et al., 1998). It is worth mentioning that, as demonstrated by a comparative peak width studies on powder and hot-pressed samples of stishovite (e.g. Li et al., 1994), starting with a hot-pressed sample in current experiment minimizes the build-up of intergranular stresses at high pressures observed in powder X-ray diffraction. Previous analysis suggested that the detecting limit of the deviatoric stress using current energy dispersive X-ray diffraction method is about 0.05 GPa (see Wang et al., 1998). At this stress level, its effect on Eos analysis was found insignificant (Wang et al., 1998). Successful examples of equation of state analysis using data from the combined acoustic and X-ray diffraction techniques can be found in previous studies on wadsleyite (Li et al., 1998a,b), magnesiowustite (Kung et al., 2002) and forsterite (Li et al., 2004).

3.2. Finite-strain analysis using combined X-ray diffraction and acoustic data

An advantage of current experiment is that both sample density and acoustic velocities at the same pressure and temperature conditions are measured in a single experiment, resulting in direct determination of the elastic moduli without using previously published density data. In addition, these measurements allow us to fit P and S wave velocity and density data simultaneously to a third-order finite-strain equation of state to obtain elastic moduli as well as their pressure derivatives. The finite-strain equations are expressed as the following (Davis and Dziewonski, 1975),

$$\rho V_P^2 = (1-2\varepsilon)^{5/2}(L_1 + L_2\varepsilon) \tag{1}$$

$$\rho V_S^2 = (1-2\varepsilon)^{5/2}(M_1 + M_2\varepsilon) \tag{2}$$

where

$$\varepsilon = 0.5\left[(1-\rho/\rho_0)^{2/3}\right] \tag{3}$$

$$L_1 = K_0 + (4/3)G_0 \tag{4}$$

$$L_2 = 5(K_0 + 4G_0/3) - 3K_0(K_0' + 4G_0'/3) \tag{5}$$

$$M_1 = G_0 \tag{6}$$

$$M_2 = 5G_0 - 3K_0G_0' \tag{7}$$

The fitted coefficients, L_1, L_2, M_1, and M_2, are used for the calculation of the bulk and shear moduli as well as their pressure derivatives at ambient conditions using Eqs. (4)–(7). The bulk modulus can be either isothermal (K_{0T}) or adiabatic (K_{0S}) depending on the conditions imposed. It is a common practice that data obtained from acoustic measurements at high pressure and room temperature are used directly in the fit to obtain the approximation of the adiabatic elastic moduli K_{0S} and G_0 and their pressure derivatives $(\partial K_{0S}/\partial P)_S$, and $(\partial G_0/\partial P)_S$ (e.g. Zha et al., 1998a, 2000; Kung et al., 2002; Mueller et al., 2002) (precisely, the derivative from such fitting is $(\partial K_{0T}/\partial P)_T$ rather than $(\partial K_{0S}/\partial P)_T$ see details in Section 3.3). Note that in these fittings, pressure is absent; therefore, the results on the elastic bulk and shear moduli are not affected by the accuracy of the pressure measurements using additional pressure markers enclosed in these experiments (see also Kung et al., 2002). Instead, the pressure can be directly calculated as discussed below.

3.3. Pressure calibration

With current data, not only we can determine the elastic moduli and their pressure derivatives without using pressure standard as described above, we can also directly determine the sample pressure using the measured density and velocities (e.g. Ruoff et al., 1973; Zha et al., 1998a, 2000). Following Birch (1952), the third-order finite-strain equations for bulk modulus and pressure can be expressed as the following,

$$K = K_0(1 - 2\varepsilon)^{5/2}(1 + (5 - 3K'_0)\varepsilon) \tag{8}$$

$$P = -3K_0\varepsilon(1 - 2\varepsilon)^{5/2}(1 + 3(4 - K'_0)\varepsilon/2) \tag{9}$$

Similar to Eqs. (1) and (2), Eqs. (8) and (9) can be applied to both isothermal and adiabatic conditions. Eq. (8) can be derived from Eqs. (1) and (2) using $K_S = \rho(V_P^2 - 4V_S^2/3)$. The adiabatic bulk modulus is related to its isothermal value by Eq. (10),

$$K_T = K_S/(1 + \alpha\gamma T) \tag{10}$$

where α is thermal expansivity, and γ is the gruneisen parameter. Combining Eqs. (8) and (10), for isothermal compressions, Eq. (8) can be written in the following form,

$$K_S/(1 + \alpha\gamma T) = K_{0T}(1 - 2\varepsilon)^{5/2}(1 + (5 - 3K'_{0T})\varepsilon) \tag{11}$$

in which the left-hand side represents the isothermal bulk modulus which is converted from the adiabatic bulk modulus K_S from ultrasonic measurement. The thermal expansivity at high pressure is obtained from the following equations,

$$\alpha = a_0(0, 300) + (d\alpha/dP)_T P \tag{12}$$

$$(d\alpha/dP)_T = (\partial K_T/\partial T)_P/K_T^2 \tag{13}$$

A least square fit to the observed K_S and strain using Eqs. (11)–(13) yields the optimized K_{0T} and K'_{0T} while α_0 and $(\partial K_T/\partial T)_P$ are constrained using previously published data. The value of $(d\alpha/dP)_T$ is calculated and refined iteratively during fitting. The gruneisen parameter γ can be constrained either by $\gamma\rho =$ constant or simply assume γ is constant. Since $\alpha\gamma T$ is small (on the order of 10^{-2} for most mantle minerals) over a wide pressure range at room temperature, no significant change in the fitted results (less than the uncertainty) is observed for a range of α_0 and $(\partial K_T/\partial T)_P$. If high temperature data are available, both α_0 and $(\partial K_T/\partial T)_P$ can be refined together with K_T and K'_{0T}. Using the fitted K_{0T} and K'_{0T}, the sample pressure can be calculated using Eq. (9). Alternatively, for room temperature data, one can use a simplified approach in which the observed K_S is used directly in Eq. (8) to constrain K_{0S} and $(\partial K_S/\partial P)_T$ at ambient conditions (e.g. Zha et al., 2000). These results, however, need to be converted to the isothermal values using thermoelastic parameters α_0, γ, $(\partial K_T/\partial T)_P$ before calculating pressure using Eq. (9).

Generally, in X-ray diffraction studies, pressures are inferred from a pressure standard positioned close to the sample in the high pressure chamber by measuring changes in unit-cell volumes (e.g. NaCl, Au, MgO, and Pt). Therefore, by enclosing NaCl pressure standard in current experiment, we can compare the sample pressures calculated using Eq. (9) with those inferred from NaCl pressure standard to evaluate the reliability of the pressure measurement in these experiments. Furthermore, conducting such experiment on NaCl itself will allow us to evaluate and calibrate the reliability of the pressure scale currently in use, and establish the absolute pressure. In the following discussion, we will refer the pressures obtained using X-ray diffraction data on NaCl and the Decker pressure scale (Decker, 1971) as the "NaCl pressure" and those calculated from velocity and density data as "sample pressure".

4. Example of Mg_2SiO_4 wadsleyite

The comprehensive nature of current technique is illustrated below using the results on a polycrystalline specimen of Mg_2SiO_4 wadsleyite. The polycrystalline sample was hot-pressed using the 1000-ton Uniaxial Split-Cylinder Apparatus (USCA-1000) in the High Pressure Lab at Stony Brook at 15 GPa, 1500 K. Acoustic velocity data on this sample up to 7 GPa and 873 K have been reported (Li et al., 1998b). In this study, the measurement has been extended to $P \sim 20$ GPa at 300 K using the T-25 apparatus at 13-ID-D at GSECARS/APS using the techniques described above. In this experiment, the pressure was first increased to a designated peak value at room temperature and then the temperature was increased to about ~673 K to relax the deviatoric stress caused by solid pressure medium (NaCl + BN) surrounding the sample. After reaching the peak P and T conditions, thc pressure was released slowly and data were also collected along decompression at room temperature. The sample recovered from this experiment remained its original length, suggesting that the surrounding NaCl provided a desired pseudo-hydrostatic stress condition; and the sample had only been subjected to elastic strain during the entire experiment. X-ray diffraction patterns of the sample showed the same peak widths during cold compression, heating as well as decompression of this experiment, suggesting that the intergranular stresses were very low. Following the procedures described in previous studies (Weidner et al., 1992; Wang et al., 1998), analysis on relative peak position shift of NaCl at ~20 GPa at room temperature indicated the stress level was about 0.02 GPa, which was within the resolution of the current set-up (Wang et al., 1998). We estimated the effect of stress on the velocity using the relation $\Delta V_{ij}/V_{ij} = L_{ij} \times \sigma/E$ (e.g. dos Santos and Bray, 2002), where L_{ij} is the acoustoelastic constant in which i represents the stress loading direction and j is the wave propagation direction, E is the Young's modulus, and σ is the imposed stress. For wadsleyite at 20 GPa, E is ~270 GPa, at a stress level of 0.02 GPa as inferred from NaCl, the maximum velocity change ($L_{11} \sim -2$ to -4) is less than 0.03%, which is far less than the reported uncertainty in the data.

Figure 5 summarizes the change of the unit-cell volume as a function of pressure from the current experiment. For comparison, pressures determined using NaCl and those using Eq. (9) as described above are both presented. The standard deviations in the calculated pressure caused by propagation of errors in the measurements of density (volume), bulk modulus, shear modulus and their pressure derivatives are 0.14 GPa at 10 GPa and 0.23 GPa at 20 GPa. First, we notice that, within the experimental uncertainties, the data collected along compression and those along decompression after heating are indistinguishable. Comparing the unit-cell volume change as a function of pressure with two previous studies, one on iron-free wadsleyite up to 4 GPa (Hazen et al., 1990), and the other on wadsleyite with 16% iron content by Fei et al. (1992) to 27 GPa, we found that all data below 5 GPa are compatible with each other if we consider the uncertainty in our pressure determination using Decker (1971) NaCl pressure scale (~3%). Beyond 5 GPa, however, the NaCl pressure deviates from sample pressure; it underestimates the sample pressure by as much as ~12.0 (1.2)% at the peak pressure. Current volume data as a

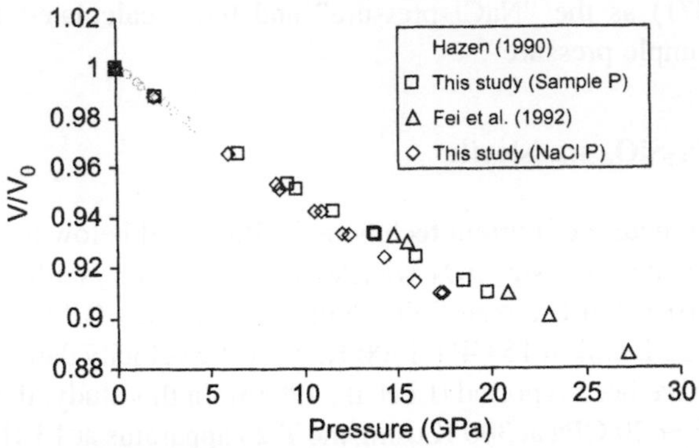

Figure 5. Volume change of wadsleyite as a function of pressure. Data from Fei et al. (1992) on iron bearing wadsleyite and iron free wadsleyite from Hazen et al. (1990) are also plotted. The squares are plotted against pressures determined from Decker's Eos for NaCl and diamonds are against pressures from finite-strain calculation. The uncertainty in volume is ~0.1% and uncertainty in pressure is about 0.05 GPa.

function of sample pressure are in complete agreement with previous studies of Hazen et al. (1990) and Fei et al. (1992) up to the peak pressure of this study.

Fitting the current data to a third-order Birch–Murnaghan Eos using pressures obtained from Decker's Eos for NaCl, we obtained a range of K_{0T} from 155 to 169 GPa for the reported V_0 values ranging from 535.6 (Hazen et al., 1990) to 538.5 A^3 (Sawamoto et al., 1984, and this study) while K_0' is fixed at 4.0. These results are in good agreement with those from a compression study by Hazen et al. (1990) in which ruby pressure scale and alcohol plus petroleum jelly were used as pressure medium. However, they are 2–10% lower than the values obtained using the ultrasonic method at ambient conditions (this study) and those from compression studies by Fei et al. (1992) using gold as pressure marker and neon as gas pressure medium. As discussed later, underestimated pressure from NaCl relative to the sample pressure is likely responsible for the lower values in bulk modulus in these Eos fittings.

Figure 6 plots travel-time data for the P and S waves as a function of sample pressure to the peak pressure of ~20 GPa. Both P and S travel times exhibit monotonic decrease as a function of pressure, reaching 10 and 12% for P and S waves, respectively. No anomalous behavior in both P and S waves travel times are obvious at this pressure range. The length change of the polycrystalline sample at high pressure needed for calculating velocities can be obtained by two methods, one is from the direct measurement using X-ray imaging, and the other is to infer sample length from the measured unit-cell volume when no plastic deformation presents in the sample. For the latter, the sample volume change in Figure 5 is used to determine sample length at high pressure using the relationship $l/l_0 = (V/V_0)^{1/3}$; where l is sample length and V is unit-cell volume; subscript zero represents value at ambient conditions. The sample lengths at high pressures derived from these two methods are compared in Figure 7. The results are consistent within 0.5–1% at all pressures except the data at ~2 GPa that the image length is about 1.8% lower than that derived from the unit-cell volume. Improved agreement between these two methods

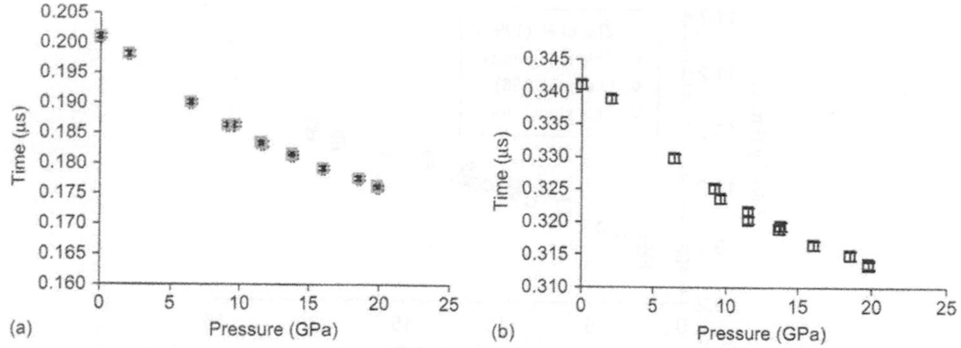

Figure 6. Travel time as a function of pressure for P wave (a) and S wave (b). The error bars are less than the size of the symbols.

can be found at higher pressures in compression and all data along decompression, presumably due to a better alignment of the sample inside the high-pressure apparatus while the flowing pressure transmitting medium is minimal. Nonetheless, compared with the travel time change observed in the pressure range of current study (Fig. 6, 10–12%), a difference of 1.8% in the determined sample length using these two methods has insignificant effect on the calculated velocities (Fig. 8).

From the travel times and the length data in Figure 7, P and S wave velocities at high pressures are obtained and plotted in Figure 8. Previous acoustic data available for comparison include polycrystalline data to 12 GPa using ultrasonic technique (Li et al., 1996) as well as the single crystal results of Brillouin scattering in a diamond anvil cell by Zha et al. (1998b). It is worth noting that in previous ultrasonic experiment of Li et al. (1996), the pressure was determined based on the observation of the phase transitions of Bi (2.55, 7.7 GPa) and ZnTe (12.0 GPa) enclosed in their acoustic

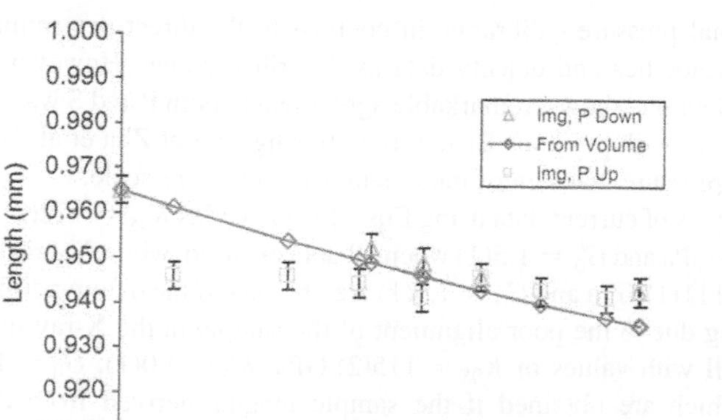

Figure 7. Comparison of the sample lengths at high pressure determined using X-ray imaging and those derived from X-ray diffraction. Squares and triangles are obtained from X-ray images along compression and decompression, respectively; diamonds are derived from X-ray diffraction.

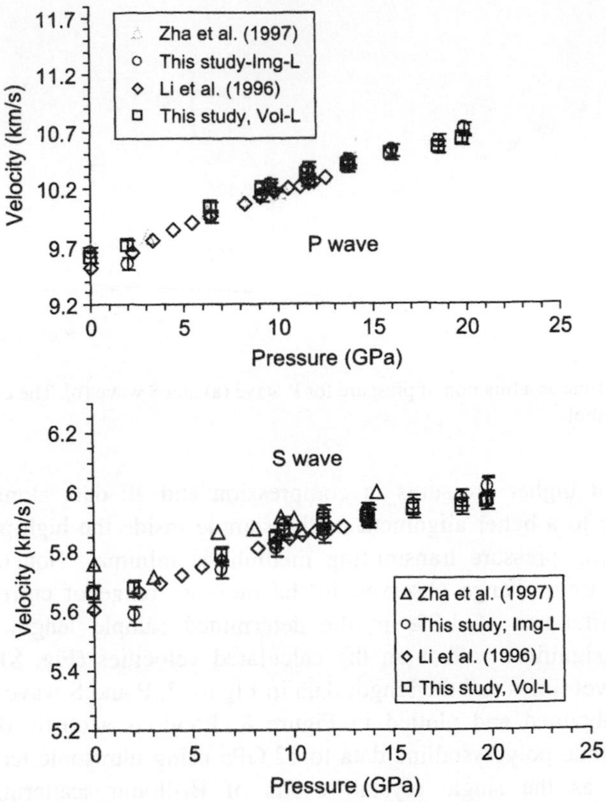

Figure 8. P and S wave velocities of wadsleyite from this study compared with previous measurements using ultrasonic method by Li et al. (1996) and Brillouin scattering techniques by Zha et al. (1998b). Squares: lengths from X-ray diffraction used; Circles: lengths from images used.

experiment as internal pressure calibrants; in contrast to the direct determination using measured acoustic velocities and density data as described above. However, as seen in Figure 8, these two data sets show a remarkable agreement in both P and S wave velocities. Current data also agree with previous Brillouin scattering data of Zha et al. (1998b) very well in which their pressure was determined using ruby pressure scale.

Finite-strain analysis of current data using Eqs. (1)–(7) yields $K_{0S} = 172(2)$ GPa, $K_0' = 4.3(1)$, $G_0 = 110(1)$ GPa and $G_0' = 1.5(1)$ when all data are used, while $K_{0S} = 174(2)$ GPa, $K_0' = 4.2(1)$, $G_0 = 111(1)$ GPa and $G_0' = 1.5(1)$ are obtained if the datum point ~2 GPa is excluded from fitting due to the poor alignment of the sample in the X-ray image. These results compare well with values of $K_{0S} = 175(2)$ GPa, $K_0' = 4.0(1)$, $G_0 = 112(1)$ GPa, and $G_0' = 1.4(1)$ which are obtained if the sample lengths derived from the unit-cell volumes are used. The values in the parenthesis indicate one standard deviation. Compare with another ultrasonic study of polycrystalline wadsleyite up to 3 GPa by Gwanmesia et al. (1990) ($K_{0S}' = 4.7$, $G_0' = 1.7$) and our previous measurements (Li et al., 1998b) on the same sample using combined ultrasonic and X-ray diffraction techniques to 7 GPa ($K_{0S}' = 4.5$, $G_0' = 1.5$), a gradual decrease in the pressure derivatives of bulk modulus is suggested. Current results on K_{0S}, K_0', G_0 and G_0' are very consistent with studies at similar pressure range using ultrasonic method (Li et al., 1996) and Brillouin scattering techniques (Zha et al., 1998b). When applying these new results to the modeling of the mantle composition, current study provides further supports to our previous discussions that the upper mantle and transitions zone favors a pyrolitic composition (Li et al., 1998b, 2001).

The discrepancy between NaCl pressure and sample pressure observed in this study is illustrated in Figure 9; the pressure estimated from the Eos of NaCl underestimates the sample pressure throughout the pressure range of this study, reaching a maximum of 2.4 GPa (12.0 (1.2)%) at 20 GPa. This discrepancy is much larger than the maximum difference in the Eos of NaCl reported in previous studies (Brown, 1999). On contrary, Zha and co-workers compared pressures calculated using Brillouin scattering and X-ray diffraction measurements on San Carlos olivine and found that the ruby pressure scale overestimates sample pressure 2–3(2)% in the pressure range 0–32 GPa (Ruby pressure scale was calibrated against NaCl scale up to 20 GPa, see details in Zha et al. (1998a) and references thererein). The disagreement between current study and Zha et al. (1998a) is clearly beyond the mutual uncertainties of the techniques used in these two studies and remains to be reconciled by further studies. On the other hand, using similar techniques as in this study, discrepancy between NaCl pressure and sample pressure similar to those observed in this study (Fig. 9) has been reported in previous studies on other materials up to 10 GPa, including MgO (Woody, 2004), Pyrope-garnet (Liu, 2000) and (MgFe)O (Kung et al., 2002). In these experiments, non-hydrostatic stresses are believed to be released since the sample assembly was heated to temperatures as high as 1273 K prior to data collection (see details in Liu, 2000; Woody, 2004). It is likely the observed difference between NaCl pressure and sample pressure in this study is a total effect of remnant stresses on the sample (rather than on NaCl) and the accuracy of NaCl pressure scale in use. Although evidence from peak width analysis as well as stress calculation using NaCl X-ray diffraction suggested that the deviatoric stress level on our sample is insignificant, quantitative investigation using X-ray diffraction data collection at multiple diffracting angles (or 2D area detector) in conjunction with an apparatus with the capability of controlling sample stress state are needed to decouple these two effects (Wang et al., 2003, Uchida et al., this volume).

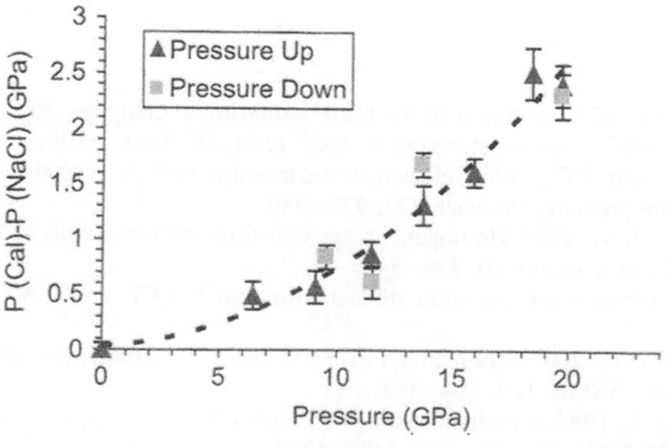

Figure 9. Observed difference between sample pressure and those determined from Decker's NaCl Eos at high pressures. Triangles: data along compression; squares: data on decompression.

5. Summary

We have demonstrated a state-of-art technique for the study of elastic properties of polycrystalline and single crystal materials using simultaneous ultrasonic and X-radiation techniques at high pressures and temperatures in Kawai type multi-anvil high pressure apparatus. Simultaneous equation of state, pressure calibration, and sound velocity measurements have been illustrated using our data for wadsleyite phase of Mg_2SiO_4 up to 20 GPa at room temperature. The bulk and shear moduli and their pressure derivatives are $K_{0S} = 174(2)$ GPa, $K_0' = 4.2(1)$, $G_0 = 111(1)$ GPa, and $G_0' = 1.5(1)$. Remarkable agreement can be found between current data and previous studies using ultrasonic interferometry and Brillouin scattering techniques at similar pressure ranges. The sample pressure determined using the combined acoustic data and density data differs from those determined using Decker NaCl scale by ~12.0 (1.2)% at the peak pressure of 20 GPa of this study. Further investigation of this discrepancy is needed in order to differentiate the stress effect from the accuracy of the Decker pressure scale. Nonetheless, successful application of these techniques to the study of the behavior and physical properties of Earth materials, including unquenchable mantle phases and multi-phase aggregates, will greatly improve our interpretation of seismic observations and our understanding about the structure, composition and dynamics of the Earth's interior.

Acknowledgements

This research was supported by National Science Foundation under grant EAR000135550 to BL. We thank the reviewers and the editor for their help improving the manuscript. These experiments were performed at GeoSoilEnviroCARS (Sector 13), Advanced Photon Source (APS), Argonne National Laboratory. GeoSoilEnviroCARS is supported by the National Science Foundation – Earth Sciences (EAR-0217473), Department of Energy – Geosciences (DE-FG02-94ER14466) and the State of Illinois. Use of the APS was supported by the US Department of Energy, Basic Energy Sciences, and Office of Energy Research, under Contract No. W-31-109-Eng-38.. Mineral Physics Institute Publication No. 351.

References

Birch, F., 1952. Elasticity and constitution of the Earth's interior. J. Geophys. Res. 54, 227–286.

Brown, J.M., 1999. The NaCl pressure standard. J. Appl. Phys. 86, 5801–5808.

Chen, G., Li, B., Liebermann, R.C., 1996. Selected elastic moduli of single-crystal olivines from ultrasonic experiments to mantle pressures. Science 272, 979–980.

Davis, G.F., Dziewonski, A.M., 1975. Homogeneity and constitution of the Earth's lower mantle and outer core. Phys. Earth Planet. Interiors 10, 336–343.

Decker, D.L., 1971. High-pressure equation of state for NaCl, KCl, and CsCl. J. Appl. Phys. 42, 3239–3244.

dos Santos, A.A., Bray, D.E., 2002. Compairson of acoustoelastic methods to evaluate stresses in steel plates and bars, Trans. ASME, 124, 354–358.

Fei, Y., Mao, H.-K., Shu, J., 1992. Simultaneous high-*P*, high-*T* X-ray diffraction study of b-$(MgFe)_2SiO_4$ to 26 GPa and 900 K. J. Geophys. Res. 97, 4489–4495.

Flesch, L.M., Li, B., Liebermann, R.C., 1998. Sound velocities of polycrystalline $MgSiO_3$-orthopyroxene to 10 GPa at room temperature. Am. Mineral. 83, 444–450.

Fujisawa, H., 1998. Elastic wave velocities of forsterite and its beta-spinel form and chemical boundary hypothesis for the 410 km discontinuity. J. Geophys. Res. 103, 9591–9608.

Fujisawa, H., Ito, E., 1985. Measurements of ultrasonic wave velocities of tungsten carbide as a standard material under high pressure. J. Appl. Phys. 24 (Suppl. 24-1), 103–105.

Gieske, J.H., Barsch, G.R., 1966. Pressure dependence of the elastic constants of single crystalline aluminum oxide. Phys. Status Solidi 29, 121–131.

Gwanmesiam, G.D., Rigden, S.M., Jackson, I., Liebermann, R.C., 1990. Pressure dependence of elastic wave velocity for β-Mg_2SiO_4 and the composition of the Earth's mantle. Science 250, 794–797.

Gwanmesia, G.D., Chen, G., Liebermann, R.C., 1998. Sound velocities in $MgSiO_3$-garnet to 8 GPa. Geophys. Res. Lett. 25, 4553–4556.

Hazen, R.M., Zhang, J.M., Ko, J., 1990. Effect of Fe/Mg on the compressibility of synthetic wadsleyite: β-$(Mg_{1-x}Fe_x)_2SiO_4$ ($x \leq 0.25$). Phys. Chem. Miner. 17, 416–419.

Jackson, I., Niesler, H., 1982. The elasticity of periclase to 3 GPa and some geophysical implications. In: Manghnani, M., Syono, Y. (Eds), High Pressure Research: Application to Earth and Planetary Sciences. Terra Scientific Publishing Co./American Geophysical Union, Tokyo/Washington, DC, pp. 93–113.

Jackson, I., Rigden, S.M., 1998. Composition and temperature of the Earth's mantle: seismological models interpreted through experimental studies of Earth materials. In: Jackson, I. (Ed.), The Earth's Mantle. Cambridge University Press, Cambridge, UK, pp. 405–460.

Knoche, R., Webb, S.L., Rubie, D.C., 1998. Measurements of acoustic wave velocities at $P-T$ conditions of the Earth's mantle. In: Manghnani, M.H., Yagi, T. (Eds), Properties of Earth and Planetary Materials at High Pressure and Temperature. American Geophysical Union, Washington, DC, pp. 119– 128.

Kumazawa, M., Anderson, O.L., 1969. Elastic moduli, pressure derivatives, and temperature derivatives of single-crystal olivine and single-crystal forsterite. J. Geophys. Res. 74, 5961–5972.

Kung, J., Li, B., Weidner, D.J., Zhang, J., Liebermann, R.C., 2002. Elasticity of $(Mg_{0.83},Fe_{0.17})O$ ferropericlase at high pressure: ultrasonic measurements in conjunction with X-radiation techniques. Earth Planet. Sci. Lett. 203, 557–566.

Kung, J., Li, B., Liebermann, R.C., Wang, Y., Uchida, T., 2004. In-situ measurement of the elasticity across the orthopyroxene → high clinopyroxene phase transition. Phys. Earth Planet. Interiors 147, 27–44.

Larson, A.C., Von Dreele, R.B., 1988. GSAS Manual, Report LAUR 86-748. Los Alamos National Laboratory, Los Alamos, NM.

Li, B., Liebermann, R.C., 2000. Sound velocities of wadsleyite β-$(Mg_{0.88}Fe_{0.12})_2SiO_4$ to 10 GPa. Am. Mineral. 85, 292–295.

Li, B., Weidner, D.J., 1999. Rheological property measurement at high pressure and high temperature using ultrasonic and X-ray diffraction techniques. In: Manghnani, M., Nellis, W.J., Nicol, M.F. (Eds), Science and Technology of High Pressure: Proceedings of AIRAPT-17, p. 135.

Li, B., Zhang, J., Wang, Y., Weidner, D.J., Liebermann, R.C., 1994. Characterization of microscopic stress in a hot-pressed polycrystalline stishovite at high P and T. Eos Trans. Am. Geophys. Union 75, 346.

Li, B., Gwanmesia, G.D., Liebermann, R.C., 1996. Sound velocities of olivine and beta polymorphs of Mg_2SiO_4 at Earth's transition zone pressures. Geophys. Res. Lett. 23, 2259–2262.

Li, B., Chen, G., Gwanmesia, G.D., Liebermann, R.C., 1998a. Sound velocity measurements at mantle transition zone conditions of pressure and temperature using ultrasonic interferometry in a multi-anvil apparatus. In: Manghnani, M.H., Yagi, T. (Eds), Properties of Earth and Planetary Materials at High Pressure and Temperature. American Geophysical Union, Washington, DC, pp. 41–61.

Li, B., Liebermann, R.C., Weidner, D.J., 1998b. Elastic moduli of wadsleyite to 7 gigapascals and 873 kelvin. Science 281, 675–677.

Li, B., Liebermann, R.C., Weidner, D.J., 2001. $P-V-V_P-V_S-T$ measurements on wadsleyite to 7 GPa and 873 K: implications for the 410 km seismic discontinuity. J. Geophys. Res. 106, 30575–30591.

Li, B., Chen, K., Kung, J., Liebermann, R.C., Weidner, D.J., 2002. Sound velocity measurement using transfer function method. J. Phys.: Condens. Matter 14, 11337–11342.

Li, B., Kung, J., Liebermann, R.C., 2004. Modern techniques in measuring elasticity of Earth materials at high pressure and high temperature using ultrasonic interferometry in conjunction with synchrotron X-radiation in multi-anvil apparatus. Phys. Earth Planet. Interiors 143–144, 559–574.

Liebermann, R.C., Li, B., 1998. Elasticity at high pressures and temperatures. In: Hemley, R. (Ed.), Ultrahigh-Pressure Mineralogy: Physics and Chemistry of the Earth's Deep Interior, Vol. 37, pp. 459–492.

Liu, J., 2000. Ultrasonic wave velocities of pyrope-majorite garnets at high pressure and high temperature. Ph.D. Thesis, State University of New York at Stony Brook, p. 167.

Liu, J., Chen, G., Gwanmesia, G.D., Liebermann, R.C., 2000. Elastic wave velocities of pyrope-majorite garnets ($Py_{62}Mj_{38}$ and $Py_{50}Mj_{50}$) to 9 GPa. Phys. Earth Planet. Interiors 120, 153–163.

McSkimin, H.J., 1950. Ultrasonic measurement techniques applicable to small solid specimens. J. Acoust. Soc. Am. 22, 413–418.

Meng, Y., Weidner, D.J., 1993. Deviatoric stress in a quasi-hydrostatic diamond anvil cell: effect on the volume-based pressure calibration. Geophys. Res. Lett. 20, 1147–1150.

Mueller, H.J., Lautherjung, J., Schilling, F.R., Lathe, C., Nover, G., 2002. Symmetric and asymmetric interferometric method for ultrasonic compressional and shear wave velocity measurements in piston cylinder and multi-anvil high pressure apparatus. Eur. J. Mineral. 14, 581–598.

Ruoff, A.L., Lincoln, R.C., Chen, Y.C., 1973. A new method of absolute high pressure determination. J. Phys. D: Appl. Phys. 6, 1295–1306.

Sawamoto, H., Weidner, D.J., Sasaki, S., Kumazawa, M., 1984. Single-crystal elastic properties of the modified-spinel (beta) phase of Mg_2SiO_4. Science 224, 749–751.

Schreiber, E., Anderson, O.L., 1966. Pressure derivatives of the sound velocities of polycrystalline alumina. J. Am. Ceram. Soc. 49, 184–190.

Spetzler, H.A., Chen, G., Whitehead, S., Getting, I.C., 1993. A new ultrasonic interferometer for the determination of equation of state parameters of sub-millimeter single crystals. In: Liebermann, R.C., Sondergeld, C.H. (Eds), Experimental Techniques in Mineral and Rock Physics, Vol. 141. PAGEOPH, pp. 341–377.

Truell, R., Elbaum, C., Chick, B., 1969. Ultrasonic Methods in Solid State Physics. Academic Press, New York, p. 464.

Uchida, T., Wang, Y., Rivers, M.L., Sutton, S.R., Weidner, D.J., Vaughan, M.T., Chen, J., Li, B., Secco, R.A., Rutter, M.D., Liu, H., 2002. A large-volume press facility at the advanced photon source: diffraction and imaging studies on materials relevant to the cores of planetary bodies. J. Phys.: Condens. Matter 44, 11517–11523.

Wang, Y., Weidner, D.J., Meng, Y., 1998. Advances in equation of state measurements in SAM-85. In: Manghnani, M.H., Yagi, T. (Eds), Properties of Earth and Planetary Materials at High Pressure and Temperature. American Geophysical Union, Washington, DC, pp. 365–372.

Wang, Y., Durham, W.B., Getting, I.C., Weidner, D.J., 2003. The deformation DIA: a new apparatus for high temperature triaxial deformation to pressures up to 15 GPa. Rev. Sci. Instrum. 74, 3002–3011.

Warren, B.E., 1990. X-ray Diffraction. General Publishing Company Ltd, Toronto, Ont., p. 378.

Webb, S.L., 1989. The elasticity of the upper mantle orthosilicates olivine and garnet to 3 GPa. Phys. Chem. Miner. 16, 684–692.

Weidner, D.J., Vaughan, M.T., Ko, J., Wang, Y., Liu, X., Yeganeh-Haeri, A., Pacalo, R.E., Zhao, Y., 1992. Characterization of stress, pressure, and temperature in SAM85, a DIA type high pressure apparatus. In: Syono, Y., Manghnani, M. (Eds), High Pressure Research: Application to Earth and Planetary Sciences. American Geophysical Union, Washington, DC, pp. 13–17.

Woody, K.M., 2004. Elasticity and pressure calibration using polycrystalline MgO. MS Thesis, State University of New York at Stony Brook.

Yoneda, A., 1990. Pressure derivatives of elastic constants of single crystal MgO and $MgAl_2O_4$. J. Phys. Earth 38, 19–55.

Yoneda, A., Morioka, M., 1992. Pressure derivatives of elastic constants of single crystal forsterite. In: Syono, Y., Manghnani, M. (Eds), High Pressure Research: Application to Earth and Planetary Sciences. American Geophysical Union, Washington, DC, pp. 206–214.

Zha, C.-S., Duffy, T.S., Downs, R.T., Mao, H.-K., Hemley, R.J., 1998a. Brillouin scattering and X-ray diffraction of San Carlos olivine: direct pressure determination to 32 GPa. Earth Planet. Sci. Lett. 159, 25–33.

Zha, C.-S., Duffy, T.S., Downs, R.T., Mao, H.-K., Hemley, R.J., Weidner, D.J., 1998b. Single-crystal elasticity of the α and β of Mg_2SiO_4 polymorphs at high pressure. In: Manghnani, M.H., Yagi, T. (Eds), Properties of Earth and Planetary Materials at High Pressure and Temperature. American Geophysical Union, Washington, DC, pp. 9–16.

Zha, C.-S., Mao, H.K., Hemley, R.J., 2000. Elasticity of MgO and a Primary Pressure Scale to 55 GPa, PNAS 97, 13494–13499.

Advances in High-Pressure Technology for Geophysical Applications
Jiuhua Chen, Yanbin Wang, T.S. Duffy, Guoyin Shen, L.F. Dobrzhinetskaya, editors

Chapter 4
Simultaneous determination of elastic and structural properties under simulated mantle conditions using multi-anvil device MAX80

Hans J. Mueller, Christian Lathe and Frank R. Schilling

Abstract

An ultrasonic interferometry high-pressure set-up for elastic wave velocity measurements under simulated Earth's mantle conditions has been developed. A DIA-type multi-anvil apparatus MAX80 permanently located at HASYLAB, Hamburg, Germany for X-ray diffraction (XRD) under in situ condition was equipped for simultaneous ultrasonic measurements. Two of the six anvils were equipped with lithium niobate P- and S-wave transducers of 33.3 MHz natural frequency. The pressure and temperature limits of the high-pressure apparatus were not reduced as a side effect of the modification. The ultrasonic configuration allows all kinds of interferometric measurements with compressional and shear waves. In addition to the classical ultrasonic interferometry the newly developed data transfer function technique (DTF), first described in [J. Phys. Condens. Matter 14 (2002) 11337], is discussed in detail to give the readers the chance to use this valuable and important new method.

The results for natural San Carlos olivine up to 3 GPa pressure are compared with published data of several authors. The data for hot-isostatic-pressed anorthite solved discrepancies between published high-pressure and normal-pressure data for polycrystalline anorthite leading to $v_p = 7.28$ km/s, $v_s = 3.93$ km/s at ambient conditions and $dv_p/dp = 0.027$ km/s GPa, $dv_s/dp = 0.001$ km/s GPa. The obtained data sets correspond to published results within the accuracy of the method.

As an example for unquenchable phase transitions we measured the elastic wave velocities at the high-pressure clinoenstatite ($MgSiO_3$, HCEn) – low-pressure (LCEn) transition at high pressure, high temperature conditions in conjunction with in situ XRD. For ultrasonic interferometry experiments LCEn powder synthesized at ambient pressure was hot-isostatic-pressed at 0.4 GPa and 1400°C for 2 h in MAX80 to obtain low-porosity samples. The elastic wave velocities v_p and v_s of the CEn sample were measured in situ using the classical interferometric technique as well as the recently developed ultrasonic data transfer function (DTF) technique for the elastic wave velocities as a function of pressure at 700°C. To compare the results, v_p and v_s were measured at 6.7 and 7.5 GPa using both interferometric techniques. The results correspond within the limits of less than 1%.

1. Introduction

During the last decade the progress of global seismology in general and of the tomographic method in particular in terms of resolution, amount of data, and quality of their processing reveals a lot of new and exciting structural details of the Earth's deep interior (van der Hilst, 1995; Li et al., 2000). Understanding and modeling of mantle dynamics, crust mantle interaction, formation of plumes, and many others require much more detailed insights into the structural and physical properties of materials relevant for great depths

(Kohlstedt et al., 1996). Ultradeep subduction, penetrating into the lower mantle, the incidence of deep earthquakes and their relation to the nature of the transition zone as well as the discussion of slab recycling in plumes require comparable effort in high-pressure research and precise geophysical observation. Different multi-anvil high-pressure devices have been used with great success in experimental mantle mineralogy for many years (Shimomura et al., 1985; Yagi, 1988; Funamori et al., 1996a,b; Suzuki et al., 1996; Oguri et al., 2000; Hirose et al., 2001; Nishiyama and Yagi, 2003). In many cases pioneering work was achieved at the Mineral Physics Institute, Stony Brook, e.g. ultrasonic interferometry in the DIA-type multi-anvil cell SAM85 (Chen et al., 1996; Li et al., 1996a,b, 1998, 2001a; Liebermann and Li, 1998; Kung and Rigden, 1999; Kung et al., 2000, 2001a, b, 2002; Li, 2003), development of the DTF technique (Li et al., 2002), X-radiography (Li et al., 2001b), and double-stage T-CUP up to 23 GPa for *in situ* X-ray diffraction (XRD) (Vaughan, 1993; Vaughan et al., 1995). The *in situ* study of complex phase relations, the understanding and description of unquenchable high-pressure phases, and the investigation of the kinetics of phase transitions and mineral reactions require time-resolved measurements. Synchrotron radiation allows *in situ* structural investigations under simulated mantle conditions. Furthermore, simultaneous measurement of compressional and shear wave velocities, especially using the DTF technique, and structural investigation might be the critical key to understand the ongoing processes in more detail.

The different elastic properties, elastic wave velocities, shear modulus, bulk modulus, Young's modulus, Poisson's ratio, provide detailed information about the mechanical behavior of samples (Kern, 1982). Elastic properties are particularly sensitive to phase transitions. The existing knowledge of the physical properties of high-pressure phases is mostly limited to equilibrium conditions. Considering the significance of non-equilibrium structures for the Earth's deep interior, reliable time-resolved measurements during transition processes are required. Simultaneously determined elastic wave velocities and structural information provides an independent way to measure compressibility and elastic moduli without pressure calibrant, independent of any standard material (Spetzler and Yoneda, 1993; Yoneda and Spetzler, 1994; Getting, 1998; Zha et al., 2000; Mueller et al., 2002) (see also Mueller et al., 2005, pages 427–449, this volume). Elastic properties are important for thermodynamic calculations and to understand the kinetics of mineral reactions (Haussuehl et al., 1980; Hoffbauer et al., 1985; Lauterjung and Will, 1985; Angel and Ross, 1996; Zinn et al., 1997).

In addition to the description of up-to-date ultrasonic interferometric techniques we present exemplary results of high-pressure interferometric measurements of compressional and shear wave velocities in single-crystal San Carlos olivine polycrystalline anorthite, and at the high-*P* (HCEn)–low-*P* (LCEn) clinoenstatite transition (Mueller et al., 2002, 2003, 2005).

Enstatite, the pure magnesium silicate end-member of pyroxene stoichiometry, $MgSiO_3$, exists in at least five polymorphs (Bowen and Andersen, 1914). Using single-crystal XRD analyses in a diamond anvil cell, Angle et al. (1992) determined the clinoenstatite (CEn) transformation from the $P2_1/c$-to the $C2/c$-polymorph to be at about 5.5–8.0 GPa and room temperature (RT) conditions. They also determined the structure of the HCEn polymorph and estimated thermodynamic data for the CEn-transition. A further important conclusion of their study was that the LCEn–HCEn transition is not quenchable, reverting to the $P2_1/c$-structure upon decompression at RT. The current understanding of phase relations in the system $MgSiO_3$ is summarized, e.g. by Presnall (1995, Fig. 8). Up to now, the position of the HCEn–LCEn transition for $MgSiO_3$ is only deduced from thermodynamic data by Angel and Hugh-Jones (1994). Additionally, large discrepancies exist between recently performed experimental studies for the OEn–HCEn transition at high pressures and temperatures. For clarification of these discrepancies and to better characterize the HCEn–LCEn transition we performed *in situ* experiments at elevated temperatures and various pressures.

2. Techniques, methods and materials description

2.1. Multi-anvil high-pressure apparatus MAX80

MAX80 is a DIA-type multi-anvil high-pressure apparatus with six tungsten carbide anvils compressing a cubic sample volume of maximum $8 \times 8 \times 8\ mm^3$ (Fig. 1). The apparatus is installed at beamline F2.1 at the HamburgerSynchrotronstrahlungsLabor (HASYLAB) for high-pressure–high-temperature synthesis and *in situ* measurements. The anvils are driven by a 2500 N uniaxial hydraulic ram. Three anvil sets with different truncations exist – 6, 5, and 3.5 mm. The corresponding cube length is 8, 6, and 5.5 mm resulting in maximum pressures of about 7, 9, and 12 GPa. The maximum attainable temperature is 2000 K produced by an internal graphite heater. One or two of the original anvil spacers had to be replaced by redesigned parts. The new spacers have a cavity in their center to keep the ultrasonic transducer free of stress from the load of the hydraulic press. The ultrasonic anvils are equipped with two P-wave and two S-wave transducers. Because of their high conversion factor and high thermal stability lithium niobate transducers overtone polished with a resonant frequency of 33.3 MHz were cemented on the polished rear side of the ultrasonic anvils using epoxy resin diluted by acetone to reduce its viscosity for a minimum thickness of the glue film. The resulting strong coupling of the transducer to the anvil is of fundamental importance for the interferometric method, because the strong coupling results in a broad band characteristics of the transducer.

Figure 1. Boron-epoxy cube of MAX80 with gaskets after the experiment. Top and front lateral anvils (6 mm truncation) are removed.

Diffraction patterns are recorded in an energy-dispersive mode using X-rays from the storage ring DORIS III at HASYLAB and a Ge-detector. NaCl is used as pressure standard. The pressure is calculated using an in-house program from the XRD data following the EoS of Decker (1971). For further details, see Mueller et al., 2005, pages 427–449, this volume and Mueller et al. (2002, 2003).

2.2. Multi-anvil apparatus cell assembly

The high-pressure cell consists of a cube made by pressing and adjacent machining of epoxy resin mixed with amorphous boron with the weight ratio 1:4 for better compressive strength containing the ultrasonic configuration, the heater, the pressure standard, and the thermocouple (Fig. 2). The interfaces between the sample and the close-fitting buffer rods/reflector bars are polished for optimal ultrasonic coupling. The sample is surrounded by a boron nitride sleeve for electrical insulation and as pressure transmitting medium inside a stepped graphite heater. The stepping results in stronger heat production at the ends of the sample which compensates the heat flow to the colder anvils resulting in a smaller temperature gradient throughout the sample (Knoche et al., 1997, 1998). For experiments, if sodium chloride is not in use as ultrasonic reflector, a sleeve made from a mixture of sodium chloride and boron nitride is used as pressure calibrant. The copper rings contact the heater at the top and bottom anvils, and the pyrophyllite rings are a quasi-hydrostatic pressure transmitting medium. The total length of the set-up (see Fig. 2) reduced from 8 mm to about 6.9 mm during the experiments by plastic deformation. Even very brittle buffer rods made from fused quartz or polycrystalline corundum did not crush during the

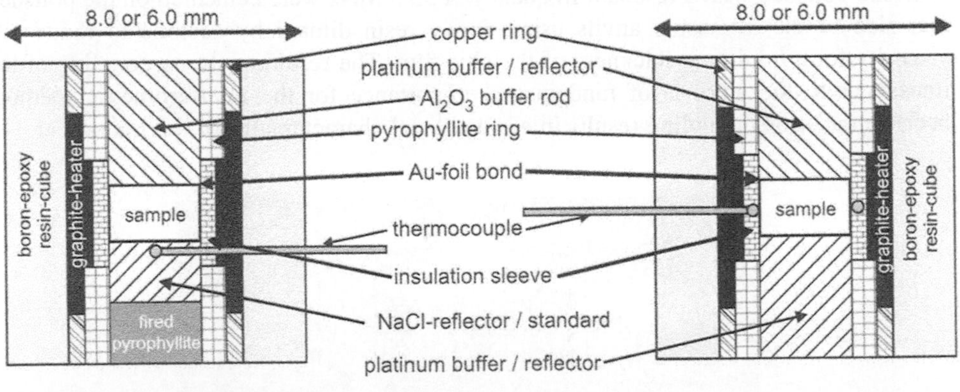

Figure 2. MAX80 set-ups for combined XRD and ultrasonic measurements. For a low-loss transmission of ultrasonic energy from the buffer rod made of brittle material to the sample the interface is covered by Au foils. The NaCl cylinder reflects the ultrasonic waves back to the transducer and is used as pressure standard at the same time. The adjacent fired pyrophyllite part prevents the blow-out of the NaCl cylinder at elevated temperatures. Opposite to set-ups for only XRD measurements the thermocouple has to be located outside the sample center to keep the ultrasonic travel path undisturbed. The sample is surrounded by a hBN-sleeve and the adjacent graphite heater. The copper rings contact the heater to the top and bottom tungsten carbide anvils. The boron-epoxy resin cube is a pressure transmitting medium highly permeable for X-rays.

experiments. Olivine and anorthite only deformed elastically as measured by a dial gauge indicator before and after the experiments. Corresponding to the anvil sets with truncations of 6, 5, and 3.5 mm cell assemblies with 8, 6, and 5 mm length of the sides were designed. A thermocouple inside the graphite heater on the sample surface, or inside the NaCl reflector close to the sample is used to control the temperature. The temperature inside the sample at its center is derived from a calibration using special cell assemblies with an additional thermocouple at the sample center or inside the NaCl reflector and a calibration of electrical current, respectively. The maximum deviation of the temperature measured during the experiments from the true sample temperature was found to be 10–25° depending on the method and on cube deformation. To minimize electromagnetically induced noise to the ultrasonic signal, a DC power source was used for heating. Even DC electrical heating requires the grounding of the anvil where the transducers are assembled to avoid interferences of the electric current with the transducers during heating.

A stress test was performed to get quantitative information about the level of non-hydrostatic stress inside the sample at cold compression, especially. A common way to do this is measuring the unit cell deformation of NaCl. Over all pressure conditions up to 8 GPa and the unit cell parameters derived from 111 to 200 were compared. We found maximum deviation of the calculated volumes of the unit cell between +0.03 and +0.25, i.e. any differential stress resulting in negative quotients were not found. As an additional indication for a high degree of hydrostatic pressure conditions at different elevated temperatures we found no shift of the phase boundary between high-*P* (HCEn)–low-*P* (LCEn) clinoenstatite (see Section 3.3) derived from the results of experiments using a powder or a hot-isostatic-pressed (HIP) sample, otherwise the last-mentioned samples should apparently cross the phase boundary at lower pressure conditions because of the higher internal stress. In this case one or two of the unit cell parameters (see Table 1 and Fig. 17) would also systematically deviate from

Table 1. Variation of CEn unit cell parameters with pressure, and temperature.

P (GPa)	*T* (°C)	*a* (Å)	*b* (Å)	*c* (Å)	β (–)	*V* (Å^3)
LCEn						
6.61(5)	300	9.438(18)	8.624(12)	5.076(9)	107.5(2)	394.1(9)
7.20(5)	550	9.438(8)	8.621(6)	5.076(4)	107.68(8)	393.5(4)
7.50(5)	700	9.452(9)	8.613(6)	5.069(4)	107.80(8)	392.8(4)
7.88(5)	900	9.405(9)	8.609(6)	5.066(4)	107.78(9)	390.6(4)
HCEn						
6.61(5)	20	9.224(9)	8.658(6)	4.915(4)	101.71(7)	384.4(4)
6.61(5)	250	9.244(8)	8.665(4)	4.925(5)	101.54(5)	386.5(9)
7.20(5)	34	9.195(8)	8.623(6)	4.919(4)	101.63(7)	382.0(4)
7.20(5)	500	9.239(5)	8.671(4)	4.927(3)	101.56(6)	386.7(4)
7.50(5)	41	9.197(8)	8.615(6)	4.906(4)	101.49(8)	380.9(4)
7.50(5)	650	9.237(9)	8.643(5)	4.925(3)	101.68(5)	385.0(9)
7.88(5)	20	9.188(10)	8.593(7)	4.896(5)	101.71(9)	378.5(5)
7.88(5)	850	9.222(5)	8.651(6)	4.910(5)	101.52(4)	383.8(9)

The 1σ uncertainties of the last digits of the lattice refinements are given in parentheses.

published data, and from results of single-crystal DAC experiments, especially. All that was not found.

2.3. Ultrasonic interferometry techniques

2.3.1. Frequency sweep method

Ultrasonic interferometry, using the interference between the incident and reflected waves inside the sample, was first described by McSkimin (1950). It allows high-precision measurements of the travel time through the sample, independent of the delay travel time and its variation with pressure and temperature in anvils and buffer rods. Piezoelectric transducers for the generation and detection of ultrasonic waves are cemented at the rear side of the piston outside the true pressure cell (see Mueller et al., 2005, pages 427–449, this volume). The amount of energy reflected or transmitted at an interface is given by the reflection factor R, where

$$R = \frac{Z_2 - Z_1}{Z_2 + Z_1} \tag{1}$$

and

$$Z = \rho v_i \tag{2}$$

with Z the acoustic impedance, ρ density, v_i the compressional or shear wave velocity and the transmission factor D

$$D = \frac{2Z_2}{Z_2 + Z_1} \tag{3}$$

with Z_1 the acoustic impedance of medium 1 and Z_2 the acoustic impedance of medium 2.

For negative R values a phase shift of 180° is observed (see Niesler and Jackson, 1989). For example, if we take an olivine sample ($v_p \approx 8.25$ km/s, $\rho \approx 3.34$ g/cm^3) between a fused quartz buffer ($v_p \approx 5.57$ km/s, $\rho \approx 2.60$ g/cm^3) and a reflector made of platinum ($v_p \approx 3.96$ km/s, $\rho \approx 21.40$ g/cm^3), the reflection factor at the interface buffer–sample becomes ≈ 0.3, and >0.5 at the transition to the reflector. This means that nearly three quarters of the energy reach the sample and half of the energy is reflected at the sample's rear side (Mueller et al., 2002).

The most popular interferometric technique is called double-pulse phase-comparison method (McSkimin, 1950). It effectively eliminates all interferences, which are not useful for further evaluation. Its precision is about 1–3 times higher (Schreiber et al., 1973; Li et al., 1998) than classical travel-time methods (Birch, 1960, 1961). The difference between several destructive and constructive interferences is used to reveal the reverberation time inside the sample. The high precision of the ultrasonic interferometry requires a highly precise sample length measurement under *in situ* conditions, because calculating the elastic wave velocities from the reverberation time requires the sample length (see Mueller et al., 2005, pages 427–449, this volume). Using a broad range of frequencies leads to the detection of a high number of constructive and destructive interferences, yielding to higher precision of the regression analysis (see Fig. 3).

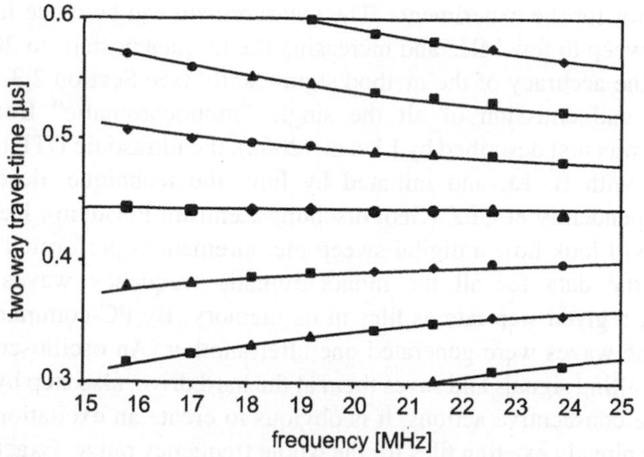

Figure 3. Travel time curves as a function of frequency. Picking all available maxima and minima as a function of frequency allows the determination of the travel time inside the sample as a regression result for the horizontal point sequence between the curves of opposite curvature.

We mostly used a slightly modified method, similar to that published by Shen et al. (1998) for diamond anvil cells, where only one single tone burst (a sinusoidal wave limited in time to few microseconds) of about three to sixfold duration of the travel time inside the sample is used. The narrower frequency band of the prolonged burst increases the sharpness of the interference pattern compared to the shorter bursts of the double-pulse method. Because of the broader frequency range most of our measurements were performed using tone bursts of 4 μs duration. Additional coupling media, e.g. gold foil as described by Niesler and Jackson (1989), are only necessary at the interface between two brittle media, e.g. corundum buffer rod and San Carlos olivine sample or for making the evaluation of X-radiographs (see Mueller et al., 2005, pages 427–449, this volume) easier (Li et al., 2001a,b).

Ultrasonic interferometry requires a special equipment for generation, superposition, and display of rf-signals. In the last two decades of the 19th century, the Australian Scientific Instruments Ultrasonic Interferometer (Rigden et al., 1988, 1992, 1994; Niesler and Jackson, 1989), became the standard equipment. Today a combination of digital generators and oscilloscopes controlled by a PC took on the task. For details of the electronic equipment used for our measurements see Mueller et al., 2005, pages 427–449, this volume.

2.3.2. Ultrasonic data transfer function technique

2.3.2.1. Pulse shaping of the excitation function

The classical digital sweep interferometry is very time consuming. A 60 MHz frequency sweep with 100 kHz steps lasts for more than 30 min. Consequently a single measurement of v_p and v_s requires more than 1 h. This is a serious limitation not only for all transient measurements, but also limits the data collection at elevated temperatures to prevent the boron-epoxy cubes and the anvils from overheating. So the ultrasonic interferometry is the limiting factor for the experiments. The measurement can be made faster by limiting the frequency sweep to few MHz and increasing the frequency steps to 200 kHz or more. But this limits the accuracy of the method significantly (see Section 2.3.1). A solution is the generation and emission of all the single "monochromatic" frequencies simultaneously. This was first described by Li et al. (2002), the ultrasonic DTF technique. Based on discussions with B. Li, and initiated by him, the technique, described here was developed independently at GFZ (GeoForschungsZentrum Potsdam), Dept. 4.

At first we will look how a digital sweep measurement is performed practically. The generator has the data for all the monochromatic frequency waves in the desired bandwidth with a given step rate as files in its memory. By PC-command the files were called-in and the waves were generated one after another. An oscilloscope receives and digitizes the resulting signals and saves them at the hard drive, also step by step. If we plan to "unify" these consecutive actions, it is obvious to create an excitation function as the sum of all these already existing files for the whole frequency range. Exactly that we did as our first approach.

$$Y(t) = \sum_{i=1}^{i=n} y_i(t) \tag{4}$$

with Y the amplitude of summarized excitation function, y_t the amplitude of sinusoidal waves between upper and lower cut-off frequency and t time (Fig. 4).

Figure 8a and b shows the result for the whole 4 μs duration and the first 200 ns with higher time magnification. The function increases very steeply from zero followed by a little less dramatic decrease and a relatively low attenuation for later points in time. If we want to apply this to an ultrasonic transducer, we have to realize that first of all the transducer is a mechanical vibratory system driven by piezoelectricity and its reversal, respectively, i.e. we have to keep in mind its inertia. The excitation function would act as a shock. The transducer would mostly respond with an attenuated oscillation in its natural frequency. The excitation must be much more intensive to make the forced oscillations of the transducer stronger, far from resonance. What we have to do is inverting the function in time and amplitude, removing the first point – 0 – and "assembling" this inverted function to the beginning of the original function. Now we have a slowly increasing oscillation

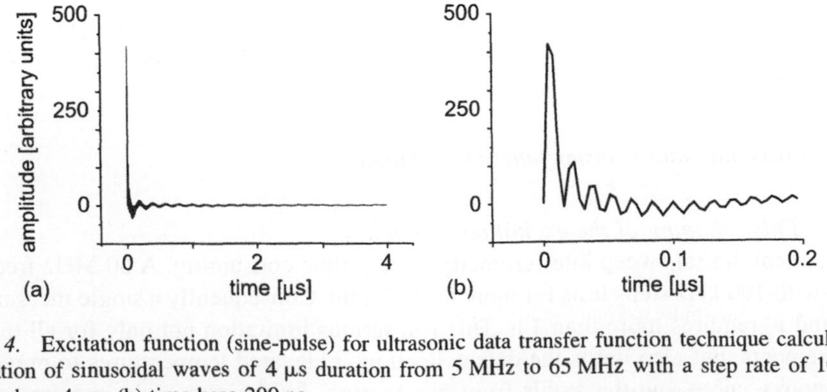

Figure 4. Excitation function (sine-pulse) for ultrasonic data transfer function technique calculated by summation of sinusoidal waves of 4 μs duration from 5 MHz to 65 MHz with a step rate of 100 kHz: (a) time base 4 μs, (b) time base 200 ns.

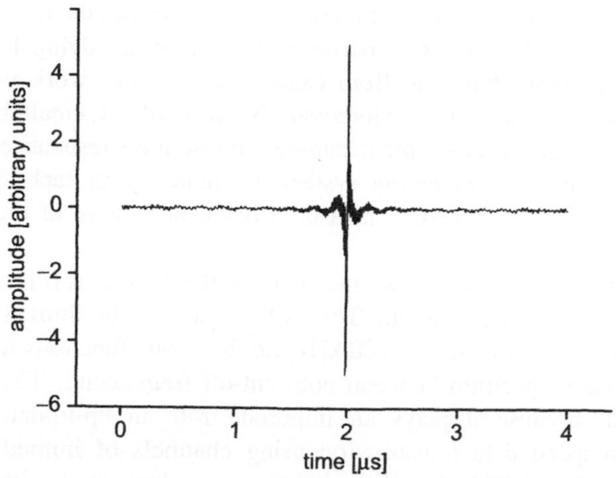

Figure 5. Completed sine-pulse excitation function created from the sine-pulse (Fig. 4) by "assembling" its copied function inverted in time and amplitude at the beginning of the original function.

culminating in two consecutive, but opposite symmetrical deflections followed by a slow tailing off. Because the function has two times the maximum duration of the used arbitrary waveform generator we have to cut-off the first and last 2 μs. This results in limited deformation of the frequency spectrum (Seidel and Myszkowski, 2004), but it is much simpler than the other option writing a sequence file, i.e. a command for using more than one file after another, because in this case we have to check very carefully that the switching time among the files is much less than the sampling rate. Otherwise we have to remove the corresponding number of points, because the function is only very effective, if it is totally symmetrical in amplitude and time. Actually this *sine-pulse* (Fig. 5) was very successfully used in many of our experiments. If we look to high-frequency engineering, pulses of this type are rarely used in spite of their effective excitation, because of their high demands on symmetry. Otherwise the spectrum deforms dramatically. This disadvantage can be limited by using a *cosine-pulse* created by adding up all single waves starting each

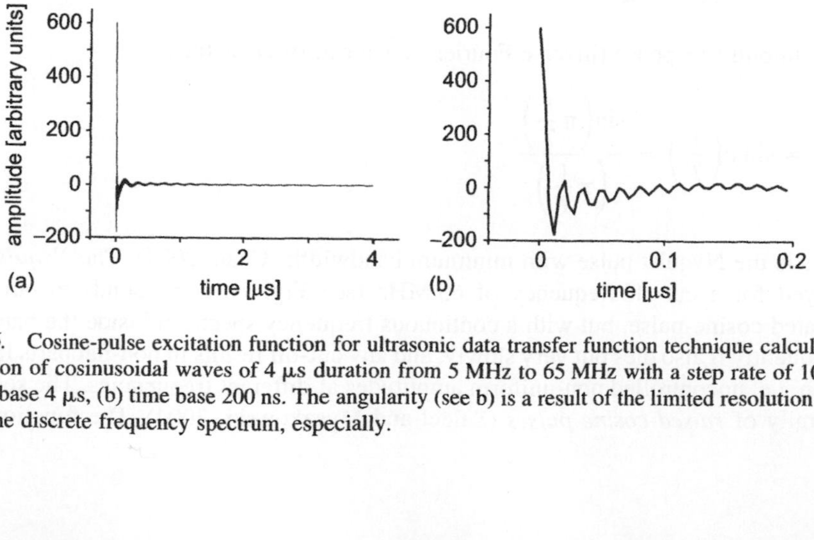

Figure 6. Cosine-pulse excitation function for ultrasonic data transfer function technique calculated by summation of cosinusoidal waves of 4 μs duration from 5 MHz to 65 MHz with a step rate of 100 kHz: (a) time base 4 μs, (b) time base 200 ns. The angularity (see b) is a result of the limited resolution in time and of the discrete frequency spectrum, especially.

of them with phase $\pi/2$, i.e. the cosine function. Figure 6 shows the result in two different time bases, analogous to Figure 8. Comparing both indicates the dying down seems to be a little faster for the cosine-function. Both excitation functions work very well and are simple to calculate by each user. Moreover, because of its creation as the sum of single wave files, it is also very simple to suppress transducers resonance very effectively, even individually for each oscillating system, by multiplying each single file with a factor corresponding to the inverted, measured resonance curve of the transducer-glue anvil system.

If we look in detail to our sine-pulse and cosine-pulse functions (Figs. 4 and 6) we find they look multi-cornered, non-smooth. This is the result of the limited sampling rate of 250 MHz and the frequency step of 100 kHz, i.e. both our functions do not represent a continuous frequency spectrum between both cut-off frequencies. The problem is very well investigated, because it plays an important role in up-to-date communication technology. High-speed data transmission using channels of limited bandwidth (cell phones, modems, digital TV, digital cameras, camcorders, etc.) without inter-symbol interference (ISI) require thorough pulse optimization. So, a channel specified by pulse response $h(t)$ is ISI free, if

$$H(e^{-j2\pi fT}) = \frac{1}{T}\sum_{n=-\infty}^{\infty} H\left(f + \frac{n}{T}\right) = 1 \tag{5}$$

This condition is called *Nyquist Criterion*. We will come back to this at the end of the section. A $h(t)$ that satisfies Nyquist criterion is called *Nyquist pulse* (Ekbal, 2004). The transfer function $H(\omega)$ of the ideal Nyquist filter is rectangular with single-sided bandwidth ω_0:

$$H(\omega)/T \begin{cases} 1 & \text{for } -1 < \omega/\omega_0 < 1 \\ 0 & \text{otherwise} \end{cases} \tag{6}$$

or

$$H(\omega) \begin{cases} T & \text{for } -17(2T) < \omega/(2\pi) < 1/(2T) \\ 0 & \text{otherwise} \end{cases} \tag{7}$$

The impulse response (inverse Fourier transform; of H) is then

$$h(t) = \sin c\left(\frac{t}{T}\right) \equiv \frac{\sin\left(\pi\frac{t}{T}\right)}{\left(\pi\frac{t}{T}\right)} \tag{8}$$

This is the Nyquist pulse with minimum bandwidth (Chan, 2004). The *Nyquist pulse*, displayed for a cut-off frequency of 65 MHz (see Fig. 7), corresponds to our above-calculated cosine-pulse, but with a continuous frequency spectrum inside the bandwidth. Consequently it also dies out very slowly, and any cut-off results in non-flat parts in spatial domain, i.e. uncontrolled non-uniform amplitudes at different frequencies. The solution is the family of *raised-cosine pulses* (Seidel and Myszkowski, 2004). The function of the

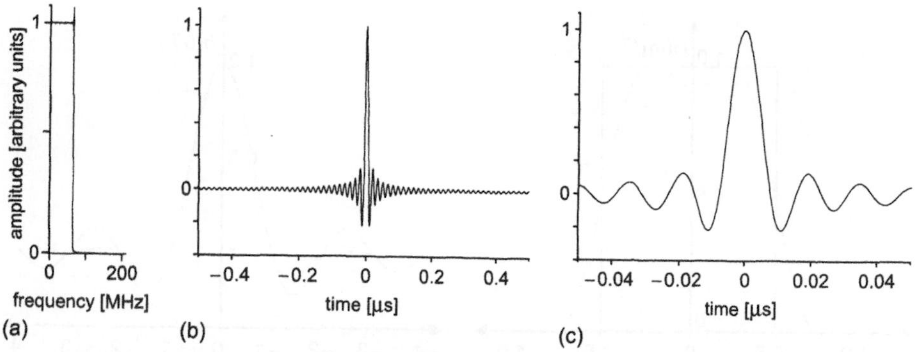

Figure 7. Sine-function for a low-pass filter with a cut-off frequency of 65 MHz, displayed in two time bases including its FFT: (a) FFT, (b) time base −0.5 to 0.5 µs, (c) time base −0.05 to 0.05 µs.

ideal Nyquist pulse – $\sin c(t/T)$ – is multiplied by an additional fall off function.

$$h(t) = \sin c\left(\frac{t}{T}\right)\left[\frac{\cos\left(\frac{\alpha\pi t}{T}\right)}{1-\left(\frac{2\alpha\pi t}{T}\right)^2}\right] \tag{9}$$

with α the roll-off factor controlling the slope steepness of frequency spectrum function, $0 < \alpha < 1$.

The raised cosine-pulse falls as $1/\alpha^2 t^3$, whereas the ideal Nyquist pulse only falls as $1/t$. The raised-cosine transfer function is the corresponding Fourier transform.

$$H(\omega) = \begin{cases} T & |\omega| \le (1-\alpha)\frac{\pi}{T} \\ \frac{T}{2}\left[1 - \sin\left(\frac{T}{2\alpha}\left(|\omega| - \frac{\pi}{T}\right)\right)\right] & (1-\alpha)\frac{\pi}{T} \le |\omega| \le (1+\alpha)\frac{\pi}{T} \\ 0 & (1+\alpha)\frac{\pi}{T} \le |\omega| \end{cases} \tag{10}$$

Figure 8 compares the Nyquist pulse and the raised-cosine pulse with different roll-off factors α in time and spatial domain (modified from Rapppaport, 2002; Chan, 2004). To complete things it should be mentioned that already further developed pulses exist as square-root raised-cosine pulse, the "better than" Nyquist-pulse (Beaulieu et al., 2001) and others. But our demands for the DTF technique are met by the raised-cosine pulse with one exception – the resonance suppression. To implement this for the available pulses we will use the fast Fourier transform (FFT), a special form of the discrete Fourier transform, available at many PC-software and installed at many digital oscilloscopes. The relation (Ekbal, 2004) between discrete-time Fourier transform $P(e^{-j2\pi fT})$ and continuous-time Fourier transform $P(f)$ is

$$P(e^{-j2\pi fT}) = \frac{1}{T}\sum_{n=-\infty}^{\infty} P\left(f + \frac{n}{T}\right) \tag{11}$$

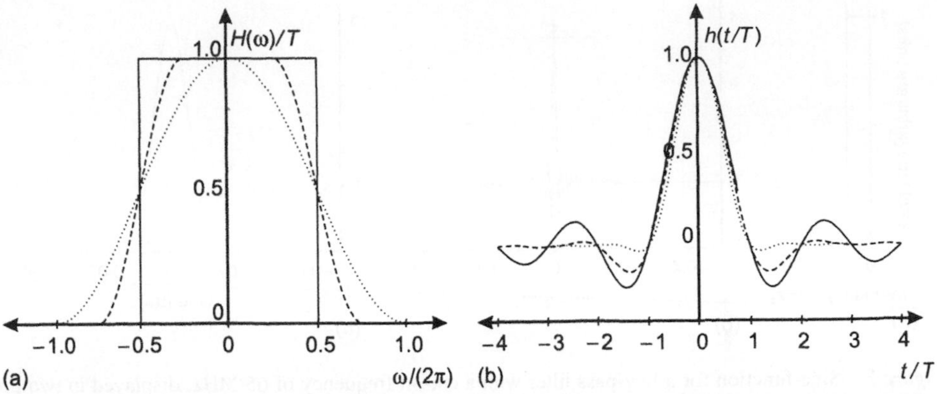

Figure 8. Comparison of transfer functions $H(\omega)$ of the ideal Nyquist pulse and the raised-cosine pulse with different roll-off factors, as well as their impulse responses (inverse Fourier transform of H) in spatial domain. A raised cosine-pulse with roll-off factor $\alpha = 0$ is an ideal Nyquist pulse (modified from Chan, 2004): —, $\alpha = 0$; ---, $\alpha = 0.5$; ·····, $\alpha = 1$.

We use our low-pass cosine-pulse with a cut-off frequency of 65 MHz and modify it with a tall resonance suppression of 85% at 33.33 MHz in shape of an inverse Gauss function and a simple rectangular cut-out inside the same frequency range −31 to 36 MHz. Figure 9 shows the results in spatial and time domain with a time base of 200 ns. The influence of the resonance suppression to the cosine-pulse in time domain is clearly visible, but a difference between the two ways of suppression is missing in this time base. Careful checking unwraps, however, there is a difference (not displayed in the figure), but it is very small and appears after about 8 μs, far outside our time span of interest. We can summarize, the implementation of a transducer resonance suppression is possible and useful with any excitation function for optimum use of the receiver sensitivity. To use the

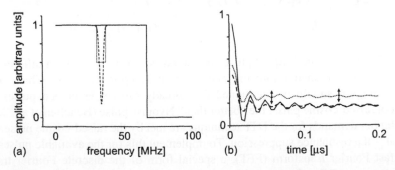

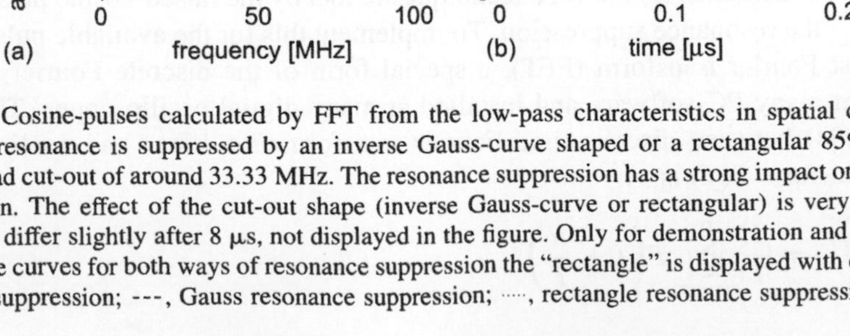

Figure 9. Cosine-pulses calculated by FFT from the low-pass characteristics in spatial domain. The transducer resonance is suppressed by an inverse Gauss-curve shaped or a rectangular 85% amplitude 5 MHz broad cut-out of around 33.33 MHz. The resonance suppression has a strong impact on the pulse in time domain. The effect of the cut-out shape (inverse Gauss-curve or rectangular) is very small. Both pulses only differ slightly after 8 μs, not displayed in the figure. Only for demonstration and to be able to compare the curves for both ways of resonance suppression the "rectangle" is displayed with offset. —, no resonance suppression; ---, Gauss resonance suppression; ·····, rectangle resonance suppression.

exact resonance curve of system might be less important, because the difference between the different shaped cut-outs is very minor. Therefore, our first simple approach, forming the excitation function from the available monochromatic waves, already worked very successfully with results comparable to those of the more sophisticated functions.

2.3.2.2. Evaluation of the data transfer function

Contrary to the sweep technique (600 files of 4 μs duration and 2048 data points) the received DTF should be saved with a much longer time base and a much higher resolution. This is plausible, because the amount of information spread out over several hundreds of files for the sweep technique is now concentrated in one single file. We saved about 40 μs with a resolution of 65,536 data points. Li et al. (2002) published 50 μs and 100,000 data points. With increased resolution the results of the evaluation of the DTF improves, i.e. the reproduction of the monochromatic signals. The received DTF is the response of the system to the excitation function containing all monochromatic frequencies between the upper and lower cut-off frequencies.

Convolving vectors u and v means, algebraically, the same operation as multiplying the polynomials whose coefficients are the elements of u and v. If $m = \text{length}(u)$ and $n = \text{length}(v)$, then w is a vector of length $m + n - 1$ whose kth element is

$$w(k) = \sum_j u(j)v(k + 1 - j) \tag{12}$$

The sum is over all the values of j which lead to legal subscripts for $u(j)$ and $v(k + 1 - j)$, specifically $j = \max(1, k + 1 - n) : \min(k, m)$. The convolution theorem says, roughly that convolving two sequences is the same as multiplying their Fourier transforms. In order to make this precise, it is necessary to pad the two vectors with zeros and ignore round-off error.

$$f \cdot g \leftrightarrow F \otimes G \tag{13}$$

That means, reproduction of all the analogues of the monochromatic signals received and saved by the sweep technique require the stepwise convolution of the DTF with each of the monochromatic frequencies. Corresponding to Eq. (12) the time axis has to be re-scaled by the factor C.

$$C = \frac{(m + n_i - 1)}{m} \tag{14}$$

with m the length of the DTF and n_i the length of the monochromatic signal amplitude.

The signal has to be strictly monochromatic. Otherwise the convolution will fail in reproducing the response of the system for this single frequency, because a non-monochromatic oscillation consists of more than one frequency peaks in spatial domain, i.e. multiplying in spatial domain would be no longer effective for selection. That also means the sampling frequency for the DTF and the signal must satisfy the Nyquist/Shannon sampling theorem (see Eq. (5)): "A signal can be properly reconstructed from its samples if the original signal is sampled at a frequency that is greater than twice

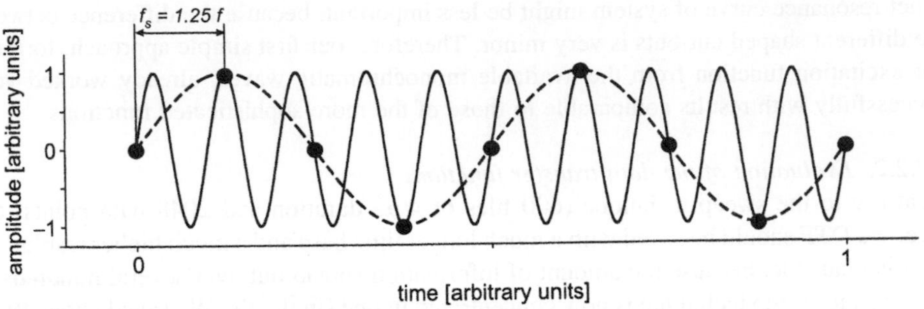

Figure 10. Violation of the Nyquist theorem, i.e. the sampling frequency has to be twice the signal frequency at the minimum, results in failing the reproduction of the original analog signal from the digitized signal. The displayed lower frequency also matches all data points. During the reproduction from the digital file this lower frequency will also appear. In computer graphics the effect results in the well-known Moiré patterns and aliasing (modified from Seidel and Myszkowski, 2004).

the highest frequency component in its spectrum."

$$f < f_{\mathrm{Ny}} \equiv \frac{1}{2} f_{\mathrm{s}} \tag{15}$$

with f the signal frequency, f_{Ny} the Nyquist frequency, and f_{s} the sampling frequency.

Otherwise the signal reproduced from the file do not represent the original signal. The data also represent another signal of lower frequency (see Fig. 10). We all know the effect from digital cameras. In digital photography and computer graphics the effect is called aliasing and results in Moiré patterns, non-existing in the original optical images.

Figure 11 compares a 36 MHz signal saved by sweep technique with the corresponding signal reproduced from the transfer function by convolution of the same experiment. The time base was adapted corresponding Eq. (14). The signals match to a great extent. Analog to the sweep technique the reproduced signals are further processed as published by Knoche et al. (1997, 1998), Li et al. (1998), Shen et al. (1998), Mueller et al. (2002, 2003, 2005). The DTF technique reduces the time for one velocity measurement from more than 30 min to a few seconds, but shifts the efforts from the measurement itself to the subsequent evaluation. Transient measurements without significant reduction of the precision are possible, if the response is measured with high resolution, because the response to all frequencies is recorded at the same time.

2.4. Sample preparation

2.4.1. San Carlos olivine

San Carlos olivine is widely used for equation-of-state investigations to derive reliable compositional constraints from observed seismic velocity profiles, because it is a good representation of the most abundant mineral in the upper mantle (Chen et al., 1996; Chang-Sheng-Zha et al., 1998). Consequently it was ideal for testing a new method, as the results can be compared with published data. The samples came from San Carlos, an ultramafic inclusion locality, about 30 km east of Globe, Arizona. A detailed petrological

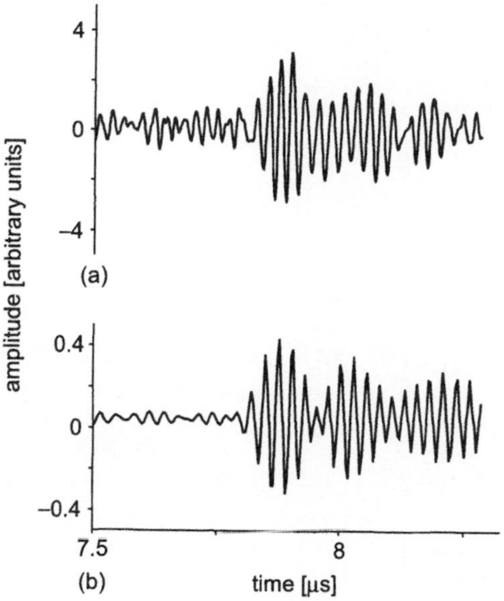

Figure 11. Comparison of a 39 MHz signal, saved during a sweep technique (b) measurement with the corresponding signal, reproduced by convolving (a) the data transfer function with a monochromatic 39 MHz oscillation.

and geochemical description is given by Frey and Prinz (1978). In the terminology of Frey and Prinz, two different types of xenoliths occur. All xenoliths investigated belong to group I. A recrystallized or annealed near-equilibrium texture is characteristic for all samples. The grain size of recrystallized olivine grains is about 30–35 μm. A few grains with wavy extinction are about several hundred micrometers in diameter (Wirth, 1996).

2.4.2. Anorthite

The anorthite samples were manufactured by E. Rybacki from crushed $CaAl_2Si_2O_8$ glass powder, which was hot-isostatically pressed and crystallized at 300 MPa confining pressure and temperatures between 900 and 1200°C in a Paterson gas pressure apparatus. The original sample size was 10 mm diameter and 20 mm length. The porosity was less than 1 vol.%, the density 2.75 g/cm^3. The grains are prismatic with an average aspect ratio of about 2.7. Twins are abundant. The (arithmetic) mean grain size is 3.4 ± 0.2 μm (for details see Rybacki et al., 1998; Rybacki and Dresen, 2000).

The grain size of the samples was measured by scanning electron microscopy. The samples of the three materials were shaped with a high-precision (±0.5 μm) cylindrical grinding machine.

2.4.3. Clinoenstatite

For the following high-pressure experiments LCEn powder was synthesized from a gel with a molar ratio of $MgO:SiO_2 = 1:1$ by heating up to 1500°C for 2 h followed by

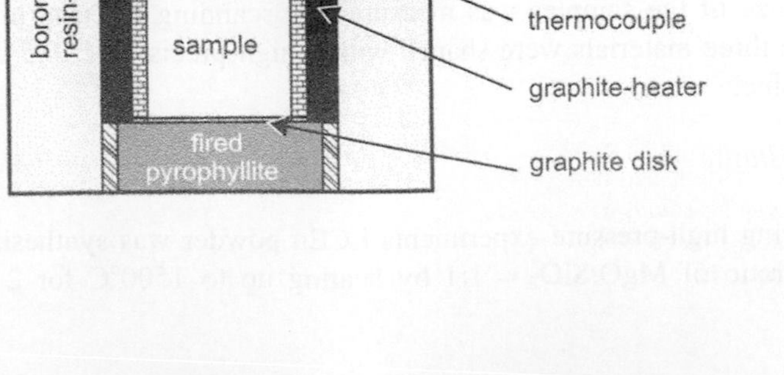
8.0 or 6.0 mm
copper ring
BN sleeve
standard (NaCl + BN)
thermocouple
graphite-heater
graphite disk
boron-epoxy resin-cube
sample
fired pyrophyllite

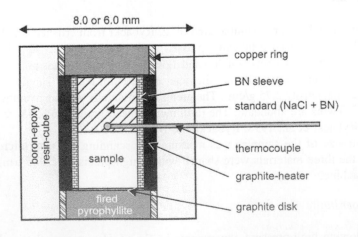

3. Results and analysis

3.1. San Carlos olivine

Figure 13 shows the results for ultrasonic compressional and shear waves of San Carlos olivine up to 3 GPa pressure using a symmetrical as well as an asymmetrical configuration.

Both elastic wave velocities were corrected for the elastic sample shortening by calculating the compressibility by successive approximation from our v_p and v_s data (Mueller et al., 2002). This approximation converges according to Banach's fixpoint theorem for meaningful v_p and v_s data. Using this method we derived velocity data independent of literature data and any sample comparison. For details of sample length measurements in multi-anvil devices, see, e.g. Mueller et al., 2005, pages 427–449, this volume. This procedure was checked by X-radiography.

In addition to our results for both cell assemblies, Figure 13 also shows the values of c_{22}, c_{33}, c_{44}, c_{55} and c_{66}, published by Webb (1989), Zaug et al. (1993) and Chen et al. (1996).

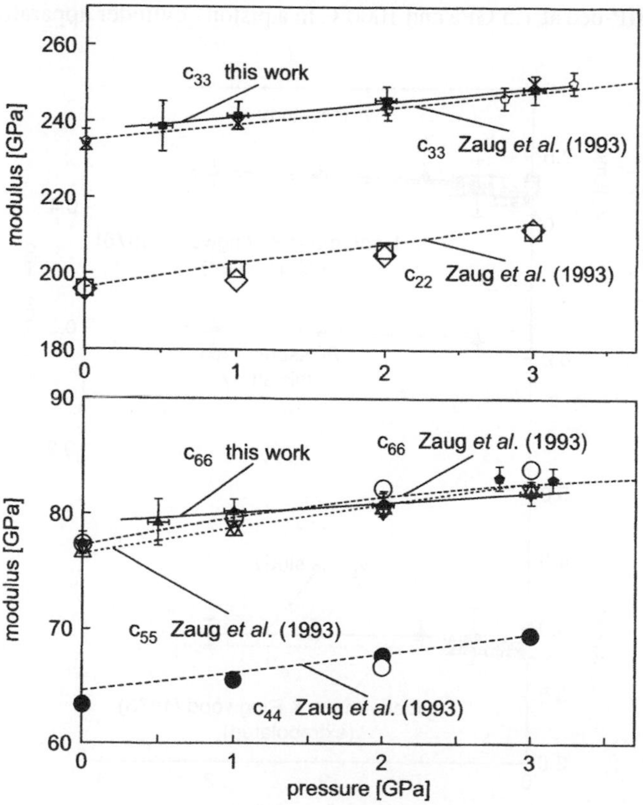

Figure 13. Elastic moduli of San Carlos olivine versus pressure up to 3 GPa measured in the multi-anvil apparatus MAX80. Published data (Webb, 1989; Zaug et al., 1993; Chen et al., 1996) are plotted for comparison. ■, c_{33} this study; ▲, c_{66} this study; ⬠, c_{33} Zaug et al. (1993); ⬟, c_{66} Zaug et al. (1993); ×, c_{33} Webb (1989); □, c_{22} Webb (1989); ●, c_{44} Webb (1989) (mode 4); △ c_{55} Webb (1989) (mode 5); ○, c_{66} Webb (1989) (mode 12); ◇, c_{22} Chen et al. (1996); ▽, c_{55} Chen et al. (1996).

The solid line is the linear least square best fit for our data. The dashed line is the second-order polynomial least square best fit for the data of Zaug et al. (1993). Our data for the [001] direction are in agreement with the published data within the limit of experimental errors (~1.5%). The measurements with the asymmetrical configuration were performed at a sample manufactured in a direction of 36° from the [001] direction at the [100] plane. Consequently the modulus derived from the compressional wave velocity (not displayed) has a value between the published data for c_{22} and c_{33}, whereas published c_{55}, c_{66} and the module, calculated from our corresponding transverse wave data correspond to each other in the limits of experimental uncertainty (~1.5%). The data were compared by an in-house program (Schilling, 1998) using a solution for the Christoffel equation. The reference data were taken from Landolt-Bömstein (Hearmon, 1984).

3.2. Anorthite

Figure 14 compares our high pressure v_p and v_s data (solid line) for polycrystalline anorthite with hydrostatic high-pressure results, measured up to 0.75 GPa at polycrystalline samples, HIP-ped at 1.5 GPa and 1000°C in a piston–cylinder apparatus, published by

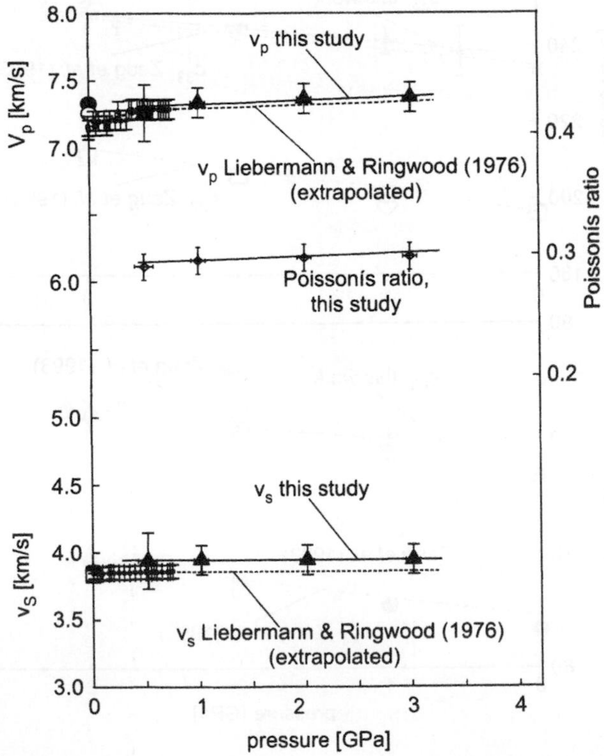

Figure 14. v_p, v_s and Poisson's ratio ν of anorthite versus pressure up to 3 GPa measured in the multi-anvil apparatus MAX80. Published data (Liebermann and Ringwood, 1976; Bass, 1995) are plotted for comparison. ●, arithmetic mean value of v_p (Bass, 1995); ○, geometric mean value of v_p (Bass, 1995); ■, arithmetic mean value of v_s (Bass, 1995); □, geometric mean value of v_s (Bass, 1995).

Liebermann and Ringwood (1976, dashed line) and with the arithmetic and geometric mean values of v_p and v_s deduced from normal-pressure elastic moduli published by Gebrande (1982) and Bass (1995). Our graphs correspond to the normal-pressure data (Bass, 1995) and to the high-pressure data up to 0.75 GPa of Liebermann and Ringwood (1976) within the limits of experimental errors (~1.3 to 1.7%). Our pressure derivatives (linear best fit) also correspond to the extrapolated high-pressure data of Liebermann and Ringwood (1976) (second-order polynomial best fit) within these limits. The unusual pressure independency of the shear wave velocity, i.e. v_s is constant with pressure within the precision of the pulse transmission technique (±0.05 km/s), described by Liebermann and Ringwood (1976) is validated by our measurements for polycrystalline samples. The slight deviation to the absolute values might be caused by the fact that the remaining pore space in our sample might be smaller and that a more uniform crystallization seems to be achieved, due to hot-isostatic pressing in a Paterson apparatus. The central graph is the Poisson's ratio, about 0.29 at room conditions with a slight increase to 0.30 at 3 GPa pressure at 20°C, determined from the presented velocity data. The sample shortening under pressure was derived from the compressibility the same way as described in Section 3.1.

3.3. Clinoenstatite

Regression analysis (Belsley et al., 1980; Holland and Redfern, 1997) was used for lattice-refining the energy-dispersive X-ray data of HCEn and LCEn determined with MAX80 at HASYLAB (see Fig. 15). The method is capable of determining lattice parameters with high resolution from the *in situ* X-ray results, i.e. powder-diffraction data from a beam of white synchrotron radiation impinging on a small sample surrounded by heater, electrical insulator and gasket material. The narrow slits between the X-ray absorbing anvils result in an observation of a limited part of the diffraction cone. The regression diagnostics refinement is based on minimization of the differences between the measured d_{hkl} and its calculated values. The modeling was performed using the program UnitCell from Department of Earth Sciences, Cambridge University.

We started the high pressure/high temperature experiments (run 3/24, see Fig. 16) using the set-up shown in Figure 12 with a sample of pure LCEn powder. By raising the pressure above 6.5 GPa at RT pure HCEn was formed. The phase transition was observed by *in situ* XRD measurements. The pressure and temperature of the first appearance of LCEn in the X-ray diffraction pattern was determined by successively raising the temperature in steps of 50 K at a given P (Fig. 16). This procedure was performed for three different P conditions, 6.61(5), 7.20(5) and 7.50(5) GPa. We used the reaction from HCEn–LCEn to determine the phase boundary, as this reaction is kinetically less hindered than the back-reaction.

Run 3/25 (see Fig. 16) is the continuation of run 3/24 at a pressure of 7.89(5) GPa. To ensure that the results are not distorted by hysteresis effects because of multiple crossing the phase boundary back and forth, we only crossed the phase boundary once by increasing temperature in this experiment. Figure 15 shows the energy-dispersive XRD spectra of HCEn and LCEn as measured in MAX80 under *in situ* conditions at 6.61(5) GPa and 250 and 300°C, respectively. During the phase transition the position of the strongest

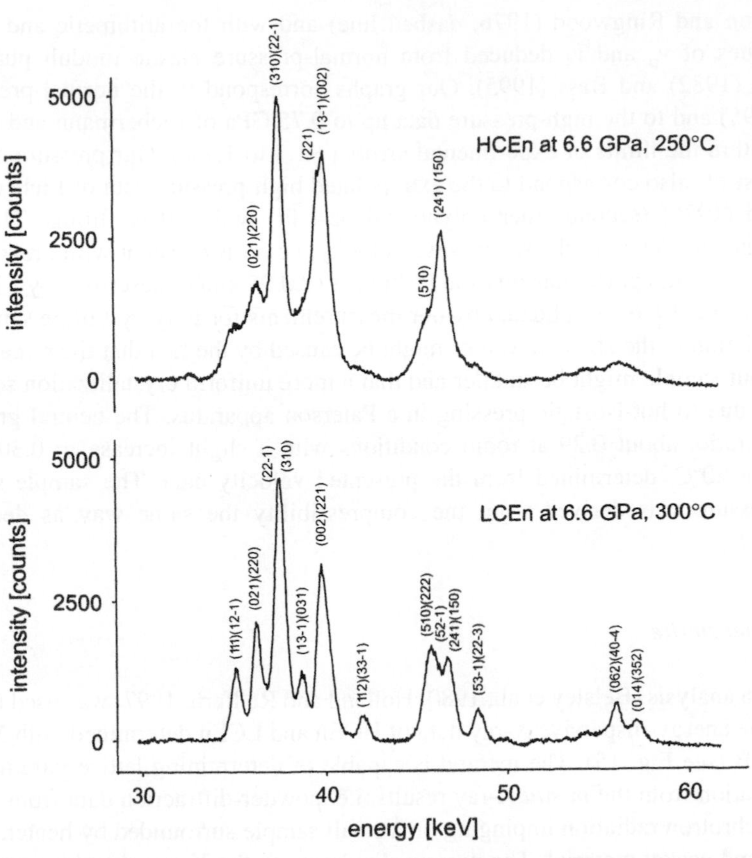

Figure 15. XRD data for HCEn and LCEn at 6.6 GPa and 250/300°C, measured in MAX80. The stronger peaks have a limited significance for phase detection because the energy shift is very small. Several smaller peaks are used for phase identification.

diffraction lines change by only a small amount. However, between four and seven diffraction lines with lower intensity could be used to securely distinguish between LCEn and HCEn.

Run 3/26 (see Fig. 16) reproduced the *P*, *T* regime of run 3/24 with a slightly higher pressure, but the sample was a 1:1 per volume mixture of HCEn and hBN. The phase boundary was crossed at pressures of 6.74(5), 6.93(5) and 7.28(5) GPa, respectively. Due to the dilution of CEn in BN, the detected intensity of the diffraction lines was less for the two-component samples in run 3/26, but still sufficient for the evaluation. The experiment with the mixed sample gave slightly higher pressures for the phase boundary. This might be a result of the different compressibility of both constituents of the mixture, resulting in lower pressure in the more compressible medium, as discussed by Dietrich and Arndt (1982) and Will et al. (1982). Consequently, the data of the first cycle of run 3/24 and run 3/25 mostly define the maximum temperature conditions of the phase boundary as shown in Figure 16. The solid line between the last existence of HCEn and the first appearance of LCEn is the best fit to our results. Our results only represent the minimum *P* conditions of the HCEn–LCEn phase boundary, which is approximated by *P* (GPa) = 0.0021 (GPa/°C)

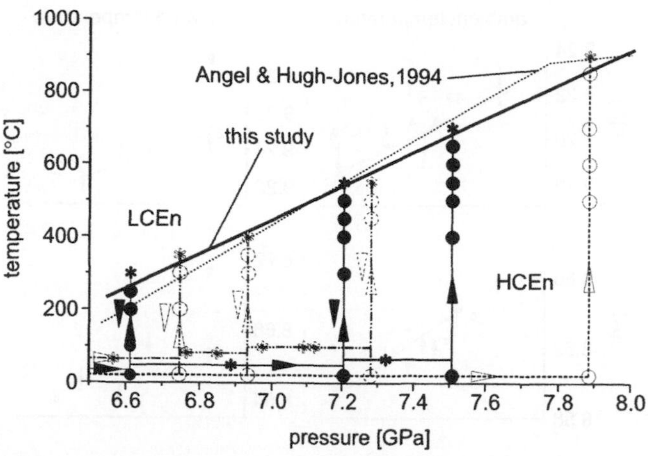

Figure 16. Scheme of experimental runs to determine the HCEn–LCEn phase boundary. Run 3/24 and 3/26 crossed the phase boundary HCEn–LCEn three times. The arrows indicate the pressure temperature path of the experiments. By raising the pressure above 6.5 GPa at RT HCEn was formed followed by increasing the temperature in steps of 50 K up to the first existence of HCEn. Then the sample was cooled by switching off the power and the pressure was increased to form HCEn again. To rule out hysteresis effects run 3/25 crossed the phase boundary only once. Run 3/26 used a sample mixture of clinoenstatite and hBN 1:1. The solid line represents our results of the maximum temperature condition of the HCEn–LCEn phase boundary. The dotted line represents the data published by Angel and Hugh-Jones (1994). —, run 3/24 (100% CEn); ····, Run 3/25 (100% CEn); ---, run 3/26 (50% CEn + 50% hBN); —, this study; ✱, LCEn; ●, HCEn.

T (°C) + 6.048 (GPa). Nevertheless, our results (see Fig. 16) fall within the pressure range determined by Angel and Hugh-Jones (1994) at ambient conditions. The invariant point defined by the intersection of the HCEn–LCEn equilibrium determined within this study is in good accordance with the invariant point deduced by OEn–LCEn reaction after Angel and Hugh-Jones (1994) which lies at about 7.9 GPa and 865°C. This is contrast to the experimental results of Kanzaki (1991) and Ulmer and Stalder (2001).

Figure 17 compares our cell parameters with the results of Angel and Hugh-Jones (1994) and Shinmei et al. (1999) for HCEn at ambient temperature and with the results of Shinmei et al. (1999) for HCEn at maximum temperature near the phase transition, see also Table 1. The results are in good agreement in the limits of the 2σ experimental uncertainty. The unit-cell volumes (Table 1) of this study correspond to those determined by Shinmei et al. (1999) within a multi-anvil press and at pressures $<$7 GPa with those by Angel and Hugh-Jones (1994) from as diamond anvil study using synthetic single crystals of CEn.

Figure 18 shows the results of our ultrasonic experiments with a HIP-ped clinoenstatite sample. Similar to run 3/24 (see Fig. 16) we targeted to transform the sample to a minimum porosity HCEn sample by raising the pressure at normal temperature up to 6.7 GPa. Because there is some indication that the sample was not completely transformed to HCEn before passing the phase boundary to LCEn between 250 and 300°C we do not report this ultrasonic results. A further temperature increase up to 700°C ensured representative ultrasonic data, because the measurements started deep inside the

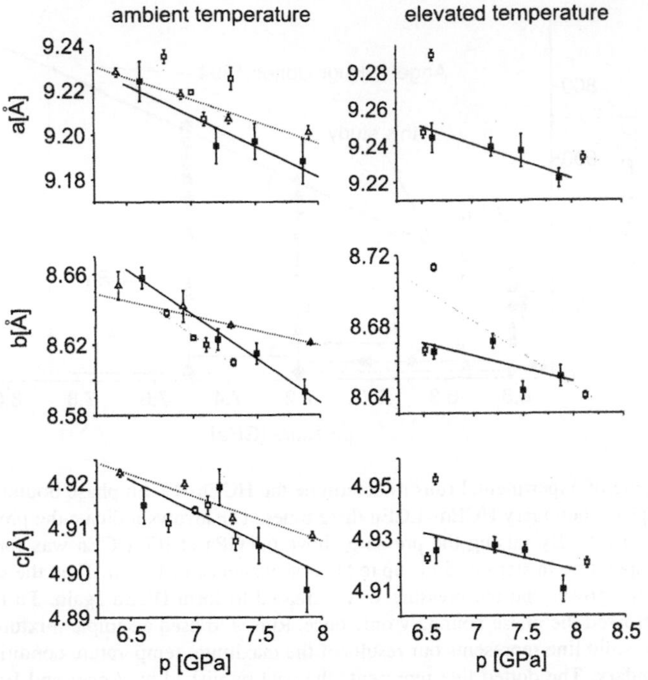

Figure 17. Variation of HCEn unit cell parameters with pressure at RT and at elevated temperatures, close to the HCEn–LCEn phase boundary. The results of this study (■) (—) (see also Table 1) are compared with the data published by Shinmei et al (1999)(□) (…) and Angel and Hugh-Jones (1994) (△) (- - -). The lines represent the least square linear best fit for the data sets of this study and of the comparative authors.

LCEn-stability field and the sample never left it during the following pressure increase up to 7.5 GPa. The gradual temperature increase at constant pressure load also targets a minimum deviatoric stress inside the sample. The displayed best fit lines represent the velocity dependence on pressure at constant temperature of 700°C for LCEn. For v_p and v_s a temperature derivative at 700°C of 0.8 and 0.7 km/(s GPa) was determined, respectively. The ultrasonic measurements were performed using the new developed ultrasonic transfer function DTF technique. To compare the results performed at 6.7 and 7.5 GPa, v_p and v_s were also measured using the classical sweep technique. The data are in good agreement.

Recently Kung et al. (2004) published the results of very thorough and innovative experiments on elastic wave velocities at the orthopyroxene–HCEn transformation, as a systematic continuation of the measurements of Flesch et al. (1998) with orthopyroxene up 10 GPa. Different from our experiments an OEn sample entered the HCEn-stability field at much higher pressures and temperatures of about 16 GPa/650°C.

The LCEn–HCEn phase transition might be an important reaction in deeper parts of cold, fast subducting slabs, where the temperature increase is retarded. Our preliminary results indicate a velocity drop of less than 0.5% within a cold, fast subducting pyrolitic mantle.

During the transformation of the 1:1 sample a strong reduction in porosity is observed. This indicates that at the phase transition the rheological behavior of the sample allows a

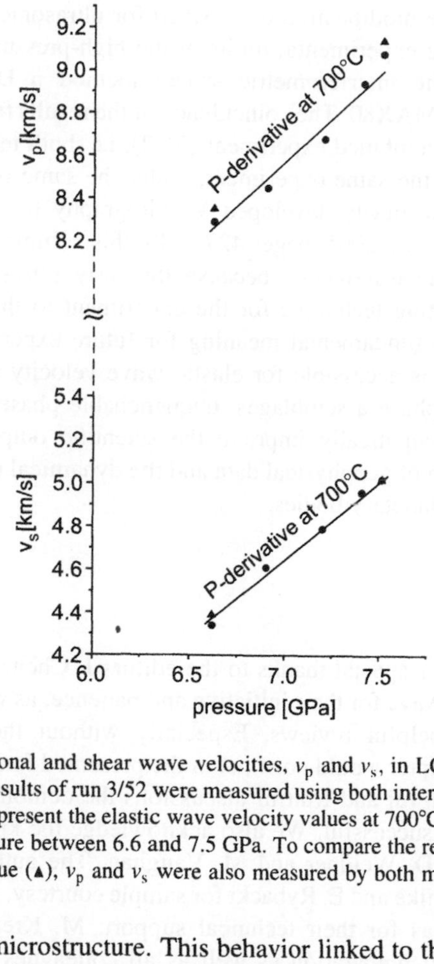

Figure 18. Compressional and shear wave velocities, v_p and v_s, in LCEn in dependence on pressure at 700°C. The displayed results of run 3/52 were measured using both interferometric techniques (see text for details). The data (●) represent the elastic wave velocity values at 700°C, measured by the DTF technique in dependence on pressure between 6.6 and 7.5 GPa. To compare the results of the DTF method with the classical sweep technique (▲), v_p and v_s were also measured by both methods at 6.6 and 7.5 GPa.

modification of its microstructure. This behavior linked to the CEn phase transformation can be explained by transformation plasticity (e.g. Poirier, 1982; Schmidt et al., 2002). Therefore, a reduced shear strength related to the CEn transition might result in a markedly reduced viscosity of CEn-bearing rocks and should influence the rheology of the lithospheric mantle of down-going slabs.

4. Conclusions

The results show the power of the demonstrated ultrasonic interferometric measurements in conjunction with XRD in multi-anvil devices under simulated Earth's mantle conditions. The results for San Carlos olivine and HIP-polycrystalline anorthite were compared with published data and illustrate the accuracy and reliability of the method. The results for clinoenstatite demonstrate the potential of simultaneous elastic and X-ray measurements to study unquenchable phase transitions.

For the optimum adaptation to different samples and experimental conditions several cell assemblies and the corresponding anvils were developed and tested under high-pressure conditions. The modification of MAX80 for ultrasonic measurements had no negative side effect on the experimental limits of the high-pressure/high-temperature apparatus. In addition to the interferometric sweep method a DTF technique was developed and optimized for MAX80. The coincidence of the results from both techniques could be demonstrated by a combined experiment (3/52), i.e. both techniques were used for the same sample, during the same experiment, under the same pressure/temperature conditions. Together with the newly developed X-radiography for *in situ* deformation measurements (see Mueller et al., 2005, pages 427–449, this volume) the DTF technique allows extensive transient measurements, because this way ultrasonic interferometry changed from the most limiting technique for the experiment to the fastest one of the applied methods. This has a fundamental meaning for future experiments, because the kinetics of phase transitions is accessible for elastic wave velocity measurements now. Experiments with complex phase assemblages, unquenchable phases, volatile-saturated and molten systems will dramatically improve the scientific output of high-pressure research for the interpretation of geophysical data and the dynamical understanding of the interior of Earth and other planetary bodies.

Acknowledgements

We would like to express our special thanks to the editors J. Chen, Y. Wang, T. Duffy, G. Shen, and L. Dobrzhinetskaya for their initiative and patience, as well as two unknown references for their very helpful reviews. Especially without the very constructive guidance of Y. Wang the chapter would not exist in the presented form. Our special thank is for B. Li. Without his helpful and fruitful discussions the demonstrated development would have been much less successful. We also acknowledge the kind guidance and the support of B. Liebermann, D. Weidner and M. Vaughan. The authors are particularly indebted to B. Wunder, R. Milke and E. Rybacki for sample courtesy, including the sample HIP by the latter, as well as for their technical support; M. Kreplin, G. Berger and S. Gehrmann for the special preparation as well as all colleagues of both mechanical workshop for their dedicated support.

References

Angel, R.J., Hugh-Jones, D.A., 1994. Equations of state and thermodynamic properties of enstatite pyroxenes. J. Geophys. Res. 99 (B10), 19777–19783.

Angel, R.J., Ross, N.L., 1996. Compression mechanisms and equation of state. Philos. Trans. R. Soc. London, A 354, 1449–1459.

Angle, R.J., Chopelas, A., Ross, N.L., 1992. Stability of high-density clinoenstatite at upper-mantle pressures. Nature 358, 322–324.

Bass, J.D., 1995. Elasticity of minerals, glasses and melts. In: Ahrens, Th.J. (Ed.), Mineral Physics and Crystallography: A Handbook of Physical Constants, AGU Reference Shelf 2, pp. 45–63.

Beaulieu, N.C., Tan, C.C., Damen, M.O., 2001. A "better than" Nyquist pulse. IEEE Commun. Lett. 5 (No. 9), 367–368.

Belsley, D.A., Kuh, E., Welsh, R.E., 1980. Regression Diagnostics: Identifying Influential Data and Sources of Colinearity. Wiley, New York.

Birch, F., 1960. The velocity of compressional waves in rocks to 10 kilobars. Part 1. J. Geophys. Res. 65, 1083–1102.

Birch, F., 1961. The velocity of compressional waves in rocks to 10 kilobars. Part 2. J. Geophys. Res. 66, 2199–2224.

Bowen, N.L., Andersen, O., 1914. The binary system $MgO-SiO_2$. Am. J. Sci. 37, 487–500.

Chan, H.A., 2004. EEE482F Telecommunication. Pulse Shaping. [online] [cited 2004-04-02]. Available from: http://www.eng.uct.edu.za/~achan/eee482f/note/eee482f-0009.ppt.

Chang-Sheng-Zha, Duffy, T.S., Downs, R.T., Ho-Kwang-Mao, Hemley, R.J., 1998. Brillouin scattering and X-ray diffraction of San Carlos olivine: direct pressure determination to 32 GPa. Earth Planet. Sci. Lett. 159 (1–2), 25–33.

Chen, G., Li, B., Liebermann, R.C., 1996. Selected elastic moduli of single-crystal olivines from ultrasonic experiments to mantle pressures. Science 272, 979–980.

Decker, D.L., 1971. High-pressure equation of state for NaCl, KCl, and CsCl. J. Appl. Phys. 42, 3239–3244.

Dietrich, P., Arndt, J., 1982. Effect of pressure and temperature on the physical behavior of mantle-relevant olivine, orthopyroxene and garnet. In: Schreyer, W. (Ed.), High-Pressure Researches in Geoscience. E. Schweizerbart'sche Verlagsbuchhandlung, Stuttgart, pp. 293–306.

Ekbal, A., 2004. Receiver SNR and Nyquist pulses, [online]. [Cited 2004-02-03]. Portable Document Format. Available from: http://www.stanford/edu/class/ee379a/handouts/lec8.pdf.

Flesch, L.M., Li, B., Liebermann, R.C., 1998. Sound velocities of polycrystalline $MgSiO_3$-orthopyroxene to 10 GPa at room temperature. Am. Mineral. 83, 444–450.

Frey, F.A., Prinz, M., 1978. Ultramafic inclusions from San Carlos, Arizona: petrologic and geochemical data bearing on their petrogenesis. Earth Planet. Sci. Lett. 38, 129–176.

Funamori, N., Yagi, T., Uchida, T., 1996a. High-pressure and high-temperature in situ x-ray diffraction study of iron to above 30 GPa using MA8-type apparatus. Geophys. Res. Lett. 23 (No. 9), 953–956.

Funamori, N., Yagi, T., Utsumi, W., Kondo, T., Uchida, T., 1996b. Thermoelastic properties of $MgSiO_3$ perovskite determined by in situ x-ray observations up to 30 GPa and 2000 K. J. Geophys. Res. 101, 8257–8269.

Gebrande, H., 1982. Elasticity and inelasticity. In: Angenheister, G. (Ed.), Physical Properties of Rocks, Landoldt-Börnstein, group V, Vol. 1, Subvol. b, Springer, Berlin, pp. 1–147.

Getting, I.C., 1998. The practical pressure scale: fixing fixed points and future prospects. Eos 79, F830.

Haussuehl, S., Wallrafen, F., Recker, K., Eckstein, J., 1980. Growth, elastic properties and phase transition of orthorhombic $Li_2Ge_7O_{15}$. Z. Kristallogr. 153, 329–337.

Hearmon, R.F.S., 1984. The elastic constants of crystals and other anisotropic materials. In: Hellwege, K.H., Hellwege, A.M. (Eds), Landolt-Börnstein Tables, III/18, 559 p. Springer, Berlin, pp. 1–154.

Hirose, K., Fei, Y., Ono, S., Yagi, T., Funakoshi, K., 2001. In situ measurements of the phase transition boundary in $Mg_3Al_2Si_3O_{12}$: implications for the nature of the seismic discontinuities in the Earth's mantle. Earth Planet. Sci. Lett. 184, 567–573.

Hoffbauer, W., Will, G., Lauterjung, J., 1985. Compressibility of forsterite up to 300 kbar measured with synchrotron radiation. Z. Kristallogr. 170, 80–81, (in German).

Holland, T.J.B., Redfern, S.A.T., 1997. Unit cell refinement from powder diffraction data: the use of regression diagnostics. Mineral. Mag. 61, 65–77.

Kanzaki, M., 1991. Ortho/clinoenstatite transition. Phys. Chem. Miner. 17, 726–730.

Kern, H., 1982. Elastic-wave velocity in crustal and mantle rocks at high pressure and temperature: the role of the high–low quartz transition and of dehydration reactions. Phys. Earth Planet. Interiors 29, 12–23.

Knoche, R., Webb, S.L., Rubie, D.C., 1997. Experimental determination of acoustic wave velocities at Earth mantle condition using a multianvil press. Phys. Chem. Earth 22, 125–130.

Knoche, R., Webb, S.L., Rubie, D.C., 1998. Measurements of acoustic wave velocities at $P-T$ conditions of the Earth's mantle. In: Manghnani, M.H., Yagi, T. (Eds), Properties of Earth and Planetary Materials at High Pressure and Temperature, Geophysical Monograph 101. AGU, Washington, DC, pp. 119–128.

Kohlstedt, D.L., Keppler, H., Rubie, D.C., 1996. Solubility of water in the α, β and γ phases of $(Mg, Fe)_2SiO_4$. Contrib. Mineral. Petrol. 123, 345–357.

Kung, J., Gwanmesia, G.D., Liu, J., Li, B., Liebermann, R.C., 2000. PV3T experiments: simultaneous measurement of sound velocities (V_p and V_s) and sample volume (V) of polycrystalline specimens of mantle minerals at high pressures (P) and temperatures (T). Eos 81 (No. 48), F1151.

Kung, J., Angel, R.J., Ross, N.L., 2001a. Elasticity of $CaSnO_3$ perovskite. Phys. Chem. Miner. 28, 35–43.

Kung, J., Weidner, D.J., Li, B., Liebermann, R.C., 2001b. Determination of the elastic properties at high pressure without pressure scale. Eos 82, F1383.

Kung, J., Li, B., Weidner, D.J., Zhang, J., Liebermann, R.C., 2002. Elasticity of (Mg0.83 Fe0.17)O ferropericlase at high pressure: ultrasonic measurements in conjunction with X-radiation techniques. Earth Planet. Sci. Lett. 203, 557–566.

Kung, J., Li, B., Uchida, T., Wang, Y., Neuville, D., Liebermann, R.C., 2004. In situ measurements of sound velocities and densities across the orthopyroxene – high pressure clinopyroxene transition in $MgSiO_3$ at high pressure. Phys. Earth Planet. Interiors 147, 27–44.

Lauterjung, J., Will, G., 1985. Investigation of olivine–spinel phase transition with time-delayed energy-dispersive X-ray diffraction using the example of Mg_2GeO_4. Z. Kristallogr. 170, 117–119, (in German).

Li, B., 2003. Compressional and shear wave velocities of ringwoodite γ-Mg_2SiO_4 to 12 GPa. Am. Mineral. 88, 1312–1317.

Li, B., Gwanmesia, G.D., Liebermann, R.C., 1996a. Sound velocities of olivine and beta polymorphs of Mg_2SiO_4 at Earth's transition zone pressures. Geophys. Res. Lett. 23 (No. 17), 2259–2262.

Li, B., Jackson, I., Gasparik, T., Liebermann, R.C., 1996b. Elastic wave velocity measurement in multi-anvil apparatus to 10 GPa using ultrasonic interferometry. Phys. Earth Planet. Interiors 98, 79–91.

Li, B., Chen, G., Gwanmesia, G.D., Liebermann, R.C., 1998. Sound velocity measurements at mantle transition zone conditions of pressure and temperature using ultrasonic interferometry in a multianvil apparatus. In: Manghnani, M.H., Yagi, T. (Eds), Properties of Earth and Planetary Materials at High Pressure and Temperature, Geophysical Monograph 101. AGU, Washington, DC, pp. 41–61.

Li, X., Kind, R., Priestley, K., Sobolev, S.V., Tilmann, F., Yuan, X., Weber, M., 2000. Mapping the Hawaiian plume conduit with converted seismic waves. Nature 405, 938–941.

Li, B., Liebermann, R.C., Weidner, D.J., 2001a. P–V–V_p–V_s–T measurement on wadsleyite to 7 GPa and 873 K: implications for the 410-km seismic discontinuity. J. Geophys. Res. 106 (B12), 30575–30591.

Li, B., Vaughan, M.T., Kung, J., Weidner, D.J., 2001b. Direct length measurement using X-radiography for the determination of acoustic velocities at high pressure and high temperature. NSLS Activity Report 2, 103–106.

Li, B., Chen, K., Kung, J., Liebermann, R.C., Weidner, D.J., 2002. Sound velocity measurement using transfer function method. J. Phys. Condens. Matter 14, 11337–11342.

Liebermann, R.C., Li, B., 1998. Elasticity at high pressure and temperatures. In: Hemley, R. (Ed.), Ultrahigh-Pressure Mineralogy, Rev. Mineral., Vol. 37, pp. 459–492.

Liebermann, R.C., Ringwood, A.E., 1976. Elastic properties of anorthite and the nature of the lunar crust. Earth Planet. Sci. Lett. 31, 69–74.

McSkimin, H.J., 1950. Ultrasonic measurement techniques applicable to small solid specimens. J. Acoust. Soc. Am. 22, 413–418.

Mueller, H.J., Lauterjung, J., Schilling, F.R., Lathe, C., Nover, G., 2002. Symmetric and asymmetric interferometric method for ultrasonic compressional and shear wave velocity measurements in piston–cylinder and multi-anvil high-pressure apparatus. Eur. J. Mineral. 14, 581–589.

Mueller, H.J., Schilling, F.R., Lauterjung, J., Lathe, C., 2003. A standard free pressure calibration using simultaneous XRD and elastic property measurements in a multi-anvil device. Eur. J. Mineral. 15, 865–873.

Mueller, H.J., Schilling, F.R., Lathe, C., 2005. Calibration based on a primary pressure scale in a multi-anvil device. In: Chen, J., Wang, Y., Duffy, T., Shen, G., Dobrzhinetskaya, L. (Eds.), Frontiers in High Pressure Research: Geophysical Applications. Elsevier Science, Amsterdam, pp. 427–449.

Niesler, H., Jackson, I., 1989. Pressure derivatives of elastic wave velocities from ultrasonic interferometric measurements on jacketed polycrystals. J. Acoust. Soc. Am. 86, 1573–1585.

Nishiyama, N., Yagi, T., 2003. Phase relation and mineral chemistry in pyrolite to 2200°C under the lower mantle pressures and implications for dynamics of mantle plumes. J. Geophys. Res. 108, 7-1–7-12.

Oguri, K., Funamori, N., Uchida, T., Miyajima, N., Yagi, T., Fujino, K., 2000. Post-garnet transition in the natural pyrope: a multi-anvil study based on in situ X-ray diffraction and transmission electron microscopy. Phys. Earth Planet. Interiors 122, 175–186.

Ohashi, Y., 1984. Polysynthetically-twinned structures of enstatite and wollastonite. Phys. Chem. Miner. 10, 217–229.

Poirier, J.P., 1982. On transformation plasticity. J. Geophys. Res. 87, 6791–6797.

Presnall, D.C., 1995. Phase diagrams of earth-forming minerals. In: Ahrens, T.J. (Ed.), Mineral Physics and Crystallography, A Handbook of Physical Constants. Am. Geophys. Union, AGU Reference Shelf 2, Washington, pp. 248–268.

Rapppaport, T.S., 2002. Wireless communications. Principles and Practice. 2/2. Chap. 6. Modulation Techniques for Mobile Radio. [online] [Cited 2002 Pearson]. Available from: http://www.utexas.edu/~wireless/EE381K11_Spring03/Presentation_6.ppt.

Rigden, S.M., Jackson, I., Niesler, H., Liebermann, R.C., Ringwood, A.E., 1988. Pressure dependence of the elastic wave velocities for Mg_2GeO_4 spinel up to 3 GPa. Geophys. Res. Lett. 15, 604–608.

Rigden, S.M., Gwanmesia, G.D., Jackson, I., Liebermann, R.C., 1992. Progress in high-pressure ultrasonic interferometry, the pressure dependence of elasticity of Mg_2SiO_4 polymorphs and constraints on the composition of the transition zone of the Earth's mantle. In: Syono, Y., Manghnani, M. (Eds), High Pressure Research: Application to Earth and Planetary Sciences. Terra Scientific Publishing Co. and AGU, Tokyo and Washington, DC, pp. 167–182.

Rigden, S.M., Gwanmesia, D., Liebermann, R.C., 1994. Elastic wave velocities of a pyrope–majorite garnet to 3 GPa. Phys. Earth Planet. Interiors 86, 35–44.

Rybacki, E., Dresen, G., 2000. Dislocation and diffusion creep of synthetic anorthite aggregates. J. Geophys. Res. 105 (B11), 26017–26036.

Rybacki, E., Dresen, D.H., Wirth, R., 1998. Creep of synthetic anorthosite at high temperature and high pressure. Eos 79, F852.

Schilling, F.R., 1998. PETROPHYSIK – Ein mineralogischer Ansatz, Habilitationsschrift, Fachbereich Geowissenschaften, Freie Universität Berlin, p. 259, (in German).

Schmidt, C., Bruhn, D., Wirth, R., 2002. Experimental evidence of transformation plasticity in silicates: minimum of creep strength in quartz. Earth Planet. Sci. Lett. 6468, 1–8.

Schreiber, E., Anderson, O.L., Soga, N., 1973. Elastic Constants and their Measurement. McGraw-Hill Book Company, New York, p. 196.

Seidel, H-P., Myszkowski, K., 2004. Computer graphics – signal processing. [online] Portable Document Format. Available from: http://mpi-sb.mpg.de/units/ag4/teaching/uebung/lecture23.pdf.

Shen, A.H., Reichmann, H.-J., Chen, G., Angel, R.J., Bassett, W.A., Spetzler, H., 1998. GHz ultrasonic interferometry in a diamond anvil cell: P-wave velocities in periclase to 4.4 GPa and 207°C. In: Manghnani, M.H., Yagi, T. (Eds), Properties of Earth and Planetary Materials at High Pressure and Temperature, Geophysical Monograph 101. AGU, Washington, DC, pp. 71–77.

Shimomura, O., Yamaoka, S., Yagi, T., Wakutsuki, M., Tsuji, K., Kawamura, H., Hamaya, N., Fukunaga, O., Aoki, K., Akimoto, S., 1985. Multi-anvil type X-ray system for synchrotron radiation. In: Minomura, S. (Ed.), Solid State Physics Under Pressure. Terra Science Publishing, Tokyo, pp. 351–356.

Shinmei, T., Tomioka, N., Fujino, K., Kuroda, K., Irifune, T., 1999. In situ X-ray diffraction study of enstatite up to 12 GPa and 1473 K and equations of state. Am. Mineral. 84, 1588–1994.

Spetzler, H., Yoneda, A., 1993. Performance of the complete travel-time equation of state at simultaneous high pressure and temperature. Pure Appl. Geophys. 141, 379–392.

Suzuki, T., Akaogi, M., Yagi, T., 1996. Pressure dependence of Ni, Co and Mn partitioning between iron hydride and olivine, magnesiowüstite and pyroxene. Phys. Earth Planet. Interiors 96, 209–220.

Ulmer, P., Stalder, R., 2001. The $Mg(Fe)SiO_3$ orthoenstatite–clinoenstatite transitions at high pressures and temperatures determined by Raman-spectroscopy on quenched samples. Am. Mineral. 86, 1267–1274.

van der Hilst, R., 1995. Complex morphology of subducted lithosphere in the mantle beneath the Tonga trench. Nature 374, 154–157.

Vaughan, M.T., 1993. In situ X-ray diffraction using synchrotron radiation at high *P* and *T* in multi-anvil device. In: Luth, R.W. (Ed.), Experiments at High Pressure and Applications to the Earth's Mantle, Short Course Handbook 21. Mineralogical Association of Canada, Edmonton, Alberta, pp. 95–130.

Vaughan, M.T., Weidner, D.J., Wang, J.H., Chen, J.H., Koleda, C.C., Getting, I.C., 1995. T-CUP: a new high-pressure apparatus for X-ray studies. NSL Activity Report, p. B140, Available from: http://www.chipr.sunysb.edu/sam85/tcup/tcup.html.

Webb, S.L., 1989. The elasticity of the upper mantle orthosilicates olivine and garnet to 3 GPa. Phys. Chem. Miner. 16, 684–692.

Will, G., Hinze, E., Nuding, W., 1982. Energy-dispersive X-ray diffraction applied to the study of minerals under pressure up to 200 kbar. In: Schreyer, W. (Ed.), High-Pressure Researches in Geoscience. Schweizerbart'sche Verlagsbuchhandlung, Stuttgart, pp. 177–201.

Wirth, R., 1996. Thin amorphous films (1–3 nm) at olivine grain boundaries in mantle xenoliths from San Carlos, Arizona. Contrib. Mineral. Pertol. 124, 44–54.

Yagi, T., 1988. MAX80: large-volume high-pressure apparatus combined with synchrotron radiation. Eos 69 (12), 18–27.

Yoneda, A., Spetzler, H., 1994. Temperature fluctuation and thermodynamic properties in Earth's lower mantle: an application of the complete travel time equation of state. Earth Planet. Sci. Lett. 126 (4), 369–377.

Zaug, J.M., Abramson, E.H., Brown, J.M., Slutsky, L.J., 1993. Sound velocities in olivine at Earth mantle pressures. Science 260, 1487–1489.

Zha, C.-S., Mao, H.K., Hemley, R.J., 2000. Elasticity of MgO and a primary pressure scale to 55 GPa. Proc. Natl Acad. Sci. 97, 13494–13499.

Zinn, P., Hinze, E., Lauterjung, J., Wirth, R., 1997. Kinetic and microstructural studies of the quartz-coesite phase transition. Phys. Chem. Earth 22, 105–111.

Advances in High-Pressure Technology for Geophysical Applications
Jiuhua Chen, Yanbin Wang, T.S. Duffy, Guoyin Shen, L.F. Dobrzhinetskaya, editors

Chapter 5

Laboratory measurement of seismic wave dispersion and attenuation at high pressure and temperature

Ian Jackson

Abstract

At sufficiently high temperatures and in the presence of fluids, the small-strain mechanical behaviour of geological materials is inevitably frequency dependent. In order to understand the nature of viscoelastic relaxation manifest in the attenuation and dispersion of seismic waves, laboratory methods suited to the seismic-frequency interrogation of geological materials are required. Torsional forced-oscillation and related microcreep methods, providing for the measurement of shear modulus and the associated strain-energy dissipation under conditions of high pressure and temperature and independently controlled pore-fluid pressure, will be reviewed. Phenomenological approaches to the modelling of high-temperature viscoelastic behaviour will be outlined. The application of these experimental and analytical methods will be illustrated by reference to recent work on polycrystalline olivine.

1. Geophysical motivation

The Earth and its constituent materials are stressed on timescales ranging from less than a second to more than a billion years by naturally occurring processes including faulting and seismic wave propagation, tides and glacial loading, and tectonic activity that is the surface expression of convective overturn of the interior (Fig. 1). The apparent paradox whereby the Earth behaves as a rigid but brittle solid on the shortest of these timescales, but as a convecting fluid on the longest timescales, is resolved by invoking a non-elastic contribution to the strain that results from the stress-induced migration of crystal defects. Partial relaxation of the shear modulus associated with such defect migration will occur for loading cycles longer than a characteristic timescale related to the mobility of the defect (e.g. vacancy, dislocation or grain-boundary) in question. Because the relevant diffusivity will be strongly temperature dependent, it is inevitable that the resulting viscoelastic behaviour will be both frequency and temperature dependent as illustrated in Figure 1. As a consequence, laboratory information on seismic properties of geological materials, typically obtained at MHz–GHz frequencies with ultrasonic and opto-acoustic techniques, cannot in general be applied without modification at the much lower (sub-Hz) frequencies of teleseismic wave propagation.

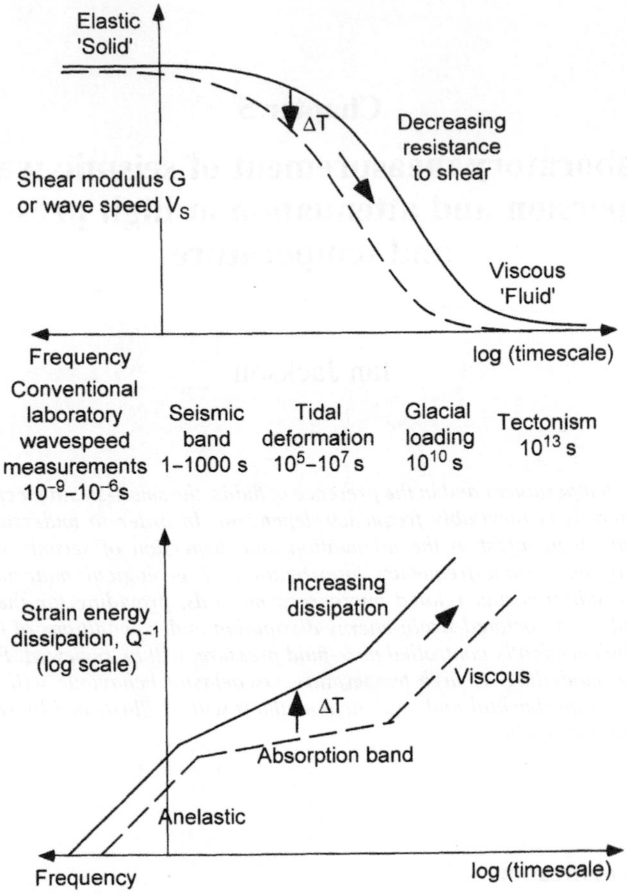

Figure 1. Schematic illustration of the inevitably frequency and temperature dependent mechanical behaviour of the Earth and its constituent materials stressed at high temperature. At sufficiently long timescales (low frequencies) for a given temperature, elastic behaviour gives way to viscoelastic deformation with reduced modulus and wave speed and associated dissipation of strain energy (from Jackson et al., 2005).

2. Experimental methods

Various experimental methods each capable of interrogation of geological and other materials over a restricted frequency range are reviewed schematically in Figure 2. At the highest of these frequencies (THz) and corresponding wavelengths approaching interatomic distances (<nm) lattice vibrations (phonons) become dispersive. Away from the origin (the centre of the Brillouin zone), the phase velocity $v = \omega/k$ and the group velocity $u = \mathrm{d}\omega/\mathrm{d}k$ diverge developing different sensitivities to the variation of frequency ω or wavevector k. Frequencies in the MHz–GHz range correspond to acoustic wavelengths of μm–mm convenient for ultrasonic and opto-acoustic wave propagation experiments on sub-mm to cm-sized laboratory specimens. Resonance experiments on

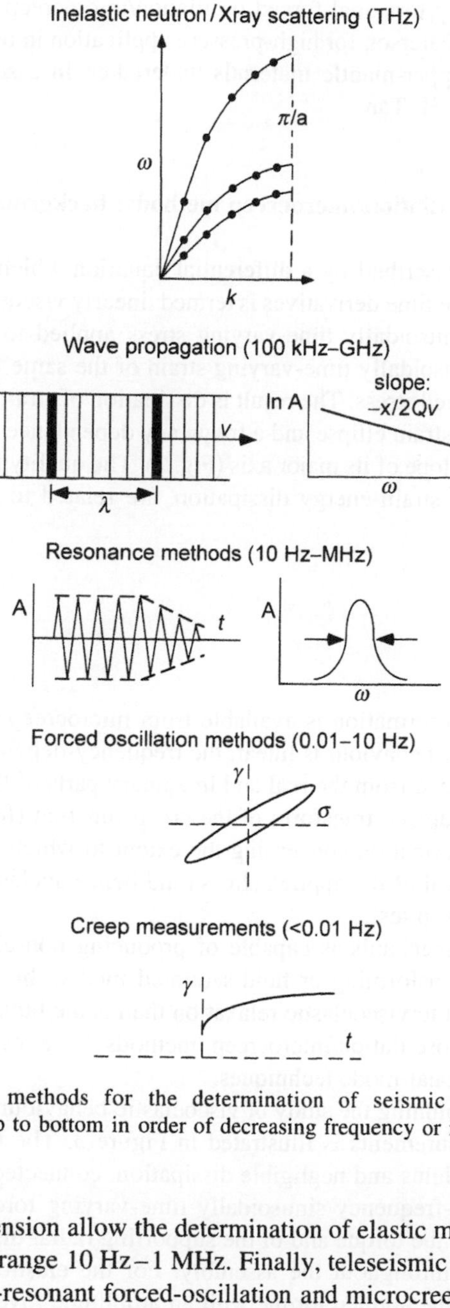

Figure 2. Experimental methods for the determination of seismic properties of geological materials arranged from top to bottom in order of decreasing frequency or increasing time scale of the mechanical test.

specimens of such dimension allow the determination of elastic moduli and quality factor Q at frequencies in the range 10 Hz–1 MHz. Finally, teleseismic frequencies (mHz–Hz) are accessible with sub-resonant forced-oscillation and microcreep methods described in detail below. In combination these techniques offer the prospect of a greatly improved understanding of the high-temperature viscoelastic relaxation responsible for the dispersion and dissipation sketched in Figure 1.

Some of the high-frequency methods and their application under conditions of high pressure and temperature are described elsewhere in this volume. In this paper I will review the low-frequency torsional forced oscillation/microcreep methods developed in collaboration with M.S. Paterson for high-pressure application in our ANU laboratory and recent applications to upper-mantle materials undertaken in association with J.D. Fitz Gerald, U.H. Faul and B.H. Tan.

3. Torsional forced oscillation/microcreep methods: background

Mechanical behaviour described by a differential equation which is linear in stress and strain and their respective time derivatives is termed linearly viscoelastic (e.g. Nowick and Berry, 1972, p. 42). Sinusoidally time-varying stress applied to a linearly viscoelastic material results in a sinusoidally time-varying strain of the same frequency with a phase lag δ relative to the applied stress. The result is dissipation of strain energy manifest in the finite area of the stress–strain ellipse and a frequency dependence of the shear modulus G given by the reciprocal slope of its major axis (Fig. 2). The quality factor Q, and its inverse Q^{-1} here referred to as strain-energy dissipation, are related to the phase lag δ by the relationship

$$Q^{-1} = \tan\delta. \tag{1}$$

In principle, the same information is available from microcreep tests (Fig. 2). Provided only that the viscoelastic behaviour is linear, the frequency-dependent shear modulus and dissipation can be obtained from the real and imaginary parts of the dynamic compliance that is essentially the Laplace transform of the creep function (formally defined below). Valuable additional information concerning the extent to which the non-elastic strain is recoverable upon removal of the applied stress (and hence anelastic) is also contained in the record of a microcreep test.

Although there are mechanisms capable of producing non-elastic dilatational strain particularly in phase-transforming or fluid-saturated media, the shear modulus is much more commonly subject to viscoelastic relaxation than is the bulk modulus. Accordingly, torsional-mode forced-oscillation/microcreep methods have been much more widely developed than extensional-mode techniques.

The principle underpinning the study of viscoelastic behaviour through torsional mode forced-oscillation measurements is illustrated in Figure 3. The unknown and an elastic standard of known modulus and negligible dissipation, connected mechanically in series are subjected to a low-frequency sinusoidally time-varying torque. At sufficiently low frequency, the phase of the torque and of the supporting radial distribution of shear stress is essentially uniform throughout the assembly. For the elastic standard, the resulting strain is in phase with the applied torque with an amplitude inversely proportional to its shear modulus. Similarly, for the unknown, except that its viscoelastic behaviour is manifest in a phase lag of strain behind stress. Thus measurement of the relative amplitudes and phases of the torsional mode displacements in standard and unknown provides for the determination of the shear modulus of the unknown relative to the standard and of the strain energy dissipation Q^{-1} for the unknown.

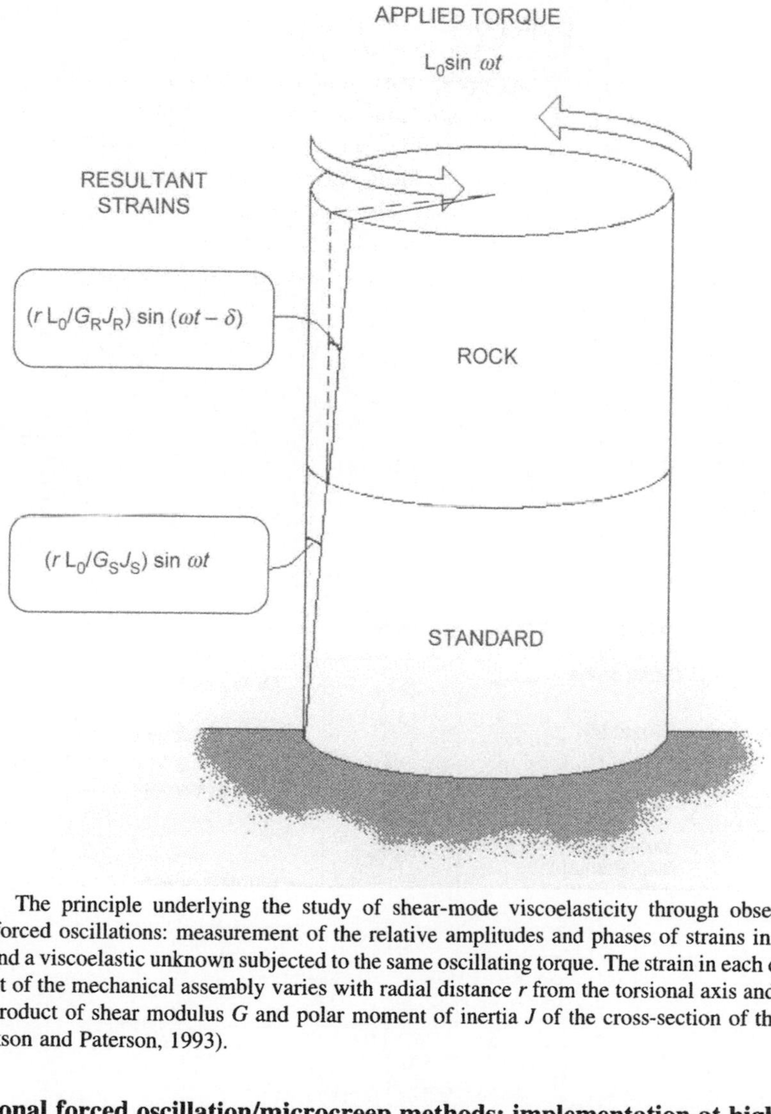

Figure 3. The principle underlying the study of shear-mode viscoelasticity through observation of torsional forced oscillations: measurement of the relative amplitudes and phases of strains in an elastic standard and a viscoelastic unknown subjected to the same oscillating torque. The strain in each cylindrical component of the mechanical assembly varies with radial distance r from the torsional axis and inversely with the product of shear modulus G and polar moment of inertia J of the cross-section of the cylinder (after Jackson and Paterson, 1993).

4. Torsional forced oscillation/microcreep methods: implementation at high temperature and pressure

The method illustrated in Figure 3 has been widely implemented for studies at high temperature and atmospheric pressure (e.g. Woirgard et al., 1981; Berckhemer et al., 1982; Gadaud et al., 1990; Getting et al., 1997; Gribb and Cooper, 1998a). In our ANU laboratory, techniques have been developed for the conduct of such measurements under conditions of *simultaneously* high pressure and temperature within an internally heated gas-charged pressure vessel. The equipment and experimental techniques have been described in detail elsewhere (Jackson and Paterson, 1987, 1993; Jackson, 2000). Accordingly, the present discussion is restricted to a brief overview highlighting the most recent developments.

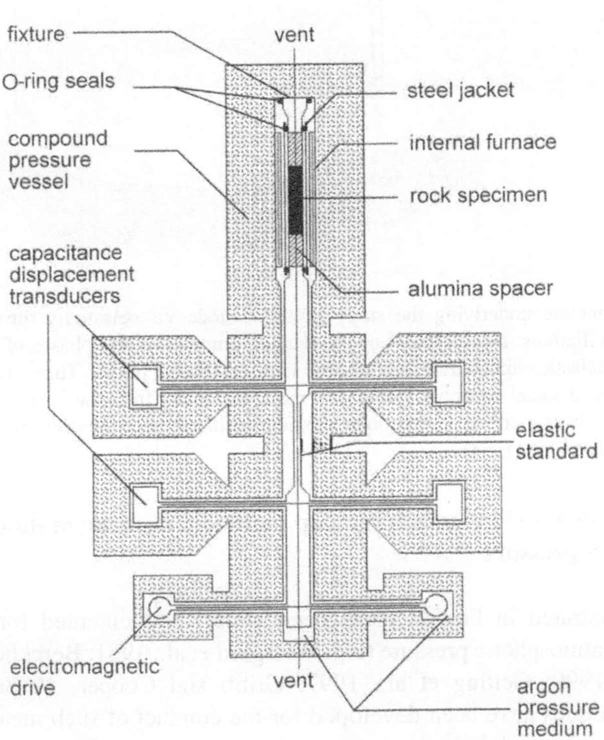
fixture
vent
O-ring seals
steel jacket
compound
pressure
vessel
internal furnace
rock specimen
capacitance
displacement
transducers
alumina spacer
elastic
standard
electromagnetic
drive
vent
argon
pressure
medium
700 mm

The entire assembly is housed within a compound hardened steel pressure vessel of minimal internal volume. Pressurisation to the usual operating pressure of 200 MPa is effected by a double-acting single-stage air-driven gas booster and a large-displacement intensifier. The upper part of the assembly, containing the specimen and alumina torsion rods, is heated by an internal electric furnace capable of controlled temperature to 1600 K uniform within 10 K across the length of the specimen.

The twist of the assembly induced by the electromagnetically applied torque is measured at stations above and below the elastic standard by pairs of parallel-plate capacitance displacement transducers (Brennan and Stacey, 1977). The two transducers within each pair are connected diagonally in parallel in order to discriminate against any flexural mode contributions. At each station the parallel combinations of capacitors are connected to the inductances on either side of the adjustable tap of a six-decade ratio transformer to form an AC bridge. The displacement transducers are calibrated under operating conditions by measuring the synchronously detected bridge out-of-balance voltage corresponding to switching the ratio transformer through a prescribed increment.

Torsional microcreep or forced-oscillation tests are performed by applying a suitable computer-synthesised voltage-time signal to the pair of matched electromagnetic drivers. The raw data describing the resulting deformation of the assembly and consisting of pairs of calibrated displacement-versus-time records are A/D converted for acquisition by the computer. Maximum shear strain amplitudes at the cylindrical surface of the specimen are typically $\sim 10^{-5}$. Forced-oscillation tests are performed sequentially at selected periods within the range 1–100 s sometimes extended to ~1000 s. Microcreep tests comprise prescribed periods of steady positive and negative torque application separated by intervals in which the torque is set to zero – to facilitate examination of the extent to which any non-elastic strain is recovered following removal of the applied torque.

Fourier analysis is used to extract the relative amplitudes and phases of the two displacement-time sinusoids at the known forcing period from the forced-oscillation records. A correction described in detail by Jackson and Paterson (1987, 1993) for the influence of the torques exerted on the twisting assembly by the gas pressure medium is routinely applied. This correction is significant at periods of 1–10 s but negligible at longer periods.

Similar tests on a reference assembly identical to the specimen assembly except for the replacement of the specimen by a dummy of polycrystalline alumina provide the information needed to eliminate the contribution to the overall compliance from the torsion rods spanning the regions of strong temperature gradient at either end of the furnace hot-zone (Fig. 5). Lucalox™ alumina (General Electric Lighting, Cleveland, Ohio, USA) is used throughout because of its exceptional creep resistance (Jackson, 2000). The chemical environment of the specimen, including oxygen fugacity, is controlled by isolating the specimen itself from the alumina torsion rods and mild-steel sleeve with high-purity foils of an appropriate metal alloy (e.g. $Ni_{70}Fe_{30}$ for Ni-bearing olivine).

The raw time-domain microcreep data for the specimen and reference assemblies are fitted to the Andrade creep function [Eq. (4), below]. The analytical Laplace transforms of the fitted creep functions [Eq. (7), below] allow further processing in the frequency domain as for the forced-oscillation data (Jackson, 1998, 2005).

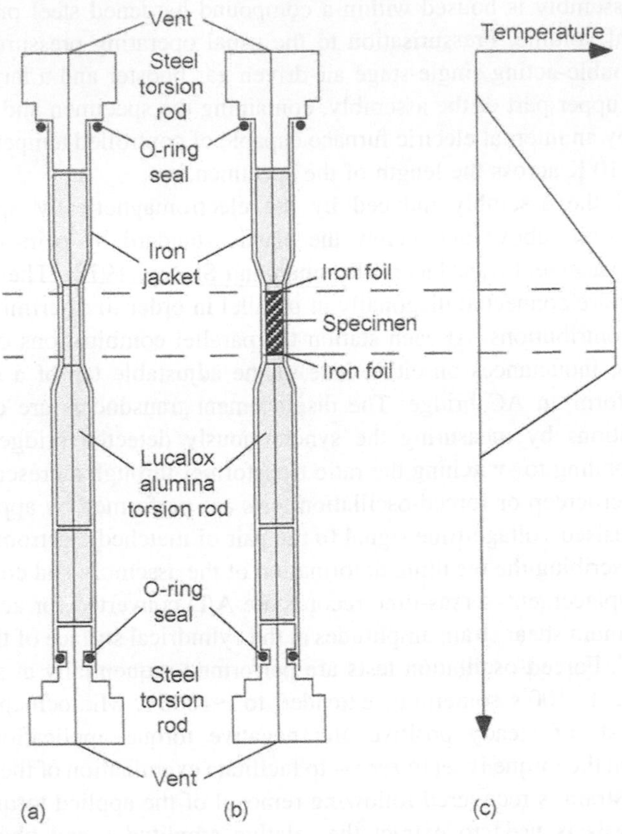

Figure 5. Schematic representation of the reference (a) and specimen (b) assemblies currently employed in high-temperature torsional forced oscillation and microcreep tests, and of the temperature profile (c) to which they are exposed during testing within the gas-medium high-pressure apparatus described by Jackson and Paterson (1993) (after Jackson, 2005).

Quantitative compatibility of the observed modulus dispersion and the dissipation (through the Kramers–Kronig relationship of linear theory) along with the insensitivity of the modulus and dissipation to variation of the maximum strain amplitude up to at least 5×10^{-5} provide strong evidence for the linearity of the stress-strain relationship as defined in Section 3. Consistency between the modulus and dissipation directly measured in forced oscillation tests and the corresponding quantities derived as described in Section 5 from microcreep records provides further evidence of linear behaviour (Jackson, 2000).

In order to facilitate studies of the viscoelastic behaviour of cracked and fluid-saturated media, pore-fluid reservoirs can be connected to the upper and lower vents as illustrated in Figure 6. These arrangements provide for the control of pore fluid pressure independently of the confining pressure and for measurements of crack porosity and permeability (Lu and Jackson, 1998, 2005).

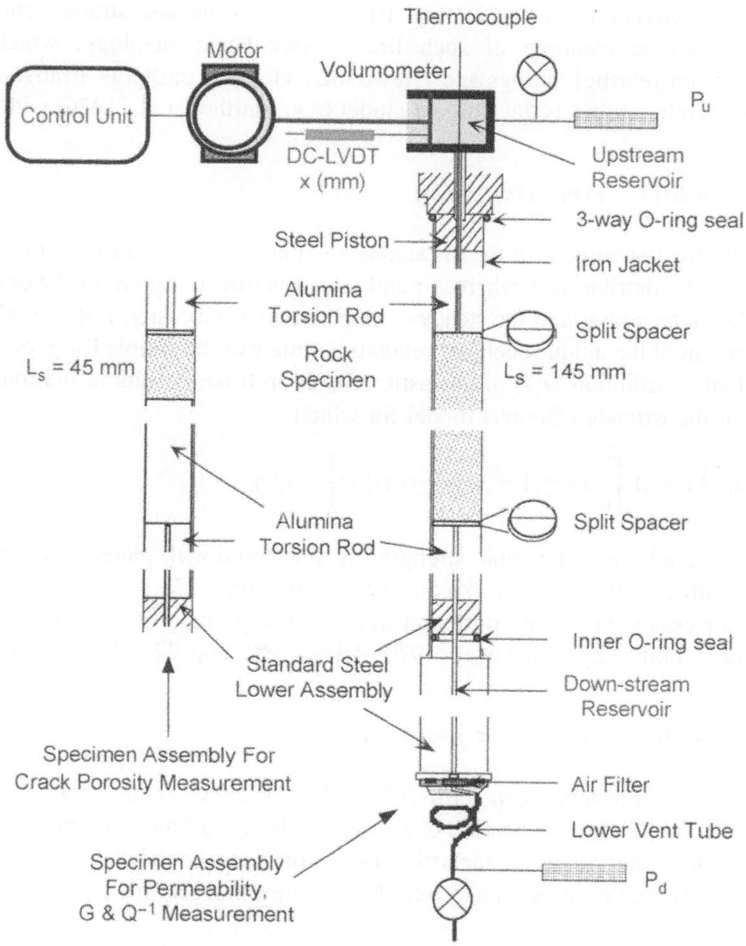

Figure 6. Experimental configuration for independent control of pore-fluid pressure and measurement of permeability and crack porosity in conjunction with torsional forced oscillation/microcreep studies of the viscoelastic behaviour of fluid-saturated rocks. (from Lu and Jackson, 2005).

5. Phenomenology of high-temperature linear viscoelastic behaviour

5.1. The creep function

For sufficiently small stresses, the stress-strain behaviour must be linear and the response is represented in the time domain by the creep function $J(t)$ which is defined as the strain resulting from the application at time $t = 0$ of unit step-function stress (e.g. Nowick and Berry, 1972). For the special case of elastic behaviour, the strain appears essentially instantaneously and thereafter remains constant for the duration of stress application. However, at high temperatures and long timescales, and in the presence of fluids, the response $J(t)$ will usually be more complicated than the elastic ideal, involving not only an instantaneous (elastic) component, but also a time-dependent contribution which may be a mixture of recoverable (anelastic) and irrecoverable (viscous) strains. The simplest moderately realistic example of such linear viscoelastic rheology, which can be constructed from (elastic) springs and (Newtonian viscous) dashpots arranged in series and parallel combinations, is the Burgers model (e.g. Findley et al., 1976) with the creep function

$$J(t) = J_U + \delta J[1 - \exp(-t/\tau)] + t/\eta. \tag{2}$$

J_U and δJ are the magnitudes of the instantaneous (elastic) and anelastic (time-dependent but recoverable) contributions, whereas τ and η are the time constant for the development of the anelastic response, and the steady-state Newtonian viscosity, respectively.

Replacement of the unique anelastic relaxation time τ of the simple Burgers model by a user-specified distribution $D(\tau)$ of anelastic relaxation times results in the much greater flexibility of the extended Burgers model for which

$$J(t) = J_U\left\{1 + \Delta\int_0^\infty D(\tau)[1 - \exp(-t/\tau)]\mathrm{d}\tau\right\} + t/\eta \tag{3}$$

Here Δ, the anelastic relaxation strength, is the fractional increase in compliance associated with complete (i.e. $t = \infty$) anelastic relaxation.

No less successful in the description of transient creep, but physically less transparent, is the Andrade model (e.g. Amin et al., 1970; Poirier, 1985, pp. 27–28) for which the creep function is

$$J(t) = J_U + \beta t^n + t/\eta \quad (1/3 < n < 1/2). \tag{4}$$

The form of the transient-creep term βt^n in the Andrade creep function implies the existence of a continuous distribution of anelastic relaxation times stretching from 0 to ∞ (Jackson, 2000), rather than an anelastic absorption band bounded by upper and lower relaxation times readily incorporated into the extended Burgers model.

5.2. The dynamic compliance

The strain $\varepsilon(t) = \varepsilon_0 \exp \mathrm{i}(\omega t - \delta)$ resulting from the application of sinusoidally time-varying stress $\sigma(t) = \sigma_0 \exp(\mathrm{i}\omega t)$ can be evaluated from the creep function provided that the behaviour is linear. This is done by superposition of the creep-function responses to each of a series of infinitesimal step-function applications of stress, that together represent the entire history $\sigma(t)$ of stress application (e.g. Nowick and Berry, 1972). Thus an expression is obtained for the dynamic compliance $J^*(\omega)$ given by

$$J^*(\omega) = J_1(\omega) - \mathrm{i}J_2(\omega) = \varepsilon(t)/\sigma(t) = \mathrm{i}\omega\int_0^\infty J(\xi)\exp(-\mathrm{i}\omega\xi)\mathrm{d}\xi \tag{5}$$

where $\omega = 2\pi f = 2\pi/T_o$ is the angular frequency corresponding to period T_o. This integral is the Laplace transform of $J(t)$ with transform variable $s = \mathrm{i}\omega$. With Laplace transforms of each of the terms in the Burgers and Andrade creep functions tabulated in standard compilations (e.g. Abramowitz and Stegun, 1972), the following expressions for the dynamic compliance are readily derived.

For the extended Burgers model the real and negative imaginary parts of $J^*(\omega)$ are

$$J_1(\omega) = J_U\{1 + \Delta \int_0^\infty D(\tau)d\tau/(1 + \omega^2\tau^2)\}, \text{ and}$$

$$J_2(\omega) = \omega J_U \Delta \int_0^\infty \tau D(\tau)d\tau/(1 + \omega^2\tau^2) + 1/\eta\omega \tag{6}$$

respectively, whereas, for the Andrade model, the corresponding expressions are

$$J_1(\omega) = J_U + \beta\Gamma(1+n)\omega^{-n}\cos(n\pi/2), \text{ and}$$

$$J_2(\omega) = \beta\Gamma(1+n)\omega^{-n}\sin(n\pi/2) + 1/\eta\omega \tag{7}$$

where $\Gamma(1+n)$ is the Gamma function (Findley et al., 1976; Gribb and Cooper, 1998b). From such expressions for $J_1(\omega)$ and $J_2(\omega)$, the shear modulus

$$G(\omega) = \left[J_1^2(\omega) + J_2^2(\omega)\right]^{-1/2}, \tag{8}$$

and the associated strain energy dissipation

$$Q^{-1}(\omega) = J_2(\omega)/J_1(\omega), \tag{9}$$

can be evaluated.

6. Alternative strategies for modelling the "high-temperature background"

6.1. Introduction

Sub-resonant forced-oscillation techniques allow measurement of the dissipation Q^{-1} as a function not only of oscillation period T_o and temperature T but also key material microstructural parameters such as average grain size d and, potentially, dislocation density ρ. At high temperatures and low frequencies Q^{-1} typically increases mildly and monotonically with increasing period and exponentially with increasing temperature – this pattern of behaviour being referred to as the high-temperature background (Nowick and Berry, 1972). The associated frequency dependence of the shear modulus is reported only in a minority of studies.

Notwithstanding the classic work on elastically and diffusionally accommodated grain-boundary sliding (Raj and Ashby, 1971; Raj, 1975) commonly invoked (e.g. Gribb and Cooper, 1998b) to explain the high-temperature viscoelastic behaviour of fine-grained material, there is as yet no satisfactory microphysical model of these processes (Jackson et al., 2002; Faul et al., 2004). Under these circumstances, experimentally observed viscoelastic behaviour is best modelled by specifying an empirical creep function from which both the dissipation and modulus dispersion can be calculated in an internally consistent way through Eqs. (8) and (9). Gribb and Cooper (1998b) showed that a rheology given by the Andrade creep function [Eq. (4)] with $n \sim 1/2$ approximates the frequency dependence of Q^{-1} measured on a fine-grained reconstituted dunite specimen. It was subsequently demonstrated that for individual polycrystalline olivine specimens tested at constant temperature, the frequency dependence of both Q^{-1} and G could be simultaneously fitted to creep functions of either the Andrade or extended Burgers type (Jackson, 2000; Tan et al., 2001).

In order to apply such creep-function approaches to Q^{-1} (and G) datasets for multiple specimens with different microstructures tested at various temperatures it is necessary, but not straightforward, to build into the creep function the appropriate dependence upon temperature and microstructural variables such as grain size (see Section 8).

6.2. *Master-variable approaches to the fitting of dissipation data*

For this reason, simpler alternative procedures have often been employed – usually for fitting Q^{-1} only. A popular approach has been to combine the key independent variables period, temperature, grain size etc. into a single master variable whose value is the sole determinant of the dissipation and possibly also modulus (e.g. Kê, 1947; Schoeck et al., 1964). For restricted ranges of oscillation period, temperature and grain size, dissipation data for fine-grained polycrystals may be adequately represented by a power-law model of the form

$$Q^{-1} = AX^{\alpha} \tag{10}$$

with the controlling or "master" variable X specified by

$$X = (T_o/d)\exp(-E/RT) \tag{11}$$

(Jackson et al., 2002).

An approach with a potentially wider domain of validity involves substitution of the master variable

$$X_B = T_o(d/d_R)^{-m} \exp[(-E_B/R)(1/T - 1/T_R)] \tag{12}$$

for period T_o (in seconds) in the Andrade expressions for $J_1(\omega)$ and $J_2(\omega)$ [Eq. (7)]. In Eq. (12) T_R and d_R are reference values of temperature and grain size respectively, E_B is an activation energy and $m > 0$ is the grain-size exponent. The master variable X_B is a pseudoperiod in the sense that $X_B = T_o$ for $d = d_R$ and $T = T_R$, but for smaller (larger) grain size and/or higher (lower) temperatures, X_B is greater (less) than T_o in order to account for the expected grain-size and temperature dependence of viscoelastic relaxation times.

This approach (see also Cooper, 2003, pp. 272–3) has been successfully tested on the dissipation data for a suite of fine-grained melt-free olivine polycrystals (Jackson et al., 2004). The Andrade-pseudoperiod approach has the advantage over the previously used power-law dependence of Q^{-1} on pseudoperiod [Eq. (10)] of describing departures from power-law behaviour encountered at the highest temperatures, smallest grain sizes and longest periods (Fig. 7, below) and associated with the progressive transition from essentially anelastic to viscous behaviour.

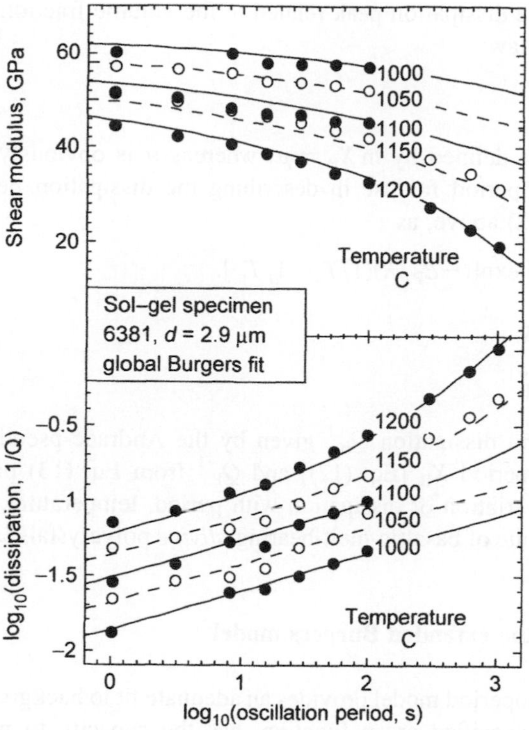

Figure 7. Representative data from torsional forced oscillation experiments on fine-grained melt-free polycrystalline olivine (Jackson et al., 2002). Data are indicated by the plotting symbols. The curves represent the extended Burgers model [Eq. (18)], fitted to G and Q^{-1} data for a suite of four specimens ranging in mean grain size from 2.9 to 165 μm, evaluated at $d = 2.9$ μm.

7. Master-variable modelling of a superimposed dissipation peak

For some materials, especially those containing a secondary phase of relatively low viscosity on the grain-boundaries or in grain-edge triple junctions, a dissipation peak that is typically substantially broader than the Debye relaxation peak of the simple Burgers (or Standard Anelastic) solid, is superimposed upon the monotonically frequency- and temperature-dependent background dissipation (e.g. Jackson et al., 2004).

Instead of seeking, in the first instance, to specify the underlying distribution $D(\tau)$ of anelastic relaxation times (see Section 8), Jackson et al. (2004) chose to model the dissipation peak Q_P^{-1} by the following Gaussian function

$$Q_P^{-1} = B \exp(-z^2/2) \tag{13}$$

with

$$z = (\ln X_P - \mu)/\sigma. \tag{14}$$

B is the height of the dissipation peak related to the volume fraction ϕ of the secondary phase by the power law:

$$B = B_0\phi^l. \tag{15}$$

The peak position is defined by $\ln X_P = \mu$, whereas σ is obviously a measure of peak width. X_P, a pseudoperiod for use in describing the dissipation peak, was defined by analogy with Eq. (12) above, as

$$X_P = T_o(d/d_r)^{-m}\exp[(-E_P/R)(1/T - 1/T_r)]. \tag{16}$$

A model of the form

$$Q^{-1} = Q_B^{-1} + Q_P^{-1} \tag{17}$$

With the background dissipation Q_B^{-1} given by the Andrade-pseudoperiod model with background pseudoperiod X_B [Eq. (12)] and Q_P^{-1} from Eq. (13) provides an excellent description of the variation of dissipation with period, temperature, grain size and melt fraction data for a suite of basaltic-melt-bearing *olivine* polycrystals (Jackson et al., 2004).

8. Application of the extended Burgers model

The Andrade-pseudoperiod model provides an adequate fit to background-only dissipation data, and with its specified creep function, has the capacity to provide an internally consistent representation of both the dissipation and modulus dispersion. Without a specified creep function, the Andrade–Gaussian-pseudoperiod model cannot similarly provide as readily for the simultaneous modelling of both shear modulus and dissipation.

The extended Burgers model is therefore preferred for its superior flexibility and also greater transparency. Firstly, it provides a cleaner separation of anelastic and viscous contributions to the viscoelastic behaviour at long period by imposing an upper cut-off for the distribution of anelastic relaxation times. Secondly, it readily accommodates both background and peak contributions through appropriate specification of the distribution $D(\tau)$ of anelastic relaxation times as follows.

The strengths and weaknesses of the various empirical approaches to the description of high-temperature viscoelastic behaviour are compared in Table 1.

8.1. High-temperature background

For the practical description of high-temperature background-only behaviour, the extended Burgers model is modified from Eq. (3) to take the form

$$J(t) = J_U(P)\left\{1 + \delta J_U(T)/J_U(P) + \Delta\int_0^\infty D(\tau)[1 - \exp(-t/\tau)]\mathrm{d}\tau + t/\tau_M\right\} \tag{18}$$

(Faul and Jackson, 2005). $J_U(P)$ is the unrelaxed compliance, given by the reciprocal of the unrelaxed shear modulus G_U with its usual anharmonic pressure dependence, and $\tau_M = \eta J_U$ is the Maxwell relaxation time. $\delta J_U(T)$ is an adjustment to J_U for the effect of

Table 1. Attributes of various phenomenological models for the representation of high-temperature viscoelastic behaviour.

Model	Creep function based?	Capacity to model anelastic Q^{-1} bkg'd?	Capacity to model broad Q^{-1} peak?	Capacity to model both Q^{-1} & G?	Comments
Power-law pseudoperiod	N	Y	N	N	Q^{-1} only; restricted T_o,T, d domain
Andrade-pseudoperiod	Y	Y	N	Y	Lacks flexibility to model peak
Andrade-Gaussian-pseudoperiod	N	Y	Y	N	No creep function: cannot model both Q^{-1} & G
Extended Burgers	Y	Y	Y	Y	Provides required flexibility

temperature explained below, and Δ is the relaxation strength associated with the distribution $D(\tau)$ of anelastic relaxation times.

At relatively low temperatures, $\delta J_U(T)$ is simply the anharmonic temperature dependence of J_U. However, for any given material tested in the laboratory there exists a threshold temperature above which the temperature dependence of G is significantly enhanced by viscoelastic relaxation (~900°C for fine-grained polycrystalline olivine). For temperatures above this threshold it is found to be necessary to account for the cumulative dispersion associated with relaxation times shorter than those responsible for the modelled dissipation within the observational window. For polycrystalline olivine, this adjustment is adequately modelled as

$$\delta J_U(T) = (\delta J_U/\delta T)(T - T_R)(d/d_R)^{-m_J} \tag{19}$$

The background dissipation occurring within the observational window is attributed to the following (normalised) distribution of anelastic relaxation times

$$D_B(\tau) = \alpha\tau^{\alpha-1}/(\tau_H^\alpha - \tau_L^\alpha) \quad \text{for } \tau_L < \tau < \tau_H \text{ and zero elsewhere} \tag{20}$$

(e.g. Minster and Anderson, 1981).

8.2. Dissipation peak

For a melt-related dissipation peak superimposed upon the high-temperature background, the distribution of anelastic relaxation times given by Eq. (20) is augmented by the separately normalised distribution

$$D_P(\tau) = \sigma^{-1}(2\pi)^{-1/2}\exp\{-[\ln(\tau/\tau_P)/\sigma]^2/2\} \tag{21}$$

(Kampfmann and Berckhemer, 1985) associated with relaxation strength Δ_P related to volume fraction ϕ of the secondary phase as in Eq. (15):

$$\Delta_P = \Delta_{P0}\phi^l. \tag{22}$$

The upper and lower limits to the distribution of anelastic relaxation times, respectively τ_H and τ_L, the peak relaxation time τ_P, and the Maxwell relaxation time τ_M are each of the form

$$\tau = \tau_R(d/d_R)^m \exp[(E/R)(1/T - 1/T_R)] \tag{23}$$

where τ_R is the value of τ at the reference temperature T_R and the reference value of grainsize d_R. Different grain-size exponents m_A for τ_L, τ_H and τ_P, and m_V for τ_M are allowed.

8.3. *Least-squares fitting of G, Q^{-1} data*

The extended Burgers rheology [Eq. (18) along with Eqs. (5, 6,[1] 8, 9)] has been fitted to the observed variation of both G and Q^{-1} with period, temperature and mean grain size through use of the iterative Levenberg–Marquardt strategy (Press et al., 1986) in order to minimize the misfit function

$$\chi^2(\underline{a}) = \sum_i \left\{ \left[\frac{G_i - G(T_{oi}, T_i, d_i, \underline{a})}{\sigma(G_i)} \right]^2 + \left[\frac{Q_i^{-1} - Q^{-1}(T_{oi}, T_i, d_i, \underline{a})}{\sigma(Q_i^{-1})} \right]^2 \right\} \tag{24}$$

where $\underline{a}$ with components a_j is the vector of model parameters, all or some of whose values are to be refined. Analytical expressions for the derivatives $\partial G/\partial a_j$ and $\partial Q^{-1}/\partial a_j$ required in calculation of the gradient vector and the Hessian matrix are derived from Eqs. (8) and (9), along with Eqs. (6[1]) and (19)–(23). In the absence of well-determined a priori errors and with the benefit of experience, the standard errors in G and Q^{-1} have been set at $\sigma(G)/G = 0.03$ and $\sigma(\log Q^{-1}) = 0.05$, the latter being equivalent to $\sigma(Q^{-1})/Q^{-1} = 0.12$. The numerical integrations with respect to τ are performed with the Romberg method.

9. Case study: Fo_{90} olivine ± basaltic melt

The experimental and analytical methods described above are well illustrated by the recent work in our ANU laboratory on fine-grained olivine polycrystals (Tan et al., 1997, 2001; Jackson et al., 2002, 2004; Faul and Jackson, 2005; Faul et al., 2004). Dense, texturally mature Fo_{90} olivine polycrystals have been prepared by hot-isostatic pressing of either natural (San Carlos) or synthetic (sol–gel) precursor powders with or without added basaltic melt glass.

[1] Suitably modified by inclusion in the expression for J_1 of the correction term involving $\delta J_U(T)$ from Eq. (18).

For temperatures below ~900°C the behaviour is essentially elastic. Both dissipation and modulus dispersion are minimal and the shear modulus closely approaches the temperature-dependent aggregate modulus calculated from single-crystal elasticity data. At higher temperatures, the behaviour becomes markedly viscoelastic.

9.1. Melt-free polycrystalline olivine

For melt-free materials the dissipation varies smoothly and monotonically with oscillation period and temperature in the way described previously as "high-temperature background" (Fig. 7). For such genuinely melt-free materials ($\phi < 10^{-4}$) there is no evidence of any significant dissipation peak superimposed upon the background (Jackson et al., 2004). Associated with the marked strain-energy dissipation is a frequency dependence (dispersion) of the shear modulus that becomes progressively more pronounced with increasing temperature (Fig. 7).

The fact that the $\log Q^{-1} - \log T_o$ trends for different temperatures are almost linear and parallel indicates a close approach to the power-law master-variable behaviour of Eq. (10). However, the frequency dependence of Q^{-1} becomes systematically stronger at the highest temperatures (1150–1200°C) and longest periods (≥100 s) for this very fine-grained specimen. This correlates with evidence from complementary torsional microcreep tests (Fig. 8) of a progressively more dominant viscous (irrecoverable) component in the observed deformation with increasing temperature beyond 1100°C. Essentially identical behaviour recorded from tests at

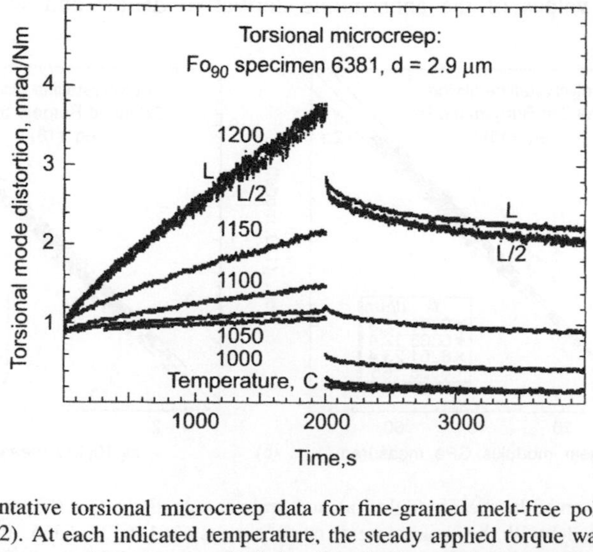

Figure 8. Representative torsional microcreep data for fine-grained melt-free polycrystalline olivine (Jackson et al., 2002). At each indicated temperature, the steady applied torque was switched off after 2000 s allowing assessment of the extent to which the viscoelastic strain is recoverable, i.e. anelastic rather than viscous. Close consistency between the 1200°C records for torque amplitudes differing by a factor of 2 (labelled "L" and "L/2" respectively) indicates a close approach to linearity of the viscoelastic behaviour.

1200°C and different amplitudes of the applied torque provides evidence for linearity of the viscoelastic behaviour.

Such data for a suite of melt-free olivine aggregates with different mean grain sizes have been used in testing the various models for viscoelastic behaviour described above. The power-law pseudoperiod model [Eq. (10)] was found to provide an adequate fit to the 1–100 s Q^{-1} data for a suite of four such melt-free specimens ranging in mean grain size from 3 to 23 μm and tested at temperatures of 1000–1200°C (or 1300°C) (Jackson et al., 2002). Subsequently, it was demonstrated that the Andrade-pseudoperiod model [Eq. (12)] provides a superior fit to an expanded dataset incorporating additional 300–1020 s dissipation data (Jackson et al., 2004). Most recently, the extended Burgers model [Eq. (18)] has been fitted simultaneously to a total of $N = 206$ dissipation and shear modulus data for a suite of four genuinely melt-free specimens ($\phi \leq 10^{-4}$) ranging in mean grain size from 2.9 to 165 μm and tested at periods of 1–1020 s and temperatures of 1000–1300°C (Faul and Jackson, 2005). The curves superimposed on the data of Figure 7 represent this model evaluated at the 2.9 μm grain size for this particular specimen. The model provides an entirely adequate and internally consistent description of the frequency and temperature dependence of both G and Q^{-1}.

The overall quality of the fit to the entire dataset, spanning the range $20 < G < 65$ GPa and $-2 < \log Q^{-1} < 0$ is evident from the comparison in Figure 9 of the measured and modelled values of G and Q^{-1}. The misfit of the data by the model is generally within the range $\pm 2\sigma$.

This model provides an internally consistent description of the impact of viscoelastic relaxation in polycrystalline olivine on both shear modulus (and hence wave speed) and attenuation. The model requires only modest extrapolation in grain-size and temperature for upper-mantle application. The significant extrapolation in pressure is achieved by using published values of the anharmonic derivative dG/dP and scaling the various

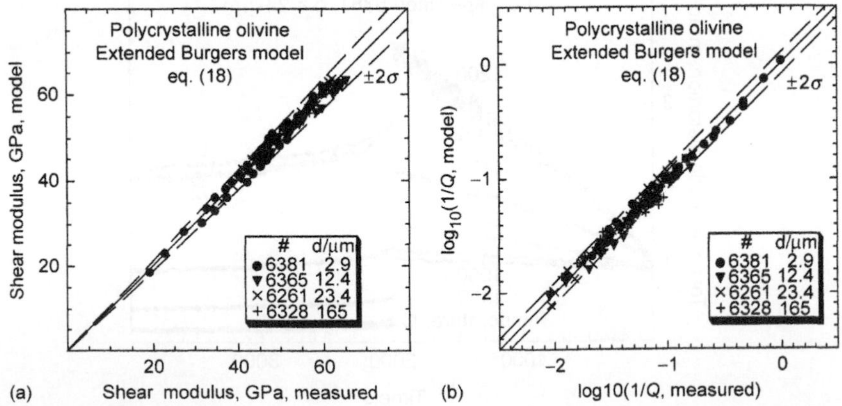

Figure 9. Comparison of the shear modulus (a), and strain energy dissipation (b) predicted by the optimal extended Burgers model with the measurements on specimens 6261, 6328, 6365, and 6381 of Tan et al. (2001) and Jackson et al. (2002). The model (with $\chi^2 = 213$) was fitted to a total of $N = 206$ G and Q^{-1} data for the four specimens ranging in mean grain size from 2.9 to 165 μm tested in forced oscillation at periods of 1–1022 s and temperatures of 1300–1000°C during slow staged cooling. The solid and broken lines correspond respectively to zero and $\pm 2\sigma$ misfit. (from Faul and Jackson, 2005).

relaxation times [Eq. (23)] by a factor exp (PV/RT) with an estimated activation volume V.

In order to demonstrate directly the dispersion of the shear modulus shown schematically in Figure 1, a single fine-grained olivine specimen has recently been measured by both ultrasonic interferometry (>10 MHz) and the torsional forced-oscillation technique (<1 Hz) (Jackson et al., 2005). The ultrasonically determined shear modulus is in close accord with the aggregate modulus calculated from single-crystal elasticity data obtained at comparably high frequencies whereas a consistently lower and more strongly temperature-dependent modulus is found at the long periods of the forced-oscillation technique (Fig. 10).

9.2. Basaltic melt-bearing olivine polycrystals

Synthetic Fo_{90} olivine polycrystals containing small fractions of melt have been similarly prepared either by deliberate addition of basaltic melt glass to the precursor powder or by inadvertent use of mildly impure San Carlos olivine. Six such specimens ranging in mean grain size from 7 to 52 μm and in maximum melt fraction (at 1240–1300°C) from 0.0001 to 0.037 have been tested with the torsional forced-oscillation and microcreep methods described above (Jackson et al., 2004).

The melt-bearing specimens display qualitatively different behaviour from their melt-free equivalents as illustrated by the representative data shown in Figure 11. This difference is most striking in Figure 11b where a plateau of almost frequency-independent dissipation is seen for periods of 1–100 s and temperatures of 1100–1200°C. When the same dissipation data are plotted against reciprocal absolute temperature for each of

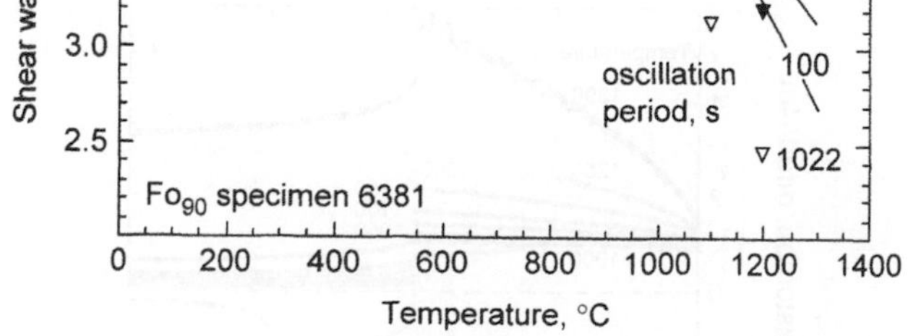

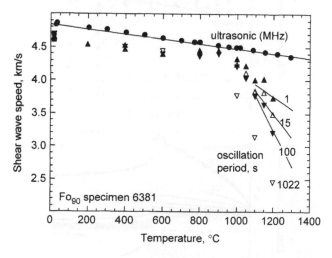

Figure 10. Comparison of the results for a fine-grained synthetic olivine polycrystal obtained at MHz frequencies by ultrasonic interferometry with those at seismic frequencies (mHz–Hz) from torsional forced-oscillation methods. The short lines superimposed upon the forced-oscillation data at the representative periods of 1, 15 and 100 s periods indicate the temperature derivatives calculated at 1200°C from the measured dissipation Q^{-1} (from Jackson et al., 2005).

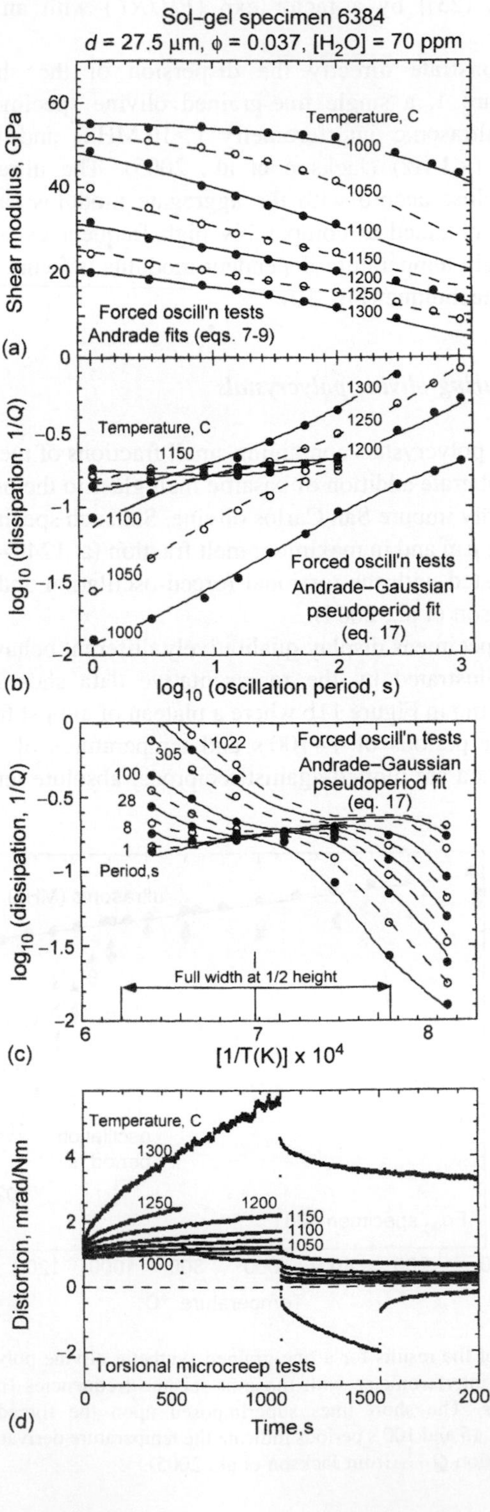
Sol–gel specimen 6384
d = 27.5 μm, ϕ = 0.037, [H2O] = 70 ppm
Shear modulus, GPa
Temperature, C
1000
1050
1100
1150
1200
1250
1300
Forced oscill'n tests
Andrade fits (eqs. 7-9)
(a)
log10 (dissipation, 1/Q)
Temperature, C
1300
1250
1200
1150
1100
1050
1000
Forced oscill'n tests
Andrade–Gaussian
pseudoperiod fit
(eq. 17)
(b)
log10 (oscillation period, s)
log10 (dissipation, 1/Q)
1022
305
100
28
8
1
Period,s
Forced oscill'n tests
Andrade–Gaussian
pseudoperiod fit
(eq. 17)
Full width at 1/2 height
(c)
[1/T(K)] x 10^4
Distortion, mrad/Nm
Temperature, C
1300
1250
1200
1150
1100
1050
1000
Torsional microcreep tests
(d)
Time,s

a series of fixed oscillation periods (Fig. 11c), it becomes clear that the dissipation consists of a broad peak superimposed upon a smoothly monotonic frequency- and temperature-dependent background. Thus the apparently complex variation with temperature of the frequency dependence of Q^{-1} in Figure 11b is explained as follows. At 1000°C, the peak centre is located near 1000 s period and its short-period side enhances the frequency dependence of Q^{-1} for periods of 1–1000 s. As the temperature increases, the peak is displaced to progressively shorter periods. At 1150°C, the peak is located near 1 s period and the nearly frequency-independent dissipation for periods of 1–100 s results from near-cancellation between the frequency dependence of the long-period side of the peak and that of the background. At 1300°C, the peak is located at such short periods as to have minimal impact within the 1–300 s range.

The peak is broader than the Debye peak of the simple Burgers solid by about two decades in period. The Andrade–Gaussian pseudoperiod model represented by the curves in Figure 11(b and c) evidently provides a satisfactory fit. Indeed, this model is capable of representing the dissipation data for the entire suite of melt-bearing specimens by taking into account the inter-specimen variations in mean grain size and maximum melt fraction (Fig. 12). The microcreep records of Figure 11(d) reveal the extent to which the non-elastic strain is recoverable upon removal of the applied torque (i.e. anelastic). The proportion of irrecoverable (viscous) strain increases systematically with increasing temperature.

Subsequently, substantial progress has been made towards the ultimate goal of an extended Burgers model capable of representing the variations of both the shear modulus and dissipation for the suite of melt-bearing specimens. The progress so far is illustrated in Figure 13 showing an adequate fit of the extended Burgers model to the G and Q^{-1} data for a single representative melt-bearing specimen.

9.3. Viscoelastic relaxation mechanisms

Alternative microphysical explanations for dissipation and associated shear modulus dispersion in fine-grained olivine polycrystals have been thoroughly canvassed (Gribb and Cooper, 1998b; Jackson et al., 2002; Cooper, 2003; Faul et al., 2004). The possibility of dislocation damping (e.g. Minster and Anderson, 1981; Karato and Spetzler, 1990) is excluded by the generally very low but somewhat variable dislocation densities in these materials. Stress-induced redistribution of melt between grain-edge tubules ("melt squirt", Mavko and Nur, 1975), although superficially an

Figure 11. Representative results of torsional forced-oscillation and microcreep tests on melt-bearing polycrystalline olivine (specimen 6384). (a) Shear modulus versus log oscillation period at each of a series of temperatures. Forced-oscillation data are indicated by plotting symbols; curves represent the optimal Andrade fits [Eqs. (7) and (8)]. (b) Log dissipation versus log period at each of a series of temperatures from torsional forced oscillation tests. Data are indicated by the plotting symbols; curves represent the Andrade–Gaussian pseudoperiod fit $Q^{-1}(T_0,T)$ for this specimen [Eq. (17)]. (c) Same data and fit as in (b) but now plotted as log dissipation versus reciprocal absolute temperature at each of a series of oscillation periods (1, 3, 8, 15, 28, 54, 100, 305, 612, and 1022 s). (d) Microcreep records plotted as angular distortion versus time. The interval during which the steady torque is applied is terminated after either 500 or 1000 s allowing observation of the partial recovery of the creep strain (from Jackson et al., 2004).

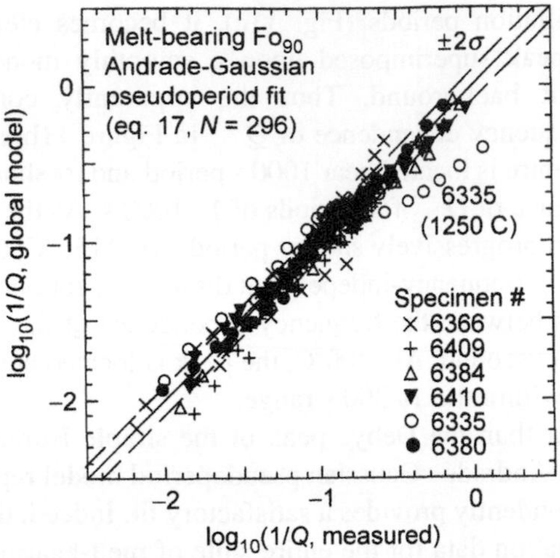

Figure 12. A comparison of measured and modelled dissipation for the melt-bearing specimens of Jackson et al. (2004). The six specimens range in mean grain size from 7 to 52 μm and in maximum melt fraction (at 1240–1300°C) from 0.0001 to 0.037 and were tested at oscillation periods of 1–1022 s and temperatures of 1300–1000°C during slow staged cooling. The different plotting symbols denote the different specimens. The continuous and broken lines correspond to zero and $\pm 2\sigma$ misfit, respectively, relative to the optimal global Andrade–Gaussian-pseudoperiod model $Q^{-1}(T_o,T,d,\phi)$ for which $\chi^2 = 899.4$ for the total of $N = 296$ data (from Jackson et al., 2004).

attractive option, does not provide a satisfactory explanation for the dissipation peak superimposed upon the high-temperature background for melt-bearing materials (Jackson et al., 2004, see also Fig. 11). Instead, the height of the peak and the mild grain-size sensitivity of its location in frequency-temperature space are suggestive of elastically accommodated grain-boundary sliding (Raj and Ashby, 1971). Accordingly, it has been suggested that the elastically accommodated sliding in these materials is facilitated by the rounding of grain edges at triple-junction melt tubules, whereas the absence of a similar dissipation peak for genuinely melt-free olivine reflects the inhibition of elastically accommodated sliding by tight grain-edge intersections (Faul et al., 2004). The monotonically frequency- and temperature-dependent background dissipation and associated dispersion for both classes of material are attributed to diffusionally accommodated grain-boundary sliding (Raj and Ashby, 1971; Raj, 1975).

This interpretation requires that grain-boundary sliding with diffusional accommodation can occur without prior elastically accommodated sliding in the melt-free materials, and that the two mechanisms of accommodation can operate simultaneously in the melt-bearing materials. These inferences concerning the behaviour of real materials are inconsistent with the classic theory of grain-boundary sliding (Raj and Ashby, 1971; Raj, 1975) which envisages sliding with elastic and diffusional accommodation as occurring sequentially. The development of a more realistic model for the transition at high temperature in fine-grained materials from elastic, through anelastic, to viscous behaviour remains a major challenge.

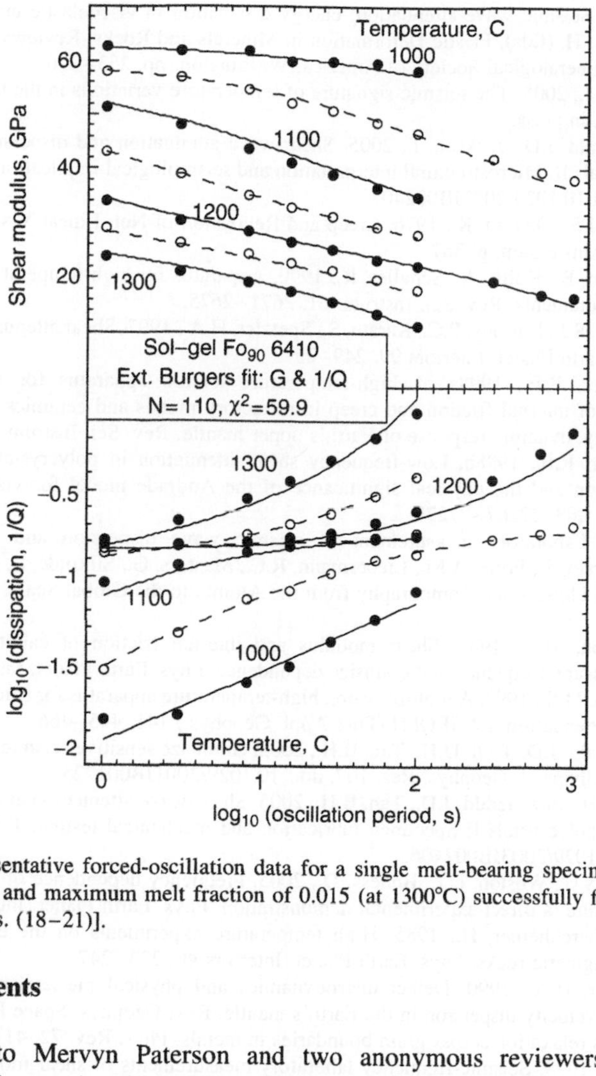

Figure 13. Representative forced-oscillation data for a single melt-bearing specimen (6410) of mean grain size 11.3 μm and maximum melt fraction of 0.015 (at 1300°C) successfully fitted to an extended Burgers model [Eqs. (18–21)].

Acknowledgements

I am grateful to Mervyn Paterson and two anonymous reviewers for thoughtful suggestions leading to significant improvements in the manuscript.

References

Abramowitz, M., Stegun, I.A., 1972. Handbook of Mathematical Functions with Formulas, Graphs, and Mathematical Tables. Dover Publications, New York, p. 1046.

Amin, K.E., Mukherjee, A.K., Dorn, J.E., 1970. A universal law for high-temperature diffusion controlled transient creep. J. Mech. Phys. Solids 18, 413–426.

Berckhemer, H., Kampfmann, W., Aulbach, E., Schmeling, H., 1982. Shear modulus and Q of forsterite and dunite near partial melting from forced oscillation experiments. Phys. Earth Planet. Interiors 29, 30–41.

Brennan, B.J., Stacey, F.D., 1977. Frequency dependence of elasticity of rock – test of seismic velocity dispersion. Nature 268, 220–222.

Cooper, R.F., 2003. Seismic wave attenuation: energy dissipation in viscoelastic crystalline solids. In: Karato, S., Wenk, H. (Eds), Plastic Deformation in Minerals and Rocks, Reviews in Mineralogy and Geochemistry. Mineralogical Society of America, Washington, pp. 253–290.

Faul, U.H., Jackson, I., 2005. The seismic signature of temperature variations in the upper mantle. Earth Planet. Sci. Lett., in press.

Faul, U.H., Fitz Gerald, J.D., Jackson, I., 2005. Shear-wave attenuation and dispersion in melt-bearing olivine polycrystals II. Microstructural interpretation and seismological implications. J. Geophys. Res. 109, B06202, doi:10.1029/2003JB002407.

Findley, W.N., Lai, J.S., Onaran, K., 1976. Creep and Relaxation of Non-Linear Viscoelastic Materials. North-Holland, Amsterdam, p. 367.

Gadaud, P., Guisolan, B., Kulik, A., Schaller, R., 1990. Apparatus for high-temperature internal friction differential measurements. Rev. Sci. Instrum. 61, 2671–2675.

Getting, I.C., Dutton, S.J., Burnley, P.C., Karato, S., Spetzler, H.A., 1997. Shear attenuation and dispersion in MgO. Phys. Earth Planet. Interiors 99, 249–257.

Gribb, T.T., Cooper, R.F., 1998a. A high-temperature torsion apparatus for the high-resolution characterization of internal friction and creep in refractory metals and ceramics: Application to the seismic-frequency, dynamic response of Earth's upper mantle. Rev. Sci. Instrum. 69, 559–564.

Gribb, T.T., Cooper, R.F., 1998b. Low-frequency shear attenuation in polycrystalline olivine: Grain boundary diffusion and the physical significance of the Andrade model for viscoelastic rheology. J. Geophys. Res. 103, 27267–27279.

Jackson, I., 2000. Laboratory measurement of seismic wave dispersion and attenuation: recent progress. In: Karato, S., Forte, A.M., Liebermann, R.C., Masters, G., Stixrude, L. (Eds), Earth's Deep Interior, Mineral Physics and Tomography from the Atomic to the Global Scale. AGU, Washington, pp. 265–289.

Jackson, I., Paterson, M.S., 1987. Shear modulus and internal friction of calcite rocks at seismic frequencies: pressure, frequency and grainsize dependence. Phys. Earth Planet. Interiors 45, 349–367.

Jackson, I., Paterson, M.S., 1993. A high-pressure, high-temperature apparatus for studies of seismic wave dispersion and attenuation. PAGEOPH (Pure Appl. Geophys.) 141, 445–466.

Jackson, I., Fitz Gerald, J.D., Faul, U.H., Tan, B.H., 2002. Grainsize sensitive seismic wave attenuation in polycrystalline olivine. J. Geophys. Res. 107, doi: 10.1029/2001JB001225.

Jackson, I., Faul, U.H., Fitz Gerald, J.D., Tan, B.H., 2005. Shear-wave attenuation and dispersion in melt-bearing olivine polycrystals I: Specimen fabrication and mechanical testing. J. Geophys. Res. 109, B06201, doi:10.1029/2003JB002406.

Jackson, I., Webb, S.L., Weston, L.J., Boness, D., 2005. Frequency dependence of elastic wave speeds at high-temperature: a direct experimental demonstration. Phys. Earth Planet. Interiors, 148, 85–96.

Kampfmann, W., Berckhemer, H., 1985. High temperature experiments on the elastic and anelastic behaviour of magmatic rocks. Phys. Earth Planet. Interiors 40, 223–247.

Karato, S., Spetzler, H.A., 1990. Defect microdynamics and physical mechanisms of seismic wave attenuation and velocity dispersion in the Earth's mantle. Rev. Geophys. Space Phys. 28, 399–421.

Kê, T., 1947. Stress relaxation across grain boundaries in metals. Phys. Rev. 72, 41–46.

Lu, C., Jackson, I., 1998. Seismic-frequency laboratory measurements of shear mode viscoelasticity in crustal rocks II: thermally stressed quartzite and granite. Pure Appl. Geophys. 153, 441–473.

Lu, C., Jackson, I., 2005. Seismic-frequency laboratory measurements of shear mode viscoelasticity in crustal rocks III: The role of pore fluid. Geophysics, submitted.

Mavko, G.M., Nur, A., 1975. Melt squirt in the asthenosphere. J. Geophys. Res. 80, 1444–1448.

Minster, J.B., Anderson, D.L., 1981. A model of dislocation-controlled rheology for the mantle. Phil. Trans. Roy. Soc. Lond. 299, 319–356.

Nowick, A.S., Berry, B.S., 1972. Anelastic Relaxation in Crystalline Solids. Academic Press, New York, p. 677.

Poirier, J.-P., 1985. Creep of Crystals. High-Temperature Deformation Processes in Metals, Ceramics and Minerals. Cambridge University Press, Cambridge, p. 260.

Press, W.H., Flannery, B.P., Teukolsky, S.A., Vetterling, W.T., 1986. Numerical Recipes: the Art of Scientific Computing. Cambridge University Press, Cambridge, p. 818.

Raj, R., 1975. Transient behaviour of diffusion-induced creep and creep rupture. Metall. Trans. A 6A, 1499–1509.

Raj, R., Ashby, M.F., 1971. On grain boundary sliding and diffusional creep. Metall. Trans. 2, 1113–1127.

Schoeck, G., Bisogni, E., Shyne, J., 1964. The activation energy of high temperature internal friction. Acta Metal. 12, 1466–1468.

Tan, B.H., Jackson, I., Fitz Gerald, J.D., 1997. Shear wave dispersion and attenuation in fine-grained synthetic olivine aggregates: preliminary results. Geophys. Res. Lett. 24, 1055–1058.

Tan, B.H., Jackson, I., Fitz Gerald, J.D., 2001. High-temperature viscoelasticity of fine-grained polycrystalline olivine. Phys. Chem. Minerals 28, 641–664.

Woirgard, J., Mazot, P., Riviere, A., 1981. Programmable system for the measurement of the shear modulus and internal friction, on small specimens at very low frequencies. J. Phys. Chem. 5, 1135–1140.

II
Rheology

Advances in High-Pressure Technology for Geophysical Applications
Jiuhua Chen, Yanbin Wang, T.S. Duffy, Guoyin Shen, L.F. Dobrzhinetskaya, editors
Published by Elsevier B.V. (2005)

Chapter 6
High-temperature plasticity measurements using synchrotron X-rays

Donald J. Weidner, Li Li, William Durham and Jiuhua Chen

Abstract

Synchrotron X-rays coupled with high-pressure deformation facilities have transformed the tools for studying the rheological properties of materials at high pressure and temperature. The new D-DIA device is capable of producing 15 GPa of pressure at 2000 K and simultaneously introducing a differential stress field that can attain steady state flow with strains reading in excess of 50%. X-ray transparent anvils enable diffraction in both the vertical and horizontal directions, providing sufficient information to define differential stress in the sample while the strain rate is monitored by X-ray images. This new tool opens the door to new studies relevant to the processes deep within the Earth, illuminating the dynamics of deep earthquakes and mantle flow.

1. Introduction

The determination of the flow properties of the Earth's materials at mantle conditions has challenged Earth scientists for decades. Flow of the solid mantle to accommodate plate tectonics and material behavior during the stress instability that results in deep earthquakes motivate the understanding of flow properties of minerals at elevated pressure and temperature.

The vision has been to build a device capable of deforming a sample at controlled rates with sustained pressure in excess of 10 GPa and temperature up to 2000 K, while providing accurate measurements of the time history of the strain and stress. Successes have come in the past 3 years that have enabled such measurements using synchrotron X-rays to define the strain and stress in the sample in a multi-anvil high pressure system with an axial stress capability. During the next 5 years, the challenge is to improve the accuracy of stress measurements to 10 MPa and expand the pressure regime for such measurements to over 20 GPa.

A series of breakthroughs have enabled this new technology. They include:

- Use of a DIA – a cubic multi-anvil high-pressure device – in conjunction with a synchrotron source that enables X-ray analysis of the sample (Shimomura et al., 1985).
- Development of D-DIA for deformation experiments (Durham et al., 2002; Wang et al., 2003).
- Analysis of stress using X-ray diffraction (Singh, 1993; Mao et al., 1998).
- Analysis of strain from X-ray images(Vaughan et al., 2000; Li et al., 2003).
- Use of X-ray transparent anvils in the multi-anvil system in order to obtain the necessary diffraction data for stress analysis (Chen et al., 2004; Li et al., 2004a).

- The understanding of the effect of plasticity on X-ray stress measurements (Li et al., 2004b).
- Implementation of conical slits to allow white energy-dispersive or 2D detectors to allow monochromatic angle-dispersive measurements (Chen et al., 2004; Li et al., 2004a; Uchida et al. in this book).

Time-dependent flow experiments are among the most demanding of mechanical property measurements. In the traditional deformation experiment, differential stress is monitored by a load cell that indicates the amount of force that is applied to the cylindrical sample. Normalizing by area, with appropriate corrections for friction between the point of measurement and the sample, yields the average differential stress applied to the sample. In a similar manner, strain is monitored by displacement of the pistons that are acting on the sample. Again, with appropriate corrections for length changes in the parts other than the sample, the relative displacement of the two ends of the sample are defined. With the aid of an intense X-ray source, the strategy changes. Diffraction, which defines the elastic changes in lattice spacings, is used to monitor stress, and radiographic images define sample length with time, yielding strain rate. In this new measurement protocol, the stress and strain are measured in the sample directly, with no need for corrections for friction or length changes in soft parts of the cell assembly (Chen et al., 2004). On the down side, experimental precision of length measurements with X-rays is limited to about one part in 10^4, still somewhat short of the highest precision deformations experiments.

2. Analysis of strain from X-ray images

The plastic strain of the sample can be monitored directly from X-ray radiographs. A powerful and non-divergent synchrotron X-ray beam, larger than the dimension of the sample, provides an ideal probe for measuring sample length. By combining relatively X-ray transparent materials as the high-pressure cell assembly, samples can be defined between heavy metal foils (such as Pt, Ni, Au, Re, etc.). These heavy metals absorb X-ray, and thus define sharp edges in the images and serve as strain markers. The image itself is produced from the fluorescence of an X-ray sensitive material, which gives off visible light when bombarded by an X-ray. The image is magnified and recorded by a digital camera. The imaging optics and two images are illustrated in Figure 1. Strain, as a manifestation of the change of sample length, can be derived from these images. Thus, an instantaneous strain measurement can be derived *in situ* at controlled experimental conditions.

These measurements allow us to define the onset of plastic behavior by comparing the image strain with the strain of the lattice planes derived from diffraction measurements. The ratio of sample length to the lattice spacing is proportional to the number of unit cells required to define the sample length. This number should remain constant, as the sample is elastically deformed. Heating, stress, pressure will all keep this value constant if the material is elastically responding. A change in this ratio (termed the Kung ratio) implies that the sample length has increased or decreased in the number of atoms – thus, it has been plastically deformed.

The confirmation of steady state deformation is made possible by these *in situ* strain measurements. Time variations in the strain rate indicate that the sample and cell assembly

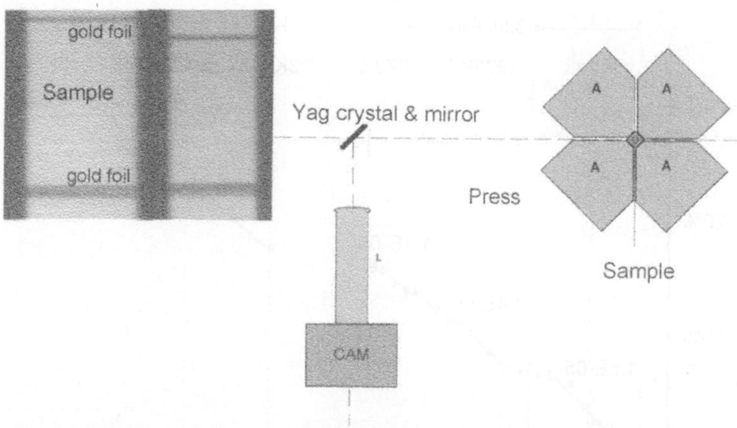

Figure 1. Imaging optics and images of sample. The X-ray beam passes through the sample and excites fluorescence in a YAG single crystal. The visible signal is guided with a mirror and focused with a microscope objective onto a CCD detector in a digital camera. Two images of a sample, one before deformation and one during deformation, delineated with gold foil above and below the sample are also indicated. Strain is manifested by the shortening of the sample between the two images.

are still responding to the changes in applied forces in a transient fashion. Only by observing a constant rate of deformation of the sample are we able to confirm steady state, and then only when the differential stress has attained a time-independent value as well.

High-precision plastic strain measurements are crucial for deformation experiments. The controlling factor for the precision is the image of the metal foils, and ultimately the diffraction limit of visible light. Metal foils have their own width and the foil itself will evolve with deformation, thus complicating the determination of absolute sample length. The main error that contributes to the strain measurement is from the detection of sample length change. Thus, even if the measured sample length (l) has an error of 100 μm, the sample length change (Δl) with 0.1 μm precision provides a precision of 0.0001 for a 1 mm sample. A standard image analysis software can detect object motion to a sub-pixel precision. Li et al. (2003) assimilated this idea into image analysis for strain detection. The current optics allows around one pixel in the image related to 1 μm for the sample with strain resolution of 10^{-4}. Since the strain markers themselves deform, images that obtain this strain need to be gathered every 100–1000 s yielding a strain rate resolution of about $10^{-6}\ s^{-1}$.

To illustrate the procedures for deducing mechanical properties, we will use, as an example, data for San Carlos olivine being deformed in the D-DIA (the apparatus is discussed below). The full experiment, SAN37, shortened the sample by 30% with active deformation at two different pressures (3.5 and 6.5 GPa) and temperatures between 1000 and 1500 K. Figure 2 illustrates strain as a function of time that is measured from the sample images. Uncertainties in strain are indicated or are smaller than the symbol. These uncertainties are based on assuming 0.5 pixel resolution of sample length change. This is probably an overestimate of the uncertainty as judged by the statistical scatter of the observations. These data represent the deformation of the sample, changes in slope reflect changes in the parameter of the driving system – the D-DIA. As indicated here, it is

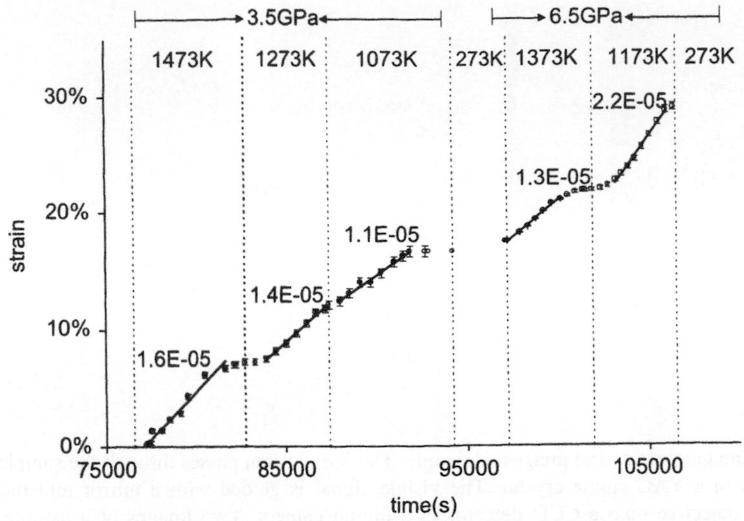

Figure 2. Strain as a function of time for sample in run San37. The San Carlos sample is deformed in the D-DIA at elevated pressure and temperature. Segments of constant strain rate are indicated by the straight lines and the strain rate is indicated in the figure. Error bars represent the error due to 0.5 pixel uncertainty of the deformation between two images.

possible to define several regions where the strain *vs* time curve is fairly linear, indicating a constant strain rate. Strain rate can be changed by changing the parameters of the D-DIA driving system. Here we see a variation of about a factor of two in strain rate.

3. Analysis of stress from diffraction spectra

The spacing between lattice planes is insensitive to the plastic history of the sample. This spacing is a state function, and thus is defined by the pressure, temperature, and stress at the point in the specimen where the lattice spacing is sampled. While volume is mostly controlled by the pressure and temperature, changes in shape reflect the deviatoric stress field. For a cylindrically symmetric stress field, the change in lattice spacing parallel and perpendicular to the unique stress axis uniquely define the differential elastic strain, which, when multiplied by the appropriate elastic moduli, yields the differential stress.

The Debye ring representing diffraction of a polycrystalline sample becomes distorted from a circle when the X-rays pass along the radial direction of this cylindrical symmetry as illustrated in Figure 3. Singh (1993) showed that the shape of the lattice spacing sampled by the Debye ring is given by:

$$\varepsilon_{hkl}(\Psi) = \Delta\sigma/3\{1 - 3\cos^2(\Psi)\}\{(S_{11} - S_{12})(1 - \Gamma(hkl)) + S_{44}\Gamma(hkl)/2\} \quad (1)$$

for a cubic crystal, where $\Delta\sigma$ is the differential stress that is felt by the subset of grains that give rise to the (hkl) reflection, $\varepsilon_{hkl}(\Psi)$ is the value of $\Delta d/d$ referenced to the hydrostatic value of d as a function of the angle, Ψ. S_{ij} are the elastic compliance values for the

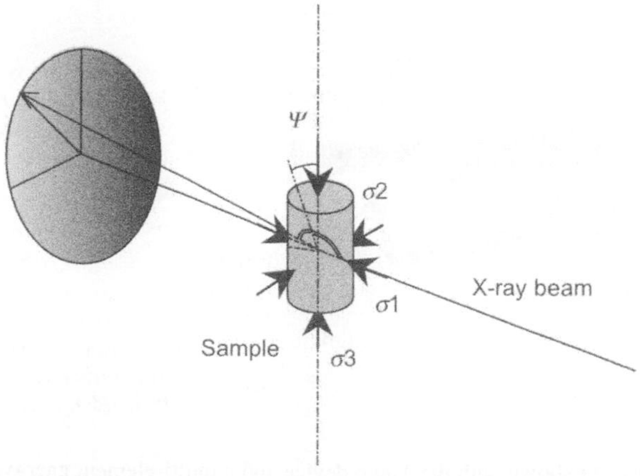

Figure 3. Diffraction geometry for stress measurement. The sample, under a cylindrical stress field has a synchrotron X-ray beam pass through and the beam is diffracted into a distorted Debye ring. The *d-spacing* is determined by the 2θ angle which is a function of Ψ (the angle between the diffraction vector and σ_3). The variation of d with Ψ results because the lattice spacings depend on their orientation relative to the applied stress field.

material and $\Gamma(hkl)$ takes on a value between 0 and 1 depending on the specific diffraction reflection.

Indeed, each subpopulation of grains that give rise to a diffraction observation may in general experience its unique stress field. In this case, if we define the differential strain as:

$$\varepsilon_{hkl}(\text{diff}) = \varepsilon_{hkl}(\Psi = 0) - \varepsilon_{hkl}(\Psi = 90) \tag{2}$$

then

$$\sigma_{hkl}(\text{diff}) = \varepsilon_{hkl}(\text{diff})/E_{hkl} \tag{3}$$

where E_{hkl} is the Young's modulus for the particular orientation that is defined by the diffraction vector. Nye (1957), Uchida et al. (1996)give expressions for E_{hkl} for all crystal systems in terms of the single crystal elastic moduli.

Measurement of stress is thus redefined as measurement of the differential elastic strain using X-ray diffraction. The strain is now measured with both monochromatic X-rays in angle-dispersive mode with a 2D detector (Duffy et al., 1999; Merkel et al., 2002), Uchida et al., in this book) and with white radiation using energy-dispersive tools with multiple detectors (located at $\Psi = 0$ and 90°) with a conical slit system illustrated in Figure 4 (Chen et al., 2004; Li et al., 2004a). On one hand, experiments using monochromatic X-rays have the advantage that the entire Debye ring is sampled, allowing departure from cylindrical symmetry of stress to be determined. In addition, with high-resolution 2D detectors, it is expected to be slightly more accurate than white radiation. On the other hand, the monochromatic exposure is generally longer than with white radiation, thus smearing out time variations in the stress field. Furthermore, at high stress, the peak broadening due to heterogeneous stresses will degrade the signals for both white and monochromatic, erasing the gains in accuracy obtained with the monochromatic beam. White radiation can also be collimated at the detector allowing only X-rays from

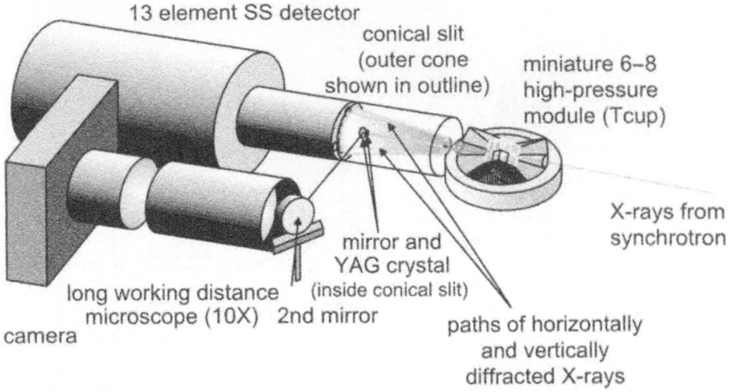

Figure 4. The conical slit shown with the T-cup device and a multi-element energy-dispersive detector. The diffracted X-ray from the sample is collimated by a conical slit system which allows diffraction at a fixed 2θ angle. The multiple-detector is placed behind the slit system to capture the diffracted beam for several values of ψ.

the sample into the detector. Monochromatic radiation will record scattering from the pressure medium and furnace as well as the sample thus causing an overly complex diffraction pattern that may possibly overwrite the sample peaks with background noise. Indeed, both systems work and the final choice will depend on the specific experiment. Realizable accuracies for differential strain measurements for both white and monochromatic studies are about 10^{-4}, which yields a stress accuracy of 20 MPa for a material, such as olivine, with a Young's modulus of about 200 GPa. In these early days of using this technique, most experiments are struggling to achieve this level of accuracy.

4. The deformation DIA

The D-DIA (Durham et al., 2002; Wang et al., 2003) is a modification of the cubic anvil device known as the DIA (e.g. Shimomura et al., 1985) a single-stage, six anvil compression device. The DIA operates up to ~15 GPa and ~2000 K with 4 mm truncations on each of the six anvils. The D-DIA modification gives independent displacement control to one pair of opposing anvils, reducing the cubic symmetry of the DIA to tetragonal, thus allowing high-strain deformation experiments to 15 GPa.

The DIA applies pressure by forcing each of the six anvils to advance on the cubic pressure medium with a single hydraulic ram. The deformation modification is made by including independent rams for both the upper and lower anvil. Once pressure is achieved by advancing the main ram, a differential stress can be produced by advancing the upper and lower anvils. Control of the three rams can provide a constant pressure and a constant strain rate of the sample. The D-DIA system is illustrated in Figure 5. Strains up to 80% have been achieved on 1 mm samples.

The four horizontal tungsten carbide anvils have been replaced with X-ray transparent cubic boron nitride or sintered diamond, both of which are available commercially. This modification allows diffraction to be observed both parallel to the maximum stress

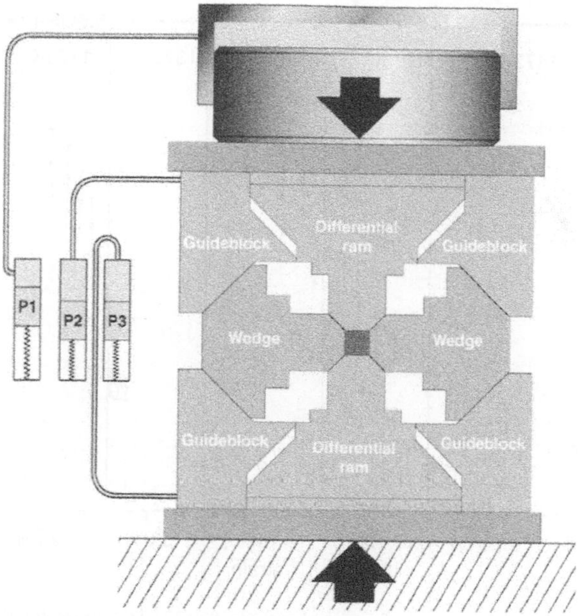

Figure 5. Schematic of D-DIA. The D-DIA high-pressure device is driven by one main ram indicated by the big arrows. This ram drives the top and bottom anvils together and the side anvils (2 of the 4 are illustrated) are driven into the sample chamber by a wedge-type effect. The upper and lower anvils have each an additional ram that can drive them independently of the main ram and hence independently of the side anvils. During deformation, the main ram may need to be backed off of the sample in order to maintain a constant pressure (which can be monitored by the diffractions observations).

direction and the minimum stress axis. This device was first installed at a synchrotron in 2002 and operation has now become nearly a routine at both the GSECARS beam line at the Advanced Photon Source and at X17 at the National Synchrotron Light Source. The details of operation, cell assemblies, and design are given by Durham et al. (2002), Wang et al. (2003).

With the San Carlos olivine example, we illustrate the stress that is measured as a function of strain in Figure 6. In this case, stress is measured in energy-dispersive mode with four solid-state detectors and white X-rays. The detectors are positioned at $\Psi = 0°$, 90°, 180°, and 270°. With the assumption that the stress is aligned with the cylindrical geometry of the deformation system, this is sufficient to define the differential stress using a single diffraction peak. For this sample, four diffraction peaks were used for calculating stress $\{(112), (130), (131), (021)\}$. The error bars indicate the range of stress obtained from these four measurements.

We use the olivine sample itself as the pressure standard using the third-order Birch–Murnaghan equation of state. The cell volume measured after the experiment is used as the reference volume. The hydrostatic *d*-spacing (a, b, c) for orthorhombic olivine are derived by the following Eq. (4):

$$d(P,T) = (d_V + 2d_H)/3 \tag{4}$$

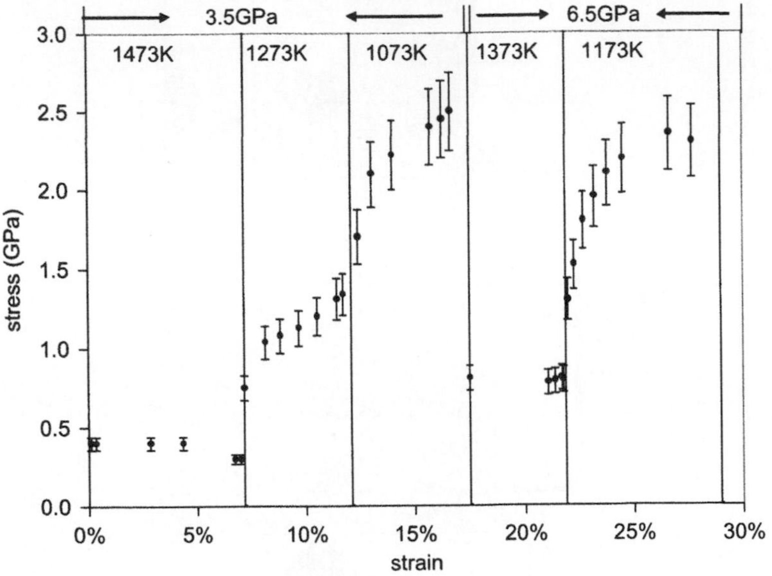

Figure 6. Stress as a function of strain for sample in run San37. The stress for the San Carlos olivine specimen is calculated using the diffraction peaks {(112), (130), (131), (021)}. Error bars indicate the variation in estimated stress from these four peaks. Strain is defined from the corresponding value indicated in Figure 2. Strain hardening can be observed in the lower temperature measurements, yet approach to steady state can be observed at most conditions by the approach to constant stress.

in which $d(P,T)$ is the hydrostatic cell parameters at P and T, d_V is the d spacing measured with vertical detector, d_H is the d spacing measured with horizontal detector. In this study, 3.5 (± 0.2) and 6.5 (± 0.2) GPa are the two pressures during the experiment.

The differential stress values indicated in Figure 6 have different variations with strain that depends on the temperature. For the highest temperature, the stress remains fairly constant as strain increases, while for the lowest temperature, the stress increases by about a factor of two during the deformation. This behavior may indicate strain hardening at low temperature. The steady-state may be identified by the achievement of constant stress.

An example of stress measurement using monochromatic X-rays is shown in Figure 7. In this case, the sample is polycrystalline rocksalt, NaCl deformed at room pressure through a total shortening strain of approximately $k = 0.30$, where we use the symbol k to distinguish plastic strain or *kung* from elastic strain ε, the state variable. Shown as insets in Figure 7 are the Debye rings imaged at two points during the run, at $k = 0.11$ and at $k = 0.27$ during two deformation steps, the first under a confining load of $30T$, the second under $50T$ following the methodology presented by Uchida et al. (2004). It can be seen that mean stress σ_m rose significantly as a result of the increased confining load, but the differential stress $\Delta\sigma$ remained nearly constant.

The D-DIA has as a feature the possibility of applying either a compressive stress along the vertical axis by pushing the differential rams closer together, or an extensional stress along this axis by retracting the rams. Figure 8 illustrates the use of both actions on a pyrope sample at 5.5 GPa and 1570 K. In the compression cycle, the anvils were driven at

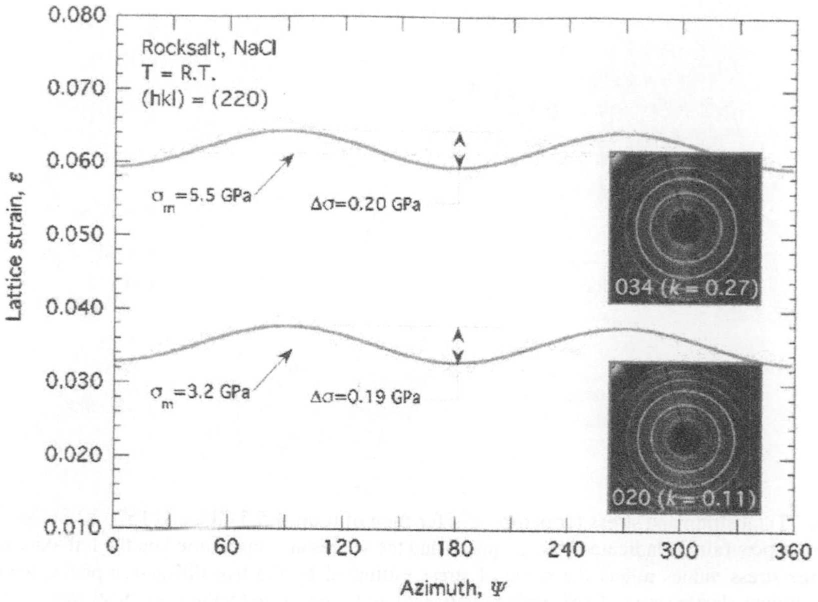

Figure 7. Illustration of stress measurement using monochromatic X-ray diffraction, for salt sample dd029, run at the GSECARS sector at APS. The raw data are the Debye rings captured by a CCD camera. Two such sets of Debye rings are shown in the insets, taken during deformation at plastic strains $k = 0.11$ and 0.27 under different confining loads. Each ring in the pattern results from diffraction by a different (*hkl*) lattice plane in the salt. Using Bragg's law, the 2θ radii around the $0 \leq \Psi \leq 360°$ of the Debye rings can be converted to lattice spacing $d(\Psi)$ and then normalized to lattice strain $\varepsilon(\Psi)$. The main plot shows $\varepsilon(\Psi)$ for only the (220) diffraction in salt, the outermost of the two most prominent Debye rings in each of the patterns. 360 individual points make up each measurement in the main plot. Those points are fit to curves of the form $1 - 3\cos^2\Psi$, which can then be used to calculate stress using Eq. (1) and the known S_{ij} for salt. The values of lattice strain at $\Psi = 37°$ gives the value of the two confining pressures (mean stress $\sigma_m = 3.2$ and 5.5 GPa), while the amplitudes measure the differential stress ($\Delta\sigma \sim 0.2$ GPa in both cases).

two different speeds which are reflected in the different slopes to the strain *vs* time curve. In this phase, the sample was shortened by almost 50%. In the second phase, the differential pistons were retracted creating a negative strain rate. The squares in this diagram indicate the stress calculated from the average elastic strain of five diffraction peaks. The error bars on the stress points represent the span of calculated stress for these five peaks. Indeed, each peak may sample different stress levels as we discuss below. Thus, the error bars may be an overestimate of the uncertainty of the mean stress. As expected, when the strain rate changes sign, the differential stress also changes sign.

5. The integrated system

The D-DIA together with the diffraction and imaging systems provides a powerful new tool for exploring many phenomena where stress is a critical issue. Quantitative

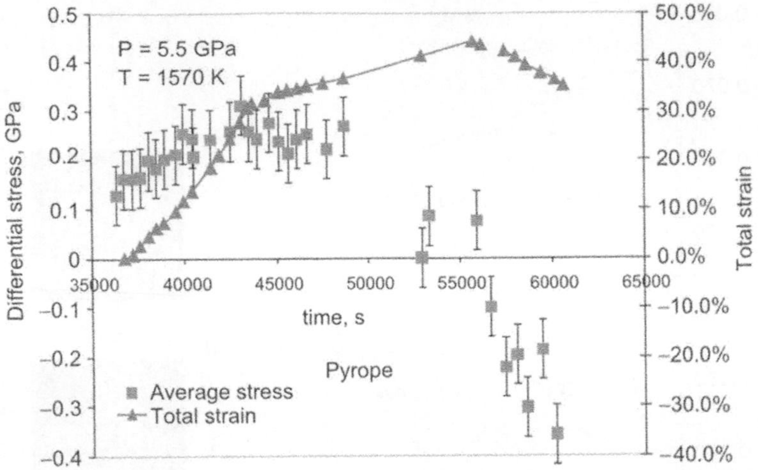

Figure 8. Total strain and stress for pyrope as a function of time at 5.5 GPa and 1570 K. Stress, deduced from the elastic strain, is indicated by the squares and the values are represented on the left axis. The error bars on the stress values reflect the range of stress estimated by the five diffraction peaks, which were used to estimate elastic strain. Total strain is indicated by triangles and values are displayed on the right-hand axis. The first break in the slope of the strain *vs* time was caused by a change in the rate that the deforming anvils were driven. The second slope change coincides with a change in the direction that the deforming anvils were driven. This created a change in sign of the slope of the strain *vs* time curve and a change in sign of the differential stress.

determination of rheological properties is certainly the central goal where stress and strain rate are determined at elevated pressure and temperature while the sample is experiencing a constant strain rate. Other new uses are already being discovered as outline below.

5.1. Study stress distribution in a polycrystalline sample

Equation (3) above allows for the measurement of stress in different subpopulations of grains, where the subpopulation is defined by the diffraction planes. Such variations will occur within a polycrystalline sample in all but the Reuss state. Indeed the Reuss (1929)–Voigt (1928) models define bounds for the subgrain stress–strain states for populations when the sample is elastically deformed. However, when the polycrystalline sample is plastically deformed, the Reuss–Voigt models no longer bound the stress distribution in the sample (Chen et al., 2004). Instead, the Taylor (1938)–Sachs (1928) bounds may control the stress state in the material (Li et al., 2004b). In the elastic case, the elastic moduli dictate the stress–strain variation among the subgrain populations, in the plastic case, it is the plastic anisotropy that governs the stress variations among the different populations. The X-ray diffraction method of defining stress allows us to explore the variations of stress among different subgrain populations of a monomineralic aggregate or among the different materials in a polyphase aggregate. This observation will be very useful in defining residual stresses in materials that have been deformed and recovered.

This new tool will also allow the study of the flow properties of multiphase aggregates. Stress in each material can be monitored along with the deformation of the entire specimen. Such new style of data will enable tests of flow models for these rock-like aggregates.

5.2. Measure elastic properties

If indeed the stress field achieves the Reuss state, then Eq. (3) allows the determination of some of the elastic properties of the sample from the elastic strain anisotropy. This approach has been widely used with diamond anvil cells and side entry diffraction (see Mao et al. (1998)). Weidner et al. (2004) explore the limitations of this approach. The main difficulty comes from the Reuss–Voigt assumptions which may be extremely wrong for a material that is being plastically deformed. However, by following a proper stress path, this method may be very useful in defining static single crystal elastic moduli. In particular, if the sample is annealed by heating after it is compressed, reducing the differential stress to near zero, then the stress–strain relations observed once a new stress field is applied, may be useful to this end.

One also has the opportunity to measure acoustic velocities in specimens with a defined non-hydrostatic stress field using techniques such as by Li et al. (1996). This provides a whole new set of material properties known as the third-order elastic moduli.

5.3. Provide hydrostatic stress

Solid media devices often subject the sample to non-hydrostatic stresses during the pressure loading cycle. Some experiments will be affected by the resulting, room temperature, yielding of the sample. The D-DIA with stress detection capability provides an opportunity to reduce such effects. Stress can be monitored during loading, and by adjusting the differential rams, it may be reduced or eliminated. This approach can allow nearly hydrostatic loading of the sample.

6. Future developments

The D-DIA, complete with stress and strain metrics, is in its infancy. Future goals means pushing the stress–strain rate metrics to deliver accuracies of 10 MPa in stress and 10^{-7} s^{-1} in strain rate. These are probably possible with current technology. Currently, the system is optimized with polycrystalline samples with grain size less than 10 μm. Future studies may be carried out using coarser samples or single crystal samples if stress proxies are developed. These proxies may be polycrystalline samples placed in series with the sample in the sample chamber. Owing to continuity of normal stress across interfaces, the unique principal stress will be continuous. However, the radial stress continuity relies on the strength of the confining material. Cells can be developed with weak confining material with robust diffraction patterns that can be used to define variation of confining stress around the sample and around the proxy. In this manner, the validity of the proxy can be assured.

These new systems will need to develop higher pressures. It may be possible to follow the design of the D-DIA with a 6–8 type device. The 6–8 system has been modified for synchrotron work with the introduction of the T-cup (Vaughan et al., 1998). This system operates to pressures of 25 GPa. The next generation deformation device may be a T-cup modified to include a differential stress system.

The data analysis software has not kept up with the hardware developments. Tools for near real time analysis of data are important. Software for processing 2D monochromatic data is lacking. Now the approach is to bin the data for small ranges of Ψ. A better approach is to model the entire 2D data set with stress as a variable, much the same way as Rietveld refinements model 1D data. This possible stress refinement approach for data processing is an advantage of the 2D monochromatic diffraction over the energy-dispersive diffraction. Converting the data from 2D to 1D inherently looses precision. It will be better to fit the raw data.

Acknowledgements

The authors thank the SAM group (especially Michael Vaughan and Liping Wang) for providing technical support, Zhong Zhong for his support at the NSLS beam line. We also thank T. Uchida and Y. Wang of GSECARS for support at the APS. This research was carried out in part at the NSLS, which is supported by the US Department of Energy, Division of Material Sciences and Division of Chemical Sciences under contract No. DE-AC02_98CH10886 and COMPRES for support of the beam lines X17 and in part at the APS in the GSECARS beamline. This research was supported by the NSF Grant EAR-9909266, EAR0135551, and EAR0229260 and the Centre National de la Recherche Scientifique (CNRS). Samples of garnet were synthesized by Gabriel Gwanmesia under grants: NSF EAR-0106528 and EAR-0135431 Work by WBD performed under the auspices of the US Department of Energy by the Lawrence Livermore National Laboratory under contract W-7405-ENG-48. MPI publication No. 353.

References

Chen, J., Li, L., Weidner, D.J., Vaughan, M., 2004. Deformation experiments using synchrotron X-rays: in situ stress and strain measurements at high pressure and temperature. Phys. Earth Planet. Interiors 143–144, 347–356.

Duffy, T.S., Shen, G.Y., Shu, J.F., Mao, H.K., Hemley, R.J., Singh, A.K., 1999. Elasticity, shear strength, and equation of state of molybdenum and gold from X-ray diffraction under nonhydrostatic compression to 24 GPa. J. Appl. Phys. 86, 6729–6736.

Durham, W., Weidner, D., Karato, S., Wang, Y., 2002. New developments in deformation experiments at high pressure. In: Wenk, R. (Ed.), Plastic Deformation of Minerals and Rocks. Mineralogical Society of America, Washington, DC, pp. 291–329.

Li, B., Gwanmesia, G.D., Liebermann, R.C., 1996. Sound velocities of olivine and beta polymorphs of Mg_2SiO_4 at Earth's transition zone pressures. Geophys. Res. Lett. 23, 2259–2262.

Li, L., Raterron, P., Weidner, D., Chen, J., 2003. Olivine flow mechanisms at 8 GPa. Phys. Earth Planet. Interiors 138, 113–129.

Li, L., Weidner, D., Raterron, P., Chen, J., Vaughan, M., 2004a. Stress measurements of deforming olivine at high pressure. Phys. Earth Planet. Interiors 143–144, 357–367.
Li, L., Weidner, D.J., Chen, J., Vaughan, M.T., Davis, M., Durham, W.B., 2004b. X-ray strain analysis at high pressure: effect of plastic deformation in MgO. J. Appl. Phys. 95, 8357–8365.
Mao, H.-k., Shu, J., Shen, G., Hemley, R.J., Li, B., Singh, A.K., 1998. Elasticity and rheology of iron above 220 GPa and the nature of the Earth's inner core. Nature 296, 741–743.
Merkel, S., Wenk, H.R., Shu, J., Shen, G., Gillet, P., Mao, H.-k., Hemley, R.J., 2002. Deformation of polycrystalline MgO at pressures of the lower mantle. J. Geophys. Res. 107, 2271.
Nye, J.F., 1957. Physical Properties of Crystals. Oxford University Press, Ely House, London.
Reuss, A., 1929. Berechnung der fließgrenze von mischkristallen auf grund den konstanten des einkristalls. Z. Angew. Math. Mech. 9, 49.
Sachs, G., 1928. Zur ableitung einer fliessbedingung. Z. Ver. Dtsch. Ing. 72, 734–736.
Shimomura, O., Yamaoka, S., Yagi, T., Wakatsuki, M., Tsuji, K., Kawamura, H., Hamaya, N., Fukuoga, O., Aoki, K., Akimoto, S., 1985. Multi-anvil type X-ray system for synchrotron radiation. In: Minomura, S. (Ed.), Solid State Physics Under Pressure. Terra Scientific Publishing, pp. 351–356.
Singh, A.K., 1993. The lattice strains in a specimen (cubic symmetry) compressed nonhydrostatically in an opposed anvil device. J. Appl. Phys. 73, 4278–4286.
Taylor, G.I., 1938. Plastic strain in metals. J. Inst. Met. 62, 307–315.
Uchida, T., Funamori, N., Yagi, T., 1996. Lattice strains in crystals under uniaxial stress field. J. Appl. Phys. 80, 739–746.
Uchida, T., Wang, Y., Rivers, M.L., Sutton, S.R., 2004. Yield strength and strain-hardening of MgO up to 8 Gpa measured in the deformation-DIA with monochromatic X-ray diffraction. Earth Planet. Sci. Lett. 226, 117–126.
Vaughan, M.T., Weidner, D.J., Wang, Y.B., Chen, J.H., Koleda, C.C., Getting, I.C., 1998. T-cup: a new high-pressure apparatus for X-ray studies. Rev. High Pressure Sci. Technol. 7, 1520–1522.
Vaughan, M., Chen, J., Li, L., Weidner, D., Li, B., 2000. Use of X-ray imaging techniques at high-pressure and temperature for strain measurements. In: Nicol, M.F. (Ed.), AIRAPT-17. Universities Press, Hyderabad, India, pp. 1097–1098.
Voigt, W., 1928. Lehrbuch der Kristallphysik. Teubner, Berlin.
Wang, Y.B., Durham, W.B., Getting, I.C., Weidner, D.J., 2003. The deformation-DIA: a new apparatus for high temperature triaxial deformation to pressures up to 15 GPa. Rev. Sci. Instrum. 74, 3002–3011.
Weidner, D.J., Li, L., Davis, M., Chen, J., 2004. Effect of plasticity on elastic modulus measurements. Geophys. Rev. Lett. 31, 19090.

Advances in High-Pressure Technology for Geophysical Applications
Jiuhua Chen, Yanbin Wang, T.S. Duffy, Guoyin Shen, L.F. Dobrzhinetskaya, editors

Chapter 7

Stress and strain measurements of polycrystalline materials under controlled deformation at high pressure using monochromatic synchrotron radiation

Takeyuki Uchida, Yanbin Wang, Mark L. Rivers and Steve R. Sutton

Abstract

We describe techniques for stress and strain measurements using monochromatic diffraction and imaging instruments developed at the GSECARS beamline of the Advanced Photon Source for controlled quantitative deformation experiments at high pressures and high temperatures. Details about the experimental procedure, including sample and cell preparation, system operation, and data analysis are discussed. Based on these techniques, examples of data obtained for a powdered MgO sample are presented, demonstrating the capability of decoupling differential stress from pressure by advancing and retracting the differential rams while adjusting the main hydraulic ram, so that both positive ($\sigma_1 > \sigma_2 = \sigma_3$) and negative ($\sigma_1 < \sigma_2 = \sigma_3$) differential stress conditions can be achieved and controlled under a given pressure. This technical capability enables us to examine sample properties within either the elastic or the plastic regimes. Differential lattice strains computed from distortion of diffraction Debye rings recorded on a 2D charge coupled device X-ray detector, can be used as a "stress gauge" with a resolution between 10^{-5} and 10^{-4}, a result of the complete 360° azimuth angle coverage. The errors in total axial strain, obtained from the sample length measurements in the vertical direction, are between 10^{-4} and 10^{-3}. Differential stresses, computed from the differential lattice strains for reflections hkl, are in general non-equivalent. Differential stress versus total strain curves provide information on hysteresis loops and the yield point. Beyond the yield point, the flow stress of MgO increases with total axial strain, with significant work hardening. This behavior is consistent with previous stress–strain measurements using a gas-medium apparatus, demonstrating the capability of quantitative deformation experiments under high pressure. Further developments are needed to better understand the correlations between X-ray stress measurements, texture observation, and flow mechanism.

1. Introduction

The deformation-DIA (D-DIA) (Wang et al., 2003) has opened new opportunities to study high-pressure, high-temperature rheological properties of mantle materials that are of fundamental importance in constraining the dynamics of the Earth's interior. However, a number of theoretical and experimental issues still need to be addressed in order to perform quantitative deformation research under conditions corresponding to the Earth's mantle. In this chapter, we present theoretical and technical developments in performing controlled deformation experiments using monochromatic synchrotron radiation at the GeoSoilEnviroCARS (GSECARS) beamlines of the Advanced Photon Source (APS).

Under non-hydrostatic stress conditions, the X-ray diffraction Debye ring observed from a deformed sample is distorted. Singh (1993) developed a theory to correlate the distortion of the diffraction Debye rings (statistic average of lattice strains in grains that contribute to diffraction) to differential stress in a polycrystalline sample, using the iso-stress (Reuss, 1929) and iso-strain (Voigt, 1928) bounds. Since, within the elastic regime, the real situation is between these two bounds (see Watt et al., 1976), lattice strains for the iso-stress and iso-strain models were combined by introducing an empirical parameter α, which represents certain arithmetic averaging of the two bounds. Because the Reuss and Voigt bounds are defined under the elastic, infinitesimal strain framework, this "stress gauge" by definition is only meaningful within the elastic regime. The effects of plastic deformation and slip systems were not considered in the theory.

Despite the limitations, this theory has been applied to lattice strain measurements under pressure in various high-pressure devices, ranging from the Drickamer cell (Funamori et al., 1994; Uchida et al., 1998) to the diamond anvil cell (DAC; Mao et al., 1998; Singh et al., 1998; Duffy et al., 1999a; Duffy et al., 1999b; Kavner and Duffy, 2001; Merkel et al., 2002a; Merkel et al., 2002b; Shieh et al., 2002; Kavner, 2003; Merkel et al., 2003; Akahama et al., 2004; Merkel et al., 2004; Shieh et al., 2004). The original lattice strain theory on cubic symmetry (Singh, 1993) has been extended to hexagonal (Singh and Balasingh, 1994) and to all other symmetries (Uchida et al., 1996), so that any polycrystalline sample can be studied to pressure and temperature conditions corresponding to the Earth's core using the DAC (e.g. Mao et al., 1998).

One of critical problems in all these studies is that the non-hydrostatic stress is closely coupled with pressure, and the amount of total sample strain (especially plastic strain) cannot be controlled or accurately determined. Therefore, quantitative rheological information (such as the dependence of sample strength on strain and strain rate) cannot be reliably obtained.

In addition to lattice strain measurements, there are several approaches for stress measurements using X-ray diffraction. Other techniques, such as using diffraction peak width as a stress gauge, can be found in Weidner et al. (1994), Chen et al. (2002), and Cordier et al. (2004) and the references therein. Chen et al. (2004) described techniques of *in situ* stress and strain measurements using X-ray imaging and diffraction previously developed at different facilities and currently available at the National Synchrotron Light Source (NSLS) for deformation experiments. Li et al. (2003) and Kung et al. (2004) present a method to process image data. In this chapter, we handle diffraction peak shifts (i.e. distortion of the Debye rings) using 2D monochromatic X-ray diffraction and radiographic images. For the development of deformation apparatus, see Wang et al. (2003), Chen et al. (2004), and Xu et al. (in this volume).

Details of the experimental technique, including sample and cell preparation, system operation, and data analysis are described for controlled deformation experiments at high pressures and high temperatures. By employing advanced anvil materials such as cubic boron nitride (cBN) and sintered diamond (SD), monochromatic X-ray diffraction can be carried out using a 2D X-ray detector, allowing measurements of lattice strain with unprecedented resolution (on the order of 10^{-5}). Examples of results obtained on MgO, which is stable up to 227 Gpa (Duffy et al., 1995), are also given to demonstrate that (1) differential stress can be controlled independently from pressure, generating a stress field that is well defined and characterized, (2) the total sample strain (mostly plastic) can be controlled and measured *in situ* throughout the deformation processes, (3) the relatively large sample height/diameter ratio (typically about 1) enables multiple deformation cycles, varying not only the magnitude, but also the sign of the differential stress and total strain (i.e. compression and extension), thus allowing examination of material behavior within the elastic and plastic regimes separately, and (4) by comparing the resultant stress–strain relations using our diffraction techniques with those obtained using a conventional gas-medium deformation apparatus at similar pressures (Paterson and Weaver, 1970), quantitative information can be obtained with our techniques.

Sections 2–4 describe procedures for sample preparation, experiment operation, as well as data analysis. Section 5 shows several examples of results obtained for MgO, demonstrating the data quality and wealth of information that can be obtained in our deformation experiments and discusses applicability of lattice strain as a "stress gauge". Section 6 discusses issues relevant to the achievable resolution of strain measurements. Section 6 also discusses issues that need to be addressed to further advance the quantitative deformation experiments.

2. Experimental

2.1. System setup and the D-DIA

The main panel of Figure 1 shows schematic configuration of the entire setup (top view), including the incident X-ray optics, the D-DIA apparatus, as well as the imaging and diffraction detection setup. The D-DIA apparatus (Fig. 1a) consists of a pair of guide blocks (only the lower guide block is shown), each with a built-in differential hydraulic ram that is controlled independently from the main hydraulic ram, which is used for pressure control in the same way as in the conventional DIA (e.g. Osugi et al., 1964). The D-DIA module is compressed in the 250 ton press (Wang et al., 1998) installed at the GSECARS 13-BM-D beamline (Rivers et al., 1998) at the APS.

A Si (111) monochromator is used to select a photon energy up to 65.0 keV, corresponding to a wavelength of 0.191 Å, with an overall beam size about $3 \times 3\ \mathrm{mm}^2$. Entrance slits, consisting of two pairs of tungsten carbide blocks, can be driven in and out of the beam path to adjust the incident beam size. The X-ray imaging system, adopted from the X-ray tomography setup at the GSECARS (Rivers et al., 1999), consists of a YAG single crystal phosphor and a mirror and is mounted on the press frame to monitor sample length during deformation (e.g. Uchida et al., 2001). Diffraction and imaging modes are interchangeable by moving both the entrance slits and the phosphor-mirror assembly.

2.2. Diffraction mode

In this mode, entrance slits are moved into the beam path to aperture the incident beam to $100 \times 100\ \mu\mathrm{m}^2$, with the phosphor-mirror assembly driven out from the beam path. The diffracted X-rays are detected by a 2D X-ray charge coupled device (CCD) detector (Bruker SMART 1500; 1024×1024 pixel^2). We use X-ray transparent cBN or SD anvils,

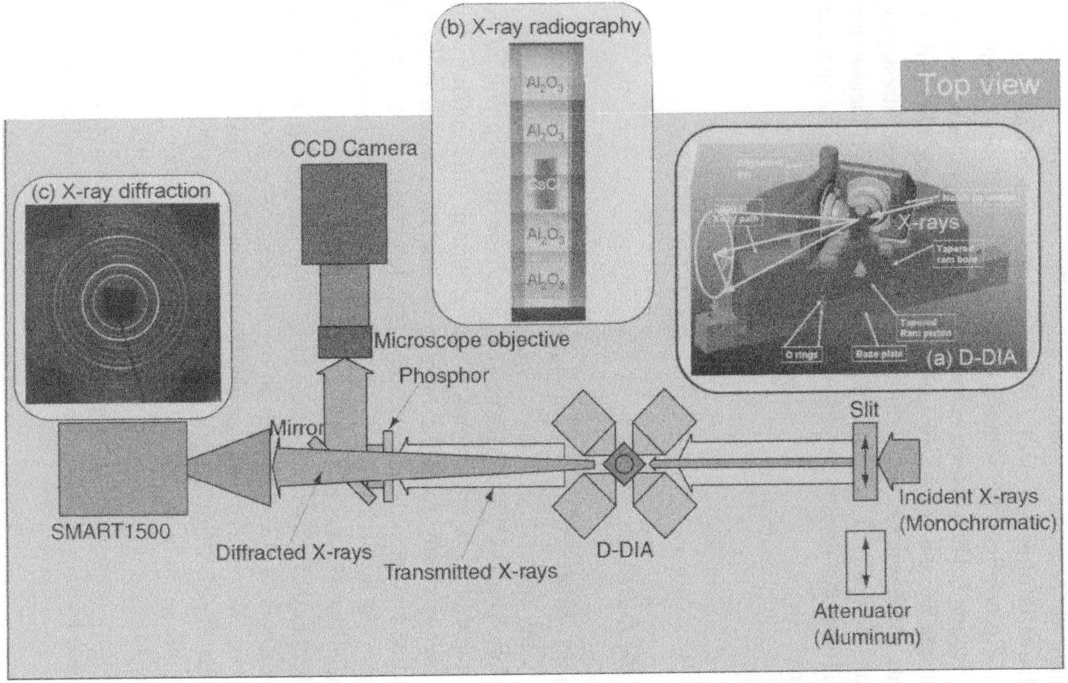

Figure 1. Schematic illustration of X-ray diffraction and imaging setup (top view). Transmitted X-rays are converted by the YAG single-crystal phosphor to visible light, which is then reflected by the mirror through the microscope objective into the PentaMax CCD camera. For the imaging mode, the phosphor–mirror assembly is in and incident slits are out of the beam path. For the X-ray diffraction, incident slits are moved in and both CCD camera and the phosphor-mirror assembly are moved out from the beam path. Diffracted X-rays travel above the phosphor-mirror assembly and are detected by a Bruker SMART 1500 CCD detector. Monochromatic beam of 65.0 keV ($\lambda = 0.191$ Å) is used for both imaging and diffraction. (a) Bottom guide block of the D-DIA and X-ray geometry. (b) X-ray image of the sample chamber. The sample is compressed by sintered Al_2O_3 pistons from top and bottom. (c) An example of X-ray diffraction.

and collect complete Debye rings over the entire 360° detector azimuth angles χ with 2θ values up to about 11° (Fig. 1c). Because of the high photon energies (low wavelengths), *d*-spacings down to about 1 Å can be recorded within this 2θ range.

The angle dispersive technique has certain advantages over a 2-azimuth detector configuration (i.e. using two solid-state detectors that are 90° apart in azimuth angle, e.g. Funamori et al., 1994; Uchida et al., 1998; Weidner et al., 2004) in that distortion of the Debye rings is more robustly determined and detector drift can be detected and corrected during a prolonged experiment (see Section 6). Note that in our angle dispersive diffraction mode, no beam collimation is performed on the detector side and thus we observe diffraction lines of all materials in the X-ray beam path. However, contamination of diffraction signal due to other materials can be minimized by using amorphous boron as the pressure medium and graphite as the heating material, and making the beam cross sectional area smaller.

2.3. *Imaging mode*

In this mode, entrance slits are driven out of the beam path to allow the entire incident beam (3×3 mm^2) to pass through the sample assembly, and the phosphor-mirror assembly is moved into the beam path. Transmitted X-rays project the absorption contrast of the high-pressure cell (the sample, deformation pistons, etc.) onto the YAG phosphor, which converts the X-rays into visible light that is reflected by the mirror and viewed by a PentaMax CCD camera (1300×1000 pixel^2) through a microscope objective. Using a 5× objective, an area of 1.9×1.5 mm^2 can be recorded, with a resolution of 1.5 μm pixel^{-1}. An example of X-ray image of the high-pressure cell is shown in Figure 1b. By analyzing these images collected under various stages during deformation, total axial strain, which contains both elastic and plastic components, and strain rate, are calculated based on the sample length measurement over time.

2.4. *Cell assembly*

A schematic of the cell assembly is shown in Figure 2. The pressure medium (6 mm edge length cubes or $6 \times 7 \times 7$ mm^3 rectangular blocks) is made of a mixture of amorphous boron and epoxy resin, with a sample chamber of 1.6 mm diameter that is stacked in the following sequence from top to bottom (Fig. 2a): crushable Al_2O_3, sintered Al_2O_3 piston, the sample surrounded by a hexagonal BN sleeve, sintered Al_2O_3 piston, and crushable Al_2O_3. For X-ray transparent samples, the interfaces between the sample and the sintered Al_2O_3 pistons are marked with thin (2 μm) gold foils to allow accurate sample length measurement by X-ray imaging. For high-temperature experiments (Fig. 2b), a graphite heater tube is used around the sample with platinum foils on top and bottom for electrical contact. Temperature is monitored by a W/Re thermocouple, which is inserted into the small drill hole made in the (110) plane (Fig. 2b).

It is important to polish the sintered Al_2O_3 pistons carefully so that the two ends are parallel to each other and both are perpendicular to the cylindrical axis, as these ends define the sample length through X-ray imaging. Tilted piston ends make the stress

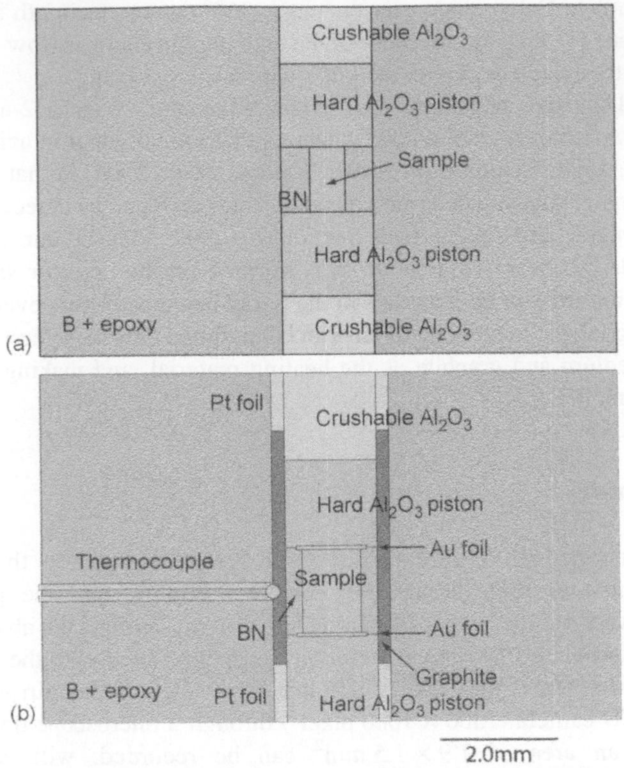

Figure 2. Schematic cross sectional view of cell assembly in the (110) perpendicular to the X-ray for (a) room-temperature and (b) high-temperature deformation experiments. The sample temperature can be increased by resistance heating of graphite and measured by a W/Re thermocouple, which is inserted into the small drill hole perpendicular to the beam. Top and bottom surfaces of Al_2O_3 pistons are polished well so that the sample parallelism remains unchanged during deformation.

distribution heterogeneous and precise determination of sample length difficult, thus compromising accuracy in total strain measurements.

The samples generally have an aspect ratio (height over diameter) of about 1. The stress state in the samples is approximated as having one unique compressive principal stress (σ_1) in the vertical direction (parallel to the compression axis) and two equal principal stresses ($\sigma_2 = \sigma_3$) in the horizontal directions. Observations on distortion of the Debye ring confirmed this stress state (Section 4).

2.5. Sample preparation

It is recommended to use pre-sintered samples in deformation experiments, as loose powders have undetermined porosity and, as the sample is deformed, the total strain measured by imaging will contain significant contribution from compaction. In our experiments, we synthesize samples beforehand in a multi-anvil press, following carefully determined pressure-temperature paths, which are specific to the stability field and grain-size characteristic of the materials to be studied. When a pre-sintered sample is used, unwanted non-hydrostatic stress may be accumulated during isotropic compression due to high density and strength of the sample and Al_2O_3 pistons, introducing a strong uniaxial compressive stress component σ_1 ($>\sigma_3$). To avoid this situation, a $6 \times 7 \times 7$ mm^3 rectangular pressure medium can be used, with the shorter dimension in the vertical direction (parallel to the sample column). The differential rams are adjusted, if necessary, during isotropic compression so that non-hydrostatic stress can be minimized before initialization of the deformation process.

In situations where the sample cannot be pre-sintered (in reality many users do not have such facility), cold-pressed polycrystalline samples may be used. In this case, additional errors may be introduced in sample length measurement. Typically we choose the reference sample length at the very beginning of the deformation cycle. There is certain arbitrariness in the total strain definition, especially when multiple deformation cycles are applied. Therefore, for low-pressure deformation experiments, we repeat deformation cycles on our way up to and down from high pressure and double-check the consistency of the results. Example results obtained for cold-pressed sample will be presented in Section 5.

3. Operation procedure

Prior to an experiment, tilt and rotation of the X-ray CCD detector relative to the incident X-ray beam are calibrated with a diffraction standard (CeO_2). The sample assembly is then loaded in the D-DIA and the sample-detector distance is determined from the diffraction pattern of the sample at ambient condition. The six anvils are advanced to touch the pressure medium gently. When the rectangular pressure medium is used, the top and bottom anvils are advanced slightly (0.5 mm) more than side anvils by driving the differential rams. The main ram load is then increased with the differential rams valved off to compress the sample isotropically to the target pressure. The stress condition is monitored during isotropic compression because a sintered sample sometimes experiences strong non-hydrostatic stress. If necessary, the differential rams are retracted or advanced to reduce the non-hydrostatic stress level. This operation procedure is important in order not to generate large differential stresses to deform the sample plastically before starting deformation.

Under a given pressure, the differential rams are advanced (retracted) at a constant speed to shorten (lengthen) the sample, while diffraction patterns and radiographic images are repeatedly recorded with typical exposure times of 300 and 100 s, respectively, at the vertical center of the sample, where frictional effects across the sample–piston interface are minimal. Strain rates for the advancing and retracting cycles are variable by changing the pumping speed of the differential rams, generally in the range of 10^{-3}–10^{-6} s^{-1}. An additional 10:1 gear reducer is available for each differential ram for even slower strain rates. The main hydraulic ram load is constantly adjusted to compensate for pressure effects due to load changes caused by the differential rams.

Typically each deformation cycle requires several hours to complete. Slower strain rate experiments (10^{-6} s^{-1}) require 10^4 s to detect 10 μm length change for 1 mm long sample. Within the limited beamtime, multiple deformation cycles are still achievable by driving the differential rams at various pressures and speeds if data collection is conducted continuously. Often these cycles are performed at elevated temperatures. Heating is achieved by a large power supply with several hundred Watts of electrical power, and temperature in the experimental station may increase during a prolonged experiment. This temperature increase causes the detector position to drift slightly, introducing additional errors in the lattice strain determination (Section 6). Thus it is important to maintain a constant hutch temperature during the experiment.

4. Data reduction

4.1. Image analysis

4.1.1. Zinger removal

Due to the scattering of high-energy X-rays inside the hutch, random photons will directly strike the CCD chip, generating isolated sharp bright spots (generally only one pixel in size) in the images and diffraction patterns recorded by the CCD detectors (Figs 3a and 4a). A numerical median filter is used to remove such zingers, which replaces intensity of each point with the median of all eight neighboring pixels as well as the pixel of interest. This filter is very effective in restoring the "true" image, as can be seen in Figure 3b (image) and Figure 4b (diffraction). The procedure is important in reducing the noise in the images for obtaining accurate sample length and distortion of the Debye rings.

4.1.2. Sample length measurement

For less absorbing samples, the sample length is defined by the gold foils located at the sample–piston interfaces (Fig. 5a). After zinger removal, each image is decomposed into a series of 1-pixel stripes along the vertical loading axis. Each 1-pixel stripe exhibits two intensity minima caused by the absorption of the gold foils (Fig. 5b). This intensity profile is flipped upside down (Fig. 5c), and an auto curve-fitting routine is applied to determine the center of the peaks in terms of pixels, for all the stripes. The distance between the two maxima (Fig. 5d) defines the sample length in terms of pixels which can be converted into real length after a calibration.

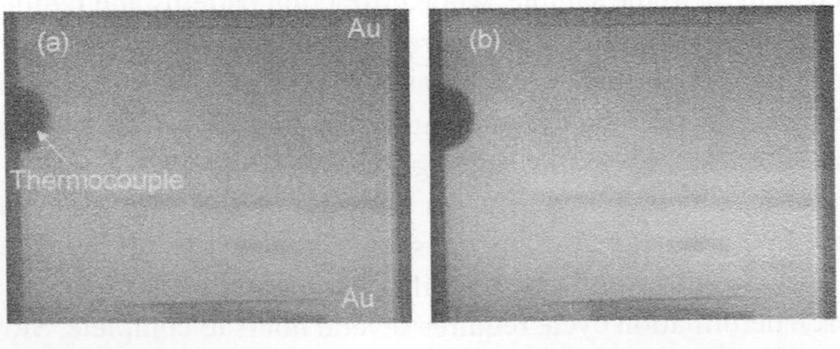

Figure 3. Representative X-ray image (a) before and (b) after median filtering. The filtering process removes most of the zingers, allowing robust fitting results for Au foils' distance. These images correspond to schematic design in Figure 2(b).

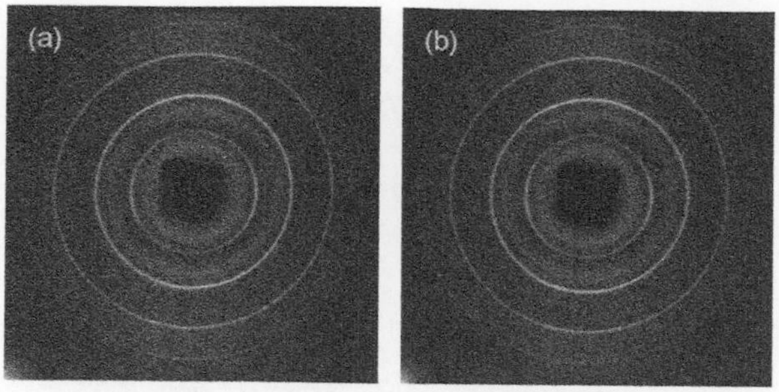

Figure 4. An example of X-ray diffraction (a) before and (b) after median filtering. Most of the zingers, including the one on a diffraction line, can be eliminated.

In cases where the sample is more X-ray absorbing (Fig. 6a), no gold foils are used. The intensity profile of the 1-pixel stripe is characterized by a U-shaped trough with two sharp edges marking the ends of the sample (Fig. 6b). By taking the absolute values of the derivatives of the intensity distribution with respect to the vertical pixels, the end positions are defined by the two peaks in Figure 6c. The auto curve-fitting routine is then applied to all the 1-pixel stripes to obtain the sample length in terms of pixels. Generally all of the available 1-pixel stripes are used (more than 300 stripes over the entire sample diameter) (Fig. 6d). These length measurements are then used to define an average sample length.

Sub-pixel resolution can be achieved for each stripe, but such resolution is difficult to accomplish for the entire sample length. Each cell part suffers certain machining errors and contains some heterogeneities in mechanical properties, hence the sample may tilt during the deformation cycles, making it difficult to define a unique sample length, based on the 2D X-ray image alone. For better resolution, precise cell parts are crucial. In our deformation experiments, all cell parts were carefully prepared using a computerized numerical control (CNC) machine, so that the top and bottom pistons remained parallel throughout the experiment. The resultant errors in sample length measurements can be within 1 pixel over the entire sample diameter, corresponding to errors in total axial strain on the order of 10^{-4}.

Although Li et al. (2003) claimed sub-pixel resolution by measuring the relative length change in two consecutive images, data processing methods for situations when the strain marker was tilted or deformed were not described. Occasionally, the sample column (the stack of crushable alumina, deformation pistons, and the sample) would buckle under compression, resulting in poor images. Most of the time, this problem requires detailed image analysis. Figure 7a shows an example of such an image. To quantify the sample length, the oval-shaped Au shadows over the diameter are analyzed and fit by a second-order polynomial function. The minimum of the curve is then defined, with its pixel position identified. Because the curvature remains unchanged during the deformation, the minimum point is used to calculate relative sample length changes. Figure 7b shows the sample length measurement thus obtained. Self-consistent sample length changes can still be obtained by this technique. Typical errors of length measurement in such samples are several pixels, corresponding to errors in total axial strain on the order of 10^{-3}. When the curvature changes, an additional error of a few pixels is expected.

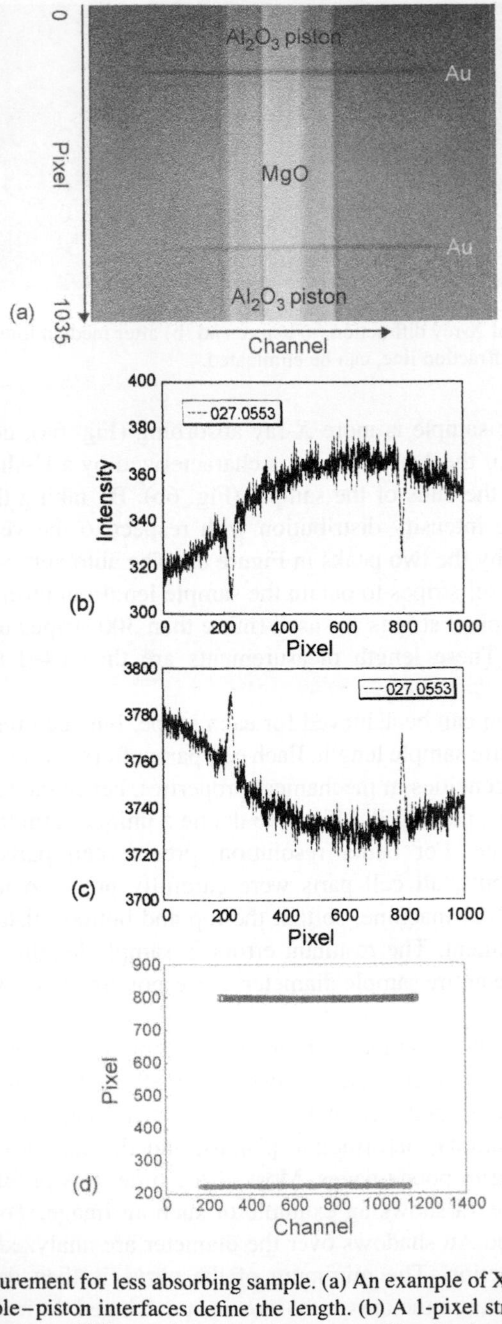

Figure 5. Length measurement for less absorbing sample. (a) An example of X-ray image. The 2 μm Au foils located at the sample–piston interfaces define the length. (b) A 1-pixel stripe decomposed from the image (a). Each 1-pixel stripe exhibits two intensity minima caused by the absorption of the gold foils. (c) Flipped intensity profile of (b). Auto curve-fitting routine determines the peak position for all 1-pixel stripes. (d) Resultant peak positions as a function of horizontal channel. The sample length is obtained from the peak positions.

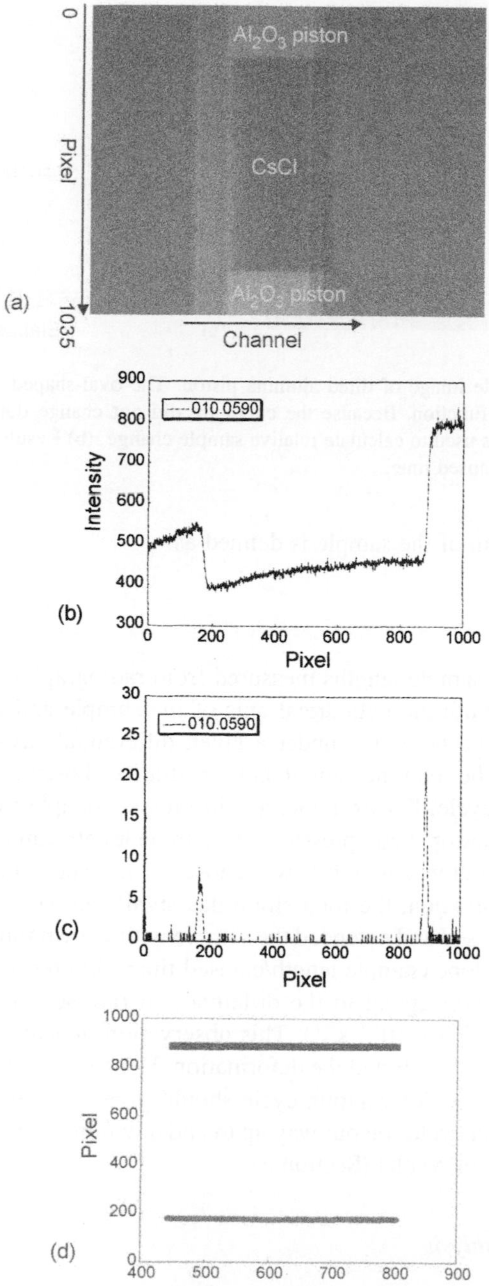

Figure 6. Length measurement for X-ray absorbing sample. (a) An X-ray image. (b) A 1-pixel stripe created from the image (a), featuring a U-shaped trough with two sharp edges. (c) Absolute values of the derivatives of the intensity distribution (b) with respect to vertical distance, showing the end positions as the two peaks. Auto curve-fitting routine determines the peak position for all 1-pixel stripes. (d) Fitting result. The sample length is obtained from the peak positions.

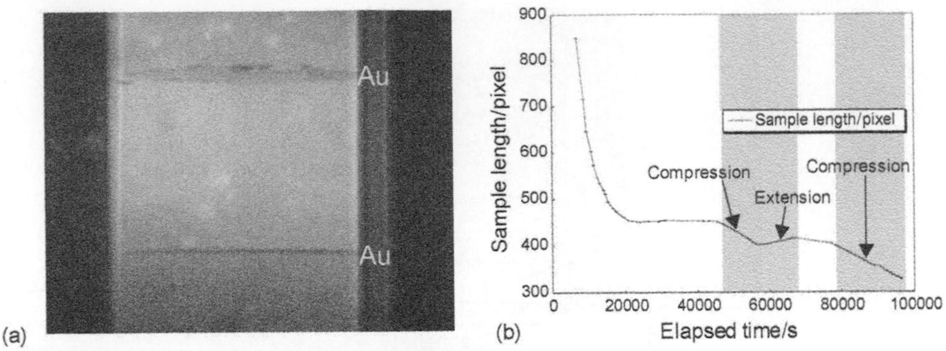

Figure 7. (a) An example image of tilted alumina piston. The oval-shaped Au absorption is fit by a second-order polynomial function. Because the curvature did not change during the deformation, the minimum of the curve was used to calculate relative sample change. (b) Result of relative sample length change as a function of elapsed time.

The total axial strain of the sample is defined as

$$\varepsilon_{\text{total}} = \frac{(l_0 - l)}{l_0}, \tag{1}$$

where l_0 and l are the sample lengths measured from radiographic images in the vertical direction (parallel to both the cylindrical axis of the sample and compressive principal stress σ_1) at a reference point and under a given differential stress during deformation cycles, respectively. The reference length l_0 is arbitrarily chosen at the very first point of the first deformation cycle. For example, a cold pressed sample may contain significant porosity. During the isotropic compression, the sample length can change by a factor of 2 (e.g. Fig. 7b). If the reference length l_0 was chosen at the very first point of experiment, before isotropic compression, the total strain has significant contribution from porosity reduction. We compressed and extended the sample with known ram-advancement speeds in Figure 7b, and the slope (sample length/elapsed time) difference between compression and extension cycles correspond to the difference in ram-advancement speeds (in this example, strain rate is $9.5 \times 10^{-6}\ \text{s}^{-1}$). This observation indicates that most of porosity had been removed when we started the deformation. Hence the reference sample length at the very beginning of the deformation cycle should give us meaningful total strain. We also repeat deformation cycles on our way up to and down from high pressure and double-check the consistency of results (Section 5).

4.2. Lattice strain analysis

Our data reduction technique is similar to that of Merkel et al. (2002a). After spatial and flat field corrections as well as zinger removal, each 2D diffraction pattern (Fig. 8a) is processed using the ESRF software package FIT2D (Hammersley, 1998). The observed Debye rings, originally in the polar coordinates (Fig. 8a), are transformed into Cartesian coordinates (Fig. 8b), in terms of detector azimuth χ with the azimuth zero arbitrarily defined in the vertical direction (parallel to the maximum principal stress axis).

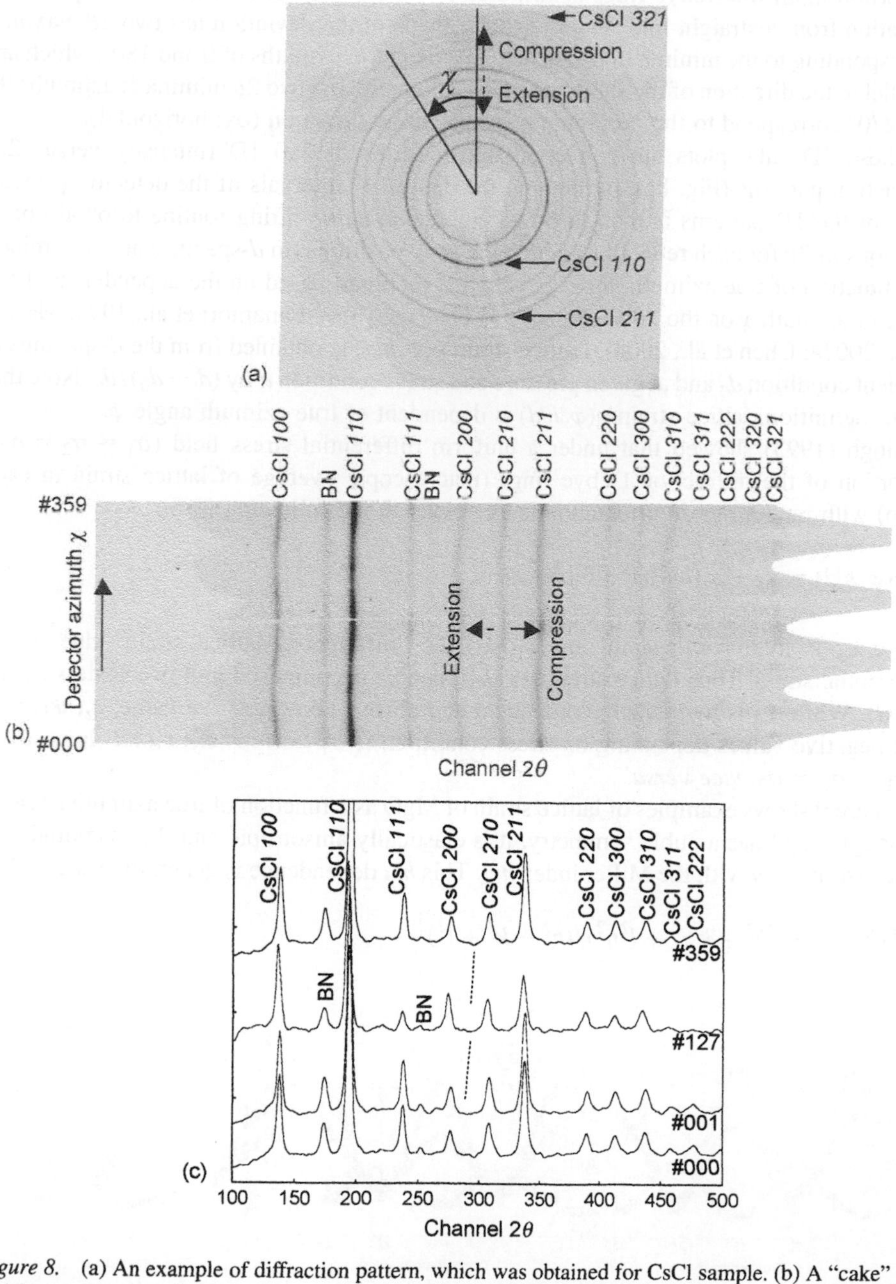

Figure 8. (a) An example of diffraction pattern, which was obtained for CsCl sample. (b) A "cake" plot, which can subsequently be saved in tiff format, converted from 2D diffraction data using FIT2D program. Data are binned at 1° steps. The vertical axis is detector azimuth angles χ, from 0 to 359° (from bottom to top), of the Debye rings, with zero degree arbitrarily defined as diffraction vector being vertical (parallel to the compression axis σ_1), and the horizontal axis is 2θ, from 0 to 11° (left to right). Distortion in the Debye rings and variety of modulation are clearly visible. (c) Series of 1D diffraction patterns that were transformed from the "cake" plot. Each diffraction pattern is fit to obtain 2θ angle of each peak. True azimuth angle φ is computed from 2θ angle and detector azimuth angle χ while lattice strains are from 2θ values at ambient and given pressure conditions.

Distortion in all the Debye rings is now clearly visible in these so-called "cake plots" as deviation from a straight line. The sinusoidal shape of the deviation has two 2θ maxima, corresponding to the minima in *d*-spacing, near detector azimuths of 0 and 180°, which are parallel to the direction of the shortening (σ_1). Likewise, the two 2θ minima at azimuths 90 and 270° correspond to the minimum principal stress direction (σ_3; horizontal).

These 2D cake plots are converted into a series of 360 1D (intensity versus 2θ) diffraction patterns (Fig. 8c), by binning the data in 1° intervals of the detector χ angle. Each of the 1D patterns is then fit by an automated curve-fitting routine to obtain peak positions in 2θ for each reflection, so that the peak positions (in *d*-spacing) are determined as a function of true azimuth angle φ, which is modified based on the dependence of the detector azimuth χ on the 2θ angle: $\cos\varphi = \cos\theta\cos\chi$ (see Funamori et al., 1994; Merkel et al., 2002a; Chen et al., 2004). Lattice strain $\varepsilon(\varphi, hkl)$ is obtained from the *d*-spacings at ambient condition d_0 and at given pressure and stress condition d by $(d - d_0)/d_0$. Note that in our definition, lattice strain $\varepsilon(\varphi, hkl)$ is dependent of true azimuth angle φ.

Singh (1993) showed that under a uniform differential stress field ($\sigma_1 \neq \sigma_2 = \sigma_3$), distortion of the diffraction Debye rings (macroscopic average of lattice strain in each grain) with random grain orientation is expressed in the following form:

$$\varepsilon(\varphi, hkl) = \varepsilon_p - \varepsilon_t(hkl)(1 - 3\cos^2\varphi), \tag{2}$$

where ε_p is hydrostatic strain and $\varepsilon_t(hkl)$ the "differential lattice strain" due to the differential stress. Thus lattice strain $\varepsilon(\varphi, hkl)$ can be decomposed into two scalars ε_p and $\varepsilon_t(hkl)$. While ε_p is hydrostatic component and always takes positive value, $\varepsilon_t(hkl)$ may take negative values depending on stress condition. If $\sigma_1 > \sigma_2 = \sigma_3$, $\varepsilon_t(hkl)$ is positive. If $\sigma_1 < \sigma_2 = \sigma_3$, *vice versa*.

Figure 9 shows examples of lattice strain of MgO as a function of true azimuth angle φ. Although MgO has a cubic symmetry, it is elastically anisotropic and the magnitudes of lattice strain vary with the Miller index *hkl*. This *hkl* dependence is generally described by

$$\Gamma(hkl) = (h^2k^2 + k^2l^2 + l^2h^2)/(h^2 + k^2 + l^2)^2, \tag{3}$$

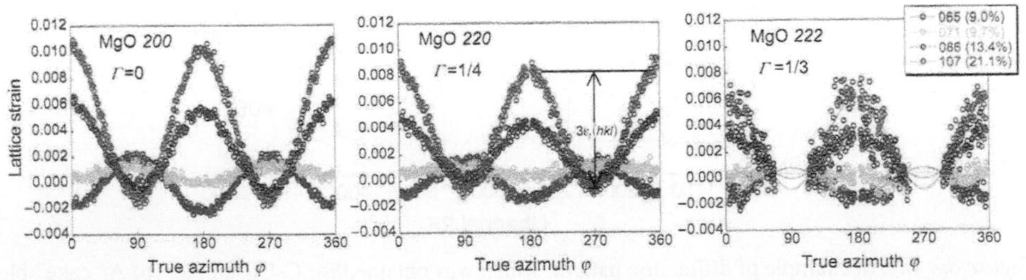

Figure 9. Azimuth dependence of lattice strain $\varepsilon_t(hkl)$ on MgO 200, 220, and 222 reflections taken from the deformation cycle at 1.0 GPa. The legend shows the data file number with total axial strain in brackets. Data #065 were collected at the onset of deformation, #071 at the yield point, #086 in the first-stage hardening, and #107 in the second-stage hardening. Depending on $\Gamma(hkl)$, differential lattice strains (magnitude of modulation) are systematically different. Note the change in the direction of lattice strain modulation, due to reversing differential ram directions (positive: compression; negative: extension).

and elastic anisotropy

$$S = S_{11} - S_{12} - (S_{44}/2). \tag{4}$$

As Funamori et al. (1994) pointed out, under a constant differential stress, when the anisotropy of the sample is positive (i.e. $S > 0$, such as MgO), reflections with the minimum $\Gamma(hkl)$ such as 200 and 400 ($\Gamma = 0$) exhibit largest modulation (Fig. 9). When the anisotropy is negative ($S < 0$, such as NaCl and CsCl), reflections with the maximum $\Gamma(hkl)$ such as 111 and 222 ($\Gamma = 1/3$) should show largest modulation. However, CsCl does not follow this rule, rather it shows similar trend with MgO data (Fig. 10). This anomaly is due to the difference in strength and work hardening along various crystalline orientations. Differential stresses in different subgroups of grains can be grossly different after yielding (for detailed discussion, see Section 5).

The magnitude of $\varepsilon_t(hkl)$, defined as differential lattice strain, is obtained by fitting lattice strain as a function of true azimuth angle φ using Eq. (2). Figure 11 shows an example of differential lattice strain $\varepsilon_t(hkl)$ of MgO as a function of total sample strain. The reference sample length l_0 is taken at the very first point of the first deformation cycle.

4.3. *Differential stress and hydrostatic stress*

As in conventional deformation experiments, the macroscopic stress field ($\sigma_1 \neq \sigma_2 = \sigma_3$) in the sample is divided into two terms at the center of the sample,

$$\sigma_{ij} = \begin{pmatrix} \sigma_1 & 0 & 0 \\ 0 & \sigma_3 & 0 \\ 0 & 0 & \sigma_3 \end{pmatrix} = \begin{pmatrix} \sigma_p & 0 & 0 \\ 0 & \sigma_p & 0 \\ 0 & 0 & \sigma_p \end{pmatrix} + \begin{pmatrix} 2t/3 & 0 & 0 \\ 0 & -t/3 & 0 \\ 0 & 0 & -t/3 \end{pmatrix}, \tag{5}$$

where σ_1 and σ_3 represent the maximum and minimum principal stresses, respectively, with compression being positive. In Eq. (5), $\sigma_p(= (\sigma_1 + 2\sigma_3)/3)$ and $t(= \sigma_1 - \sigma_3)$ are hydrostatic pressure and differential stresses, respectively.

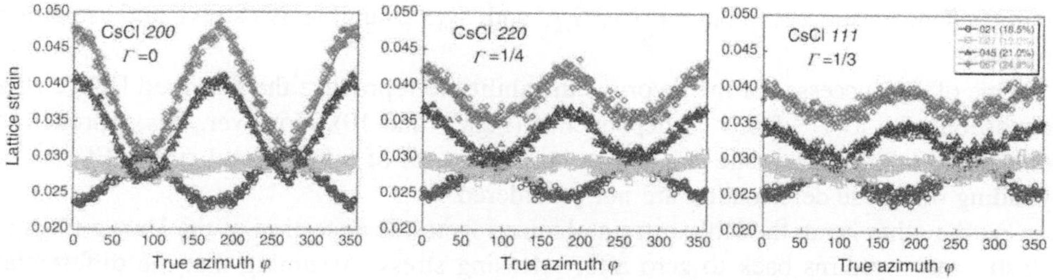

Figure 10. An example of the azimuth dependence of lattice strain $\varepsilon_t(hkl)$ on CsCl 200, 220, and 111 reflections taken from the deformation cycle at 2.5 GPa. The legend shows the data file number with total axial strain in brackets. From data #045 to #067, the magnitude of strain modulation does not increase for 220 and 111 reflections, indicating that grains contributing these reflections has yielded, while it still increase for 200 reflection.

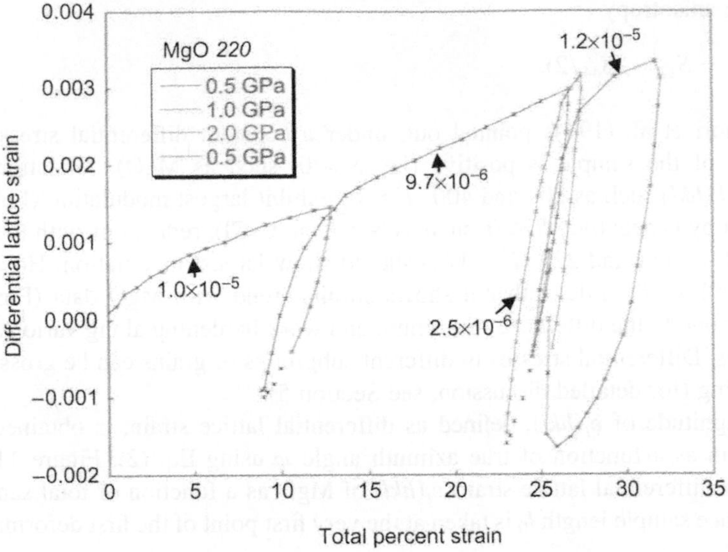

Figure 11. An example of differential lattice strain versus total axial percent strain plot, obtained for MgO 220 reflection in the first run. The reference sample length of total strain was arbitrarily chosen at the very beginning of the first deformation cycle. For each deformation cycle, the linear relationship between lattice strain and total strain is limited to very small total strains (<1%). Above the linear (elastic) regime, the slope changes with increasing total strain, indicating significant work hardening. The sample strain rates (in s^{-1}) is labeled for sample shortening.

The lattice strain $\varepsilon(\varphi, hkl)$ of any *d*-spacing in a polycrystalline sample with random grain orientation is expressed as

$$\varepsilon(\varphi, hkl) = \varepsilon_p - \left(\frac{t}{3}\right)(1 - 3\cos^2\varphi)\left(\frac{1}{2G(hkl)}\right), \tag{6}$$

where $G(hkl)$ is the shear modulus (see Nye, 1957 for each crystal symmetry). Singh (1993) showed in detail the term that corresponds to $G(hkl)$ for cubic materials with three independent elastic compliances S_{11}, S_{12}, and S_{44}, assuming iso-stress and iso-strain models.

One of the successes of the theory is the ability to reproduce the observed Debye ring distortion with a $(1 - 3\cos^2\varphi)$ dependence (Figs 9 and 10). However, this approach is only valid within the elastic regime in the framework of infinitesimal strain. Effects of yielding or plastic deformation are not considered.

On the other hand, the lattice strain obtained from the distortion of the Debye rings is elastic and it returns back to zero after releasing stress. Assuming that the differential lattice strain $\varepsilon_t(hkl)$ has a linear relation with differential stress, we have a simple elastic relation from Eqs. (2) and (6):

$$\varepsilon_t(hkl) = \left(\frac{t(hkl)}{3}\right)\left(\frac{1}{2G(hkl)}\right). \tag{7}$$

For cubic systems,

$$\frac{1}{2G(hkl)} = S_{11} - S_{12} - 3S\Gamma(hkl), \tag{8}$$

where $\Gamma(hkl)$ and S are defined by Eqs. (3) and (4), respectively. Note that here we do not assume either iso-stress or iso-strain models and differential stresses $t(hkl)$ are not necessarily identical. For example, $t(hkl)$ are expressed for cubic systems as

$$t(222) = \frac{3\varepsilon_t(222)}{(S_{44}/2)}, \tag{9}$$

$$t(200) = \frac{3\varepsilon_t(200)}{(S_{11} - S_{12})}, \tag{10}$$

and

$$t(220) = \frac{3\varepsilon_t(220)}{\left[\frac{1}{4}(S_{11} - S_{12}) + \frac{3}{8}S_{44}\right]}. \tag{11}$$

Using these differential stresses, we discuss the similarity between our lattice strain observations and stress measurements using conventional deformation apparatus, and consider correlation between lattice strain and differential stress in Section 5. For crystals with any symmetry other than cubic, Eq. (7) is still applicable, but the content of shear modulus $G(hkl)$ varies (Uchida et al., 1996).

For infinitesimal strain, hydrostatic stress (pressure) is obtained from the hydrostatic strain ε_p,

$$\sigma_p = K_T \times 3\varepsilon_p = \left(\frac{1}{S_{11} + 2S_{12}}\right)\varepsilon_p. \tag{12}$$

where K_T is isothermal bulk modulus. Since strain is not infinitesimal in high-pressure experiments, Eq. (12) needs to be modified according to the third-order Birch–Murnaghan equation of state (EOS),

$$\sigma_p = \left(\frac{3}{2}\right)K_T((1-\varepsilon_p)^{-7} - (1-\varepsilon_p)^{-5})\left[1 - \left(\frac{3}{4}\right)(4 - K'_T)((1-\varepsilon_p)^{-2} - 1)\right], \tag{13}$$

where K'_T is pressure derivative of isothermal bulk modulus K_T. We calculate pressure based on known equations of state of the samples from the hydrostatic volumetric strain obtained from the curve-fit at the "magic" azimuth angle $\varphi = \sin^{-1}\left(\sqrt{1/3}\right) = 35.3°$, where the second term in Eqs. (2) and (6) vanishes. For MgO, both K_T and S_{ij} are taken from Spetzler (1970).

The differential strain $\varepsilon_t(hkl)$ is treated as the infinitesimal strain, as the finite hydrostatic strain has been taken into account by Eq. (13). This involves a redefinition of strain, by replacing d_0 with d_p, the hydrostatic d-spacing under pressure,

$$\varepsilon_t(\varphi, hkl) = (d_p - d)/d_p. \tag{14}$$

We apply the infinitesimal strain method to handle the small amount of distortion shown in Eq. (14). Note that while X-ray lattice strains are elastic and are a manifestation of volumetric (pressure) and differential stresses, the total sample axial strain contains a significant plastic component. This will be denoted as "total strain" to be distinguishable from the elastic lattice strains.

5. Example: deformation of MgO

In this section we illustrate the information one can obtain by employing the technique outlined above, using the results from a powdered MgO sample. Another purpose in this section is to demonstrate a baseline of "calibration", so to speak, by comparing our results with previous conventional deformation experiments at comparable pressures, using a gas-medium apparatus (Paterson and Weaver, 1970). Scientific results and discussions can be found in Uchida et al. (2004).

The cold pressed MgO samples were deformed in two runs. In the first run, the sample was driven through four deformation cycles: (1) at 0.5 GPa hydrostatic pressure: compression with differential lattice strains increased to about 0.2%, followed by extension, where lattice strain was reversed to −0.15%, (2) at 1 GPa hydrostatic pressure: compression with lattice strains to 0.4%, followed by extension, where lattice strains were reversed to −0.25%, (3) at 2 GPa, compression with lattice strains up to 0.5%, followed by extension, where lattice strains were reversed to −0.3%, and (4) back to 0.5 GPa, to reproduce the lattice–total strain curve in step (1). In the second run, pressure was first increased to 4.9 GPa and then the sample was shortened and lengthened under an average pressure of about 6.4 GPa (4.9–7.9 GPa). After the deformation cycle, the sample temperature was increased.

5.1. Stress–strain curves and hysteresis loops

Figure 11 plots the differential lattice strain $\varepsilon_t(220)$ against total axial strain $\varepsilon_{\text{total}}$ in the first run. For clarity, data from other reflections 200 and 222 are not plotted. Note that multiple deformation curves have been obtained from a single experiment, for both compression and extension; at various pressures while conventional deformation devices generally cannot conduct extension measurements under pressure. The hysteresis loops identify the total stress and strain threshold for the elastic regime, i.e. the initial linear segment after the differential rams are reversed. Note that after the first compression–extension cycle, the differential lattice strain returns to zero at the different total strain level, due to plastic deformation (Fig. 11). If one examines the hysteresis behavior of all three reflections, differential lattice strains are different at a given total strain, because of the build up of residual stress in the sample after plastic deformation (Uchida et al., 2004).

For better understanding, Figure 12 shows schematics of a hysteresis loop in the differential stress versus total strain curve, where differential stress is computed from the differential lattice strain using Eqs. (9)–(11). When the differential rams reversed the direction from extension to compression (marked as reference point: R_0), the sample is in the elastic regime. Depending on the history of the previous deformation cycle (dashed line), residual stress t_0 may vary. Then differential stress linearly increases with total strain up to initial yield point with the slope of $1/6G(hkl)$ because in the elastic regime, relative change in total strain is identical to relative change in lattice strain. Beyond the yield point,

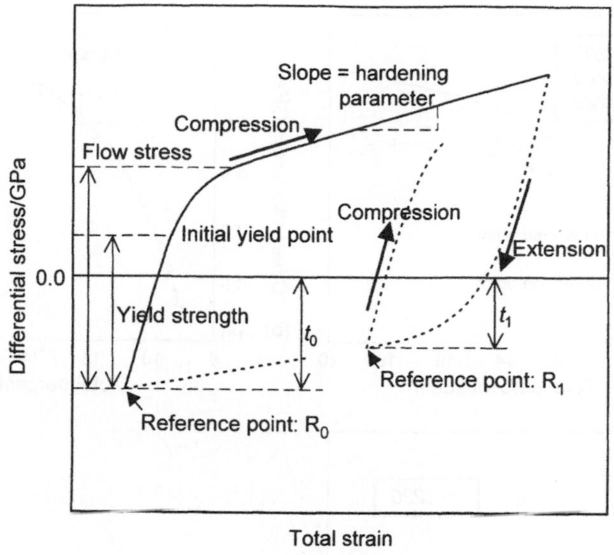

Figure 12. Schematics of a hysteresis loop in the differential stress, which is computed from the differential lattice strain using Eqs. (9)–(11), versus total strain curve. When the differential rams reversed the direction from extension to compression (marked as reference point: R_0), the sample is in the elastic regime. Depending on the history of previous deformation cycle (dashed line), residual stress t_0 may vary. In the elastic regime, differential stress linearly increases with total strain up to initial yield point with the slope of $1/6G(hkl)$. Beyond the yield point, the slope becomes shallower and differential stress sometimes shows no total strain dependence, but another time gradually increases with total strain as in Figure 12. This is referred to as "hardening". Hardening parameter is defined as slope in the second hardening. If the sample further experiences compression and extension cycle as shown in Figure 12, residual stress t_1 for the next extension–compression reference point R_1 is smaller than t_0. The magnitude of residual stress depends on the deformation history. On the other hand, magnitudes of linear segments from R_0 and R_1 are identical. Yield strength is the difference of differential stress at reference and initial yield points. Flow stress is differential stress after the sample yields and therefore flow stress > yield strength.

the differential stress saturates, sometimes showing no total strain dependence, but another time gradually increasing with total strain as in Figure 12. This is referred to as "hardening". If the sample experiences the compression and extension cycle as shown in Figure 12, residual stress t_1 for the next extension–compression reference point R_1 is smaller than t_0. The magnitude of residual stress depends on the deformation history. On the other hand, magnitudes of linear segments from R_0 and R_1 are identical. In the following sections, we verify these observations.

5.2. Differential stress

Differential stress is computed from the differential lattice strain shown in Figure 11, using Eqs. (9)–(11). Figure 13a and b shows the differential lattice strain and differential stress change, respectively, during the deformation cycle at 1.0 GPa shown in Figure 11. Differential stresses from 200 and 220 are almost identical throughout the elastic and plastic regimes and differential stress from 222 may be largest in the elastic regime (Fig. 13b).

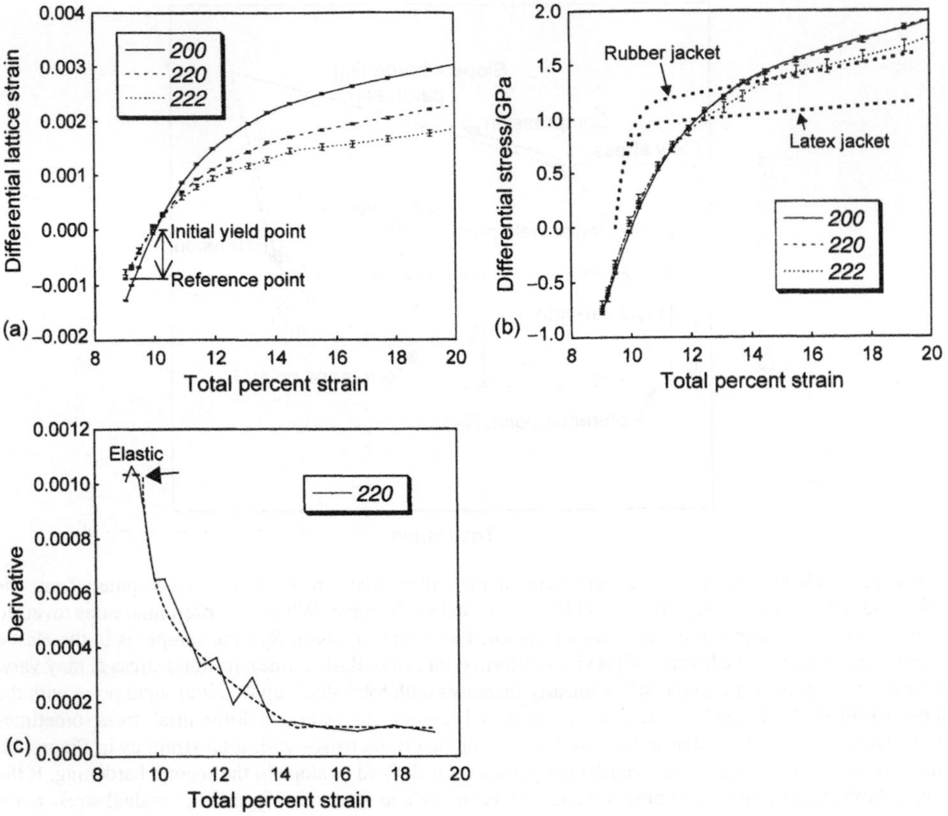

Figure 13. An example of (a) detail lattice strain change of MgO during the deformation cycle at 1.0 GPa shown in Figure 11. (b) Differential stress change obtained from (a) using Eqs. (9)–(11). For comparison, stress data of Paterson and Weaver (1970) are plotted by normalizing the horizontal scale with jacketing material (rubber or latex) indicated. (c) Derivative of lattice strain versus total strain curve (a) in 220 reflection, taken from the deformation cycle at 1.0 GPa. Yield point is defined as the initial deviation from linearity in (a) or abrupt change (indicated by arrow) from elastic to elasto-plastic regimes in the derivative in (c). Note that the difference in lattice strain between 200 and 222 increases with total strain. The sample reaches steady state at around 14% total strain, where lattice strain linearly increases with total strain. The dashed curves in (c) are guide to the eye.

Figure 13b shows that the differential stress calculated from differential lattice strain data on MgO is consistent with results obtained in a gas-medium apparatus at essentially the same pressure (~1 GPa) (Paterson and Weaver, 1970) in that the sample reaches initial yield point within 1% of total strain, and the sample shows strong strain-hardening. The principle of our stress measurement is different from previous conventional deformation apparatus, where stress measurement was based on force/area at the ends of the sample, and strain measurement relied on strain-gauge data typically near the vertical center of the sample. Consistent results attest the capability of our stress gauge for high-pressure deformation experiments.

Our stress gauge obtained from different reflection may give varying estimates within about 10–15% in the plastic regime (Fig. 13b). This difference is actually smaller than the data presented by Paterson and Weaver (1970), using different jacketing materials in their experiments.

5.3. Detection of the yield point

The initial yield point is identified at each loop as the deviation from the linear differential lattice strain versus total strain relation in Figure 11. The yield point is more clearly and accurately detected by examining the derivative of the curve. Figure 13c shows the derivative of the differential lattice strain versus total strain curve (only for 220 reflection). In an ideal elasto-plastic material, the derivative curve should start out as a flat line (a constant slope that corresponds to the elastic constant for the given crystallographic direction), and then become zero beyond the yield point (ideal plasticity, with no strain hardening). In Figure 13c, the abrupt change in the derivative, corresponding to the initial yield point, is indicated by the arrow. Beyond the initial yield point, however, the derivative first decreases rapidly and then becomes a non-zero constant (Fig. 13c), while the lattice strain increases linearly with total strain (Fig. 13a), showing strain-hardening.

To distinguish differential stresses at different stages during deformation, we will refer to the differential stress of the initial yield point as "the yield strength", and the differential stress beyond the yield point as "flow stress" (Fig. 12).

5.4. Yield strength

To determine the yield strength, we first measure the magnitude of the linear segment (i.e. difference of differential lattice strains between the reference point of each compression cycle and the yield point), as indicated by the arrow in Figure 13a. The same process is also applied to the linear segment for each extension cycle. There is no systematic difference in the magnitude of the linear segment of differential lattice strain for both compression and extension cycles, since extensive yield is achieved by compression of surrounding material in the radial direction ($\sigma_3 > \sigma_1 \neq 0$), which is one of the critical differences between high-pressure deformation and conventional extension test ($\sigma_3 = 0 > \sigma_1$).

In our measurement, the magnitude of the linear segment includes both extensive and compressive components, and the extensive strength should be equal to compressive strength under pressure. Thus, the yield strength is computed by halving the magnitude of the linear segment and multiplying by the shear modulus for each *hkl* using Eqs. (9)–(11). The yield strength thus calculated is plotted in Figure 14, together with previous strength data for comparison. Our yield stress data are rather insensitive to pressure and, within the resolution, can be regarded as a constant at pressures from 0.5 to 7 GPa. On the other hand, all previous differential stress measurements in the DAC and multi-anvil apparatus, which could not quantify the total sample strain, yielded much greater values (Fig. 14). Our results indicate that these previous high-pressure deformation data do not reflect the yield strength, but rather certain flow stress at some unknown total strain. This explains the wide scatter in those measurements: At room temperature, flow stress of MgO depends strongly

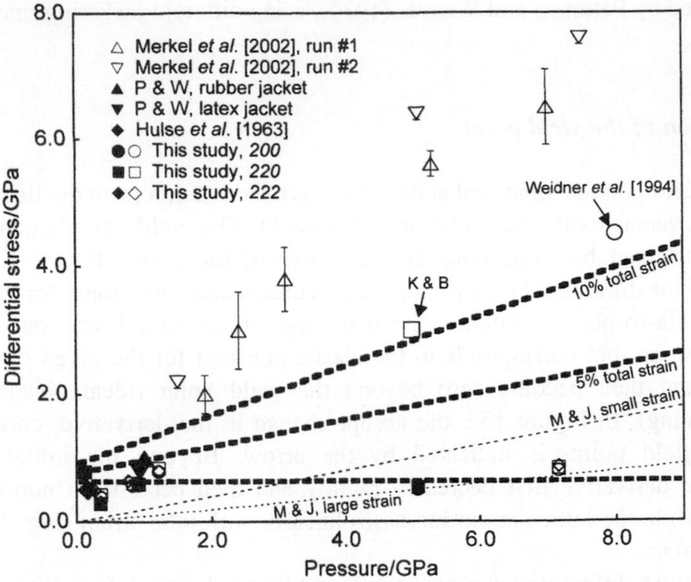

Figure 14. Comparison of differential stress with the data from previous studies. Only the initial yield strengths from our data are shown, as the flow stress varies greatly with total axial strain. Solid and open symbols in our data indicate the data collected during compression and extension cycles, respectively. Heavy dashed horizontal line is a fit to our yield stress data. Two other heavy dashed lines indicate differential stress level at 5 and 10% total strains. Abbreviations: K&B: Kinsland and Bassett (1977), M&J: Meade and Jeanloz (1988), P&W: Paterson and Weaver (1970).

on total strain, due to strain-hardening (Fig. 14). Based on our results, we may estimate flow stresses at various total strain levels (dashed line in Fig. 14) in the previous studies. Over 10% total strain was likely to have been achieved.

5.5. Hardening parameter

Figures 11 and 13a show that strain-hardening of MgO is characterized by two different stages. In the first stage, from immediately after the yield point at 10% to about 14% total strain, the flow stress rapidly increases with total strain. The second stage, where lattice strain linearly increases with increasing total strain, may be quantitatively characterized by a hardening parameter $\partial\varepsilon_t(hkl)/\partial\varepsilon_{\text{total}}$ (or $\partial t(hkl)/\partial\varepsilon_{\text{total}}$ in Fig. 13b). The hardening parameter is insensitive to pressure below 2 GPa (Fig. 11), and in good agreement with that in Paterson and Weaver (1970) (Fig. 13b). Even after the sample went through compression and extension cycles at several different pressures, lattice strain in the second stage hardening increases almost linearly with increasing total strain, following a unique trend for each *hkl* (Fig. 11). This second-stage hardening has similarities with the elastic regime in that differential stress linearly increases with increasing total strain and that the magnitude of lattice strain is greatest for 200 and least for 222. Without careful examination of the relationship between lattice strain and total strain, the elastic regime cannot be identified with confidence.

6. System and technique evaluation

6.1. Lattice strain resolution

The uncertainties of peak positions obtained by fitting the 1D diffraction data are generally less than the scatter shown in Fig. 9. The azimuth dependence of lattice strain is fit by Eq. (2). The error bars reported in Figure 11, representing one standard deviation in the least-squares fit, reflect the magnitude of scatter in Figure 9 as well as the errors in curve-fitting to obtain peak position. The intensity of the 111 reflection is fairly weak compared to 200, and 220 reflections, resulting in larger scatter of lattice strain (peak position) in azimuth dependence (Fig. 9). Thus the error bar of the 111 reflection is larger than that of other two reflections in Figure 11. Intrinsic lattice strain change is more significant than the size of the errors for all reflections.

In two azimuth-angles measurement (e.g. Funamori et al., 1994; Uchida et al., 1998; Weidner et al., 2004), the errors in the diffraction peak fit are directly transferred to the error in azimuth angle fit (2 equations for 2 parameters). Therefore, typical uncertainties are similar to the uncertainties in energy-dispersive diffraction of 10^{-4} (Chen et al., 2004). In our case, we fit the data using Eq. (2) for the entire 360 azimuth angles. Typical fitting errors in Eq. (2) are in the order of 10^{-5}, which is better than the errors in the diffraction peak fitting. This is because we have 360 peak position data for the entire azimuth range and these data were obtained using consistent peak and background fitting parameters. In addition, the fit for lattice strain is sensitive to random errors, but not to systematic errors, because it is the relative variation in peak position that determines the strain. This is one of the advantages of using a CCD detector and monochromatic diffraction. As a result, it is possible to measure lattice strains with resolution between 10^{-5} and 10^{-4} (for weak peaks such as the MgO 111 reflection) corresponding to a differential stress level of 0.02 GPa, for an elastic constant of 200 GPa, typical of most mantle minerals.

6.2. Detector drift during heating cycle

The accuracy in lattice strain measurement is also affected by several factors. One of the important factors is detector position drift due to temperature variation in the hutch, when the sample is heated for a prolonged period and the hutch temperature is not controlled. Figure 15 shows the "raw" lattice strain change with increasing temperature from 298 to 800 K. The asymmetry of the lattice strain variation with azimuth angle becomes more pronounced as temperature increases. This is caused by the change in the detector height relative to the incident beam, due to thermal expansion of the detector stand. When the hutch temperature increased by 1–2 K, the detector height changed by ca. 14 μm. The detector center is held at 1.4 m from the floor by the detector support; thus the inferred height change corresponds to a linear thermal expansion coefficient of 10^{-5} K^{-1} for the support frame.

It is essential to construct better mechanical support and anvil cooling system, keeping room temperature constant to avoid such drift, although the asymmetry can be corrected by curve-fitting from the entire 360° azimuth coverage in our data collection. Figure 16 compares raw lattice strain (open circles) and corrected lattice strain (solid circles)

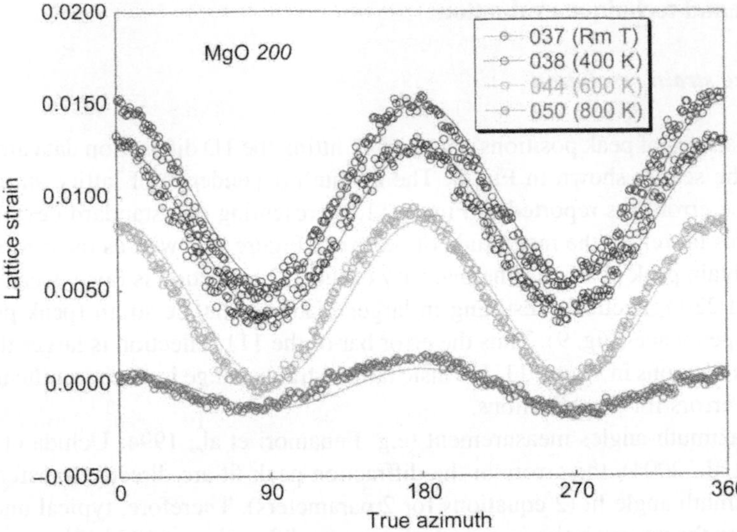

Figure 15. Lattice strain change (MgO 200 reflection) with increasing the sample temperature from room temperature to 800 K. Each solid line is a fit to the data using Eq. (2). The asymmetry of the lattice strain variation with azimuth angle becomes larger with temperature increase.

at 800 K. Note that the curve-fitting results (solid line) based on Eq. (2) are identical for both raw and corrected data. Because we observe the entire 360 azimuth angles, small distortions can be corrected even when the detector position is drifted. The error of course becomes larger (between 10^{-4} and 10^{-3}) because of systematic peak shift, but this error

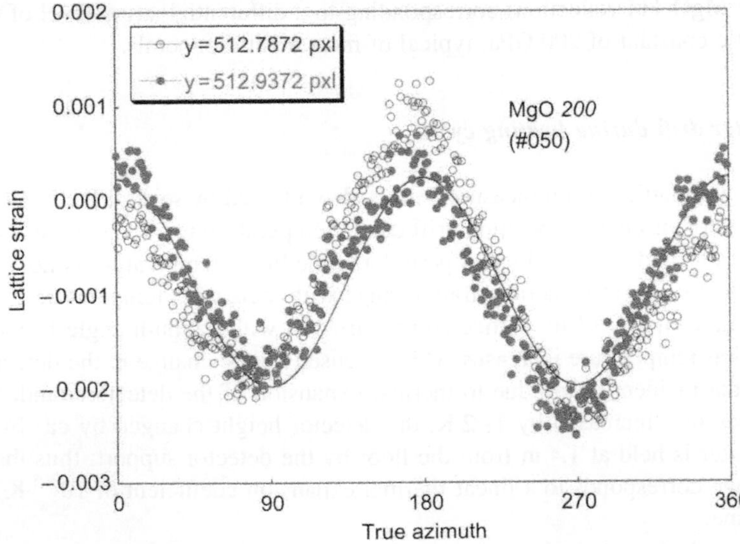

Figure 16. Comparison of raw and corrected lattice strains at 800 K. The detector vertical center was inferred to move 0.15 pixels, corresponding to 14 μm. However, fitting results (solid lines) are identical for both raw and corrected data.

can be detected in the differential stress–total strain curves. This result demonstrates the power of 2D monochromatic diffraction with 360° azimuth coverage. If diffraction data are collected at only two azimuth angles (e.g. Funamori et al., 1994; Uchida et al., 1998; Li et al., 2004; Weidner et al., 2004), such small lattice strains would be undetectable, resulting in systematic errors.

6.3. Further development to advance quantitative experiments

Texture observation has been extensively performed on both natural and recovered samples from high-pressure experiments to infer flow mechanism such as active slip system (e.g. Wenk et al., 2000; Merkel et al., 2002a; Cordier et al., 2004). Merkel et al. (2002a) reported {110}⟨110⟩ slip system is the most active slip system for MgO. In our X-ray observation, on the other hand, differential lattice strain for 200 reflection is greatest below 2 GPa and 111 reflection is expected to be greatest over 6 GPa (Uchida et al., 2004). Because sample yielding happens via slip system, the {110}⟨110⟩ slip system should be correlated to the greatest 200 differential lattice strain.

Even after exposed to 35% total strain, the MgO sample showed no systematic X-ray intensity variation. On the other hand, CsCl sample (in our different study) exhibited the development of intensity variation as deformation cycles are repeated up to 23% total strain (Fig. 17), but did not show any systematic anomaly in lattice strain modulation. Because X-ray intensity is proportional to the number of diffraction planes that satisfy the Bragg's law, non-uniform intensity variation observed in CsCl sample implies the development of preferred orientation. The intensity variation is dependent of the sample position as well. At the center of the sample, the intensity variation is maximal. As the sample position, where X-ray diffraction data were collected, moves up in the vertical direction, the intensity variation becomes smaller. Minimum intensity variation was observed at the position very close to Al_2O_3 piston, where the sample has more constraint by solid Al_2O_3 piston. These observations in X-ray diffraction need to be correlated to texture analysis.

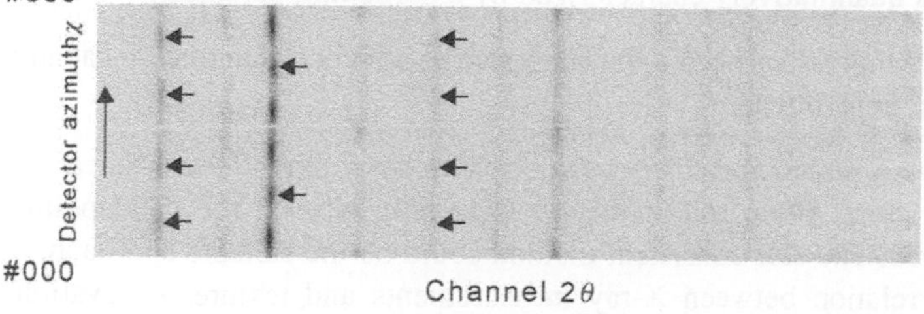

Figure 17. An example of intensity distribution in lattice strain on CsCl collected after several deformation cycles. X-ray intensity variation is observed (indicated by arrows for some stronger intensity peaks) in addition to that caused by absorption of cBN anvils (see the intensity distribution for BN peak). Depending on the Miller index, the intensity distribution is different.

As can be seen in Figure 13, differential stress is computed from each reflection *hkl*. However, the average differential stress is unknown. A method to average differential stresses needs to be established. A self-consistent method (Hutchinson, 1970; Clausen et al., 1999) could be an approach to average differential stresses using Taylor (1938) and Sachs (1928) bounds. Li et al. (2004) applied a self-consistent approach to the data obtained for polycrystalline MgO deformed in the D-DIA, adopting a 2-azimuth detector configuration (e.g. Funamori et al., 1994). However, all of their models (Table 1 in Li et al., 2004) show that the magnitudes of differential lattice strains are in reversed order when compared to our observation (Fig. 13), independent of which slip system is activated. In addition, Li et al. (2004) did not present any averaged stress. This unsuccessfulness of the models suggests that additional careful application of the self-consistent method to experimental results will be essential.

7. Summary

1. We described the capability to decouple stress from pressure by advancing and retracting differential rams with adjusting main ram. This operation allows the examination of the properties of samples within either elastic or plastic regimes.
2. Using a CCD detector, distortion of diffraction Debye rings is observed for complete 360° azimuth angles, enabling typical resolution in lattice strain measurements to be between 10^{-5} and 10^{-4}. The accuracy in total strain measurement from X-radiography is between 10^{-4} and 10^{-3}.
3. Although lattice strain theory collapses in the plastic regime, differential lattice strain computed from distortion of diffraction Debye rings can still be used as a stress gauge. Assuming elastic constants are still applicable beyond the yield point, differential stress can be computed from the differential lattice strain.
4. Yield point can be detected as deviation from the linearity between total strain and differential lattice strain.
5. Strain-hardening is characterized by two different stages. In the first stage hardening, immediately after the yield point, the flow stress rapidly increases with total strain. The second stage hardening, where lattice strain linearly increases with increasing total strain, is quantitatively characterized by a hardening parameter

The following issues need to be addressed to advance quantitative capabilities of these deformation experiments.

1. There is no satisfactory theory to correlate X-ray stress measurements to the average sample stress. More robust theory, especially beyond the yield point, needs to be developed. The self-consistent method could be the suitable approach.
2. The correlation between X-ray measurements and texture observation needs to be clarified. While X-ray measurement cannot infer flow mechanism, such as slip systems involved, texture observations can only be performed on quenched samples. Because flow mechanism is not necessarily the same at all pressure and temperature conditions, systematic examination of the correlation is of fundamental importance.

Acknowledgements

We thank A. Hammersley and ESRF personnel for the public use of FIT2D. We also thank T. Duffy, S. Merkel, M. Jessell, J. Chen, and a reviewer for useful discussion and critical review. The synchrotron work was performed at GeoSoilEnviroCARS (GSECARS), Sector 13, Advanced Photon Source at Argonne National Laboratory. We thank N. Lazarz, F. Sopron, M. Jagger, C. Pullins, N. Nishiyama, and GSECARS personnel for their support during the design and commissioning of the D-DIA. GSECARS is supported by the National Science Foundation – Earth Sciences, Department of Energy – Geosciences, W. M. Keck Foundation, and the U.S. Department of Agriculture. Use of the advanced photon source was supported by the US Department of Energy, Basic Energy Sciences, Office of Energy Research, under Contract No. W-31-109-Eng-38. Work partly supported by the NSF grant EAR001188.

References

Akahama, Y., Kawamura, H., Singh, A.K., 2004. A comparison of volume compressions of silver and gold up to 150 GPa. J. Appl. Phys. 95, 4767–4771.

Chen, J., Weidner, D.J., Vaughan, M.T., 2002. The strength of $Mg_{0.9}Fe_{0.1}SiO_3$ perovskite at high pressure and temperature. Nature 419, 824–826.

Chen, J., Li, L., Weidner, D., Vaughan, M., 2004. Deformation experiments using synchrotron X-rays: in situ stress and strain measurements at high pressure and temperature. Phys. Earth Planet. Inter. 143-144, 347–356.

Clausen, B., Lorentzen, T., Bourke, M.A.M., Daymond, M.R., 1999. Lattice strain evolution during uniaxial tensile loading of stainless steel. Mater. Sci. Eng. A259, 17–24.

Cordier, P., Ungár, T., Zsoldos, L., Tichy, G., 2004. Dislocation creep in $MgSiO_3$ perovskite at conditions of the Earth's uppermost lower mantle. Nature 428, 837–840.

Duffy, T.S., Hemley, R.J., Mao, H.K., 1995. Equation of state and shear strength at multimegabar pressures: magnesium oxide to 227 GPa. Phys. Rev. Lett. 74, 1371–1374.

Duffy, T.S., Shen, G., Heinz, D.L., Shu, J., Ma, Y., Mao, H.K., Hemley, R.J., Singh, A.K., 1999a. Lattice strains in gold and rhenium under nonhydrostatic compression to 37 GPa. Phys. Rev. B60, 15061–15073.

Duffy, T.S., Shen, G., Shu, J., Mao, H.K., Hemley, R.J., Singh, A.K., 1999b. Elasticity, shear strength, and equation of state of molybdenum and gold from x-ray diffraction under nonhydrostatic compression to 24 GPa. J. Appl. Phys. 86, 6729–6736.

Funamori, N., Yagi, T., Uchida, T., 1994. Deviatoric stress measurement under uniaxial compression by a powder x-ray diffraction method. J. Appl. Phys. 75, 4327–4331.

Hammersley, A.P., 1998. Fit2d: V9.129 reference manual v3.1. Internal Rep. ESRF98HA01, ESRF, Grendole, France.

Hutchinson, J.W., 1970. Elastic-plastic behaviour of polycrystalline metals and composites. Proc. R. Soc. London A319, 247–272.

Kavner, A., 2003. Elasticity and strength of hydrous ringwoodite at high pressure. Earth Planet. Sci. Lett. 214, 645–654.

Kavner, A., Duffy, T.S., 2001. Strength and elasticity of ringwoodite at upper mantle pressure. Geophys. Res. Lett. 28, 2691–2694.

Kinsland, G.L., Bassett, W.A., 1977. Strength of MgO and NaCl polycrystals to confining pressures of 250 kbar at 25°C. J. Appl. Phys. 48, 978–984.

Kung, J., Li, B., Uchida, T., Wang, Y., Neuville, D., Liebermann, R.C., 2004. In situ measurements of sound velocities and densities across the orthopyroxene → high-pressure clinopyroxene transition in $MgSiO_3$ at high pressure. Phys. Earth Planet. Inter. 147, 27–44.

Li, L., Paterron, P., Weidner, D., Chen, J., 2003. Olivine flow mechanisms at 8 GPa. Phys. Earth Planet. In. 138, 113–129.

Li, L., Weidner, D.J., Chen, J., Vaughan, M.T., Davis, M., Durham, W.B., 2004. X-ray stress analysis in deforming materials. J. Appl. Phys. 95, 8357–8365.
Mao, H.K., Shu, J., Shen, G., Hemley, R.J., Li, B., Singh, A.K., 1998. Elasticity and rheology of iron above 220 GPa and the nature of the Earth's inner core. Nature 396, 741–743.
Meade, C., Jeanloz, R., 1988. Yield strength of MgO to 40 GPa. J. Geophys. Res. 93, 3261–3269.
Merkel, S., Wenk, H.R., Shu, J., Shen, G., Gillet, P., Mao, H.K., Hemley, R.J., 2002a. Deformation of polycrystalline MgO at pressures of the lower mantle. J. Geophys. Res. 107 (B11), 2271, 10.1029/2001JB000920.
Merkel, S., Jephcoat, A.P., Shu, J., Mao, H.K., Gillet, P., Hemley, R.J., 2002b. Equation of state, elasticity, and shear strength of pyrite under high pressure. Phys. Chem. Miner. 29, 1–9.
Merkel, S., Wenk, H.R., Badro, J., Montagnac, G., Gillet, P., Mao, H.K., Hemley, R.J., 2003. Deformation of $(Mg_{0.9},Fe_{0.1})SiO_3$ perovskite aggregates up to 32 GPa. Earth Planet. Sci. Lett. 209, 351–360.
Merkel, S., Wenk, H.R., Gillet, P., Mao, H.K., Hemley, R.J., 2004. Deformation of polycrystalline iron up to 30 GPa and 1000 K. Phys. Earth Planet. Lett. 145, 239–251.
Nye, J.F., 1957. Physical Properties of Crystals. Oxford University Press, Oxford.
Osugi, J., Shimizu, K., Inoue, K., Yasunami, K., 1964. A compact cubic anvil high pressure apparatus. Rev. Phys. Chem. Jpn. 34, 1–6.
Paterson, M.S., Weaver, C.W., 1970. Deformation of polycrystalline MgO under pressure. J. Am. Ceram. Soc. 53, 463–471.
Reuss, A., 1929. Berechnung der fließgrenze von mischkristallen auf grund der plasticitätsbedingung für einkristalle. Z. Angew. Math. Mech. 9, 49.
Rivers, M.L., Duffy, T.S., Wang, Y., Eng, P.J., Sutton, S.R., Shen, G., 1998. A new facility for high-pressure research at the Advanced Photon Source. In: Manghnani, M.H., Yagi, T. (Eds), Properties of Earth and Planetary Materials at High Pressure and Temperature, Geophysical Monograph Series, Vol. 101. AGU, Washington, DC, pp. 79–86.
Rivers, M.L., Sutton, S.R., Eng, P.J., 1999. Geoscience applications of x-ray computed microtomography. In: Bonse, U. (Ed.), Proceedings of SPIE, Developments in X-ray Tomography II, Vol. 3722. The International Society for Optical Engineering, Washington, DC, pp. 78–86.
Sachs, G., 1928. Zur Ableitung einer Fließbedingung. Z. Ver. Deu. Ing. 72-22, 734–736.
Shieh, S.R., Duffy, T.S., Li, B., 2002. Strength and elasticity of stishovite across the stishovite-$CaCl_2$-type structural phase boundary. Phys. Rev. Lett. 89, 255507.
Shieh, S.R., Duffy, T.S., Shen, G., 2004. Elasticity and strength of calcium silicate perovskite at lower mantle pressure. Phys. Earth Planet. Int. 143-144, 93–105.
Singh, A.K., 1993. The lattice strains in a specimen (cubic system) compressed nonhydrostatically in an opposed anvil device. J. Appl. Phys. 73, 4278–4286.
Singh, A.K., Balasingh, C., 1994. The lattice strains in a specimen (hexagonal system) compressed nonhydrostatically in an opposed anvil high pressure setup. J. Appl. Phys. 75, 4956–4962.
Singh, A.K., Balasingh, C., Mao, H.K., Hemley, R.J., Shu, J., 1998. Analysis of lattice strains measured under nonhydrostatic pressure. J. Appl. Phys. 83, 7567–7575.
Spetzler, H., 1970. Equation of state of polycrystalline and single-crystal MgO to 8 kilobars and 800 °K. J. Geophys. Res. 75, 2073–2087.
Taylor, G.I., 1938. Plastic strain in metals. J. Inst. Met. 62, 307–315.
Uchida, T., Funamori, N., Yagi, T., 1996. Lattice strains in crystals under uniaxial stress field. J. Appl. Phys. 80, 739–746.
Uchida, T., Yagi, T., Oguri, K., Funamori, N., 1998. Analysis of powder X-ray diffraction data obtained under uniaxial stress field. Rev. High Pressure Sci. Technol. 7, 269–271.
Uchida, T., Wang, Y., Rivers, M.L., Sutton, S.R., 2001. Stability field and thermal equation of state of ε-iron determined by synchrotron X-ray diffraction in a multi-anvil apparatus. J. Geophys. Res. 106, 21799–21810.
Uchida, T., Wang, Y., Rivers, M.L., Sutton, S.R., 2004. Yield strength and strain-hardening of MgO up to 8 GPa measured in the deformation-DIA with monochromatic X-ray diffraction. Earth Planet. Sci. Lett. 226, 117–126.
Voigt, W., 1928. Lehrbuch der Krystalphysik. Teubner, Berlin.
Wang, Y., Rivers, M., Sutton, S., Eng, P., Shen, G., Getting, I., 1998. A multi-anvil, high-pressure facility for synchrotron radiation research at GeoSoilEnviroCARS at the Advanced Photon Source. Rev. High Pressure Sci. Technol. 7, 1490–1495.

Wang, W., Durham, W.B., Getting, I.C., Weidner, D.J., 2003. The deformation-DIA: a new apparatus for high temperature triaxial deformation to pressures up to 15 GPa. Rev. Sci. Instrum. 74, 3002–3011.
Watt, J.P., Davies, G.F., O'Connell, R.J., 1976. The elastic properties of composite materials. Rev. Geophys. Space Phys. 14, 541–563.
Weidner, D.J., Wang, Y., Vaughan, M.T., 1994. Yield strength at high pressure and temperature. Geophys. Res. Lett. 21, 753–756.
Weidner, D.J., Li, L., Davis, M., Chen, J., 2004. Effect of plasticity on elastic modulus measurements. Geophys. Res. Lett. 31, L06621, 10.1029/2003GL019090.
Wenk, H.R., Matthies, S., Hemley, R.J., Mao, H.K., Shu, J., 2000. The plastic deformation of iron at pressures of the Earth's inner core. Nature 405, 1044–1047.

Advances in High-Pressure Technology for Geophysical Applications
Jiuhua Chen, Yanbin Wang, T.S. Duffy, Guoyin Shen, L.F. Dobrzhinetskaya, editors

Chapter 8

Development of a rotational Drickamer apparatus for large-strain deformation experiments at deep Earth conditions

Yousheng Xu, Yu Nishihara and Shun-ichiro Karato

Abstract

In this chapter we describe a new type of torsion apparatus – the rotational Drickamer apparatus (RDA). Large-strain deformation experiments have been performed in the RDA at pressures and temperatures up to ~15 GPa and ~1700 K, respectively. The apparatus consists of opposing tungsten carbide anvils that are supported by a pyrophyllite gasket. The sample is sandwiched between the two anvils, and alumina or YAG insulating disks. The sample space is heated by two disk heaters made of a mixture of TiC and diamond. Deformation is achieved by rotating one anvil relative to the other through a Harmonic Drive™ gear box. Deformation experiments with shear strains of up to ~2 have been conducted for samples of (Mg,Fe)O, $(Mg,Fe)_2SiO_4$ olivine, and wadsleyite at rotation rates of $(3-7)\times 10^{-4}$ rpm corresponding to shear strain rates of $(0-5)\times 10^{-4}$ s^{-1}. However, the actual shear strain rates are lower presumably due to deformation of portions of the sample assembly other than the sample itself. A conical window is made in the confining cylinder for in situ X-ray stress and strain measurements. In this apparatus, both uniaxial compression and shear deformation occur. To determine the uniaxial stress and the shear stress separately, X-ray diffraction measurements were done at five different angles with respect to the rotation (compression) axis (i.e. 0, 45°, 90°). The sample thickness change and shear deformation were monitored by an imaging system during the synchrotron experiments. This apparatus allows quantitative studies of plastic deformation and microstructural development at a prescribed strain rate at pressure and temperature conditions equivalent to the deep mantle (500 km).

1. Introduction

Although knowledge of the rheological properties of Earth materials is critical to our understanding of the dynamics of the Earth's interior, high-resolution quantitative studies of rheological properties have been limited to low pressures. Challenges in conducting high-pressure rheological measurements include: (i) the controlled generation of deviatoric stress (strain) and (ii) the measurements of deviatoric stress. At low pressures, deviatoric stress (strain) can be generated in a controlled fashion by moving a piston through pressure seals, and the deviatoric stress can be measured by a load-cell (e.g. Tullis and Tullis, 1986). The most reliable rheological measurements can be made using a gas-medium apparatus with an internal load-cell (Paterson, 1990). However, the maximum pressure attainable by this apparatus is limited to ~0.5 GPa. With a modification to a sample assembly, quantitative rheological experiments can be performed up to ~3–4 GPa using a solid-medium, Griggs-type deformation apparatus (Green and Borch, 1987; Stöckhert and Renner, 1998).

Experimental studies on rheological properties beyond this pressure range are difficult because an unsupported deformation piston tends to fail when the deviatoric stress exceeds the strength of the piston (~ a few GPa; Getting et al., 1993). Piston failure can be avoided by moving the piston within a high-pressure assembly. Fujimura et al. (1981), Bussod et al. (1993) and Karato and Rubie (1997) used this approach to conduct deformation experiments at pressures exceeding 5 GPa. In these cases, the sample assembly is designed such that the strength is anisotropic. This causes deviatoric stress to be applied upon isotropic compression. Therefore, the stress generation is not well controlled and the magnitude of stress decreases significantly during a single run, making it difficult to interpret the results.

Wang et al. (2003) recently developed a new type of high-pressure deformation apparatus by modifying a cubic anvil apparatus in which the motion of two opposed anvils is controlled independently from the other four anvils to generate a deviatoric stress in a controlled fashion. In this apparatus, called D-DIA, deformation anvils are more effectively supported than in the Griggs apparatus; therefore, deformation experiments can be conducted at significantly higher pressures. D-DIA has been tested to ~10 GPa at high temperatures using tungsten carbide anvils (Wang et al., 2003; Li et al., 2004a). However, beyond 10 GPa, the anvils tend to fail because of the limitations of the lateral support. To achieve pressures beyond 10 GPa with a DIA type apparatus, one needs to use a stronger anvil material, such as sintered diamond (Utsumi et al., 1992). Recognizing the limitation of the D-DIA apparatus, Yamazaki and Karato (2001) developed a new apparatus by adding a rotational actuator to a Drickamer apparatus (RDA: rotational Drickamer apparatus). In the RDA, the deformation piston is supported by a gasket and a cylinder as in static high-pressure experiments with a conventional Drickamer apparatus. Consequently, deformation experiments can be performed at conditions similar to those of static high-pressure experiments (~20 GPa with tungsten carbide anvils). Additionally since the sample shape does not change with strain, high-strain deformation experiments are possible, which is essential to studies of deformation microstructures.

Another important challenge in high-pressure rheology experiments is the accurate determination of the deviatoric stress applied to the sample. A load-cell, such as those used in low-pressure apparatuses, is impractical because of the limitations of the sample volume and the pressures induced. Some microstructures, such as dislocation density, can be used to estimate the stress (e.g. Karato and Jung, 2003), but these relationships require calibrations, which are available only for low-pressure minerals such as olivine (Kohlstedt and Weathers, 1980). X-ray diffraction-based methods of stress determination are the most promising, particularly the method based on the measurements of the orientation dependence of lattice spacing (e.g., Singh, 1993).

The RDA was designed such that X-ray measurements can be made during an experiment. A brief account of the development of this apparatus was made by Yamazaki and Karato (2001). In this chapter, we report the results of further technical developments, including modifications to the hydraulic system, addition of a rotational actuator, and modifications to the sample assembly including the heater design, and the thermocouple lead through. We also report details of the operation of this apparatus at synchrotron facilities, and describe some preliminary results from deformation experiments on (Mg,Fe)O, $(Mg,Fe)_2SiO_4$ olivine and wadsleyite, in which stress and strain were measured *in situ* up to ~15 GPa and ~1600 K.

2. Experimental techniques

2.1. The rotational Drickamer apparatus (RDA)

The RDA (shown in Fig. 1) is a modification of the Drickamer apparatus, one of the opposed anvil-type high-pressure apparatuses in which high pressure is generated by squeezing a thin, disk-shaped sample between two anvils. We use a 70-ton hydraulic press. The pistons are inserted in a cylinder and are supported by a gasket. This support makes it possible to operate this apparatus at higher pressures than the Bridgman apparatus (e.g. Ringwood, 1975) which has no lateral support. Torque is applied by connecting the top anvil to a rotational actuator. The top anvil can rotate relative to the bottom anvil, which is fixed to the hydraulic press frame. A Harmonic Drive™ gearbox with a reduction ratio of 1/3360 is used as a reduction gear, and an AC servomotor is used to drive the gear. Large strain can be generated by continuously rotating the anvil. The direction of shear deformation is normal to that of compression and therefore the effects of compressional and torsional deformation can be separated. This is critical for studies of deformation-induced microstructures.

2.2. Sample assembly

Figure 2a shows a schematic diagram of the sample assembly for deformation experiments at high pressure and high temperature. A sample (typically ~1.6 mm diameter and ~0.2 mm thickness) is sandwiched between two alumina or YAG disks, two heater disks made of TiC + diamond, and two zirconia disks. Thin foils of W3%Re and W25%Re are inserted between two halves of samples, acting as a thermocouple as well as a strain

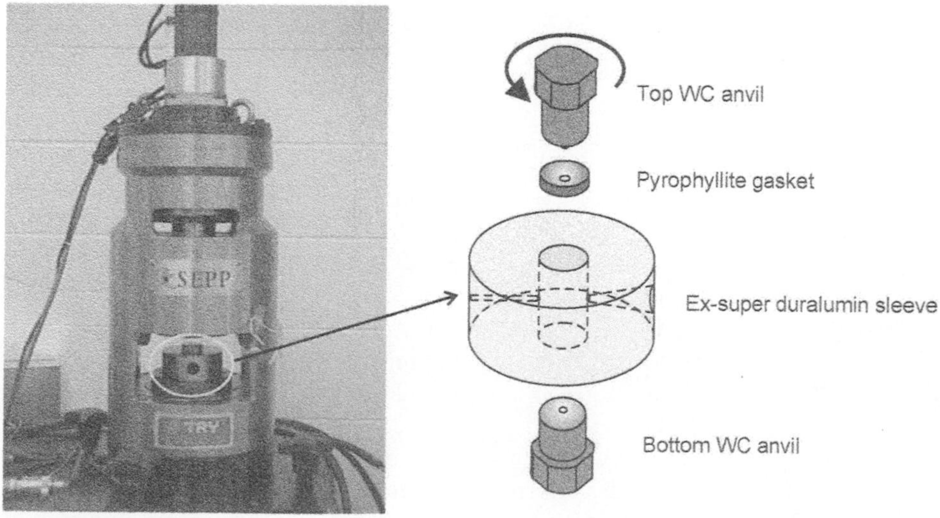

Figure 1. A rotational Drickamer apparatus for rheology and deformation microstructure studies under high *P* and *T*.

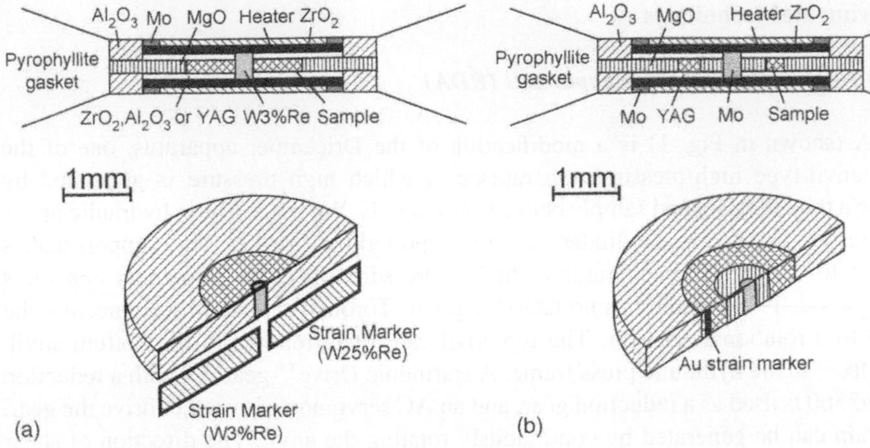

Al2O3 Mo MgO Heater ZrO2
Pyrophyllite gasket
ZrO2,Al2O3 or YAG W3%Re Sample
1mm
Strain Marker (W25%Re)
Strain Marker (W3%Re)
(a)
Al2O3 MgO Heater ZrO2
Pyrophyllite gasket
Mo YAG Mo Sample
1mm
Au strain marker
(b)

experiments because of the high X-ray absorption of thermocouple materials. The temperature in these experiments is estimated from a temperature–power calibration curve. Including the radial variation of temperature estimated from finite element calculations, the uncertainties in temperature are about ±100 K. In synchrotron *in situ* experiments, pressure can be estimated using a temperature inferred from an off-line calibration or from the equation of state of the sample material (wadsleyite: Fei et al., 1992; Jackson and Rigden, 1996; Li and Liebermann, 2000, etc.).

2.4. Stress and strain measurements using synchrotron radiation

Stress measurements were obtained using the X-ray diffraction technique (Weidner, 1998; Chen et al., 2004; Li et al., 2004a) with high-intensity white X-rays at the X17B2 beamline of the National Synchrotron Light Source (NSLS). A 13-element solid-state detector is used to record the energy dispersive X-ray diffraction patterns. The incident beam was collimated to $0.05 \times 0.05\ \mathrm{mm}^2$. The diffracted X-rays are collimated using a conical slit (0.05 mm gap) and detected by five of the 13 detector elements (one top, two side and two 45° orientated elements). ^{57}Co γ-rays (14 and 122 keV) and characteristic X-rays of Pb ($K\alpha_1$ and $K\alpha_2$) were used for energy calibration. The diffraction angle 2θ, which is mechanically fixed by the dimensions of the conical slit, is ~6.5° and was calibrated using a standard material (Al_2O_3 powder). A typical data collection time is 600 s.

We measure the lattice spacings of a given material as a function of the orientation of lattice planes with respect to the vertical (direction of force for pressurization). One complication in the stress measurements with torsion tests is that the stress in a sample is not constant because of the radial variation in strain rate. In the current X-ray diffraction geometry, the effective volume from which X-ray diffraction occurs has an elongated parallelpiped shape. Consequently, X-ray diffraction occurs in a region that contains grains experiencing different magnitudes of stress (Fig. 3). In order to minimize the stress variation in the effective sample volume from which X-ray diffraction occurs, we currently use a donut-shaped sample with OD = ~1.6 mm and ID = ~1.0 mm. With this sample geometry, the variability of stress in a sample is ±8–13% for $n = 3$–5 (n: stress exponent) $((\sigma(r_1)/\sigma(r_2) = (r_1/r_2)^{1/n})$ where $\sigma(r_{1,2})$ is stress at distance from the rotation axis $r_{1,2}$).

Unlike the multi-anvil apparatus and the D-DIA apparatus, the stress state of a sample in the RDA can be more complicated. With the assumption that the material motion occurs parallel to the compression axis (for pressurization) and/or parallel to the rotation direction (for torsion), the deviatoric stress state in a sample can be decomposed into two components as:

$$\sigma = \begin{bmatrix} \frac{2}{3}t & 0 & 0 \\ 0 & -\frac{1}{3}t & 0 \\ 0 & 0 & -\frac{1}{3}t \end{bmatrix} + \begin{bmatrix} 0 & \frac{\sigma_s}{2} & 0 \\ \frac{\sigma_s}{2} & 0 & 0 \\ 0 & 0 & 0 \end{bmatrix} = \begin{bmatrix} \frac{2}{3}t & \frac{\sigma_s}{2} & 0 \\ \frac{\sigma_s}{2} & -\frac{1}{3}t & 0 \\ 0 & 0 & -\frac{1}{3}t \end{bmatrix} \tag{1}$$

where a coordinate in which the x_1 direction is the compression direction (the rotational axis), the x_2 direction is the shear direction, and the x_3 direction is the normal to both of

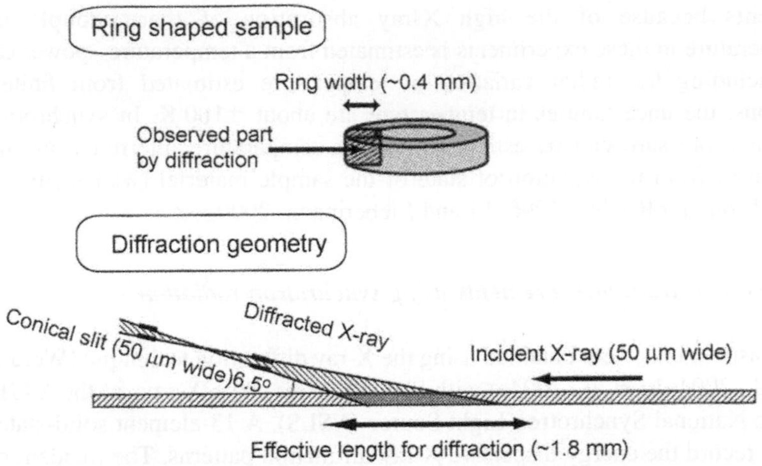

Figure 3. A schematic diagram showing the sample area through which the X-ray beam passes.

them, and t and σ_s represent the two deviatoric stress components respectively. In this case, the variation in d-spacing with orientation ($d_{hkl}(\Psi)$) is given by

$$\frac{d_{hkl}(\psi)}{d^0_{hkl}} = I + \frac{t}{6M}(1 - 3\cos^2\psi) - \frac{\sigma_s}{2M}\sin\psi\cos\psi \tag{2}$$

where d^0_{hkl} is the d-spacing corresponding to the hydrostatic stress, Ψ is the orientation of the plane normal to a specific lattice plane relative to the orientation of compression, and M is an appropriate elastic (shear) modulus that depends on the crystallographic orientation of the plane (hkl). Therefore, from measurements of the orientation (Ψ) dependence of d-spacing, we can determine both t and σ_s (these two values provide a measure of creep strengths of a sample for two deformation geometries from which plastic anisotropy can also be determined). Equation (2) contains three unknowns to be determined (d^0_{hkl}, t and σ_s). Therefore, we need to have data from at least three different Ψ angles. The validity of the assumption of no radial flow is not supported by our observations, but as long as the X-ray diffraction is measured at a point where radial material motion is parallel to the X-ray beam, this deviation has only small effects on the stress measurements.

We used the shear elastic modulus of an aggregate obtained by ultrasonic measurements (wadsleyite: Li and Liebermann, 2000; Li et al., 2001; Mayama et al., 2004). The actual value of M to be used here depends on the stress–strain distribution in deforming material and is not very obvious. For further discussion on this issue, see Li et al. (2004b) and Weidner et al. (2004).

The sources of errors in stress measurements include the alignment error of the X-ray for different diffraction angles and the drift of the amplifier to determine the d-spacing (i.e. energy). With current set-up for X-ray diffraction, the error of lattice spacing measurement is 10^{-3} that leads to an error of ~100 MPa. The error in strain measurement is ~10% due to the limitation of resolution of image of a strain marker.

Because the stress state of a sample changes with distance from the axis of rotation, it is important to get the X-ray diffraction from the same spots at different angles with respect to the rotation (compression) direction. In the configuration of our experiments, the effective length for X-ray diffraction is relatively large (~1.8 mm) compared to the small sample size (OD, 1.6 mm; ring width, 0.3 mm) (Fig. 3). Consequently, alignment of incident X-ray beam and the conical slit must be accurate to determine the stress with high precision.

A synchrotron facility also allows us to monitor deformation of each component of the sample assembly *in situ*. The progression of shear strain in a sample was monitored using an X-ray image of a strain marker (Au or Re). Consequently, the strain rate (deformation geometry) of a sample can be determined *in situ* from X-ray images even in cases where deformation occurs in other portions of sample assembly. The total strain of a sample can also be measured after an experiment from the geometry of a strain marker in a recovered sample.

2.5. Deformation geometry and microstructural characterization

Deformation geometry was determined by the measurements of sample thickness and the rotation of a strain marker, or through *in situ* X-ray radiography. Both uniaxial compression and simple shear deformation occurred. Deformation microstructures were observed both by optical microscopy and scanning electron microscopy on recovered samples. Scanning electron microscopy observations include observations of grain-boundary morphology, dislocation microstructures (by decoration for olivine) and measurements of crystallographic orientations by EBSD (electron backscatter diffraction) technique.

3. Experimental procedure

Deformation experiments with the RDA are conducted as follows: first, the sample is pressurized at room temperature. During this process significant deformation occurs in the sample as well as surrounding materials. Then temperature is raised to ~1600–1800 K. The sample is kept at ~1600–1800 K for 1–2 h to anneal the deformation microstructures introduced during room-temperature pressurization. Annealing of the sample is important for several reasons. First, a number of dislocations are generated during room-temperature deformation. If a high-temperature deformation experiment is conducted without annealing, then the high dislocation density in the sample will affect rheology through "work-hardening." Second, when stress is applied at high temperatures to a sample that contains a large number of dislocations with low-temperature slip system(s), then the fabric produced by deformation (i.e. lattice preferred orientation) would reflect low-temperature slip system(s) even though deformation occurred at high temperatures.

After annealing the motor is turned on at a constant rotational rate. After rotation to a certain angle is completed, the motor is turned off and the heater current is turned off, effectively quenching the sample. Both stress and strain (displacement) are monitored *in situ* during the experiments conducted at the synchrotron facility. In off-line experiments, deformation geometry and microstructures of quenched samples are studied after each experiment and stress is estimated from the rheology of the sample determined during synchrotron experiments.

4. Preliminary results

Deformation experiments were performed up to ~15 GPa and ~1700 K to shear strains of up to ~ 2. Samples of $(Mg_{0.75}Fe_{0.25})O$ magnesiowüstite, and $(Mg_{0.9}Fe_{0.1})_2SiO_4$ olivine and wadsleyite were deformed and their microstructures were observed.

4.1. Deformation history

During each experiment or for each sample, we need to make sure that initial stage deformation does not significantly affect subsequent rheological and microstructural observations. In experiments that do not make use of synchrotron X-rays, deformation history is inferred mostly through the analyses of microstructures after the run. However, at a synchrotron radiation facility, some aspects of deformation history can be tracked *in situ*. For instance, using the synchrotron facility, we noted that significant compressional deformation of the porous heater occurs during room-temperature compression, while subsequent deformation of the heater is minimal. Similarly, the majority of the uniaxial compression of a sample occurs during pressurization, although appreciable compression also occurs during torsion tests.

For olivine, annealing at ~8 GPa, ~1500 K for 2 h without rotation (Run#D-42) resulted in a dislocation density of $\sim 6 \times 10^{13}\ m^{-2}$ corresponding to a stress (it is considered to be uniaxial deviatoric stress *t*) of ~2 GPa (Karato and Jung, 2003). This relatively high deviatoric stress was effectively relaxed by increasing the annealing temperature. Annealing at ~8 GPa, ~1800 K for 1 h (Run#D-48) resulted in a dislocation density of $\sim 2 \times 10^{12}\ m^{-2}$ corresponding to a deviatoric stress of ~0.2 GPa. These results are largely consistent with the results of a previous study by Karato et al. (1993).

For wadsleyite, a series of experiments at different pressures and/or temperatures have been performed. After high-*P* and high-*T* annealing, the grain-boundaries of the recovered sample (Run#D33) are essentially in morphological equilibrium (Fig. 4a) and the fabrics of the samples are weak (Fig. 5a). Therefore, the fabrics formed during room-temperature pressurization will be mostly annealed and consequently the initial stage of deformation

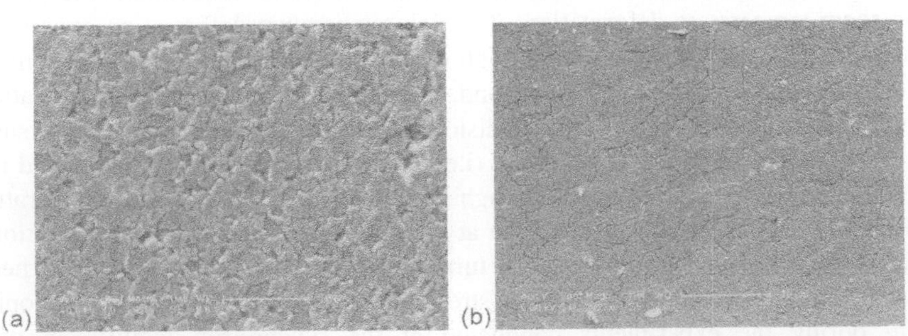

Figure 4. Photographs of wadsleyite samples. (a) After annealing but before deformation (Run#D33) and (b) after deformation (Run#D28). The scale bars are 10 μm.

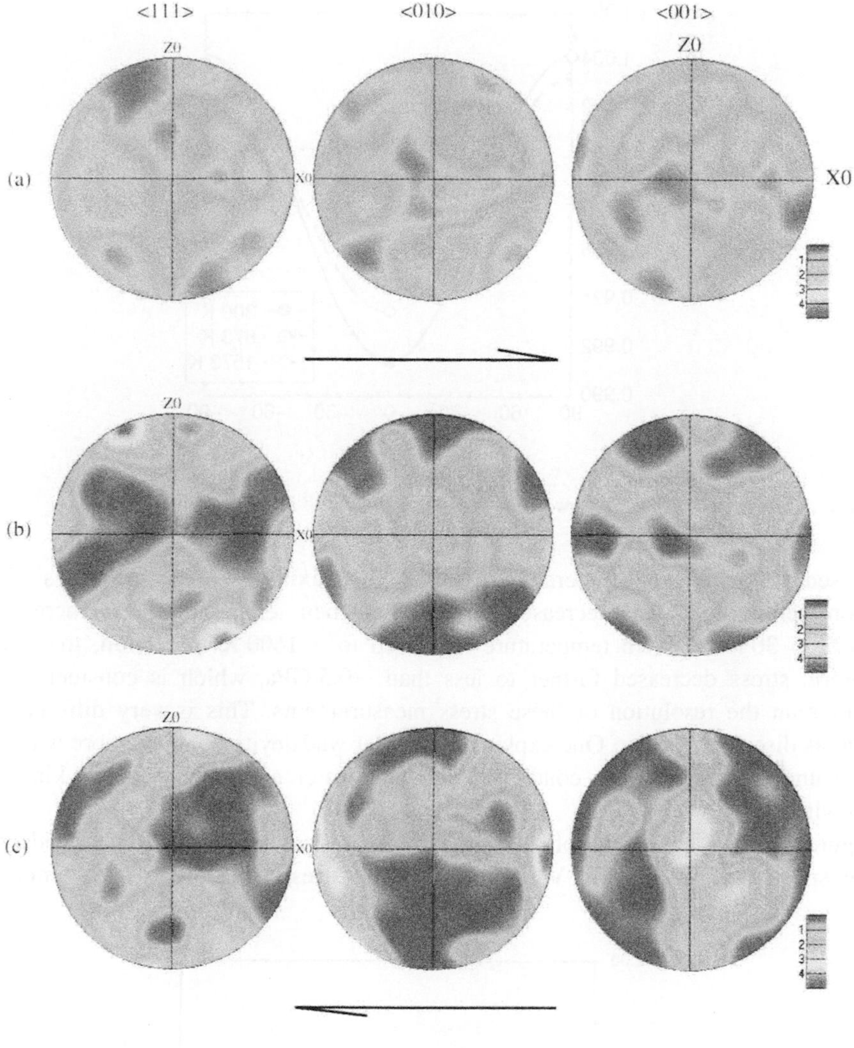

Figure 5. Pole figures of wadsleyite samples. (a) Annealed sample (Run#D33). 482 grains were measured. X0 is normal to the plane including uniaxial compression and radius directions and Z0 is uniaxial compressional direction. All the data points have been corrected for the sectioning effects. (b) Deformed sample (Run#bet19). 286 grains were measured. (c) Deformed sample (Run#D37). 146 grains were measured. For (b) and (c), X_0 is shear direction and Z_0 is shear plane normal. For (a)–(c), equal area projections (Schmid net) were used (a half-width of 15°); the numbers in the legend represents the density of points relative to the random distribution (1 for the random distribution).

will have a minimal effect on the fabrics of sheared wadsleyites. Once deformation starts, a fabric related to shear deformation and uniaxial compression develops.

X-ray measurements also confirmed that annealing could reduce the differential stress generated during pressurization. In an *in situ* X-ray observation experiment, the uniaxial deviatoric stress in wadsleyite was determined by X-ray diffraction data collected at $\Psi = 0$ and $\pm 90°$. The results show that the uniaxial deviatoric stress dramatically

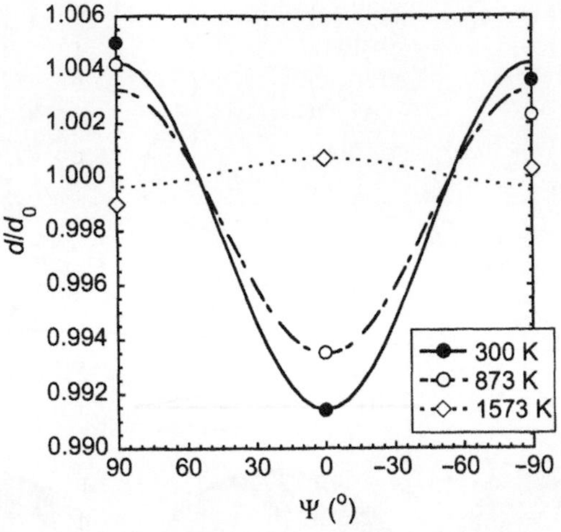

Figure 6. Normalized *d*-spacing of wadsleyite (240) plane vs. Ψ at different temperatures.

decreased with increasing temperature (Fig. 6). The uniaxial deviatoric stress was ~3 GPa at room temperature and decreased to ~2 GPa when temperature was increased to ~900 K in 30 min. When temperature increased to ~1600 K in 30 min, the uniaxial deviatoric stress decreased further to less than ~0.5 GPa, which is considered to be smaller than the resolution of these stress measurements. This is very different from olivine as discussed earlier. One explanation is that wadsleyite contains more water than olivine under nominally dry conditions and that water enhances recovery kinetics in wadsleyite.

Figure 7 shows the results of X-ray measurements of the orientation dependence of lattice spacing of wadsleyite. We found that the stress state after room-temperature

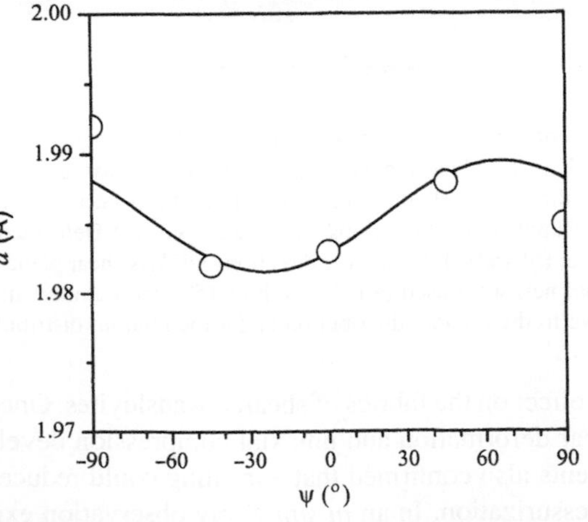

Figure 7. Results of X-ray diffraction measurements of orientation dependence of *d*-spacing of wadsleyite (240) plane during rotation deformation at ~14 GPa, ~1600 K, $\dot{\gamma} \sim 6 \times 10^{-5}\,s^{-1}$.

pressurization is dominated by uniaxial compression (~3 GPa), whereas the stress state after heating and applying a torque contains both components ($t \sim 0.6$ GPa; $\sigma_s \sim 0.7$ GPa). The stress state at high temperature could not be determined precisely due to the poor alignment of the detectors. We anticipate that the relative magnitude of two components changes with temperatures and annealing conditions and in future experiments will seek conditions under which the simple shear component dominates.

4.2. Magnesiowüstite

A magnesiowüstite sample was deformed at ~10 GPa and ~1500 K to a rotation angle of ~20° with a rotation rate of $\sim 4 \times 10^{-4}$ rpm without synchrotron radiation (Table 1). The recovered sample shows that the deformation of magnesiowüstite was particularly successful in terms of achieving large strain in the sample and nearly uniform deformation (Fig. 8). The deformation geometry determined by the thickness change and the strain-marker rotation showed that deformation was not truly simple shear. Shortening of ~50% occurred in addition to the shear strain of ~2.1. The shear strain-rate at the outer edge of the sample is estimated to be $\dot{\gamma} \sim 4 \times 10^{-4}\ \mathrm{s^{-1}}$. The magnitude of the shear strain is ~70% of what is expected from the rotation angle of anvils, indicating that most of the deformation occurred in the sample and not in the surrounding materials.

Figure 9 shows pole figures of $(Mg_{0.75}Fe_{0.25})O$ compared with those of Yamazaki and Karato (2002). The shear deformation experiments of Yamazaki and Karato (2002) were conducted at confining pressures of 300 MPa using a gas-medium deformation apparatus. Both our samples and their samples were deformed at ~1500 K and the strain rate of $\sim 4 \times 10^{-4}\ \mathrm{s^{-1}}$ of our sample is similar to that of $\sim 5 \times 10^{-4}\ \mathrm{s^{-1}}$ of their samples. Yamazaki and Karato (2002) pointed out that, with increasing strain, the position of

Table 1. Experimental conditions and results.

Run#	Sample	P (GPa)	Annealing T (K)	Annealing t (min)	Deformation T (K)	Deformation t (min)	$\theta_{rotation}$ (°)	$\dot{\theta}_{rotation}$ (°min^{-1})	γ	$\dot{\gamma}$ ($10^{-5}\ \mathrm{s^{-1}}$)
D15	mw	10	1500	60	1500	120	20	0.17	2.1	44
D42	ol	8	1500	120						
D43	ol	8	1600	120	1500	400	40	0.10	1.2	5
D48	ol	8	1800	60						
D33	wads	15	1700	60						
D30	wads	15	1500	60	1500	150	38	0.25	0.5	6
D28	wads	15	1500	60	1500	155	40	0.26	0.7	7
D38	wads	15	1600	60	1600	325	40	0.12	0.7	3
D44	wads	15	1600	60	1600	300	30	0.10	*	*
bet19	wads	15	1600	240	1600	300	30	0.10	1.0	6
D37	wads	15	1700	60	1700	155	40	0.26	*	*

Note: The pressure error is ±1 GPa and temperature error is ±100 K; strain (rate) was measured from a strain marker shape at the outer edge of each sample; * indicates no strain marker was used.

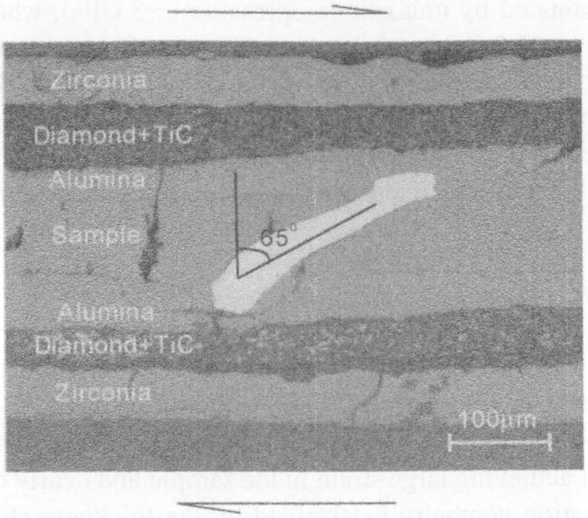

Figure 8. A photomicrograph of the deformed $(Mg_{0.75}Fe_{0.25})O$ assembly. The center horizontal layer is the sample and the white region is the strain marker.

concentration of ⟨100⟩ axes rotates toward to the direction of the shear plane normal. The strain of our sample ($\gamma = 2.1$) is between their two samples ($\gamma = 1.4$ and 4.0). The fact that the strength of the deformation fabrics in our sample is intermediate to the strength of the deformation fabrics of their two samples is consistent with the magnitude of strains in the samples. The similarity between the deformation fabrics in these studies suggests that the uniaxial component of deformation does not have major effects on deformation fabrics and the dominant slip systems are similar for these conditions.

4.3. Wadsleyite

Several deformation experiments have been performed on wadsleyite (Table 1). Among the total of seven experiments, six were conducted without synchrotron X-rays and one experiment (Run#bet19) was conducted at NSLS.

After deformation grains show clear elongation (Fig. 4b). All experiments except Run#bet19 were conducted by using the first assembly shown in Figure 2a with alumina disks between the sample and heater disks. In these cases, the strains are very small compared to the expected total strains (~70% of the expected total strain for magnesiowüstite but only ~15% of the expected total strain for wadsleyite, Fig. 10). This is presumably due to the fact that wadsleyite is significantly stronger than magnesiowüstite and significant deformation occurred in other portions of the sample assembly. The recovered assemblies show that the alumina disks, which were between the sample and the heater disks, deformed more than the sample. To minimize the deformation of the surrounding disks, we also tried YAG disks instead of alumina disks in Run#bet19. The results show that this sample deformed more than samples from earlier experiments and the strain of the sample is ~25% of the expected total strain (Fig. 10). Although this is an improvement, more progress will be needed to obtain better rheological and microstructural data.

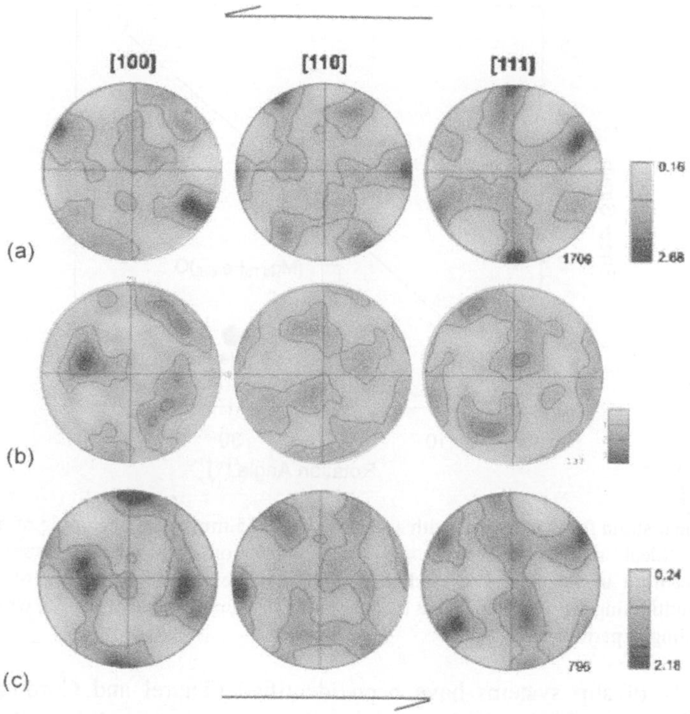

Figure 9. Comparison of pole figures of $(Mg_{0.75}Fe_{0.25})O$ with those of Yamazaki and Karato (2002). (a) $\gamma = 1.4$, $\dot{\gamma} = 5.0 \times 10^{-4}\,s^{-1}$, deformed at 300 MPa and 1473 K (Yamazaki and Karato, 2002); (b) $\gamma = 2.1$, $\dot{\gamma} = 2.9 \times 10^{-4}\,s^{-1}$, deformed at ~10 GPa and 1500 K; (c) $\gamma = 4.0$, $\dot{\gamma} = 5.2 \times 10^{-4}\,s^{-1}$, deformed at 300 MPa and 1473 K (Yamazaki and Karato, 2002). Equal area projections (Schmid net) were used (a half-width of 15° and the cluster size of 3° are used in the plotting). The number below each set of pole figure shows the number of grains for which orientations are measured. The numbers in the legend represents the density of points relative to the random distribution (1 for the random distribution). X0 is shear direction and Z0 is shear plane normal. All our data points have been corrected to their own shear directions.

Figure 5b shows the pole figure of the sample Run#bet19 deformed at ~15 GPa, ~1600 K, and strain rate of $\sim 6 \times 10^{-4}\,s^{-1}$. The pole figure is characterized by [100] axes sub-parallel to the shear plane normal. [010] forms two point maxima: one is concentrated around the shear plane normal Y0 generated by uniaxial compression (the deviation from girdle generated by uniaxial compression is due to the correction made for each point to its own shear direction. See the caption of Figure 5); another is sub-parallel to the shear direction X0. The fact that some of the b axes cluster in the [010] shear normal direction suggests that uniaxial compression has an effect on the fabrics introduced during deformation. In another experiment Run#D37, the sample was deformed at ~15 GPa and ~1700 K. The pole figure is characterized by a [010] point maximum near the shear direction. The effects of uniaxial compression are not seen, probably due to the high strain in this experiment (Fig. 5c). Fabrics similar to this were also observed in samples deformed by stress relaxation tests (Lawlis et al., 1999). However, stress relaxation tests also show a variety of fabrics depending on the conditions of deformation. In wadsleyite,

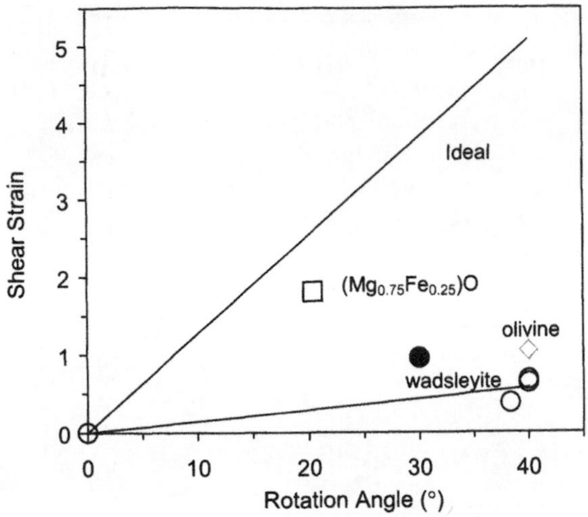

Figure 10. Shear strain for the samples with a thickness of 0.15 mm at the beginning of deformation vs. rotation angle. "Ideal" represents the case where no deformation occurred in pressure media and all deformation occurred in samples. The thickness of the samples at the beginning of deformation is calculated by multiplying the initial thickness by the sample shortening value of ~30%, which is estimated from the annealing experiments.

a wide variety of slip systems have been identified (Thurel and Cordier, 2003) and therefore deformation fabrics likely change with deformation conditions (for olivine see Carter and Avé Lallemant, 1970; Jung and Karato, 2001; Katayama et al., 2004). Consequently, a systematic study must be made to define the variation of deformation fabrics with deformation conditions (e.g. stress, temperature and water content) before experimental results can be applied to interpret seismic anisotropy.

5. Discussions and perspectives

We have presented a progress report on our development of a new type of deformation apparatus, the RDA. The initial tests are encouraging inasmuch as they demonstrate that this apparatus can be used for quantitative deformation studies to at least 15 GPa. However, we have identified a number of points that need to be improved. (1) Deformation occurs not only in the samples but also in other surrounding components of the sample assembly. This is more severe for samples such as wadsleyite. Obviously, this problem is related to the strength contrast between the sample and its surrounding components. (2) Extensive flattening occurs during the initial stage pressurization as well as during torsion tests. (3) Thus far, the achievable temperature range is limited to ~1700 K. Above this temperature, anvils (made of WC) tend to fail (at ~15 GPa) presumably due to the reduction of their strength at higher temperatures. (4) The resolution of stress measurements in our initial tests was not high enough due to a number of complications including absorption of X-rays by various components of the sample assembly, less than satisfactory alignment of the X-ray beam, insufficient thickness of the samples, etc. In addition, there are uncertainties as to which elastic constants should be used as *M* in Equation (2) due to the issue of grain-scale stress/strain distribution (e.g. Li et al., 2004b). (5) Plastic deformation depends on the chemical environment, for example oxygen fugacity and water content (water fugacity). In our experiments the chemical environment of deformation has not been controlled. The use of diamond presumably sets the oxygen fugacity between the iron-wüstite and magnetite–wüstite buffers and the open system sets the water content in wadsleyite as high as 1000 wt ppm. A more detailed analysis of the effects of the chemical environment on plastic deformation is needed to further pin down the microscopic mechanisms of deformation.

Acknowledgments

We thank T. Yagi, T. Uchida and S. Ozaki for the discussion on the design of the apparatus and D. Weidner, Y. Wang, J. Chen for the discussion on the design of the synchrotron experiments. Z. Wang and D. Yamazaki made the major contribution in the early stage of apparatus development. The synchrotron experiments were carried out at National Synchrotron Light Source (NSLS) and technically supported by C. Koleda, L. Wang, M. Vaughan. Z. Jing, I. Katayama, W. Landuyt, K. N. Matsukage, P. Skemer helped us do the synchrotron experiments. Z. Jiang helped us to do preferred orientation measurements. K. Leinenweber kindly offered us the Thermal Modeling Program to calculate temperature distribution in the RDA assembly. M. Long kindly donated the magnesiowüstite sample to us. TiC + diamond disk heaters used in some experiments are kindly provided by M. Akaishi (National Institute for Materials Science, Japan). This work is supported by the grant from National Science Foundation and the U.S. Department of Energy, Division of Material Sciences and Division of Chemical Sciences.

References

Bussod, G.Y., Katsura, T., Rubie, D.C., 1993. The large volume multi-anvil press as a high $P-T$ deformation apparatus. Pure Appl. Geophys. 141, 579–599.

Carter, N.L., Avé Lallemant, H.G., 1970. High temperature flow of dunite and peridotite. Geol. Soc. Am. Bull. 81, 2181–2202.

Chen, J., Li, L., Weidner, D., Vaughan, M., 2004. Deformation experiments using synchrotron X-rays: *in situ* stress and strain measurements at high pressure and temperature. Phys. Earth Planet. Int. 143–144, 347–356.

Fei, Y., Mao, H.-K., Shu, J., Parthasarthy, G., Bassett, W.A., Ko, J., 1992. Simultaneous high-P, high-T X ray diffraction study of β-$(Mg,Fe)_2SiO_4$ to 26 GPa and 900 K. J. Geophys. Res. 97, 4489–4495.

Fujimura, A., Endo, S., Kato, M., Kumazawa, M., 1981. Preferred orientation of b-Mn_2GeO_4. Programme and Abstracts, The Japan Seismological Society, p. 185.

Getting, I.C., Chen, G., Brown, J.A., 1993. The strength and rheology of commercial tungsten carbide cermets used in high-pressure apparatus. PAGEOPH 141, 545–577.

Green, H.W. II, Borch, R.S., 1987. The pressure dependence of creep. Acta Metall. 36, 1301–1305.

Jackson, I., Rigden, S.M., 1996. Analysis of $P-V-T$ data: constraints on the thermoelastic properties of high-pressure minerals. Phys. Earth Planet. Int. 96, 85–112.

Jung, H., Karato, S., 2001. Water-induced fabric transitions in olivine. Science 293, 1460–1463.

Karato, S., Jung, H., 2003. Effects of pressure on high-temperature dislocation creep in olivine. Phil. Mag. 83, 401–414.

Karato, S., Rubie, D.C., 1997. Toward experimental study of plastic deformation under deep mantle conditions: a new multianvil sample assembly for deformation experiments under high pressures and temperatures. J. Geophys. Res. 102, 20111–20122.

Karato, S., Rubie, D.C., Yan, H., 1993. Dislocation recovery in olivine deep upper mantle conditions: implications for creep and diffusion. J. Geophys. Res. 98, 9761–9768.

Katayama, I., Jung, H., Karato, S., A new type of olivine fabric at modest water content and low stress. Geology, 32, 1045–1048.

Kohlstedt, D.L., Weathers, M.S., 1980. Deformation-induced microstructures, paleopiezometers, and differential stresses in deeply eroded fault zones. J. Geophys. Res. 85, 6269–6285.

Lawlis, J.D., Frost, D., Rubie, D., Cordier, P., Karato, S., 1999. Deformation and fabric development in transition zone minerals. EOS 80, F975.

Li, B., Liebermann, R.C., 2000. Sound velocities of wadsleyite β-$(Mg_{0.88}Fe_{0.12})_2SiO_4$ to 10 GPa. Am. Miner. 85, 292–295.

Li, B., Liebermann, R.C., Weidenr, D.J., 2001. *P–V–*Vp–Vs–*T* measurements on wadsleyite to 7 GPa and 873 K: implications for the 410-km seismic discontinuity. J. Geophys. Res. 106, 30575–30591.

Li, L., Weidner, D.J., Raterron, P., Chen, J., Vaughan, M., 2004a. Stress measurements of deforming olivine at high pressure. Phys. Earth Planet. Int. 143–144, 357–367.

Li, L., Weidner, D.J., Chen, J., Vaughan, M.T., Davis, M., Durham, W.B., 2004b. X-ray strain analysis at high pressure: effect of plastic deformation in MgO. J. Appl. Phys. 95, 8357–8365.

Mayama, N., Suzuki, I., Saito, T., Ohno, I., Katsura, T., Yoneda, A., 2004. Temperature dependence of elastic moduli of β-$(Mg,Fe)_2SiO_4$. Geophys. Res. Lett. 31, 10.1029/2003GL019247.

Paterson, M.S., 1990. Rock deformation experimentation. In: Duba, A.G., Durham, W.B., Handin, J.W., Wang, H.F. (Eds), The Brittle–Ductile Transition in Rocks, The Heard Volume, American Geophysical Union Geophysical Monograph, Vol. 56. AGU, Washington, DC, pp. 187–194.

Ringwood, A.E., 1975. Composition and Petrology of the Earth's Mantle. McGraw-Hill, New York.

Singh, A.K., 1993. The lattice strains in a specimen (cubic system) compressed nonhydrostatically in an opposed anvil device. J. Appl. Phys. 73, 4278–4286.

Stöckhert, B., Renner, J., 1998. Rheology of crustal rocks at ultrahigh pressure. In: Hacker, B.R., Liou, J.G. (Eds), When Continents Collide: Geodynamics and Geochemistry of Ultrahigh-Pressure Rocks. Kluwer Academic Publishers, Princeton, pp. 57–95.

Thurel, E., Cordier, P., 2003. Plastic deformation of wadsleyite: I. High-pressure deformation in compression. Phys. Chem. Miner. 30, 256–266.

Tullis, T.E., Tullis, J., 1986. Experimental rock deformation. In: Hobbs, B.E., Heard, H.C. (Eds), Mineral and Rock Deformation. American Geophysical Union, Washington, DC, pp. 297–324.

Utsumi, W.T., Yagi, T., Leinenweber, K., Shimomura, O., Taniguchi, T., 1992. High pressure and high temperature generation using sintered diamond anvils. In: Syono, Y., Manghnani, M.H. (Eds), High-Pressure Research: Application to Earth and Planetary Sciences. Centre for Academic Press, Tokyo, pp. 37–42.

Wang, Y., Durham, W.B., Getting, I.C., Weidner, D.J., 2003. The deformation-DIA: a new apparatus for high temperature triaxial deformation to pressures up to 15 GPa. Rev. Sci. Instrum. 74, 3002–3011.

Weidner, D.J., 1998. Rheological studies at high pressure. In: Hemley, R.J. (Ed.), Ultrahigh-Pressure Mineralogy: Physics and Chemistry of the Earth's Deep Interior, Vol. 37. Mineralogical Society of America, Washington, DC, pp. 493–524.

Weidner, D.J., Li, L., Davis, M., Chen, J., 2004. Effect of plasticity on elastic modulus measurements. Geophys. Res. Lett. 31, 10.1029/2003GL019090.

Yamazaki, D., Karato, S., 2001. High pressure rotational deformation apparatus to 15 GPa. Rev. Sci. Instrum. 72, 4207–4211.

Yamazaki, D., Karato, S., 2002. Fabric development in (Mg,Fe)O during large strain, shear deformation: implication for seismic anisotropy in Earth's lower mantle. Phys. Earth Planet. Int. 131, 251–267.

III
Melt and glass properties

Advances in High-Pressure Technology for Geophysical Applications
Jiuhua Chen, Yanbin Wang, T.S. Duffy, Guoyin Shen, L.F. Dobrzhinetskaya, editors

Chapter 9

Density measurements of molten materials at high pressure using synchrotron X-ray radiography: melting volume of FeS

Jiuhua Chen, Donald J. Weidner, Liping Wang, Michael T. Vaughan and Christopher E. Young

Abstract

A new technique for density measurement of molten materials in a multi-anvil press using synchrotron X-ray radiography is described. This technique takes advantage of a linear conversion of X-ray intensity to radiograph brightness, and records two-dimensional variations in transmitted X-ray intensities across a reference sphere in the sample on a single exposure. Comparing with the existing technique of one-dimensional scan using a small beam of X-rays for the melt density measurement at high pressure, this method gains a shorter data collection time and larger data coverage (two-dimensional). Melting volume of FeS at 4.1 GPa is determined to be 0.28 cm^3/mol and slope of the melting curve is estimated to be 41°C/GPa. This experiment demonstrates an accuracy of 1% for the density measurement with respect to X-ray diffraction method for crystalline phase.

1. Introduction

Physical properties (e.g. density and elasticity) of minerals, combined with geophysical evidence (e.g. seismology and gravitational observations), are very important for understanding a planet's structure. While density measurements of crystalline minerals at high pressure and temperature can be routinely carried out using *in situ* diffraction techniques, experiments to determine the density of melts at static high pressure are rare (Katayama, 1996; Sanloup et al., 2000; Eggert et al., 2002; Shen et al., 2002). Traditionally, spectroscopic methods (e.g. extended X-ray absorption fine structure – EXAFS, Nuclear magnetic resonance – NMR, vibrational spectroscopy – infrared and Raman) and X-ray pair-distribution function (PDF) analysis are used to characterize local structure of non-crystalline materials (Bell, 1969; Desbat and Vanhuong, 1983; Di Cicco and Minicucci, 1999; Katayama, 2001; Katayama et al., 2001). However, none of these techniques can provide direct density information for non-crystalline materials. Limitation of experimental techniques has hindered us from studying equations of state of molten minerals, which is essential for understanding the Earth, especially the Earth's core.

Katayama et al. (1993, 1996) pioneered direct density measurement of liquid under high pressure and temperature by using X-ray absorption spectroscopy at the synchrotron beamline of the Photon Factory. The technique was soon adopted at ESRF – European Synchrotron Radiation Facilities (Katayama et al., 1998) and applied to the equation of state studies of Earth Science related materials (Sanloup et al., 2000). In these experiments, a small X-ray beam (typically 50 μm × 50 μm or greater in cross section) passes through a liquid sample maintained at high pressure and temperature in a large-volume press. The sample density is derived through the Beer–Lambert absorption law ($I = I_0 e^{-\mu\rho t}$, where I and I_0 are the intensities of the transmitted and incident beams, respectively, μ and ρ are the mass absorption coefficient and density of the sample, and t is the sample length in the X-ray path) and by measuring the intensities of the incident and transmitted beams. For samples with known chemical composition, i.e. mass absorption coefficient, determining the sample length becomes crucial for deriving an accurate density. In previous experiments, either a small reference sphere is embedded in the sample or a rigid cylindrical sample capsule is used to contain the sample so that the sample length can be derived from a scan of the X-ray beam across the reference sphere or cylindrical capsule. Nevertheless, scanning across the reference sphere or cylindrical sample increases the period of time for maintaining the sample at high temperature above its melting point, and one-dimensional scan across a sphere may introduce additional experimental error when the scan path is off the sphere center. To improve the technique, we have explored a method to measure melt density at high pressure using X-ray radiography (Chen et al., 2000). Shen et al have applied the radiographic method to measure molar volume of molten indium at high pressures in a diamond anvil cell (Shen et al., 2002).

In this chapter, we report experimental details for measuring the density of molten materials in a large-volume press using an X-ray radiography technique and apply this to the melting volume of iron sulfide (FeS). Iron is believed to be the principal constituent of terrestrial planets' cores, but some light elements are required to match the densities of the cores extracted from gravimetric and/or seismic (for Earth) data to the experimental data of candidate materials (Birch, 1952). Sulfur is a very important candidate for the light element in the Earth's core and the Martian core (Mason, 1966; Rama Murthy and Hall, 1970; Lewis, 1971; Rama Murthy and Hall, 1972). Therefore, experimental measurements of the density and elasticity of the molten iron–sulfur system with different compositions at high pressure and temperature becomes critical for understanding compositions of the planets' cores. Phase stability and density of solid FeS have been extensively studied at high pressures (Taylor and Mao, 1970; King and Prewitt, 1982; Fei et al., 1995; Kusaba et al., 2000; Kavner et al., 2001; Kobayashi et al., 2001; Urakawa et al., 2002). Density measurement of molten Fe–FeS system at high pressure has only been challenged recently by using both *in situ* (Sanloup et al., 2000) and *ex situ* (Balog and Secco, 2003) techniques. Knowledge of the high pressure behavior of molten FeS is certainly essential for understanding terrestrial planets' cores.

2. Experimental setup

The experimental system is installed at the superconductor wiggler beam line, X17B, of the National Synchrotron Light Source (NSLS). Figure 1 shows a sketch of the system layout. The upstream aperture defines the source beam size down to 2 mm (horizontal) by 3 mm (vertical). The energy of the X-ray is selected by a Si(220) double-crystal monochromator. A set of motor-driven incident slits is used after the monochromator to further reduce the beam size to 0.1 mm by 0.1 mm for X-ray diffraction measurements. High pressures and temperatures are generated by using a DIA-type cubic-anvil press

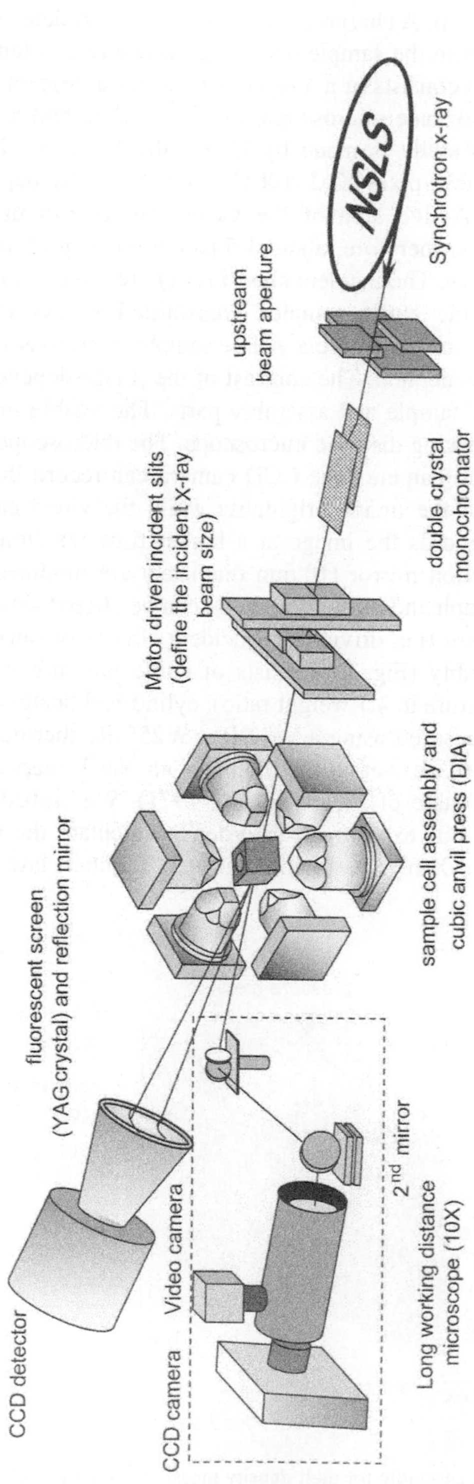
Synchrotron x-ray
NSLS
upstream
beam aperture
double crystal
monochromator
Motor driven incident slits
(define the incident X-ray
beam size)
sample cell assembly and
cubic anvil press (DIA)
fluorescent screen
(YAG crystal) and reflection mirror
CCD detector
CCD camera
Video camera
2nd mirror
Long working distance
microscope (10X)

SAM85 (Weidner et al., 1992). A charge-coupled device (CCD) detector is used to collect X-ray diffraction signals from the sample under high pressure and temperature.

The radiography system consists of a YAG crystal as a fluorescent screen, two optical mirrors, a long working distance microscope, a CCD camera and a video camera. The CCD camera used in this study is made by Diagnostic Scientific Instrument (Model: SPOT 2E) with 1315 × 1035-pixel KAF-1401E chip, pixel bit depth = 12 bits, pixel size = 6.8 μm × 6.8 μm. Active area of the CCD chip is 8.98 mm × 7.04 mm. The highest spatial resolution is, therefore, about 1.5 μm, after magnifying a 2 mm × 1 mm image to full size of the chip. The incident slits (Fig. 1) are removed for the radiographic measurement to capture the entire sample. Transmitted X-rays through the sample impinge on the fluorescent screen, where a visible sample image based on the intensity of the transmitted X-rays is generated. The contrast of the image depends on the difference of absorption between the sample and assembly parts. The visible image is reflected by two mirrors into a long working distance microscope. The microscope splits the image to a CCD camera and a video camera. The CCD camera can record the image at a preset exposure time to optimize the image brightness while the video camera monitors the image in real-time and records the image at a higher time resolution. The fluorescent screen and the first reflection mirror (10 mm diameter) are mounted beneath the CCD area detector. The radiograph and diffraction mode can be altered simply by changing the size of incident X-ray beam (i.e. driving the incident slits in or out).

The sample cell assembly (Fig. 2) consists of cubic pressure transmitting medium (made of boron:epoxy mixture in 4:1 weight ratio), cylindrical heater and sample capsule. Sample temperature is measured using a W3%Re–W25%Re thermocouple, and sample pressure is calculated based on measured volume of an NaCl internal pressure calibrant and Decker's equation of state of NaCl (Decker, 1971). We embed a reference sphere (Al_2O_3) in the sample in our experiment in order to calculate the density through the intensity of the transmitted X-rays, based on the Beer–Lambert law.

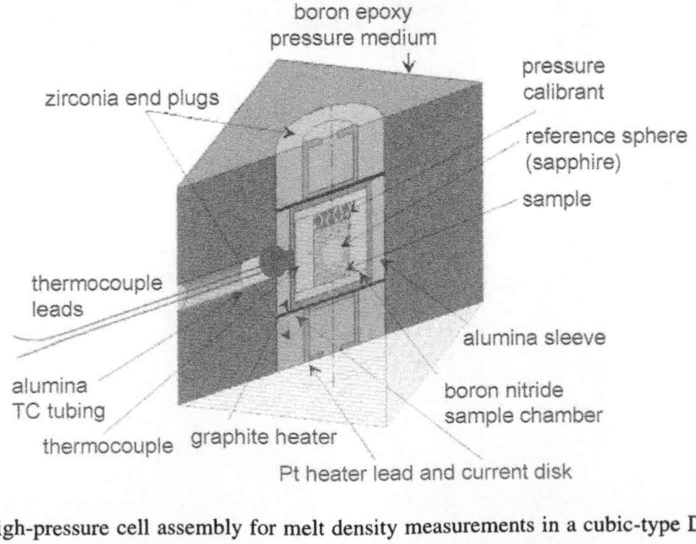

Figure 2. High-pressure cell assembly for melt density measurements in a cubic-type DIA apparatus.

3. System calibration

Deriving the sample density from the Beer–Lambert law requires measurement of transmitted X-ray intensities. Instead of scanning the sample across a reference sphere with a small X-ray beam and measuring intensities of incident beam and transmitted beam, the radiographic method records brightness distribution across the entire sample on an optical image. Variation of the brightness on the image, converted from X-rays impinging on the YAG fluorescent screen, is proportional to the X-ray intensity variation, which, in turn, is due to the difference in X-ray absorption of the sample and reference sphere. It is essential to confirm the linear relation between the brightness of the radiographic image and the X-ray intensity within the experimental X-ray intensity range. Tests are conducted using a step-absorber with identical thickness increment for each step. A radiograph image of the step-absorber is shown in Figure 3a. The brightness of the image is plotted as a function of pixel number in the horizontal axis in Figure 3b. The superimposed solid line in Figure 3b represents the calculated intensities of X-rays transmitted through the step-absorber after a linear normalization. This indicates a good linear correspondence between the brightness and the X-ray intensity. This linear relation is also illustrated in the plot of image brightness against X-ray intensity in Figure 4, where the data are nicely fit into a straight line ($R^2 = 0.9999$).

Uniformity of the X-ray intensity across the sample is another factor that can affect the experiment accuracy. Before each experiment, an image of the incident beam without sample in the path is taken to confirm the beam uniformity in intensity.

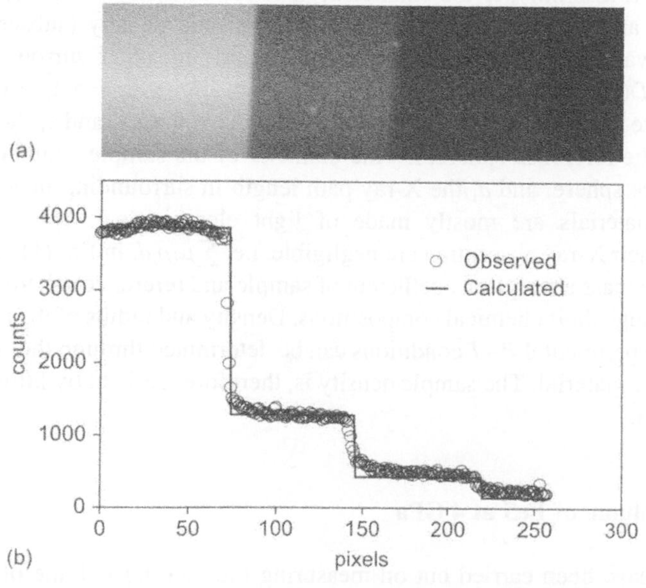

Figure 3. (a) Radiograph image of a step-absorber consisting of three equal thickness layers of copper foil; (b) Comparison of observed brightness and calculated X-ray intensity transmitted through the step-absorber after linear normalization.

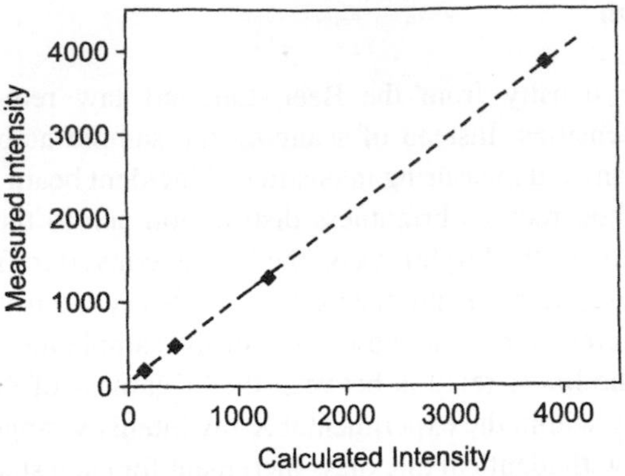

Figure 4. Relation of image brightness and X-ray intensity. A linear fit to the data is shown by the broken line ($R^2 = 0.9999$).

4. Date processing

Based on the linear conversion from X-ray intensity to image brightness and the Beer–Lambert law, the image brightness $B(x, y)$ in the reference sphere area for a cylindrical-symmetry cell assembly (Fig. 2) is expressed as

$$B(x, y) = I_0(x, y)Ke^{-\left[\mu_s\rho_s(D(x,y)-l(x,y))+\mu_r\rho_r l(x,y)+\sum\mu_i\rho_i d_i\right]} \quad (1)$$

where x and y are coordinates on the image, $I_0(x, y)$ is the intensity of the X-ray before entering the cell assembly, K the coefficient of X-ray intensity to radiograph brightness conversion, μ and ρ the mass absorption coefficient and density (subscripts s, r, and i represent the value for sample, reference sphere and any other surrounding materials, respectively), $D(x, y) = \left[D_0^2 - (x - x_c)^2\right]^{1/2}$, $l(x, y) = 2\left[R^2 - (x - x_r)^2 - (y - y_r)^2\right]^{1/2}$, x_c the x coordinate of the axis of the cylindrical sample chamber, x_r and y_r the coordinates of the center of the reference sphere, D_0 the diameter of the sample chamber, R the radius of the reference sphere, and d_i the X-ray path length in surrounding materials. Since the surrounding materials are mostly made of light element (e.g. boron and nitrogen), variations of their X-ray absorption are negligible, i.e. $\sum\mu_i\rho_i d_i$ in Eq. (1) can be treated as a constant. The mass absorption coefficient of sample and reference sphere material can be calculated through their chemical compositions. Density and radius of the reference sphere at any given experimental P–T conditions can be determined through the equation of state of the reference material. The sample density is, therefore, derived by fitting Eq. (1) to the brightness data.

5. Melting volume of FeS at 4 GPa

Experiments have been carried out on measuring the melting volume of FeS using this technique. A 0.5 mm-diameter ball of single crystal Al_2O_3 is used as the reference sphere. The sample is loaded in the high-pressure cell with NaCl as a pressure calibrant (Fig. 2) and compressed at room temperature; then the sample temperature is increased. During the

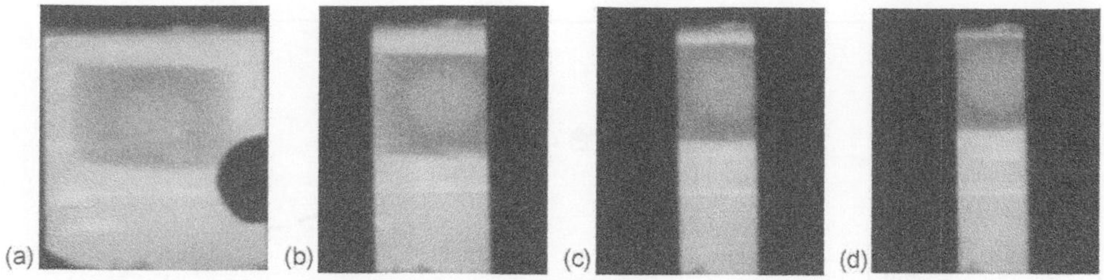

Figure 5. X-ray radiograph images of high-pressure cell between the WC anvils at (a) ambient condition; (b) 1.7 GPa and room temperature; (c) 4.1 GPa and room temperature; and (d) 4.1 GPa and 1300°C. The round shape dark area on the right of plate (a) is the image of the thermocouple junction next to the sample. As pressure increases, the anvil gap becomes tighter and, therefore, the visible image becomes narrower. The round shape image is from the reference sphere embedded in the sample.

heating, the sample pressure is monitored by taking diffraction patterns of the pressure calibrant using the CCD detector and the hydraulic load is adjusted to keep the pressure constant. A sequence of radiographic data are collected during heating while the pressure is kept at 4.1 ± 0.2 GPa. X-ray diffraction patterns from the sample are also collected after each radiographic exposure. Figure 5 shows some representative radiographs recorded during pressure loading and heating. The energy of the monochromatic X-ray beam is 39.77 ± 0.01 keV. The mass X-ray absorption coefficients of FeS and Al_2O_3 are calculated based on the element mass absorption coefficients from the International Tables for Crystallography (Creagh and Hubbell, 1992). A typical fitting of the brightness data is shown in Figure 6. Below the melting temperature, the sample density is measured through both radiographic method and X-ray diffraction after the sample transforms from FeS IV to FeS V, a NiAs-type structure (Fei et al., 1995). As shown in Figure 7, before the sample melts, the densities derived from the radiograph agree with the X-ray densities very well. The experimental error is estimated within 1%. A density drop is observed at 1300°C indicating melting of the sample. Loss of crystalline diffraction peaks in the X-ray diffraction pattern confirms the melting.

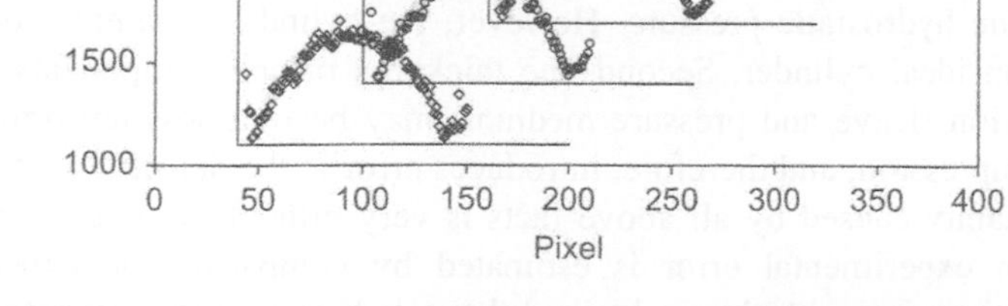

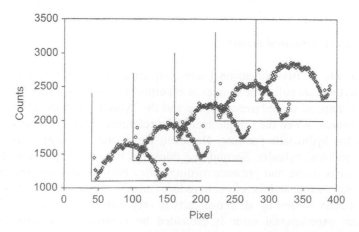

Figure 6. Representative fitting of observed brightness (open symbols) and calculated intensity (lines) for five consecutive scans across the reference sphere on the radiograph image.

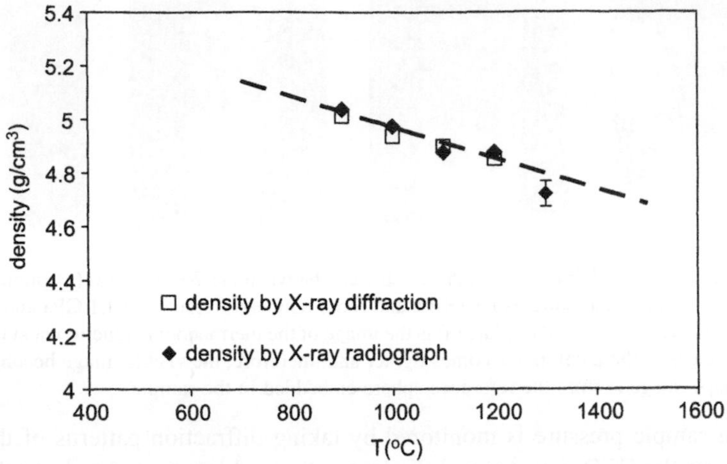

Figure 7. Measured densities of FeS as function of temperature at 4.1 GPa. Bar attached to the symbol represents an estimated experimental error. The straight line represents a linear fit to the density data of crystalline phase from radiograph method. Melting occurs at 1300°C.

Extrapolating the density of solid phase to 1300°C, we get the melting volume of FeS at 4.1 GPa, 0.28 cm^3/mol. If we adopt the relation between entropy change and melting volume of simple substances: $\Delta S_m = R\ln(2) + \alpha K \Delta V_m$ (where R is the molar gas constant, α and K are the volume thermal expansion coefficient and the bulk modulus of the solid phase, respectively, and ΔV_m the melting volume), suggested by Tallon (1979), we estimate the entropy change during melting to be 6.79 J/K/mol using $\alpha K = 3.7 \times 10^{-3}$ (Fei et al., 1995). From Clausius–Clapeyron equation: $dT/dP = \Delta V_m/\Delta S_m$ the slope of the melting curve at 4.1 GPa is, therefore, estimated to be 41°C/GPa. This value lies between two distinct DAC experimental results of ~30°C/GPa by Boehler (1992) and ~60°C/GPa by Ryzhenko and Kennedy (1973) and Williams and Jeanloz (1990) in this pressure region.

6. Sources of experimental errors

Although the use of two-dimensional data can improve the accuracy of the density measurement, there are still some sources of experimental error. First, the imperfectness of the spherical shape of the reference sphere and the cylindrical shape of the sample is a major error source. When the sample melts, the reference sphere probably is not easily distorted by the hydrostatic pressure. However, the cylindrical sample sometimes may depart from an ideal cylinder. Second, the thickness of cell components (e.g. capsule, heater, insulation sleeve and pressure medium) may become less uniform in the X-ray path upon compression, and therefore, introduces error in the density calculation. Experimental uncertainty caused by all above facts is very difficult to assess. In this chapter, the maximum experimental error is estimated by comparing the experimental data of radiograph imaging with those obtained through X-ray diffraction before the sample is molten.

As shown in Figure 2, the thermocouple junction is placed outside the sample capsule. The temperature measured by the thermocouple may actually be lower than the real sample temperature. No correction is made for this possible discrepancy; neither is for pressure effect on thermocouple EMF.

7. Concluding remarks

The brightness of the X-ray radiograph image from a YAG single-crystal fluorescent screen has a linear response to the intensity of incident X-rays. The use of an X-ray radiograph in measuring the density of the non-crystalline material avoids point-by-point scanning in data collection, increases the number of data from one-dimension to two-dimension, and therefore, improves the data statistics and precision. With reference to the result from X-ray diffraction in the case of the crystalline sample, this technique offers experimental accuracy of 1% at high pressures. Measurements of melting volume of FeS supply accurate data, which yield more reasonable result on melting curve calculation. This density measurement technique has a great potential in EOS studies of molten minerals.

Acknowledgements

Research carried out at the National Synchrotron Light Source, Brookhaven National Laboratory, which is supported by the US Department of Energy, Division of Materials Sciences and Division of Chemical Sciences, under Contract No. DE-AC02-98CH10886. The study is also supported by NSF research grant (EAR039879). We would like to thank Dr Z. Zhong for his technical support at the X17 beamline. MPI Pub. No. 351.

References

Balog, P.S., Secco, R.A., 2003. Equation of state of liquid Fe-10 wt%S: implications for the metallic cores of planetary bodies. J. Geophys. Res. 108 (B2), 2124, doi:10.1029/2001JB001646.

Bell, R.J., 1969. Pair distribution function for a 2-dimensional liquid. Nature 221 (5175), 50.

Birch, F., 1952. Elasticity and constitution of the Earth's interior. J. Geophys. Res. 57, 227–286.

Boehler, R., 1992. Melting of the Fe–FeO and the Fe–FeS systems at high-pressure – constraints on core temperatures. Earth Planet. Sci. Lett. 111 (2–4), 217–227.

Chen, J., Weidner, D.J., Vaughan, M.T., 2000. Density measurements of molten minerals at high pressure using synchrotron x-ray radiography. EOS, Trans. AGU 81 (48), 1195.

Creagh, D.C., Hubbell, J.H., 1992. X-ray absorption (or attenuation) coefficients. In: Prince, E. (Ed.), International Tables for Crystallography C, Kluwer Academic, Dordrecht, pp. 220–235.

Decker, D.L., 1971. High-pressure equation of state for NaCl, KCl and CsCl. J. Appl. Phys. 42, 3239–3244.

Desbat, B., Vanhuong, P., 1983. Structure of liquid-hydrogen fluoride studied by infrared and Raman-spectroscopy. J. Chem. Phys. 78 (11), 6377–6383.

Di Cicco, A., Minicucci, M., 1999. Solid and liquid short-range structure determined by EXAFS multiple-scattering data analysis. J. Synchrotron Radiat. 6, 255–257.

Eggert, J.H., Weck, G., Loubeyre, P., Mezouar, M., 2002. Quantitative structure factor and entity measurements of high-pressure fluids in diamond anvil cells by x-ray diffraction: argon and water. Phys. Rev. B 65, 172105.

Fei, Y., Prewitt, C.T., Mao, H.-k., Bertka, C.M., 1995. Structure and density of FeS at high pressure and high temperature and the internal structure of Mars. Science 268, 1892–1894.
Katayama, Y., 1996. Density measurements of non-crystalline materials under high pressure and high temperature. High Pressure Res. 14, 383–391.
Katayama, Y., 2001. XAFS study on liquid selenium under high pressure. J. Synchrotron Radiat. 8, 182–185.
Katayama, Y., Tsuji, K., Chen, J., Koyama, N., Kikegawa, T., Yaoita, K., Shimomura, O., 1993. Density of liquid tellurium under high-pressure. J. Non-Cryst. Solids 156, 687–690.
Katayama, Y., Tsuji, K., Shimomura, O., Kikegawa, T., Mezouar, M., Martinez-Garcia, D., Besson, J.M., Hausermann, D., Hanfland, M., 1998. Density measurements of liquid under high pressure and high temperature. J. Synchrotron Radiat. 5, 1023–1025.
Katayama, Y., Mizutani, T., Utsumi, W., Shimomura, O., Tsuji, K., 2001. X-ray diffraction study on structural change in liquid selenium under high pressure. Physica Status Solidi B-Basic Res. 223 (2), 401–404.
Kavner, A., Duffy, T.S., Shen, G.Y., 2001. Phase stability and density of FeS at high pressures and temperatures: implications for the interior structure of Mars. Earth Planet. Sci. Lett. 185 (1–2), 25–33.
King, H.E., Prewitt, C.T., 1982. High-pressure and high-temperature polymorphism of iron sulfide (FeS). Acta Crystallogr. Sect. B-Struct. Sci. 38 (Jul), 1877–1887.
Kobayashi, H., Takeshita, N., Mori, N., Takahashi, H., Kamimura, T., 2001. Pressure-induced semiconductor–metal–semiconductor transitions in FeS. Phys. Rev. B 63, 115203.
Kusaba, K., Utsumi, W., Yamakata, M., Shimomura, O., Syono, Y., 2000. Second-order phase transition of FeS under high pressure and temperature. J. Phys. Chem. Solids 61 (9), 1483–1487.
Lewis, J.S., 1971. Consequences of the presence of sulfur in the core of the Earth. Earth Planet. Sci. Lett. 11, 130–134.
Mason, B., 1966. Composition of the Earth. Nature 211, 616–618.
Rama Murthy, V., Hall, H.T., 1970. The chemical composition of the Earth's core: possibility of sulfur in the core. Phys. Earth Planet. Interiors 2, 276–282.
Rama Murthy, V., Hall, H.T., 1972. The origin and chemical composition of the Earth's core. Phys. Earth Planet. Interiors 6, 123–130.
Ryzhenko, B., Kennedy, G.C., 1973. The effect of pressure on the eutectic in the system Fe–FeS. Am. J. Sci. 273, 803–810.
Sanloup, C., Guyot, F., Gillet, P., Fiquet, G., Mezouar, M., Martinez, I., 2000. Density measurements of liquid Fe–S alloys at high pressure. Geophys. Res. Lett. 27, 811–814.
Shen, G., Sata, N., Mewille, M., Rivers, M.L., Sutton, S.R., 2002. Molar volumes of molten indium at high pressures measured in a diamond anvil cell. Appl. Phys. Lett. 81 (8), 1411–1413.
Tallon, J.L., 1979. The entropy change on melting of simple substances. Phys. Lett. 76A (2), 139–142.
Taylor, L.A., Mao, H.K., 1970. A high-pressure polymorph of trolite, FeS. Science 170 (3960), 850–851.
Urakawa, S., Hasegawa, M., Yamakawa, Y., Funakoshi, K.I., Utsumi, W., 2002. High-pressure phase relationships for FeS. High Pressure Res. 22 (2), 491–494.
Weidner, D.J., Vaughan, M.T., Ko, J., Wang, Y., Liu, X., Yeganeh-Haeri, A., Pacalo, R.E., Zhao, Y., 1992. Characterization of stress, pressure, and temperature in SAM85, a DIA type high pressure apparatus. In: Synono, Y., Manghnani, M.H. (Eds), High-pressure research: application to Earth and planetary sciences, Terra Scientific Publishing Company/AGU, Tokyo/Washington, DC, pp. 13–17.
Williams, Q., Jeanloz, R., 1990. Melting relations in the iron–sulfur system at ultra-high pressures: implications for the thermal state of the Earth. J. Geophys. Res. 95 (B12), 19,299–19,310.

Advances in High-Pressure Technology for Geophysical Applications
Jiuhua Chen, Yanbin Wang, T.S. Duffy, Guoyin Shen, L.F. Dobrzhinetskaya, editors

Chapter 10

Viscosity and density measurements of melts and glasses at high pressure and temperature by using the multi-anvil apparatus and synchrotron X-ray radiation

E. Ohtani, A. Suzuki, R. Ando, S. Urakawa, K. Funakoshi and Y. Katayama

Abstract

This chapter summarizes the techniques for the viscosity and density measurements of silicate melts and glasses at high pressure and temperature by using the X-ray radiography and absorption techniques in the third generation synchrotron radiation facility, SPring-8, Japan. The falling sphere method using in situ X-ray radiography makes it possible to measure the viscosity of silicate melts to pressures above 6 GPa at high temperature. We summarize the details of the experimental technique of the viscosity measurement, and the results for some silicate melts such as albite and diopside–jadeite melts. X-ray absorption method is applied to measure the density of silicate glasses such as basaltic glass and iron sodium disilicate glass up to 5 GPa at high temperature. A diamond capsule, which is not reactive with the glasses, is used for the density measurement of the glasses. The present density measurement of the glasses indicates that this method is useful for measurement of the density of silicate melts at high pressure and temperature.

1. Introduction

Magma plays key roles in the dynamic and fractionation processes in the evolving Earth. In order to evaluate the role of magmas quantitatively, we need information on physical properties of magmas. Viscosity and density are the fundamental physical properties of magmas, which controls transportation and crystallization of magmas in the Earth's interior.

Various procedures have been applied for determination of viscosity and density of magmas. In order to measure these properties at high pressure and temperature, the falling sphere technique combined with the quenching method using a piston cylinder apparatus has been applied first to measure the density and viscosity of the basaltic melt and various silicate melts such as diopside, jadeite, and albite melts up to 2.5 GPa (e.g. Fujii and Kushiro, 1977). Various materials such as boron nitride, graphite, and platinum spheres were used as a density marker. The quenching method was extended to about 7 GPa by using a diamond crystal as a density marker (Mori et al., 2000). *In situ* X-ray viscometry of silicate melts at high pressure and temperature combined with the synchrotron X-ray radiography has been developed by Kanzaki et al. (1987)

and this method enabled us to determine the viscosity up to the pressures exceeding 10 GPa (e.g. Suzuki et al., 2002; Reid et al., 2003). In this chapter we will summarize the detailed techniques of this method.

Density is one of the fundamental physical properties of magma, which controls the generation, transportation, and eruption of magma in the Earth. Density measurement of melts and glasses has been made using several methods, such as falling sphere method, sink–float method, shock wave experiment, and X-ray absorption method. The falling sphere method has been applied to measure the density of the basaltic melt up to 1.5 GPa by quenching method (Fujii and Kushiro, 1977). The sink–float method was applied to determine the density of peridotite and basaltic melts at high pressure using various crystals such as olivine and diamond as density markers (e.g. Agee and Walker, 1993; Ohtani et al., 1993; Suzuki et al., 1995; Ohtani and Maeda, 2001). Suzuki and Ohtani (2003) confirmed existence of the density crossover between the peridotite magma and olivine at the base of the upper mantle, and discussed its effect for the differentiation in the deep magma. Balog et al. (2003) measured the density of the Fe–FeS melt up to 20 GPa using a corundum–metal composite density marker in the multi-anvil apparatus. Density measurements of silicate melts above 20 GPa were carried out by the shock compression method (e.g. Rigden et al., 1984; Miller et al., 1991).

Katayama et al. (e.g. Katayama et al., 1996, 1998) developed X-ray absorption technique to measure the density of liquid metal at high pressures using a large volume press such as the cubic apparatus or the Toroidal press (Paris–Edinburgh press) combined with the synchrotron radiation. Katayama et al. (1996, 1998) applied the method to the density measurements of mono-atomic liquid metals, such as Bi and Te at high pressures up to a few GPa. Sanloup et al. (2000, 2004) and Chen et al. (2000, 2005) also applied this method for determining the density of Fe–FeS, FeS, and Fe–FeSi liquids at high pressures. The advantage of this method is to measure the density of liquid at desired pressure and temperature conditions without density markers. This method has a potential to provide the reliable density data of silicate melts within 1% error. We have developed a method to measure the density of silicate glass by using this method, and found that it is promising to determine the density of magmas at high pressure and temperature.

In this chapter, we summarize the techniques for the viscosity and density measurements of silicate melts and glasses at high pressure and temperature by using the X-ray radiography and absorption techniques in the third generation synchrotron radiation facility, SPring-8, Japan.

2. *In situ* X-ray viscometry at high pressure and temperature

2.1. *Experimental procedure for viscosity measurement using high pressure and temperature* in situ *X-ray viscometry*

Viscosity can be measured by the falling sphere method using the X-ray radiography (e.g. Suzuki et al., 2002). In the viscosity measurement, the Stokes' equation with the

Faxen correction (Faxen, 1925) is applied to correct the wall effect for the terminal velocity:

$$\eta = \frac{2gr^2\Delta\rho}{9v}\left\{1 - 2.104\left(\frac{r}{r_c}\right) + 2.09\left(\frac{r}{r_c}\right)^3 - 0.95\left(\frac{r}{r_c}\right)^5\right\}$$

v = terminal velocity, g = acceleration due to gravity, r = sphere radius, $\Delta\rho$ = density difference between the sphere and the melt, η = viscosity, and r_c is the inner radius of container.

Falling spheres were made of platinum or rhenium with 65 to 135 μm in diameter. The size of spheres was measured by the scanning electron microscopic images before the experiments. The schematic figure of the X-ray radiography system of the beam-line is schematically shown in Figure 1. The density of platinum or rhenium at high pressure and temperature was calculated from the equation of state (e.g. Jamieson et al., 1982; Ahrens and Johnson, 1995).

The experiments are conducted at the bending magnet beam line site of BL04B1 in SPring-8 (Utsumi et al., 1998). A white X-ray beam (20 ~ 50 keV) with a beam size of 4 mm × 4 mm is introduced from the bending magnet to the Kawai multi-anvil apparatus SPEED1500 (Kawai and Endo, 1970). The first stage anvil is made of steel for the SPEED 1500 apparatus. DIA type configuration is adopted for the guide-block system to drive the second stage multi-anvil made of tungsten carbide. The radiographic image is obtained by a fluorescence screen made of a Ce doped-YAG crystal. The fluorescence image is collected by a CCD camera (Hamamatsu Photonix, Co., Japan, C4880 high speed camera). The resolution of the X-ray imaging is 6.80 ± 0.02 μm/pixel.

The *in situ* high pressure and temperature X-ray radiography are made at pressures up to 6 GPa by using the furnace assembly given in Figure 2. The truncated edge length (TEL) of the tungsten carbide anvil is 12 mm for the experiments to the pressure of 6 GPa. The experiment for higher pressures to 13 GPa is made using the anvils with the TEL of 8 mm (e.g. Reid et al., 2003).

The pressure medium was a Cr-doped semi-sintered magnesia. The edge length of the octahedral pressure medium is 14 mm for TEL of the anvil of 12 mm and 11 mm

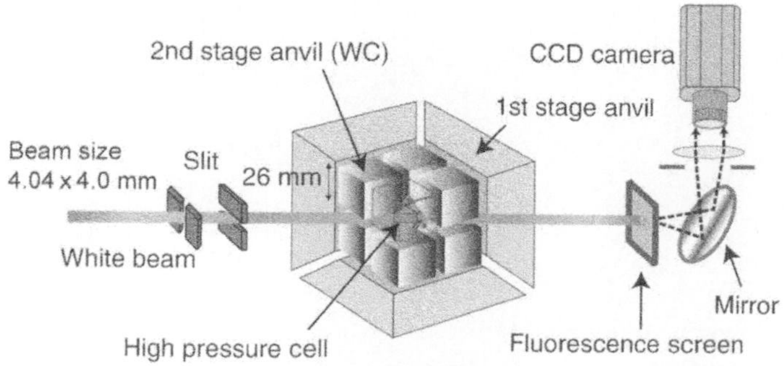

Figure 1. The schematic figure of the X-ray optical system for the falling sphere viscometry using the *in situ* X-ray radiography at high pressure and temperature.

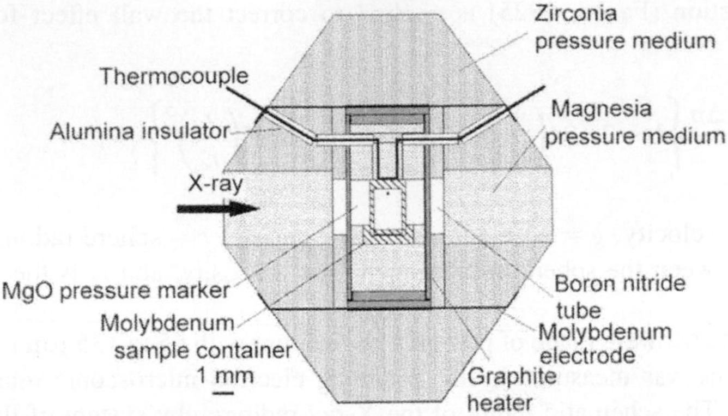

Zirconia
pressure medium
Thermocouple
Magnesia
pressure medium
Alumina insulator
X-ray
MgO pressure marker
Boron nitride
tube
Molybdenum
sample container
Molybdenum
electrode
Graphite
heater
1 mm

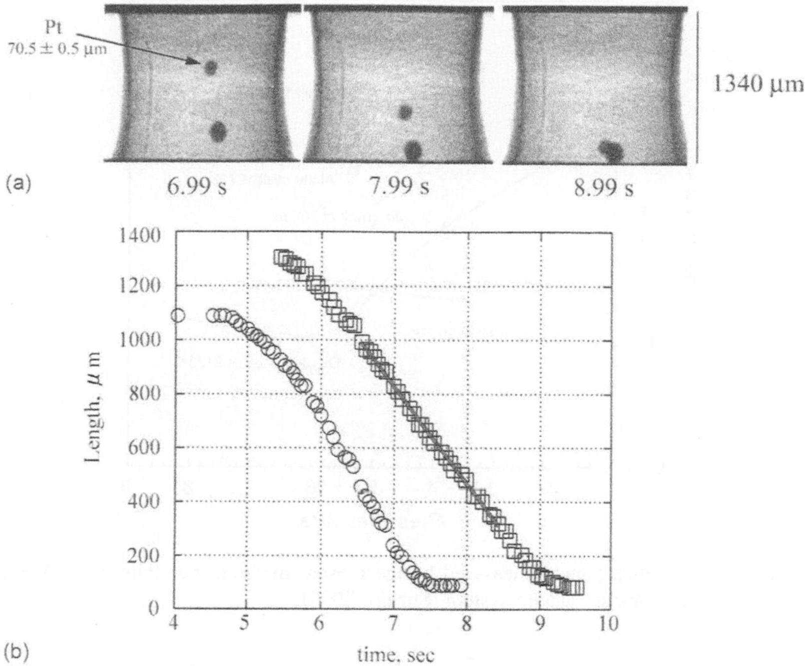

Figure 3. (a) The radiographic image of the falling platinum spheres placed in the model basaltic (MORB) melt in a molybdenum capsule. (b) The falling distance with time for the two spheres. We can determine the falling velocity as 360 ± 3 μm/s for a smaller falling sphere with a diameter of 70.5 ± 0.5 μm.

The molecular dynamic calculation is also available to estimate the liquid density (Matsui, 1998). Since the density difference between the Pt and silicate liquid is very large, the uncertainty associated with the ambiguity of the silicate liquid density can be negligible in the viscosity measurement.

Based on this technique, we clarified the viscosity at high pressure and temperature for some silicate systems including the albite, diopside, diopside–jadeite compositions. Small amount of water may affect the viscosity of the silicate melts. We carefully dried the furnace and the glass samples before the experiments. Therefore, the effect of water will be negligible, although we did not check the water content after the experiments. Our result on the albite viscosity is consistent with the previous works made by the quenching method (e.g. Fujii and Kushiro, 1977; Mori et al., 2000). Therefore, we can safely ignore the effect of water in our experiments.

Figure 4 summarizes the example of the change of viscosity with pressure for the albite melt (Suzuki et al., 2002) and the diopside–jadeite melt (Suzuki, 2000). The structure of the silicate melt can be expressed by NBO/*T*, i.e. the ratio of nonbridging oxygens (NBO) to tetrahedrally coordinated cations (Si^{4-} and Al^{3-}) at ambient pressure (Mysen et al., 1980). This figure implies that the pressure dependency of the viscosity is related to the liquid structure, NBO/*T*, i.e. the viscosity decreases with

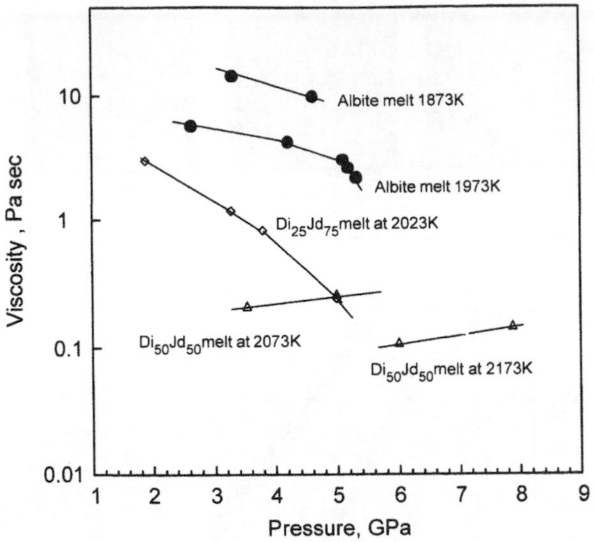

Figure 4. Viscosity of silicate melts measured by the present method; the albite melt (Suzuki et al., 2002), and the melts of diopside–jadeite system (Suzuki, 2000).

pressure in the albite melt with NBO/$T = 0$ and the $Di_{25}Jd_{75}$ melt with NBO/$T \sim 0.36$, whereas it increases with pressure in the $Di_{50}Jd_{50}$ melt with a larger NBO/$T \sim 0.8$.

3. Measurement of the density of silicate glass at high pressure using X-ray absorption method

3.1. X-ray absorption method for the density measurement of liquid metals

Intensity of transmitted X-ray through material with density ρ is expressed as $I = I_0 \exp(-\mu\rho t)$, where I_0 is intensity of incident X-ray, μ is mass absorption coefficient, and t is thickness of the material. When μ and t are known, the density can be evaluated by measuring I_0 and I.

In high-pressure experiments, the sample is enclosed by the sample container, heater, and pressure medium, therefore, the intensity of the transmitted X-ray depends on the absorption of these materials. It is difficult to evaluate density of the sample without consideration of the change of thickness and densities of materials surrounding the sample with pressure. Katayama et al. (1996, 1998) solved these difficulties by using synchrotron radiation and a hard sample container such as a sapphire (Al_2O_3) capsule for the density measurement of liquid metals, and measured their densities up to 5 GPa. When the sample is the most absorbent material along the X-ray path, a highly intense synchrotron X-ray reduces the effect of the variation of thickness and densities of the surrounding materials on estimation of the absorption by the sample. If the sample container deforms elastically with compression, the thickness of the sample can be evaluated by calculating the container size based on its equation of state. A sapphire (Al_2O_3) capsule used by Katayama et al. (1998) as a sample container for the density measurement of liquid metals is highly incompressible with a bulk modulus of 252 GPa.

3.2. *Experimental technique for the density measurement of the silicate glass*

When this method is applied to silicate melt or glass, which are composed of light elements such as silicon and oxygen, it is difficult to use sapphire for the capsule material since the

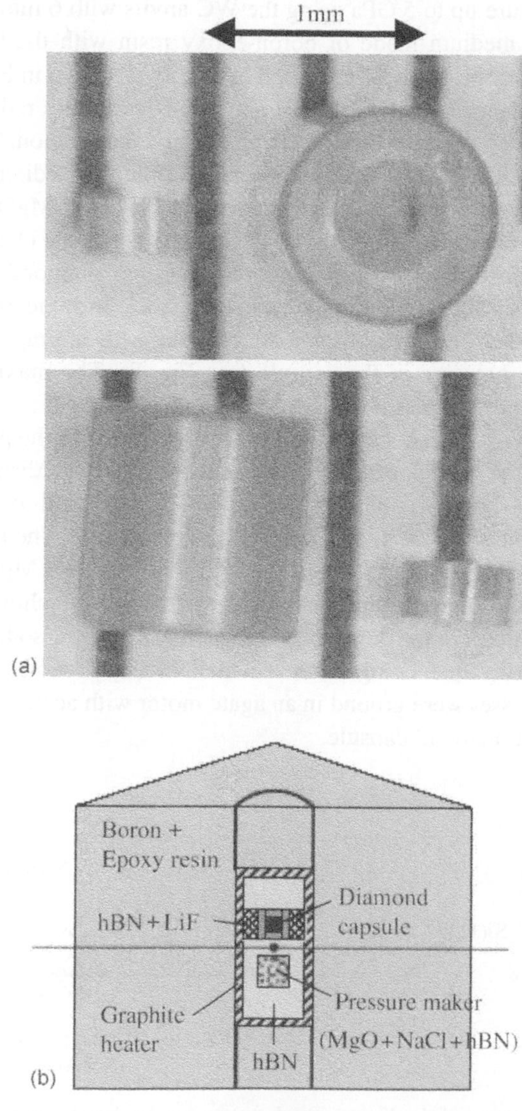

Figure 5. The cell assembly for the X-ray absorption method at high pressure and temperature. (a) The single crystal diamond capsule and lid used in this study, (b) A cross-section of the cell assembly.

X-ray absorption by silicates is as low as sapphire. Sapphire is also reactive with silicates at high temperature. We choose a single crystal diamond as a sample container for the silicate samples. We used type-IB single crystal diamond synthesized by Sumitomo Electric Hardmetal Co. Ltd. The diamond capsule is composed of one ring and two lids (Fig. 5), which are processed by Syntech Co. The diamond ring has a dimension of 1.0 mm in outer diameter, 0.5 mm in inner diameter, and 1.0 mm in length. The photograph of the diamond capsule and the detail of the cell assembly are given in Figure 5a and b.

High-pressure and high-temperature X-ray absorption experiments of silicates have been carried out at BL22 of SPring-8, Japan, where the cubic press, SMAP-I, is installed. We generate the pressure up to 5 GPa using the WC anvils with 6 mm truncation and the cubic shape pressure medium made of boron-epoxy resin with the 9 mm edge length. Cross-section of an example of the sample cell assembly is shown in Figure 5. A graphite tube is used as a resistance heater. A ring of LiF and BN mixture reduces the deviatoric stress in the diamond capsule to avoid fracturing during compression. Temperature of the sample is measured by a W3%Re–W25%Re thermocouple with a diameter of 0.15 mm at the center of the heater. Pressure is determined by NaCl or MgO pressure markers placed close to the diamond capsule based on the equations of state (Jamieson et al., 1982; Brown, 1999). The absorption profile indicates that two diamond lids proceed with increasing pressure to compress the sample from both sides. In some experiments, we also placed the MgO marker inside the diamond capsule with a graphite lid in order to determine the precise pressure in the capsule (see Fig. 8). The maximum pressure and temperature of the present experiments are 5 GPa and 1273 K.

We tested two kinds of silicate glass samples for application of the present method. The silicate samples are the model basaltic glass and iron sodium disilicate glass with a composition of $Na_{2.21}Fe_{1.00}Si_{1.95}O_{6.00}$. We simplified the mid oceanic ridge basalt composition by Melson et al. (1976) to a six component mixture. The model basaltic glass has the composition with SiO_2 51.81 wt%, Al_2O_3 15.95, FeO 9.97, MgO 7.86, CaO 11.69, Na_2O 2.72. The basaltic glass sample was synthesized by quenching of the melt from 1200°C at 1 atm without controlling oxygen fugacity. The iron sodium disilicate glass sample was prepared by quenching from the melt at fO_2 around magnetite–hematite buffer. The silicate glasses were ground in an agate motor with acetone, and the powdered glass was packed in a diamond capsule.

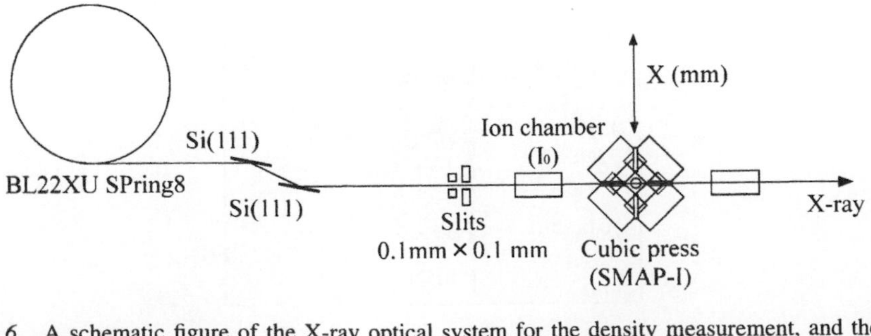

Figure 6. A schematic figure of the X-ray optical system for the density measurement, and the cubic press, SMAP-I. A monochromatized X-ray beam (0.1 mm × 0.1 mm) is introduced to the cubic press for the absorption measurement.

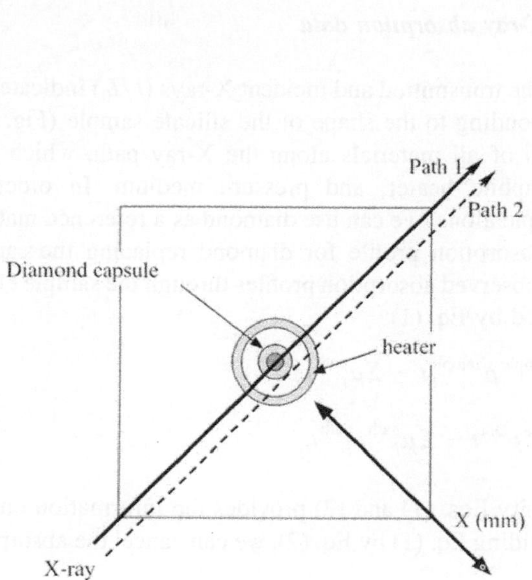
Path 1
Path 2
Diamond capsule
heater
X (mm)
X-ray

the upstream and downstream of the sample (Fig. 6). The synchrotron X-ray is fixed in position, but the press position is moved three-dimensionally by 1 μm step. Intensity measurement is conducted with 20 μm step perpendicular to the X-ray path, so that the X-ray absorption profile can be acquired in the radial direction of the cylindrical sample.

Determination of the pressure in the run is made by measuring the diffracted X-ray profiles of the pressure markers and the glass samples by an imaging plate, which is replaced for the downstream ion chamber, using angle-dispersive X-ray diffraction method. The diffraction pattern contains all peaks diffracted by the materials in the X-ray path, because receiving slit systems are not used. Therefore, the diffraction profile is too complex to recognize the pattern of glass sample, but we can readily detect new peaks appearing with crystallization of the glass.

3.3. Analysis of the X-ray absorption data

The intensity ratio of the transmitted and incident X-rays (I/I_0) indicates a clear parabolic shaped profile corresponding to the shape of the silicate sample (Fig. 8). These profiles include the absorption of all materials along the X-ray path, which are the sample, a diamond capsule, graphite heater, and pressure medium. In order to evaluate the absorption by the sample alone, we can use diamond as a reference material and draw the background for the absorption profile for diamond replacing the sample as shown in Figure 8. In this case, observed absorption profiles through the sample (1) and the diamond lid (2) can be expressed by Eq. (1):

$$I^{S}/I_0^{S} = \exp(-\mu^{\text{Sample}}\rho^{\text{Sample}}t - \Sigma\mu_i^{\text{path}}\rho_i^{\text{path}}t_i) \quad (1)$$

$$I^{D}/I_0^{D} = \exp(-\mu^{\text{Dia}}\rho^{\text{Dia}}t - \Sigma\mu_i^{\text{path}}\rho_i^{\text{path}}t_i) \quad (2)$$

The ratio of the intensity Eqs. (1) and (2) provides the information on the density of the sample ρ^{Sample}, i.e. dividing Eq. (1) by Eq. (2), we can cancel the absorption by the path as follows:

$$\frac{I^{S}/I_0^{S}}{I^{D}/I_0^{D}} = \exp[(\mu^{\text{Dia}}\rho^{\text{Dia}} - \mu^{\text{Sample}}\rho^{\text{Sample}})t] \quad (3)$$

The sample thickness t can be written as $t = 2\sqrt{R^2 - (x - x_0)^2}$ where R is the inner radius of the sample container, x is the coordinate, and x_0 is the center of parabolic shaped profile. We can calculate the density of the sample from the observed absorption profile by fitting these equations to the absorption data. Here, the density of diamond is calculated based on the Birch–Murnaghan equation of state using the bulk modulus, its pressure and temperature derivatives (Bass, 1995) and the thermal expansion coefficient (Saxena et al., 1993) of diamond, and mass absorption coefficient μ is calculated by Victoreen's equation (MacGillavry and Rieck, 1983).

Parabola of the absorption profile slightly undulates (Fig. 8) due to the effect of absorption of materials surrounding diamond and a finite size of the X-ray beam. Edges of the parabolic absorption profile incline outward, because X-ray we used has a finite size

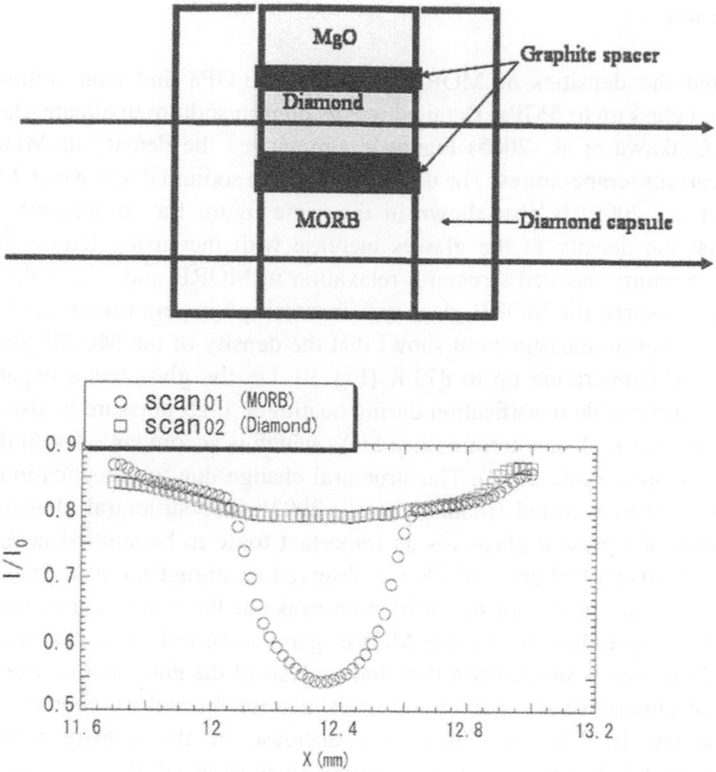

Figure 8. Intensity ratio of the transmitted and incident X-rays, I/I_0. A parabolic shaped profile corresponds to the cylindrical shape of the silicate sample. The density of the sample (MORB) can be determined by comparison of the absorption along the path through the sample (1) with that through the diamond lid (2). X (mm) indicates the position of the X-ray beam across the pressure medium as shown in Figure 7.

(100 μm in *x*-direction). In spite of these difficulties, the error of the density produced by the least square fitting is within 1%, which means that the X-ray absorption method can determine the relative density with very high precision (less than 1%).

The accuracy of the present density measurements may be evaluated by comparison of the density of the glass measured directly and estimated by this method. The density of the iron sodium disilicate glass was measured to be 2.961 g/cm^3 directly by the sink–float method with the solutions of the mixture of methylene and benzene. Whereas the density was determined as 2.877 g/cm^3 at the ambient condition by the X-ray absorption method, which is smaller by about 3% than that of the direct measurement by the sink–float method. This discrepancy may imply the accuracy of the present X-ray absorption method. Although the quality of the absorption profile may affect the accuracy, the present discrepancy might be mainly caused by the reliability of the mass absorption coefficients estimated by the Victoreen's equation (MacGillavry and Rieck, 1983). This problem can be solved by the direct measurement of the mass absorption coefficient of the relevant silicate samples.

3.4. Discussion

We measured the densities of MORB glass up to 3 GPa and iron sodium disilicate ($Na_2FeSi_2O_6$) glass up to 5 GPa. Detailed results on iron sodium disilicate glass are given elsewhere (Urakawa et al., 2005) Figure 9 summarizes the density of MORB glass at 3 GPa and various temperatures. The density of the iron sodium disilicate at 2.5 and 5 GPa (Urakawa et al., 2005) is also shown in the same figure for comparison. This figure indicates that the density of the glasses increase with increasing temperatures, which shows a temperature-induced structural relaxation in MORB and iron sodium disilicate glasses. We measured the MORB glass with increasing temperature to 1273 K at 3 GPa. The X-ray absorption measurement shows that the density of the MORB glass increases with increasing temperature up to 873 K (Fig. 9), i.e. the glass has a negative thermal expansivity. Irreversible densification during heating at high pressure is also observed in some oxide glasses such as vitreous silica SiO_2, which is accompanied with the structural relaxation (Inamura et al., 2002). The structural change due to densification of vitreous silica has also been reported (Inamura et al., 2003). The structural change due to the densification of the present glasses is an important topic to be studied in future. At the temperatures between 873 and 1073 K we observed an abrupt increase in the density of the glass and an appearance of the diffraction peaks at the temperatures above 1073 K, indicating that crystallization of the MORB glass occurred at temperatures between 873 and 1073 K. The *in situ* X-ray diffraction analysis of the glass in this condition clearly indicates that clinopyroxene, garnet, and coesite were formed by crystallization of the glass. At above 1073 K, we observed a decrease of the density with increasing temperature, which is caused by the thermal expansion of the minerals formed by

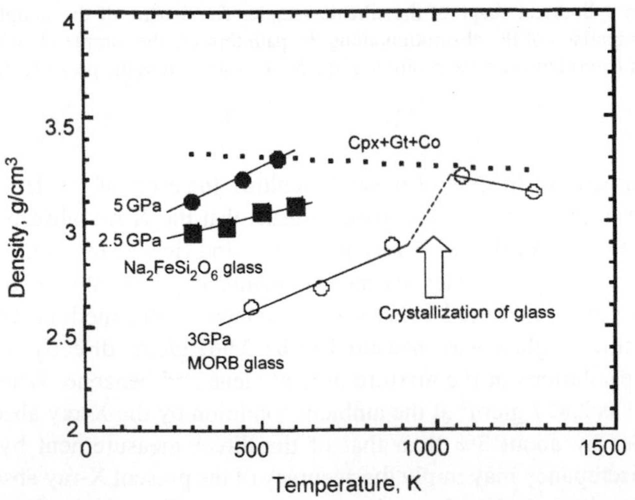

Figure 9. Densification of glasses at high pressure and temperature. The densities of the MORB glass and $Na_2FeSi_2O_6$ glass increase with increase of the temperature. A drastic increase in the density of the MORB glass indicates crystallization of the glass (see the text). The calculated density of the crystallized assemblage of clinopyroxene, garnet, and coesite is also shown in this figure, and it is consistent with the density measured in this study.

crystallization of the MORB glass. The calculated density of the crystallized assemblage, a mixture of clinopyroxene, garnet, and coesite is also shown in Figure 9. The density of the sample at 1073 and 1273 K at 3 GPa is consistent with those calculated by the equation of state of the minerals within the error of 3%, which is consistent with the accuracy of the density measurement of iron sodium disilicate glass made at the ambient condition.

The present results on the density measurements in the silicate glass starting materials provide reliable density values at high pressure and temperature, which indicates that the X-ray absorption method is a promising new method to determine the silicate melt density at high pressure and temperature, although the present measurement is still preliminary and is made only for determination of the density of silicate glasses at high pressure and temperature.

Acknowledgements

The authors are grateful to O Shimomura, Y. Inamura, W. Utsumi, H. Kaneko, H. Saito, T. Watanuki, A. Machida, T. Okada, T. Kubo, T. Kondo, and H. Terasaki for their help and encouragement in the present experiments using SPring-8 facilities. *In situ* X-ray radiography for viscosity measurement was conducted at the SPring-8 of the Japan Synchrotron Research Institute (Proposal number, 1999A0137-ND-np; 1999B0156-ND-np, 2000A0220-ND-np). High-pressure X-ray absorption experiments were performed under contract with Japan Atomic Energy Research Institute. This work was partially supported by the Grant-in-Aid of Scientific Research of Priority Area (no.12126201) from the Ministry of Education, Culture, Sport, Science, and Technology, and Scientific Research S (no.14102009) from Japan Society for the Promotion of Science to E. Ohtani., and Scientific Research B (No.13440163) to S. Urakawa.

References

Agee, C.B., Walker, D., 1993. Olivine flotation in mantle melt. Earth Planet. Sci. Lett. 114, 315–324.

Ahrens, T.J., Johnson, M.L., 1995. Shock wave data for minerals. In: Ahrens, T.J. (Ed.), Mineral Physics and Crystallography: A Handbook of Physical Constants. American Geophysical Union, Washington, DC, pp. 143–183.

Balog, P.S., Secco, R.A., Rubie, D.C., Frost, D., 2003. Equation of state of liquid Fe-10 wt% S: implications for the metallic cores of planetary bodies. J. Geophys. Res. 108 (B2), 2124, doi:10.1029/2001JB001646.

Bass, J.D., 1995. Elasticity of minerals, glass, and melts. In: Ahrens, T.J. (Ed.), Mineral Physics and Crystallography: A Handbook of Physical Constants. American Geophysical Union, Washington, DC, pp. 45–63.

Brown, M.J., 1999. The NaCl pressure standard. J. Appl. Phys. 86, 5801–5808.

Chen, J., Weidner, D.J., Vaughan, M.T., 2000. Density measurements of molten minerals at high pressure using synchrotron X-ray radiography. EOS, Trans. AGU 81 (48), 1195.

Chen, J., Weidner, D.J., Wang, L., Vaughan, M.T., 2005. Density measurements of molten materials at high pressure using synchrotron X-ray radiography: melting volume of FeS. In: Chen, J., Wang, Y., Duffy, T.S., Shen, G., Dobrzhinetskaya, L.F. (Eds), Advances in High-Pressure Technology for Geophysical Applications. Elsevier, Amsterdam, pp. 185–194.

Faxen, H., 1925. Gegenseitige Einwirkung zweier Kugeln, die in einer Zaehen Fluessigkeit fallen. Ark Mat Asron Fys 19, 1–8.

Fujii, T., Kushiro, I., 1977. Density, viscosity, and compressibility of basaltic liquid at high pressures. Carnegie Inst. Year Book 79, 419–424.

Inamura, Y., Katayama, Y., Funakoshi, K., 2002. The Temperature and Pressure Dependence of the Structure for Vitreous Silica. SPring-8 User Experiment Report No. 9 (2002A), 49.

Inamura, Y., Katayama, Y., Kohira, S., 2003. High Energy X-Ray Diffraction Measurement of Densified Vitreous Silica. SPring-8 User Experiment Report No. 10 (2002B), 47.

Jamieson, J.C., Fritz, J.M., Manghnani, M.H., 1982. Pressure measurement at high temperature in X-ray diffraction studies: gold as a primary standard. In: Akimoto, S., Manghnani, M.H. (Eds), High Pressure Research in Geophysics. Center for Academic Publication, Tokyo, pp. 27–48.

Kanzaki, M., Kurita, K., Fujii, T., Kato, T., Shimomura, O., Akimoto, S., 1987. A new technique to measure the viscosity and density of silicate melts at high pressure. In: Manghnani, M.H., Syono, Y. (Eds), High Pressure Research in Mineral Physics. American Geophysical Union, Washington, DC, pp. 195–200.

Katayama, Y., Tsuji, K., Kanda, H., Nosaka, H., Yaoita, K., Kikegawa, T., Shimomura, O., 1996. Density of liquid tellurium under pressure. J. Non-Cryst. Solids 207, 451–454.

Katayama, Y., Tsuji, K., Shimomura, O., Kikegawa, T., Mezouar, M., Martinez-Garcia, D., Besson, J.M., Hausermann, D., Hanfland, M., 1998. Density measurements of liquid under high pressure and high temperature. J. Synchrotron Radiat. 5, 1023–1025.

Kawai, N., Endo, S., 1970. The generation of ultrahigh hydrostatic pressures by a sprit sphere apparatus. Rev. Sci. Instrum. 41, 1178–1181.

Kress, V.C., Williams, Q., Carmichael, I.S.E., 1988. Ultrasonic investigation of melts in the system Na_2O–Al_2O_3–SiO_2. Geochim. Cosmochim. Acta 52, 283–293.

Lange, R.L., Carmichael, I.S.E., 1987. Densities of Na_2O–K_2O–CaO–MgO–FeO–Fe_2O_3–Al_2O_3–TiO_2–SiO_2 liquids: new measurements and derived partial molar properties. Geochim. Cosmochim. Acta 51, 2931–2946.

MacGillavry, C.H., Rieck, G.D., 1983. International table for X-ray crystallography: volume III. In: Physical and Chemical Table. International Union of Crystallography, Reidel, Dordrecht, p. 362.

Matsui, M., 1998. Computational modeling of crystals and liquids in the system Na_2O–CaO–MgO–Al_2O_3–SiO_2. In: Manghnani, M.H., Yagi, T. (Eds), Properties of the Earth and Planetary Materials at High Pressure and Temperature. American Geophysical Union, Washington, DC, pp. 145–151.

Melson, W.G., Wright, T.L., Byery, G., Nelen, J., 1976. Chemical diversity of abyssal volcanic glass erupted along Pacific, Atlantic, and Indian Ocean sea-floor spreading centers. In: Sutton, G.H., Manghnani, M.H., Moberly, R. (Eds), The Geophysics of the Pacific Ocean Basin and its Margin, Vol. 19. AGU, Washington, DC, Geophysics Monograph Series, pp. 351–367.

Miller, G.H., Stolper, E.M., Ahrens, T.J., 1991. The equation of state of a molten komatiite 1. Shock wave compression to 36 GPa. J. Geophys. Res. 96, 11831–11848.

Mori, S., Ohtani, E., Suzuki, A., 2000. Viscosity of the albite melt to 7 GPa at 2000 K. Earth Planet. Sci. Lett. 175, 87–92.

Mysen, B.O., Birgo, D., Scarf, C.M., 1980. Relations between the anionic structure and viscosity of silicate melts. A Raman spectroscopic study. Am. Mineral. 65, 690–710.

Ohtani, E., Maeda, M., 2001. Density of basaltic melt at high pressure and stability of the melt at the base of the lower mantle. Earth Planet. Sci. Lett. 193, 69–75.

Ohtani, E., Suzuki, A., Kato, T., 1993. Flotation of olivine in the peridotite melt at high pressure. Proc. Jpn Acad. 69, 23–28.

Reid, J.E., Suzuki, A., Funakoshi, K., Terasaki, H., Poe, B.T., Rubie, D.C., Ohtani, E., 2003. The viscosity of $CaMgSi_2O_6$ liquid at pressures up to 13 GPa. Phys. Earth Planet. Interiors 139, 45–54.

Rigden, S.M., Ahrens, T.J., Stolper, E.M., 1984. Density of liquid silicates at high pressures. Science 226, 1071–1074.

Sanloup, C., Guyot, F., Gillet, P., Fiquet, G., Mezouar, M., Martinez, I., 2000. Density measurement of liquid Fe–S alloys at high-pressure. Geophys. Res. Lett. 27, 811–814.

Sanloup, C., Fiquet, G., Gregoryanz, E., Morad, G., Mesouar, M., 2004. Effect of Si on liquid Fe compressibility: implication for sound velocity in core materials. Geophys. Res. Lett. 31, L07604, doi:10.1029/2004GL019526.

Saxena, S.K., Chatterjee, N., Fei, Y., Shen, G., 1993. Thermodynamic data on oxides and silicates. Springer, Berlin, p. 428.

Suzuki, A., 2000. Physical properties of silicate melt at high pressure and high temperature: applications to the planetary interior. Ph.D thesis, Tohoku University.

Suzuki, A., Ohtani, E., 2003. Density of peridotite melts at high pressure. Phys. Chem. Miner. 30, 449–456.

Suzuki, A., Ohtani, E., Kato, T., 1995. Flotation of diamond in mantle melt at high pressure. Science 269, 216–218.

Suzuki, A., Ohtani, E., Funakoshi, K., Terasaki, H., 2002. Viscosity of albite melt at high pressure and high temperature. Phys. Chem. Miner. 29, 159–165.

Utsumi, W., Funakoshi, K., Urakawa, S., Yamakata, M., Tsuji, K., Konishi, H., Shimomura, O., 1998. SPring-8 beamline for high pressure science with multi-anvil apparatus. Rev. High Pressure Sci. Technol. 7, 1484–1486.

Urakawa, S., Ando, R., Ohtani, E., Suzuki, A., Katayama, Y., 2005. New approach to density of silicate melts at high pressures: application of X-ray absorption technique. Geochim. Cosmochim. Acta, in press.

Advances in High-Pressure Technology for Geophysical Applications
Jiuhua Chen, Yanbin Wang, T.S. Duffy, Guoyin Shen, L.F. Dobrzhinetskaya, editors

Chapter 11

The effect of composition, compression, and decompression on the structure of high-pressure aluminosilicate glasses: an investigation utilizing ^{17}O and ^{27}Al NMR

Jeffrey R. Allwardt, Jonathan F. Stebbins,
Burkhard C. Schmidt and Daniel J. Frost

Abstract

This contribution reviews recent work on the determination of the structure of high-pressure aluminosilicate glasses and presents new ^{17}O and ^{27}Al magic-angle spinning (MAS) and ^{17}O triple-quantum (3Q) MAS NMR spectra that investigate the structure of $K_3AlSi_3O_9$ (KAS) and $Ca_3Al_2Si_6O_{18}$ (CAS) glasses quenched at 5 GPa. Comparison of the ^{27}Al MAS spectra for the high-pressure and ambient-pressure CAS and KAS glasses shows that the average Al-coordination increases with increasing pressure and that there is a strong compositional dependence to the amount of high-coordinated Al, as the CAS glass contains significantly more $^{[5]}Al$ and $^{[6]}Al$ than the KAS glasses. The spectra of the 5 GPa and the ambient pressure glasses were compared and show that high-coordinated Al are formed by the mechanisms: $^{[4]}Si{-}NBO + 4^{[4]}Si{-}O{-}^{[4]}Al \rightarrow 5^{[4]}Si{-}O{-}^{[5]}Al$ and $2^{[4]}Si{-}NBO + 4^{[4]}Si{-}O{-}^{[4]}Al \rightarrow 6^{[4]}Si{-}O{-}^{[6]}Al$, which are similar to the mechanisms for the generation of $^{[5]}Si$ and $^{[6]}Si$ in high-pressure silicate glasses. Decompression times were varied (14 h and 1 s) and longer times are shown to reduce the average Al-coordination. Additionally, the reproducibility of glass making in a multi-anvil apparatus was investigated by comparing the ^{27}Al MAS spectra of two glasses made in the same high-pressure experiment, but in separate capsules. This suggests that the reproducibility of the average Al-coordination is about 0.1 (e.g. 5.0 ± 0.1) for glasses with an Al/(Al + Si) ratio of 0.25.

1. Introduction

Pressure has been shown to drastically affect macroscopic properties of melts, such as density, viscosity, and diffusivity (Rigden et al., 1984; Scarfe et al., 1987; Reid et al., 2001). A fundamental understanding of the pressure effect on the melt structure is required to formulate structurally realistic models for the pressure dependence of macroscopic melt properties. For this reason, there has been considerable effort to understand structural transitions in high-pressure glasses and melts. However, results have been somewhat controversial because of uncertainties in the structural interpretations of *in-situ* data from Raman spectroscopy and the apparent structural relaxation that occurs during decompression of quenched glasses. The purpose of this paper is to review aspects of what is known and suspected about the pressure effects on the structure of aluminosilicate melts and glasses and to provide new data on several aspects of these important questions. These include: the large and obvious effects of composition, the effect of temperature and thermal history, and the effect of decompression on quenched high-pressure samples. First, we briefly review some of the recent work on the pressure dependence of physical properties.

1.1. Pressure dependence of viscosity and density of silicate melts

Melt density (and hence buoyancy) and viscosity have direct consequences for magma ascent and therefore the chemical evolution of the Earth and other planetary bodies (Stolper et al., 1981). There has been a significant amount of work to determine the pressure dependence on pressure of both properties, which can be quite significant (Kushiro, 1976, 1978; Rigden et al., 1984, 1988; Scarfe et al., 1987; Agee and Walker, 1993; Reid et al., 2001, 2003; Suzuki et al., 2002; Lange, 2003). The pressure dependence of viscosity has been shown to vary strongly with composition and can be either positive or negative. For example, the viscosity of $NaAlSi_3O_8$ melts seems to decrease with increasing pressure (Wolf and McMillan, 1995; Suzuki et al., 2002). However, two recent studies have observed a minimum in the viscosity of albite and MORB melts at 8 and 3 GPa, respectively (Poe and Rubie, 2000; Ando et al., 2003), while Reid et al. (2003) reports a maximum in viscosity of diopside melt at 10 GPa. These studies suggest that important structural changes occur in these melts as a function of pressure.

The qualitative pressure dependence of the density seems to be much more straightforward, as it has been shown to increase with pressure. At ambient pressure, silicate melts have long been known to be less dense than the corresponding crystals, however, melts are also much more compressible. This means that the densities of melts and source rocks become more similar as pressure increases, which could lead to neutral buoyancy and a higher probability that the melt becomes trapped and solidifies in the surrounding mineral assemblage (Stolper et al., 1981). Additionally, through the use of shock wave measurements and olivine floatation experiments, it has been shown that at pressures corresponding to the Earth's upper mantle, melts can become more dense than typical mantle minerals, which would cause olivine and other crystalline materials to float (Rigden et al., 1984; Agee and Walker, 1993; Ohtani et al., 1993). These and other experiments have experimentally documented the increasing melt density with increasing pressure (Agee and Walker, 1988; Miller et al., 1991; Suzuki et al., 1995; Knoche and Luth, 1996; Ohtani et al., 1998; Suzuki et al., 1998; Ohtani and Maeda, 2001; Suzuki and Ohtani, 2003). Additionally, studies have measured the densities of glasses quenched from high-pressure liquids and confirmed that they, like the densities of the melts, increase with increasing pressure of synthesis (Kushiro, 1978; Poe et al., 2001). For example, very recent data (Allwardt et al., in press-a) indicates that rapidly (1 s) and conventionally (~14 h) decompressed $Ca_3Al_2Si_6O_{18}$ glasses quenched at 5 GPa (CAS5R2 and CAS5S in Table 1 of this study) are densified by 14 and 11% relative to the ambient-pressure glass (CAS0), which is similar to that suggested by the pressure–density relationships shown for basaltic composition melts (Ohtani and Maeda, 2001). At least some of the high-pressure melt structure is thus preserved in the decompressed glasses; especially when the decompression is rapid (~1 s). This suggests that although these measurements cannot be directly applied to the density of high-pressure melts, density determination of glasses at

Table 1. Experimental conditions for the high-pressure samples.

Sample	Composition	*P* (GPa)	*T* (°C)	Decompression
KAS0	$K_3AlSi_3O_9$	0.0001	1450	N/A
KAS5R	$K_3AlSi_3O_9$	5	1960	Rapid
KAS5S	$K_3AlSi_3O_9$	5	1960	Conventional
KAS6	$K_3AlSi_3O_9$	6	25	Conventional
CAS0	$Ca_3AlSi_6O_{18}$	0.0001	1450	N/A
CAS5R1	$Ca_3AlSi_6O_{18}$	5	1900	Rapid
CAS5R2	$Ca_3AlSi_6O_{18}$	5	1900	Rapid
CAS5S	$Ca_3AlSi_6O_{18}$	5	1960	Conventional

ambient conditions can be used as at least a first approximation for the percentage of densification that occurs. Very recently, densities of some high-pressure melts have been directly determined by the X-ray absorption method using synchrotron radiation (Urakawa et al., 2003; Ohtani et al., 2005), which may represent a real breakthrough in this field. Knowledge of how the structure of a melt changes with pressure is essential to obtaining a fundamental understanding of why these macroscopic properties are significantly affected by pressure and how this might affect the buoyancy and ascent rate of mantle melts.

1.2. Aluminosilicate glass and melt structure: background and terminology

To understand the structure of high-pressure glasses and melts, one must first describe their structure at ambient pressure, which begins with defining the network forming cations (e.g. Si and Al) and network modifying cations (e.g. Mg, Ca, Na, K, etc.). The network formers provide the structural framework of the glass and are predominantly tetrahedrally coordinated in aluminosilicate glasses when the molar percentage of modifier oxides (e.g. CaO, K_2O, etc.) are greater than or equal to the molar percentage of alumina (Al_2O_3) (Mysen, 1988). Modifier oxides in excess of the alumina break up this framework and create non-bridging oxygen (NBO). A NBO is an oxygen bonded to only one network former, leaving the network modifiers to balance the remaining valence charge, while bridging oxygen (BO) are bonded to two network formers (e.g. Si–O–Si, Si–O–Al, Al–O–Al). The mixing of Si and Al (and its accompanying charge balancing modifier) is not random in aluminosilicate glasses, because there is an energetic penalty for oxygen bonded to a pair of aluminum atoms (aluminum avoidance, Navrotsky et al., 1985). Al–O–Si linkages are generally more energetically favorable than a combination of Si–O–Si and Al–O–Al linkages (Loewenstein, 1954), although the penalty is small enough that the higher energy species can exist in silicate glasses (Stebbins et al., 1997; Lee and Stebbins, 1999, 2000, 2002). Raman and nuclear magnetic resonance (NMR) spectroscopies have illustrated that Si is almost exclusively the host for NBO in aluminosilicate melts (Mysen et al., 1981, 1985; Stebbins et al., 2001; Allwardt et al., 2003a). Therefore, the structural framework of "geologically relevant" depolymerized aluminosilicate glasses (for $SiO_2 > 40$ mol%) generally consists mostly of Si–O–Si and Si–O–Al linkages in tetrahedral coordination where the NBO are located on the Si.

The average NBO/tetrahedron ratio (NBO/*T*) is often taken as a measure of the average network connectivity in melts and glasses. Most natural magmas have an average NBO/*T* ratio between 0 and 1.4 (Mysen, 1988), where lower values represent rhyolites and higher values represent more alkali basalt type compositions. Another way to convey the connectivity of a glass is through the use of the Q^n notation, where *n* is the number of BO bonded to the tetrahedral network former (Mysen, 1988). This notation concentrates on the immediate environment of the tetrahedrally coordinated network former (i.e. Si, Al) and its average value can be determined by the equation: $n = 4 - (\mathrm{NBO}/T)$. Even for integer average values of *n*, however, multiple Q-species have been shown to exist. For instance, if $n = 3$, Q^3 will not be the only Q-species present in the structure, but there will also be Q^4 and Q^2 species (Brawer and White, 1975, 1977; Murdoch et al., 1985; Brandriss and Stebbins, 1988; Zhang et al., 1997). As more network modifiers are added to the melt, the NBO/*T* ratio increases and in general the viscosity decreases due to a decrease in the network connectivity of the melt (Hess et al., 1995, 1996). However, recent work has shown that the use of NBO/*T* and mean Q-species notations is not always appropriate for aluminosilicates as $^{[5]}$Al have been detected in both charge balanced (Stebbins et al., 2000; Toplis et al., 2000; Neuville et al., 2004) and depolymerized glasses (Allwardt et al., 2003b). Additionally, triclusters, which are oxygen bonded to three instead of two tetrahedrally coordinated network formers, have been suggested as a structural species to explain anomalous ambient pressure viscosity data near the $NaAlO_2-SiO_2$ join (Toplis et al., 1997) and $CaAl_2O_2-SiO_2$ (Toplis and Dingwell, 2004).

1.3. Pressure effects on network cation coordination

Known effects of pressure on the structures of crystalline silicates and aluminosilicates led to the suggestion that the most obvious effect of pressure on molten silicates would be an increase in the coordination numbers of network forming cations (Al and Si), although such structural transitions would be expected to be continuous because melts lack the long-range order of crystalline materials (Waff, 1975; Stolper and Ahrens, 1987). This would also suggest that there are not likely to be first-order discontinuities in macroscopic properties with increasing pressure resulting from network former coordination changes in aluminosilicate melts (Stolper and Ahrens, 1987). Recent evidence from experimental work and molecular dynamics simulations confirms that such changes occur in high-pressure silicate melts, which has led to proposing several types of mechanisms for the generation of high-coordinated Si (Wolf et al., 1990; Xue et al., 1991; Diefenbacher et al., 1998). Much of this work is based on data from glasses that were quenched at high pressures and subsequently decompressed. However, one must realize that at best, such results record the structure at the glass transition temperature (T_g) and pressure, not the experimental temperature and pressure, because the large configurational component to the heat capacity (C_P) necessitates that the structure changes with temperature (see below). Also, another potential pitfall of measuring the structure at ambient conditions for glasses quenched from high-pressure melts is the problem of structural relaxation during decompression, (see below). However, even with these experimental difficulties, these studies can help us understand the mechanisms of "permanent" (i.e. quenchable) densification, which must be an important part of the densification of melts at high pressure and temperature. Additionally, they are also likely to identify important, qualitative effects of compositional changes on densification mechanisms that will be useful in comparing different types of magmas.

Most early experimental work on the structure of high-pressure aluminosilicate glasses concentrated on glasses along the $NaAlO_2$–SiO_2 join due to their relatively good glass-forming ability and the low melting temperatures relative to other aluminosilicates (Kushiro, 1976, 1978; Sharma et al., 1979; Mysen et al., 1980, 1983; McMillan and Graham, 1980; Hochella and Brown, 1985). One early study used infrared spectra and aluminum X-ray absorption lines to suggest that the average Al-coordination of $NaAlSi_3O_8$ and $NaAlSi_2O_6$ increases between 1 atm and 3 GPa (Velde and Kushiro, 1978); however, further work in these systems with Raman and X-ray spectroscopies observed no evidence to support this increasing Al-coordination at pressures to 4 GPa (Sharma et al., 1979; Mysen et al., 1980, 1983; McMillan and Graham, 1980; Hochella and Brown, 1985). More recently, Al–K edge XANES studies of several high-pressure (4.4 GPa) glasses along the $NaAlSi_3O_8$–$NaAlSi_2O_6$ join has been used to suggest that high-coordinated Al exist in these glasses (Li et al., 1996). Al-27 magic-angle spinning (MAS) nuclear magnetic resonance (NMR) spectra (9.4 and 11.7 T) of glasses quenched from a liquid at 8 and 10 GPa consisted of a tetrahedral Al peak that appeared to contain intensity in the "tail" at approximately 25 and 0 ppm which were used to suggest the presence of $^{[5]}$Al and $^{[6]}$Al species, respectively (Stebbins and Sykes, 1990). A recent ^{27}Al triple-quantum (3Q) MAS spectrum (7.1 T) confirmed the presence of $^{[5]}$Al in Ab glasses quenched at 8 GPa (Lee, 2004) while a very recent ^{27}Al MAS (18.8 T) study has quantified the amount of $^{[5]}$Al (15%) and $^{[6]}$Al (6%) in the 10 GPa glass (Allwardt et al., in press-b). Additionally, results from molecular dynamics simulations have also long suggested that the average coordination of network formers increases in high-pressure melts with increasing pressure (Angell et al., 1982, 1987; Nevins and Spera, 1998; Suzuki et al., 2002).

Five- and six-coordinated silicon ($^{[5]}$Si and $^{[6]}$Si) were first definitively observed in alkali silicate glasses quenched from high-pressure melts by Stebbins and McMillan (1989) and Xue et al. (1989). In subsequent Raman and ^{29}Si NMR studies, high-coordinated Si have been suggested to be generated by the mechanisms:

$$Q^3 + Q^4 = {}^{[5]}Si + Q^{4*} \tag{1}$$

$$2Q^3 + Q^4 = {}^{[6]}Si + 2Q^{4*} \tag{2}$$

(Wolf et al., 1990; Xue et al., 1991), where Q^{4*} is a $^{[4]}$Si with three $^{[4]}$Si and one $^{[5]}$Si or $^{[6]}$Si neighbor. In this model, NBO associated with a Q^3 species are consumed to form high-coordinated silicon ($^{[5]}$Si, $^{[6]}$Si) and Q^3 and Q^4 species are transformed to Q^{4*} species during this process. The work of Xue et al. (1991) and Stebbins and McMillan (1993) showed that glass compositions with an average Q-species of 3.5 (NBO/$T = 0.5$) contain the most high-coordinated Si, presumably because these compositions contain large amounts of Q^3 and Q^4 species. If mechanisms 1 and 2 are put in terms of the oxygen environment, they become:

$$\mathrm{NBO} + 4\mathrm{BO} \rightarrow 5\ {}^{[4]}\mathrm{Si{-}O{-}}{}^{[5]}\mathrm{Si} \tag{3}$$

$$2\mathrm{NBO} + 4\mathrm{BO} \rightarrow 6{}^{[4]}\mathrm{Si{-}O{-}}{}^{[6]}\mathrm{Si} \tag{4}$$

This approach recently confirmed that mechanisms 3 and 4 create the $^{[5]}Si$ and $^{[6]}Si$ in potassium tetrasilicate composition ($K_2Si_4O_9$) by combining ^{17}O 3QMAS and ^{29}Si MAS NMR spectra of quenched high-pressure and ambient-pressure glasses, where the loss of NBO corresponded to the increase in the fraction of $^{[5]}Si$ and $^{[6]}Si$ (Allwardt et al., 2004). Consequently, the decrease in the percentage of NBO would also likely increase the coordination number of the modifier cations due to the replacement of higher charge density bonds to NBO with lower charge density bonds to BO. Five- and six-coordinated Si were also observed in a high-pressure $CaSi_2O_5$ glass (Stebbins and Poe, 1999). Additionally, a very recent study used *in-situ* high-temperature and -pressure X-ray diffraction of $CaSiO_3$ and $MgSiO_3$ melts up to 6 GPa to show that changes occur in Ca^{2+} and Mg^{2+} environments and the intermediate range order (Funamori et al., 2004). These authors then stated that these were likely the dominant compression mechanisms, while the increase in the Si-coordination is the secondary mechanism for melts of this composition and experimental conditions.

In aluminosilicate melts an additional complication is introduced as network Al and Si can both respond to pressure by increasing coordination and may, in a sense, "compete" with each other. Poe et al. (2001) used Raman and XANES spectroscopies to investigate compression mechanism(s) in aluminosilicate glasses and found evidence to suggest that there are two separate mechanisms. A loss of intensity in the low frequency region of the Raman spectra for the $44CaO-12Al_2O_3-44SiO_2$ glasses was used to suggest that at pressures below 6 GPa the oxygen was increasing its average coordination by either forming triclusters or having Ca^{2+} cations inhibiting the vibrations of the T–O–T species. At higher pressures, the increase in intensity in the 1570 eV region of the XANES spectra was used to suggest that either the T–O–T angle was narrowing or there was an increase in the average Al-coordination. Known effects of pressure on crystal structures led to the suggestion that AlO_6 appears at lower pressures than SiO_6 species in high-pressure melts (Waff, 1975), which was later confirmed by ^{27}Al and ^{29}Si MAS NMR of $Na_3AlSi_7O_{17}$ glasses (Yarger et al., 1995). The data from the glasses of Yarger et al. (1995), when compared to the results from a metaluminous composition ($NaAlSi_3O_8$) of Stebbins and Sykes (1990), clearly demonstrates that more $^{[5]}Al$ and $^{[6]}Al$ occur in depolymerized NAS aluminosilicate glasses, which suggests that the presence of NBO somehow facilitates the generation of $^{[5]}Al$ and $^{[6]}Al$. As mentioned previously, this agrees with previous studies of silicates where NBO have been shown to be important for the generation of $^{[5]}Si$ and $^{[6]}Si$ (Xue et al., 1991; Allwardt et al., 2004). Si–NBO (Allwardt et al., 2003a) and Si–O–Al (Loewenstein, 1954; Lee and Stebbins 1999, 2000) species are the energetically favorable structural species in depolymerized aluminosilicate glasses, which, combined with the observation that the generation of high-coordinated Al is dependent on the number of NBO in the system, led Allwardt et al. (2004) to propose that the mechanisms that create high-coordinated Al are most likely:

$$\text{Si–NBO} + 4\text{Si–O–Al} \rightarrow 5\,^{[4]}\text{Si–O–}^{[5]}\text{Al} \quad (5)$$

$$2\text{Si–NBO} + 4\text{Si–O–Al} \rightarrow 6\,^{[4]}\text{Si–O–}^{[6]}\text{Al} \quad (6)$$

As in the silicate glasses, these mechanisms would likely increase the average coordination number of the modifer cations. Even though mechanisms 5 and 6 are reasonable for depolymerized glasses, they may not be valid for charge-balanced aluminosilicate glasses due to the lack of an adequate percentage of NBO to drive the mechanism (Stebbins et al., 1997; Lee and Stebbins, 2000). Recently, two different mechanisms were proposed for charge-balanced systems (Allwardt et al., in press-b). The first involves the formation of triclusters by the mechanism proposed by Toplis et al. (1997) for ambient-pressure glasses to create the NBO that could then be used in mechanisms 5 and 6. The other possibility forces a BO into the coordination sphere of an AlO_4 species to form the high-coordinated Al. Both mechanisms yield a $^{[3]}O$, but the $^{[3]}O$ in the "tricluster mechanism" is bonded to three tetrahedral network formers while the other mechanism results in a $^{[3]}O$ bonded to one or more high-coordinated Al.

Despite being possibly the most applicable system to geologic melts due to the percentage of Al and the NBO/*T* ratio (Mysen, 1988), relatively little work has been done on determining the structure of high-pressure aluminosilicate glasses containing significant concentrations of NBO. An early study used changes in a portion (850–1300 cm^{-1}) of Raman spectra to confirm that the reaction: $2Q^3 = Q^2 + Q^4$, favored more disorder (right) at high pressures (Mysen, 1990) as suggested previously by ^{29}Si NMR on a Na-silicate glass quenched at pressure (Xue et al., 1989). The NBO/*T* ratio at ambient conditions has been shown to be an important factor in the average Al-coordination in high-pressure glasses where more NBO in the ambient-pressure glass seems to yield more high-coordinated Al (Yarger et al., 1995; Lee, 2004; Allwardt et al., 2004). However, more studies are needed on high-pressure aluminosilicate glasses to determine whether an initial average Q-species of 3.5 (NBO/$T = 0.5$) yields a maximum in the amount of high-coordinated Al as is the case for Si in Al-free compositions (Wolf et al., 1990; Xue et al., 1991). A very recent ^{27}Al MAS NMR (18.8 T) study (Allwardt et al., 2003c) managed to isolate the effect of the type of modifier cation in high-pressure $M^{x+}_{(9-3x)}Al_2Si_6O_{18}$ glasses ($M = Ca^{2+}$, Na^+, or K^+) and clearly showed that the cation type significantly affects the average Al-coordination (Fig. 1), where the alkaline earth aluminosilicates contain more high-coordinated Al than the alkali aluminosilicates. Additionally, this study directly showed that the Al-coordination begins to increase at pressures as low as 3 GPa, which was the minimum pressure explored. These findings suggest that actual mantle melts may contain significant amounts of $^{[5]}Al$ and $^{[6]}Al$ species at the modest pressures relevant to typical mantle melting, which would mean that density predictions based on values for the bulk modulus and its first pressure derivative measured at ambient conditions are problematic as they would not account for this structural change and could systematically underestimate the density.

1.4. Effects of temperature and thermal history on structure

Because melting temperatures of melts can be over 1000°C higher than the glass transition temperature in high-pressure aluminosilicate melts, determining the effect of temperature on melt structure is essential to extrapolating data from glasses to the structure of silicate liquids. NMR and molecular dynamics simulations of ambient pressure, NBO-containing aluminosilicate glasses and melts have suggested that the average Al-coordination increases with increasing temperature (Stebbins and Farnan, 1992; Poe et al., 1993, 1994). Recently, this work was extended to high-pressure (10 GPa) $NaAlSi_3O_8$ (Ab) and

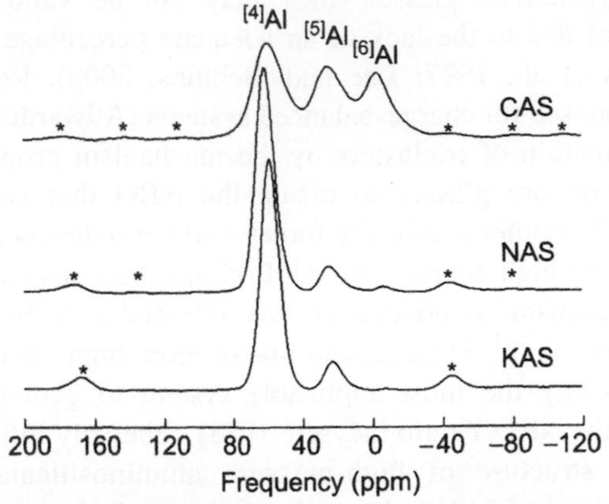

Figure 1. Background-subtracted ^{27}Al MAS NMR (18.8 T) showing how the choice of modifier cation affects the percentage of high-coordinated Al in $M^{x+}_{(9-3x)}Al_2Si_6O_{18}$ glasses quenched from the melt at 5 GPa (Allwardt et al., 2003c). The sample names indicate the modifier cation used in the aluminosilicate glass (e.g. CAS = calcium aluminosilicate). * denotes a spinning sideband.

$Na_3AlSi_7O_{17}$ (NAS) glasses (Fig. 2; Allwardt et al., in press-b) by synthesizing glasses that represent slightly different temperatures. The spectra for the $Na_3AlSi_7O_{17}$ glasses showed that the amount of $^{[5]}$Al and $^{[6]}$Al increases with increasing temperature, which is in agreement with previous ambient-pressure studies of depolymerized melts and glasses.

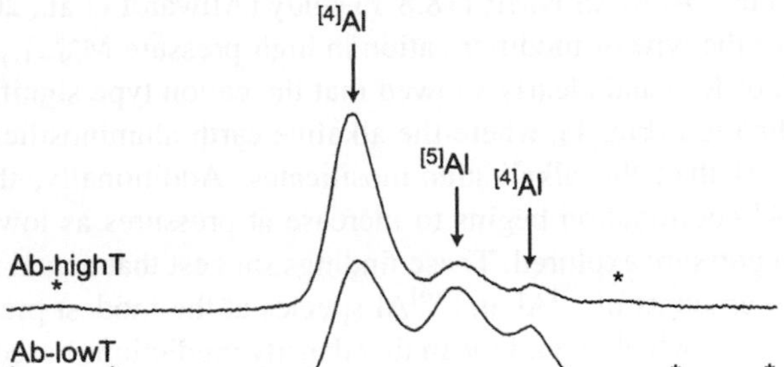

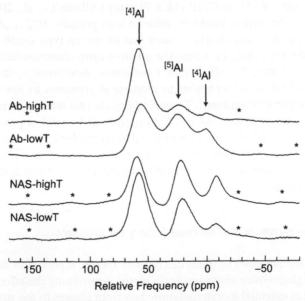

Figure 2. Background-subtracted ^{27}Al MAS NMR spectra (18.8 T) showing the effect of temperature on the average Al-coordination in $NaAlSi_3O_8$ (Ab) and $Na_3AlSi_7O_{17}$ (NAS) glasses at 10 GPa (Allwardt et al., in press-b). The high-temperature samples were quenched from the liquid (2200°C) while the low-temperature glasses were annealed for 12 h near the glass transition temperature (630 and 475°C, respectively).

Since the T_g is below the melting temperature and the Al-coordination of depolymerized glasses increases with temperature, it appears that depolymerized glasses quenched from the high-pressure liquid represent a minimum value for the percentage of high-coordinated Al species present in high-temperature melts. However, the NMR spectra indicate that $NaAlSi_3O_8$ glasses show the opposite effect at 10 GPa, as the amount of $^{[5]}Al$ and $^{[6]}Al$ drastically decreases with increasing temperature (Fig. 2). Further experiments are required in this area to determine if this is an anomalous system or if other charge-balanced glasses behave similarly to $NaAlSi_3O_8$.

1.5. The effect of decompression on the structure of high-pressure glasses

Another important factor to consider when relating the structure of high-pressure glasses to the structure of high-pressure melts is whether relaxation occurs during the decompression of high-pressure glasses, which would affect the structure that can be measured at ambient conditions. Two different approaches have utilized Raman spectroscopy, combined with the diamond anvil cell, to measure the effect of compression and decompression in silicate glasses by collecting spectra of the sample at the pressure of interest and collecting another spectrum after decompression. The first method involved compressing an ambient pressure glass to the desired pressure and measuring the spectrum without ever heating the sample (Hemley et al., 1986; William and Jeanloz, 1988; Wolf et al., 1990; Sharma et al., 1996). The results of this method have been extensively reviewed by Wolf and McMillan (1995) and, for the most part, are beyond the scope of this brief review. The second method involves annealing the glass at T_g (and pressure) and collecting a spectrum (Farber and Williams, 1996; Robinson, McMillan and Wolf, in McMillan and Wolf, 1995, Figure 25). The compression mechanisms occurring during the two types of experiments seem to be similar (Wolf et al., 1990; Xue et al., 1991). However, the NMR data (Yarger et al., discussed in Wolf and McMillan, 1995, p. 543) and Raman spectra of decompressed glasses clearly show that glasses that were annealed at high pressure or quenched from a high pressure melt (Farber and Williams, 1996; Robinson, McMillan and Wolf, in McMillan and Wolf, 1995, Figure 25; Yarger et al., discussed in Wolf and McMillan, 1995, pp. 543) are structurally distinct from those that were compressed at room temperature only. This is not surprising as the increased temperature of the latter method would provide additional energy to allow structural transitions to occur. Although the exact amount of structural relaxation that occurs during decompression is still uncertain and debated, it is obvious that the measured structures of decompressed high-pressure glasses are different from ambient pressure glasses (before densification) indicating that at least some of the high-pressure melt structure is retained. Additionally, it is extremely difficult to quantify the amount of high-coordinated species with Raman and infrared (IR) spectroscopies due to the strong overlapping character of bands and the uncertainties in peak assignments. For this purpose NMR spectroscopy is much more robust in the quantification of the structure of high-pressure glasses than Raman or IR spectroscopies, but has the disadvantage that it currently cannot be applied *in-situ* at high pressure to study inorganic glasses (or melts at high temperature).

In a recent ^{29}Si NMR and Raman spectroscopic study on high-pressure $K_2Si_4O_9$ glasses, Allwardt et al. (2004) approached the problem of determining the decompression effects on the glass structure by performing multi-anvil experiments at 5.7 GPa using two different decompression times (14 h and 1 s). The ^{29}Si MAS NMR spectra of the recovered samples suggest that a small fraction of $^{[6]}Si$ reverts to $^{[5]}Si$ (~1% of the total Si) during the slower decompression. The effect of varying decompression rates was also visible in the Raman spectra where the spectrum of the rapidly decompressed glass resembled that of a conventionally decompressed glass synthesized at 8 GPa. More importantly, the differences between the Raman spectra of the conventionally and rapidly decompressed glasses (5.7 GPa) mimicked the differences between the Raman spectra collected *in-situ* and that collected on the decompressed glass (Robinson, McMillan, and Wolf, in McMillan and Wolf, 1995, Figure 25). Structural determination at ambient conditions offers the opportunity for a much more robust quantitative approach with fewer assumptions as there are fewer spectroscopic uncertainties compared to measurements at extreme conditions, such as temperature and pressure effects on IR absorption coefficients and Raman scattering cross-sections (McMillan and Wolf, 1995), and anharmonic vibrations (Brown et al., 1995; Daniel et al., 1995; McMillan and Wolf, 1995).

1.6. NMR and structural studies of high-pressure glasses

Most recent work on the structure of high-pressure glasses has used solid-state NMR as it is quantitative and nuclide specific, so it can measure percentages of specific structural species surrounding specific types of atoms (e.g. Al, Si, O, etc.). One difficulty of early NMR investigations of high-pressure aluminosilicate glass structure was that ^{27}Al is a quadrupolar nuclide, so peak widths and positions are dependent on the external magnetic field used. Al-27 MAS spectra from lower fields are difficult to quantify due to this effect as their interpretation can depend on accurately representing the unconstrained, overlapping peaks and their low-frequency tails for the different Al-coordinations (Stebbins et al., 2000; Gan et al., 2002). However, technological advances in high-field NMR spectroscopy have greatly increased the available external magnetic fields and sample spinning rates, which allows a more reliable approach to quantifying ^{27}Al MAS NMR spectra as the peaks are more Gaussian in shape and contain a significantly smaller "tail" that is not as prone to overlap as those observed in spectra from lower fields (Stebbins et al., 2000). Figure 3 shows how the external magnetic field affects the spectra as even the peaks in the relatively high field spectrum collected at 14.1 T are overlapping and difficult to quantify compared to the spectrum from 18.8 T. Aluminum speciation can also be quantified at lower fields by utilizing the narrower peak widths of satellite transitions (Jäger, 1994). However, as can be seen in Figure 3, the relative intensity of the spinning sidebands is significantly smaller than that of the central transition, which makes collecting spectra with an adequate signal to noise ratio much more time consuming, especially for small, high-pressure samples. Another approach to quantifying Al-species is to use ^{27}Al triple-quantum (3Q) MAS NMR as this technique produces a two-dimensional spectrum where the second dimension is free of second-order quadrupolar broadening, which allows peak fitting of the projections of the second dimension. The quantification of 3QMAS NMR spectra, however, is dependent on the quadrupolar coupling constant (C_Q) of the structural

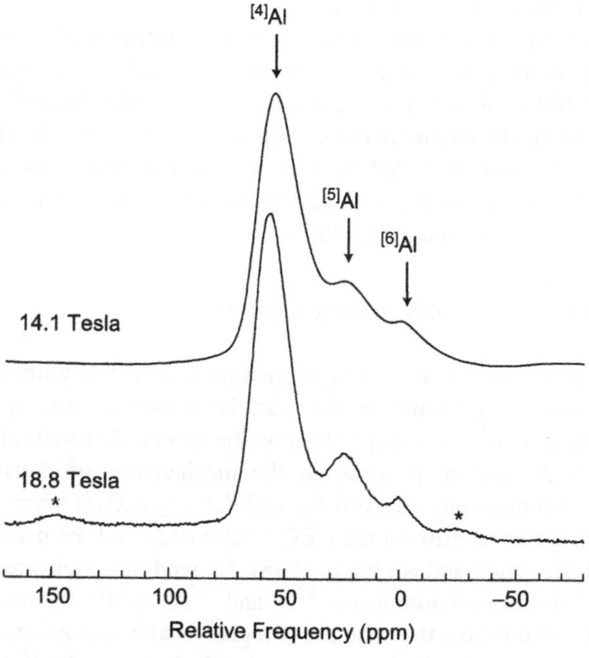

Figure 3. Background-subtracted ^{27}Al MAS NMR spectra illustrating how the external magnetic field affects the spectra for a high-pressure 10 GPa $NaAlSi_3O_8$ glass that was quenched from a melt (Allwardt et al., in press-b).

species, which is dependent on the local charge symmetry (Baltisberger et al., 1996). This does not appear to be a significant issue for ^{27}Al NMR as the C_Q of the relevant species ($^{[4]}Al$, $^{[5]}Al$, $^{[6]}Al$) are similar. However, more work is required to verify that small differences in the C_Q of the different Al species is not introducing systematic quantification errors for ^{27}Al 3QMAS spectra. The major advantage of 3QMAS is that it offers an alternative to high magnetic field spectrometers and allows at least a semi-quantitative treatment of ^{27}Al spectra with "low" field spectrometers; however, it takes significantly more time to collect a spectrum with an adequate signal to noise ratio to detect small percentages of high-coordinated Al.

O-17 NMR has been used extensively to investigate the structure of ambient-pressure glasses; however, the use of ^{17}O NMR to determine the structure of high-pressure glasses is still in its infancy (Xue et al., 1994; Lee et al., 2003; Lee, 2004; Allwardt et al., 2004). As a result, relatively little is known about peak assignments of species that are not present in ambient-pressure glasses. Xue et al. (1994) used a $K_2Si_4O_9$–wadeite crystalline model compound to aid in peak assignments for high-pressure glasses and found that the peak location of the $^{[4]}Si–O–^{[6]}Si$ species overlaps with the peak for $^{[4]}Si–O–^{[4]}Si$ species. This peak location was, however, well resolved in the ^{17}O 3QMAS spectrum (Allwardt et al., in 2004). Additionally, to further constrain peak assignments in ^{17}O NMR for high-pressure silicate glasses, a 3QMAS spectrum of the crystalline $CaSi_2O_5$-triclinic phase was collected to determine how oxygen bonded to $^{[5]}Si$ are represented in ^{17}O NMR spectra. This spectrum clearly demonstrated that the $^{[4]}Si–O–^{[5]}Si$ species do not show a distinct signature, but appear to be either "$^{[4]}$Si–O–$^{[4]}$Si-like" or "$^{[4]}$Si–O–$^{[6]}$Si-like," depending on the local environment of the oxygen (Allwardt et al., 2004). This strongly indicates that Si-coordination, and presumably Al-coordination, cannot be reliably inferred from ^{17}O NMR, but must be measured directly with ^{29}Si or ^{27}Al NMR.

We also note that the development of new types of small volume NMR probes, coupled with the higher sensitivity provided by higher field magnets, have greatly improved obtainable signal to noise ratios for the small samples typically made in multi-anvil high-pressure apparatus (Stebbins and Poe, 1999).

1.7. Motivation and methodology of the present study

Here we present new data on several high-pressure Ca- and K-aluminosilicate glasses, to illustrate the type of information that can be obtained from quenched samples. We demonstrate large effects of composition on the extent of structural changes, explore decompression effects, and help constrain the mechanisms of densification in high-pressure glasses. Compositions ($K_3AlSi_3O_9$ and $Ca_3Al_2Si_6O_{18}$) were chosen as simple models of mafic magmas in terms of the NBO/*T* ratio (0.5) and the percentage of alumina (Mysen, 1990; McDonough and Rudnick, 1998). By studying high-pressure glasses with different rates of decompression using ^{17}O and ^{27}Al NMR, it may be possible to mechanistically determine how the structure of a glass is affected by longer decompression times, although not necessarily the total extent of such changes. The ^{17}O 3QMAS and ^{27}Al MAS NMR spectra of these glasses can then be used to quantitatively test proposed mechanisms of decompression. Finally, by comparing the ^{17}O 3QMAS and ^{27}Al MAS spectra of the rapidly decompressed sample to those of the ambient pressure sample, models of the mechanism for the generation of five- and six-coordinated Al in melts at high pressure can be compared to the previously proposed structural compression mechanisms.

2. Experimental procedures

Ambient pressure $K_3AlSi_3O_9$ (KAS) and $Ca_3Al_2Si_6O_{18}$ (CAS) glasses were used as the starting materials for the high-pressure samples. K_2CO_3, CaO, Al_2O_3, and SiO_2 were used and all oxides and carbonates were dried at 1000 and 250°C, respectively, prior to mixing. In addition, 0.2 wt% cobalt oxide was added to the oxide powders to speed spin-lattice relaxation in the NMR experiments. The powder for the KAS glass was held at 750°C for about 16 h to decarbonate the sample prior to melting. The ambient-pressure glasses were made in 400 mg batches by dipping the bottom of the Pt crucible in water to quench the melt into a glass. The glasses were also isotopically enriched by replacing conventional SiO_2 with 70% ^{17}O-enriched SiO_2, which was synthesized as has been previously reported in Stebbins et al. (1997). All steps that involved the heating of ^{17}O-enriched materials were done in argon to minimize ^{17}O exchange with the environment.

The high-pressure glasses were synthesized in a multi-anvil "Kawai-type" apparatus (1000 ton) at Bayerisches Geoinstitut, Bayreuth, Germany (Rubie, 1999). A 25/17 assembly (edge lengths of the MgO octahedron/truncated edge lengths of the tungsten carbide anvils) with a $LaCrO_3$ stepped furnace was used. This assembly is similar to that shown in Figure 2 of Frost et al. (2004) where the main difference is that a Re-capsule was used in place of MgO. The 25/17 assembly was chosen to maximize sample volume and allow the synthesis of two glasses per multi-anvil experiment. Two rhenium-foil capsules were stacked on top of each other, each with an outer diameter of 4 mm and a length of about 2 mm. The "high" experimental temperatures required the use of Re-capsules as 1960°C is beyond the useful range of Pt at 5 GPa. Each experiment consisted of an initial compression to the desired pressure (at room temperature) followed by heating at a rate of 200–300°C min^{-1} to the final experimental temperature (Table 1), which was controlled using a D-type thermocouple located at the top of the sample capsule. The experiments were held at the experimental temperatures for 1–5 min until the heater was turned off, which isobarically quenched the melt to a glass within a few seconds. The P calibration was done at 1200°C using the quartz-coesite (3.23 GPa, Hemingway et al., 1998) and the $CaGeO_3$ garnet–perovskite (5.94 GPa, Susaki et al., 1985) transitions. Two or more 5 GPa glasses were synthesized for each composition (KAS and CAS), where the rate of decompression was varied (~14 h-standard and ~1 s-rapid). Rapid decompression resulted from opening an electric valve in the oil pressure system (Langenhorst et al., 2002) after the temperature had dropped below 100°C. To investigate the best-case reproducibility of the experiments, two CAS glasses (CAS5R1 and CAS5R2) were made in the same run to ensure that the experimental conditions were nearly identical. Sample names and experimental conditions are shown in Table 1 and consist of the composition (KAS or CAS), the pressure (1 atm or 5 GPa) and the rate of decompression (rapid or standard). For instance, a KAS glass that was synthesized at 5 GPa and rapidly decompressed is referred to as KAS5R. The exception is a glass that was compressed to 6 GPa, but never heated, which was named KAS6 (Table 1). All high-pressure glasses were found to be entirely amorphous when examined with a 400X petrographic microscope, which means that any changes in the ^{27}Al MAS spectra can thus be attributed to structural changes in the glass.

The ^{17}O MAS and 3QMAS NMR spectra were collected on a Varian Unity/Inova 600 spectrometer (14.1 T) while the ^{27}Al MAS spectra were collected on a Varian 18.8 T spectrometer operated by the Stanford Magnetic Resonance Laboratory. Both used a Varian/Chemagnetics "T3" probe with 3.2 mm zirconia rotors. The ^{17}O and ^{27}Al frequencies are reported relative to H_2O and $Al(NO_3)_3$, respectively. The ^{27}Al MAS experiments used a single pulse acquisition with pulse widths corresponding to approximately 30° "solid" radiofrequency (rf) tip angles (approximately 1 μs) and 0.1 s delays between pulses were used to optimize the signal to noise ratio. No differential relaxation was observed in the reported spectra relative to experiments with longer delay times. A typical ^{27}Al MAS NMR experiment consisted of about 15,000 transients while samples were spun at 22 kHz to separate the first spinning sideband of the tetrahedral Al peak from the octahedral Al peak. The ^{17}O MAS and 3QMAS experiments were conducted at 14.1 T with samples spinning at 20 kHz. The MAS experiments used pulse widths of 0.25 μs and delays of 1 s between pulses. The 3QMAS spectra were collected with the same experimental equipment as the ^{17}O MAS experiments and a rf power of 145 kHz. A shifted echo pulse sequence was used, consisting of two hard pulses (3.0 and 1.0 μs, respectively) followed by a soft pulse (21 μs) (Massiot et al., 1996). Delay times of 8 s were used between acquisitions and the ^{17}O 3QMAS data were processed using the software package, RMN (FAT) (P.J. Grandinetti, The Ohio State University).

The resulting 3QMAS spectra are two-dimensional plots in which the isotropic dimension is free of second order quadrupolar broadening and the projection of the MAS dimension yields a spectrum similar to that measured by conventional 1-D MAS NMR.

High-pressure glass samples are small by MAS NMR standards so differentiating between the signal resulting from the glass and that from the rotor becomes critical to accurately interpreting the spectra. The zirconia rotors used here contain trace amounts of $^{[6]}Al$ impurities that create peaks in the ^{27}Al MAS spectra. The signal and intensity is rotor-specific so only two rotors were used to minimize the time devoted to collecting rotor backgrounds. Signal resulting from the rotor ranges from -5 to 20 ppm in the ^{27}Al MAS spectra with the majority of the intensity centered at about 0 ppm. To remove this signal and not introduce additional noise to the glass spectra, 75,000 transients were collected for both rotors. These were subtracted from the glass spectra to yield background subtracted ^{27}Al MAS NMR spectra of the high-pressure glasses. These spectra were then fit with 2 Gaussian peaks for each of the Al-coordinations (4, 5, and 6) to approximate the slight non-Gaussian peak shapes, which are apparent even at 18.8 T. Oxygen-17 MAS NMR does not have the same difficulties as ^{27}Al NMR because ^{17}O is of low natural abundance and the zirconia rotor background (385 ppm, Stebbins, 1995) is well outside the typical chemical shift ranges for aluminosilicate glasses (0–150 ppm). However, for experiments where there is either little sample or low ^{17}O-enrichment, one should keep in mind that the first spinning sideband of this narrow peak can lie inside the range of interest. For instance, spinning the sample at 20 kHz with a 14.1 T spectrometer results in the first spinning sideband of the zirconia rotor at about 130 ppm, which is approximately the location of a Si–NBO coordinated by Sr^{2+} (Allwardt, unpublished data).

3. New results

The background-subtracted ^{27}Al MAS NMR spectra (18.8 T) of the four $Ca_3Al_2Si_6O_{18}$ glasses are shown in Figure 4 and the results of peak fitting can be seen in Table 2. The spectrum of the ambient-pressure glass (CAS0) is dominated by a peak centered at 58 ppm, which coincides with the region of the spectrum that is commonly assigned to fully polymerized (Q^4) tetrahedral Al ($^{[4]}Al$) in aluminosilicates (isotropic chemical shifts of approximately 56–64 ppm, Stebbins, 1995). A small unresolved shoulder peak (~2%) centered near 25 ppm is present. This suggests that there is $^{[5]}Al$ present at ambient pressures as chemical shifts for five- or six-coordinated Al are in the ranges of 30–45 ppm and 0–16 ppm, respectively, which would yield expected peak maxima a few ppm lower for the external magnetic fields (18.8 T) used in this study (Du and Stebbins, 2005). Three CAS glasses were synthesized at 5 GPa. Two were decompressed rapidly (CAS5R1 and CAS5R2) while the other was decompressed more slowly over 14 h (CAS5S). Pressure obviously affects the aluminosilicate glass structure as the spectra show that the $^{[4]}Al$ peak shifts to 62 ppm and there is an increases in the intensity of the five- (25 ppm) and six-coordinated peaks (-2 ppm) (Fig. 4). Peak areas derived from the two spectra show that the rapidly decompressed glasses contain between 36–38% $^{[5]}Al$ and 28–31% $^{[6]}Al$, which indicates that the average Al-coordination increases from 4.0 to about 5.0 (Table 2). If all Si are assumed to be tetrahedrally coordinated, as suggested by the ^{29}Si MAS spectrum of a $Ca_3AlSi_6O_{18}$ glass synthesized at an even higher pressure (8 GPa, Allwardt et al., in press-a),

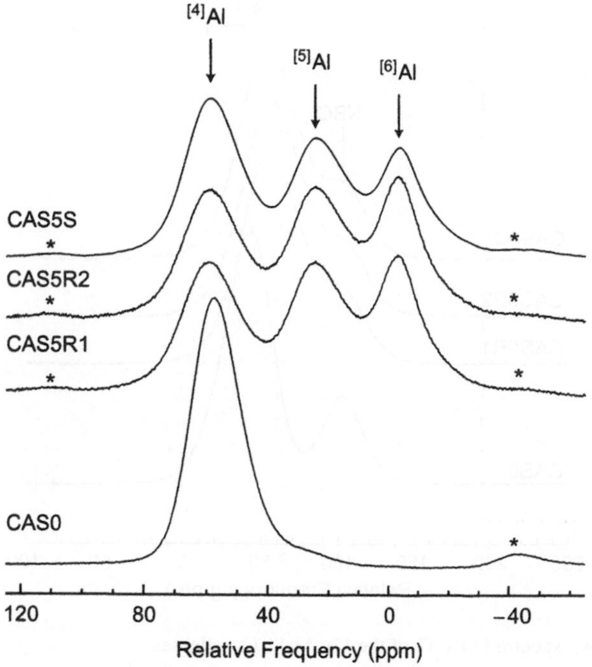

Figure 4. Background-subtracted ^{27}Al MAS spectra (18.8 T) of the $Ca_3Al_2Si_6O_{18}$ glasses of this study. Experimental conditions and compositions are shown in Table 1.

the average coordination number of the network formers increases from 4.0 to 4.2, respectively (Table 2). Figure 4 also shows that less $^{[5]}Al$ and $^{[6]}Al$ are present in the CAS5S glass (31 and 22%, respectively) than the CAS5R glasses, suggesting that some high-coordinated Al are lost during the slower decompression.

The ^{17}O MAS NMR spectra (14.1 T) of the four CAS glasses are shown in Figure 5. Previous ^{17}O NMR work on ambient pressure Ca-aluminosilicate glasses indicates that the

Table 2. Results from peak fitting the ^{27}Al MAS NMR spectra. Columns show the relative percentages of Al species and mean coordination numbers for both Al and total network formers (Al + Si).

Sample	$^{[4]}Al$	$^{[5]}Al$	$^{[6]}Al$	Average Al	Average (Si + Al)
KAS0	99.7	0.3	0	4	4
KAS5R	93.6	6.4	<0.4	4.09	4.02
KAS5S	87.1	12.2	0.7	4.14	4.03
KAS6	94.5	5.5	<0.4	4.06	4.01
CAS0	98	2	0	4	4
CAS5R1	33	37	30	5	4.2
CAS5R2	33	39	28	4.9	4.2
CAS5S	47	31	22	4.8	4.2

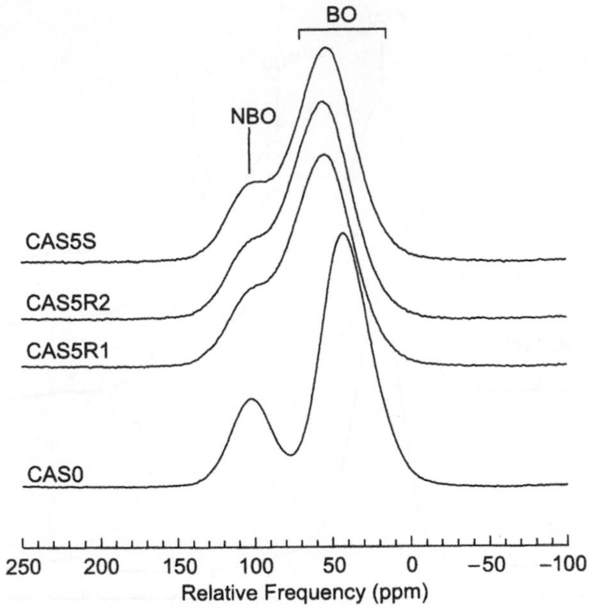

Figure 5. O-17 MAS spectra (14.1 T) of the $Ca_3Al_2Si_6O_{18}$ glasses.

peak at 102 ppm is due to NBO bonded to both Si and Ca (Stebbins et al., 1997; Allwardt et al., 2003a). The spectrum of the CAS0 glass also contains a composite bridging oxygen (BO) peak, which can be separated into contributions from Si–O–Al and Si–O–Si peaks (Stebbins et al., 1997; Lee and Stebbins, 2002), although they are severely overlapping in the MAS spectrum and hard to resolve. By comparing the spectrum of CAS0 with the spectra of the CAS5R glasses, it is obvious that as pressure is increased the amount of NBO decreases, the BO composite peak decreases and shifts to higher frequency, and additional intensity appears in the region between the BO and NBO peaks (Fig. 5). This additional intensity is likely related to oxygen bonded to high-coordinated Al, but the exact structural interpretation is not obvious. O-17 3QMAS NMR spectra were collected for the CAS glasses (Fig. 6) in an attempt to resolve the BO peaks and the new high-pressure peak(s). Previous work using ^{17}O 3QMAS NMR has shown that there are two predominant, distinct BO species in the ambient-pressure CAS glass: Si–O–Al and Si–O–Si (Stebbins et al., 1997; Lee and Stebbins, 2002). A comparison of the spectra for the CAS0 and CAS5R2 glasses shows that NBO and Si–O–Al species becomes less abundant at high pressure while there is no obvious change in the intensity of the Si–O–Si peak. As is also observed in the MAS spectra, additional intensity appears in the region between the NBO and Si–O–Al peaks in the spectrum of the CAS5R2 glass, but still cannot be resolved with 3QMAS. Figure 6 also shows that when the standard decompression is used, more Si–O–Al species are present in the structure than is seen in the rapidly decompressed glass. The peak position of the NBO observed in the spectra of high-pressure glasses appears similar to that seen at ambient pressure although it is possible that this may be due to overlap with the high-pressure peak(s).

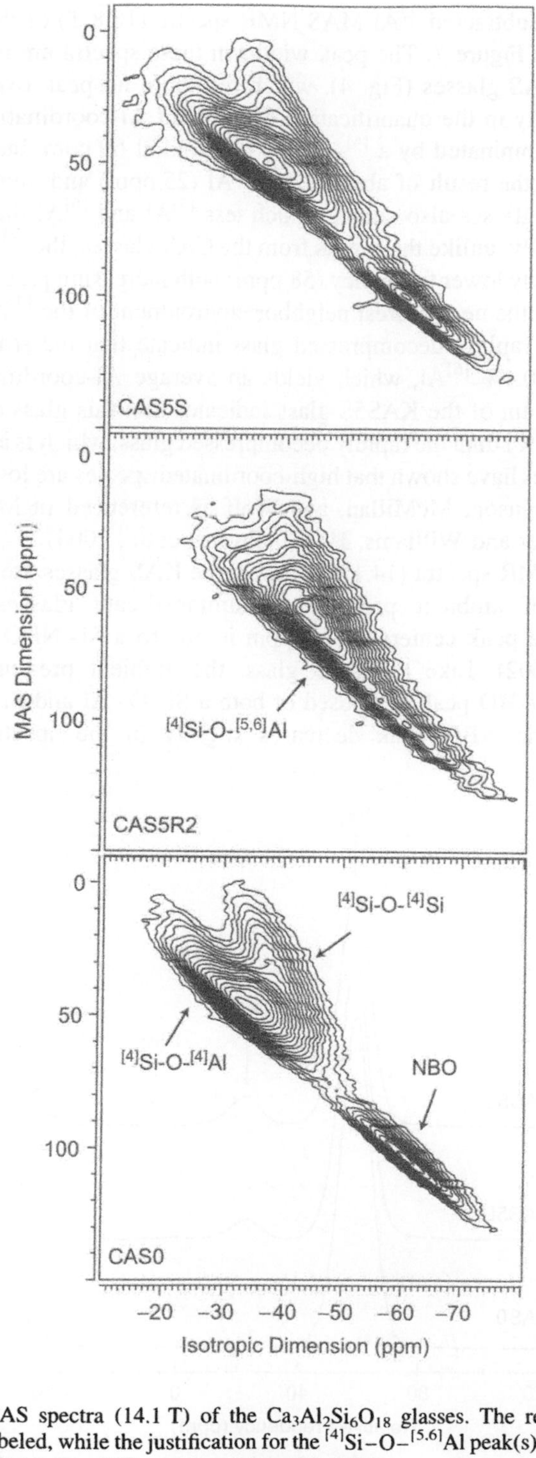

Figure 6. O-17 3QMAS spectra (14.1 T) of the $Ca_3Al_2Si_6O_{18}$ glasses. The regions for the relevant structural species are labeled, while the justification for the $^{[4]}Si{-}O{-}^{[5,6]}Al$ peak(s) is discussed in the text. Contours are drawn at intervals of 5% and range from 8 to 98%.

The background-subtracted ^{27}Al MAS NMR spectra (18.8 T) of the three $K_3AlSi_3O_9$ glasses are shown in Figure 7. The peak widths in these spectra are narrower than those measured for the CAS glasses (Fig. 4), which results in no peak overlap and therefore yields less uncertainty in the quantification of different Al-coordinations (Table 2). The KAS0 spectrum is dominated by a $^{[4]}Al$ peak centered at 60 ppm, but also contains two other peaks that are the result of about 0.3% $^{[5]}Al$ (25 ppm) and about 0.6% crystalline impurity. The KAS glasses also contain much less $^{[5]}Al$ and $^{[6]}Al$ than the CAS glasses (Table 2). Additionally, unlike the results from the CAS glasses, the $^{[4]}Al$ peak of the KAS glass shifts to a slightly lower frequency (58 ppm) with increasing pressure, which is likely due to the change in the next nearest neighbor environment of the $^{[4]}Al$. Peak areas from the spectrum of the rapidly decompressed glass indicate that the structure contains 6% $^{[5]}Al$ and less than 0.4% $^{[6]}Al$, which yields an average Al-coordination of only 4.09. However, the spectrum of the KAS5S glass indicates that this glass contains more $^{[5]}Al$ and $^{[6]}Al$ (12 and 0.7%) than the rapidly decompressed glass, which is an unexpected result as all previous studies have shown that high-coordinated species are lost, not gained during decompression (Robinson, McMillan, and Wolf as referenced in McMillan and Wolf, 1995, pp. 305; Farber and Williams, 1996; Allwardt et al., 2004).

The ^{17}O MAS NMR spectra (14.1 T) of the three KAS glasses are shown in Figure 8. Previous studies of ambient pressure K-aluminosilicate glasses with ^{17}O NMR have shown that the peak centered at 75 ppm is due to a Si–NBO surrounded by K^+ (Oglesby et al., 2002). Like the CAS glass, the ambient pressure KAS glass also contains a composite BO peak composed of both a Si–O–Al and Si–O–Si component. The intensity of the NBO peak decreases slightly in the spectra for the KAS5S

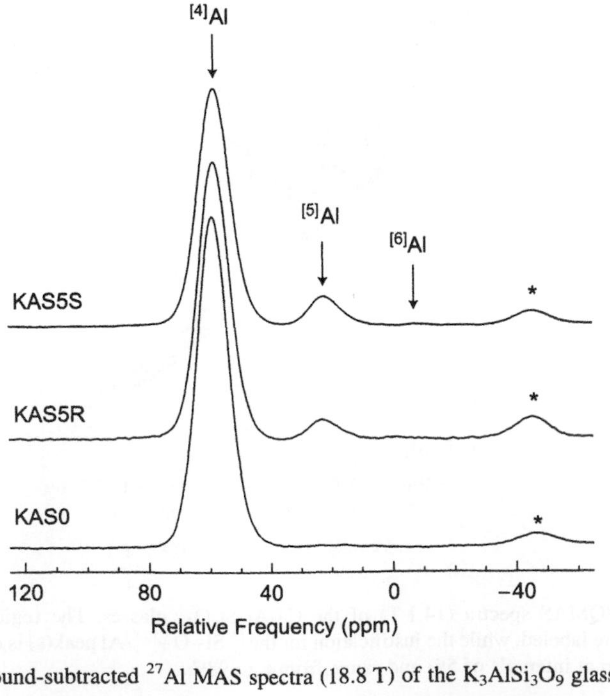

Figure 7. Background-subtracted ^{27}Al MAS spectra (18.8 T) of the $K_3AlSi_3O_9$ glasses of this study.

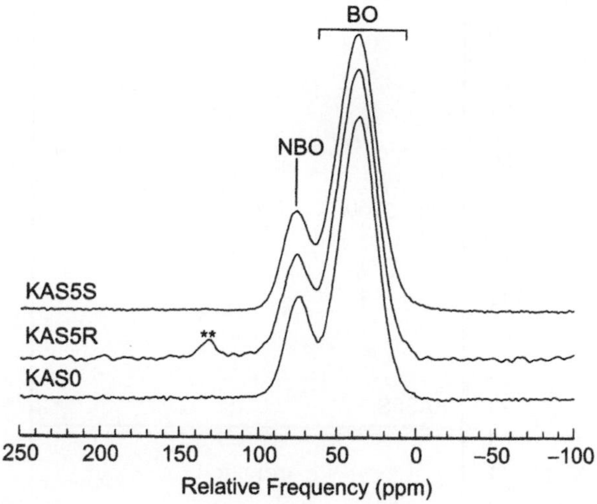

Figure 8. O-17 MAS spectra (14.1 T) of the $K_3AlSi_3O_9$ glasses. ** denotes a spinning sideband of the zirconia rotor.

and KAS5R glasses relative to that for KAS0. A small amount of additional intensity appears between 45 and 70 ppm, presumably due to the presence of oxygen bonded to high-coordinated Al, as was also suggested for the ^{17}O NMR spectra of the CAS glasses. The 3QMAS spectra show that there are two distinct bridging oxygen peaks and an NBO peak of the KAS0 spectrum (Fig. 9) and the spectra of the high-pressure glasses show little change relative to the KAS0 glass. The KAS5R glass lost some of its ^{17}O-enrichment either during synthesis or storage as the signal to noise ratio of the MAS spectrum is significantly reduced relative to the KAS5S glass, even though it represents twice the number of transients. This has been observed to varying degrees in all high-pressure glasses of this study and those of our previous study (Allwardt et al., 2004).

4. Discussion

4.1. *Approximating percentages for unresolved ^{17}O MAS peaks*

The high signal-to-noise ratio and quantitative nature of conventional MAS NMR, combined with the additional resolution available from the second dimension of the 3QMAS spectra can be used to better approximate changes in structure in the CAS and KAS glasses. Although there is overlap of the NBO peak with the high-pressure peak(s) in both the MAS and 3QMAS spectra, it seems reasonable to approximate the percentage of NBO by fitting a single Gaussian to this region (Table 3), as the high-frequency side of the peak is fairly well resolved and the 3QMAS spectra (Figs 6 and 9) suggest that the peak position and width are not different at high pressure. Beyond this, there has not been an adequate amount of work with ^{17}O NMR with crystalline model compounds to reliably assign the additional peak(s) that are exclusive to high-pressure systems. It is apparent that

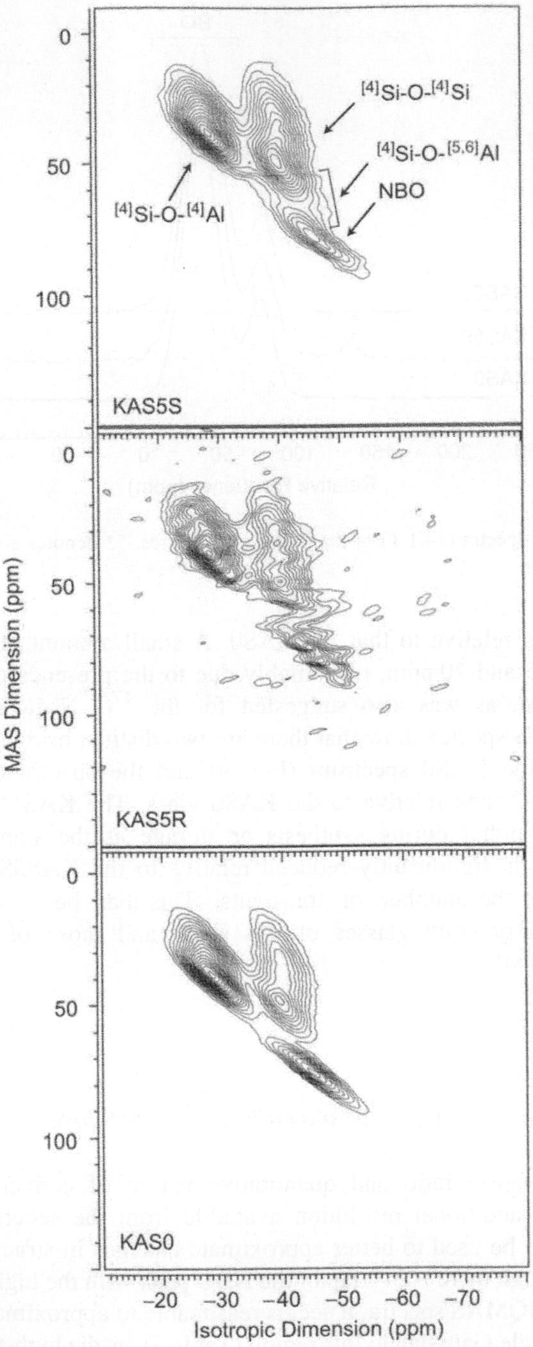

Figure 9. O-17 3QMAS spectra (14.1 T) of the $K_3AlSi_3O_9$ glasses. The $^{[4]}Si-O-^{[5,6]}Al$ region is approximated based on increased intensity in this region and it similarity to that seen in the high-pressure CAS glasses. Contours are drawn at intervals of 5% and range from 8 to 98%.

Table 3. Measured and predicted percentages of oxygen species.

Sample	Measured		Predicted from ^{27}Al data	
	NBO	Minimum Si–O–$^{[5,6]}$Al	$^{[4]}$Si–O–$^{[5]}$Al	$^{[4]}$Si–O–$^{[6]}$Al
KAS0	22	0	0	0
KAS5R	22	4	3	0
KAS5S	22	5	6	0.5
CAS0	23	0	1	0
CAS5R1	15	25	20	21
CAS5R2	14	25	21	19
CAS5S	16	21	17	15

the unresolved intensity between the NBO and BO peaks is related in some way to the introduction of high-coordinated Al into the glass structure, but the exact peak assignment and peak location are not obvious from the available data. To determine whether the high-pressure peak(s) is solely due to $^{[4]}$Si–O–$^{[6]}$Al or if this also includes some or all of the $^{[4]}$Si–O–$^{[5]}$Al species, the ^{17}O MAS spectra of the high-pressure samples were subtracted from the spectra for the 1 atm glasses (Fig. 10). This approach will not yield the percentage of oxygen associated with high-coordinated Al as it does not account for the decreasing intensities of the partially overlapping NBO and $^{[4]}$Si–O–$^{[4]}$Al peaks. Since the percentage of these other peaks decrease with the increase in the percentages of high-coordinated Al,

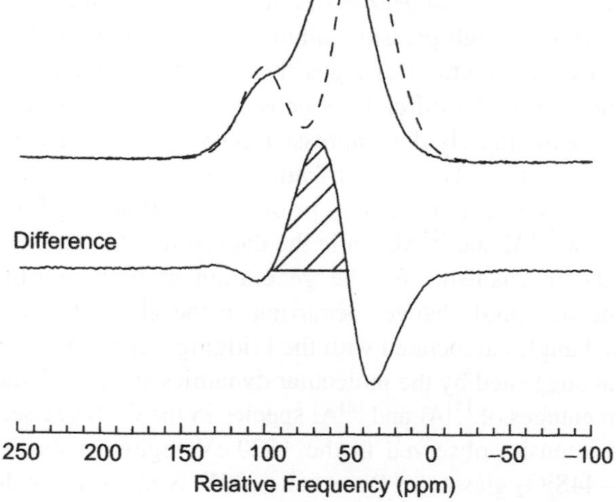

Figure 10. The ^{17}O MAS NMR spectra of CAS1 (dashed) and CAS5R2 (solid) shown to emphasize the differences between the spectra. The spectrum labeled "difference" is the result of subtracting CAS1 from CAS5R2 and the shaded region was used to approximate the minimum value of the $^{[4]}$Si–O–$^{[5,6]}$Al present in the spectrum.

the approximate value obtained from subtracting the spectra likely represents a minimum value for the $^{[4]}Si-O-^{[5,6]}Al$ species. These difference spectra were then fit with one Gaussian peak to approximate the peak area due to oxygen bonded to high-coordinated Al (Table 3). This value can be compared to the value expected from the ^{27}Al MAS results (Table 3). For example, the ^{27}Al MAS spectrum of the CAS5R2 glass shows that 38 and 28% of the aluminum in this glass are five- and six-coordinated, respectively (Table 2) and if one assumes that there are minimal amounts of oxygen that bridge two high-coordinated Al, as suggested by Lee (2004), the ^{17}O NMR spectrum should contain 21% $^{[4]}Si-O-^{[5]}Al$ (38% (percentage of $^{[5]}Al$) $\times$ 5 (coordination number) $\times$ 2/18 (aluminum/oxygen ratio) = 21%) and 19% $^{[4]}Si-O-^{[6]}Al$ (28% $\times$ 6 $\times$ 2/18 = 19%). The minimum percentage of $^{[4]}Si-O-^{[x]}Al$ measured from the difference spectrum is 25%. The predicted values for the percentage of $^{[4]}Si-O-^{[5]}Al$ and $^{[4]}Si-O-^{[6]}Al$ species in the other glasses are also shown in Table 3, along with the approximated percentages of $^{[4]}Si-O-^{[x]}Al$. Since the predicted values of $^{[4]}Si-O-^{[6]}Al$ are less than the measured minimum values for the high-pressure peak for all samples, it seems very likely that the high-pressure peak is due at least to some degree, to the presence of both $^{[4]}Si-O-^{[5]}Al$ and $^{[4]}Si-O-^{[6]}Al$ species. Once again, the ^{17}O spectra measure the immediate environment of the oxygen so it is not directly dependent on the Al-coordination. O-17 spectra for high-pressure crystalline materials like Na- and K-hollandite, ($NaAlSi_3O_8$ and $KAlSi_3O_8$, Gillet et al., 2000; Finger and Hazen, 2000) and andalusite (Al_2SiO_5) as well as quantum mechanical modeling of structures with varying Al–O distances and coordinations (Lee, 2004) will be extremely useful in the refinement of peak assignments for ^{17}O MAS and 3QMAS NMR spectra of high-pressure aluminosilicate glasses.

4.2. Compression mechanisms in high-pressure aluminosilicate melts

Although there are uncertainties in peak assignments and difficulties with peak overlap in the ^{17}O NMR spectra of high-pressure aluminosilicate glasses, they contain useful information for assessing structural changes that occur with increasing pressure. More specifically, by determining the differences between the spectra of the ambient-pressure glasses with those for the rapidly decompressed glasses, it is possible to determine how pressure affects the structure. The results from the KAS glasses may better enable the determination of how the structure changes prior to the formation of high-coordinated Al as it contains very few $^{[5]}Al$ and $^{[6]}Al$, while the data from CAS glasses is more favorable for investigating the mechanisms for the generation of high-coordinated Al in high-pressure melts. The structural changes occurring in the glass, if any, are likely to be a decrease in the bond angles associated with the bridging oxygen and/or an increase in the K^+ coordination, as suggested by the molecular dynamics studies of Matsui (1996, 1998). Also, the large percentages of $^{[5]}Al$ and $^{[6]}Al$ species in the CAS glasses (Fig. 4) indicates that the increased intensity observed in the 1570 eV region in the XANES spectra of $44CaO-12Al_2O_3-44SiO_2$ glasses of Poe et al. (2001) is more likely due to the presence of high-coordinated Al than a narrowing of the T–O–T angle.

O-17 MAS and 3QMAS NMR can measure (or at least approximate) the percentages of Si–O–Al, Si–O–Si, Si–NBO, and $^{[4]}Si-O-^{[5,6]}Al$ species (Table 3) and the comparison of these percentages provides an excellent opportunity to directly test potential compression mechanisms. Analysis of the 3QMAS spectra from the KAS glass suggests that the percentages of Si–O–Al and Si–O–Si show little variation with pressure suggesting that Al-avoidance, at least for glasses in this compositional range, is not drastically affected by pressure. This may indicate that the decreasing T–O–T angle suggested from the Raman spectra of Sykes et al. (1993) is due to the collapse of the random network (Sykes et al., 1993). However, the depolymerized composition of this study is not an ideal system to investigate the pressure dependence of Al-avoidance due to the low percentages of Al. This indicates that the variation in the amount of Si–O–Al and Si–O–Si species resulting from the pressure dependence of Al-avoidance is negligible for glasses with similar Al/(Al + Si) ratios so changes in the ^{17}O spectra for the high-pressure CAS glasses are due to the formation of $^{[4]}Si–O–^{[5,6]}Al$. Due to the uncertainty in the peak assignments and peak overlap, we used the ^{27}Al MAS results to predict the percentages of $^{[4]}Si–O–^{[5]}Al$ and $^{[4]}Si–O–^{[6]}Al$ species, which allows us to avoid the inherent problems associated with estimating the percentages of these species without adequate constraints. The ^{27}Al MAS data predicts that the CAS5R2 glass contains 21% $^{[4]}Si–O–^{[5]}Al$ and 19% $^{[4]}Si–O–^{[6]}Al$ species and this, combined with mechanisms 5 and 6, would predict that the percentages of the total oxygen that are $^{[4]}Si–O–^{[4]}Al$ and Si–NBO should decrease by 29% (21% × 4/5 + 19% × 4/6) and 11% (21% × 1/5 + 19% × 2/6), respectively. Considering the uncertainties in the quantification due to overlapping peaks, the difference of 9% measured for the amount of NBO seems consistent with mechanisms 5 and 6. Qualitatively speaking, the predicted percentages for the loss of $^{[4]}Si–O–^{[4]}Al$ also seems reasonable for the differences observed in the 3QMAS spectra, although, once again, severe peak overlap prevents a reliable quantification. Additionally, the NBO percentages in the spectra of the KAS1 and KAS5R glasses should stay approximately the same (6% × 1/5 + 0.5% × 2/6 = 1%), which is observed in the minor differences between the ^{17}O NMR spectra. These two pieces of evidence support mechanisms 5 and 6, which were suggested by Allwardt et al. (2004) to be responsible for the creation of high-coordinated Al in high-pressure aluminosilicate glasses.

4.3. Effect of slower decompression on the structure of high-pressure glasses

It is clear that at least some of the high-pressure structural changes of aluminosilicate melts are captured in glasses that are measured at ambient conditions. However, a determination of the effect of decompression on the glass structure is an essential component in understanding how studies measuring structure at one bar relate to the glass structure present at high pressure. Although studies of ambient-pressure samples alone may not observe the full extent of the changes that occur during decompression, the quantitative nature of NMR can constrain the manner in which the structure changes. A comparison of the ^{27}Al MAS spectra of the CAS5R glasses with that for the CAS5S glass shows that 15% of the aluminum (representing about 4% of the total network formers) reverts to tetrahedral coordination during the slower decompression (Table 2). In our previous study on high-pressure $K_2Si_4O_9$ glasses, there was much less of a difference between glasses with differing decompression rates (Allwardt et al., 2004), which may suggest that Al polyhedra are more susceptible to local relaxation during decompression than Si groups. In addition to the decrease in the average Al-coordination, the ^{17}O NMR data show that there is a higher percentage of NBO (~2% of the oxygen) in CAS5S relative to CAS5R. Based on the reversal of mechanisms 5 and 6, one would predict that 2% of the total oxygen would become NBO from the conversion of $^{[4]}$Si–O–$^{[5,6]}$Al species to NBO and $^{[4]}$Si–O–$^{[4]}$Al. This indicates that the $^{[5]}$Al and $^{[6]}$Al that are lost during decompression are simply due to the reversal of mechanisms 5 and 6, although this is not the only way to explain the structural differences between the glasses.

An unexpected result of this study is that the KAS5R contains less high-coordinated Al than the KAS5S sample (Table 2). One explanation for this could be that the average Al-coordination increases as the sample is being held at room temperature during the gradual decompression. To test this potential explanation, a KAS glass was compressed to 6 GPa and decompressed without heating (KAS6, Table 1). The peak fitting of the ^{27}Al MAS spectrum of this glass (not shown) indicates that it contains 5.5% $^{[5]}$Al and less than 0.4% $^{[6]}$Al (approximate detection limit), which is significantly less than the 12.2 and 0.7%, respectively, measured for the KAS5S (Table 2). This suggests that this explanation cannot completely account for the increase in $^{[5]}$Al and $^{[6]}$Al suggested to occur during the slower decompression. However, due to structural differences, glasses quenched at pressure would likely behave somewhat differently to room temperature compression/decompression than glasses that were only compressed and decompressed without annealing/melting. Since there were no obvious experimental differences between the experiments, an alternate explanation for the differences between the 5 GPa KAS glasses is that it simply indicates problems with the reproducibility in the synthesis of high-pressure glasses, however further work remains to determine whether problems with reproducibility are the reason for this unexpected result.

4.4. Reproducibility of the structure of high-pressure glasses using a multi-anvil apparatus

To directly test the reproducibility of high-pressure aluminosilicate glasses in an ideal case, two CAS5R samples were synthesized in separate capsules during the same multi-anvil experiment (CAS5R1 and CAS5R2). Figure 4 shows that CAS5R1 contains more $^{[6]}$Al and less $^{[5]}$Al than CAS5R2. Quantifying the spectra reveals that the variation in the different aluminum coordinations is about 3% of the total aluminum for any given Al-coordination (Table 2), which translates into a variation in the average Al-coordination of about 0.1. This may result from subtle differences in the cooling rates of different regions of the multi-anvil assembly, which would produce portions of the glass with slightly different fictive temperatures. A better way to consider these structural changes is to think about them in terms of the percentage of network formers. Realistically, the actual reproducibility of high-pressure glasses is less than that reported here as small compositional variations, temperature and pressure calibrations, and cooling characteristics of different assembly sizes can all be factors in reproducing experimental data, which could be especially true for samples synthesized at different multi-anvil facilities. These data, combined with the differences observed in the KAS5R and KAS5S glasses, could suggest that high-pressure glasses are less reproducible than previously thought and the main limitation to accurately determining the structure of high-pressure glasses lies in the reproducibility of the glasses, not in the quantification of the NMR spectra. Once again, this stresses the importance of collecting all structural measurements (e.g. ^{17}O, ^{27}Al, ^{29}Si, Raman, etc.) on the same glass and not inferring structural speciation from previous work in similar systems.

5. Conclusions

From this and other recent studies, it is clear that there is still much to do in determining the structure of "geologically relevant" high-pressure melts. With this said, there has been significant progress toward this goal in recent years as spectroscopic methods have improved over those used in earlier studies. We conclude, based on the data presented in this study:

1. As has been observed in previous studies, the average coordination of the network formers increases with increasing pressure.
2. The field strength of the modifier cation plays a key role in the amount of high-coordinated Al present in a high-pressure aluminosilicate glass where a higher field strength (e.g. CAS) yields more $^{[5,6]}Al$ than a lower field strength (KAS). This correlation between the Al-coordination and the field strength of the modifier cation is consistent with previous work on alkaline-earth boroaluminates (Bunker et al., 1991).
3. Our data supports the mechanisms proposed by Allwardt et al. (2004) for creating $^{[5]}Al$ and $^{[6]}Al$ in high-pressure aluminosilicate melts: $Si{-}NBO + 4Si{-}O{-}Al \rightarrow 5^{[4]}Si{-}O{-}{}^{[5]}Al$ and $2Si{-}NBO + 4Si{-}O{-}Al \rightarrow 6^{[4]}Si{-}O{-}{}^{[6]}Al$.
4. The structure of glasses with drastically different decompression rates is not the same, which suggests that decompression affects the structure of high-pressure aluminosilicate glasses.
5. More work remains to determine how reproducible high-pressure glass structure is in multi-anvil apparatuses, but it seems as though the limitation in determining the structural speciation lies in the reproducibility of the glass structure, not the detection limits of NMR.

Acknowledgements

We are indebted to Drs Brent Poe and Marc Hirschmann for valuable discussions on high-pressure glass/melt structure and assistance in the synthesis of high-pressure glasses. We would also like to thank Drs Eiji Ohtani, Mike Toplis, and an anonymous reviewer for helpful comments on the original manuscript and Drs Jiuhua Chen, Yanbin Wang, Tom Duffy, Guoyin Shen, and Larissa Dobrzhinetskaya for organizing this review volume and offering us the opportunity to contribute. JRA and JFS thank the National Science Foundation (USA) for grant number EAR-0104926 and BCS and DJF acknowledge funding from the Bayerisches Geoinstitut. We are especially grateful to the Bayerisches Geoinstitut visitor's program for additional funding and for facilitating the visit of JRA to their laboratories.

References

Agee, C.B., Walker, D., 1988. Static compression and olivine floatation in ultrabasic silicate liquid. J. Geophys. Res. – Solid Earth Planets 93, 3437–3449.

Agee, C.B., Walker, D., 1993. Olivine floatation in mantle melts. Earth Planet. Sci. Lett. 114, 315–324.

Allwardt, J.R., Lee, S.K., Stebbins, J.F., 2003a. Bonding preferences of non-bridging oxygen: evidence from ^{17}O MAS and 3QMAS NMR on calcium aluminate and low-silica Ca-aluminosilicate glasses. Am. Mineral. 88, 949–954.

Allwardt, J.R., Poe, B.T., Schmidt, B.C., Stebbins, J.F., 2003b. Compression mechanisms and temperature effects in high-pressure glasses investigated by ^{27}Al and ^{17}O 3QMAS NMR. Geophy. Res. Abstr. 5, 07757.

Allwardt, J.R., Stebbins, J.F., Frost, D.J., Schmidt, B.C., 2003. The effect of modifier cations on the amount of five- and six-coordinated aluminum present in high-pressure aluminosilicate glasses, Bayerisches Forschunsintitut für Experimentelle Geochemie und Geophysik, Universität Bayreuth, Annual Report.

Allwardt, J.R., Schmidt, B.C., Stebbins, J.F., 2004. Structural mechanisms for the generation of high-coordinated Si and the effect of decompression on the structure of high-pressure $K_2Si_4O_9$ melts and glasses. Chem. Geol. 213, 137–151.

Allwardt, J.R., Stebbins, J.F., Schmidt, B.C., Frost, D.J., Withers, A.C., and Hirschmann, M.M., in press-a. Correlating the average aluminum coordination and density in high-pressure aluminosilicate glasses: Implications for the structural changes and densification in basaltic magmas. Am. Mineral.

Allwardt, J.R., Poe, B.T., and Stebbins, J.F., in press-b. The effect of fictive temperature on Al-coordination in high-pressure (10 GPa) sodium aluminosilicate glasses. Am. Mineral.

Ando, R., Ohtani, E., Suzuki, A., Rubie, D.C., Funakoshi, K., 2003. Pressure dependency of viscosities of MORB melts. Eos Trans. AGU 84 (46), Fall Meet. Suppl., Abstract V31D-0958.

Angell, C.A., Cheeseman, P.A., Tammadon, S., 1982. Pressure enhancement of ion mobilities in silicate liquids from computer simulation studies to 800 kbar. Science 218, 885–887.

Angell, C.A., Cheeseman, P.A., Kadiyala, R., 1987. Diffusivity and thermodynamic properties of diopside and jadeite melts by computer simulation studies. Chem. Geol. 62, 85–95.

Baltisberger, J.H., Xu, Z., Stebbins, J.F., Wang, S.H., Pines, A., 1996. Triple-quantum two-dimensional ^{27}Al magic-angle spinning nuclear magnetic resonance spectroscopic study of aluminosilicate and aluminate crystals and glasses. J. Am. Chem. Soc. 118, 7209–7214.

Brandriss, M.E., Stebbins, J.F., 1988. Effects of temperature on the structure of silicate liquids: ^{29}Si NMR results. Geochim. Cosmochim. Acta 52, 2659–2670.

Brawer, S.A., White, W.B., 1975. Raman spectroscopic investigation of structure of silicate glasses, I. The binary alkali silicates. J. Chem. Phys. 63, 2421–2432.

Brawer, S.A., White, W.B., 1977. Raman spectroscopic investigation of structure of silicate glasses (II). Soda-alkaline earth-alumina ternary and quaternary glasses. J. Non-Cryst. Solids 23, 261–278.

Brown, G.E. Jr., Farges, F., Calas, G., 1995. X-ray scattering and X-ray spectroscopy studies of silicate melts. In: Stebbins, J.F., McMillan, P.F., Dingwell, D.B. (Eds), Structure, Dynamics and Properties of Silicate Melts. Mineralogical Society of America, Washington, DC, pp. 317–410.

Bunker, B.C., Kirkpatrick, R.J., Brow, R.K., Turner, G.L., Nelson, C., 1991. Local structure of alkaline-earth boroaluminate crystals and glasses: II, ^{11}B and ^{27}Al MAS NMR spectroscopy of alkaline-earth boroaluminate glasses. J. Am. Ceram. Soc. 74, 1430–1438.

Daniel, I., Gillet, P.H., Poe, B.T., McMillan, P.F., 1995. In-situ high temperature Raman spectroscopic studies of aluminosilicate liquids. Phys. Chem. Mineral. 22, 74–86.

Diefenbacher, J., McMillan, P.F., Wolf, G.H., 1998. Molecular dynamics simulations of $Na_2Si_4O_9$ liquid at high pressure. J. Phys. Chem. B 102, 3003–3008.

Du, L.-S., Stebbins, J.F., 2005. Site connectivities in sodium aluminoborate glasses: multinuclear and multiple quantum NMR results. Solid State NMR, 27, 37–49.

Farber, D.L., Williams, Q., 1996. An in situ Raman spectroscopic study of $Na_2Si_2O_5$ at high pressures and temperatures: structures of compressed liquids and glasses. Am. Mineral. 81, 273–283.

Frost, D.J., Poe, B.T., Trønnes, R.G., Libske, C., Duba, A., Rubie, D.C., 2004. A new large volume multianvil system. Phys. Earth Planet. In. 143–144, 507–514.

Funamori, N., Yamamoto, S., Yagi, T., Kikegawa, T., 2004. Exploratory studies of silicate melt structure at high pressures and temperatures by in situ X-ray diffraction. J. Geophys. Res. 109, B03203.

Gan, Z., Gor'kov, P., Cross, T.A., Samoson, A., Massiot, D., 2002. Seeking higher resolution and Sensitivity for NMR of quadrupolar nuclei at ultrahigh magnetic fields. J. Am. Chem. Soc. 124, 5634–5635.

Hemingway, B.S., Bohlen, S.R., Hankins, W.B., Westrum, E.F. Jr., Kuskov, O.L., 1998. Heat capacity and thermodynamic properties for coesite and jadeite, reexamination of the quartz–coesite equilibrium boundary. Am. Mineral. 83, 409–418.

Hemley, R.J., Mao, H.K., Bell, P.M., Mysen, B.O., 1986. Raman spectroscopy of SiO_2 glass at high pressure. Phys. Rev. Lett. 57, 747–750.

Hess, K.-U., Dingwell, D.B., Webb, S.L., 1995. The influence of excess alkalis on the viscosity of a haplogranitic melt. Am. Mineral. 80, 297–304.

Hess, K.-U., Dingwell, D.B., Webb, S.L., 1996. The influence of alkaline-earth oxides (BeO, MgO, CaO, SrO, BaO) on the viscosity of a haplogranitic melt: systematics of non-Arrhenian behaviour. Eur. J. Mineral. 8, 371–381.

Hochella, M.H., Brown, G.E., 1985. The structures of albite and jadeite composition glasses quenched from high pressure. Geochim. Cosmochim. Acta 48, 2631–2640.

Jäger, C., 1994. Satellite transition spectroscopy of quadrupolar nuclei. In: Diehl, P., Fluck, E., Günther, H., Kosfeld, R., Seelig, J. (Eds), Solid State NMR III: Inorganic Matter. Springer-Verlag, Berlin, pp. 135–170.

Knoche, R., Luth, R.W., 1996. Density measurements on melts at high pressure using the sink/float method: limitations and possibilities. Chem. Geol. 128, 229–243.

Kushiro, I., 1976. Changes in viscosity and structure of melt of $NaAlSi_2O_6$ composition at high pressures. J. Geophys. Res. 81, 6347–6350.

Kushiro, I., 1978. Viscosity and structural changes of albite ($NaAlSi_3O_8$) melt at high pressures. Earth Planet. Sci. Lett. 41, 87–90.

Lange, R.A., 2003. The fusion curve of albite revisited and the compressibility of $NaAlSi_3O_8$ liquid with pressure. Am. Mineral. 88, 109–120.

Langenhorst, F., Boutsie, M., Deutsch, A., Hornemann, U., Matignon, C., Migault, A., Romain, J.P., 2002. Experimental techniques for the simulation of shock metamorphism: a case study on calcite. In: Davison, L., Horie, Y., Sekine, T. (Eds), High-Pressure Shock Compression of Solids V- Shock Chemistry with Applications to Meteroite Impacts. Springer, New York, pp. 1–27.

Lee, S.K., 2004. Structure of silicate glasses and melts at high pressure: quantum chemical calculations and solid-state NMR. J. Phys. Chem. B 108, 5889–5900.

Lee, S.K., Stebbins, J.F., 1999. The degree of aluminum avoidance in aluminosilicate glasses. Am. Mineral. 84, 937–945.

Lee, S.K., Stebbins, J.F., 2000. Al–O–Al and Si–O–Si sites in framework aluminosilicate glasses with Si/Al = 1: quantification of framework disorder. J. Non-Cryst. Solids 270, 260–264.

Lee, S.K., Stebbins, J.F., 2001. The extent of intermixing among framework units in silicate glasses and melts. Geochim. Cosmochim. Acta 66, 303–309.

Lee, S.K., Fei, Y., Cody, G., Mysen, B.O., 2003a. Disorder in silicate glasses and melts at 10 GPa. Geophys. Res. Lett. 30, 1845.

Li, D., Secco, R.A., Bancroft, G.M., Fleet, M.E., 1996. Pressure induced coordination change of Al in silicate melts from Al K edge XANES of high pressure $NaAlSi_2O_6$–$NaAlSi_3O_8$ glasses. Geophys. Res. Lett. 22, 3111–3114.

Loewenstein, W., 1954. The distribution of aluminum in the tetrahedra of silicates and aluminates. Am. Mineral. 39, 92–96.

Massiot, D., Touzo, B., Tremeau, D., Coutures, J.P., Virlet, J., Florian, P., Grandinetti, P.J., 1996. Two-dimensional magic-angle spinning isotropic reconstruction sequences for quadrupolar nuclei. Solid State Nucl. Magn. Reson. 6, 73–83.

Matsui, M., 1996. Molecular dynamics simulations of structures, bulk moduli and thermal expansivities of silicate liquids in the system CaO-MgO-Al_2O_3-SiO_2. Geophys. Res. Lett. 23, 395–398.

Matsui, M., 1998. Computational modeling of crystals nad liquids in the system Na_2O-CaO-MgO-Al_2O_3-SiO. In: Manghani, M., Yagi, T. (Eds), Properties of Earth and Planetary Materials at High Pressure and Temperature. American Geophysical Union, Washington D.C., 145–151.

McDonough, W.F., Rudnick, R.L., 1998. Pressure effects on silicate melt structure and properties. In: Hemley, R.J. (Ed.), Ultrahigh-pressure mineralogy. Mineralogical Society of America, Washington, DC, pp. 139–164.

McMillan, P., Graham, C.M., 1980. The Raman spectra of quenched albite and orthoclase glasses from 1 atm to 40 kb. In: Ford, C.E. (Ed.), Progress in Experimental Petrology. Eaton Press, Boston, pp. 112–115.

McMillan, P.F., Wolf, G.H., 1995. Vibrational spectroscopy of silicate liquids. In: Stebbins, J.F., McMillan, P.F., Dingwell, D.B. (Eds), Structure, Dynamics and Properties of Silicate Melts. Mineralogical Society of America, Washington, DC, pp. 247–315.

Miller, G.H., Stolper, E.M., Ahrens, T.J., 1991. The equation of state of a molten komatite: 1. shock-wave compression to 36 GPa. J. Geophys. Res. – Solid Earth Planets 96, 11831–11848.

Murdoch, J.B., Stebbins, J.F., Carmichael, I.S.E., 1985. High-resolution ^{29}Si NMR study of silicate and aluminosilicate glasses: the effect of network-modifying cations. Am. Mineral. 70, 332–343.

Mysen, B.O., 1988. Structure and properties of silicate melts. Elsevier, Amsterdam.

Mysen, B.O., 1990. Effect of pressure, temperature, and bulk composition on the structure and species distribution in depolymerized alkali aluminosilicate melts and quenched melts. J. Geophys. Res. 95, 15733–15744.

Mysen, B.O., Virgo, D., Scarfe, C.M., 1980. Relations between the anionic structure and viscosity of silicate melts- a Raman spectroscopic study. Am. Mineral. 65, 690–710.

Mysen, B.O., Virgo, D., Kushiro, I., 1981. The structural role of aluminum in silicate melts- a Raman spectroscopic study at 1 atmosphere. Am. Mineral. 66, 678–701.

Mysen, B.O., Virgo, D., Danckwerth, P., Seifert, F.A., Kushiro, I., 1983. Influence of pressure on the structure of melts on the joins $NaAlO_2$–SiO_2, $CaAl_2O_4$–SiO_2, and $MgAl_2O_4$–SiO_2. Neues Jahrbuch Für Mineralogie-Abhandlungen 147, 281–303.

Mysen, B.O., Virgo, D., Seifert, F.A., 1985. Relationship between properties and structure of aluminosilicate melts. Am. Mineral. 70, 88–105.

Navrotsky, A., Geisinger, K.L., McMillan, P., Gibbs, G.V., 1985. The tetrahedral framework in glasses and melts- inferences from molecular orbital calculations and implications for structure, thermodynamics, and physical properties. Phys. Chem. Miner. 11, 284–298.

Neuville, D.R., Cormier, L., Masssiot, D., 2004. Al environment in tectosilicate and peraluminous glasses: a ^{27}Al MQ-MAS NMR, Raman, and XANES investigation. Geochim. Cosmochim. Acta. 68, 5071–5079.

Nevins, D., Spera, F.J., 1998. Molecular dynamics simulations of molten $CaAlSi_2O_8$: dependence of structure and properties on pressure. Am. Mineral. 83, 1220–1230.

Oglesby, J.V., Zhao, P.D., Stebbins, J.F., 2002. Oxygen sites in hydrous aluminosilicate glasses: the role of Al–O–Al and H_2O. Geochim. Cosmochim. Acta 66, 291–301.

Ohtani, E., Maeda, M., 2001. Density of basaltic melt at high pressure and stability of the melt at the base of the lower mantle. Earth Planet. Sci. Lett. 193, 69–75.

Ohtani, E., Suzuki, A., Kato, T., 1993. Flotation of olivine in the peridotite melt at high-pressure. Proc. Jpn Acad. B-Phys. Biol. Sci. 69, 23–28.

Ohtani, E., Suzuki, A., Kato, T., 1998. Flotation of olivine and diamond in mantle melt at high pressure: implications for fractionation in the deep mantle and ultradeep origin of diamond. In: Manghnani, M.H., Yagi, T. (Eds), Properties of Earth and Planetary Materials at High Pressure and Temperature, Geophysical Monograph, Vol. 101. American Geophysical Union, Washington, DC, pp. 227–239.

Ohtani E., Susuki, A., Ando R., Urakawa, S., Funakoshi, K., Katayama, Y., 2005. Viscosity and density measurements of melts and glasses at high pressure and temperature by using the multianvil apparatus and synchrotron X-ray radiation. In: Chen, J., Wang, Y., Duffy, T., Shen, G., and Dobrzhinetskaya, L., (Eds), Advances in High Pressure Technology for the Geophysical Applications. Elsevier, Amsterdam.

Poe, B.T., Rubie, D.C., 2000. Transport properties of silicate melts at high pressure. In: Aoki, H., Syono, Y., Hemley, R.J. (Eds), Physics Meets Mineralogy. Cambridge University Press, Cambridge, pp. 340–353.

Poe, B.T., McMillan, P.F., Coté, B., Massiot, D., Coutures, J.P., 1993. Magnesium and calcium aluminate liquids: in situ high-temperature ^{27}Al NMR spectroscopy. Science 259, 786–788.

Poe, B.T., McMillan, P.F., Coté, B., Massiot, D., Coutures, J.P., 1994. Structure and dynamics in calcium aluminate liquids: high-temperature ^{27}Al NMR and Raman spectroscopy. J. Am. Ceram. Soc. 77, 1832–1838.

Poe, B.T., Romano, C., Zotov, N., Cibin, G., Marcelli, A., 2001. Compression mechanisms in aluminosilicate melts: Raman and XANES spectroscopy of glasses quenched from pressures up to 10 GPa. Chem. Geol. 174, 21–31.

Reid, J.E., Poe, B.T., Rubie, D.C., 2001. The self-diffusion of silicon and oxygen in diopside ($CaMgSi_2O_6$) liquid up to 15 GPa. Chem. Geol. 174, 77–86.

Reid, J.E., Suzuki, A., Funakoshi, K.I., Terasaki, H., Poe, B.T., Rubie, D.C., Ohtani, E., 2003. The viscosity of $CaMgSi_2O_6$ liquid at pressures up to 13 GPa. Phys. Earth Planet. In. 139, 45–54.

Rigden, S.M., Ahrens, T.J., Stolper, E.M., 1984. Densities of liquid silicates at high pressures. Science 226, 1071–1074.

Rigden, S.M., Ahrens, T.J., Stolper, E.M., 1988. High-pressure equation of state of molten anorthite and diopside. J. Geophys. Res. 94, 9508–9522.

Rubie, D.C., 1999. Characterising the sample environment in multi-anvil high-pressure experiments. Phase Transitions 68, 431–451.

Scarfe, C.M., Mysen, B.O., Virgo, D., 1987. Pressure dependence of the viscosity of silicate melts. In: Mysen, B.O. (Ed.), Magmatic Processes, pp. 59–67.

Sharma, S.K., Virgo, D., Mysen, B.O., 1979. Raman study of the coordination of aluminum in jadeite melts as a function of pressure. Am. Mineral. 64, 779–787.

Sharma, S.K., Wang, Z., Van der Laan, S., 1996. Raman spectroscopy of oxide glasses at high pressure and high temperature. J. Raman Spectrosc. 27, 739–746.

Stebbins, J.F., 1995. Nuclear magnetic resonance spectroscopy of silicates and oxides in geochemistry and geophysics. In: Ahrens, T.J. (Ed.), Handbook of Physical Constants. American Geophysical Union, Washington, DC, pp. 303–332.

Stebbins, J.F., McMillan, P.F., 1989. Five and six-coordinated Si in $K_2Si_4O_9$ glass quenched from 1.9 GPa and 1200°C. Am. Mineral. 74, 965–968.

Stebbins, J.F., Sykes, D., 1990. The structure of $NaAlSi_3O_8$ liquid at high pressure: new constraints from NMR spectroscopy. Am. Mineral. 75, 943–946.

Stebbins, J.F., Farnan, I., 1992. Effects of high-temperature on silicate liquid structure: a multinuclear NMR study. Science 255, 586–589.

Stebbins, J.F., McMillan, P., 1993. Composition and temperature effects of 5-coordinated silicon in ambient pressure silicate glasses. J. Non-Cryst. solids 160, 116–125.

Stebbins, J.F., Poe, B.T., 1999. Pentacoordinate silicon in high-pressure crystalline and glassy phases of calcium disilicate ($CaSi_2O_5$). Geophys. Res. Lett. 26, 2521–2523.

Stebbins, J.F., Oglesby, J.V., Xu, Z., 1997. Disorder among network modifier cations in silicate glasses: new constraints from triple-quantum oxygen-17 NMR. Am. Mineral. 82, 1116–1124.

Stebbins, J.F., Kroeker, S., Lee, S.K., Kiczenski, T.J., 2000. Quantification of five- and six-coordinated aluminum ions in aluminosilicate and fluoride-containing glasses by high-field, high-resolution ^{27}Al NMR. J. Non-Cryst. Solids 275, 1–6.

Stebbins, J.F., Oglesby, J.V., Kroeker, S., 2001. Oxygen triclusters in crystalline $CaAl_4O_7$ (grossite) and in calcium aluminosilicate glasses: ^{17}O NMR. Am. Mineral. 86, 1307–1311.

Stolper, E.M., Ahrens, T.J., 1987. On the nature of pressure-induced coordination changes in silicate melts and glasses. Geophys. Res. Lett. 14, 1231–1233.

Stolper, E.M., Walker, D., Hager, B.H., Hays, J.F., 1981. Melt segregation from partially molten source regions: the importance of melt density and source regions size. J. Geophys. Res. 86, 6261–6271.

Susaki, J., Akaogi, M., Akimoto, S., Shimomura, O., 1985. Garnet-perovskite transformation in CaGeO3-insitu X-ray measurements using synchrotron radiation. Geophys. Res. Lett. 12, 729–732.

Suzuki, A., Ohtani, E., 2003. Density of peridotite melts at high pressure. Phys. Chem. Mineral. 30, 449–456.

Suzuki, A., Ohtani, E., Kato, T., 1995. Flotation of diamond in mantle melt at high-pressure. Science 269, 216–218.

Suzuki, A., Ohtani, E., Kato, T., 1998. Density and thermal expansion of a peridotite melt at high pressure. Phys. Earth Planet. In. 107, 53–61.

Suzuki, A., Ohtani, E., Funakoshi, K., Terasaki, H., Kubo, T., 2002. Viscosity of albite at high-pressure and high temperature. Phys. Chem. Miner. 29, 159–165.

Sykes, D., Poe, B.T., McMillan, P.F., Luth, R.W., Sato, R.K., 1993. A spectroscopic investigation of anhydrous $KAlSi_3O_8$ and $NaAlSi_3O_8$ glasses quenched from high pressure. Geochim. Cosmochim. Acta 57, 1753–1759.

Toplis, M.J., Dingwell, D.B., 2004. Shear viscosities of $CaO-Al_2O_3-SiO_2$ and $MgO-Al_2O_3-SiO_2$ liquids: implications for the structural role of aluminum and the degree of polymerization of aluminosilicate melts. Geochim. Cosmochim. Acta, 68, 5169–5188.

Toplis, M.J., Dingwell, D.B., Lenci, T., 1997. Peraluminous viscosity maxima in $Na_2O-Al_2O_3-SiO_2$ liquids: the role of triclusters in tectosilicate melts. Geochim. Cosmochim. Acta 61, 2605–2612.

Toplis, M.J., Kohn, S.C., Smith, M.E., Poplett, I.J.F., 2000. Fivefold-coordinated aluminum in tectosilicate glasses observed by triple quantum MAS NMR. Am. Mineral. 85, 1556–1560.

Urakawa, S., Ando, R., Ohtani, E., Katayama, Y., 2003. Density measurement of silicate glass by in-situ absorption method. Eos Trans. AGU 84 (46), Fall Meet. Suppl., Abstract V41B-07.

Velde, B., Kushiro, I., 1978. Structure of sodium aluminum-silicate melts quenched at high-pressure: infrared and aluminum K-radiation data. Earth. Planet. Sci. Lett. 40, 137–140.

Waff, H.S., 1975. Pressure-induced coordination changes in magmatic liquids. Geophys. Res. Lett. 2, 193–196.

William, Q., Jeanloz, R., 1988. Spectroscopic evidence for pressure-induced coordination changes in silicate glasses and melts. Science 239, 902–905.

Wolf, G.H., McMillan, P.F., 1995. Pressure effects on silicate melt structure and properties. In: Stebbins, J.F., McMillan, P.F., Dingwell, D.B. (Eds), Structure, Dynamics and Properties of Silicate Melts. Mineralogical Society of America, Washington, DC, pp. 506–561.

Wolf, G.H., Durben, D.J., McMillan, P.F., 1990. High-pressure spectroscopic study of sodium tetrasilicate ($Na_2Si_4O_9$) glass. J. Chem. Phys. 93, 2280–2288.

Xue, X., Stebbins, J.F., Kanzaki, M., Trønnes, R.G., 1989. Silicon coordination and speciation change in a silicate liquid at high pressures. Science 245, 962–964.

Xue, X., Stebbins, J.F., Kanzaki, M., McMillan, P.F., Poe, B., 1991. Pressure induced silicon coordination and tetrahedral structural changes in alkali oxide-silica melts up to 12 GPa: NMR, Raman, and infrared spectroscopy. Am. Mineral. 76, 8–26.

Xue, X., Stebbins, J.F., Kanzaki, M., 1994. Correlations between ^{17}O NMR parameters and local structure around oxygen in high-pressure silicates: Implications for the structure of silicate melts at high pressure. Am. Mineral. 79, 31–42.

Yarger, J.L., Smith, K.H., Nieman, J., Diefenbacher, J., Wolf, G.H., Poe, B.T., McMillan, P.F., 1995. Al coordination changes in high-pressure aluminosilicate liquids. Science 270, 1964–1967.

Zhang, P., Grandinetti, P.J., Stebbins, J.F., 1997. Anionic species determination in $CaSiO_3$ glass using two-dimensional ^{29}Si NMR. J. Phys. Chem. 101, 4004–4008.

Advances in High-Pressure Technology for Geophysical Applications
Jiuhua Chen, Yanbin Wang, T.S. Duffy, Guoyin Shen, L.F. Dobrzhinetskaya, editors

Chapter 12

The application of ^{17}O and ^{27}Al solid-state (3QMAS) NMR to structures of non-crystalline silicates at high pressure

Sung Keun Lee, Yingwei Fei, George D. Cody and Bjorn O. Mysen

Abstract

The recent development in two-dimensional solid-state triple quantum (3Q) magic angle spinning (MAS) NMR offers much improved resolution compared with conventional one-dimensional MAS NMR, allowing the structural details of amorphous silicates to be revealed. Since the first experimental application of 3QMAS NMR to the silicate glasses quenched from melts at high pressure in a multi-anvil apparatus [Lee, S. K., Fei, Y., Cody, G. D., Mysen, B. O., 2003a. Geophys. Res. Lett. 30, 1845], there has been continued progress in the understanding of the structures of silicate melts at high pressure. Here, we present the recent progress and insights made by ^{17}O and ^{27}Al 3QMAS NMR spectra of silicate glasses quenched from melts at pressures up to 10 GPa in a multi-anvil apparatus, revealing new details of melt structures at high pressure.

The atomic structure of sodium silicate and aluminosilicate glasses at high pressure is significantly different from that at ambient pressure. There is evidence of extensive cation ordering among highly coordinated network polyhedra, such as $^{[5,6]}Al$ and $^{[5,6]}Si$, that affect corresponding thermodynamic and transport properties. New oxygen sites are observed at high pressure in a series of silicate glasses with varying degree of polymerization, e.g. $^{[5,6]}Al{-}O{-}^{[4]}Si$, $^{[5,6]}Si{-}O{-}^{[4]}Si$, and $Na{-}O{-}^{[5,6]}Si$. The fractions of these clusters tend to increase with pressure with corresponding reduction of non-bridging oxygen ($Na{-}O{-}^{[4]}Si$), thus increasing the net degree of polymerization. The fraction of $^{[5,6]}Al$ in aluminosilicate glasses increases with pressure, consistent with previous studies, but decreases with increasing degree of polymerization of melts from depolymerized to fully polymerized melts at an isobaric condition. The bond angle and length distribution, as well as the range in distortion of framework units increase with increasing pressure, which increases the topological entropy. These results shed light on a new opportunity for studying structures of silicate glasses and melts at high pressure and help to provide microscopic constraints for the melt properties in the Earth's interior.

1. Introduction

Silicate melts are the ubiquitous components of igneous processes in the Earth's crust and mantle and serve as the key transport agents for physico-chemical differentiation and evolution of the Earth (e.g. Mysen, 1988). Understanding the thermodynamic and transport properties of silicate melts is crucial to provide better understanding of the relevant processes including melting, dynamics (migration and emplacement of melts), and the distribution of elements in the Earth's interior (Kushiro, 1976; 1983; Dunn, 1982; Navrotsky et al., 1982; Shimizu and Kushiro, 1984; Neuville and Richet, 1991;

Toplis et al., 1997; Tinker et al., 2003). It is well known that molecular structures and the extent of disorder of silicate melts controls these macroscopic properties and the corresponding processes in the Earth's interior (Angell et al., 1982; Mysen et al., 1982; Mysen, 1990; Yarger et al., 1995; Poe et al., 2001).

While the atomic structures of crystalline Earth materials or molecules at high pressure are relatively well known due to the progress in high-pressure technologies in mineral physics, much less is known about the structures of amorphous silicates including glasses and melts. This is mostly because X-ray scattering or vibrational spectroscopy provide limited information (e.g. coordination environment or topological variation) on the specific nature of atomic disorder: pronounced chemical and topological disorder with lack of periodicity in the oxide glasses make diffraction patterns or the modes in the vibrational spectra further broadened. (See also Wolf and McMillan, 1995 for extensive review of structure and properties of silicate melts and glasses at high pressure).

Solid-state NMR has particular advantages in studying structures of silicate glasses at high pressure. Pioneering one-dimensional ^{29}Si and ^{27}Al magic angle spinning (MAS) NMR studies of silicate glasses quenched from melts at high pressure provided clear evidence of pressure-induced coordination changes in Si and Al in that the fraction of highly coordinated framework units (e.g. $^{[5,6]}Si$ and $^{[5,6]}Al$) increases with pressure, which can partly account for the anomalies in pressure dependence of diffusivity and viscosity (Xue et al., 1989, 1991; Yarger et al., 1995; Poe et al., 1997). The first ^{17}O MAS NMR for binary silicate glasses at high pressure suggested the pressure-induced changes in connectivity of silicate melts (Xue et al., 1994).

Recent advances in solid-state NMR allow us to explore the distributions of cations and anions in the non-crystalline silicates. Most notable progress has been the development of two-dimensional triple quantum (3Q) MAS NMR that provides much improved resolution over conventional one-dimensional MAS NMR (Frydman and Harwood, 1995; Amoureux et al., 1996; Baltisberger et al., 1996; Lee and Stebbins, 2000b). Various aspects of disorder (e.g. chemical and topological) in silicate glasses and melts have been recently quantified using this technique (Lee and Stebbins, 2000b; 2002; 2003a,b; Lee et al., 2001; 2003b). The ^{17}O 3QMAS NMR of the silicate glasses quenched from melts at the high pressure provided the first new structural details of anion environment in simple binary silicates (Lee et al., 2003a) and subsequently the oxygen environment in ternary aluminosilicate glasses and melts have been revealed (Lee, 2004b; Lee et al., 2004). These results show that the connectivity at high pressure is significantly different from that at ambient pressure. In particular, in the silicate and aluminosilicate melts, the proportion of non-bridging oxygen (NBO) generally decreases with increasing pressure, leading to the formation of new oxygen clusters that connect $^{[5,6]}Si$ and $^{[5,6]}Al$ (in addition to $^{[4]}Al$ and $^{[4]}Si$), such as $^{[4]}Si\text{-}O\text{-}^{[5,6]}Si$, $^{[4]}Si\text{-}O\text{-}^{[5,6]}Al$ and $Na\text{-}O\text{-}^{[5,6]}Si$ (Lee, 2004b; Lee et al., 2004). Quantum chemical calculations were also used to better understand the relationship between structurally relevant NMR parameters and atomic configurations at high pressure (Lee et al., 2003a; Lee, 2004b).

Here, these recent results using two-dimensional solid-state NMR (3QMAS) and quantum chemical calculations on the pressure-induced changes in atomic structures and the properties of silicate melts with varying pressure and composition are reviewed (Lee et al., 2003a; 2004b; Lee, 2004). It should be noted that we did not attempt to provide an extensive review of the structure of silicate glasses and melts obtained from many other experimental and theoretical methods. Much more extensive general reviews on the structure of glasses and melts at ambient and high pressure can be found in the other, excellent previous reviews (e.g. Wolf and McMillan, 1995). Relevant review can also be found in the recent contribution (Allwardt et al., 2004). Instead, we rather focus on a review about the studies of structures of silicate glasses and melts using "two-dimensional solid-state NMR" of quadrupolar nuclides (e.g. ^{17}O and ^{27}Al). This contribution is composed of two main parts. In Section 2, the methodology for pressure generation, experimental conditions, and spectroscopic advances in high-resolution solid-state NMR are summarized. After describing multi-anvil equipment for the sample synthesis (Sections 2.1 and 2.2), the recent advances in high-resolution solid-state NMR, in particular, 3QMAS NMR and its application to $NaAlSiO_4$ glass are discussed in Section 2.3. We also briefly describe the insights from quantum chemical calculations on the atomic structures at high pressure and their relationship with NMR parameters (e.g. isotropic chemical shift, Section 2.4). In Section 3, we summarize how the quantification of disorder in silicate glasses and melts is obtained from these solid-state NMR techniques. These results help to establish the classification scheme of disorder in non-crystalline silicates (Section 3.1). We then present the recent results on the structures of silicate glasses and melts at high pressure. Finally, we explore the pressure-induced changes in the coordination environment of framework units and the corresponding connectivity reflected in BO and NBO environment at the high pressure. The variation of the extent of disorder with pressure will also be discussed in Section 3.2.

2. Techniques, methods, and materials description

2.1. Pressure generation: multi-anvil equipments

Pressure generation was achieved by compressing eight tungsten carbide cubes and octahedron pressure medium (+sample assembly) with hydraulic ram. Sample assemblies are defined by ratio between octahedron edge length (OEL) and truncated edge length (TEL) of the cubes (e.g. OEL/TEL). Two assemblies (18/11 and 10/5) were used to generate pressure up to 10 GPa. (18/11 for below 8 GPa and 10/5 for 10 GPa, respectively, for the experimental results given here). The graphite and rhenium heaters were used for each assembly. Pressure calibrations and temperature gradient experiments were performed previously and can be found elsewhere (Bertka and Fei, 1997). Figure 1 shows the schematic diagram for the 18/11 assembly where the glass starting materials were sealed inside the platinum tube. The Pt tube was inserted in an Al_2O_3 tube for insulation. This assembly is heated by graphite furnace. The furnace was further surrounded by zirconia ceramics.

2.2. Starting materials and glass sample synthesis

Various silicate glasses quenched from melts at high pressure were synthesized in the 1000 and 1500 ton multi-anvil equipments at the Geophysical Laboratory. Among the systems of investigation, the compositions of the glasses that will be presented here are

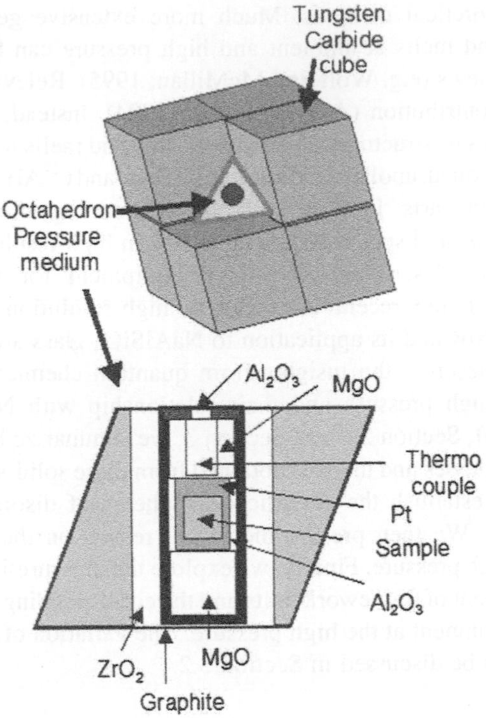

Figure 1. Schematic diagram of multi-anvil cube assembly (up) and cross section of 18/11 assembly (bottom).

$[(Na_2O)_{1-x}(Al_2O_3)_x]3SiO_2$ (where $x = 0$, 0.25, and 0.5). The starting materials are ^{17}O enriched sodium silicate glasses that were synthesized from carbonates (Na_2CO_3) and oxides such as Al_2O_3 and ^{17}O enriched SiO_2 by melting the mixtures above the melting temperature after decarbonation and then quenching them. These starting glasses were heated above the melting temperature at varying pressure conditions (up to 10 GPa) in the multi-anvil apparatus and then quenched (Lee et al., 2004). The initial quench rate is estimated about >500 K/s. The quenched sample from the 18/11 assembly is about 20 mg and the diameter is slightly larger than the inner diameter of the 2.5 mm rotor used for the NMR experiments. The sample obtained from the 10/5 assembly is smaller than that of the rotor with about 5–6 mg.

2.3. High resolution solid-state NMR: two-dimensional 3QMAS NMR

2.3.1. Application of high-resolution solid-state NMR to non-crystalline silicate at ambient and high pressure

Solid-state NMR is an element-specific, usually quantitative probe of atomic environments of nuclides of interest and has been experiencing major experimental improvements for the last decade. The notable progress includes high-spinning

MAS probe at high static magnetic field and 3QMAS NMR for quadrupolar nuclides (e.g. ^{23}Na, ^{27}Al, ^{11}B, ^{17}O, etc.) (Frydman and Harwood, 1995). Thanks to these recent advances, the detailed information of distributions of cations and anions in complex crystalline and non-crystalline solids has been obtained (Amoureux et al., 1996, 1998; Dirken et al., 1997; Stebbins and Xu, 1997; Wang and Stebbins, 1998; Xu et al., 1998; Wu and Dong, 2001; Lee et al., 2003c). The advantage of 3QMAS NMR (in particular for ^{17}O) is briefly demonstrated below.

Figure 2 shows comparison between liquid and solid-state NMR (and also magic angle spinning vs. static). ^{17}O static NMR spectrum for liquid water reveals sharp peak, which is one of the characteristic features of liquid-state NMR where the anisotropy of the nuclear spin interactions are averaged out by the rapid molecular motions. The ^{17}O static NMR spectrum for a crystalline solid, however, exhibits broad peaks for multiple oxygen sites, e.g. in the case of the layer silicates (pyrophyllite + kaolinite mixture). This broad spectrum does not provide explicit information of the oxygen sites for the clay minerals and is mainly due to the anisotropic nuclear spin interactions in the solids (e.g. nuclear dipolar interaction, quadrupolar interaction, and chemical shift anisotropy) (Mehring, 1983). Furthermore, quadrupolar nuclides, such as ^{17}O with nuclear spin larger than $\frac{1}{2}$ have additional broadening due to quadrupolar interactions between quadrupole moment of nuclei and electric field gradients generated by surrounding electrons. Broadening caused by these anisotropic interactions can be substantially reduced by spinning the sample at the magic angle (54.74°) with respect to the static external magnetic field (MAS). MAS decreases the spectral broadening and results in a high-resolution

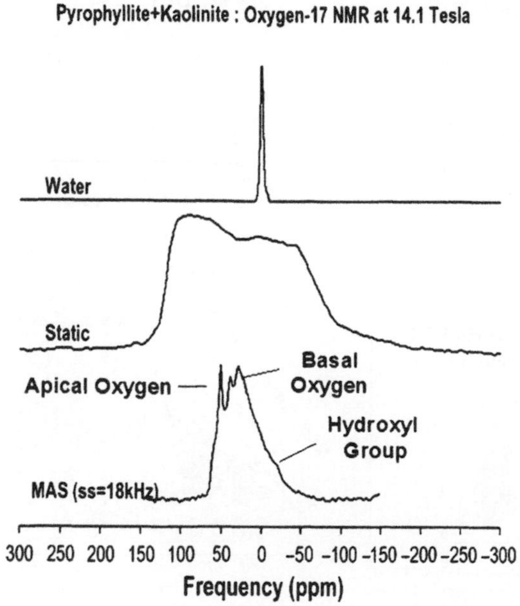

Figure 2. Comparison between liquid and solid-state NMR. ^{17}O liquid-state NMR for H_2O (up) vs. solid-sate NMR spectra for pyrophyllite + kaolinite mixture (static: middle, MAS with spinning speed of 18 kHz: bottom) (see Lee et al., 2003c for experimental conditions for ^{17}O MAS NMR spectrum for the layer silicate mixture).

spectrum of oxygen sites in solids, e.g. clay minerals where distinct oxygen sites, such as apical ($2^{[6]}Al-O-^{[4]}Si$), basal oxygens ($^{[4]}Si-O-^{[4]}Si$), and hydroxyl groups ($2^{[6]}Al-O-OH$) were well resolved (Fig. 2, bottom) (Lee et al., 2003c). While rapid spinning MAS yields relatively narrow spectrum for the solids, not all of these broadenings due to anisotropy are fully averaged out. This is particularly troublesome for non-crystalline solids including oxide glasses. For example, in Figure 3a the ^{17}O MAS NMR spectrum for $NaAlSiO_4$ glass exhibits broad features and does not resolve individual oxygen environments in the glasses (Lee and Stebbins, 2000a,b).

Recent advances and progress in 3QMAS NMR greatly improved resolution by removing residual anisotropy due to the second-order quadrupolar interactions. 3QMAS NMR provides two-dimensional spectra where one isotropic dimension provides a spectrum free from quadrupolar broadening and the other MAS dimension provides similar spectrum with one-dimensional MAS NMR (Frydman and Harwood, 1995; Amoureux et al., 1996; Baltisberger et al., 1996). Figure 3b shows the ^{17}O 3QMAS NMR spectrum for $NaAlSiO_4$ glasses where the three distinct peaks for oxygen sites such as $^{[4]}Si-O-^{[4]}Si$, $^{[4]}Si-O-^{[4]}Al$, and $^{[4]}Al-O-^{[4]}Al$ (as shown in the ring of aluminosilicate cluster, Fig. 3, up left) are clearly resolved (Lee and Stebbins, 2000a,b). The spectrum is often shown as a two-dimensional contour plot (bottom left) and also represented as a projection on the isotropic dimension (isotropic projection), which has much higher resolution than conventional MAS NMR spectrum, allowing us to quantify the distribution of framework units ($^{[4]}Si$, $^{[4]}Al$) and other aspects of disorder in aluminosilicate glasses

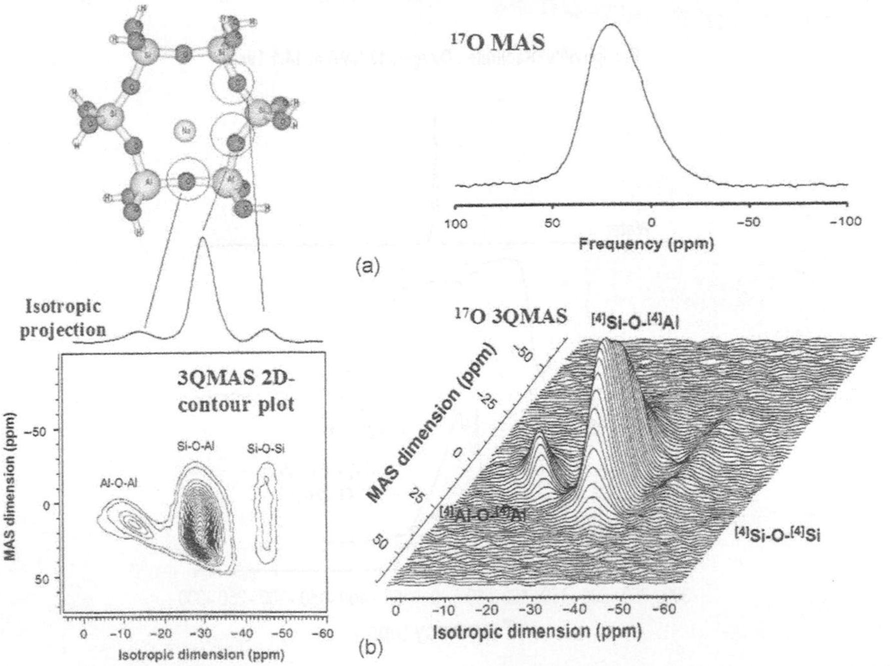

Figure 3. Comparison of ^{17}O MAS (a) and 3Q (triple quantum) MAS NMR (b) for sodium aluminosilicate ($NaAlSiO_4$) glasses (Lee and Stebbins, 2000a,b). The model aluminosilicate ring cluster with $^{[4]}Si-O-^{[4]}Si$, $^{[4]}Si-O-^{[4]}Al$, and $^{[4]}Al-O-^{[4]}Al$ are shown (left, up).

and melts (Lee and Stebbins, 2000a,b; 2003b; Lee et al., 2003b). In the following section, these advances in the quantification of the extent of disorder and the structures of non-crystalline solids at ambient (Section 3.1) and high pressure (Section 3.2) using mainly ^{17}O and ^{27}Al 3QMAS NMR are summarized. Finally, it should be noted that the 3QMAS efficiency depends primarily on the magnitude of quadrupolar interactions and partly on the other experimental conditions (Frydman and Harwood, 1995; Amoureux et al., 1996; Baltisberger et al., 1996; Lee and Stebbins, 2000b). These effects were carefully investigated here and were used to calibrate the experimental intensity.

2.3.2. NMR spectroscopy

The ^{17}O and ^{27}Al 3QMAS NMR spectra for sodium silicate and aluminosilicate glasses and melts were collected at 7.1 T (300 MHz) on CMX infinity 300 spectrometer with 2.5 mm MAS probe (spinning speed of 19 kHz) at a Larmor frequency of 78.2 MHz for ^{27}Al and 40.7 MHz for ^{17}O. The detailed experimental conditions were reported previously (Lee, 2004b). ^{17}O and ^{27}Al 3QMAS NMR spectra were collected using a shifted-echo pulse sequence, with two hard pulses (3.3 and 0.75 μs) and one soft pulse (11 μs) (Baltisberger et al., 1996; Lee, 2004b). Tap water and 0.1 M $AlCl_3$ *(aq)* were the external references for ^{17}O and for ^{27}Al, respectively. ^{17}O and ^{27}Al 3QMAS NMR data were processed with a shear transformation, with the isotropic and MAS dimension frequencies scaled as described in Baltisberger et al. (1996) (Frydman and Harwood, 1995; Amoureux et al., 1996 also given below): for a spin-5/2 nuclei such as ^{27}Al and ^{17}O, the observed peak positions (resonance frequencies) in the isotropic (δ_{3QMAS}) and MAS dimensions (δ_{MAS}) are given by:

$$\delta_{3QMAS} = -17/31\delta_{iso}^{CS} + 10/31\delta_{iso}^{2Q} \tag{1}$$

$$\delta_{MAS} = \delta_{iso}^{CS} + \delta_{iso}^{2Q} \tag{2}$$

where δ_{iso}^{CS} and δ_{iso}^{2Q} are the isotropic chemical shift and the second-order quadrupolar shift for spin-5/2 nuclei that is $6000P_q^2/\omega_0^2$ (ω_0 is the Larmor frequency and P_q is quadrupolar coupling product, respectively). Mean values of P_q and δ_{iso}^{CS} of each oxygen or aluminum site can be estimated from the above relations. In this chapter, most of the two-dimensional NMR spectra (except Fig. 2) are presented in a way that the isotropic and MAS dimensions are plotted in opposite directions to some of the previously reported spectra (e.g. Lee and Stebbins, 2000a,b) but the interpretation of the spectra should be identical to these studies as the same relations given above are used.

2.4. Insights from quantum chemical calculations

Quantum chemical calculations have been one of the complementary tools to explore the relationship between molecular structures of silicate glasses and melts and their spectroscopic responses, particularly for isotropic chemical shift (δ_{iso}) from NMR (Tossell and Lazzeretti, 1988; Tossell, 1998). Here, these methods were used to give insight into the relationship between atomic configurations around oxygen sites in silicate melts at high pressure and structurally relevant NMR parameters (isotropic chemical shift, δ_{iso}).

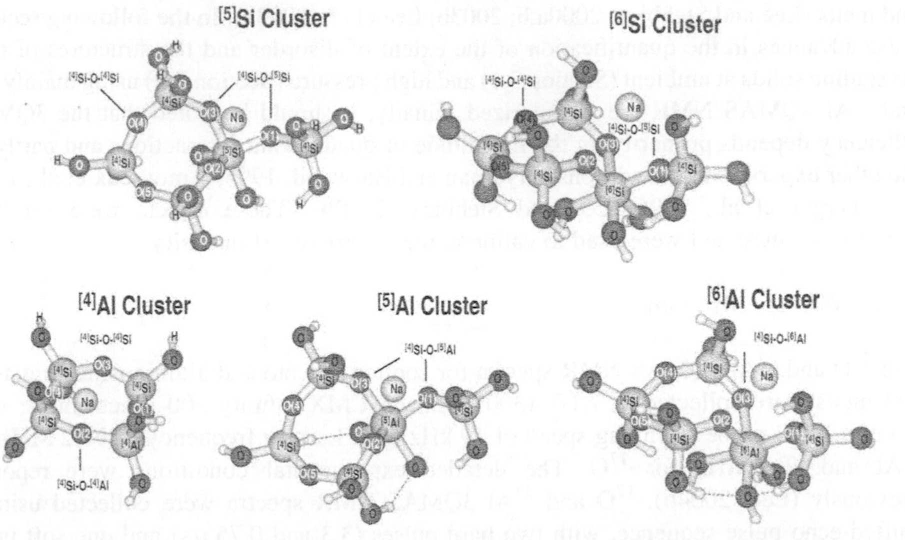

Figure 4. The optimized geometries of sodium silicate clusters with $^{[6]}Si$ and $^{[5]}Si$ (up) and optimized aluminosilicate clusters with $^{[4]}Al$, $^{[5]}Al$, and $^{[6]}Al$ that are confined in four-member ring. The geometry optimizations were performed at the HF level of theory with 6-311G(d) basis sets for the all clusters. Dotted lines refer to hydrogen bonds (these figures are modified from Lee et al., 2003c; Lee, 2004b).

Quantum chemical calculations were performed with Gaussian98 for model silicate and aluminosilicate clusters (Fig. 4) (Foresman and Frisch, 1996; Frisch et al., 1998). Some of the peak assignments of ^{17}O NMR spectra presented here are based on the correlation established by quantum chemical calculations: detailed discussions about methods of quantum chemical calculations including geometry optimization, the peak assignment, and comparison between experimental data and calculated isotropic chemical shift are given in our previous reports (Lee, 2004b).

Figure 4 shows the equilibrium geometries for model sodium silicate and aluminosilicate clusters (Lee et al., 2003a; Lee, 2004b). $^{[n]}$(Si,Al)–O bond length increases with increasing coordination number n, and there is rough inverse correlation between $^{[n]}$(Si,Al)–O and bond $^{[n]}$(Si,Al)–O–$^{[4]}$Si bond angle. Quantum chemical calculations for model silicate clusters showed that ^{17}O isotropic chemical shift (δ_{iso}) for $^{[n]}$(Si,Al)–O–$^{[4]}$Si mainly increases with increasing coordination number (n). While ^{17}O δ_{iso} seems to be affected by longer range interactions including second and third coordination spheres, it shows relatively moderate positive correlation with $^{[n]}$(Si,Al)–O bond length (Lee, 2004b).

3. Results, analyses, and discussion

3.1. Quantification of disorder in silicate glasses and melts at ambient pressure: a view from 3QMAS NMR

The structure of silicate glasses and melts can be described by quantifying the various aspects of short to medium-range disorder among framework and non-framework cations.

The recent reviews and monographs of the progress in the structures of silicate glasses and melts from other spectroscopic tools including vibrational spectroscopies (Mysen et al., 1982; McMillan et al., 1992; McMillan and Wolf, 1995) and X-rays (conventional or synchrotron radiation) can be found elsewhere (Brown et al., 1995). There have also been a significant amount of studies of glass structures using mainly one-dimensional conventional MAS NMR that were previously reviewed (Kirkpatrick et al., 1986; Engelhardt and Michel, 1987; Kirkpatrick, 1988; Eckert, 1992; 1994; Stebbins, 1995). While these methods have yielded many indirect implications for the extent of disorder in silicate melts and glasses, it should be noted that X-ray-based techniques and conventional spectroscopy including one-dimensional NMR provide *only limited information* about the chemical order among framework units (e.g. Al avoidance) or network-modifying cations, the nature of phase separation in *complex* silicates glasses, and atomic environments around NBO and BO due primarily to the limited resolution in these techniques. Modern solid-state NMR techniques, particularly, 3QMAS NMR provided a unique opportunity to obtain the information on the extent of disorder in the silicate melts (Lee, 2004a). Here, we summarize the progress made *using 3QMAS NMR of quadrupolar nuclei for silicate glasses and melts at 1 atm.*

The distribution of cations and anions in silicate glasses and melts determines various aspects of disorder. The scheme of disorder on a length scale less than 1 nm is divided into (1) chemical and (2) topological disorder. The former describes the distributions of framework and non-framework cations and the latter refers to the distribution of internal variables such as bond angle and length. Chemical order can be further divided into (1.1) connectivity and (1.2) non-framework disorder. This classification is based on the types of mixing units (e.g. framework or non-framework units). For instance, the connectivity here denotes both (1.1.1) framework disorder that describes the mixing between framework units (e.g. Si and Al) and (1.1.2) the degree of polymerization describing the mixing between framework and non-framework units (Na^+, Ca^{2+}, and K^+ etc.), and thus the formation of NBO. Non-framework disorder accounts only for the distribution of non-framework cations (network-modifying or charge-balancing cations).

Framework disorder in silicate glasses and melts has recently been quantified by the introduction of the *degree of Al avoidance* that varies from chemical order ($Q = 1$, where $^{[4]}Al{-}O{-}^{[4]}Al$ is not allowed if Si/Al > 1) to random distribution ($Q = 0$, where $^{[4]}Al{-}O{-}^{[4]}Al$ exists) and the *extent of phase separation* that varies from complete phase separation ($P = 1$) to random distribution ($P = Q = 0$) (Lee and Stebbins, 1999; 2002). Previous results have shown that Q in aluminosilicate glasses and melts ranges from about 0.85 to 0.95, meaning that the $^{[4]}Si$ and $^{[4]}Al$ distribution show tendency for the chemical order toward a Al avoidance but also allows for significant fractions of $^{[4]}Al{-}O{-}^{[4]}Al$ even if Si/Al = 1 (Lee and Stebbins, 1999; 2000a), demonstrating violation of Al-avoidance rule. The detailed degree of disorder also depends on the types of charge-balancing and network-modifying cations (Lee and Stebbins, 2000b). In binary borosilicate glasses and melts, it is shown that the distribution $^{[3]}B$ and $^{[4]}Si$ were described with P value of about 0.65, which corresponds to an ordering scheme between phase separation and random distribution (Lee and Stebbins, 2002).

The degree of polymerization here refers to the mixing between non-framework cations and framework cations and thus determines formation of NBO and BO in the oxide melts and glasses (Lee and Stebbins, 2003b; Lee et al., 2003b). This aspect of disorder can also be directly determined by probing BO and NBO using ^{17}O NMR. Non-randomness in NBO distribution was manifested in the recent studies of calcium aluminosilicate glasses and melts where selective depolymerization occurs at the silicate network rather than at the aluminate network, leading to the selective formation of (Ca, Na)–O–Si instead of (Ca, Na)–O–Al (Allwardt et al., 2003; Lee et al., 2003b), consistent with the prediction given from earlier Raman studies (Mysen, 1988).

Non-framework disorder was recently quantified by introduction of the *degree of intermixing* (Qm) that varies from chemical order (Qm = − 1, e.g. Ca–Na–Ca–Na) to random distribution of non-framework cations (Qm = 0), to a clustering of the these cations (Qm = 1) (Lee and Stebbins, 2003b; Lee et al., 2003b). Detailed analysis of NBO environment using ^{17}O NMR shows that chemical ordering among network-modifying cations is prevalent in the various mixed cation silicate glasses. The first evidence of moderate degree of chemical order was reported in Na–Ca silicate glasses and melts (Lee and Stebbins, 2003b). Ba–Mg and K–Mg silicates show strong preference for NBO formation either with Ba (in Ba–Mg silicates) or Mg (in K–Mg silicates) (Farnan et al., 1992; Lee et al., 2003b; Allwardt and Stebbins, 2004). The extent of disorder among network-modifying cations in the mixed cation silicate glasses is certainly dependent on the types of network-modifying cations. For instance, previous study using two-dimensional dynamic angle spinning (DAS) NMR of series of Na–K silicate glasses reported random distribution of Na^+ and K^+ (Florian et al., 1996).

Topological disorder due to bond angle and length distribution has been also extensively studied using two-dimensional solid-state NMR including 3QMAS and DAS (Grandinetti et al., 1995; Angeli et al., 1999; 2000). There is pronounced non-randomness in bond angle and length distribution. For instance, Na–O distribution in sodium silicate and aluminosilicate glasses was described by *perturbed non-framework cation distribution model* where the Na is homogenously distributed (contrary to modified random network model, Gaskell et al., 1991; Greaves and Ngai, 1994) but shows strong compositional effect (Lee and Stebbins, 2003a). These spectroscopic and conceptual advances for quantification of the extent of disorder and structure of silicate glasses and melts at 1 atm using solid-state NMR have been extended to silicate glasses quenched from melts at high pressure. The main results from our recent progress are given below.

3.2. Structure and properties of silicate glasses and melts at high pressure

3.2.1. Pressure-induced changes in coordination environment of framework cations in silicate and aluminosilicate glasses and melts

The previous studies of silicate glasses and melts at high pressure reported that the coordination number of framework units increases with increasing pressure (Hemley et al., 1986). As previously mentioned, in the binary silicate glasses, earlier studies using ^{29}Si MAS NMR showed that the fraction of highly coordinated Si ($^{[5,6]}Si$) increases with increasing pressure (Xue et al., 1989; 1991). In aluminosilicate glasses and melts, the fraction of highly coordinated Al also increases with increasing pressure as shown by ^{27}Al MAS NMR spectra of partially depolymerized glasses ($(Na_2O)_{0.75}(Al_2O_3)_{0.25}4SiO_2$) with pressure (Yarger et al., 1995; Poe et al., 1997). Here, some of the progress on

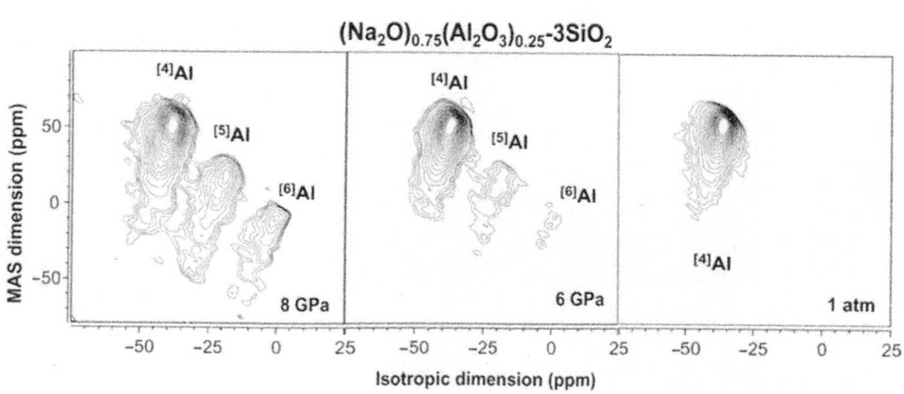

(Na2O)0.75(Al2O3)0.25-3SiO2
[4]Al
[5]Al
[6]Al
8 GPa
[4]Al
[5]Al
[6]Al
6 GPa
[4]Al
1 atm
MAS dimension (ppm)
50
0
−50
−50
−25
0
25
Isotropic dimension (ppm)

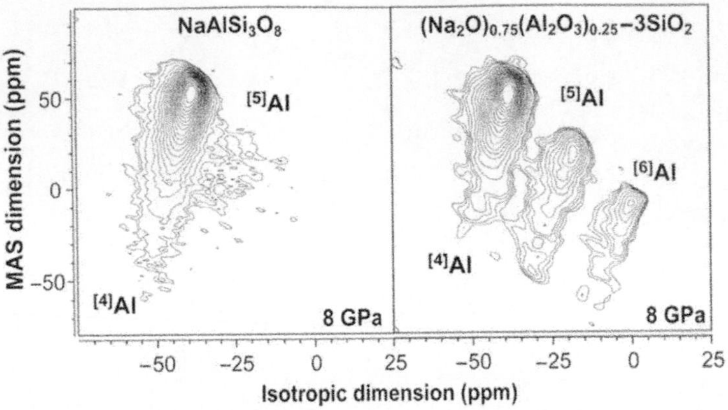

NaAlSi3O8
(Na2O)0.75(Al2O3)0.25–3SiO2
[5]Al
[5]Al
[6]Al
[4]Al
[4]Al
8 GPa
8 GPa
MAS dimension (ppm)
50
0
−50
−50
−25
0
25
−50
−25
0
25
Isotropic dimension (ppm)

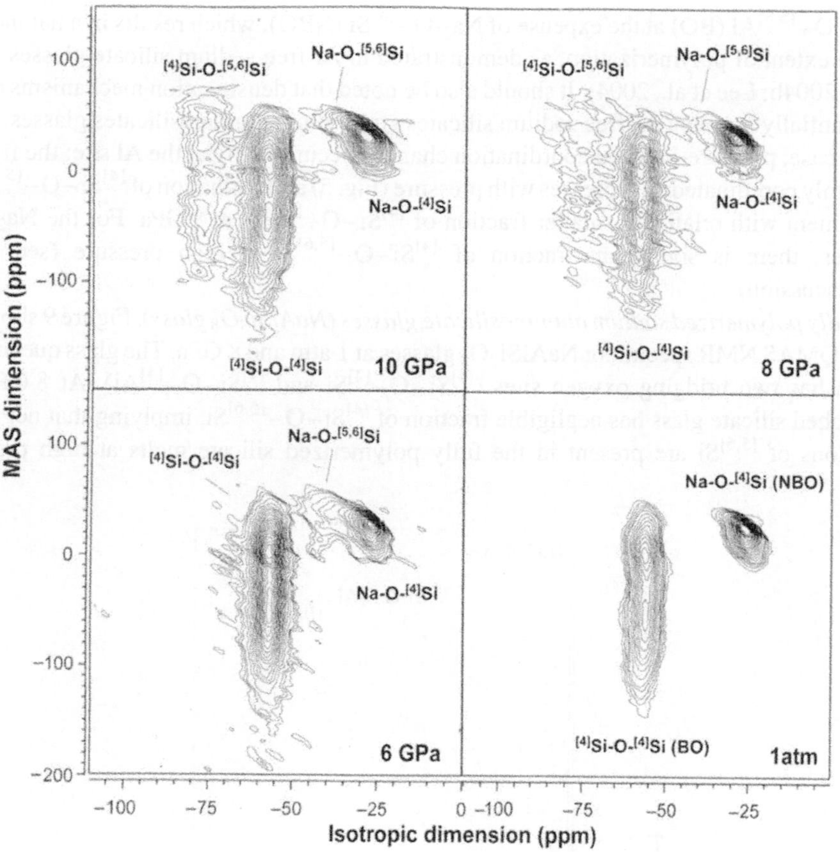

[4]Si-O-[5,6]Si
Na-O-[5,6]Si
Na-O-[4]Si
[4]Si-O-[4]Si
10 GPa
[4]Si-O-[5,6]Si
Na-O-[5,6]Si
Na-O-[4]Si
[4]Si-O-[4]Si
8 GPa
[4]Si-O-[4]Si
Na-O-[5,6]Si
Na-O-[4]Si
6 GPa
Na-O-[4]Si (NBO)
[4]Si-O-[4]Si (BO)
1atm
MAS dimension (ppm)
Isotropic dimension (ppm)

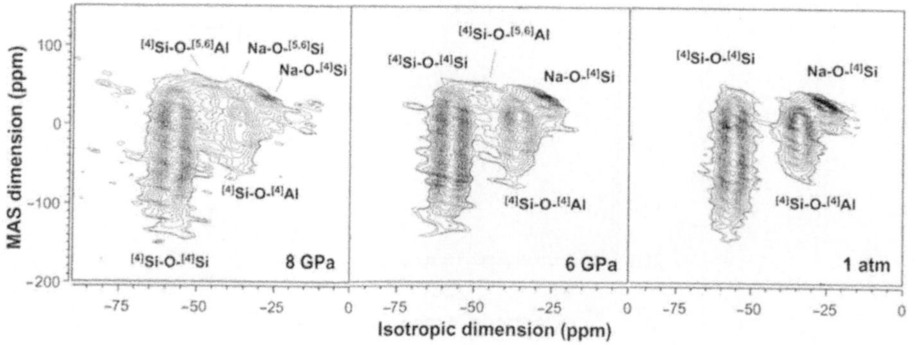

[4]Si-O-[5,6]Al
Na-O-[5,6]Si
Na-O-[4]Si
[4]Si-O-[4]Al
[4]Si-O-[4]Si
8 GPa
[4]Si-O-[5,6]Al
[4]Si-O-[4]Si
Na-O-[4]Si
[4]Si-O-[4]Al
6 GPa
[4]Si-O-[4]Si
Na-O-[4]Si
[4]Si-O-[4]Al
1 atm
MAS dimension (ppm)
Isotropic dimension (ppm)

[4]Si–O–[5,6]Al (BO) at the expense of Na–O–[4]Si (NBO), which results in a net increase in the extent of polymerization, as demonstrated in Al-free sodium silicate glasses (NS3) (Lee, 2004b; Lee et al., 2004). It should also be noted that densification mechanisms differs substantially between Al-free sodium silicates and sodium aluminosilicates glasses. In the latter case, pressure-induced coordination changes occur mainly at the Al site: the fraction of highly coordinated Al increases with pressure (Fig. 5) and formation of [4]Si–O–[5,6]Al is prominent with relatively smaller fraction of [4]Si–O–[5,6]Si at 8 GPa. For the $Na_2Si_3O_7$ glasses, there is significant fraction of [4]Si–O–[5,6]Si at high pressure (see below for discussion).

Fully polymerized sodium aluminosilicate glasses ($NaAlSi_3O_8$ glass). Figure 9 shows the ^{17}O 3QMAS NMR spectra for $NaAlSi_3O_8$ glasses at 1 atm and 8 GPa. The glass quenched at 1 atm has two bridging oxygen sites ([4]Si–O–[4]Si and [4]Si–O–[4]Al). At 8 GPa, the quenched silicate glass has negligible fraction of [4]Si–O–[5,6]Si, implying that negligible fractions of [5,6]Si are present in the fully polymerized silicate melts at high pressure.

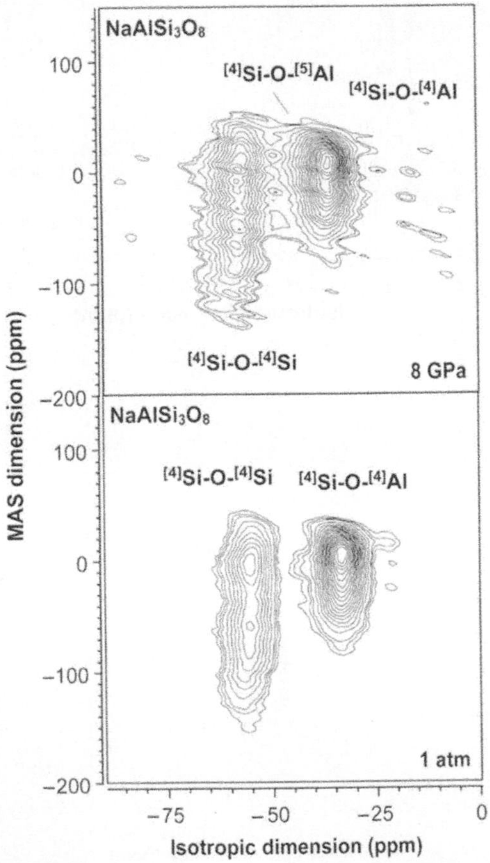

Figure 9. ^{17}O 3QMAS NMR spectra for $NaAlSi_3O_8$ glasses with varying pressure at 7.1 T. Contour lines are drawn at 5% increments from 8 to 88% of relative intensity, with added lines at the 4 and 6% levels to better show low-intensity peaks (this figure is modified from Figure 10 of Lee, 2004b).

Whereas the presence of [4]Si–O–[5,6]Al oxygen or NBO is not clear because of possible peak overlap with [4]Si–O–[4]Al in the ^{17}O 3QMAS NMR spectra at 8 GPa, it is likely that there is small fraction of [4]Si–O–[5]Al as evidenced by a few percentage of [5]Al in ^{27}Al 3QMAS NMR spectrum for the albite glasses (Fig. 5). Note that peak positions of each oxygen clusters and thus the structurally relevant NMR parameters vary systematically with pressure (The pressure-induced changes in NMR parameters have been discussed in detail elsewhere (Lee, 2004b) and pertinent aspects of these results will be discussed below.)

Effect of composition on the degree of polymerization. Speciation in silicate glasses quenched from melts at high pressure and the pressure-dependent densification mechanisms are largely dependent on composition (e.g. Na/Si, Si/Al, and the types of non-framework cations). For example, Figure 10 shows the ^{17}O 3QMAS NMR spectra for NS3, NAS150560, and NAS6 glasses quenched from melts at 8 GPa, which demonstrates the effect of composition on the structure (connectivity) of melts. In Al-free sodium silicate glasses, major pressure-induced changes involve the formation of [5,6]Si and thus oxygens (BO and NBO) connecting these highly coordinated framework units and [4]Si or alkali metal cations. Partially depolymerized aluminosilicate melts at high pressure exhibit complex connectivity involving both highly coordinated Al and [5,6]Si where [4]Al is proportionally more highly coordinated than [4]Si. Finally, the formation of highly coordinated Si and Al appears to be largely prohibited in fully polymerized silicate melts at high pressure, at least up to the pressure range studied here. The densification of fully polymerized silicate melts at high pressure involves mostly topological changes such as reduction of bond angle and an increase in bond length to the pressure ranges studied here (Lee, 2004b; Lee et al., 2004).

3.2.3. Pressure-induced changes in the extent of disorder, transport, and thermodynamic properties of silicate melts and glasses

The connectivity in silicate glasses melts at high pressure obtained from the ^{17}O NMR data suggests a non-random distribution of network polyhedra ([4,5,6]Al and [4,5,6,]Si).

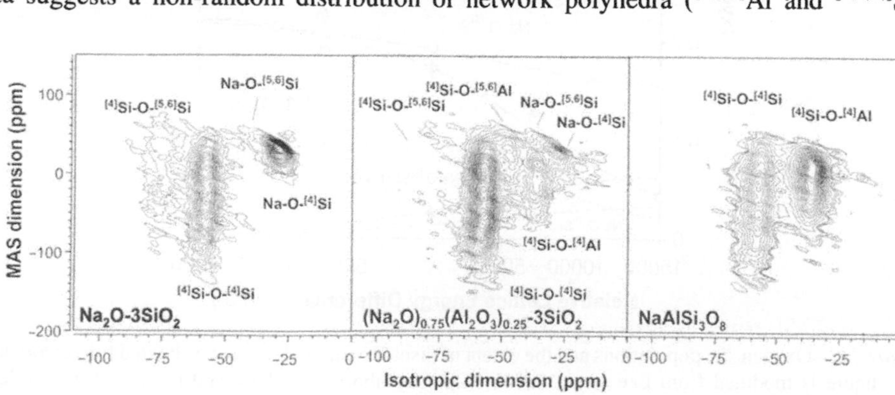

Figure 10. ^{17}O 3QMAS NMR spectra for $Na_2O-3SiO_2$ (NS3), $(Na_2O)_{0.75}(Al_2O_3)_{0.25}$ $3SiO_2$ (NAS150560), and $NaAlSi_3O_8$ (NAS6 glasses) quenched from melts at 8 GPa. The contour lines are drawn from 8 to 98% with 5% increments and additional lines at 4 and 6% (this figure is modified from Figure 10 of Lee et al., 2004).

The chemical order appears prominently in the distribution of network polyhedra, i.e. at high pressure where the formation of oxygen clusters, such as $^{[4]}Al-O-^{[4]}Al$, $^{[5]}Al-O-^{[5]}Al$, $^{[6]}Al-O-^{[6]}Al$, and $^{[5]}Si-O-^{[5]}Si$ is significantly suppressed. This suggests that a clustering of these cations seems unlikely. For example, as shown in Figure 11, numerical calculations of populations of BO and NBO with varying cluster energy difference for sodium silicate melts at 10 GPa show that the experimental oxygen site population (from ^{17}O 3QMAS NMR spectra of sodium silicate glasses quenched from melts at 10 GPa) falls into the extent of framework disorder between random distribution and chemical order (Lee et al., 2003a). The possible complication would be the presence of the edge-shared oxygen that is not likely to be the dominant oxygen species as oxygen species concentration is relatively well accounted for the mixing among corner-shared framework units, implying that the pronounced mixing model is correct. On the other

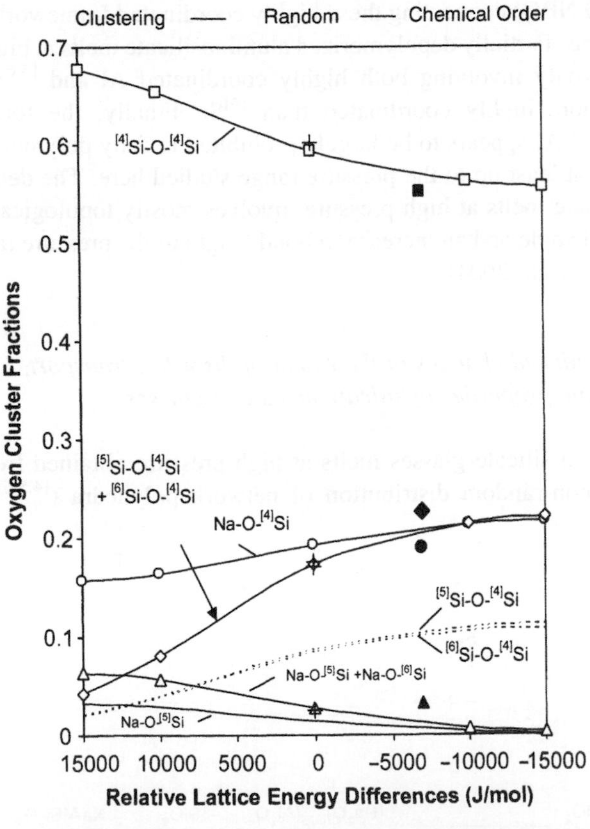

Figure 11. Oxygen site populations and the extent of disorder among highly coordinated framework units (this figure is modified from Lee et al., 2003a). Open symbols with thick and thin solid lines refer to the calculated results, and closed symbols denote experimental populations from ^{17}O 3QMAS NMR (square: $^{[4]}Si-O-^{[4]}Si$, diamond: $^{[5]}Si-O-^{[4]}Si + ^{[6]}Si-O-^{[4]}Si$, circle: $Na-O-^{[4]}Si$, and triangle: $Na-O-^{[5]}Si + Na-O-^{[6]}Si$). Squares refer the oxygen site populations with an assumption of random distribution of framework and non-framework sites (without specific preference for the formation of NBO with framework units).

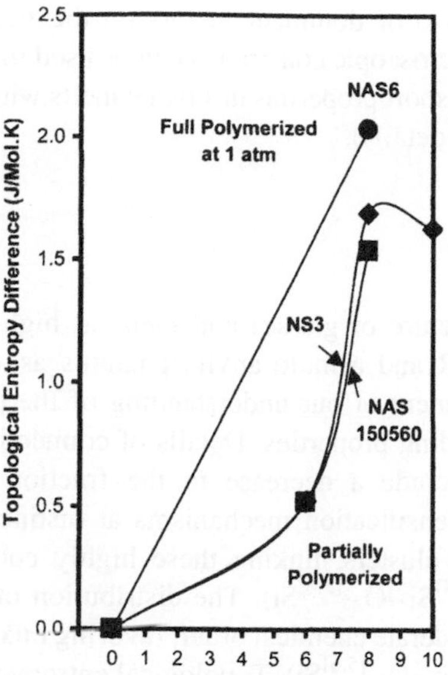

Figure 12. Variation in topological entropy of silicate resulting from the change in Si–O bond length in the $^{[4]}$Si–O–$^{[4]}$Si angle with pressure (this figure is modified from Lee, 2004a). Here, the relative entropy difference was calculated from the experimental measure of the width of Si–O length distribution and Eq. (2) from Lee, 2004b.

hand, it should be noted that the $^{[4]}$Si–O–2$^{[6]}$Al peak (edge-shared oxygen in the tri-octahedral layer silicates) could overlap with $^{[4]}$Si–O–$^{[4]}$Si, making the detection of this species difficult (Lee et al., 2003c).

The bond angle and length distribution widens with increasing pressure, resulting in an increase in topological disorder (Lee, 2004b). Figure 12 shows the variation of topological entropy with pressure due to an increase in bond length distribution in $^{[4]}$Si–O–$^{[4]}$Si, for silicate glasses with varying degrees of polymerization (Lee, 2004a). As discussed previously (Sections 3.2.1 and 3.2.2), the topological entropy increase is more prominent in the fully polymerized silicate melts than partially polymerized silicate melts.

Both chemical aspects of disorder (distribution of network polyhedra) and topological disorder significantly affect the configurational entropy and enthalpy of these systems. While it may be difficult to obtain quantitative value of configurational entropy, it appears that above the threshold pressure where the highly coordinated framework units form, some aspect of configurational entropy increases with pressure as a result of the formation of new mixing units (whose distribution deviates from that of complete chemical order as shown above) (Diefenbacher et al., 1998) and an increase in topological disorder. Below a critical pressure, the topological contribution to the total configurational entropy is likely dominating (Lee et al., 2004).

These aspects of disorder affect the total transport properties of melts, as discussed in our previous report (Lee et al., 2004). In addition to these contributions, the fraction of NBO and its pressure dependence ($(\partial X_{NBO}/\partial P)_T$, note that NBO fraction decreases with increasing pressures) are two of dominant factors on the transport properties of melts (Lee et al., 2004). These microscopic constraints can be used to account for the anomalous pressure dependence of transport properties in silicate melts with pressure (see Lee, 2004a; Lee et al., 2004 for further details).

4. Conclusions

Recent studies on the structure of glasses and melts at high-pressure, combining two-dimensional 3QMAS NMR and a multi-anvil apparatus as well as quantum chemical analyses, result in a refinement of our understanding of the pressure-induced structural changes and the corresponding properties. Details of connectivity in silicate glasses and melts at high pressure include a decrease in the fraction of NBO with increasing pressure due to multiple densification mechanisms at distinct pressure ranges, and the formation of new oxygen clusters, linking these highly coordinated framework units (e.g. $^{[4]}$Si–O–$^{[5,6]}$Al and $^{[4]}$Si–O–$^{[5,6]}$Si). The distribution of these units deviates from random, i.e. there exists moderate chemical order, favoring mixing between different types of framework units (e.g. $^{[4]}$Si–O–$^{[5,6]}$Si). Topological entropy increases with pressure, i.e. the bond angle and length, and the distortion of framework units increases with pressure. These experimental results highlight a new opportunity to investigate the molecular structures of silicate melts at high pressure and reveal connections between the microscopic signatures of pressure-induced changes with the macroscopic properties of melts (e.g. anomalous pressure-dependent viscosity and diffusivity as well as thermodynamic properties) (Lee et al., 2003a, 2004; Lee, 2004b).

Acknowledgements

This research is supported by a Carnegie Postdoctoral Fellowship to L.S.K. He thanks Drs K. Mibe and G. Gudfinnsson for help with the multi-anvil experiments at the Geophysical Laboratory and Prof. P. Grandinetti at Ohio State University for RMN software for two-dimensional NMR data processing. Part of the progress made for the silicate glasses and melts at 1 atm (Section 2.3) was from L.S.K's Ph D. thesis guided by Professor J.F. Stebbins at Stanford, supported by the Stanford Graduate Fellowship. We also thank two anonymous reviewers for helpful comments and constructive suggestions and Dr Chen and Dr Dobrzhinetskaya for help.

References

Allwardt, J., Stebbins, J.F., 2004. Ca–Mg and K–Mg mixing around non-bridging O atoms in silicate glasses: an investigation using ^{17}O MAS and 3QMAS NMR. Am. Mineral. 89, 777–784.

Allwardt, J., Lee, S.K., Stebbins, J.F., 2003. Bonding preferences of non-bridging oxygens in calcium aluminosilicate glass: evidence from O-17 MAS and 3QMAS NMR on calcium aluminate glass. Am. Mineral. 88, 949–954.

Allwardt, J.R., Stebbins, J.F., Schmidt, B.C., Frost, D.J., 2004. The effect of composition, compression, and decompression on the structure of high pressure aluminosilicate glasses: an investigation utilizing ^{17}O and ^{27}Al NMR, this volume.

Amoureux, J.P., Fernandez, C., Frydman, L., 1996. Optimized multiple-quantum magic-angle spinning NMR experiments on half-integer quadrupoles. Chem. Phys. Lett. 259, 347–355.

Amoureux, J.-P., Bauer, F., Ernst, H., Fernandez, C., Freude, D., Michel, D., Pingel, U.-T., 1998. ^{17}O multiple-quantum and 1H MAS NMR studies of zeolite ZSM-5. Chem. Phys. Lett., 285.

Angeli, F., Charpentier, T., Faucon, P., Petit, J.C., 1999. Structural characterization of glass from the inversion of Na-23 and Al-27 3Q-MAS NMR spectra. J. Phys. Chem. B 103, 10356–10364.

Angeli, F., Delaye, J.M., Charpentier, T., Petit, J.C., Ghaleb, D., Faucon, P., 2000. Investigation of Al–O–Si bond angle in glass by Al-27 3Q-MAS NMR and molecular dynamics. Chem. Phys. Lett. 320, 681–687.

Angell, C.A., Cheeseman, P.A., Tamaddon, S., 1982. Pressure enhancement of ion mobilities in liquid silicates from computer simulations studies to 800 kbar. Science 218, 885–887.

Baltisberger, J.H., Xu, Z., Stebbins, J.F., Wang, S., Pines, A., 1996. Triple-quantum two-dimensional ^{27}Al magic-angle spinning nuclear magnetic resonance spectroscopic study of aluminosilicate and aluminate crystals and glasses. J. Am. Chem. Soc. 118, 7209–7214.

Bertka, C.M., Fei, Y., 1997. Mineralogy of the Martian interior up to core-mantle boundary pressures. J. Geophys. Res. 102, 5251–5264.

Brown, G.E. Jr., Farges, F., Calas, G., 1995. In: Stebbins, J.F., McMillan, P.F., Dingwell, D.B. (Eds), X-ray Scattering and X-ray Spectroscopy Studies of Silicate Melts, Vol. 32. Mineralogical Society of America, Washington, DC, pp. 317–410.

Diefenbacher, J., McMillan, P.F., Wolf, G.H., 1998. Molecular dynamics simulations of $Na_2Si_4O_9$ liquid at high pressure. J. Phys. Chem. B 102, 3003–3008.

Dirken, P.J., Kohn, S.C., Smith, M.E., Vaneck, E.R.H., 1997. Complete resolution of Si–O–Si and Si–O–Al fragments in an aluminosilicate glass by O-17 multiple-quantum magic-angle-spinning NMR-spectroscopy. Chem. Phys. Lett. 266, 568–574.

Dunn, T., 1982. Oxygen diffusion in three silicate melts along the join diopside-anorthite. Geochim. Cosmochim. Acta 46, 2293–2299.

Eckert, H., 1992. Structural characterization of noncrystalline solids and glasses using solid state NMR. Prog. Nucl. Mag. Reson. 24, 159–293.

Eckert, H., 1994. Structural studies of non-crystalline solids using solid state NMR. In: Blümich, B. (Ed.), New Experimental Approaches and Results. Springer, Berlin, pp. 127–198.

Engelhardt, G., Michel, D., 1987. High-Resolution Solid-State NMR of Silicates and Zeolites. Wiley, New York.

Farnan, I., Grandinetti, P.J., Baltisberger, J.H., Stebbins, J.F., Werner, U., Eastman, M., Pines, A., 1992. Quantification of the disorder in network modified silicate glasses. Nature 358, 31–35.

Florian, P., Vermillion, K.E., Grandinetti, P.J., Farnan, I., Stebbins, J.F., 1996. Cation distribution in mixed alkali disilicate glasses. J. Am. Chem. Soc. 118, 3493–3497.

Foresman, J.B., Frisch, A., 1996. Exploring Chemistry with Electronic Structure Methods. Gaussian Inc., Pittsburgh.

Frisch, M.J., Trucks, G.W., Schlegel, H.B., Scuseria, G.E., Robb, M.A., Cheeseman, J.R., Zakrzewski, V.G., Montgomery, J.A. Jr., Stratmann, R.E., Burant, J.C., Dapprich, S., Millam, J.M., Daniels, A.D., Kudin, K.N., Strain, M.C., Farkas, O., Tomasi, J., Barone, V., Cossi, M., Cammi, R., Mennucci, B., Pomelli, C., Adamo, C., Clifford, S., Ochterski, J., 1998. Gaussian 98. Revision A.11.

Frydman, I., Harwood, J.S., 1995. Isotropic spectra of half-integer quadrupolar spins from bidimensional magic-angle-spinning NMR. J. Am. Chem. Soc. 117, 5367–5368.

Gaskell, P.H., Eckersley, M.C., Barnes, A.C., Chieux, P., 1991. Medium-range order in the cation distribution of a calcium silicate glass. Nature 350, 675–677.

Grandinetti, P.J., Baltisberger, J.H., Farnan, I., Stebbins, J.F., Werner, U., Pines, A., 1995. Solid-state ^{17}O magic-angle and dynamic-angle spinning NMR study of the SiO_2 polymorph coesite. J. Phys. Chem. 99, 12341–12348.

Greaves, G.N., Ngai, K.L., 1994. Ionic transport properties in oxide glasses derived from atomic structure. J. Non-Cryst. Solids 172, 1378–1388.

Hemley, R.J., Mao, H.K., Bell, P.M., Mysen, B.O., 1986. Raman spectroscopy of SiO_2 glass at high pressure. Phys. Rev. Lett. 57, 747–750.

Kirkpatrick, R.J., 1988. In: Hawthorne, F.C. (Ed.), MAS NMR Spectroscopy of Minerals and Glasses. Mineralogical Society of America, Washington, DC, pp. 341–403.

Kirkpatrick, R.J., Dunn, T., Schramm, S., Smith, K.A., Oestrike, R., Turner, G., 1986. In: Walrafen, G.E., Revesz, A.G. (Eds), Magic-Angle Sample-Spinning Nuclear Magnetic Resonance Spectroscopy of Silicate Glasses: A Review. Plenum Press, New York, pp. 302–327.

Kushiro, I., 1976. Changes in viscosity and structure of melt of $NaAlSi_2O_6$ composition at high pressures. J. Geophys. Res. 81, 6347–6350.

Kushiro, I., 1983. Effect of pressure on the diffusivity of network-forming cations in melts of jadeite composition. Geochim. Cosmochim. Acta 47, 1415–1422.

Lee, S.K., 2004a. Microscopic origins of macroscopic properties of silicate melts: implications for melt generation and dynamics. Geochim. Cosmochim. Acta, in press.

Lee, S.K., 2004b. The structure of silicate melts at high pressure: quantum chemical calculations and solid state NMR. J. Phys. Chem. B 108, 5889–5900.

Lee, S.K., Stebbins, J.F., 1999. The degree of aluminum avoidance in aluminosilicate glasses. Am. Mineral. 84, 937–945.

Lee, S.K., Stebbins, J.F., 2000a. Al–O–Al and Si–O–Si sites in framework aluminosilicate glasses with Si/Al = 1: quantification of framework disorder. J. Non-Cryst. Solids 270, 260–264.

Lee, S.K., Stebbins, J.F., 2000b. The structure of aluminosilicate glasses: high-resolution ^{17}O and ^{27}Al MAS and 3QMAS NMR study. J. Phys. Chem. B 104, 4091–4100.

Lee, S.K., Stebbins, J.F., 2002. The extent of inter-mixing among framework units in silicate glasses and melts. Geochim. Cosmochim. Acta 66, 303–309.

Lee, S.K., Stebbins, J.F., 2003a. The distribution of sodium ions in aluminosilicate glasses: a high field Na-23 MAS and 3QMAS NMR study. Geochim. Cosmochim. Acta 67, 1699–1709.

Lee, S.K., Stebbins, J.F., 2003b. Nature of cation mixing and ordering in Na–Ca silicate glasses and melts. J. Phys. Chem. B 107, 3141–3148.

Lee, S.K., Musgrave, C.B., Zhao, P., Stebbins, J.F., 2001. Topological disorder and reactivity of borosilicate glasses: ab initio molecular orbital calculations and ^{17}O and ^{11}B NMR. J. Phys. Chem. B 105, 12583–12595.

Lee, S.K., Fei, Y., Cody, G.D., Mysen, B.O., 2003a. Order and disorder of sodium silicate glasses and melts at 10 GPa. Geophys. Res. Lett. 30, 1845.

Lee, S.K., Mysen, B.O., Cody, G.D., 2003b. Chemical order in mixed cation silicate glasses and melts. Phys. Rev. B 68, 214106.

Lee, S.K., Stebbins, J.F., Weiss, C.W., Kirkpatrick, R.J., 2003c. O-17 and Al-27 MAS and 3QMAS NMR study of synthetic and natural layer-silicates. Chem. Mater. 15, 2605–2613.

Lee, S.K., Cody, G.D., Fei, Y., Mysen, B.O., 2004. The nature of polymerization and properties of silicate glasses and melts at high pressure. Geochim. Cosmochim. Acta 68, 4203–4214.

McMillan, P.F., Wolf, G.H., 1995. In: Stebbins, J.F., McMillan, P.F., Dingwell, D.B. (Eds), Vibrational Spectroscopy of Silicate Liquids, Vol. 32. Mineralogical Society of America, Washington, DC, pp. 247–316.

McMillan, P.F., Wolf, G.H., Poe, B.T., 1992. Vibrational spectroscopy of silicate liquids and glasses. Chem. Geol. 96, 351–366.

Mehring, M., 1983. High Resolution NMR in Solids. Springer, Berlin.

Mysen, B.O., 1988. Structure and Properties of Silicate Melts. Elsevier, Amsterdam.

Mysen, B.O., 1990. Effect of pressure, temperature and bulk composition on the structure and species distribution in depolymerized alkali aluminosilicate melts and quenched glasses. J. Geophys. Res. 95, 15733–15744.

Mysen, B.O., Virgo, D., Seifert, F.A., 1982. The structure of silicate melts: implications for chemical and physical properties of natural magma. Rev. Geophys. Space Phys. 20, 353–383.

Navrotsky, A., Peraudeau, G., McMillan, P., Coutures, J.P., 1982. A thermochemical study of glasses and crystals along the joins silica–calcium aluminate and silica–sodium aluminate. Geochim. Cosmochim. Acta 46, 2039–2047.

Neuville, D.R., Richet, P., 1991. Viscosity and mixing in molten (Ca, Mg) pyroxenes and garnet. Geochim. Cosmochim. Acta 55, 1011–1019.

Poe, B.T., McMillan, P.F., Rubie, D.C., Chakraborty, S., Yarger, J.L., Diefenbacher, J., 1997. Silicon and oxygen self-diffusivities in silicate liquids measured to 15 Gigapascals and 2800 Kelvin. Science 276, 1245–1248.

Poe, B.T., Romano, C., Zotov, N., Civin, G., Marecelli, A., 2001. Compression mechanisms is aluminosilicate melts: Raman and XANES spectroscopy of glasses quenched from pressures up to 10 GPa. Chem. Geol. 174, 21–31.

Shimizu, N., Kushiro, I., 1984. Diffusivity of oxygen in jadeite and diopside melts at high pressures. Geochim. Cosmochim. Acta 48, 1295–1303.

Stebbins, J.F., 1995. In: Stebbins, J.F., McMillan, P.F., Dingwell, D.B. (Eds), Dynamics and Structure of Silicate and Oxide Melts: Nuclear Magnetic Resonance Studies, Vol. 32. Mineralogical Society of America, Washington, DC, pp. 191–246.

Stebbins, J.F., Xu, Z., 1997. NMR evidence for excess non-bridging oxygen in aluminosilicate glass. Nature 390, 60–62.

Tinker, D., Lesher, C.E., Hutcheon, I.D., 2003. Self-diffusion of Si and O in diopside–anorthite melt at high pressures. Geochim. Cosmochim. Acta 67, 133–142.

Toplis, M.J., Dingwell, D.B., Hess, K.U., Lenci, T., 1997. Viscosity, fragility, and configurational entropy of melts along the join SiO_2–$NaAlSiO_4$. Am. Mineral. 82, 979–990.

Tossell, J.A., 1998. Quantum mechanical calculation of Na-23 NMR shieldings in silicates and aluminosilicates. Phys. Chem. Miner. 27, 70–80.

Tossell, J.A., Lazzeretti, P., 1988. Calculation of NMR parameters for bridging oxygens in $H_3T–O–T'H_3$ linkages (T,T = Al, Si, P) for oxygen in SiH_3O^-, SiH_3OH and SiH_3OMg^+ and for bridging fluorine in $H_3SiFSiH_3^+$. Phys. Chem. Mineral. 15, 564–569.

Waff, H.S., 1975. Pressure-induced coordination changes in magmatic liquids. Geophys. Res. Lett. 2, 193–196.

Wang, S., Stebbins, J.F., 1998. On the structure of borosilicate glasses: a triple-quantum magic-angle spinning ^{17}O NMR study. J. Non-Cryst. Solids 231, 286–290.

Wolf, G.H., McMillan, P.F., 1995. In: Stebbins, J.F., McMillan, P.F., Dingwell, D.B. (Eds), Pressure Effects on Silicate Melt Structure and Properties, Vol. 32. Mineralogical Society of America, Washington, DC, pp. 505–562.

Wu, G., Dong, S., 2001. Two-dimensional ^{17}O multiple quantum magic-angle spinning NMR of organic solids. J. Am. Chem. Soc. 123, 9119–9125.

Xu, Z., Maekawa, H., Oglesby, J.V., Stebbins, J.F., 1998. Oxygen speciation in hydrous silicate glasses – an oxygen-17 NMR study. J. Am. Chem. Soc. 120, 9894–9901.

Xue, X., Stebbins, J.F., Kanzaki, M., Tronnes, R.G., 1989. Silicon coordination and speciation changes in a silicate liquid at high pressures. Science 245, 962–964.

Xue, X., Stebbins, J.F., Kanzaki, M., McMillan, P.F., Poe, B., 1991. Pressure-induced silicon coordination and tetrahedral structural changes in alkali silicate melts up to 12 GPa: NMR, Raman, and infrared spectroscopy. Am. Mineral. 76, 8–26.

Xue, X., Stebbins, J.F., Kanzaki, M., 1994. Correlations between O-17 NMR parameters and local structure around oxygen in high-pressure silicates and the structure of silicate melts at high pressure. Am. Mineral. 79, 31–42.

Yarger, J.L., Smith, K.H., Nieman, R.A., Diefenbacher, J., Wolf, G.H., Poe, B.T., McMillan, P.F., 1995. Al coordination changes in high-pressure aluminosilicate liquids. Science 270, 1964–1967.

IV
Structural and magnetic properties

Advances in High-Pressure Technology for Geophysical Applications
Jiuhua Chen, Yanbin Wang, T.S. Duffy, Guoyin Shen, L.F. Dobrzhinetskaya, editors

Chapter 13
Decompression of majoritic garnet: an experimental investigation of mantle peridotite exhumation

Larissa F. Dobrzhinetskaya, Harry W. Green, Alex P. Renfro and Krassimir N. Bozhilov

Abstract

Garnet peridotites containing relict majoritic garnet with abundant lamellae and interstitial blebs of pyroxene were recently discovered within mantle xenoliths and ultra high-pressure metamorphic terranes, but experimental reproduction of such microstructures has not yet been carried out. This chapter presents preliminary experimental results of our studies of the microstructures associated with supersilicic (majoritic) garnet high-temperature decompression ($T = 1400°C$) from 14 to 13, 12 and 7 GPa, and from 8 to 5 GPa. Experiments were performed in a Walker-style multi-anvil apparatus on a natural mineral mix corresponding to the bulk chemistry of garnet peridotite. We produced exsolution lamellae of Mg_2SiO_4 and interstitial blebs of diopside in the run products decompressed from 14 to 13 GPa and from 14 to 12 GPa at $T = 1400°C$. Blebs of both interstitial diopside and enstatite were formed during a slow high-temperature decompression from 14 to 7 GPa. Only blebs of interstitial enstatite were exsolved from majoritic garnet during decompression from 8 to 5 GPa at the same temperature. Mg_2SiO_4 has not been reported as an exsolution product in natural ultra high-pressure garnet peridotite, nor did we observe it in our experimental charges at pressures consistent with those inferred from studies of mantle xenoliths and orogenic belt garnet peridotites (7–10 GPa). The observation that Mg_2SiO_4 and diopside may be exsolved during decompression at higher pressures is consistent with expansion of the garnet field at the expense of wadsleyite at $P > 13$ GPa in Ringwood's diagram. The new experimental results may serve as a template for interpretation of microstructures in natural garnet peridotites, providing a better understanding of the depth from which such rocks have been exhumed.

1. Introduction

The concept of mantle convection suggests that some peridotites occurring as xenoliths in kimberlites and lamproites or as lenses within ultra high-pressure metamorphic (UHPM) belts related to continental collisions were previously at depths corresponding to the mantle transition zone or even in the lower mantle (e.g. Moore and Gurney, 1985; Dobrzhinetskaya et al., 1996; Harte et al., 1999; Stachel, 2001; Gillet et al., 2002). Their rock-forming minerals, equilibrated at high pressures and temperatures, break down during decompression when tectonic processes or explosive kimberlitic and/or lamproitic events transport them to Earth's surface.

Because uplift of mantle xenoliths occurs very quickly, in many cases no more than a few days, these rocks partially preserve the mineral chemistry and microstructural relations formed at great depth and record obvious disequlibrium features such as hydrous-phase-bearing alteration rims around garnet, or some subtle ones such as diffusion profiles (e.g. Smith, 1999) and odd isotopic systematics (e.g. Shimizu, 1999). It is known that evaluations of the depths at which mantle peridotites originate from are always uncertain. On the one hand, a comparison of chemical elements partitioning between co-existing mineral pairs in natural peridotites and high P and T experimental studies demonstrates that such xenoliths were residing at depths as great as ~150–200 km when they were extracted by their volcanic hosts. On the other hand, some inclusions in diamonds (e.g. Moore and Gurney, 1985; Harte et al., 1999) and microstructural relations between intracrystalline lamellae of pyroxenes and their host-garnet (Haggerty and Sautter, 1990; Sautter et al., 1991) strongly suggest that these rocks retain microstructural features corresponding to decompression due to their uplift from depths 300 to 800 km. Discoveries of microdiamonds within metamorphic terranes related to continental collisions have demonstrated that continental material has been subducted to depths of about 120–150 km (Sobolev and Shatsky, 1990; Xu et al., 1992; Dobrzhinetskaya et al., 1995; Nasdala and Massonne, 2000; van Roermund et al., 2002; Yang et al., 2003) and returned to the Earth's surface by a tectonic exhumation.

Moreover, several recent studies now suggest that peridotites tectonically incorporated into crustal lithologies within UHPM terranes may carry microstructural patterns indicating an even deeper recrystallization history. These microstructural patterns include intracrystalline exsolution lamellae of pyroxenes in former majoritic (supersilicic) garnet in garnet peridotites from the Norwegian Caledonides, from Sulu and Qaidam regions of Central Chinese Orogenic Belt (COB), suggesting exhumation of these rocks from 185 km to >250 km (van Roermund and Drury, 1998; Ye et al., 2000; Song et al., 2004), high-pressure clinoenstatite lamellae in clinopyroxene and a high concentration of oriented rods of ilmenites and plates of chromite in olivine from Alpe Arami garnet peridotites from the Swiss Alps and Qaidam from COB, suggesting exhumation from depths >250 km (Dobrzhinetskaya et al., 1996; Bozhilov et al., 1999, 2003; Song et al., 2004), and magnetite lamellae in olivine in garnet peridotite from the Sulu region of COB suggesting exhumation from >400 km (Zhang et al., 1999). Importantly, the peridotites and eclogites of these orogenic belts are enveloped in highly deformed continental gneisses, often showing extensive migmatization. These gneisses have been thought to contain no evidence of a high-pressure history and a controversy has raged for some time about whether or not they have also experienced deep subduction. However, recent careful studies of zircons extracted from these felsic gneisses have shown them to contain inclusions of coesite (e.g. Ye et al., 2000; Liu et al., 2002, 2003). This observation confirms that the crustal gneisses have, indeed, experienced deep subduction but have otherwise lost the microstructural evidence of the high-pressure events that have been preserved in the more refractory mafic and ultramafic rocks. Although the pyroxene exsolutions from garnet in some eclogite intercalated with continental metasediments from COB terranes preserve evidence of subduction to >200 km (Ye et al., 2000), a discussion that such rocks might originate from much shallower depth (~80 km) still continues (e.g. Hirajima and Nakamura, 2004). Such considerable conflicts in interpretation arise from a lack of experimental studies of what kind of microstructures are formed during decompression of the majoritic garnet at varying pressures and temperatures.

We present here experimental data on majoritic garnet synthesized from a mineral mix of garnet peridotite bulk chemistry that has been decompressed at high temperatures and now features exsolution of Mg_2SiO_4 and pyroxenes from majoritic garnet. A brief description of the results of this experimental program has been recently reported (Dobrzhinetskaya et al., 2004). The microstructural patterns of garnet peridotites that we have produced in the laboratory during high-temperature decompression from 14 to 13, 12, or 7 GPa and from 8 to 5 GPa are consistent with chemical changes of the re-equilibrated mineral phases. Therefore, they may be valuable for establishing a standard against which to evaluate microstructures in natural rocks, where chemical characteristics related to their earlier high-pressure history are obliterated by superimposed shallow thermal events.

2. Background on majoritic garnet

Majoritic (supersilicic) garnet is the most common mineral that preserves evidence of its decompression features. This garnet remains stable at $P > 5$ GPa (e.g. Akaogi and Akimoto, 1977; Irifune, 1987) and its composition is given by the complex solid solution $M_3^{\mathrm{viii}}\,(Al_{2-2n}M_nSi_n)^{\mathrm{vi}}Si_3O_{12}$, where $M = Mg^{2+}$, $Fe^{2+}$and Ca^{2+}, $0 \leq n \leq 1$ and superscripts indicate cations oxygen coordination. With increasing pressure, the "regular" garnet becomes Al^{3+}(and Cr^{3+}) deficient due to their replacement by M and Si^{4+}ions, thereby producing majoritic (supersilicic garnet) with Si $(Si^{\mathrm{iv}} + Si^{\mathrm{vi}}) > 3$ cations per formula unit (cpfu) (e.g. Smith and Mason, 1970). Experimental studies suggest that the content of Si in the octahedral site of majoritic garnet increases with increasing pressure (Akaogi and Akimoto, 1977; Gasparik, 1990; Irifune, 1998). A temperature effect was also recently recognized (Gasparik, 1990; Fei and Bertka, 1999), although no special experimental program has been undertaken to systematically explore this effect.

Majoritic garnets were found in meteorites (e.g. Smith and Mason, 1970; Kimura et al., 2000) and as inclusions in kimberlitic diamonds (e.g. Moore and Gurney, 1985; Stachel, 2001; Gillet et al., 2002). Examples of preservation of pyroxene exsolution lamellae in naturally decompressed majoritic garnets come from both xenoliths in kimberlites (Haggerty and Sautter, 1990; Sautter et al., 1991) and from subduction-zone peridotites (van Roermund and Drury, 1998; Ye et al., 2000; van Roermund et al., 2001; Song et al., 2004). Their interpretation is supported by the following reaction:

$$\text{supersilicic garnet} \rightarrow \text{exsolved pyroxenes} + \text{less-silicic garnet} \qquad (1)$$

which, if quantified in terms of the relative volume percentage of exsolved pyroxene versus "normal" garnet, may serve as a theoretical foundation for the qualitative interpretation of the exhumation path of garnet peridotites.

However, many other garnet peridotites from subduction zones contain amoeboid-like garnet surrounded by single crystals of pyroxenes and are accompanied by olivine clustered around the curved garnet boundaries (e.g. Dobrzhinetskaya et al., 1996; Green et al., 2000; Green and Dobrzhinetskaya, 2003) or interstitial orthopyroxenes between recrystallized domains of former majoritic garnet (van Roermund and Drury, 1998). Such garnets also may contain rounded non-oriented inclusions of each of these minerals, or all three (olivine, diopside and enstatite) together. The latter microstructural combinations are not yet understood, though they are very important since they might record the decompression history of mantle peridotite.

3. Experimental procedures

3.1. The goal of experiments

The objective of our experimental program is to produce high-pressure and high-temperature suites of mineral assemblages and microstructures representative of those observed in peridotite upwelling from Earth's deep upper mantle. The chemical and microstructural features of mineral assemblages formed during experimental high-temperature decompression of a model peridotite will yield quantitative knowledge of the kinetics and mechanism of dissolution and exsolution of solid phases. Therefore, such data will provide a framework for interpretations of similar phenomena recorded in natural rocks uplifted to Earth's surface during continental collisions.

3.2. Starting materials

To create a garnet peridotite bulk composition we used the following natural mineral mix (Ward's Natural Science Establishment, Inc.): 40 wt% San-Carlos olivine (Arizona) and 20 wt% each of pyropic garnet (Arizona), diopside (Yakutia, Russia) and enstatite (Norway). Chemical compositions of starting materials are provided in Table 1.

Starting powders of two different grain sizes were prepared – a "coarse-grained" mixture of 3–5 μm (70 wt%) and 10–25 μm (30 wt%) in diameter, and a "fine-grained" mixture with all particles of 3–5 μm in diameter. The coarser powder was used initially to acquire an understanding of the starting grain size necessary to obtain homogeneous run product compositions. The minerals for all starting mixes were powdered by grinding separately to the desirable grain size in an agate mortar with an agate pestle under ethanol, then dried for 24 h at 100°C in a vacuum oven. Grain size of the final mix was determined by mechanical sieving and evaluated using an optical microscope. The starting powder was loaded into a self-sealing MgO-capsule (Dobrzhinetskaya et al., 2000) and/or a welded platinum capsule (Fig. 1). We have made all our analytical measurements and microstructural observations of the run products far away from the walls of the MgO capsule and as close as possible to the thermocouple (tc) to avoid a thermal gradient effect, which commonly exists in multi-anvil sample assemblies (Fig. 1). Positive control of oxygen fugacity was performed by placing a Ni foil at one end of the starting material, which consumes oxygen from the interstices of the starting powder by reaction into NiO and buffers fO_2 at a known level (about −5 log units).

3.3. Apparatus and assembly

The experiments were conducted in a Walker-style multi-anvil apparatus (Walker et al., 1990; Walker, 1991) using a 6 mm truncated edge octahedron and two rolls of 0.0254 mm thick Re-foil as a heater. The spaces between heater and MgO-capsule, which contains the

Table 1. Chemical composition in (wt%) of minerals used for our experiments as starting material.

Oxides	Olivine	Enstatite	Diopside	Garnet
SiO_2	41.12	57.99	56.42	42.18
TiO_2	0.02	0.08	0.17	0.50
Al_2O_3	0.01	0.83	2.58	22.38
MgO	48.52	35.64	15.58	21.15
CaO	0.03	0.29	22.02	4.96
MnO	0.10	0.13	0.06	n.m
FeO	9.35	5.43	2.21	7.85
Na_2O	0.15	0.19	0.81	0.07
Cr_2O_3	n.m	0.21	1.10	1.48
Sum	99.36	100.81	100.95	100.57
Cations	4 Oxygens	6 Oxygens	6 Oxygens	12 Oxygens
Si^{4+}	1.0104	1.9757	2.0094	2.9790
Ti^{4+}	0.0015	0.0200	0.0046	0.0266
Al^{3+}	0.0003	0.0333	0.1083	1.8629
Mg^{2+}	1.7773	1.8101	0.8272	2.2268
Ca^{2+}	0.0008	0.0106	0.8403	0.3753
Mn^{2+}	0.0021	0.0038	0.0018	0.0000
Fe^{2+}	0.1921	0.1553	0.0658	0.4637
Na^{+}	0.0104	0.0131	0.0559	0.0480
Cr^{3+}	0.0000	0.0057	0.0310	0.0826
Sum	2.9948	4.0096	3.9443	8.0127

starting material were filled with semi-sintered MgO powder and disks (Fig. 1). Temperature was measured with a W3%Re–W25%Re thermocouple placed in the middle of the starting material to avoid uncertainties due to the thermal gradient associated with the geometry of the assembly (Dobrzhinetskaya et al., 2000). We have calculated that at given geometry of the assembly the temperature of the sample is ~80°C less at the ends of the capsule than in its central part (Fig. 1). Octahedral pressure media were made from semi-sintered $MgO-Al_2O_3$ powdered ceramic material of Aremco Products (584-OS). Other materials and devices placed within the internal space of the multi-anvil apparatus included eight 25 mm tungsten carbide cubes each with a 6 mm octahedral truncation on one corner. They are arranged in the form of a larger cube with truncations facing inwards to form an octahedral cavity, into which the $MgO-Al_2O_3$ octahedron is placed (Fig. 2). To minimize friction and to provide electrical insulation, sheets of the fiberglass laminate (G-10), coated with Dry Film Vydax Mold Release, are placed between the WC cubes and the six anvils.

Temperature was maintained constant to ± 3°C; uncertainties of pressure calibration combined with small variations in the oil pressure behind the ram corresponded nominally to a total uncertainty in pressure less than 0.1 GPa. Temperature was not corrected for the effect of pressure on the thermocouple emf. The 6 mm truncated edge length (TEL) assembly was calibrated at room temperature against the phase transition of bismuth. A high-temperature calibration curve was determined by the coesite–stishovite equilibrium at 9.3 GPa and 1200°C and Mg_2SiO_4 $\alpha-\beta$ transition at 14.5 GPa and 1400°C.

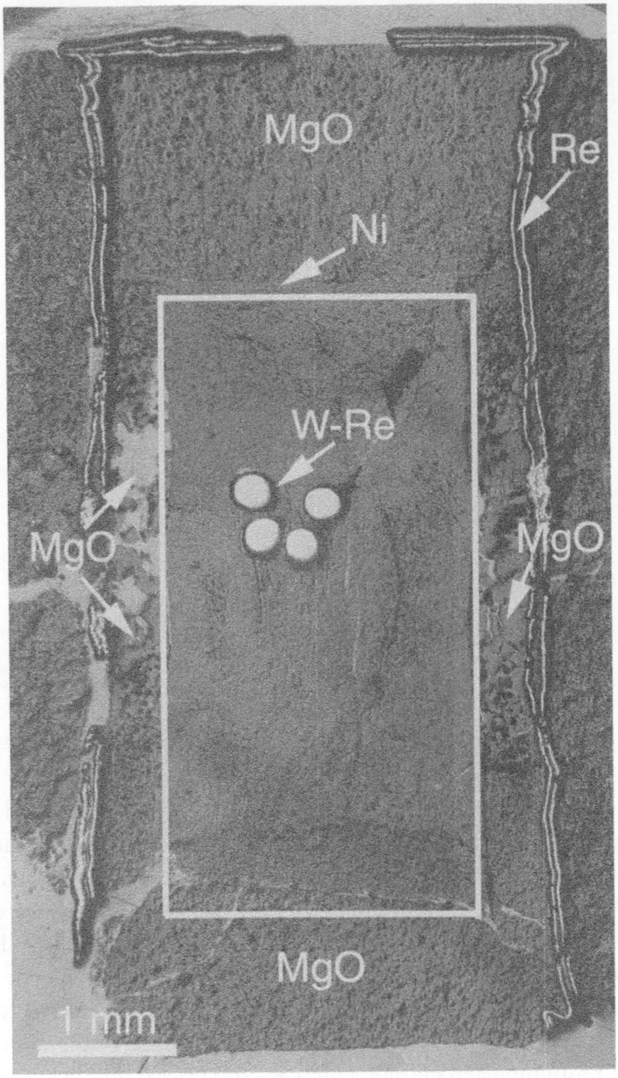

Figure 1. Optical microscope micrograph obtained with a digital camera: polished experimental sample (ma-147) recovered from the multi-anvil apparatus, then cut longitudinally perpendicular to the thermocouple (tc). Boxed area outlines the run product; W–Re (Tungsten–Rhenium) – thermocouple wires placed in the middle part of the starting material; MgO-capsule separated the run product and the Rhenium (Re) heater.

The apparatus has been also calibrated for the decompression curve with the transition of stishovite to coesite and the Mg_2SiO_4 β to α transition using the assembly described above.

We have conducted "single-cycle" and "two-cycle" experiments (Table 2). The "single-cycle" experiments (ma-140, 145, 147, 149, 216 and 217) were conducted to approach the equilibrium concentration of majoritic component due to the dissolution of

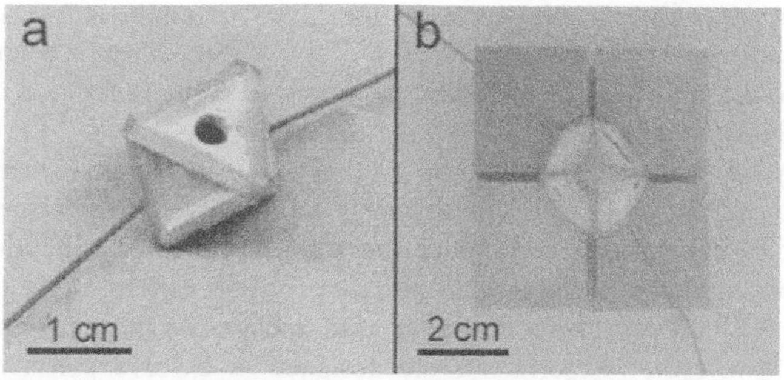

Figure 2. $MgO-Al_2O_3$ octahedron with a drilled hole and the thermocouple wires (tc) before experiment is shown on panel (a). The Rhenium heater and the starting material are placed inside of the hole. Panel (b) shows the octahedron after an experiment. The octahedron is surrounded by WC (Tungsten–Carbide) cubes with a 6 mm octahedral truncation on one corner.

Table 2. Experimental conditions and run product characterization.

Run number	Stage	Pressure (GPa)	Temperature (°C)	Time (hours)	Garnet Si (cpfu)
ma-135	1	14	1400	0.3	Equilibrium is not reached
ma-136**	1	14	1400	0.5	Equilibrium is not reached
ma-139	1	14	1400	1	Equilibrium is not reached
ma-140	1	14	1400	5	3.31 ($n = 5$)
ma-141	1	14	1400	3.5	3.30 ($n = 5$)
ma-144**	1	14	1400	26	3.32 ($n = 9$)
ma-144a	2	13	1400	26	3.27 ($n = 15$)
ma-145**	1	14	1400	26	3.34 ($n = 18$)
ma-147	1	14	1400	17	3.32 ($n = 15$)
ma-148	1	14	1400	30	3.32 ($n = 9$)
ma-148a	2	12	1400	46	3.19 ($n = 3$)
ma-149	1	8	1400	17	3.09 ($n = 6$)
ma-150	1	8	1400	17	3.09 ($n = 6$)
ma-150a	2	5	1400	55	2.98 ($n = 3$)
ma-213*	1	14	1400	32	3.32 ($n = 5$)
ma-213a	2	7	1400	5(&0.2)	3.34–3.05 ($n = 3$)
ma-216	1	14	1200	3.3	3.30 ($n = 3$)
ma-217**	1	14	1400	3	3.30 ($n = 6$)
ma-218	1	14	1400	24	3.32 ($n = 4$)
ma-219	2	7	1400	15	3.05 ($n = 5$)

Asterisk is put after the specimen number, which was depressurized from 14 to 7 GPa at $T = 1400$°C slowly (at a speed of about -1 GPa/h). Two asterisks are put after specimens with "coarse-grained" starting materials; specimen numbers with no asterisks – "fine-grained" starting material. Si = ($Si^{iv} + Si^{vi}$) content in garnet is given as an average of 3–18 analyses. Numbers ($n = 5, 6, \ldots$) reflect the number of chemical analyses used for calculation of Si content.

pyroxenes into primarily pyropic garnet at the given pressure, temperature and time conditions. Experiments were quenched to $T < 150°C$ within several seconds by shutting off the power to the apparatus, followed by programmed depressurization to atmospheric pressure over 24–35 h with RPM = −5 (−0.5 GPa/h).

The "two-cycle" experiments (ma-144, 148 and 150) are those, which include both Stage-1 and Stage-2. Stage-1 of such experiments was achieved in the same manner as for the "single-cycle" experiments to produce majoritic garnet with a certain amount of a majoritic component. Stage-1 was then followed by Stage-2, where we decreased pressure slowly (−1 to −1.5 GPa/h) to the desirable conditions while the temperature remained constant. The "two-cycle" experiments are designed to exsolve pyroxenes back from majoritic garnet during high-temperature decompression.

Three more runs (ma-213, 218, 219) were performed according to the following procedures. Experiment ma-213 was annealed at $P = 14$ GPa and $T = 1400°C$ for 32 h duration, then it was automatically depressurized to 7 GPa at the same T at speed of decompression −1.3 GPa/h. The depressurizing took 5 h followed by annealing at 7 GPa and 1400°C for 0.2 h, after which, the power was shut off to the apparatus and the sample was depressurized to atmospheric pressure at room temperature with the speed of decompression −0.5 GPa/h. Experiment ma-218 was conducted at 14 GPa and $T = 1400°C$ for 24 h as a "single-cycle" run. After its recovery from the apparatus the run product was drilled out from the assembly by microdrill; a thin disk cut perpendicular to the longer axis of the core-sample was used for scanning electron microscopy (SEM) research, whereas the remainder of the core-sample was used as a starting material for the experiment ma-219. The ma-218 run product was re-equilibrated at 7 GPa and $T = 1400°C$ for 15 h as the run ma-219.

3.4. Sample preparation

After recovery of experimental runs from the apparatus, each sample was cut longitudinally perpendicular to the thermocouple wires. Two polished sections were prepared for each experiment. One of them is used for optical microscopy, scanning electron microscopy, focused ion beam (FIB) micro-sectioning and transmission electron microscopy (TEM), the second one is stored for future research and reference. Polished sections were coated by carbon film then dried in a vacuum oven at 100°C for 2 h before each electron microscope session to avoid the effect of electron beam charging, and thereby, obtain high quality images and chemical analyses.

4. Instrumental measurements

4.1. Scanning electron microscopy

The run products were studied with the Philips XL30-FEG SEM equipped with the EDAX energy dispersive X-ray microanalysis system (EDS) with ultra-thin window Si(Li) detector, and a resolution of 137 eV at MnKα. The imaging and analysis of the chemical compositions of run products were performed with both secondary and backscattered electron modes at 15 and 20 kV. The spectral data were acquired at 1500–2000 counts/sec with a dead time below 25%, a beam current of about 1 nA, an effective spot size about 1.5 μm, and a 100 s counting time; the data were corrected using the eDXi software from EDAX and the ZAF program.

4.2. *Focused ion beam preparation of TEM samples*

The FIB of Ga^+ ions is a relatively new technique in geosciences, which has been successfully used for thin foil preparations (Heaney et al., 2001; Dobrzhinetskaya et al., 2001, 2003; Seydoux-Guillaume and Wirth, 2003). This technique allows the preparation of a foil from any point of interest recognized with a scanning electron microscope in polished rock sections. The scar of the FIB cutting remains visible on the polished surface of the sample, thereby facilitating repeated FIB milling and additional TEM examinations at a precise area on the specimen in cases where this is necessary to resolve controversial or confusing data. We have prepared several foils from majoritic garnets adjacent to the thermocouples. The foils were prepared with the FEI Strata DB 235 dual beam system using a high current density beam of Ga^+(30 kV, 20 nA). The FIB technique allowed us to mill deep trenches adjacent to areas of interest and to exhume a thin foil of the sample (10 × 5 × 0.1–0.05 μm) using a manual *ex situ* "lift-out" technique. Prepared foils were transferred onto a 3 mm diameter carbon coated standard Cu grid to be used further for TEM research.

4.3. *Transmission electron microscopy*

TEM studies were performed with a Philips CM300 instrument, operated at 300 kV. Bright Field (BF) and Dark Field (DF) images and electron diffraction patterns were obtained to study exsolution lamellae and blebs, which formed in majoritic garnet and pyroxenes during decompression experiments. The compositions of the crystalline phases were determined utilizing the EDAX energy dispersive X-ray microanalysis system with the Si(Li) detector.

5. Results

5.1. *Higher pressure experiments*

5.1.1. *"Single-cycle" experiments at $T = 1400°C$ and $P = 14$ GPa*

Exploratory experiments (ma-135, 136, 139, 140, 141, 145, 147, 218) at 14 GPa and $T = 1400°C$ were conducted with duration ranging from 0.3 to 26 h with "coarse" and "fine" grained materials to understand the kinetics of reactions and time required for an equilibration of run products. By individual spot chemical analyses as well as elemental mapping and profiling of garnet using the SEM we have determined that no equilibrium was established after 0.3–1 h in the "single-cycle" experiments. However, we established that after 17–26 h an equant microstructure of 5–7 μm grains was developed in runs with

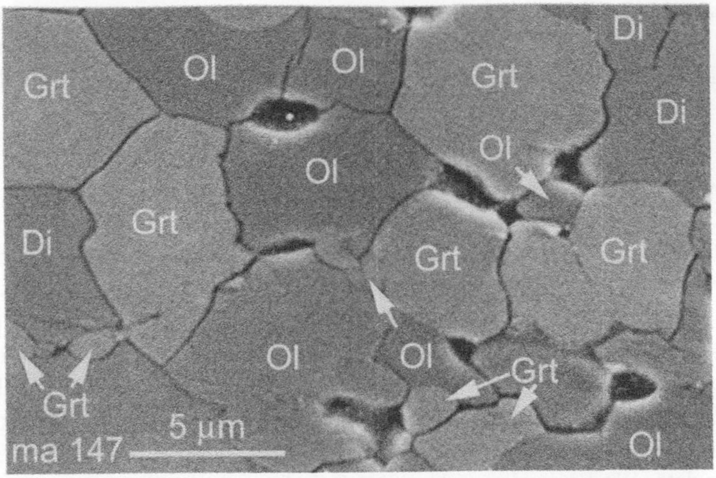

Figure 3. Secondary electron image (SEM) of an equal grain texture of the run product after "single-cycle" experiment (ma-147) annealed at $P = 14$ GPa, $T = 1400$°C for 17 h.

the "fine"-grained starting material (ma-145, 147, 218) (Fig. 3) and the chemical equilibrium was approached. Furthermore, after 26 h in the "coarse"-grained run product (ma-145), the finer grained fraction closely resembled that of ma-147 and 218, but the larger grained fraction showed zoned garnet and a small amount of relict diopside. No (clino)enstatite was found in any experiments of this series; the entire starting enstatite fraction was dissolved into garnet.

Relict diopside (ma-147) has the composition $(Mg_{0.91}Fe_{0.05}Ca_{0.74}Al_{0.14}Cr_{0.03}Na_{0.13})Si_2O_6$; its larger crystals contain lamellae of garnet, the significance of which will be discussed later. Garnet in the matrix of the run product became silica-enriched, with an average composition $(Mg_{2.45}Fe_{0.30}Ca_{0.49})(Al_{1.36}Cr_{0.07})Si_{3.32}O_{12}$. Si cation proportions range from 3.17 to 3.36 cpfu, and the average Si content is equal to 3.32 cpfu (Table 2). In contrast, the run products in ma-140, 141 and 217 (3–5 h), contain many non-equilibrated, often zoned garnets as well as abundant non-reacted diopside. Although the Si content varies in such garnets, the maximal Si = 3.30–3.31 cpfu is consistent with the well equilibrated run products of long-term experiments. The results of these experiments demonstrate that our finer-grained starting material was satisfactorily equilibrated at 1400°C and 14 GPa during 17 h and therefore, such run products are a reproducible starting material for decompression experiments.

5.1.2. *"Single-cycle" experiments equilibrated at 1200°C and 14 GPa*

Experiment ma-216 was conducted at 1200°C to evaluate the temperature effect on the Si content in majoritic garnet at the given bulk chemistry of our starting materials, because a slight temperature dependence of octahedral Si^{vi} concentrations in high-pressure garnets has been briefly mentioned in a few publications (e.g. Gasparik, 1990; Fei and Bertka, 1999). Although garnet synthesized in this run contains varying amounts of Si (3.17–3.30 cpfu), the maximum content of Si (3.30 cpfu) is close to the Si content

(3.34–3.32 cpfu) in garnet from the higher temperature experiments at 14 GPa (Table 2). Such a comparison suggests that the effect of temperature on the majoritic component is negligible at the given bulk chemistry. However, this suggestion is tentative as only one short-term experiment was performed and chemical equilibrium with co-existing minerals was not established.

5.1.3. "Two-cycle" experiments equilibrated at 1400°C and 14 GPa followed by re-annealing at 12 or 13 GPa

Experiments ma-144, 144a ("coarse"-grained) and ma-148, 148a ("fine"-grained) were performed to observe the microstructures produced upon the reaction accompanying the decompression of material previously equilibrated at 1400°C and 14 GPa. These experiments were designed to determine the mechanisms by which the system pursues a re-equilibration when majoritic garnet is decompressed to 13 or 12 GPa, conditions, under which it will exsolve significant octahedral Si in the form of pyroxene.

The run products of ma-144a and ma-148a consist of 2–5 μm equant grains of olivine and garnet with rare relict diopside accompanied by numerous tiny (<1–2 μm) interstitial grains of diopside located preferentially at garnet triple junctions (Fig. 4). No (clino)enstatite was found in the run product of either ma-144a or ma-148a. These tiny interstitial diopside crystals (ma-144a) have composition $(Mg_{0.97}Fe_{0.03}Ca_{0.73}Al_{0.13}Na_{0.12}Cr_{0.02})Si_2O_6$, whereas diopside in sample ma-147 (equilibrated at 14 GPa) has composition $(Mg_{0.91}Fe_{0.05}Ca_{0.74}Al_{0.14}Na_{0.13}Cr_{0.03})Si_2O_6$. The exsolved interstitial diopside has less Al and Na and significantly more Mg compared to that equilibrated with majoritic garnet at 14 GPa. After re-equilibration at 13 GPa (ma-144a), garnet has a composition of $(Mg_{2.47}Fe_{0.29}Ca_{0.46})(Al_{1.44}Cr_{0.07})Si_{3.27}O_{12}$ and contains less Si (3.27 cpfu) than garnet quenched to room temperature directly from $T = 1400°C$ and $P = 14$ GPa (Si = 3.32 cpfu). In the run product re-annealed at 12 GPa (ma-148a), garnet $(Mg_{2.42}Ca_{0.45}Fe_{0.27})(Al_{1.55}Cr_{0.08})Si_{3.19}O_{12}$ contains even less Si (3.19 cpfu). We consider the interstitial grains of diopside as products of exsolution from garnet with higher majoritic component because all of them have size (<1–2 μm) smaller than that of the starting material (5–7 μm) and the garnet composition reflects significant decrease in Ca and Si from specimens equilibrated at 14 GPa.

The diagram (Fig. 5) of Si (cpfu) versus $Al^{3+} + Cr^{3+}$(cpfu) of neoblastic garnets shows that garnet compositions in the runs ma-147 and ma-148 vary as expected for the course of the high-temperature decompression; the sum of highly charged cations of $Al^{3+} + Cr^{3+}$(cpfu) increases and Si cpfu decreases with decreasing pressure. Moreover, three grains of garnet situated at a distance of ~500 μm from the thermocouple in ma-148 displayed relict compositions in their cores that is close to garnet compositions obtained at 14 GPa in the ma-147 run. SEM studies of run products revealed that some large grains of garnet re-equilibrated from 14 to 13 GPa contain micron-size exsolution lamellae of Mg_2SiO_4 (ma-144a) and large grains of non-equilibrated diopside systematically include lamellae of garnet (Fig. 6). The latter features are similar to those, which we also observed in the "single-cycle" experiment at 14 GPa (ma-147).

During the slow decompression from 14 to 7 GPa (ma-213a) at 1400°C, majoritic garnet was gradually broken down, producing the interstitial exsolutions of both diopside and enstatite along adjacent garnet–garnet grain boundaries and within embayments at

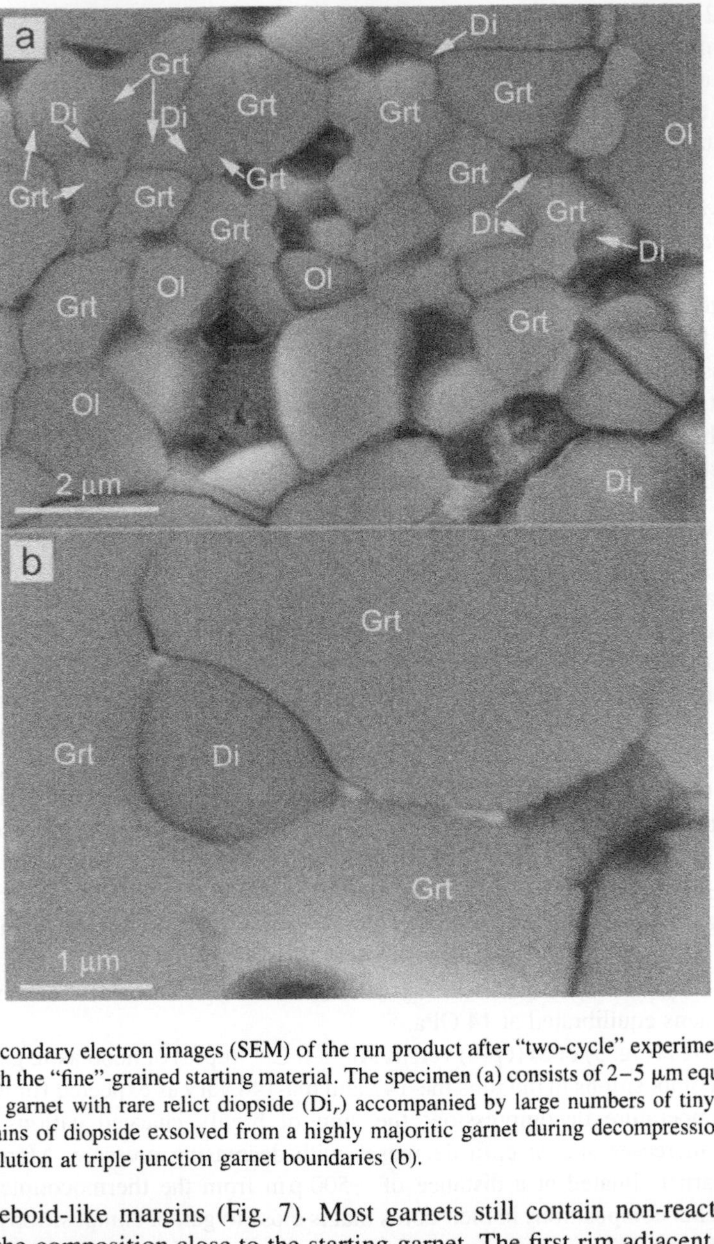

Figure 4. Secondary electron images (SEM) of the run product after "two-cycle" experiment (ma-148a) conducted with the "fine"-grained starting material. The specimen (a) consists of 2–5 μm equant grains of Mg_2SiO_4 and garnet with rare relict diopside (Di_r) accompanied by large numbers of tiny (<1–2 μm) interstitial grains of diopside exsolved from a highly majoritic garnet during decompression. Interstitial diopside exsolution at triple junction garnet boundaries (b).

garnet amoeboid-like margins (Fig. 7). Most garnets still contain non-reacted pyropic cores with the composition close to the starting garnet. The first rim adjacent to the core (Zone 1) contains majoritic garnet, $(Mg_{2.46}Fe_{0.31}Ca_{0.47})(Al_{1.32}Cr_{0.06})Si_{3.34}O_{12}$, with Si = 3.34 cpfu, whereas the outer rim (Zone 3) also is garnet, $(Mg_{2.39}Fe_{0.27}Ca_{0.43})(Al_{1.80}$-$Cr_{0.06})Si_{3.05}O_{12}$, but with a smaller majoritic component (Si = 3.05 cpfu) than the first rim; the intermediate rim (Zone 2) has composition $(Mg_{2.51}Fe_{0.27}Ca_{0.41})$ $(Al_{1.61}Cr_{0.08})$-$Si_{3.12}O_{12}$ with Si = 3.12 cpfu Such retrogressive compositional zoning of garnet is in good agreement with the microstructural evidence of diopside blebs exsolved within higher pressure rims and enstatite exsolution blebs in the outer rim. This selective exsolution

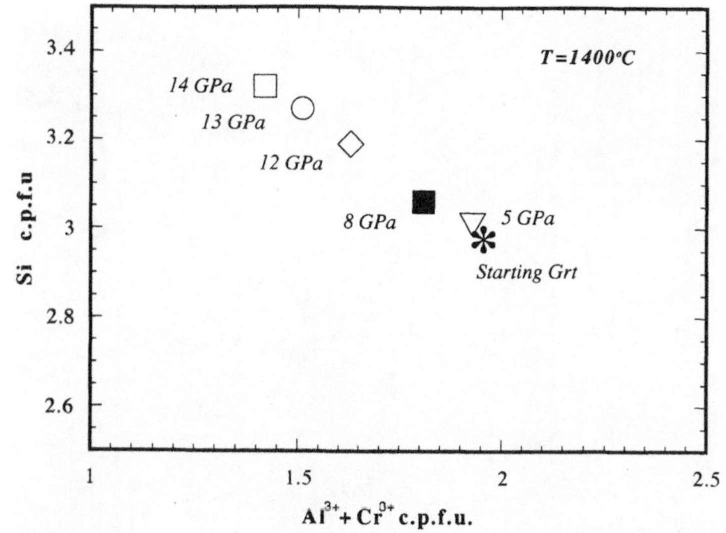

Figure 5. Diagram of average Si = (Si^{iv} + Si^{vi}) content versus Al^{3+} + Cr^{3+} (all are expressed as cpfu) in garnets from experimental samples and the starting material. Number of chemical analyses from each experiment is shown in Table 2.

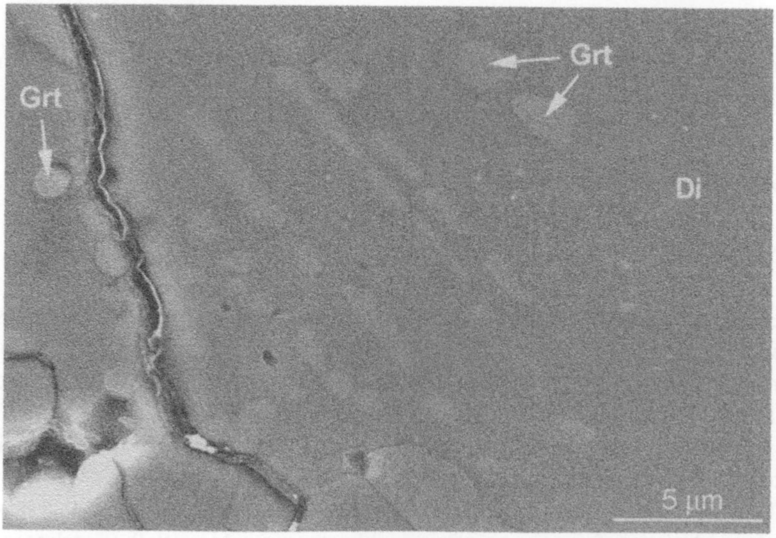

Figure 6. High-majoritic garnet nuclei and lamellae (light-gray contrast) inside of diopside (ma-144 and 144a, secondary electron image, SEM).

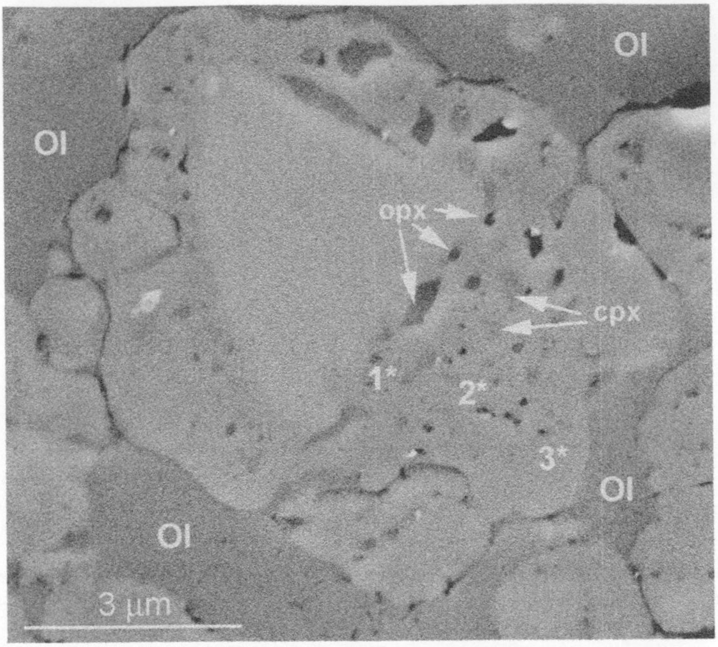

Figure 7. Secondary electron image (SEM) of zoned majoritic garnet with exsolution blebs of enstatite (darker contrast) and diopside (light-gray contrast) precipitated around a non-reacted pyropic core. The content of Si in the core is 2.92 cpfu. The recrystallyzed garnet is zoned such that the content of Si decreases gradually from 3.34 (zone 1) towards 3.05 (zone 3); the highest Si content is in the zone 1 adjacent to the non-reacted core (lighter contrast). Such zoning reflects the process of high temperature (1400°C) slow decompression from the garnet with a highest majoritic component, achieved at 14 GPa (ma-213), to the lowest at 7 GPa (ma-213a). Numbers: 1^* – zone 1, 2^*– zone 2 and 3^* – zone 3.

accompanies the chemical re-equilibration of supersilicic garnet during high-temperature decompression.

5.2. Experiments at lower pressures

5.2.1. "Single- and two-cycle" experiments

Having established a satisfactory grain size, time, and temperature of specimen synthesis and having observed the recrystallization mode of majoritic garnet growth at 14 GPa and its re-equilibration at 7, 12, and 13 GPa, we also carried out experiments on the re-equilibration of garnets with a small majoritic component at lower pressures (Table 2). These experiments, equilibrated at 1400°C and 8 GPa (ma-150) and re-annealed at 5 GPa (ma-150a), are close to the *P* and *T* conditions estimated by van Roermund and Drury (1998) for ultra-deep peridotite from Norway containing intracrystalline and interstitial exsolutions of pyroxenes. Once again, the "single-cycle" experiment, ma-149 (1400°C and 8 GPa), yielded an equant microstructure of homogeneous grains, consisting of (vol%) 40Ol + 40Grt + 13–15Di + 5–7En (Fig. 8). As expected, these

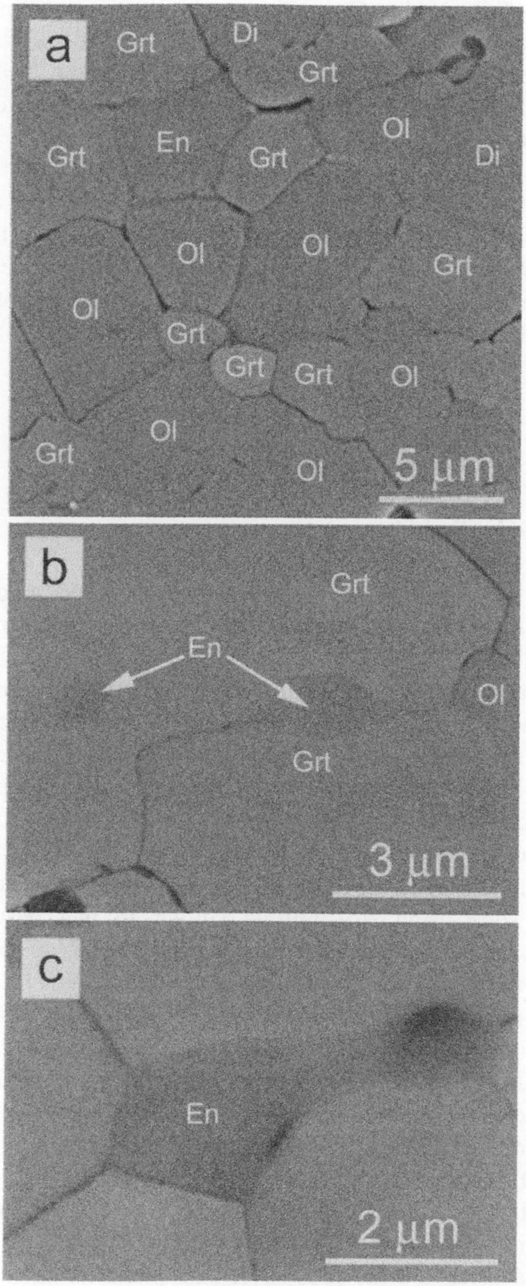

Figure 8. Secondary electron images (SEM) of lower pressure run products. Equal-grain texture (a) is composed of olivine, diopside, garnet and enstatite produced during single-cycle experiment (ma-149, $T = 1400°C$; $P = 8$ GPa). Interstitial enstatite blebs precipitated between two garnet grains (b) and at a garnet triple junction in "two-cycle" experiment (ma-150) (c).

garnets [$(Mg_{2.39}Fe_{0.22}Ca_{0.43})(Al_{1.85}Cr_{0.06})Si_{3.05}O_{12}$] contain a small majoritic component with $Si^{vi} = 0.05$ cpfu. The "two-cycle" experiment (ma-150) was held at 8 GPa and 1400°C for 17 h, then was decompressed to 5 GPa at constant $T = 1400$°C and it was annealed at these conditions for 55 more hours. The run product consisted of (vol%) 40Ol + 20Grt + 20Di + 20En. Garnet annealed at 5 GPa was found to have a composition $(Mg_{2.28}Fe_{0.28}Ca_{0.44})(Al_{1.92}Cr_{0.06})Si_{3.01}O_{12}$, with the minimal majoritic component (Si = 3.01 cpfu). Many fine-grained crystals of enstatite appeared interstitial to garnet (Fig. 8b and c), representing their exsolution from the small majoritic component. No small interstitial diopside grains similar to those present after annealing at 12 and 13 GPa were observed, suggesting that the small amount of diopside dissolved in garnet at 8 GPa had probably overgrown on relict diopside crystals remaining from the first cycle.

6. Remarkable microstructures produced in the experiments

6.1. Garnet lamellae in diopside

Garnet lamellae in diopside (Fig. 6) were observed in both "single- and two-cycle" experiments (ma-147 and ma-144/144a, respectively). In both experiments, such garnet is characterized by $Si^{vi} = 0.32–0.34$ cpfu. There is no doubt that in the "two-cycle" experiment (ma-144a) the large grain of diopside with garnet nuclei and lamellae is a relict crystal in which garnet nucleated and grew with topotactic control from the host pyroxene. We interpret that this diopside with intracrystalline lamellae of garnet reflects a "frozen" microstructure, which formed during Stage-1 of experiment ma-144, because we observed similar diopside microstructures in the "single-cycle" experiment (ma-147). To understand whether the garnet lamellae and the host diopside have a topotactic relationship we have studied the sample (ma-144a) in TEM. The foils for TEM research were prepared from the polished slide using FIB milling similar to that described in detail by Dobrzhinetskaya et al. (2003). The FIB technique allowed us to excavate several foils along the chosen cross-section through the fine-scale garnet lamella with great precision (Fig. 9a and b). The foils were cut perpendicular to the garnet lamellae. The conventional amplitude contrast imaging was performed in the TEM at an accelerating voltage of 300 kV. The structure of mineral phases and their crystallographic orientations were determined by selected area electron diffraction (SAED) measurements. Our studies revealed that the garnet lamellae are topotactically oriented with respect to the diopside host, where the (100) planes in diopside are parallel to the (121) planes in garnet (Fig. 9c and d). The orientation relationship is presumably controlled by the best-fit, minimum energy interface configuration. Across the interface the mismatch between the garnet (121) interplanar distances and the (200) pyroxene planes is only 0.1 Å. Such a small mismatch allows the creation of low-energy semi-coherent interfaces, which then control the growth shape of the inclusions. The difference in the symmetry between the host and the inclusions causes a development of the unusual elongation of the garnet crystals parallel to one set of the {121} trapezohedral faces. In the cubic garnet crystals, the 24-fold symmetrically equivalent trapezohedral faces cannot achieve the minimum energy configuration for all symmetrically equivalent faces since the matching (200) interplanar

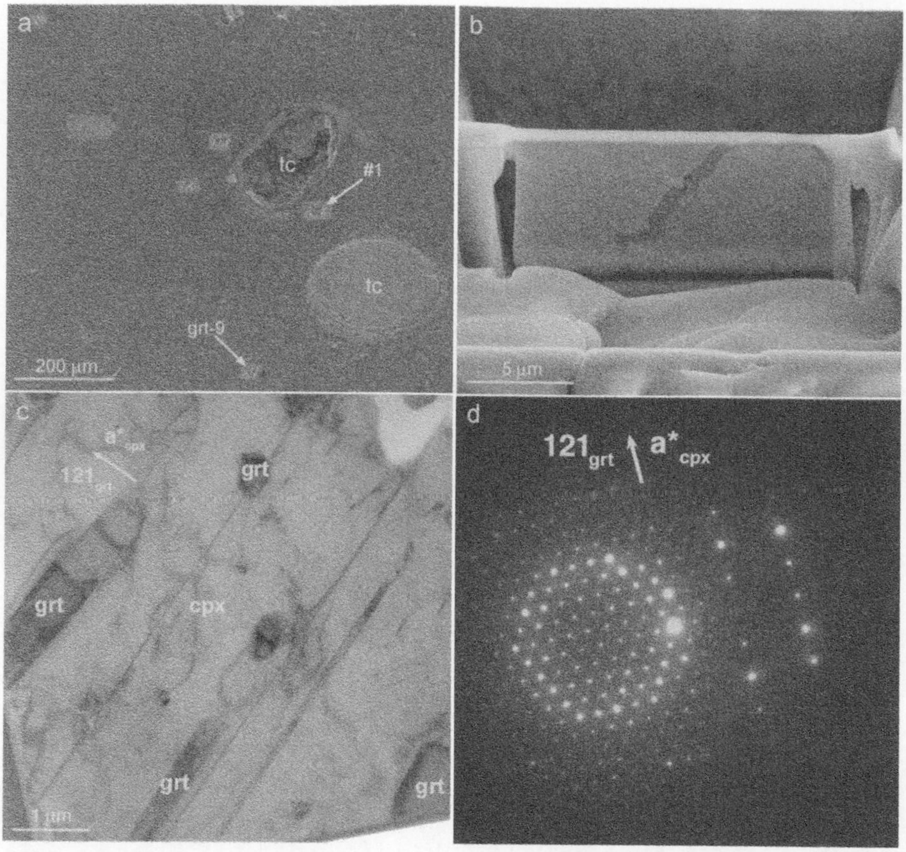

Figure 9. Electron microscope images of experimental run product containing diopside with garnet lamellae: (a) secondary electron image (SEM) of the surface of the experimental sample with traces of FIB excavations; grt-9 indicates the position of the TEM foil containing the diopside host with garnet lamellae; (b) *in situ* TEM foil preparation by FIB: in the middle of the foil there is a crack filled by epoxy during the sample preparation; garnet lamellae (lighter contrast) are faintly seen on the either side of the foil; (c) TEM Bright Filed image of the diopside host (cpx) containing topotactically oriented garnet (grt) crystals. The long interface is parallel to the trapezohedral {121} face of garnet and is subparallel to the {100} plane of the diopside host. The garnet lamellae are surrounded by mismatch dislocations that have developed in the diopside host to accommodate the slight lattice misfit across the interface; (d) selected area electron diffraction pattern taken at the interface of a garnet lamella and its diopside host. The garnet pattern, which is on the left side of the image, is parallel to zone axis [111]; the diopside pattern on the right side of the image is in [012] zone axis orientation. Both zone axes are inclined $\sim 3°$ with respect to each other. The (121) planes in garnet are parallel to the (200) planes in diopside. The long thin lamellae that intersect the pyroxene host in SW–NE direction are thin (100) diopside reflection twin domains.

distance of the host pyroxene has only 2-fold symmetry; the low symmetry of the diopside host is the controlling factor, which imposes its restrictions on the developing garnet. Therefore, the garnet crystal grows in an elongated shape to minimize the surface energy with semi-coherent interface contacts with the diopside host. Simultaneous growth of diopside and garnet would not produce such topotactically oriented garnet; rather it would produce a structure where garnet is localized mostly at the grain boundaries because of

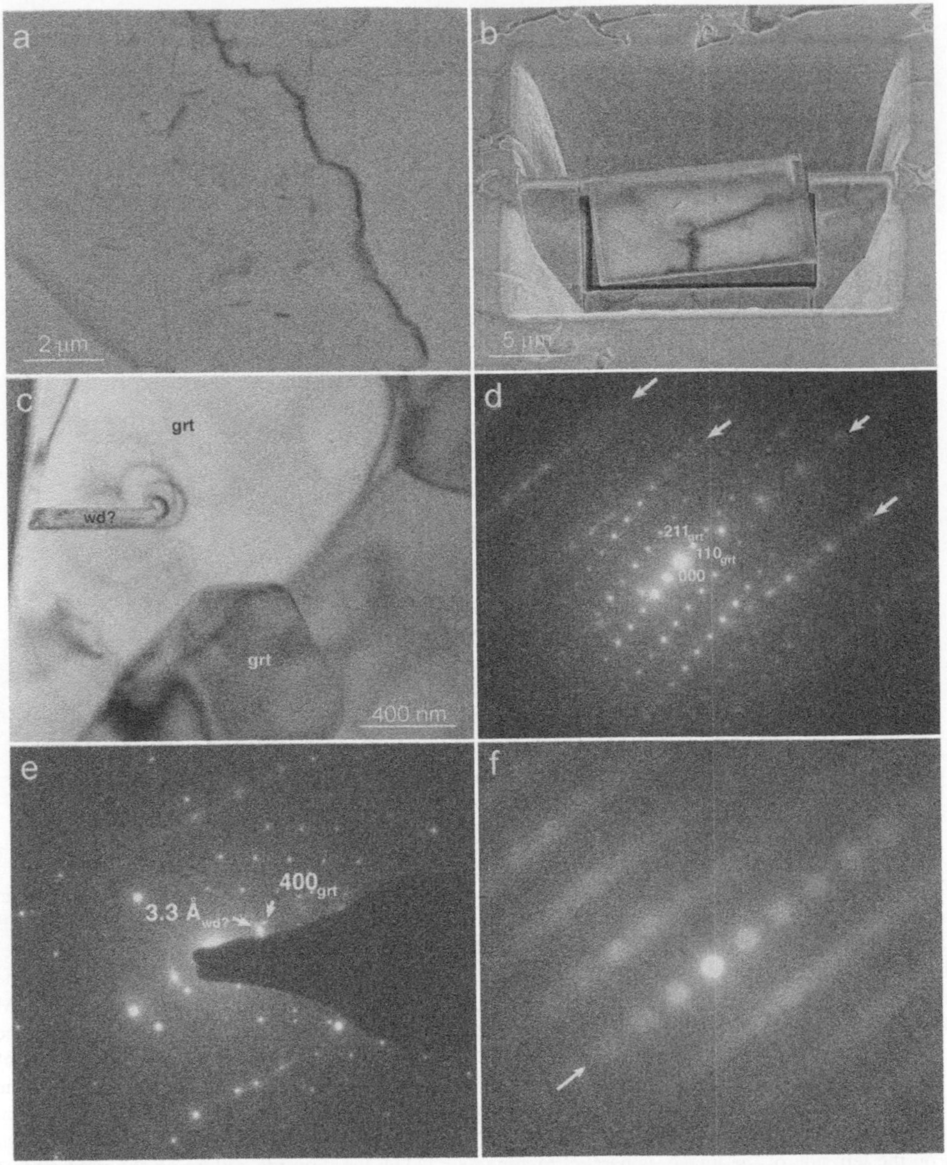

Figure 10. Electron microscope images of the garnet host and Mg_2SiO_4 lamellae exsolution: (a) secondary electron image (SEM) of garnet containing Mg_2SiO_4 lamellae (darker contrast); (b) foil preparation by FIB, in the middle of the foil there is a crack filled by epoxy; Mg_2SiO_4 lamellae are situated just above and to the left of the crack; this foil was cut from the site marked #1 on Figure 9a; (c) Bright field TEM micrograph of Mg_2SiO_4 lamella in garnet matrix. The lamella (marked as wd) is localized at grain boundaries between garnet crystals; (d) Selected area electron diffraction pattern showing the typical random orientation of the lamella with respect to the surrounding garnet grains, which is not close to any rational low-index direction in most of the garnet grains except one. The rows of reflections of the Mg_2SiO_4 phase, are marked with arrows. The streaking of the reflections suggests presence of stacking disorder; (e) Selected area electron diffraction pattern from a garnet grain in epitaxial orientation

the lattice mismatch of the two phases. Similarly, the growth of garnet prior to diopside formation would not produce the observed elongated lamellar habit of garnet inside of the diopside host, under the mentioned conditions of recrystallization.

6.2. Exsolution microstructures

6.2.1. Mg_2SiO_4 exsolution in garnet

Mg_2SiO_4 exsolution microstructures (Fig. 10a) were observed in garnet crystals only during high-temperature decompression from 14 to 13 GPa followed by long-term re-annealing at 13 GPa and $T = 1400°C$. Plates of Mg_2SiO_4 are nucleated on low-angle boundaries in garnet (Fig. 10a–c). The Mg_2SiO_4 lamellae composition is $(Mg_{1.81}Fe_{0.14}Al_{0.02}Ca_{0.01}Ni_{0.01})_{1.99}Si_{1.00}O_4$, whereas the host-garnet is $(Mg_{2.47}Fe_{0.29}Ca_{0.45})(Al_{1.44}Cr_{0.07})Si_{3.26}O_{12}$ with Si = 3.26 cpfu. The composition of the garnet host is consistent with that of other garnet crystals (Si = 3.27 cpfu) completely re-equilibrated at 13 GPa. Therefore, we have no doubt that the Mg_2SiO_4 plates were exsolved from majoritic garnet during high-temperature decompression from 14 to 13 GPa.

TEM study (Fig. 10c–f) revealed that the Mg_2SiO_4, exsolution product consists of prismatic-like lamellae and plates varying from 300 to 500 nm in length and less than 100 nm in width. A BF TEM micrograph of such a lamella incorporated within the garnet matrix is shown in Figure 10c. The lamella is localized along a garnet subgrain boundary. The orientation of the lamella with respect to the surrounding garnet subgrains is not close to any rational low-index direction in most of the garnet grains (see the diffraction pattern shown in Figure 10d). The exception is that there is a definite epitaxial orientation relationship of the Mg_2SiO_4 lamella to a contacting garnet grain, where the (100) garnet is parallel to planes in the Mg_2SiO_4 phase with 6.6 Å interplanar spacing (Fig. 10d). A significant streaking of the reflections along the 6.6 Å periodicity is observed, suggesting the presence of stacking disorder within the studied lamella. The lack of the 6.6 Å periodicity in the basal line of reflections (Fig. 10e) suggests that there is a symmetry reduction compared to the 6.6 Å (011) reflection of wadsleyite, which is not a forbidden reflection by the extinction rules. Although we also observed that the symmetry of the electron diffraction pattern (Fig. 10e) from the lamella resembles that of olivine, the interplanar spacing of 6.6 Å is inconsistent with the olivine structure. Despite the mentioned uncertainties in the structural definition of the Mg_2SiO_4 lamellae phase, we can tentatively conclude that they are comprised of a mixture of olivine and wadsleyite domains stacked randomly at the unit cell level. Such an assumption is consistent with

---

relationship to the Mg_2SiO_4 phase, where the (100) garnet is parallel to the planes in the Mg_2SiO_4 phase with 6.6 Å interplanar spacing; (f) Convergent beam electron diffraction pattern from the Mg_2SiO_4 lamella. The lack of the 6.6 Å periodicity in the basal line of reflections (marked with an arrow) suggest that there is symmetry reduction compared to the 6.6 Å (011) reflection of wadsleyite, which is not a forbidden reflection by the extinction rules. The symmetry of the diffraction pattern resembles that of olivine but the interplanar spacing of 6.6 Å rejects the possibility of olivine structure. We suggest that the lamellar Mg_2SiO_4 phase consists of a mixture of randomly stacked wadsleyite and olivine domains of less than 20 nm in size.

the *P* and *T* conditions of the experimental re-annealing of the ma-144a sample. On the olivine equilibrium phase diagram (Akaogi et al., 1989), these conditions correspond to the narrow field of the *P* and *T* where both olivine (α-phase of Mg_2SiO_4) and wadsleyite (β-phase of Mg_2SiO_4) coexist.

6.2.2. Interstitial exsolution of pyroxenes

Small interstitial diopside crystals are abundant in our higher pressure "two-cycle" experiments depressurized to 7, 12, and 13 GPa. They are distinctly smaller than the residual diopside in the "single-cycle" experiment with the equant microstructure (Fig. 4). The reduction of Si^{vi} content of majoritic garnet in our "two-cycle" experiments (Fig. 5) requires an exsolution of pyroxene and a reduction of Ca^{2+}content, and thus, requires that some or all of that exsolution be diopside. Nevertheless, no enstatite component of any kind was observed – nor were any intracrystalline lamellae of diopside. Therefore, it follows that the tiny interstitial diopside crystals are only the manifestation of pyroxene exsolution under these conditions.

However, during re-annealing at 5 GPa after equilibration at 8 GPa, small crystals of enstatite appear interstitial to garnet (Fig. 8b and c), representing exsolution of the small majoritic component. No small interstitial diopside grains similar to those of enstatite present, indicating that only a very small amount of diopside had been dissolved in garnet at 8 GPa. No intracrystalline enstatite exsolutions were observed at the scale of the SEM and TEM at this stage of our studies. van Roermund and Drury (1998) and van Roermund et al. (2000, 2001) showed that under similar conditions, natural enstatite exsolution occurred as interstitial blebs between garnets with exsolution lamellae of both pyroxenes visible only in cores of the garnet grains in excess of 4 mm in diameter (*note difference in scale of Figure* 9a and b *is a factor of 1000*). The ratio of bleb size to lamellae width in their natural material was greater than a factor of 50. If such lamellae exist in our run products, they would be expected to be ~20 nm across and be present only in the cores of grains in excess of 4–5 μm diameter. Further studies will clarify this issue.

7. Discussion

Our experiments were designed to simulate microstructural changes attending an upward migration of mantle peridotite from conditions of the mantle transition zone to shallower conditions of the lowermost part upper mantle (up to about 300–180 km).

We produced exsolution of Mg_2SiO_4 and diopside in the run products decompressed from 14 GPa to 13 GPa at $T = 1400°C$, and blebs of interstitial diopside due to high-temperature re-annealing at 12 GPa. Similar blebs of interstitial enstatite appeared only at lower pressures (5–8 GPa). Exsolution of interstitial diopside at higher pressures and interstitial enstatite at lower pressure is in agreement with mineral chemistry changes. At higher pressure (14 GPa) almost all diopside is dissolved in garnet, therefore, providing a higher value of $Si/Al^{3+} + Cr^{3+}$ratio in the newly recrystallized supersilicic garnet. During high-temperature decompression to 12 or 13 GPa followed by long-term re-annealing, some portion of diopside exsolved, precipitating as blebs at the low-energy garnet crystal junctions. Such a process correlates with the decreasing of the $Si/Al^{3+} + Cr^{3+}$ratio in re-equilibrated garnet. Enstatite exsolution from garnet becomes significant only at lower pressures (5–8 GPa).

As shown above in our experiments, exsolution of diopside from highly majoritic garnets occurs preferentially by incoherent nucleation at low energy grain-boundary sites rather than by coherent nucleation on intracrystalline defects, leading to development of conventional exsolution lamellae. This has significant implications for microstructures in natural garnet peridotites containing amoeboid-like garnets surrounded by single crystals of pyroxenes and accompanied by olivine clustered around the curved garnet boundaries (e.g. Dobrzhinetskaya et al., 1996; Green et al., 2000; Green and Dobrzhinetskaya, 2003) or interstitial orthopyroxenes occurring between recrystallized domains of former majoritic garnet (van Roermund et al., 2000, 2001).

Mg_2SiO_4 has not been reported as an exsolution product in natural deep-seated (>200–300 km) garnet peridotites (Haggerty and Sautter, 1990; van Roermund and Drury, 1998), nor did we observe it in our experimental charges at pressures consistent with those inferred from these studies (7–10 GPa). However, our observation that Mg_2SiO_4 may be exsolved during decompression at higher pressures is consistent with the expansion of the garnet field at the expense of wadsleyite at $P > 13$ GPa reported by Ringwood (1991).

Our experiments provide insight into the processes by which the garnet peridotite chemical system equilibrates at very high pressure. Rather than progressive dissolution of pyroxene components into existing pyropic garnets by diffusion, we observed that the dominant mechanism was recrystallization of garnets by chemically induced grain-boundary migrations (c.f. Hay and Evans, 1987) and also by nucleation and growth of majoritic garnets in the form of lamellae within diopside. The latter process is reminiscent of the way that the aluminosilicate polymorphs replace each other with increasing pressures and/or temperatures during crustal metamorphism, in which they nucleate on other phases rather than undergoing a direct isochemical transformation.

Acknowledgements

We thank Herman L.M. van Roermund and Martin Drury for interesting discussions of our experimental results and their implication to natural rocks. We are grateful to Frank Forgit who provided technical support to our experimental program. Efforts by Matthew Weschler and Mark Darus of the FEI Company are appreciated for cooperation in TEM foil preparations by the FEI Strata DB 235 dual beam system. A.P.R. thanks the University of California at Riverside for fellowship support.

References

Akaogi, M., Akimoto, S., 1977. Pyroxene-garnet solid solution equilibria in the system $Mg_4Si_4O_{12}$–$Mg_3Al_2Si_3O_{12}$ and $Fe_4Si_4O_{12}$–$Fe_3Al_2Si_3O_{12}$ at high pressures and temperatures. Phys. Earth Planet. Interiors 15, 90–106.

Akaogi, M., Ito, E., Navrotsky, A., 1989. Olivine-modified spinel–spinel transitions in the system Mg_2SiO_4–Fe_2SiO_4: calorimetric measurements, thermochemical calculation, and geophysical application. J. Geophys. Res. 94, 15671–15685.

Bozhilov, K.N., Green, H.W., Dobrzhinetskaya, L.F., 1999. High-pressure clinoenstatite in the Alpe Arami Peridotite. Science 284, 128–132.

Bozhilov, K.N., Green, H.W., Dobrzhinetskaya, L.F., 2003. Quantitative 3D measurement of ilmenite abundance in Alpe Arami olivine: confirmation of high-pressure origin. Am. Mineral. 88, 596–603.

Dobrzhinetskaya, L.F., Eide, E., Korneliussen, A., Larsen, R., Millege, J., Posukhova, T., Smith, D.S., Sturt, B., Taylor, W.R., Tronnes, R., 1995. Diamond in metamorphic rocks of the Western Gneiss Region in Norway. Geology 23, 597–600.

Dobrzhinetskaya, L.F., Green, H.W., Wang, S., 1996. Alpe Arami: a peridotite massif from depths of more than 300 kilometers. Science 271, 1841–1845.

Dobrzhinetskaya, L.F., Bozhilov, K.N., Green, H.W., 2000. The solubility of TiO_2 in olivine: implications for the mantle wedge environment. Chem. Geol. 163, 325–338.

Dobrzhinetskaya, L.F., Green, H.W., Mitchell, T., Dickerson, R.M., 2001. Metamorphic diamonds: mechanism of growth and inclusion of oxides. Geology 29, 263–266.

Dobrzhinetskaya, L.F., Green, H.W., Weschler, M., Darus, M., Wang, Y.-C., Massonne, H.-J., Stöckhert, B., 2003. Focused ion beam technique and transmission electron microscope studies of microdiamonds from the Saxonian Erzgebirge, Germany. Earth Planet. Sci. Lett. 210, 399–410.

Dobrzhinetskaya, L.F., Green, H.W. II, Renfro, A.P., Spengler, D., van Roermund, H., 2004. Olivine and pyroxene precipitations from majoritic garnet: examples from natural and experimental garnet peridotites. Terra Nova, 16 (6), 325–330.

Fei, Y., Bertka, C., 1999. Phase transitions in the Earth's mantle and mantle mineralogy. In: Fei, Y., Bertka, C., Mysen, B. (Eds), Mantle Petrology: Field observations and High Pressure Experimentation: A Tribute to Francis R. (Joe) Boyd. The Geochemical Society, University of Houston, Houston, USA, pp. 189–207.

Gasparik, T., 1990. Phase relations in the transition zone. J. Geophys. Res. 95, 15751–15769.

Gillet, P., Sautter, V., Harris, J., Reynard, B., Harte, B., Kunz, M., 2002. Raman spectroscopic study of garnet inclusions in diamonds from the mantle transition zone. Am. Mineral. 87 (2–3), 312–317.

Green, H.W., Dobrzhinetskaya, L.F., 2003. Garnet peridotite: wanderers of the mantle. In: Eider, E. (Ed.), Abstract Volume: Alice Wain Memorial West Norway Eclogite Field Symposium, NGU Report 2003.055. Geological Survey of Norway, Trondheim, Norway, p. 56.

Green, H.W., Dobrzhinetskaya, L.F., Bozhilov, K.N., 2000. Mineralogical and experimental evidence for very deep exhumation from subduction zones. J. Geodyn. 30, 61–76.

Haggerty, S.E., Sautter, V., 1990. Ultra-deep (> 300 km) ultramafic, upper mantle xenoliths. Science 248, 993–996.

Harte, B., Harris, J.W., Hutchison, M.T., Watt, G.R., Wilding, M.C., 1999. Lower mantle mineral associations in diamonds from Sao Luiz, Brazil. In: Fei, Y., Bertka, C., Mysen, B. (Eds), Mantle Petrology: Field observations and High Pressure Experimentation: A Tribute to Francis R. (Joe) Boyd. The Geochemical Society, University of Houston, Houston, USA, pp. 125–153.

Hay, R.S., Evans, B., 1987. Chemically induced grain boundary migration in calcite; temperature dependence, phenomenology, and possible applications to geologic systems. Contrib. Mineral. Petrol. 97 (1), 127–141.

Heaney, P.J., Vicenzi, E.P., Giannuzzi, L.A., Livi, K.J., 2001. Focused ion beam milling: a method of site-specific sample extraction for microanalysis of Earth and planetary materials. Am. Mineral. 86, 1094–1099.

Hirajima, T., Nakamura, D., 2004. Concordant and discordant UHP conditions from Yankou ultra-high pressure (UHP) unit, Sulu belt, China. 32nd International Geological Congress. Florence, Italy, August 20–28, Abstracts on CD-rom:18-2, T36.04.

Irifune, T., 1987. An experimental investigation of the pyroxene-garnet transformation in a pyrolite composition and its bearing on the constitution of the mantle. Phys. Earth Planet. Interiors 45, 324–336.

Irifune, T., Isshiki, M., 1998. Iron partitioning in a pyrolite mantle and the nature of the 410-km seismic discontinuity. Nature 392 (6677), 702–705.

Kimura, M., Suzuki, A., Kondo, T., Ohtani, E., El Goresy, A., 2000. Natural occurrence of high-pressure phases jadeite, hollandite, wadsleyite, and majorite-pyrope garnet in an H chondrite, Yamato 75100. Meteor. Planet. Sci. 35 (Suppl. S), A87–A88.

Liu, F.L., Xu, Z.Q., Liou, J.G., Katayama, I., Masago, H., Maruyama, S., Yang, J.S., 2002. Ultrahigh-pressure mineral inclusions in zircons from gneissic core samples of the Chinese Continental Scientific Drilling Site in eastern China. Eur. J. Mineral. 14 (3), 499–512.

Liu, F.L., Zhang, Z.M., Katayama, I., Xu, Z.Q., Maruyama, S., 2003. Ultrahigh-pressure metamorphic records hidden in zircons from amphibolites in Sulu Terrane, eastern China. Island Arc 12 (3), 256–267.

Moore, R.O., Gurney, J.J., 1985. Pyroxene solid solution in garnets included in diamonds. Nature 318, 553–555.

Nasdala, L., Massonne, H.-J., 2000. Microdiamonds from the Saxonian Erzgebirge, Germany: *in situ* micro-Raman characterization. Eur. J. Mineral. 12, 495–498.

Ringwood, A.E., 1991. Phase transformations and their bearing on the constitution and dynamics of the mantle. Geochim. Cosmochim. Acta 55, 2083–2110.

Sautter, V., Haggerty, S.E., Field, S., 1991. Ultra-deep (>300 km) ultramafic xenolith: new petrologic evidence from the transition zone. Science 252, 827–830.

Seydoux-Guillaume, A.-M., Wirth, R., 2003. Transmission electron microscope study of polyphase and discordant monazites: site-specific specimen preparation using the focused ion beam technique. Geology 31, 973–976.

Shimizu, N., 1999. Young geochemical features in cratonic peridotites from southern Africa and Siberia. In: Fei, Y., Bertka, C., Mysen, B. (Eds), Mantle Petrology: Field observations and High Pressure Experimentation: A Tribute to Francis R. (Joe) Boyd. The Geochemical Society, University of Houston, Houston, USA, pp. 47–56.

Smith, D., 1999. Temperature and pressures of mineral equilibration in peridotite xenoliths: review, discussion and implications. In: Fei, Y., Bertka, C., Mysen, B. (Eds), Mantle Petrology: Field Observations and High Pressure Experimentation: A Tribute to Francis R. (Joe) Boyd. The Geochemical Society, University of Houston, Houston, USA, pp. 171–188.

Smith, J.V., Mason, B., 1970. Pyroxene-garnet transformation in Coorara meteorite. Science 168, 832–833.

Sobolev, N., Shatsky, V., 1990. Diamond inclusions in garnets from metamorphic rocks: a new environment of diamond formation. Nature 343, 742–746.

Song, S., Zhang, L., Niu, Y., 2004. Ultra-deep origin of garnet peridotite from the North Qaidam ultrahigh-pressure belt, Northern Tibetan Plateau, NW China. Am. Mineral. 89, 1330–1336.

Stachel, T., 2001. Diamonds from the asthenosphere and the transition zone. Eur. J. Mineral. 13 (5), 883–892.

van Roermund, H.L.M., Drury, M.R., 1998. Ultra-high pressure ($P > 6$ GPa) garnet peridotites in Western Norway: exhumation of mantle rocks from >185 km depth. Terra Nova 10, 295–301.

van Roermund, H.L.M., Drury, M.R., Barnhoorn, M.R., De Ronde, A., 2000. Non-silicate inclusions in garnet from an ultra-deep orogenic peridotite. Geol. J. 35, 209–229.

van Roermund, H.L.M., Drury, M.R., Barnhoorn, M.R., De Ronde, A., 2001. Relict majoritic garnet microstructures from ultra-deep orogenic peridotites in western Norway. J. Petrol. 42, 117–130.

van Roermund, H.L.M., Carswell, D.A., Drury, M.R., Heijboer, T.C., 2002. Microdiamonds in megacrystic garnet websterite pod from Bardane on the island of Fjortoft, western Norway: evidence for diamond formation in mantle rocks during deep continental subduction. Geology 30, 959–962.

Walker, D., 1991. Lubrication, gasketing, and precision in multianvil experiments. Am. Mineral. 76, 1092–1100.

Walker, D., Carpenter, M.A., Hitch, C.M., 1990. Some simplification to multianvil devices for high pressure experiments. Am. Mineral. 75, 1020–1028.

Xu, S.T., Okay, A.I., Ji, S.Y., Sengor, A.M.C., Wen, S., Liu, Y.C., Jiang, L.L., 1992. Diamond from the Dabie-Shan metamorphic rocks and its implication for tectonic setting. Science 256, 80–82.

Yang, J., Xu, Z., Dobrzhinetskaya, L.F., Green, H.W., Pei, X., Shi, R., Wu, C., Wooden, J.L., Zhang, J., Wan, J., Li, H., 2003. Discovery of metamorphic diamonds in Central China: an indication of a >4000 km-long-zone of deep subduction resulting from multiple continental collisions. Terra Nova 15, 370–379.

Ye, K., Cong, B., Ye, D., 2000. The possible subduction of continental material to depths greater than 200 km. Nature 407, 734–736.

Zhang, R.Y., Shu, J.F., Mao, H.K., Liou, J.G., 1999. Magnetite lamellae in olivine and clinohumite from Dabie UHP ultramafic rocks, central China. Am. Mineral. 84, 564–569.

Advances in High-Pressure Technology for Geophysical Applications
Jiuhua Chen, Yanbin Wang, T.S. Duffy, Guoyin Shen, L.F. Dobrzhinetskaya, editors
Published by Elsevier B.V. (2005)

Chapter 14
Chemistry at extreme conditions: approaching the Earth's major interface

Leonid Dubrovinsky, Natalia Dubrovinskaia,
Alexei Kuznetsov and Innokenty Kantor

Abstract

Core–mantle boundary (CMB) is one of the most inaccessible and enigmatic regions on the Earth. Clearly distinct chemical nature of the mantle (dominated by silicates and oxides) and the outer core (liquid iron–nickel alloy) suggests a possibility of multiple and complex chemical reactions at CMB. Due to limitations of the diamond anvil cell (DAC) technique (small sample size, pressure and temperature gradients, potential contamination by carbon, risk of loss of materials on recovery, etc.), the study of chemical reactions at extreme conditions of CMB is a difficult task. Combination of different modern analytical techniques (synchrotron-based X-ray powder diffraction, Mössbauer and Raman spectroscopy, SEM, ATEM, etc.) allows the elucidation of major trends in the behavior of the geophysically and geochemically important metal-oxide (Fe–SiO_2, Fe–Al_2O_3, MgO–FeO, for example) systems at pressures and temperatures of the Earth's deep interior. Methodological aspects of investigation of chemical reactions in DACs are considered.

1. Introduction

The boundary between the Earth's mantle and core (so-called D'' layer) has drawn significant attention due to the vast contrast in properties across this layer (Gurnis et al., 1998; Jeanloz and Williams, 1998). The seismologically observed changes in density and sound wave velocities, for example, are significantly (2–3 times) greater than those across the air–rock (or air–seawater) interface at the Earth's surface. Moreover, the difference in materials across the core–mantle boundary (CMB), with predominantly crystalline rock above and liquid iron alloy below, is among the most profound in the Earth. In this sense, the CMB can be considered the primary "surface" of the planet, and it is simply due to its remote nature that it has attracted less attention than the Earth's crust. However, processes at CMB can affect us both directly or indirectly. There are a number of geophysical, geochemical, and seismological arguments, which link processes at the D'' layer and at the surface of the Earth. For example, hot-spot volcanoes on Hawaii could originate from super-plumes linked to CMB.

Recent investigations have revealed a very complex texture and/or heterogeneity of the Earth's CMB (Gurnis et al., 1998; Breger et al., 2000; Carnero and Jeanloz, 2000; Vidale and Earle, 2000a; Vidale et al., 2000b), and the importance of properties of iron for modeling composition and dynamics of the core and its interaction with the lower mantle (Knittle and Jeanloz, 1989; Poirier et al., 1998; Stixrude and Brown, 1998). Appearance of post-perovskite $CaIrO_3$-structured (Mg,Fe)SiO_3 phase in the bottom of the Earth's lower mantle (Aganov and Ono, 2004; Murakami et al., 2004) could elucidate some of peculates of the D$''$ layer (for example, its seismic anisotropy, the strongly undulating shear-wave discontinuity at its top, and possibly the anticorrelation between shear and bulk sound velocities) but many other properties (like high electrical conductivity or nature of low speed velocity zone) remain unexplained. One of the key problems in interpretation of composition and processes in the D$''$ layer at the border between the core and the lower mantle is possible chemical reactions between iron and complex Mg–Fe–Si–Al-oxides, which form the bulk of the Earth's lower mantle.

Another problem, which involves studies of chemical reactions at extreme conditions, is elucidation of the composition of the Earth's core. Composition of the Earth's outer core is a geochemical parameter crucial for understanding the evolution and current dynamics of our planet (Allegre et al., 1995; Li and Agee, 1996, 2001; Anderson and Isaak, 2002; Lin et al., 2002). Since it was recognized that the liquid metallic outer core is about 10% less dense than pure iron (Birch, 1952), different elements, lighter than iron, including Si, S, O, C, and H, were proposed as major, or at least significantly abundant elements in the Earth's core (Badding et al., 1991; Allegre et al., 1995; Hillgren et al., 2000; Gessmann et al., 2001; Li and Agee, 2001). However, the combination of experimental results with theoretical and geochemical considerations shows that it is unlikely that any one of these elements alone can account for the density deficit (Hillgren et al., 2000).

The question of the chemical composition of the Earth's core is closely related to the chemical history of the planet, particularly to the processes of accretion and segregation of the planetary body. In fact, the segregation of iron-rich metal from silicate was a major physical and chemical event in the early development of the Earth and must have involved the separation of liquid metal from either liquid or crystalline silicates (Stevenson, 1990). One consequence of the formation of a metallic core is that elements should have been strongly partitioned in the metallic core. This is exemplified in the Earth by siderophile element depletions in mantle rocks relative to primitive solar system abundances (Newsom, 1990; Drake, 2000), and is part of a long-standing problem known as the "siderophile element anomaly": the pattern of the depletion of metals in the observable mantle cannot be explained by low-pressure segregation and was first noted by A.E. Ringwood (Ringwood, 1966; O'Neill and Palme, 1998). A first-order model for the formation of the Earth's core assumes equilibrium conditions, where metal equilibrates with molten silicate in a deep magma ocean that could have extended well into the top of the Earth's lower mantle (pressures range from 28 to 45 GPa) in the early Earth assuming a homogenous accretion scenario (Li and Agee, 1996; O'Neill and Palme, 1998; Gessmann et al., 2001; Bouhifd and Jephcoat, 2003). Although element partitioning between liquid metals and complex oxides has been successfully studied at pressures below 25 GPa in multi-anvil apparatus, investigations of chemical reactions at higher pressures require diamond anvil cells (DAC) experiments.

Studying of chemical reactions in the megabar pressure range, and temperatures exceeding 2000 K is not a trivial task. The amount of reacted material is very small (in the order of a few wt%, or, in absolute values, in the order of 10^{-10} g). Despite high temperatures, spatial temperature distribution across the pressure chamber is not homogeneous. The material can be partially, or completely lost when the cell, containing the recovered sample, is opened. There are a number of important details, which are crucial for the success of such experiments. Below some methodological principles for investigation of chemical reactions in DACs at extreme conditions are discussed. Their application to studies of chemical reactions between iron and corundum, silica, and periclase, as well as to investigations of the stability of ferripericlase at conditions of the Earth's lower mantle is illustrated.

2. Methodological aspects of studies of chemical reactions in DACs

2.1. Samples preparation

Due to relatively low cost, ease of operation and compatibility with different physical techniques, a DAC becomes a standard device for evaluation of material properties at elevated pressure. Methodology of DAC operation has been described in detail in many text books (see, for example, Eremets, 1996; Holzapfel and Isaacs, 1997; Hemley, 1998), and criteria for choice of different types of cells, sizes and types of diamonds or anvil seats are not described here. Note only that the cells, diamonds, seats, etc. should be compatible with those physical methods that will be employed to study chemical transformations (for example, if ^{57}Fe spectroscopy will be used, cells and seats should be made from a non-magnetic Fe-free alloy; for X-ray diffraction studies seats with wide opening angles are preferable; if chemical reaction is to be observed using the appearance of fluorescence, low-luminescence anvils should be used, etc.)

The gaskets used in DAC experiments usually serve two purposes: they are pierced with a hole, which provides the high-pressure chamber for the experimental volume, and also give lateral support to the conical faces of the anvils. If chemical reactions are possible at high pressures and temperatures, the gasket also defines the reaction volume and can chemically interact with the materials being studied. Possible contamination by the gasket materials should always be considered when external electrical heating is used, as the diamonds, the gasket and the pressure chamber are kept at the same temperature. Beryllium and beryllium–copper alloys start to flow at moderate temperatures 250–300°C and can not support the pressure. Molybdenum, tungsten, tungsten–tantalum and iron-based alloys remain hard at higher temperatures, but at 450–500°C they start to react with diamond, and at temperatures above 400°C, and high pressure, both Mo and W actively react with (Mg, Al, Fe)-oxides and silicates (especially in the presence of water or oxygen, even in trace amounts). Re and Ir gaskets are much more chemically stable with respect to most mineralogically important oxides and silicates and could be used (in a protective external atmosphere, for example, Ar–H_2 mixture) up to at least 1200°C at pressures above 10 GPa. Datchi et al. (2000) noted that H_2O and H_2 react easily with rhenium, even at moderate temperatures and proposed the use of a gold inset in the pressure chamber to isolate the sample from the gasket. However, the use of X-ray powder diffraction has shown that at pressures up to 37 GPa, and temperature of 1000 K, not detectable reaction between rhenium and liquid water takes place (Dubrovinskaia and Dubrovinsky, 2003).

In a laser-heated DAC gaskets remain cold during the experiment and usually reactivity of the gasket material is not a problem. In this case the choice of gasket material is dictated by the usual factors, such as the necessity to form a thermal isolation layer between the diamond and the heated sample and the desire to generate homogeneous heating, etc.

In some cases non-metallic gaskets are desirable (Eremets, 1996; Holzapfel and Isaacs, 1997; Hemley, 1998; Solli and Jeanloz, 2001).

In experiments intended to study chemical reactions in DACs at high pressures and temperatures, there are several requirements for a pressure gauge. In addition to usual requirements for a pressure calibrant (see for details Eremets, 1996; Holzapfel and Isaacs, 1997) it should be (a) chemically inert at high pressure and temperature, (b) not poisoned by reagents, (c) homogeneously distributed across the pressure chamber, and (d) it should not complicate interpretation of the chemical and phase composition data (for example, it should have just a few diffraction lines, not contribute to Raman spectra, etc.). In this sense, fluorescence pressure sensors are not the best choice. While ruby is relatively stable chemically, at very high pressure and temperature it still can react with other oxides, silicates and even metals (see below). The relatively complex diffraction pattern of ruby complicates interpretation of phase compositions if X-ray powder diffraction is employed. Other popular fluorescence sensors, such as doped borates or YAG, are not stable at high pressure and temperature. Moreover, fluorescence sensors do not work at high temperatures (above about 1000 K). It is desirable to use the same material as both an internal pressure standard and a pressure medium (for example, halides (NaCl, CsCl, LiF) may be used in experiments with metals, silicates, and oxides). In the case of external electrical heating information can then be provided for pressure, exactly at the point of measurements, even at high temperature, if the thermal equations of state are known. For internally (electrically or laser) heated DACs, pressure determination in the hot spot is problematic (Kavner and Duffy, 2001) and only the so-called "cold pressure" (the pressure measured at ambient temperature before and/or after a heating cycle) is sometimes characterized.

Trace amounts of atmospheric gases (O_2, N_2, H_2O) or contaminants (for example, traces of H_2O or CO_2 for MgO; H_2O for NaCl, etc.) could be tolerated in experiments at ambient and low temperatures. The restrictions for high-temperature experiments (especially those studying chemical reactions) are more complicated. If a sample is loaded in an ambient atmosphere, the amount of oxygen trapped in the pressure chamber is sufficient to significantly affect the oxygen fugacity (McCammon et al., 1998), and may even be enough to result in partial oxidation of the iron in a laser-heated DAC, for example (Dubrovinsky et al., 1998a, 1999). In order to prevent such undesired effects, it is necessary to load samples in inert atmospheres using glove boxes. Proper treatment of materials prior to loading into the DAC for high-temperature experiments is also important. Figure 1 shows sections of diffraction images collected using a MAR345 detector from a Fe and MgO mixture at 28(1) GPa before heating, and at 2250(100) K (double-side laser heating at the ID30 beam line, ESRF). Before loading into the DAC, the MgO was heated during 12 h at 1000°C. The diffraction pattern collected at ambient temperature (Fig. 1a) contains reflections for ε-Fe and periclase only. However, during heating a number of unidentified reflections appeared (Fig. 1b). These reflections were traced on quench to ambient conditions. The additional reflections suggest that a chemical reaction(s) occurred at high pressure and temperature. However, if the MgO was heated to 1500°C for 12 h, before being loaded into the DAC, no reaction was observed up to 90 GPa (see below, Fig. 7). It is, therefore, suggested that, heating to 1000°C is not sufficient to eliminate traces of water or carbon dioxide from MgO.

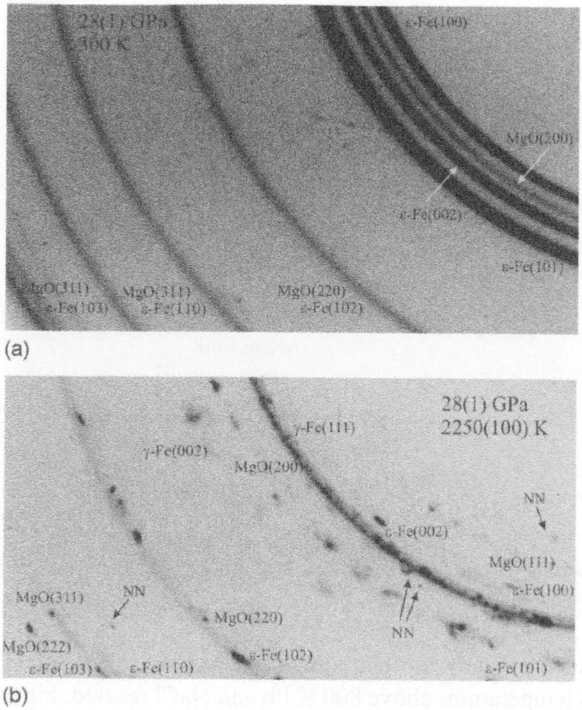

Figure 1. Parts of diffraction images collected using MAR345 detector from Fe and MgO mixture at 28(1) GPa and (a) at 300 K (before heating) and (b) at 2250(100) K (double-side laser heating at ID30 beam line at ESRF). NN marks not-identified reflections. Before loading in to DAC MgO was heated during 12 h at 1000°C, apparently not sufficient to eliminate traces of water or carbon dioxide which led to a reaction with iron. If MgO was treated before loading in DAC during 12 h at 1500°C it did not react with iron at least to 90 GPa (see text for more details).

The laser-heated DAC provides a unique opportunity to study chemical reactions in supercritical fluids – nitrogen, water, or methane, for example (Benedetti et al., 1999; Yusa, 2000; Zerr et al., 2003). Gases can be loaded into a DAC either under high pressure, or cryogenically, but in both cases great care should be taken to ensure that gases are of the highest purity. This is especially important in the case of cryogenic loading, as risk of contamination by condensed water or CO_2 is high. In the case of laser-heated DACs, the sample should not touch the surface of anvils. If contact is made, the sample will not be heated due to extremely high thermal conductivity of diamonds. In order to position the sample in the center of the pressure chamber, and avoid contact with the diamond, a hole with a step can be used (Fig. 2). Drilling the gasket to approximately half its indentation thickness from both sides, using drill-bits with different diameters will produce such a hole.

At high pressures, chemical properties of materials can change dramatically and the choice of a pressure medium (assumed to be chemically inert under experimental conditions) becomes difficult. Sodium chloride (NaCl), for example, is quite stable and (if dry) it does not react with transition metals even in a molten state at ambient pressure. Kuznetsov et al. (2002) used NaCl as a pressure medium and a pressure calibrant in

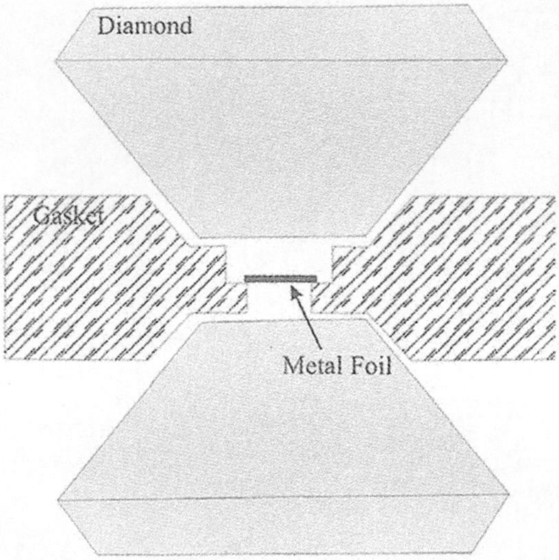

Figure 2. Schematic diagram of the hole with a step for loading the sample for laser-heating in a solidified gas pressure medium.

experiments intended to study the phase diagram of lead. They found that at pressures above 10 GPa and temperatures above 800 K Pb and NaCl reacted. Figure 3 shows Raman spectra collected from a sample recovered after heating of Pb and NaCl at 950(25) K and 16(1) GPa for 6 h. A comparison with Raman spectra of pure α-$PbCl_2$ confirms that lead and sodium chloride indeed react at high pressures and temperatures. One of the products of the reaction is lead chloride. This example shows that an *a priori* choice of pressure medium for high-P, T experiments is difficult, and it is reasonable to examine compatibility of different compounds in a series of test experiments.

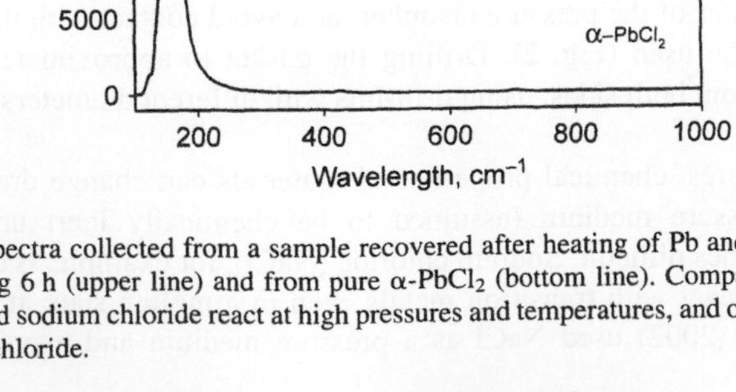

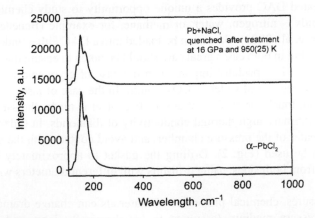

Figure 3. Raman spectra collected from a sample recovered after heating of Pb and NaCl at 950(25) K and 16(1) GPa during 6 h (upper line) and from pure α-$PbCl_2$ (bottom line). Comparison of the spectra confirms that lead and sodium chloride react at high pressures and temperatures, and one of the products of the reaction is lead chloride.

2.2. Detection of chemical reactions in DAC

Generally speaking, any physical method that is sensitive to chemical and phase composition could be applied to examine and study chemical reactions at high pressure and temperature. However, in experiments with DACs, small amounts of reagents (almost negligible), complex spatial distribution of reaction products, restrictions dictated by the properties of diamonds and gasket materials surrounding the reaction volume, limit application of most of the physical–chemical methods common for investigation of chemical reactions at ambient pressures. In this section, methods relating to chemical processes in DACs are discussed.

Diamonds are transparent to visible light, and historically, the most simple way to study processes in DACs is direct observation using an optical microscope. Alteration of colors, movement of particles or changes in shape, changes in reflectivity, etc. could suggest a reaction. Figure 4 shows changes in color due to decomposition of olivine to (Mg,Fe)SiO_3 perovskite and magnesiowüstite at 47(2) GPa after laser heating at 1900–2200 K. Knittle and Jeanloz (1989) observed a reaction between silicates and molten iron, even at megabar pressures. However, very often phase transformations (particularly melting), texturing or re-crystallization at high temperature, or even pressure-induced gradual variations in bandgap of semiconductors can mimic chemical reactions, and qualitative visual observations must be verified by quantitative methods.

Raman and infrared (IR) spectroscopy are easily coupled with DACs and are sensitive to changes in local atomic arrangements. Infrared spectroscopy in externally heated hydrothermal DACs was used, for example, to study such a problem as complex as water speciation at high pressures and temperatures (Sowerby and Keppler, 1999). Confocal Raman spectroscopy is useful as a local probe, and could be applied, not only to analyze materials in a reacting volume (see Fig. 3), but also to evaluate spatial distribution of phases, with a resolution of 3–4 μm. Application of Raman and IR spectroscopic methods

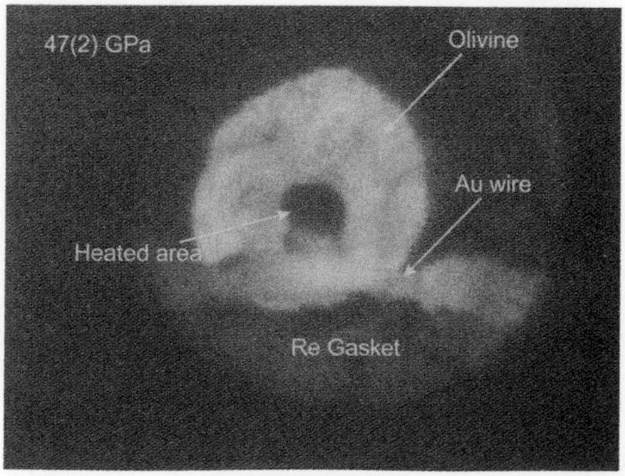

Figure 4. Changes in color due to decomposition of olivine on (Mg,Fe)SiO_3 perovskite and magnesiowüstite at 47(2) GPa after laser heating at 1900–2200 K.

is limited by phases with active vibrational modes. Note that IR spectroscopy requires special sample preparation and type IIa diamonds (see Eremets (1996) for more details). The intensity of the Raman signal depends on many factors (crystal orientation, size of crystallites, stress conditions, etc.) and the absence (or disappearance) of Raman bands cannot be considered as proof of the absence of the corresponding substance.

Unlike the Raman signal, the observation of fluorescence in DACs is easier, as the amount of doping element necessary is typically small (for example, for chromium in corundum, the level of a few hundreds of 1%), and the intensities are (usually) higher. Figure 5 shows spectra collected from a 10-μm diameter $Fe_{0.80}Ni_{0.15}Cr_{0.05}$ wire placed inside high-purity (99.9995%, Goodfellow Inc.) corundum powder at about 56 GPa, before and after internal electrical heating at 2450(50) K. Strong ruby fluorescence appeared around the heated portion of the wire, clearly showing incorporation of Cr into Al_2O_3 (i.e. a reaction between the alloy and corundum). Serghiou et al. (1998) used a Cr fluorescence marker to investigate the stability filed of $Mg_3Al_2Si_3O_{12}$ pyrope under hydrostatic pressure conditions in a CO_2-laser-heated DAC at pressures above 40 GPa and temperatures up to 3200 K. Spectra were collected from the temperature-quenched samples. From observations of fluorescence spectra resembling those of ruby, small amounts of Al_2O_3 co-existing with perovskite were followed to a pressure of 43 GPa. Above this pressure, according to Serghiou et al. (1998), entire Al_2O_3 content of the pyrope starting material is accommodated in the pyrovskite structure. Note, that monitoring of chemical reactions with fluorescence markers involves the assumption that (a) the fluorescence signature of each phase contributing to the spectra is unique and (b) the doping element is partitioning into the phase(s) of interest.

Time-resolved resistivity measurements are common methods for studying electrochemical reactions at ambient pressure. For obvious reasons, the resistivity measurements in DACs are challenging, but experiments with internally electrically heated DACs

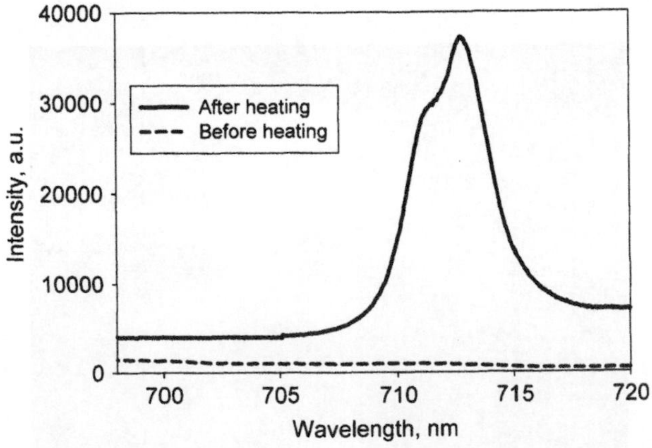

Figure 5. Spectra collected at 56(1) GPa from the $Fe_{0.80}Ni_{0.15}Cr_{0.05}$ wire placed into high purity (99.9995%, Goodfellow Inc.) corundum powder before and after internal electrical heating at 2450(50) K. Strong ruby fluorescence appeared around heated portion of the wire that manifested incorporation of Cr in Al_2O_3 (i.e. a reaction between the alloy and corundum).

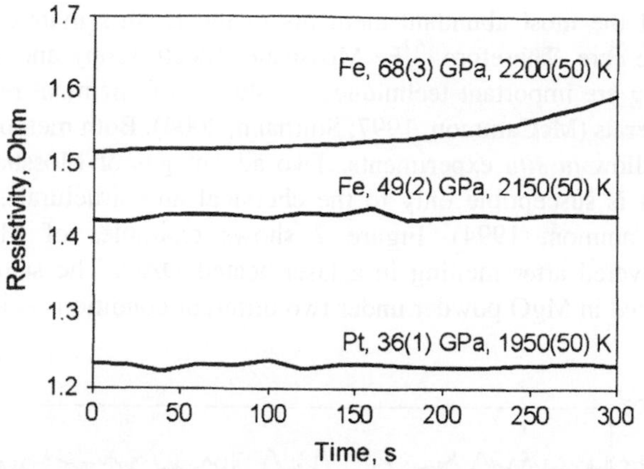

Figure 6. Variation of resistivity of platinum and $Fe_{0.80}Ni_{0.20}$ wires heated in corundum pressure medium at different pressures and temperatures. While for the period of 300 s there were no changes in resistivity of Pt wire (at 36(3) GPa and 1950(50) K) and iron wire (at 49(1) GPa and 2150(50) K), resistivity of $Fe_{0.80}Ni_{0.20}$ wire heated at 68(3) GPa and 2200(50) K changed significantly, indicating a reaction between the iron-based alloy and corundum.

provide an inexpensive (in terms of necessary equipment) and fast method for the observation of chemical reactions. Figure 6 shows the variation of resistivity of platinum and $Fe_{0.80}Ni_{0.20}$ wires, heated in a corundum pressure medium at different pressures and temperatures. While there is no change in the resistivity of the Pt wire at 36(3) GPa and 1950(50) K or the iron alloy at 49(1) GPa and 2150(50) K for the period of 300 s, the resistivity of the $Fe_{0.80}Ni_{0.20}$ wire, heated at 2200(50) K and 68(3) GPa, changes significantly (Fig. 6) indicating a reaction between the iron-based alloy and corundum.

X-ray powder diffraction is one of the most popular methods in the study of solid-state chemical reactions. There are a number of examples of successful coupling of *in situ* laser-heating with synchrotron radiation facilities, including work done at the European Synchrotron Radiation Facilities (ESRF) (Dubrovinsky et al., 1999; Andrault et al., 2000), the Advanced Photon Source (APS) (Shen et al., 1998), and at SPring-8 (Watanuki et al., 2001). However, *in situ* laser-heating experiments coupled with X-ray diffraction are still demanding, especially as temperature gradients within the laser-heated spot are large, and a great care should be taken to ensure perfect alignment of the X-ray and laser beams. Although X-ray diffraction analysis (XRDA) does not provide direct information about the chemical composition of the reacted materials, an example of an elegant solution to this problem has been given by Mao et al. (1997). In order to study the maximum solubility of $FeSiO_3$ in $(Mg,Fe)SiO_3$ perovskite in a laser-heated DAC at pressures up to 50 GPa and temperatures over 2500 K, Mao et al. (1997) used information on the dependence of the molar volume of perovskite on the Mg/Fe ratio. The latter was calibrated from an X-ray powder diffraction investigation of the samples recovered in large volume presses.

In the following sections, more examples of the application of X-ray diffraction for the analysis of chemical reactions in DACs are given. Note, that detection by X-ray powder diffraction requires that products should be well-crystallized and present in significant amounts (a few wt% at least).

Iron is one of the most abundant elements in the Earth's mantle, and a dominant component of the core. Therefore, ^{57}Fe Mössbauer spectroscopy and nuclear resonance forward scattering are important techniques for studies of chemical reactions involving iron-bearing minerals (McCammon, 1997; Sturhanh, 2004). Both methods are compatible with DAC and allow *in situ* experiments. Two advantages of Mössbauer spectroscopy are selectivity (it is susceptible only to the chemical and structural state of iron) and sensitivity (McCammon, 1994). Figure 7 shows examples of Mössbauer spectra for samples recovered after melting in a laser-heated DAC. The samples consisted of ^{57}Fe powder melted in MgO powder under two different conditions – at 85–90 GPa and

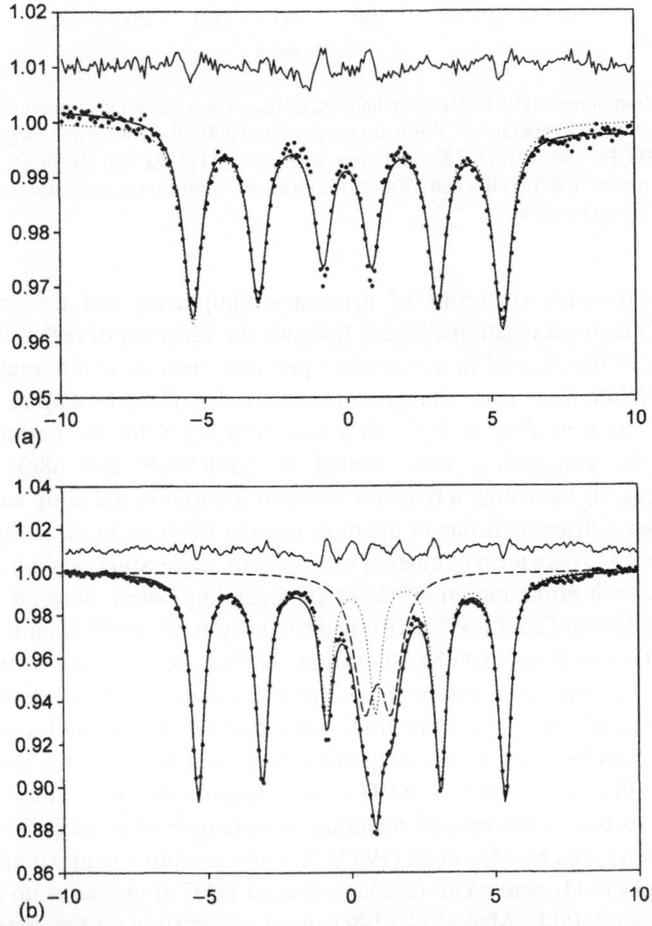

Figure 7. Mössbauer spectra of the samples of iron in MgO, recovered after laser heating at (a) 85–90 GPa and 3400–3650 K and (b) 98–105 GPa and 3450–3650 K. Spectrum (a) shows one sextet (CS 0.0(1) mm/s relative to α-Fe, B 33.4(1) T) corresponding to pure iron. Spectrum (b) consists of a sextet (CS 0.1(6) mm/s, B 32.1(1) T) (some decrease in the hyperfine field in comparison with pure α-Fe can relate to formation of Fe–Mg alloy) and a quadrupole doublet (CS 0.92(2) mm/s, QS 0.92(3) mm/s) characteristic for wüstite. Note, that sample (b) was "enriched" in reaction products, because some part of not-reacted material was lost on opening the cell.

3400–3650 K and at 98–105 GPa and 3450–3650 K. The first spectrum shows only one sextet (CS 0.0(1) mm/s relative to α-Fe, B 33.4(1) T) corresponding to pure iron. But the second consists of a sextet (CS 0.1(6) mm/s, B 32.1(1) T) and a quadrupole doublet (CS 0.92(2) mm/s, QS 0.92(3) mm/s), characteristic for wüstite. Note, that the last sample was "enriched" with reaction products, as some unreacted material was lost while opening the cell.

Quantitative chemical analysis of the material recovered from a DAC experiment is difficult, and such examples are not common in the literature. Knittle and Jeanloz (1989, 1991) pioneered the application of the X-ray microprobe for the study of products of chemical reactions between molten iron and $(Mg,Fe)SiO_3$ perovskite at conditions approaching parameters of the Earth's D'' layer, recovered after laser-heating in a DAC. SEM and/or microprobe analysis were recently employed in the investigation of iron–silicon alloy disproportionation (Lin et al., 2002; Dubrovinsky et al., 2003), the behavior of ferropericlase and magnesiowüstite at high pressure and temperature (Dubrovinsky et al., 2000a, Lin et al., 2003; Dubrovinsky et al., 2004), chemical interactions between iron and corundum or silica (Dubrovinsky et al., 2001a, 2003), and the investigation of the partitioning of Co and Ni between silicate and iron-rich metal liquids at pressures up to 42 GPa and 2500 K (Bouhifd and Jephcoat, 2003). Accurate chemical analysis with SEM or microprobe requires a flat sample surface. This can be achieved by placing the whole gasket and sample into epoxy resin and polishing it until the reaction zone is exposed. However, this procedure has serious problems associated with it, namely: (a) the sample can be partially or completely lost; (b) the reaction zone is often difficult to recognize visually and it can be polished out in the course of sample preparation; (c) due to the small initial thickness of the samples recovered from DAC experiments (in the order of 10–25 μm), even an insignificant inclination of the gasket when mounted in the epoxy resin can cause uneven thinning, which, in turn, results in an excitation volume, which is too small for certain elements (not sufficient to be analyzed). In some cases the recovered samples can contain relatively flat areas and can be analyzed using microprobe analysis without polishing (Bouhifd and Jephcoat, 2003). Strong criteria concerning the quality of the analyses should be imposed and clearly described in the results (particularly, small deviations from the sum of the total weight percent 100%). Note, that the excitation volume in SEM and microprobe analyses is rather large (in the order of 4 μm^3) and is comparable with, or bigger than the mean crystallite size of new phases produced by chemical reactions in laser- or electrically heated DACs. Therefore, the intrinsic problems of quantitative chemical analysis of materials recovered after high-P,T experiments in DACs may appear.

Analytical transmission electron microscopy (ATEM) is a prospective, method for studying the chemical composition of the samples recovered from DAC experiments (Goarant et al., 1992; Kesson et al., 1994, 1998; Dubrovinsky et al., 2003), due to high resolution achieved (down to nm scale), the sensitivity to most chemical elements, and the almost negligible amount of a material required. However, the problems associated with sample preparations for ATEM are the same as for SEM or microprobe analysis. Moreover, thinning of experimental material for the probe (by ion bombardment or mechanically) often results in changes of phase state. Additional difficulties arise due to the possible destruction of high-pressure phases (for example, amorphization of silicate perovskite) under a strong electron beam.

Tschauner et al. (1998) successfully employed SIMS techniques in the study of nickel and cobalt partitioning between silicate perovskite and metal at pressures up to 80 GPa in laser-heated DAC. SIMS provides high analytical sensitivity and a sub-micron depth resolution. Quantification of the chemical composition requires the conversion of measured mass ratios to concentrations of elements using standards that have a similar matrix. In practice, this means that the standards have to be closely related to the samples both in a structure and composition. This conversion is the main source of systematic errors. Tschauner et al. (1998) solved this problem by producing standards of various compositions by converting orthopyroxene to perovskite in a CO_2-laser-heated diamond cell.

There is one methodological problem common to all studies relying on the analysis of recovered samples following DAC experiments at high pressure and temperature. Namely, the chemical integrity of the materials may not be preserved after the release of pressure. Figure 8 shows Raman spectra collected from temperature-quenched sample after direct oxidation of chromium in oxygen at 36(3) GPa and 1550(100) K, and after complete decompression to ambient conditions (Kuznetsov et al., 2004). While the high-pressure pattern corresponds to the $CaCl_2$-type structure CrO_2 oxide, the recovered sample has the corundum type structure and composition Cr_2O_3 (as revealed by microprobe analysis).

Third-generation synchrotron facilities provide new possibilities for *in situ* studies of chemical processes at high pressure and temperature. Sanchez-Valle et al. (2003) measured the degree of dissolution of carbonate minerals in externally heated DAC at high pressure (up to 3.6 GPa) and temperature (up to 523 K). The extent of dissolution of strontionite ($SrCO_3$) has been followed as a function of time by monitoring the fluorescence of Sr cations in the fluid surrounding the crystal. Sanchez-Valle et al. (2003) have shown that Sr^{2+} concentrations as low as 1000 ppm can be detected and

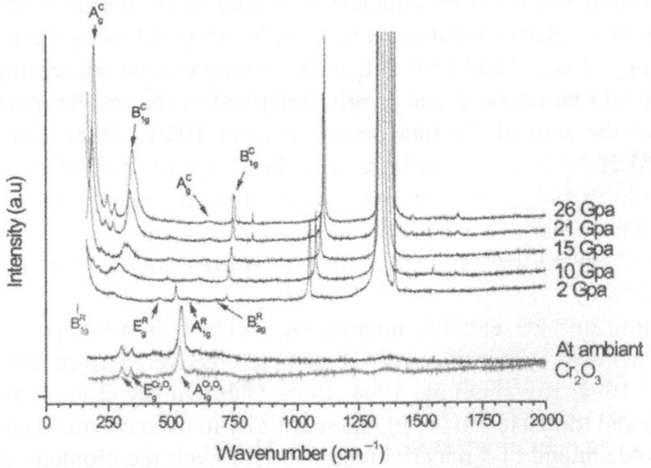

Figure 8. Raman spectra collected from temperature-quenched sample after direct oxidation of chromium in oxygen at 36(3) GPa and 1550(100) K and after complete decompression to ambient conditions (Kuznetsov et al., 2004). While high-pressure pattern corresponds to $CaCl_2$-structured CrO_2 oxide, the recovered sample has the corundum structure and Cr_2O_3 composition (as revealed by microprobe analysis).

quantitatively measured *in situ* in a DAC. The ID 22 beam line at ESRF, where Sanchez-Valle and co-authors conducted the experiments, is equipped, not only with a Si(Li) detector (used for X-ray fluorescence measurements), but also with a CCD detector allowing diffraction data collection, and an X-ray camera for sample visualization. The X-ray beam can also be focused down to a few micrometers. All this equipment provides the opportunity to enhance the methodology for *in situ* studies of chemical processes at conditions of the Earth's lower mantle.

2.3. Carbon transport in DAC

As mentioned above, small amounts of impurities in the pressure chamber (e.g. oxygen, water, nitrogen, etc.) can influence the chemical and physical properties of compounds studied at high pressure and high temperature. As a result, experimental data from DAC studies, such as pressure and temperature of phase transformations and chemical reactions, partitioning and melting temperatures, could be significantly affected by such impurities. While the appropriate choice and treatment of starting materials is optimized and the necessary precautions are taken during loading of the pressure chamber allow one to exclude practically all possible contaminations, carbon from the diamond anvils can still pollute the samples. Prakapenka et al. (2004) conducted a series of experiments to study carbon distribution inside the pressure chamber of a DAC via detection of the formation of various carbon phases after the samples were subjected to high pressures and temperatures. Prakapenka et al. (2004) used Raman spectroscopy as one of the most sensitive techniques for analysis of different carbon phases. Micro-Raman spectroscopy has high spatial resolution, with an analyzed area of 5 μm in diameter and allows "mapping" of the distribution of Raman-active substances *in situ* at high $P-T$ conditions in a small pressure chamber DAC. Prakapenka et al. (2004) demonstrated that there were carbon phases distributed in the sample chamber of DACs, and that the diamond anvils were the source of carbon in high pressure–high temperature experiments. Compression at room temperature to pressures above 40 GPa resulted in the formation of a disordered graphite phase(s) on the surface of samples (at the diamond–sample interface). The temperature-induced graphite phase formation has been observed *in situ* at high $P-T$ in DACs at pressure of about 15 GPa. In the case of laser heating at pressures above 6 GPa, a pure diamond or microcrystalline diamond and graphite phases have been obtained on the surface of samples and inside the pressure medium.

Figure 9 shows an example of Raman spectra collected from different points of a quenched sample of iron, laser-heated at about 2000 K between silica layers at 90 GPa. The bands at 1330–1335 cm^{-1} are from diamond and the band at 1590 cm^{-1} is from graphite. Figure 10 shows the element map collected from the recovered sample of FeO, compressed in an Ar pressure medium to 48(2) GPa and scanned by a Nd:YLF laser at 1500–1700 K. There is clear evidence of transport of carbon through the Ar and of C deposition on the surface of FeO sample. These examples clearly show that in experiments aiming to study chemical reactions in DACs, special attention should be paid to the possible role of carbon as an active component in the system.

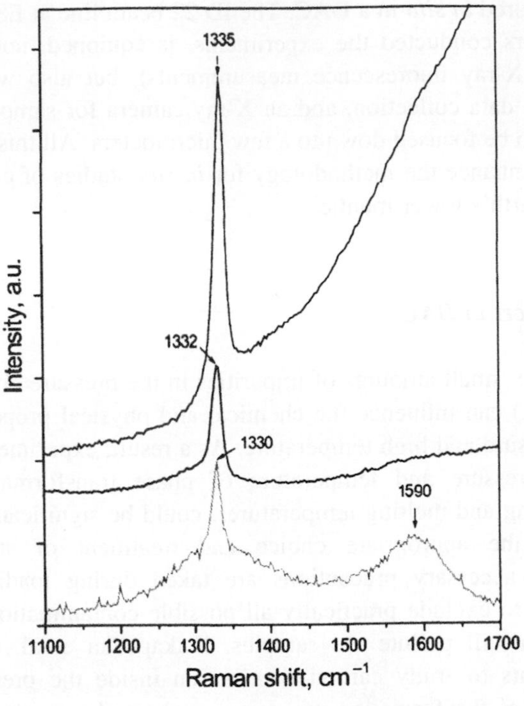

Figure 9. Examples of the Raman spectra collected from different points of a quenched sample of iron, laser-heated to about 2000 K between silica layers at 90 GPa. The peaks at 1330–1335 cm^{-1} are from diamond and at 1590 cm^{-1} – from graphite.

3. Examples of studies concerning chemical reactions in DACs at high pressure and temperature

3.1. Stability of ferropericlase in the Earth's lower mantle

There is growing geophysical, geochemical, and mineral physics evidence to suggest that the Earth's lower mantle may have a complex structure (Anderson, 2001, 2002; Kellogg et al., 1999; Shim et al., 2002; Badro et al., 2003; Wentzcovitch et al., 2004). The dominant components of the lower mantle are (Mg,Fe)SiO_3 perovskite and ferripericlase (Mg,Fe)O, and knowledge of their properties and behavior at high pressures and temperatures is essential for understanding mineralogy and dynamics of the whole Earth.

At ambient conditions, the end members of the MgO–FeO solid solution – periclase (MgO) and wüstite (FeO) – have the same halite NaCl (B1) cubic structure with a difference in lattice parameters of just ~2.1%. It is therefore, not surprising that at ambient pressure they form a complete solid solution. Periclase retains the B1 structure, at least to 227 GPa at 300 K (Duffy et al., 1995) and at temperatures above 2100 K at a megabar pressure range (Dubrovinsky et al., 1998b). On the contrary, at pressures above 17 GPa and ambient temperature wüstite transforms to a phase with a rhombohedral structure (Mao et al., 1996). At pressures above 100 GPa at 300 K it transforms to the NiAs (B8) or the anti-NiAs (a-B8) structure (Fei and Mao, 1994; Fang et al., 1999).

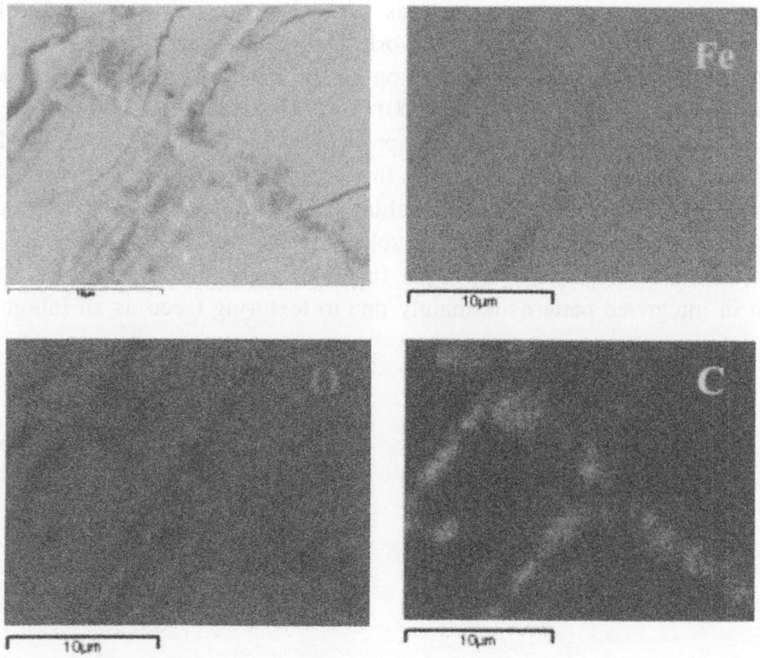

Figure 10. The elemental map collected using SEM from the recovered sample of FeO compressed in Ar pressure medium to 48(2) GPa and scanned by Nd:YLF laser at 1500–1700 K. There are clear evidences of transport of carbon through Ar and its deposition on the surface of the FeO sample.

The NaCl structure is based on cubic close packing of the anions, but the B8 (or the a-B8) structure is formed by hexagonal close packing of the anions (or cations). Based on the pioneering shock wave work on FeO by Jeanloz and Ahrens (1980), and on the thermodynamic analysis of the MgO–FeO solid solution, McCammon et al. (1983) proposed that topological difference between the B8 and the B1 structure at high pressure could lead to a miscibility gap.

In electrically externally heated DAC experiments, Dubrovinsky et al. (2000a, 2001b) found the decomposition of $(Mg_{0.80}Fe_{0.20})O$, $(Mg_{0.60}Fe_{0.40})O$, and $(Mg_{0.50}Fe_{0.50})O$ to the magnesium-rich and almost magnesium-free components at pressures above 80 GPa and temperatures 1000–1100 K. In a limited temperature range, externally heated DACs provide stable heating with homogeneous temperature distribution for a duration of 6 h in the experiments described by Dubrovinsky et al. (2000a). Lin et al. (2003) reported results on the study of iron-rich magnesiowüstite, $(Mg_{0.39}Fe_{0.61})O$ and $(Mg_{0.25}Fe_{0.75})O$, in laser-heated DACs. In contrast to previous studies, the Lin et al. (2003) experiments indicate that Mg-rich magnesiowüstite, with the rocksalt structure, is stable in the lower mantle. The conclusions from both studies (Dubrovinsky's et al. (2000a, 2001b); Lin's et al (2003)) are based on *in situ* X-ray powder diffraction data and electron microprobe analysis of recovered materials. The reason(s) for the discrepancy between the two studies is not clear. Moreover, the recent discovery of a high-spin to low-spin transition (Badro et al., 2003) in ferripericlase $(Mg_{0.83}Fe_{0.17})O$ in the 60- to 70-GPa pressure range has inferred possible complex behavior of this mineral at conditions of the Earth's lower mantle, and formed the basis for further work (Dubrovinsky et al., 2004).

Figure 11a shows the X-ray diffraction image collected during laser heating of $(Mg_{0.80}Fe_{0.20})O$ at 88(2) GPa and 1950(100) K. All diffraction lines (including the (002) line) are split. Splitting remains on decompression and it is clearly visible in diffraction patterns from quenched samples (Fig. 11b). In contrast, integrated images (Fig. 11c) do not show changes upon heating, or after quenching. This difference in X-ray diffraction data (images compared to integrated forms) is related to the well-known effect of the higher resolution 2D data compared to 1D data. In the example described above, the loss of resolution in integrated patterns is mainly due to texturing (seen as an inhomogeneous

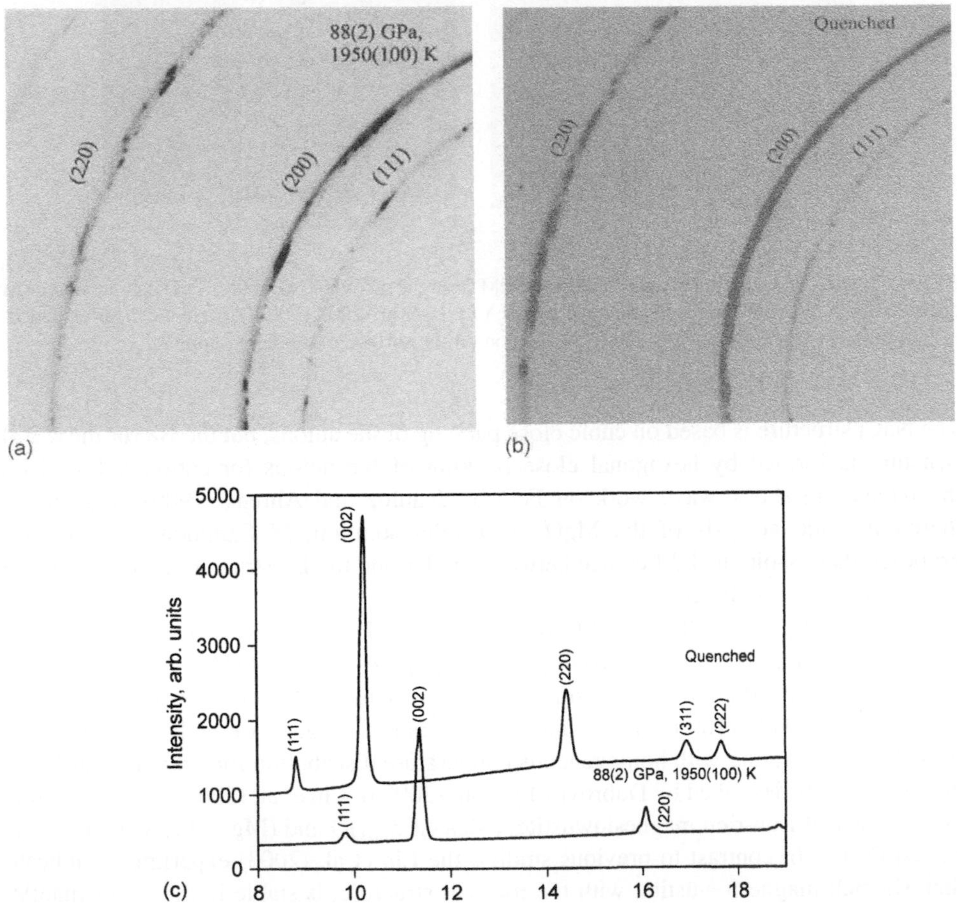

Figure 11. X-ray diffraction images of $(Mg_{0.8}Fe_{0.2})O$ collected (a) during laser-heating at 88(2) GPa and 1950(100) K; (b) at ambient temperature and 67(2) GPa; (c) from quenched sample. (c) Integrated diffraction patterns collected at 88(2) GPa and 1950(100) K (bottom line) and from the quenched sample (upper line).

distribution of intensity along the diffraction rings), which is unavoidable in laser-heated materials in the DAC.

At high pressures and temperatures, the temperature gradients in laser-heated DACs could contribute to splitting of the reflections. However, Dubrovinsky et al. (2004) observed extra lines in the patterns from both the temperature-quenched (still under pressure) and the recovered samples. For recovered samples a set of "low" *d*-spacings corresponds to an *fcc*-lattice with the cell parameter $a = 4.232(3)$ Å, which is close to the lattice parameter of the initial ferropericlase ($a = 4.2389(5)$ Å) within the uncertainty of the measurements. A set of "high" *d*-spacings corresponds to an *fcc*-lattice with the parameter $a = 4.295(8)$ Å. Such a lattice parameter could correspond to magnesiowüstite with a small amount (less than 5 mol%) of MgO. Therefore, X-ray diffraction experiments strongly suggest decomposition of ferropericlase at high pressure and temperature. This hypothesis is strongly supported by chemical analysis of the recovered samples. Figure 12 shows a back-scattered electron image of part of the recovered sample after heating for 45 min at 2350(100) K and 94(2) GPa. The appearance of a complex pattern of Fe and Mg distribution is obvious. Due to the fine intergrowth of Mg-rich and Fe-rich components, it is difficult to do accurate chemical analysis. X-ray microprobe analysis indicates that the iron content in ferropericlase may be below 7 mol% FeO, while the magnesium content in magnesiowüstite may be below 5 mol% MgO. Step-by-step polishing of the recovered sample shows that the "reaction zone" is about 2 μm thick. This thickness corresponds to the absorption length of infrared laser radiation in ferropericlase.

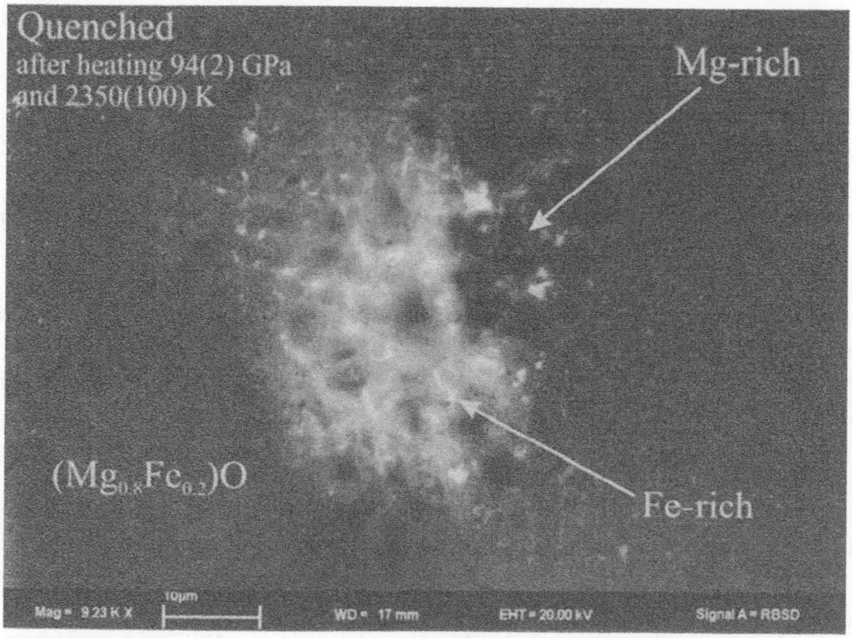

Figure 12. Back-scattered electron image of a part of the recovered sample of $(Mg_{0.8}Fe_{0.2})O$ after heating at 94(2) GPa and 2350(100) K.

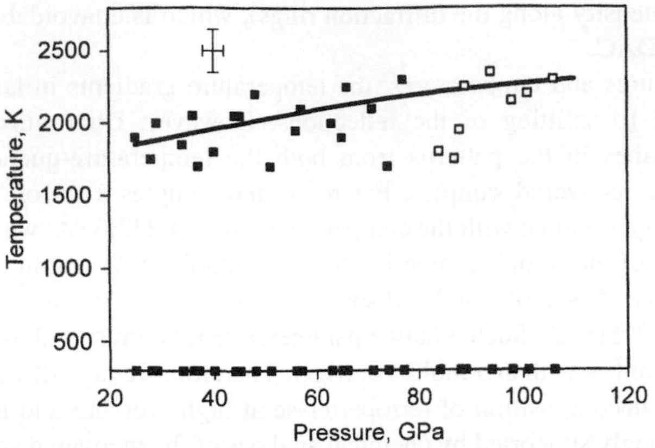

Figure 13. Phases observed in experiments on ferripericlase ($(Mg_{0.80}Fe_{0.20})O$) (Dubrovinsky et al., 2004). Closed symbols – single phase is stable; open symbols – decomposition to two phases (Mg-rich and Mg-poor) takes place; error bars show estimated uncertainties in pressures and temperatures. Continuous line is the mantle geotherm.

Dubrovinsky et al. (2004) demonstrated that ferripericlase, of mantle composition ($(Mg_{0.80}Fe_{0.20})O$), at least partially dissociates to the phases with lower and higher iron content at pressures above 80 GPa and temperatures corresponding to the Earth lower mantle geotherm (Fig. 13). Such dissociation of ferripericlase at depths in excess of 1700–2000 km should be taken into account when modeling the heterogeneity of the Earth's lower mantle. However, questions, such as, the width of the miscibility gap in the MgO–FeO system at high pressure, and, whether magnesiowüstite could dissociate completely to give MgO and FeO after a sufficiently long time of heating, have not yet been answered.

3.2. Chemical interaction of iron and corundum at high pressure and temperature

Aluminum oxide (Al_2O_3) together with SiO_2, MgO, and FeO are likely to be the most abundant components in the Earth's lower mantle, according to arguments based on cosmic abundance of chemical elements (Allégre et al., 1995). Moreover, the D$''$ layer is possibly enriched in refractories such as Al_2O_3 and CaO (Ruff and Anderson, 1980). The aluminum content of $(Mg,Fe)(Si,Al)O_3$ perovskite significantly affects its crystal chemistry, elastic properties, and the relative proportion of Fe^{3+} (McCammon, 1997). Consequently, the chemistry of the Fe–Al–O system is important for the whole mantle, and especially for the CMB region. Knowledge of exchange reactions at high pressure and high temperature between a metal from one side and refractory oxides and silicates from another side is important for understanding the early Earth differentiation (Hillgren et al., 1994).

Thermodynamic calculations (Fabrichnaya and Sundman, 1997), in agreement with available experimental data from multi-anvil apparatus, suggest that at pressures below 25 GPa, and high temperature, iron and corundum do not react. In a series of experiments suggest that at pressures above 65 GPa and high temperatures (above 2000 K) chemical reaction between iron and corundum can be described in general as

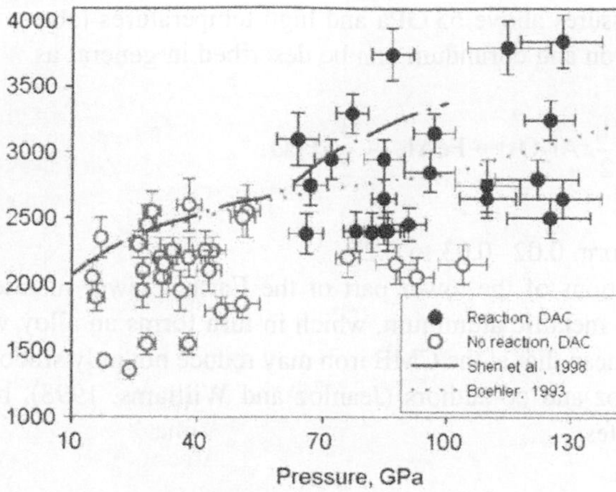

[illegible]

where x is varied form 0.02–0.1 [illegible]

Hence, at conditions of the lower part of the Earth's lower mantle iron can reduce aluminum oxide to metallic aluminum, which in turn forms an alloy with the remaining iron. It could also mean that liquid iron may reduce not only silicon, as proposed for example, by Jeanloz and co-authors (Jeanloz and Williams, 1998), but also aluminum from aluminosilicates.

1.3. Chemical interaction of iron and silica at high pressure and temperature

Although iron (with ~5 wt% Ni) is a dominant component of the Earth's core, Fe–Ni alloys are ~10% too dense for the outer liquid core and 2–5% too dense for the solid inner core, if the observed density along any reasonable geotherm is to be satisfied (Dubrovinsky et al., 2000b). Thus, on the basis of cosmochemistry, it has been proposed that the core also contains one or more light elements, and silicon is often considered as one of the main candidates (Lin et al., 2002). It is likely that such light elements dissolved from the magma ocean into the liquid metal during core formation in the early history of the Earth. At the temperature of the magma ocean [illegible] wt% Si can be dissolved in liquid Fe at 25 GPa, [illegible] the basis of [illegible] proposed that the [illegible] Dubrovinsky [illegible] between Fe and [illegible] conditions [illegible] the CMB.

The reaction between iron and silica was studied [illegible] (Dubrovinsky et al., [illegible]) [illegible]

[illegible]

suggest that at pressures above 65 GPa and high temperatures (above 2000 K) chemical reaction between iron and corundum can be described in general as

$$\left(\frac{2+3x}{2}\right)\mathrm{Fe} + \frac{1}{2}x\mathrm{Al_2O_3} \rightarrow \mathrm{FeAl}_x + \frac{3}{2}x\mathrm{FeO}$$

where x is varied form 0.02–0.03 to 0.25.

Hence, at conditions of the lower part of the Earth's lower mantle, iron can reduce aluminum oxide to metallic aluminum, which in turn forms an alloy with the remaining iron. It could also mean that at the CMB iron may reduce not only silicon, as proposed, for example, by Jeanloz and co-authors (Jeanloz and Williams, 1998), but also aluminum from aluminosilicates.

3.3. Chemical interaction of iron and silica at high pressure and temperature

Although iron (with ~5 wt% Ni) is a dominant component of the Earth's core, Fe–Ni alloys are ~10% too dense for the outer liquid core and 2–5% too dense for the solid inner core, if the observed density along any reasonable geotherm is to be satisfied (Dubrovinsky et al., 2000b). Thus, on the basis of cosmochemistry, it has been proposed that the core also contains one or more light elements, and silicon is often considered as one of the main candidates (Lin et al., 2002). It is likely that such light elements dissolved from the magma ocean into the liquid metal during core formation in the early history of the Earth. At the temperature of the magma ocean (~2800 K), 2–6 wt% Si can be dissolved in liquid Fe at 25 GPa, when an appropriate oxygen fugacity is considered (Gessmann et al., 2001). On the basis of a simple thermodynamic model, Gessmann et al. (2001) proposed that the solubility of Si at core conditions is close to zero. In order to test this hypothesis Dubrovinsky et al. (2003) examined chemical reactions between Fe and SiO_2 at conditions of the Earth's mantle down to the CMB.

The reaction between iron and silica was studied *in situ* in a DAC (Dubrovinsky et al., 2003). Iron foil, 5 μm thick, was sandwiched between two layers of amorphous silica, compressed to high pressure and laser-heated from both sides. Amorphous silica was chosen to increase the reactivity of the oxide. Figure 16 shows representative angle-dispersive X-ray diffraction patterns. At 26(1) GPa, and before heating, only reflections due to ε-Fe are present (Fig. 16a). However, after heating at 2100(100) K, silica crystallized as stishovite, and new reflections of wüstite appeared (Fig. 16 b). The lattice parameter of the iron, quenched after heating at 22–30 GPa, decreased from 2.8624(8) to 2.860–2.858 Å, which corresponds to an Fe–Si alloy with 3.6–4.5 wt% Si. However, heating of the iron and amorphous SiO_2 (or re-heating of the Fe–Si alloy and stishovite) at 3750(200) K and 87(5) GPa (Fig. 16c) (i.e. well above the melting temperature of iron for the corresponding pressure) resulted simply in the crystallization of silica phases ($CaCl_2$- and α-PbO_2-structured); no trace of wüstite was observed in the quenched sample at high pressure, and the lattice parameter of iron (2.8629(9) Å) after decompression was the same as for the starting material. Analytical transmission electron microscopy (ATEM with nanometer scale resolution) of the sample after heating at 2500(150) K and 125(10) GPa shows only a silica phase and iron (Fig. 16d). Within the analytical detection limit

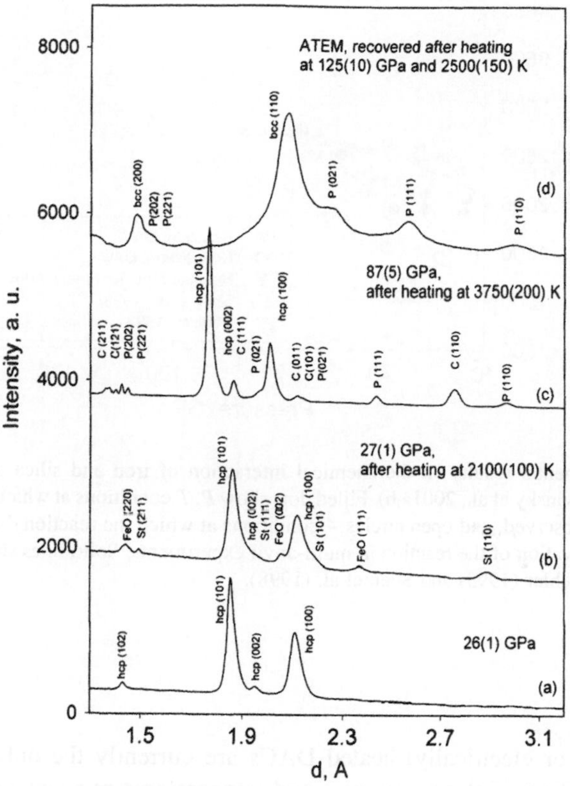

Figure 16. Representative angle-dispersive X-ray (a–c) and electron (d) diffraction patterns obtained in experiments with iron and amorphous silica as starting materials. (a) At 26(1) GPa, before heating of the sample, only the diffraction lines of ε-Fe (marked as *hcp*) are present. (b) After heating at 2100(100) K silica crystallized in stishovite (St) structure, and new lines of wüstite (FeO) appeared. (c) Heating of iron and amorphous SiO_2 mixture at 85(5) GPa and 2400(150) K resulted just in crystallization of silica phases ($CaCl_2$- and α-PbO_2-structured, marked (C) and (P), correspondingly), but wüstite was absent. (d) Selected area electron diffraction spectrum of the material recovered after heating at 125(10) GPa and 2500(150) K (d) shows only the presence of α-PbO_2-type (P) silica phase and iron (*bcc*).

(0.2 wt% Si) there is no silicon in the metallic portion of the sample. This suggests that the reaction between iron and silicon at pressures of 87 and 125 GPa was not observed.

Figure 17 summarizes the experimental results of Dubrovinsky et al. (2003) for the Fe–SiO_2 reaction. At pressures between 15 and 40 GPa iron reduces silicon, as described by the reaction:

$$x\mathrm{SiO_2} + (1 + 2x)\mathrm{Fe} \rightarrow 2x\mathrm{FeO} + \mathrm{FeSi}_x \qquad (0 < x < 0.1)$$

At higher pressures, above 80 GPa, Dubrovinsky et al. (2003) did not observe any reaction. It is suggested that iron and silica do not react at high pressure, due to a decrease in solubility of Si in ε-Fe (*hcp*-structure) with increasing pressure, as was proposed by Guyot et al. (1997) and confirmed by *ab initio* calculations and experimental observations (Dubrovinsky et al., 2003).

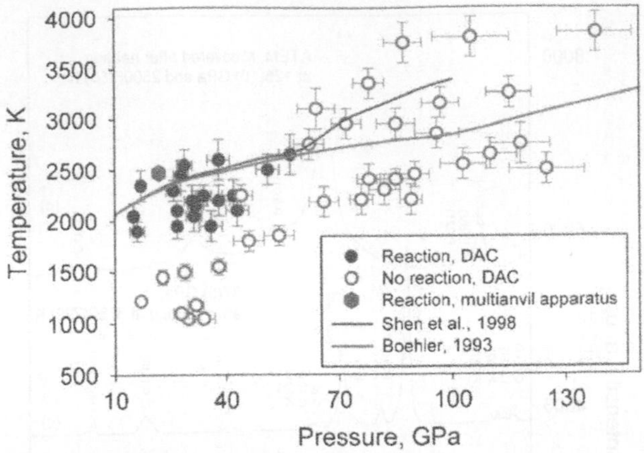

Figure 17. Experimental results on the chemical interaction of iron and silica at high pressures and temperatures (Dubrovinsky et al., 2001a,b). Filled dots show *P*, *T* conditions at which the reaction between iron and silica was observed, and open circles – conditions at which the reaction did not occur. Hexagon corresponds to observation of the reaction in multi-anvil experiments. Solid lines shows melting curve of iron according to Boehler (1993) and Shen et al. (1998).

4. Conclusions

Internally (laser- or electrically) heated DACs are currently the only technique which allows the generation of static pressures and temperatures relevant to the conditions of the Earth's lower mantle and the CMB. Due to limitations of the technique (small sample size, pressure and temperature gradients, potential contamination by carbon, risk of loss of materials on recovery, etc.) studies of chemical reactions at such extreme condition is a difficult, but exciting task. A combination of different modern analytical techniques (synchrotron based X-ray powder diffraction, Mössbauer and Raman spectroscopy, SEM, ATEM, etc.) allows the elucidation of major trends in the behavior of the geophysically and geochemically important metal-oxide systems at pressures and temperatures of the Earth's deep interior. It is remarkable to note that the chemical interactions of the most commonly studied systems (iron–silicate perovskite, iron–corundum, iron–silica, magnesium–iron monoxide) occur higher or lower of approximately the same pressure range (70–90 GPa) at high temperatures. For example, iron and corundum do not react at low pressures (below 70 GPa) and start to react at high-*P* (above 70 GPa), whereas the iron–silica system behaves reversibly: iron and silica react at pressures lower than 70 GPa and do not react at pressures above 70 GPa. In each case the nature of the chemical modifications is different. The driving force can come about from different parameters: changes in chemical activity of components, density changes due to phase transformations in the compounds involved, spin transitions in certain elements, etc. Information available now allows hypotheses to be made concerning the complex chemical reactions that occur in the D$''$ layer, which form chemical heterogeneities composed of silicon- and/or aluminum-rich and iron-rich regions.

Acknowledgements

We appreciate comments of A.J. Campbell. Help by F. Bromley allow greatly improve the manuscript. Work was financially supported by DFG grants.

References

Aganov, A., Ono, S., 2004. Theoretical and experimental evidence for a post-perovskite phase of $MgSiO_3$ in Earth's D″ layer. Nature 430, 445–448.

Allégre, C.J., Poirier, J.-P., Humles, E., Hofmann, A.W., 1995. The chemical composition of the Earth. Earth Planet. Sci. Lett. 134, 515–526.

Anderson, D.L., 2001. Top–down tectonics? Science 293, 2016–2018.

Anderson, D.L., 2002. The case for irreversible chemical stratification of the mantle. Int. Geol. Rev. 44, 97–116.

Anderson, O.L., Isaak, D.G., 2002. Another look at the core density deficit of Earth's outer core. Phys. Earth Planet. Interiors 131, 19–27.

Andrault, D., Fiquet, G., Charpin, T., Le Bihan, T., 2000. Structure analysis and stability field of beta-iron at high *P* and *T*. Am. Mineral. 85, 364–371.

Badding, J.V., Hemley, R.J., Mao, H.K., 1991. High-pressure chemistry of hydrogen in metals: in situ study of iron hydride. Science 253, 421–424.

Badro, J., Fiquet, G., Guyot, F., Rueff, J.-P., Struzhkin, V.V., Vanko, G., Monaco, G., 2003. Iron partitioning in Earth's mantle: toward a deep lower mantle discontinuity. Science 300, 789–791.

Benedetti, L.R., Nguyen, J.H., Caldwell, W.A., Jeanloz, R., 1999. Dissociation of CH_4 at high pressures and temperatures: diamond formation in giant planet interiors? Science 286, 100–102.

Birch, F., 1952. Elasticity and constitution of the Earth's interior. J. Geophys. Res. 69, 227–286.

Boehler, R., 1993. Temperatures in the Earth's core from melting-point measurements of iron at high static pressures. Nature 363, 534–536.

Bouhifd, M.A., Jephcoat, A.P., 2003. The effect of pressure on partitioning of Ni and Co between silicate and iron-rich metal liquids: a diamond-anvil cell study. Earth Planet. Sci. Lett. 209, 245–255.

Breger, L., Romanowicz, B., Rousset, S., 2000. New constraints on the structure of the inner core from P′P′. Geophys. Res. Lett. 27, 2781–2784.

Carnero, E.J., Jeanloz, R., 2000. Fuzzy patches on the Earth's core–mantle boundary? Geophys. Res. Lett. 27, 2777–2780.

Datchi, F., Loubeyre, P., LeToullec, R., 2000. Extended and accurate determination of the melting curves of argon, helium, ice (H_2O), and hydrogen (H_2). Phys. Rev. B 61, 6535–6546.

Drake, M.J., 2000. Accretion and primary differentiation of the Earth. Geochim. Cosmochim. Acta 64, 2363–2369.

Dubrovinskaia, N., Dubrovinsky, L., 2003. Whole-cell heater for the diamond anvil cell. Rev. Sci. Instrum. 74, 3433–3477.

Dubrovinsky, L.S., Saxena, S.K., Lasor, P., Weber, H.-P., 1998a. Structure of β-iron at high temperature and pressure. Science 281, 5373.

Dubrovinsky, L.S., Saxena, S.K., Lazor, P., 1998b. High-pressure and high-temperature *in situ* X-ray diffraction study of iron and corundum to 68 GPa using internally heated diamond anvil cell. Phys. Chem Miner. 25, 434–441.

Dubrovinsky, L.S., Lazor, P., Saxena, S.K., Häggkvist, P., Weber, H.-P., LeBihan, T., 1999. Study of laser heated iron using third generation synchrotron X-ray radiation facility with imaging plate at high pressure. Phys. Chem. Miner. 26, 539–545.

Dubrovinsky, L.S., Dubrovinskaia, N.A., Saxena, S.K., Annersten, H., Hålenius, E., Harryson, H., Tutti, F., Rekhi, S., Le Bihan, T., 2000a. Stability of ferropericlase in the lower mantle. Science 289, 430–432.

Dubrovinsky, L.S., Saxena, S.K., Tutti, F., Le Bihan, T., 2000b. X-ray study of thermal expansion and phase transition of iron at multimegabar pressure. Phys. Rev. Lett. 84, 1720–1723.

Dubrovinsky, L., Annersten, H., Dubrovinskaia, N., Westman, F., Harryson, H., Fabrichnaya, O., Carlson, S., 2001a. Chemical interaction of iron and corundum as a source of heterogeneity at the core–mantle boundary. Nature 412, 527–529.

Dubrovinsky, L.S., Dubrovinskaia, N.A., Annersten, H., Halenius, E., Harryson, H., 2001b. Stability of $(Mg_{0.5}Fe_{0.5})O$ and $(Mg_{0.8}Fe_{0.2})O$ magnesiowüstites in the lower mantle. Eur. Miner. J. 13, 857–861.

Dubrovinsky, L., Dubrovinskaia, N., Langenhorst, F., Dobson, D., Rubie, D., Geßmann, C., Abrikosov, I., Johansson, B., 2003. Iron–silica interaction at extreme conditions and the nature of the electrically conducting layer at the base of Earth's mantle. Nature 422, 58–61.

Dubrovinsky, L., Dubrovinskaia, N., Kantor, I., McCammon, C., Crichton, W., Urusov, V., 2004. Decomposition of ferropericlase $(Mg_{0.80}Fe_{0.20})O$ at high pressures and temperatures. J. Phys. Chem. Solids, submitted for publication.

Duffy, T.S., Hemley, R.J., Mao, H.K., 1995. Equation of state and shear strength at multimegabar pressures: magnesium oxide to 227 GPa. Phys. Rev. Lett. 74, 1371–1374.

Eremets, M., 1996. High Pressure Experimental Methods. Oxford University Press, New York.

Fabrichnaya, O.B., Sundman, B., 1997. The assessment of thermodynamic parameters in the Fe–O and Fe–Si–O systems. Cosmochim. Acta 61, 4539–4555.

Fang, Z., Solovyev, I., Sawada, H., Terakura, K., 1999. First-principal study on electronic structures and phase stability of MnO and FeO under high pressure. Phys. Rev. B 59, 762–774.

Fei, Y., Mao, H.K., 1994. In situ determination of the NiAs phase of FeO at high pressure and temperature. Science 266, 1678–1681.

Gessmann, C.K., Wood, B.J., Rubie, D.C., Kilburn, M.R., 2001. Solubility of silicon in liquid metal at high pressure: implications for the composition of the Earth's core. Earth Planet. Sci. Lett. 184, 367–376.

Goarant, F., Guyot, F., Peyroneau, J., Poirier, J.P., 1992. High-pressure and high-temperature reactions between silicates and liquid iron alloys in the diamond anvil cell, studied by analytical electron microscopy. J. Geophys. Res. 97, 4477–4487.

Gurnis, M., Wysession, M.E., Knittle, E., Buffett, B.A., (Eds), 1998. The Core–Mantle Boundary Region. AGU, Washington, DC.

Guyot, F., Zhang, J., Martinez, I., Matas, J., Ricard, Y., Javoy, M., 1997. *P–V–T* measurements of iron silicide (ε-FeSi) – implications for silicate–iron interactions in the early Earth. Eur. J. Miner. 9, 277–285.

Hemley, R.J., (Ed.) 1998. Ultrahigh-Pressure Mineralogy: Physics and Chemistry of the Earth's Deep Interior. Rev. Miner., Vol. 37, p. 671.

Hillgren, V.J., Drake, M.J., Rubie, D.C., 1994. High-pressure and high-temperature experiments on core–mantle segregation in the accreting Earth. Science 264, 1441–1445.

Hillgren, V.J., Gessmann, C.K., Li, J., 2000. An experimental perspective on the light element in Earth's core. In: Canup, R.M., Righter, K. (Eds), Origin of the Earth and Moon. University of Arizona, Tucson, pp. 245–263.

Holzapfel, W.B., Isaacs, N.S., (Eds), 1997. High-Pressure Techniques in Chemistry and Physics. Oxford University Press, Oxford, p. 388.

Jeanloz, R., Ahrens, T., 1980. Equations of state of FeO and CaO. Geophys. J.R. Astr. Soc. 62, 505–528.

Jeanloz, R., Williams, Q., 1998. The core–mantle boundary region. In: Hemley, R.J. (Ed.), Ultrahigh-Pressure Mineralogy: Physics and Chemistry of the Earth's Deep Interior, Rev. Mineral., Vol. 37, pp. 241–259.

Kavner, A., Duffy, T.S., 2001. Pressure–volume–temperature paths in the laser-heated diamond anvil cell. J. Appl. Phys. 89, 1907–1914.

Kellogg, L.H., Hager, B.H., van der Hilst, R.D., 1999. Compositional stratification in the deep mantle. Science 283, 1881–1884.

Kesson, S.E., Fitz Gerald, J.D., Shelley, J.M.G., 1994. Mineral chemistry and density of subducted basaltic crust at lower-mantle pressures. Nature 372, 767–769.

Kesson, S.E., Fitz Gerald, J.D., Shelley, J.M.G., 1998. Mineralogy and dynamics of a pyrolite lower mantle. Nature 393, 252–255.

Knittle, E., Jeanloz, R., 1989. Simulating the core–mantle boundary: an experimental study of high-pressure reactions between silicates and liquid iron. Geophys. Res. Lett. 16, 609–612.

Knittle, E., Jeanloz, R., 1991. Earth's core–mantle boundary: results of experiments at high pressures and high temperatures. Science 251, 1438–1443.

Kuznetsov, A., Dmitriev, V., Dubrovinsky, L., Prakapenka, V., Weber, H.P., 2002. FCC–HCP phase boundary in lead. Solid State Commun. 122, 125–127.

Kuznetsov, A., Dubrovinsky, L.S., Kantor, I., Kantor, A., 2004. J. Appl. Phys., submitted for publication.

Li, J., Agee, C.B., 1996. Geochemistry of mantle–core differentiation at high pressure. Nature 381, 686–689.

Li, J., Agee, C.B., 2001. Element partitioning constraints on the light element composition of the Earth's core. Geophys. Res. Lett. 28, 81–84.

Lin, J.-F., Heinz, D.L., Campbell, A.J., Devine, J.M., Shen, G., 2002. Iron–silicon alloy in the Earth's core? Science 295, 313–317.

Lin, J.-F., Heinz, D.L., Mao, H.K., Hemley, R.J., Devine, J.M., Li, J., Shen, G., 2003. Stability of magnesiowüstite in Earth's lower mantle. Proc. Natl Acad. Sci., 2003.

Mao, H.K., Shu, J., Fei, Y., Hu, J., Hemley, R.J., 1996. The wüstite enigma. Phys. Earth Planet. Interiors 96, 135–145.

Mao, H.K., Shen, G., Hemley, R.J., 1997. Multivariable dependence of Fe–Mg partitioning in the lower mantle. Science 278, 2098–2100.

McCammon, C.A., 1994. A Mössbauer milliprobe: practical consideration. Hyper. Int. 92, 1235–1239.

McCammon, C.A., 1997. Perovskite as a possible sink for ferric iron in the lower mantle. Nature 387, 694–696.

McCammon, C.A., Ringwood, A.E., Jackson, I., 1983. Thermodynamics of the system Fe–FeO–MgO at high pressure and temperature and a model for formation of the Earth's core. Geophys. J. R. Astr. Soc. 72, 577–595.

McCammon, C., Peyronneau, J., Poirier, J.P., 1998. Low ferric iron content of (Mg,Fe)O at high pressures and temperatures. Geophys. Res. Lett. 25, 1589–1592.

Murakami, M., Hirose, K., Kawamura, K., Sata, N., Ohishi, Y., 2004. Post-perovskite phase transition in $MgSiO_3$. Science 304, 855–858.

Newsom, H.E., 1990. Origin of the Earth. Oxford University Press, New York, pp. 273–288.

O'Neill, H.St.C., Palme, H., 1998. Composition of the silicate Earth: implications for accretion and core formation. In: Jackson, I. (Ed.), The Earth's Mantle. Cambridge University Press, Cambridge.

Poirier, J.P., Malavergne, V., Le Mouël, J.L., 1998. Is there thin electrically conducting layer at the base of the mantle? In: Gurnis, M., Wysession, M.E., Knittle, E., Buffett, B.A. (Eds), The Core–Mantle Boundary Region. AGU, Washington, DC, pp. 131–137.

Prakapenka, V.B., Shen, G., Dubrovinsky, L.S., 2004. Carbon transport in diamond anvil cell. High-Pressure High-Temp. 35/36, 127–151.

Ringwood, A.E., 1966. Chemical evolution of the terrestrial planets. Geochim. Cosmochim. Acta 30, 41–104.

Ruff, L.J., Anderson, D.L., 1980. Core formation, evolution, and convection; a geophysical model. Phys. Earth Planet. Interiors 21, 181–201.

Shen, G., Mao, H.K., Hemley, R.J., Duffy, T.S., Rivers, M.L., 1998. Melting and crystal structure of iron at high pressures and temperatures. Geophys. Res. Lett. 25, 373–376.

Shim, S.-H., Jeanloz, R., Duffy, T.S., 2002. Tetragonal structure of $CaSiO_3$ perovskite above 20 GPa. Geophys. Res. Lett. 29, 16148–16152.

Serghiou, G., Zerr, A., Chopelas, A., 1998. The transition of pyrope to perovskite. Phys. Chem. Miner. 25, 193–196.

Solli, D., Jeanloz, R., 2001. Nonmetallic gaskets for ultrahigh pressure diamond-cell experiments. Rev. Sci. Instrum. 72, 2110–2113.

Sowerby, J.R., Keppler, H., 1999. Water speciation in rhyolitic melt determined by in situ infrared spectroscopy. Am. Mineral. 84, 1843–1849.

Stevenson, D.J., 1990. The Origin of the Earth. Oxford Press, London, pp. 231–249.

Stixrude, L., Brown, J.M., 1998. The Earth's core. In: Hemley, R.J. (Ed.), Ultrahigh-Pressure Mineralogy: Physics and Chemistry of the Earth's Deep Interior, Rev. Mineral., Vol. 37, pp. 262–283.

Sturhahn, W., 2004. Nuclear resonant spectroscopy. J. Phys.: Condens. Matter. 16, S497–S530.

Sxanchez-Valle, C., Martinez, I., Daniel, I., Philippot, P., Bohic, S., Simionovici, A., 2003. Dissolution of strontianite at high $P–T$ conditions: an in situ synchrotron X-ray fluorescence study. Am. Mineral. 88, 978–985.

Tschauner, O., Zerr, A., Specht, S., Rocholl, A., Boehler, R., Palme, H., 1998. Partitioning of nickel and cobalt between silicate perovskite and metal at pressures up to 80 GPa. Nature 398, 604–607.

Vidale, J.E., Earle, P.S., 2000. Fine-scale heterogeneity in the Earth's inner core. Nature 404, 273–275.

Vidale, J.E., Dodge, D.A., Earle, P.S., 2000. Slow differential rotation of the Earth's inner core indicated by temporal changes in scattering. Nature 405, 445–448.

Yusa, H., 2000. Direct conversion to diamond and BN nanotube growth in nitrogen fluid: nanoscopic observation of laser-heated samples in diamond anvil cell. New Diamond Front. Carbon Technol. 10, 301–312.

Watanuki, T., Shimomura, O., Yagi, T., Kondo, T., Isshiki, M., 2001. Construction of laser-heated diamond anvil cell system for in situ X-ray diffraction study at SPring-8.

Wentzcovitch, R.M., Karki, B.B., Cococcioni, M., de Gironcoli, S., 2004. Thermoelastic properties of $MgSiO_3$–perovskite: insights on the nature of the Earth's lower mantle. Phys. Rev. Lett. 92, 18501–18504.

Zerr, A., Miehe, G., Riedel, R., 2003. Synthesis of cubic zirconium and hafnium nitride having Th_3P_4 structure. Nature Mater. 2, 185–189.

Advances in High-Pressure Technology for Geophysical Applications
Jiuhua Chen, Yanbin Wang, T.S. Duffy, Guoyin Shen, L.F. Dobrzhinetskaya, editors

Chapter 15
Pressure dependence on the magnetic properties of titanomagnetite using the reversible susceptibility method

Stuart A. Gilder and Maxime LeGoff

Abstract

This chapter describes an alternating current susceptibility system, constructed for the diamond anvil cell, that facilitates the acquisition of reversible susceptibility hysteresis loops as a function of pressure. The technique is particularly adapted to studying the effects of stress on the magnetic properties of multidomain grains because the contribution to the full (differential) susceptibility is equivalent to reversible susceptibility for such grain sizes. Here, we apply the method to synthetic multidomain grains in the titanomagnetite solid-solution series with phases containing 20, 40, and 60% Ti, and compare the results with those previously obtained from pure (0% Ti) magnetite. We show that initial susceptibility (X_{max}) and saturation moment (M_s) decrease as a function of increasing pressure and Ti concentration. The decrease is ascribed to the rotation of domains into alignment perpendicular with the maximum stress direction, conversion of a part of the reversible susceptibility into irreversible susceptibility, and an increase in the demagnetization energy via the creation of lamellar domains as well as the disappearance of closure domains. We define a constant called the median destructive stress, which is the pressure needed to remove one-half of the original susceptibility, and we show that this value is proportional to the inverse of the magnetostriction constant (λ_s), consistent with established theory. Another suite of experiments is designed to estimate the amount of reversible susceptibility that is converted into irreversible susceptibility and to determine the effect that the conversion has on $M_s(P)$. Changes in X_{max} and M_s occur fairly systematically across titanomagnetite species, with M_s/X_{max} initially increasing with pressure, before reaching a constant value where M_s becomes proportional to 4/3 X_{max}. Such behavior is attributed to systematic variations in domain wall motion in relation to stress and λ_s.

1. Introduction

Although the influence of temperature on magnetic remanence has long been a subject of research, much less is known about how pressure modifies the magnetic properties of materials. Advances in this field have been largely inhibited by experimental constraints. Early work employed large piston presses situated in proximity to astatic or flux gate magnetometers (e.g. Pozzi, 1973). Detection limits of these systems required large samples because of their higher magnetic intensities; however, the large sample sizes prevented applied pressures from exceeding a few kilobars, e.g. a few tenths of a gigapascal. Moreover, the early experimental apparatuses could only measure magnetizations along a single axis, and the applied stresses were commonly non-hydrostatic. As high-pressure investigation advanced, it became apparent that magnetic properties were sensitive to the applied stress direction and the type of imposed stress (Kean et al., 1976; Hamano, 1983; Kapicka, 1990; Valeev and Absalyamov, 2000). For stress demagnetization (i.e. stress applied in the absence of an applied magnetic field), magnetization intensity decreases faster parallel to the maximum stress direction rather than perpendicular to it. When pressure is applied in the presence of a magnetic field, both magnetic susceptibility and magnetization increase in intensity perpendicular to the maximum stress direction and decrease parallel to the maximum stress direction.

Other research has focused on how stress influences the magnetic moment at the atomic level. For example, the Mössbauer resonance technique yields important information about the existence and type of magnetic network in materials and about how those networks respond to applied stresses and magnetic fields. Although Mössbauer resonance has made important advances in understanding high-pressure phase transitions, the method does not quantify magnetic parameters such as intensity and direction of the moment, nor their stress dependencies, which are essential parameters in the Earth and planetary sciences.

More recently, BeCu and NiCrAl alloys and ceramics have been exploited in the construction of diamond or sapphire anvil cells that can achieve pressures well into the gigapascal range (e.g. Yamamoto et al., 1991; Eremets et al., 1995). The slightly magnetic nature and relatively high strength of the alloys has opened new avenues of high-pressure studies on magnetism (Tozer, 1993; Timofeev et al., 1999). Using this new technology, and adapting a technique that was originally used in superconductivity research, we developed a system that measures reversible susceptibility hysteresis loops at high pressures, which allows one to quantify the stress dependency on magnetic hysteresis parameters. Other workers have used the reversible susceptibility technique to measure tensor hysteresis loops in pipe steel (Mao and Atherton, 2000), temperature dependence on magnetic moment (Ishizuka et al., 1995) and magnetic phase transitions in invar alloy (Endo et al., 1999). It is the ability to measure hysteresis loops and to derive the associated magnetic hysteresis parameters that distinguishes our method from others using the reversible susceptibility technique. In this chapter, we describe this new method and apply it to titanomagnetite, which is an important carrier of magnetic remanence in rocks.

2. Sensor design and cell description

The measuring system we used was described in Kim et al. (1994) (see also Gilder et al., 2002; Struzhkin et al., 2002). The principle is based on the Lenz/Faraday law of electromotive force (emf) $= -\mathrm{d}\phi/\mathrm{d}t$, where ϕ is the magnetic flux passing through an electric circuit. With a coil of surface S immersed in a variable field B, the emf $= -S(\mathrm{d}B/\mathrm{d}t)$, and with a magnetic moment M placed near a coil system, the emf $= -S(\mathrm{d}(MG)/\mathrm{d}t)$, with $G = B/i$ being related to the coil geometry. Our system employed two unequal pick-up coils of 351 (detection coil) and 195 turns (compensation coil) with diameters of 3 and 4.5 mm, respectively, that were wound in opposition around a diamond (culet diameter = 370 μm), resulting in a virtually null magnetic surface (Fig. 1). This configuration has the advantage that varying magnetic field fluxes passing through both coils provoke no induction, whereas, if one coil senses a different flux than the other, the system becomes unbalanced and a voltage is generated. It follows that

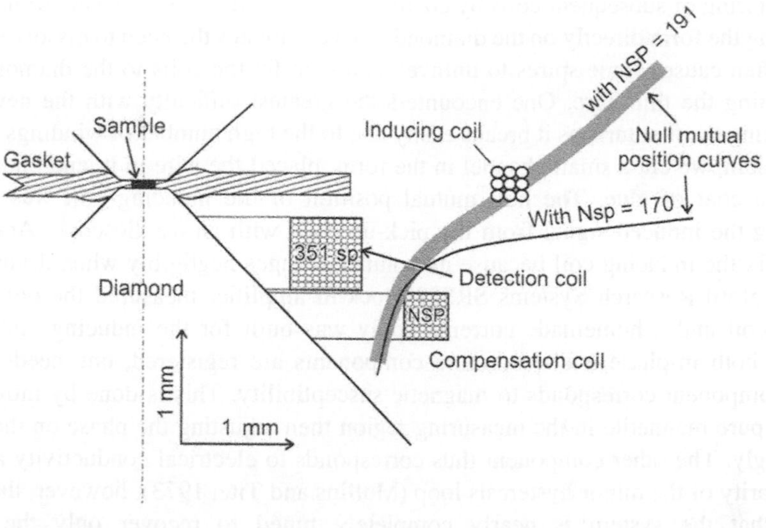

Figure 1. **Diagram of the coaxial AC susceptibility measuring system. The thick gray and thin black lines satisfy the null mutual position of the induction coil when the compensation coil has 191 and 170 turns (Nsp) of wire, respectively. The inducing coil is glued to a support fixed to the compensation and detection coils.**

the greater the number of turns of wire, and the closer the detection coil is to the magnetic source, the greater the sensitivity.

Fixed above the pick-up coils is an inducing coil, mounted in null mutual inductance, which produced a peak alternating current (AC) field of 2×10^{-4} T over the sample region. Achieving an exact surface equilibrium of both pick-up coils is not too critical when using a digital signal processor (DSP) lock-in amplifier working at frequencies higher than 10 kHz, in a magnetically quiet environment. On the contrary, the position of the induction coil requires that both pick-up coils sense as close to the same flux as possible because thermal effects on the coils are difficult to compensate with electronic circuitry. Thus, it is preferable to minimize the near null mutual induction before mounting the inducing coil. In a co-axial configuration for all three coils, which we employed (Fig. 1), the radius of the inducing coil and its distance from the detection coil have several solutions. Figure 1 shows the range of possibilities where to place the inducing coil, with the thick line representing the null mutual solution. We designed the system so that the detection coil has the smallest possible radius and sits as close as possible to the sample, and that the inducing coil lies within the plane of the sample (Fig. 1). The inducing coil must be large enough to allow a gasket to be placed within its confines, while the detection coil must be positioned low enough on the diamond so that when the diamond indents into the gasket, it does not press on the coils. The final sensitivity, given the radius and number of turns, is about 10^{-9} A m^2/μV.

We originally constructed the coils following the description of Kim et al. (1994) by machining a lucite form, then winding a 25 μm-diameter copper wire, insulated with 5 μm-thick isolation (total diameter = 35 μm), around it. We fixed the windings with a non-magnetic, water-soluble glue (Elmer's Glue-All) so they would not unbind when dissolving the lucite in methylene chloride. Although successful, we improved the manufacturing of subsequent coils by coating the diamond in epoxy resin (Araldite), then machining the form directly on the diamond. This eliminates the need to dissolve the form, which often causes some spires to unravel, and then fix the coils to the diamond, which risks cutting the thin wire. One encounters the greatest difficulty with the new method when laying the first turn, as it breaks easily due to the high number of windings. To avoid this problem, we cut a small channel in the form, placed the wire in it, and blanketed the wire in a coat of glue. The null mutual position of the inducing coil was found by observing the induced signal from the pick-up coils with an oscilloscope. Araldite was used to fix the inducing coil because its volume changes negligibly when drying.

A Stanford Research Systems SRS830 lock-in amplifier measured the output of the sensing coil and a homemade current supply was built for the inducing coil (Fig. 2). Because both in-phase and quadrature components are registered, one needs to verify which component corresponds to magnetic susceptibility. This is done by introducing a piece of pure magnetite in the measuring region then adjusting the phase on the SRS830 accordingly. The other component thus corresponds to electrical conductivity and to the non-linearity of the minor hysteresis loop (Mullins and Tite, 1973); however, the phase is set so that the system is nearly completely tuned to recover only the in-phase (susceptibility) signal while neglecting the quadrature (conductivity) signal. It is important to note that this system could be developed in the opposite sense, i.e. to measure the electrical conductivity of material under pressure. Although one attempts to fix the induction coil as close to the null position as possible, there always exists a non-perfectly null component, which can be compensated by electronic circuitry. This is done

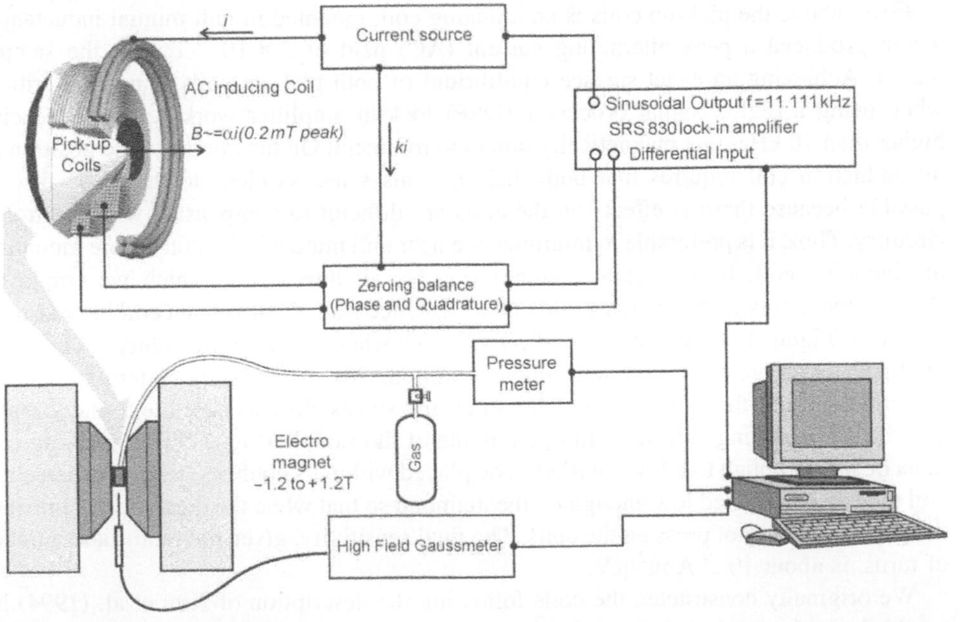

Figure 2. Cartoon of the experimental set-up. The coils are housed in a membrane-type BeCu diamond anvil cell that is attached to a helium gas tank via a capillary tube. The cell is placed in the electromagnet with the axis of the AC field parallel to the DC field generated by the electromagnet. The diameter of the pole pieces is 15 cm with a gap of 5 cm. See text for more details.

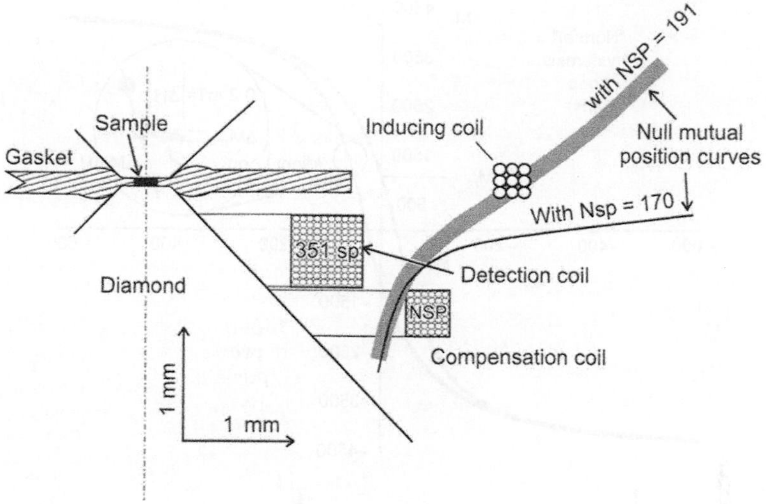
Sample
Gasket
Diamond
351 sp
NSP
Inducing coil
with NSP = 191
Null mutual
position curves
With Nsp = 170
Detection coil
Compensation coil
1 mm
1 mm

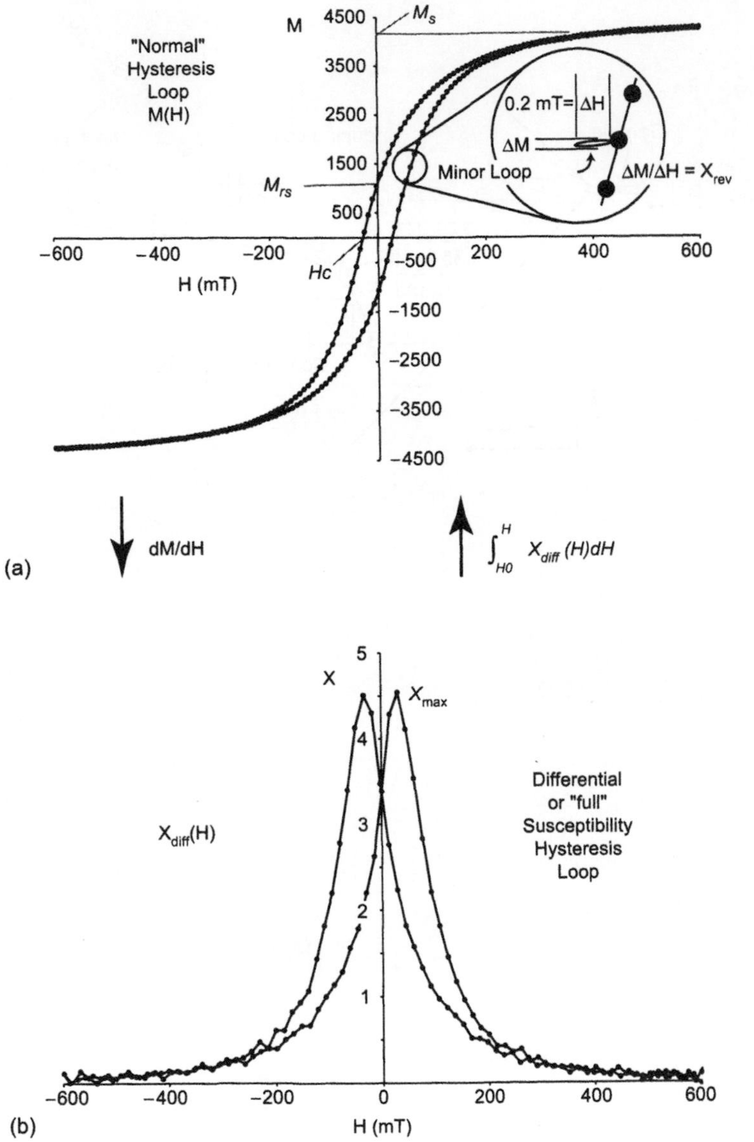

Figure 3. Explanation of the AC susceptibility method. (a) A full hysteresis loop [$M(H)$] of single domain magnetite (chiton teeth). The sinusoidal AC field forms minor loops within the interior of the full hysteresis loop. The width of the loop (ΔH) is the amplitude of the AC field (0.2 mT in this case) while the height of the loop (ΔM) is the amount of magnetization gained or lost under the 0.2 mT AC field. Reversible susceptibility (X_{rev}) equals $\Delta M/\Delta H$. (b) The derivative of the full hysteresis loop (dM/dH) is the differential, or full, susceptibility hysteresis loop [$X_{diff}(H)$]. Put another way, if one measured $X_{diff}(H)$, then took its integral, it would yield $M(H)$.

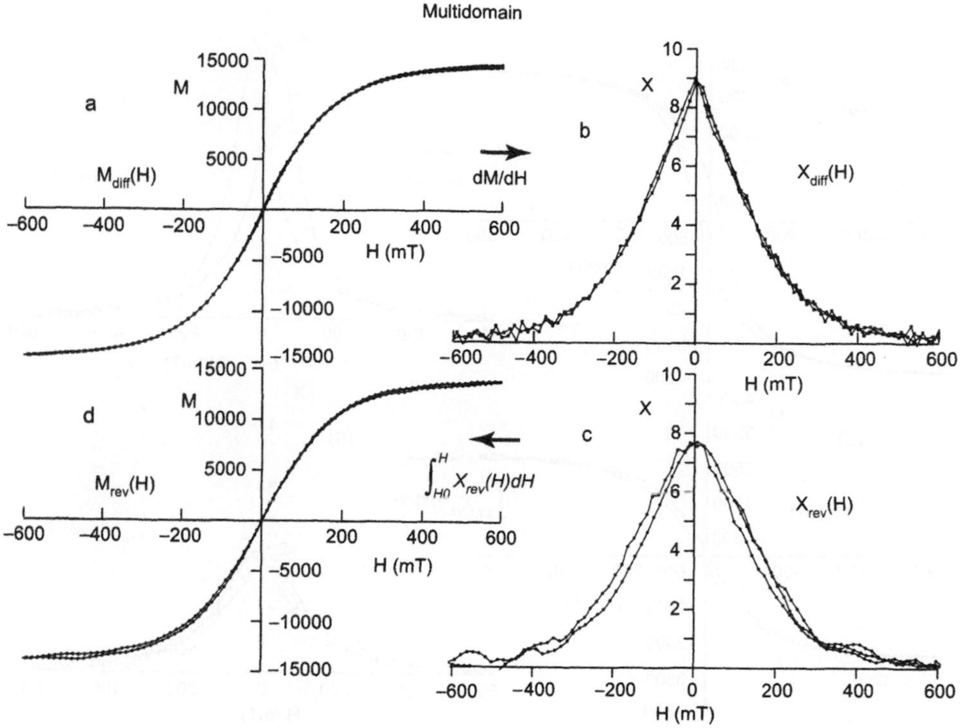

Figure 4. Comparison of the differential (full) and reversible hysteresis loops for the same sample of multidomain magnetite (TM0, magnetite extracted from the Mt Givens [California] pluton). (a) The measured differential [$M(H)$ or $M_{\text{diff}}(H)$] hysteresis loop and its derivative (b) that yields $X_{\text{diff}}(H)$. (c) The measured reversible susceptibility loop $X_{\text{rev}}(H)$ and its integral (d) that yields $M_{\text{rev}}(H)$. Note the resemblance between $M_{\text{diff}}(H)$ and $M_{\text{rev}}(H)$ ($M_s \approx 12,500$ [arbitrary units] and $H_c < 2$ mT in both cases) and between $X_{\text{diff}}(H)$ and $X_{\text{rev}}(H)$.

multidomain magnetite sample and its integral ($\int_{H_0}^{H} X_{\text{rev}}(H)\text{d}H$) is plotted in Figure 4d. One observes a good correlation between X_{rev} and X_{diff} and between $M_{\text{rev}}(H)$ and $M_{\text{diff}}(H)$. We then performed the same exercise for single domain magnetite (Fig. 5a to d), where one notes a much less than ideal fit between X_{rev} and X_{diff} and between $M_{\text{rev}}(H)$ and $M_{\text{diff}}(H)$.

4. Titanomagnetite

Titanomagnetite ($Fe_{3-x}Ti_xO_4$) [also written $xFe_2TiO_4(1-x)Fe_3O_4$] forms a complete solid solution series between end members of magnetite (Fe_3O_4) and ulvöspinel (Fe_2TiO_4). They are cubic minerals with inverse spinel structures that commonly carry the magnetic remanence of rocks; i.e. TM60 ($x = 0.6$) is the dominant magnetic mineral in rapidly quenched submarine basaltic lavas. Details concerning their structural and magnetic properties can be found in Syono (1965), O'Reilly (1984), Lindsley (1991).

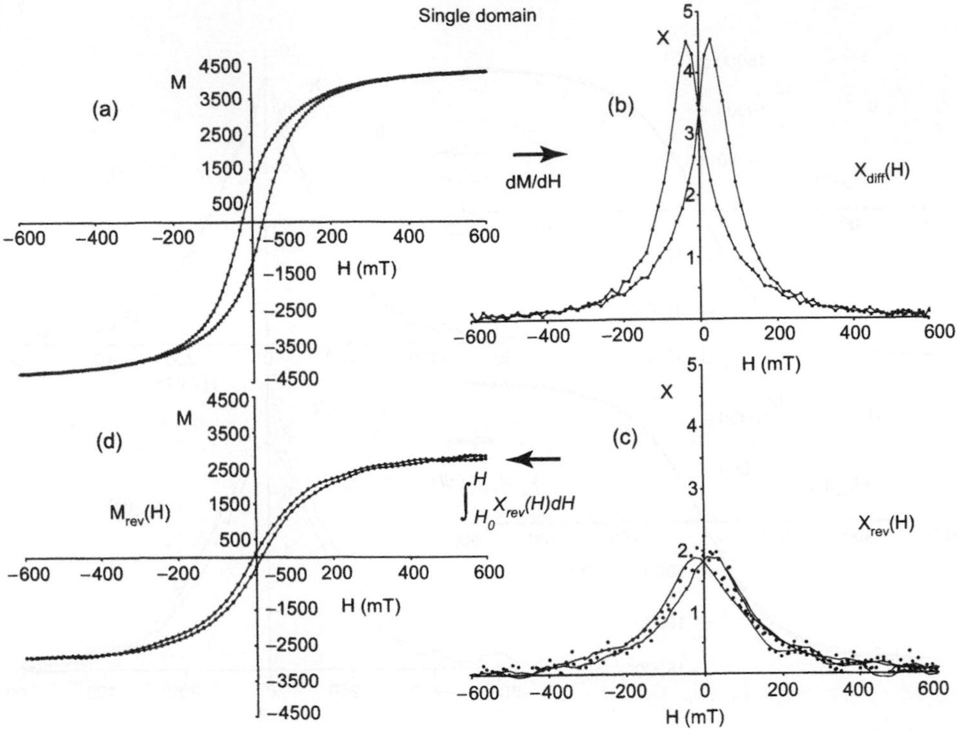

Figure 5. Comparison of the differential (full) and reversible hysteresis loops for the same sample of single domain magnetite (chiton teeth). (a) The measured differential [$M(H)$ or $M_{diff}(H)$] hysteresis loop and its derivative (b) that yields $X_{diff}(H)$. (c) The measured reversible susceptibility loop $X_{rev}(H)$ and its integral (d) that yields $M_{rev}(H)$. Note the differences between $M_{diff}(H)$ and $M_{rev}(H)$ and between $X_{diff}(H)$ and $X_{rev}(H)$. $M_{s_{diff}}$ and $H_{c_{diff}} \approx 3900$ (arbitrary units) and 27 mT; $M_{s_{rev}}$ and $H_{c_{rev}} \approx 2800$ (arbitrary units) and 11 mT.

The samples in our experiments were single titanomagnetite crystals synthesized by B.J. Wanamaker using the floating zone technique (Wanamaker and Moskowitz, 1994). Optical, Laue back-reflection and scanning electron microscopy, electron microprobe, and thermomagnetic analyses indicate that the TM60 sample is a single crystal, single phase, and chemically homogeneous except for a 3% gradient in Ti concentration over a 4 cm length of the crystal. We verified the Ti concentration and homogeneity of the samples via our own thermomagnetic analyses using an AGICO CS-3 furnace linked to a KLY-3 Kappabridge susceptibility meter (Fig. 6). Curie temperatures on heating correspond to titanium concentrations of TM19, TM39, and TM60 (Akimoto, 1962), which, given the uncertainties, are close to the intended values of $Fe_{2.8}Ti_{0.2}O_4$ (TM20), $Fe_{2.6}Ti_{0.4}O_4$ (TM40), and $Fe_{2.4}Ti_{0.6}O_4$ (TM60). Although the TM60 sample is relatively pure, TM40 and TM20 exhibit evidence for inhomogeneous or incomplete Ti diffusion during synthesis resulting in multiphase spinels. TM20 has a phase with a Curie point at 523°C (TM11) while TM40 possesses two phases with Curie points at 393°C (TM31) and 485°C (TM18). Unlike the Curie points of the major phases, those of the minor

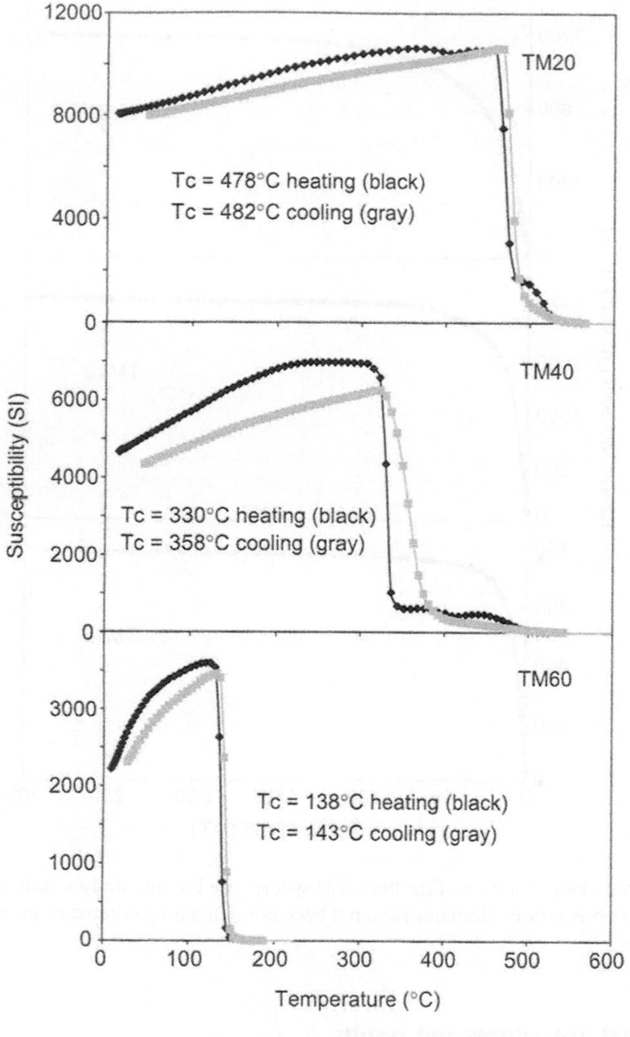

TM20
Tc = 478°C heating (black)
Tc = 482°C cooling (gray)
TM40
Tc = 330°C heating (black)
Tc = 358°C cooling (gray)
TM60
Tc = 138°C heating (black)
Tc = 143°C cooling (gray)
Susceptibility (SI)
Temperature (°C)
12000
8000
4000
0
6000
4000
2000
0
3000
2000
1000
0
0
100
200
300
400
500
600

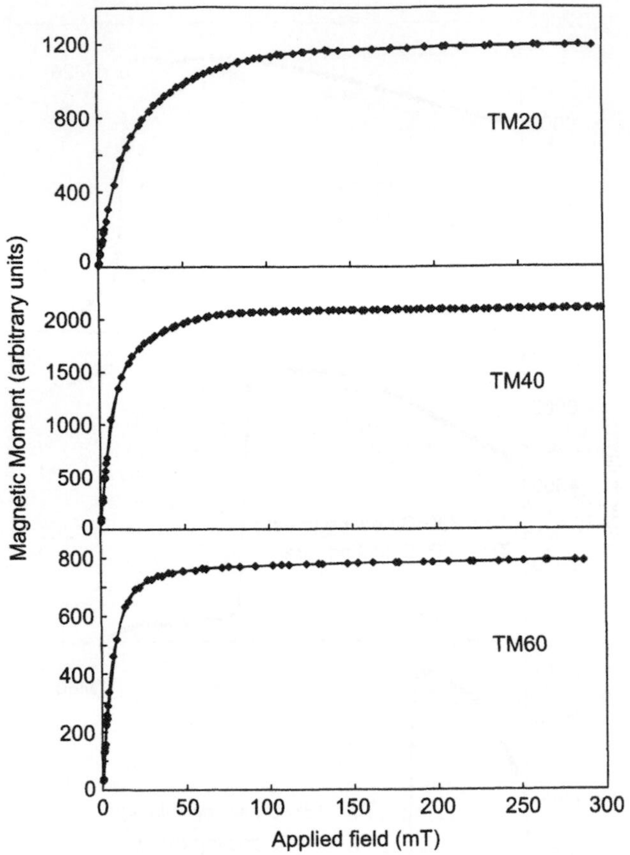

Figure 7. Hysteresis loops for each of the three TM species used in this study – only one half of the loop is shown for better observation. Moments have not been normalized by volume or mass. H_c in all cases is <2 mT.

5. Experimental procedures and results

Experimental procedures were described in Gilder et al. (2002) and are repeated briefly here. An empty gasket was loaded in the cell, the cell was placed into the electromagnet, and the gasket was slightly compressed. Reversible susceptibility of the gasket was measured while making at least three complete cycles though applied fields of ±1.2 T to obtain $X(H)$. After correcting for thermal drift, we stacked the curves corresponding to the three cycles and calculated the average X at each H (Fig. 8a). The cell was then removed from the electromagnet, a sample was loaded in the empty gasket, the cell was put back into the electromagnet, then $X(H)$ of the sample + gasket was measured using the same procedure previously applied to the gasket alone (Fig. 8b). The $X(H)$ of the gasket was subtracted from $X(H)$ of the sample + gasket to obtain $X(H)$ of the sample (Fig. 8b). Pressure was increased, $X(H)$ of the sample + gasket was measured at the new pressure, the $X(H)$ of the gasket was subtracted, etc. X(H) of the gasket varies negligibly as a

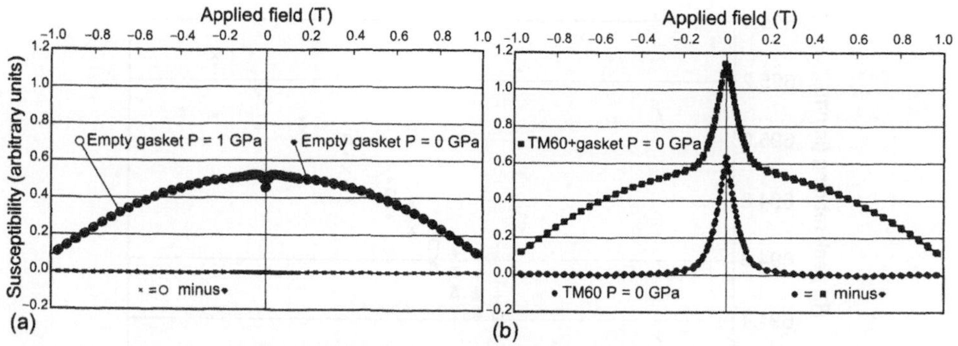

Figure 8. (a) Reversible susceptibility versus applied field of an empty gasket at high pressure (about 1 GPa) (open circles) and at ambient pressure (small solid diamonds). The difference between the two curves is negligible (small x's). (b) Reversible susceptibility versus applied field of the TM60 sample plus the gasket at $P = 0$ GPa (solid squares) and when subtracting the signal from the gasket (open circles or small solid diamonds from [a]) yielding TM60 at $P = 0$ (solid circles).

function of pressure, as seen by comparing $X(H)$ for the same gasket at $P \approx 1$ GPa and $P = 0$ GPa and their difference (Fig. 8a).

Samples consist of several tens of TM fragments, with an average size of about 20 μm/fragment, saturated in silica gel, the latter serving as a pressure medium. Although the exact amount of TM introduced in the gasket is unknown, the relative proportion of sample:gel is similar for each TM species and fairly constant across species. Knowing the precise mass of the material, by weighing beforehand, is difficult because some TM grains stay on top of the gasket during loading, then must be carefully removed to assure that no stray grains remain. We have not found a simple way to recover and then weigh those stray grains, yet this part of the procedure can be eventually improved. By filling the ~150 μm-diameter hole (made by electrolysis) in the ~100 μm-thick gasket to capacity, sample volumes should be similar and thus comparable.

Pressure was calibrated with ruby fluorescence (Adams et al., 1976; Chervin et al., 2001) by mimicking the pressure path of each experiment three to eight times, using the same gasket material and geometry and similar ruby:silica gel ratios as the TM:silica gel ratios (Fig. 9). Distinct $R1$ and $R2$ peaks in the spectra suggest dominantly hydrostatic pressure conditions were maintained in the cell. At 3 GPa, pressure measured on a ruby near the edge of the gasket hole is ~10% less than the pressure on a ruby at the center. Standard deviations are less than 0.2 GPa except at the highest pressures achieved in the TM20 experiments where they reach 0.3 GPa. While we used the mean value of the measurements, one can see in Figure 9 that pressure can vary by 0.5 GPa between two independent runs. Thus the greatest source of uncertainty in our experiments is not the relative pressure in an experiment but rather when comparing absolute pressures from one experiment to the next. Reducing this uncertainty is difficult because already the gaskets are uniform and the center of the hole can be routinely placed in the center of the diamond. Conversely, one cannot extract the cell from the electromagnet to make a pressure measurement during the course of an experiment so we see no way around the calibration method.

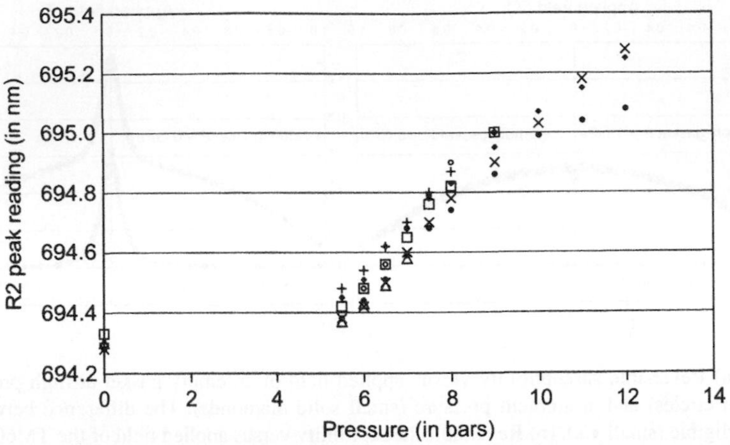

Figure 9. Pressure calibration of a BeCu gasket filled with ruby and silica gel following the experimental procedures of the type-one experiments. Each symbol represents an independent calibration. Pressure in bars is the pressure of helium gas applied to the membrane of the diamond cell (accurate to one-tenth of a bar). The R2 peak (in nanometers) is determined via ruby fluorescence then translated into pressure (Adams et al., 1976; Chervin et al., 2001).

We performed two separate experiments on each titanomagnetite phase. The first was to sequentially increase pressure until ~95% of the original signal was lost, then sequentially decrease pressure back to ambient conditions. The second was to compress to a pressure ($P = P1$), make a measurement, fully decompress ($P = 0$), make a measurement, raise the pressure above the previous value ($P = P2$), make a measurement, fully decompress ($P = 0$), etc. Reversible susceptibility hysteresis curves and their integrals for the type-one experiments are shown in Figure 10.

Because the samples are purely multidomain, two useful parameters are obtained from the experiments: the measured maximum reversible susceptibility (X_{max}) and the calculated saturation moment (M_s). Both are listed in Table 1 for each pressure step. M_{rs} and H_c are too small to be accurately determined at any pressure. The fact that X_{max} occurs at zero applied field (Figs. 4 and 10) and is measured using a 0.2 mT AC field, X_{max} should represent the initial susceptibility of the material. Fig. 11a and b plot X_{max} and M_s as a function of pressure for the type-one experiments for the three TM species together with the TM0 data from Gilder et al. (2002); the same data, normalized by the starting, pre-compressed values, are shown in Fig. 11c and d. One quickly notes the pressure sensitivity as a function of Ti concentration, with X_{max} and M_s decaying faster as a function of pressure as Ti increases.

6. Discussion

So what controls the decrease in X_{max} and M_s? For X_{max}, the response appears to lie in the change in domain structure. Bogdanov and Vlasov (1966) observed that, for pure magnetite, the number of domains increases as a function of stress, noting that the change in domain structure varies according to the crystallographic axis and the applied

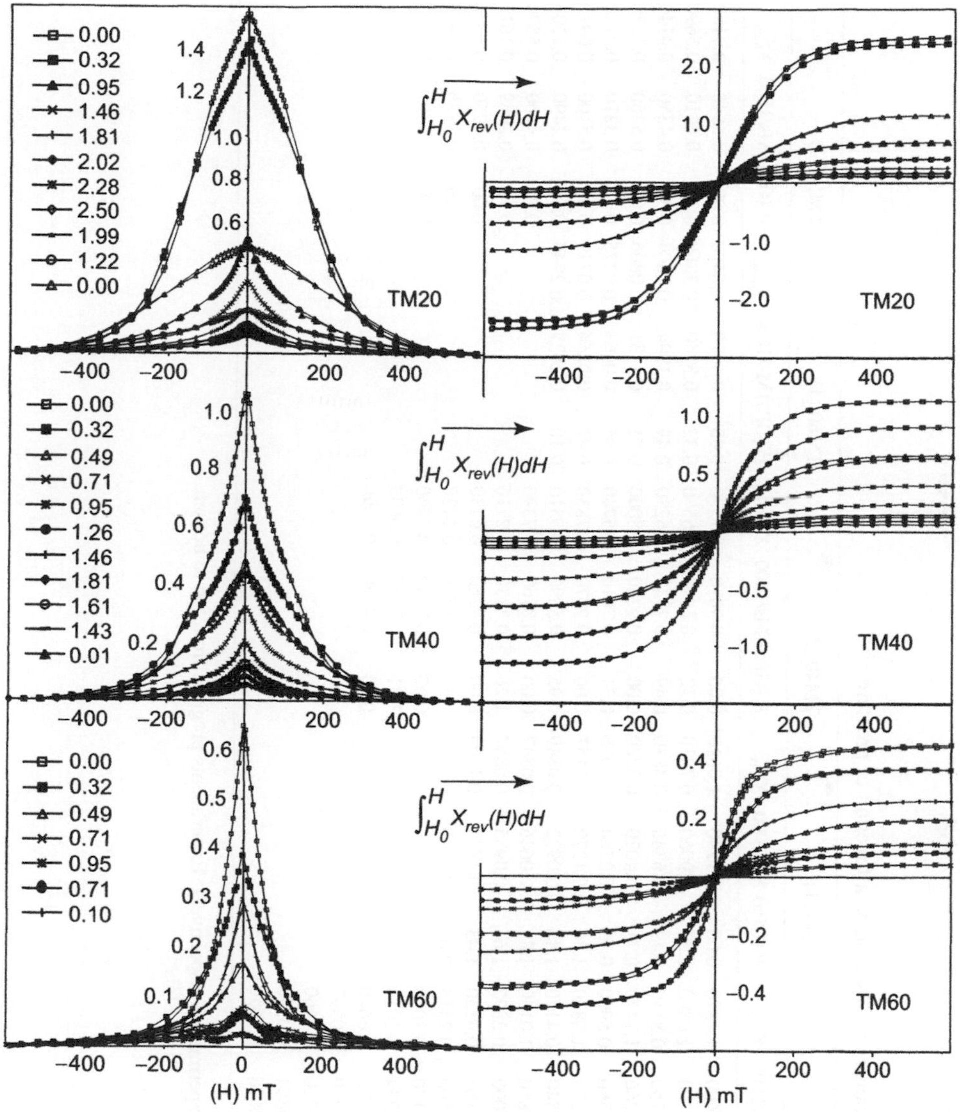

Figure 10. Measured $X_{rev}(H)$ loops for each of the TM species for the type-one experiments (left) and the integrals for each loop $[M_{rev}(H)]$ (right). Pressure shown on left is in GPa, with the experiment progressing from top to bottom.

stress direction. Because the demagnetization factor (N) increases as the number of lamellar domains increases (Dunlop and Özdemir, 1997, p. 111; Dunlop, 1983) and because X varies as $1/N$ in multidomain magnetite (Néel, 1955; O'Reilly, 1984, p. 75), the measured decrease in X_{max} can be at least partly attributed to an increase in N. Moreover, Bogdanov and Vlasov (1966) found that closure domains disappear with increasing stress, which will increase the surface pole density and the magnetostatic energy of the grain (Özdemir et al., 1995), which will in turn lead to higher N.

Table 1. Reversible susceptibility parameters measured as a function of pressure.

TM20 (1)			TM20 (2)			TM40 (1)			TM40 (2)			TM60 (1)			TM60 (2)		
P (GPa)	Ms/1000	X_{max}	*P* (GPa)	Ms/1000	X_{max}	*P* (GPa)	Ms/1000	X_{max}	*P* (GPa)	Ms/1000	X_{max}	*P* (GPa)	Ms/1000	X_{max}	*P* (GPa)	Ms/1000	X_{max}
0.00	2.5050	1.5660	0.00	1.5600	1.1870	0.00	1.1290	1.0610	0.00	0.9070	0.7370	0.00	0.4570	0.6490	0.00	0.5640	0.9180
0.32	2.3910	1.4460	0.32	1.5300	1.0880	0.32	0.9060	0.7030	0.32	0.8440	0.6640	0.32	0.3740	0.3860	0.29	0.4410	0.5690
0.95	0.6960	0.5230	0.49	1.4750	0.8110	0.49	0.6340	0.4830	0.49	0.7930	0.6210	0.49	0.1994	0.1664	0.32	0.4350	0.4140
1.46	0.4130	0.3290	0.00	1.5260	1.1470	0.71	0.4020	0.3220	0.00	0.9230	0.8000	0.71	0.1131	0.0806	0.00	0.5360	0.7830
1.81	0.2529	0.2029	0.71	1.1360	0.5900	0.95	0.2364	0.2056	0.71	0.7010	0.5740	0.95	0.0463	0.0275	0.49	0.3210	0.2784
2.02	0.1645	0.1376	0.00	1.5220	1.0830	1.26	0.1279	0.1151	0.00	0.9070	0.7830	0.71	0.0854	0.0719	0.00	0.5100	0.6970
2.28	0.1244	0.1011	0.95	0.7540	0.4110	1.46	0.0932	0.0849	0.95	0.5990	0.5310	0.10	0.2627	0.2945	0.71	0.2490	0.2106
2.50	0.1103	0.0821	0.00	1.4810	1.0150	1.81	0.0646	0.0547	0.00	0.8760	0.7390				0.00	0.4660	0.5500
1.99	0.1738	0.1154	1.26	0.5000	0.2892	1.61	0.0935	0.0857	1.26	0.3130	0.2716				0.95	0.2069	0.1610
1.22	0.4040	0.1980	0.00	1.4160	0.9130	1.43	0.1475	0.1409	0.00	0.8260	0.6730				0.00	0.4750	0.6370
0.00	1.1580	0.4880	1.46	0.3590	0.2413	0.01	0.6600	0.4410	1.46	0.1518	0.1428				1.26	0.1986	0.1475
			0.00	1.3170	0.8030				0.00	0.7430	0.5960				0.00	0.4800	0.6190
			1.81	0.2334	0.2010				1.81	0.0588	0.0693						
			0.00	1.1620	0.6640				0.00	0.4950	0.3890						
			2.15	0.1200	0.0993												
			0.00	0.9310	0.4660												
			2.50	0.0884	0.0756												
			0.00	0.7690	0.3650												

(1) and (2) refer to type-one and two experiments, respectively. Experiments progress from top to bottom.

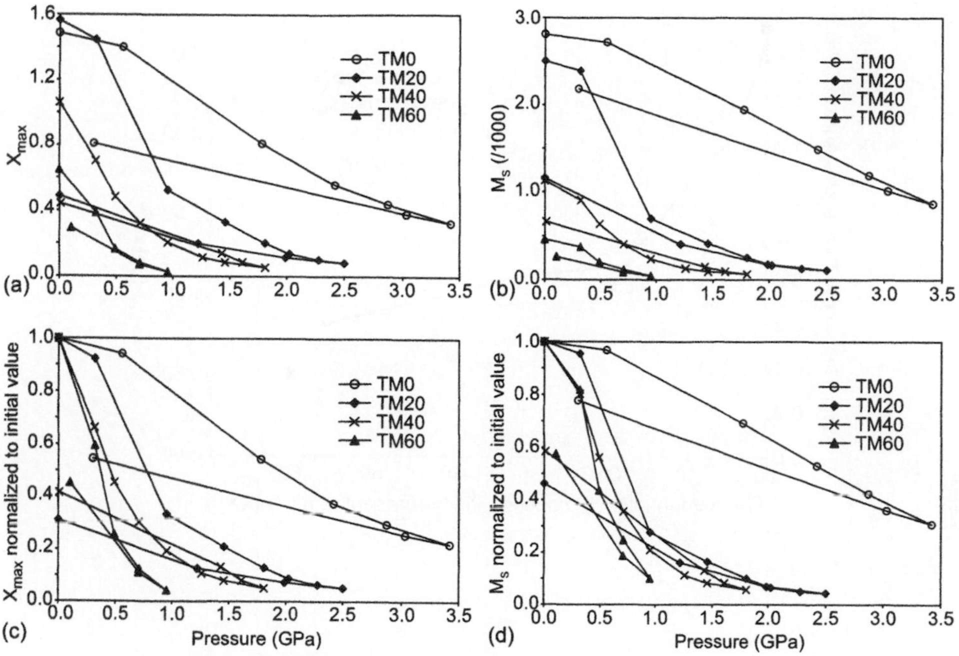

Figure 11. (a) X_{max} and (b) M_s (multiplied by 1/1000) measured at each pressure in the type-one experiments. (c) and (d) are the same data normalized by the initial (starting) X_{max} and M_s values, respectively. Data for TM0 were reported in Gilder et al. (2002).

Soffel et al. (1982) observed that titanomagnetites with higher ulvöspinel concentrations have curved domain walls and fewer closure domains. Thus, the decrease in $X_{max}(P)$ is partially related to the creation of new lamellar domains and the disappearance of closure domains.

The change in domain state results from an increase in strain, due to a rise in the strain anisotropy energy (E_{anis}), where E_{anis} is expressed as $-(3/2)\lambda_s \sigma \cos^2 \theta$, with λ_s being the net magnetostriction constant, σ, the applied stress, and θ, the angle between the magnetization vector of the grain relative to the applied stress direction (Kittel, 1949; Shive and Butler, 1969; Hodych, 1976; O'Reilly, 1984, p. 50). Although this parameter assumes a uniaxial stress, yet the stresses in our experiments are dominantly hydrostatic, the fact that λ_s and the compressibility of TM are anisotropic should lead to a non-null θ. This means that pressure acts to augment E_{anis}, and thus N. The reason for TM species with higher Ti to be more pressure sensitive, e.g. lose X_{max} faster, should therefore be related to λ_s. In other words, λ_s should vary with N, and X_{max} should be proportional to $1/\lambda_s$. Indeed, magnetostriction constants do rise as a function of titanium concentration, being 2.9, 5.0, 5.9, 8.4, 8.9 and 12.3 ($\times 10^{-5}$) for TM values of 0, 4, 10, 18, 31, and 56, respectively (Syono, 1965). This implies that, not only is X_{max} related to the spontaneous moment (decreasing effective magnetization with increasing Ti), but for multidomain grains, X_{max} also depends on λ_s, which is consistent with theory (Kern, 1961).

As TM species become enriched in Ti, so too should they become more pressure sensitive. In order to verify this, we defined a parameter called the median destructive

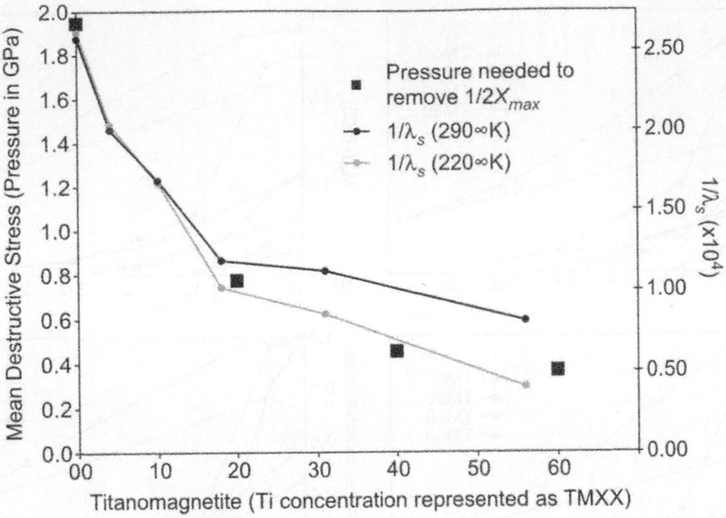

Mean Destructive Stress (Pressure in GPa)
2.0
1.8
1.6
1.4
1.2
1.0
0.8
0.6
0.4
0.2
0.0
00
10
20
30
40
50
60
Titanomagnetite (Ti concentration represented as TMXX)
2.50
2.00
1.50
1.00
0.50
0.00
$1/\lambda_s$ (x10^4)
Pressure needed to remove 1/2X_{max}
$1/\lambda_s$ (290∞K)
$1/\lambda_s$ (220∞K)

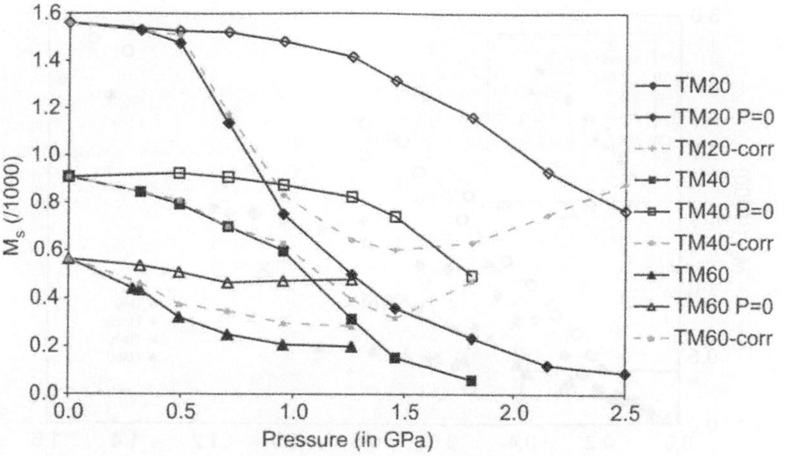

M_S (/1000)
Pressure (in GPa)
TM20
TM20 P=0
TM20-corr
TM40
TM40 P=0
TM40-corr
TM60
TM60 P=0
TM60-corr

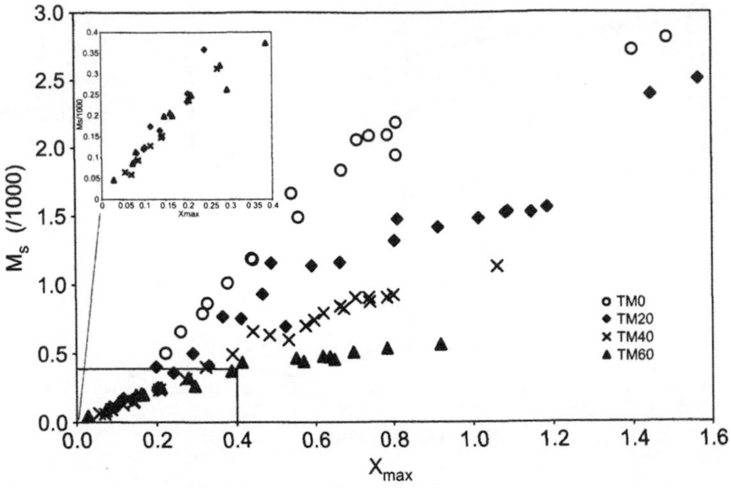

Figure 14. M_s (multiplied by 1/1000) versus X_{max} for all experiments. M_s/X_{max} increases with decreasing Ti content. With increasing pressure, M_s/X_{max} increases for all species except at the highest pressures where M_s becomes proportional to 4/3 X_{max} (inset). Data for TM0 were reported in Gilder et al. (2002).

suggest that the domain production/re-orientation process occurs differently under discrete strain states, e.g. that there are distinct strain regimes until M_s ultimately becomes linearly proportional to N, with all TM phases following the same proportionality law once a given threshold strain state is reached. Systematic changes in the domain configuration brought about by the application of stress was proposed by Kean et al. (1976) to explain differential changes in susceptibility acquired perpendicular or parallel to applied uniaxial stresses.

7. Conclusions

In this chapter we describe the theory behind the AC susceptibility method and the details surrounding the construction of an AC susceptibility system for the diamond anvil cell. By placing the cell within an electromagnet and measuring reversible susceptibility as a function of applied DC field, our system can quantify magnetic hysteresis parameters while being able to detect thermal perturbations or spurious effects from mechanical interactions that potentially occur during an experiment. When a non-electrically conducting substance can be found to serve as a gasket, our system can be adapted to measure the electrical conductivity of material under pressure.

Here, we show that pressure modifies the magnetic properties of titanomagnetite relative to the Ti concentration, with $M_s(P)$ and $X_{max}(P)$ being inversely proportional to stress and magnetostriction (λ_s). Because λ_s increases with increasing Ti, titanomagnetite species become increasingly pressure sensitive. Thus, by merely knowing the magnetostriction constants of a given magnetic mineral, one should be able to predict its sensitivity to an imposed stress. With this in mind, one can revisit the relevance of piezo-remanent magnetization and stress demagnetization to earthquake prediction or the build up or relaxation of stress in the crust due to tectonic motion (Stacey, 1964; Davis and Stacey, 1972; Smith and Johnston, 1976; Zlotnicki and Cornet, 1986), which has proved elusive. Terrains containing Ti-rich titanomagnetite phases, such as oceanic crust, should be considered the best targets, while the strain history of the rocks should be accounted for by identifying places where the rocks have been previously subjected to the lowest possible stresses.

Acknowledgements

Special thanks are given to Barbara (B.J.) Wanamaker and Bruce Moskowitz for providing the titanomagnetite. We are grateful for the helpful reviews of Gunther Kletetschka, Özden Özdemir and an anonymous reviewer. Thanks also to Denis Andrault for providing open access to his lab. The diamond cell described in this chapter was designed by Jean-Claude Chervin and his team at the Université Pierre et Marie Curie. This work was funded by INSU ("Intérieur de la Terre" program) and by the IPGP Bonus Qualité de Recherche (IPGP publication #2008).

References

Adams, D., Appleby, R., Sharma, S., 1976. Spectroscopy at very high pressures: part X. Use of ruby R-lines in the estimation of pressure at ambient and at low temperatures. J. Phys. E 9, 1140–1144.

Akimoto, S., 1962. Magnetic properties of $FeO-Fe_2O_3-TiO_2$ system as a basis of rock magnetism. J. Phys. Soc. Jpn 17 (Suppl. B-1), 706–710.

Appel, E., Soffel, H.C., 1984. Model for the domain state of Ti-rich titanomagnetites. Geophys. Res. Lett. 11, 189–192.

Appel, E., Soffel, H.C., 1985. Domain state of Ti-rich titanomagnetites deduced from domain structure observations and susceptibility measurements. J. Geophys. 56, 121–132.

Bertotti, G., 1998. Hysteresis in Magnetism. Academic Press, San Diego.

Bogdanov, A.A., Vlasov, A.Ya, 1966. On the effect of elastic stresses on the domain structure of magnetite. Izv. Phys. Solid Earth 1, 24–26.

Chervin, J.C., Canny, B., Besson, J., Pruzan, P., 1995. A diamond cell for IR microspectroscopy. Rev. Sci. Instrum. 66, 2595–2598.

Chervin, J.C., Canny, B., Mancinelli, M., 2001. Ruby-spheres as pressure gauge for optically transparent high pressure cells. High Pressure Res. 21, 305–314.

Davis, P.M., Stacey, F.D., 1972. Geomagnetic anomalies caused by a man-made lake. Nature 240, 348–349.

Dunlop, D.J., 1983. On the demagnetizing energy and demagnetizing factor of a multidomain ferromagnetic cube. Geophys. Res. Lett. 10, 79–82.

Dunlop, D.J., Özdemir, Ö., 1997. Rock Magnetism. Cambridge University Press, Cambridge.

Dunlop, D.J., Ozima, M., Kinoshita, H., 1969. Piezomagnetization of single domain grains: a graphical approach. J. Geomag. Geoelec. 21, 513–518.

Endo, S., Yamada, J., Imada, S., Ishizuka, M., Kindo, K., Miyamoto, S., Ono, F., 1999. Development of magnetization measurement under pressure and pulsed high magnetic field. Rev. Sci. Instrum. 70, 2445–2447.

Eremets, M.I., Struzhkin, V.V., Utjuzh, A.N., 1995. Miniature high pressure cells for high magnetic field applications. Physica B 211, 369–371.

Gilder, S.A., LeGoff, M., Peyronneau, J., Chervin, J.C., 2002. Novel high-pressure magnetic measurements with application to magnetite. Geophys. Res. Lett. 29, 10.1029/2001GL014227.

Gilder, S.A., LeGoff, M., Peyronneau, J., Chervin, J.C., 2004. Magnetic properties of single and multi-domain magnetite under pressures from 0 to 6 GPa. Geophys. Res. Lett. 31, L10612, doi:10.1029/2004GL019844.

Hamano, Y., 1983. Experiments on the stress sensitivity of natural remanent magnetization. J. Geomag. Geoelec. 35, 155–172.

Hodych, J., 1976. Single-domain theory for the reversible effect of small uniaxial stress upon the initial magnetic susceptibility of rock. Can. J. Earth Sci. 13, 1186–1200.

Ishizuka, M., Amaya, K., Endo, S., 1995. Precise magnetization measurements under high pressures in the diamond-anvil cell. Rev. Sci. Instrum. 66, 3307–3310.

Kapicka, A., 1990. Variations of the mean susceptibility of rocks under hydrostatic and non-hydrostatic pressure. Phys. Earth Planet. Interiors 63, 78–84.

Kean, W., Day, R., Fuller, M., Schmidt, V., 1976. The effect of uniaxial compression on the initial susceptibility of rocks as a function of grain size and composition of their constituent titanomagnetites. J. Geophys. Res. 81, 861–872.

Kern, J.W., 1961. The effect of stress on the susceptibility and magnetization of a partially magnetized multidomain system. J. Geophys. Res. 66, 3807–3816.

Kim, C., Reeves, M., Osofsky, M., Skelton, E., Liebenberg, D., 1994. A system for in situ pressure and ac susceptibility measurements using the diamond anvil cell: $T_c(P)$ for $HgBa_2CuO_{4+\delta}$. Rev. Sci. Instrum. 65, 992–997.

Kittel, C., 1949. Physical theory of ferromagnetic domains. Rev. Mod. Phys. 21, 541–583.

Lindsley, D.H., 1991. Oxide Minerals: Petrologic and Magnetic Significance, Reviews in Mineralogy, Vol. 25. Mineralogical Society of America, Washington.

Mao, W.H., Atherton, D.L., 2000. Effect of compressive stress on the reversible and irreversible differential magnetic susceptibility of a steel cube. J. Mag. Mag. Matter 214, 69–77.

Mullins, C.E., Tite, M.S., 1973. Magnetic viscosity, quadrature susceptibility, and frequency dependence of susceptibility in single domain assemblies of magnetite and maghemite. J. Geophys. Res. 78, 804–809.

Nagata, T., 1966. Magnetic susceptibility of compressed rocks. J. Geomag. Geoelec. 18, 73–80.

Nagata, T., Kinoshita, H., 1964. Effect of release of compression on magnetization of rocks and assemblies of magnetic minerals. Nature 204, 1183–1184.

Nagata, T., Kinoshita, H., 1965. Studies on Piezo-magnetization (I) magnetization of titaniferous magnetite under uniaxial compression. J. Geomag. Geoelec. 17, 121–135.

Néel, L., 1955. Some theoretical aspects of rock magnetism. Adv. Phys. 4, 191–243.

O'Reilly, W., 1984. Rock and Mineral Magnetism. Chapman & Hall, New York.

Özdemir, Ö., Xu, S., Dunlop, D., 1995. Closure domains in magnetite. J. Geophys. Res. 100, 2193–2209.

Pozzi, J.P., 1973. Effets de Pression en Magnétisme des Roches. Ph.D. Thesis, Université Paris VI, Paris, France.

Shive, P.N., Butler, R.F., 1969. Stresses and magnetostrictive effects of lamellae in the titanomagnetite and ilmenohematite series. J. Geomag. Geoelec. 21, 781–796.

Smith, B.E., Johnston, M.J.S., 1976. A tectonomagnetic effect observed before a magnitude 5.2 earthquake near Hollister, California. J. Geophys. Res. 81, 3556–3560.

Soffel, H.C., Deutsch, E.R., Appel, E., Eisenach, P., Peterson, N., 1982. The domain structure of synthetic stoichiometric TM10–TM75 and Al-, Mg-, Mn- and V-doped TM62 titanomagnetites. Phys. Earth Planet. Interiors 30, 336–346.

Stacey, F.D., 1962. Theory of the magnetic susceptibility of stressed rock. Philos. Mag. 7, 551–556.

Stacey, F.D., 1964. The seismomagnetic effect. Pure Appl. Geophys. 58, 5–22.

Struzhkin, V.V., Timofeev, Yu., Gregoryanz, E., Mao, H.K., Hemley, R.J., 2002. New methods for investigating superconductivity at very high pressures. In: Hemley, R.J., Bernasconi, M., Ulivi, L., Chiarotti, G. (Eds), High-Pressure Phenomena, Proceedings of the International School of Physics, "Enrico Fermi" Course CXLVII. IOS press, Amsterdam, pp. 275–296.

Syono, Y., 1965. Magnetocrystalline anisotropy and magnetostriction of Fe_3O_4–Fe_2TiO_4 series with special application to rock magnetism. Jpn. J. Geophys. 4, 71–143.

Timofeev, Y.A., Mao, H.K., Struzhkin, V.V., Hemley, R.J., 1999. Inductive method for investigation of ferromagnetic properties of materials under pressure. Rev. Sci. Instrum. 70, 4059–4061.

Tozer, S.W., 1993. Miniature diamond-anvil cell for electrical transport measurements in high magnetic fields. Rev. Sci. Instrum. 64, 2607–2609.

Valeev, K.A., Absalyamov, S.S., 2000. Remanent magnetization of magnetite under high pressures and shear loads. Izv. Phys. Solid Earth 36, 241–245.
Wanamaker, B.J., Moskowitz, B.M., 1994. Effect of nonstoichiometry on the magnetic and electrical properties of synthetic single crystal $Fe_{2.4}Ti_{0.6}O_4$. Geophys. Res. Lett. 21, 983–986.
Yamamoto, K., Endo, S., Yamagishi, A., Mikami, H., Hori, H., Date, M., 1991. A ceramic-type diamond anvil cell for optical measurements at high pressure in pulsed high magnetic fields. Rev. Sci. Instrum. 62, 2988–2995.
Zlotnicki, J., Cornet, F.H., 1986. A numerical model of earthquake-induced piezomagnetic anomalies. J. Geophys. Res. 91, 709–718.

V
Diffraction and spectroscopy

Advances in High-Pressure Technology for Geophysical Applications
Jiuhua Chen, Yanbin Wang, T.S. Duffy, Guoyin Shen, L.F. Dobrzhinetskaya, editors
Published by Elsevier B.V. (2005)

Chapter 16
High-pressure angle-dispersive powder diffraction using an energy-dispersive setup and white synchrotron radiation

Yanbin Wang, Takeyuki Uchida, Robert Von Dreele, Akifumi Nozawa, Ken-ichi Funakoshi, Norimasa Nishiyama and Hiroshi Kaneko

Abstract

A new powder diffraction technique has been applied to collect high-pressure angle-dispersive data using a solid-state detector (SSD) and white synchrotron radiation, with the multi-anvil apparatus SPEED-1500 at SPring-8. By scanning a well-calibrated SSD over a given 2θ range at a predetermined step size, a series of one-dimensional (1D) energy dispersive data (intensity, Int, versus energy, E) are obtained as a function of 2θ. The entire intensity dataset $Int(2\theta, E)$ consists of 4048 energy bins, covering photon energies (E) up to ~ 160 keV at 600 2θ steps, forming a large two-dimensional (2D) array. These intensity data are regrouped according to photon-energy bins, which are defined by individual channels in the multi-channel analyzer, yielding a large number of intensity-versus-2θ (angle-dispersive) datasets, $Int(E = const., 2\theta)$, each of which corresponds to a given photon energy or wavelength. Experimental data obtained on a mixture of MgO and Au to 20 GPa are used to demonstrate the feasibility of this technique. The entire dataset, selected subsets or composite scans can be used for Rietveld refinement. Data subsets are selected to simulate coarse step scans. Our analysis indicates that data within certain energy bins (up to $\sim 10\%$) may be binned together to improve counting statistics and to permit 2θ scans at 0.1–0.2° steps, without losing angular resolution. This will allow much faster data collection within about 10 min. Our test results indicate that at photon energy of ~ 80 keV, the angular resolution $\Delta\theta/\theta$ (in terms of refined FWHM) is about 0.006 within a 2θ range from 3 to 10°. This new technique is useful for high-pressure and general-purpose powder diffraction studies that have limited X-ray access to the sample using synchrotron radiation. Several advantages are discussed.

1. Introduction

X-ray diffraction is an important technique in determining structures of solids, liquids, and amorphous materials under high-pressures and high-temperatures. However, limited X-ray access in a pressure vessel often imposes technical difficulties so that high-resolution monochromatic diffraction techniques are difficult to apply. Previous high-pressure studies were mostly limited to the energy-dispersive diffraction (EDD) technique, carried out at a fixed 2θ angle. The data collection is fast (on the order of minutes or seconds); incident and diffracted beams can be carefully collimated so that clean diffraction data can be obtained without contamination from either the components of the pressure vessel or the solid pressure media surrounding the sample. However, although EDD information is useful in phase identification and the determination of unit-cell parameters, it does not provide reliable crystal chemical information such as atomic positions or bond lengths and angles. Various efforts have been made to extract quantitative crystallographic information, with limited success (e.g. Yamanaka and Ogata, 1991; Yamanaka et al., 1992; Ballirano and Caminiti, 2001); the difficulties primarily arise from the lack of appropriate descriptions of variations in the incident intensity and absorption within the experimental environment, both of which are strongly energy-dependent. EDD at multiple 2θ angles has been performed, mostly in studies of liquids and amorphous materials, in order to increase reciprocal space coverage for more accurate determination of radial distribution functions (e.g. Urakawa et al., 1996).

Many attempts have been made to perform monochromatic diffraction using various high-pressure devices. For example, step-scan angle-dispersive diffraction (ADD) studies have been carried out using the multi-anvil press (MAP), where a monochromator is used to select a narrow band of energies from the incoming radiation, and a point detector (sometimes with a crystal analyzer) is scanned at a given step size in 2θ to collect diffracted intensity (e.g. Zhao et al., 1994). The higher data quality is achieved at the cost of throwing away a vast majority (>99.9%) of the photons at other energies and at angles not covered by the detector scan path. Because of the relatively weak monochromatic beam intensity, each scan may take hours to complete. In addition, the limited X-ray access imposed by the pressure vessel often restricts the 2θ range. This especially is a problem for the MAP, in which 2θ is commonly limited to 15° or less. For a typical photon energy of 50 keV, the coverage in *d*-spacing is then limited to >0.95 Å, insufficient for a robust structure refinement. Higher-energy monochromators are costly and technically more challenging. Virtually no monochromators are currently available for photon energies above ~100 keV.

In order to speed up data collection and overcome problems due to sample graininess, 2D X-ray detectors are used in both the MAP and the diamond anvil cell (DAC). However, an immediate compromise is the loss of diffracted beam collimation. Diffraction by the pressure media, gaskets, pressure marker, etc., as well as Compton scattering in the diamonds, all contribute to the background, reducing signal to noise ratio, sometimes severely overlapping with sample diffraction signal, affecting data quality (e.g. see Angel et al., 2000). In some cases, background problems in the MAP may be circumvented using special techniques (e.g. Chen et al., 1997). The advantage of the 2D detectors is that one can observe and determine texturing in the powdered sample and improve statistics by integrating intensities over a wide azimuth angle. However, angular resolution may be degraded somewhat when intensities at various azimuth angles are integrated.

Soller slits can be used to define the scattering volume and thus overcome the background problem (e.g. Yaoita et al., 1997). By oscillating the Soller slits that lie between the sample and the 2D detector, the background scattering can be significantly reduced. However, this technique can only be applied to ADD with a single wavelength. It suffers from the same problem as the step-scan technique in the MAP in that coverage in *d*-spacing is limited. In addition, accurate alignment of the Soller slits is known to be difficult. It is also generally difficult to quantify the degree of misalignment, and small inaccuracies in the alignment can make significant differences to the measured intensities.

The advent of synchrotron radiation has provided with us a brilliant X-ray source with a wide range of photon energies. Thus, an ideal diffraction technique for high-pressure studies should take full advantage of the wide energy spectrum of a synchrotron source with simultaneous multiple ADD data collection. To this end, Wang et al. (2004) proposed a new technique for Combined Angle- and Energy-dispersive Structural Analysis and Refinement (CAESAR), which employs simultaneous ADD and EDD. By scanning a carefully calibrated multi-channel solid-state detector (SSD), a large number of EDD patterns are obtained. The entire dataset can be re-arranged as ADD patterns by plotting intensities of each channel (a given photon energy or wavelength). Subsets of the data covering narrow energy bands may be binned to give ADD patterns at a single wavelength with improved counting statistics; these may be combined in a multi-pattern Rietveld analysis to efficiently utilize the entire data set. While high photon-energy data provides coverage at the low *d*-spacing end, low photon-energy data covers the high *d*-spacing end. Test scans on NIST standard α-Al_2O_3 powder (SRM 674a) reported by Wang et al. (2004) under ambient conditions have demonstrated the usefulness of this technique in refining atomic positions that can be directly compared with conventional powder diffraction techniques.

In this chapter, we apply the CAESAR technique to high pressure, with a specially designed sample assembly to permit 2θ scan in a plane perpendicular to the cylindrical axis of the sample. We use diffraction data collected on a mixture of MgO and Au in the 1500 ton MAP (SPEED-1500) at SPring-8, Japan, to demonstrate the successfulness of this technique. The double-stage system of SPEED-1500 is capable of generating simultaneous P–T conditions of 35 GPa and 2800 K, when sintered diamond anvils are used for the second stage (Ito et al., 2004); the CAESAR technique will make routine structural refinement possible at these conditions.

2. Experimental setup

The high-pressure experiments were performed at the bending magnet beamline (BL04B1) at SPring-8, using the 1500 ton MAP, SPEED-1500 (Utsumi et al., 1998). This facility has a virtual 2θ arm consisting of two curved rails, lying in the horizontal plane, with rotation centers that coincide with the vertical axis of the press. A Ge SSD slides on the rails and is equipped with a collimator pointing towards the sample, which was located in the center of the press. The rails were optically encoded to allow an angular reproducibility in positioning of 0.01°. The multi-channel Ge SSD was calibrated against radioactive sources Co^{57}, Cd^{109}, and Am^{204}, with weighted average of Ag $K_{\alpha1}$ and $K_{\alpha2}$ (22.104 keV), Ag $K_{\beta1}$(24.942 keV) $K_{\beta2}$(25.456 keV), Am γ_1 (59.5370 keV), Cd^{109} γ_1 (88.040 keV), Co^{57} γ_2 (122.061 keV), and Co^{57} γ_3 (136.474 keV). A linear fit of channel numbers to these energies resulted in an accurate fit with typical uncertainties less than 10 eV. An amplifier with a shaping time of 4 μs was used, in order to optimize the detector for sufficient dynamic range while minimizing pile-up rejection.

SPEED-1500 is a double-stage multi-anvil apparatus (Utsumi et al., 1998). The first stage is a large DIA-type cubic anvil apparatus, employing WC anvils with a truncation

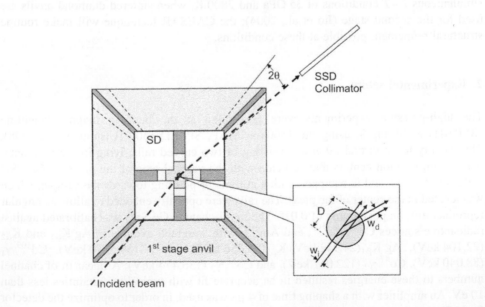

2θ
SSD
Collimator
SD
D
w_d
w_i
1st stage anvil
Incident beam

3. Results

3.1. Overview of the data

The entire EDD dataset is combined to form a 2D array of intensities, Int$(E, 2\theta)$, with each intensity value in the array corresponding to a given E (photon energy) and 2θ value. Figure 2 summarizes such 2D data. The dataset can be viewed at a fixed E or 2θ, corresponding to ADD (rows in Figure 2) or EDD (columns) spectra, respectively. By choosing the intensities at various 2θ values for certain fixed energies (wavelengths), a series of ADD patterns, Int$(E = \text{const.}, 2\theta)$, are obtained. Such a fixed-wavelength subset of the data can be fit using the Rietveld refinement software package GSAS (Larson and Von Dreele, 2001; Toby, 2001). The scans were in the horizontal plane; an initial polarization parameter of 0.5 is used in the refinement. However, the refinement results are not sensitive to the parameter, because the 2θ range was limited to 9°.

The MCA channels plotted in Figure 2 correspond to energies from 36 to 141.5 keV. The ring current varied from 100 to 80 mA between injections, which occurred at

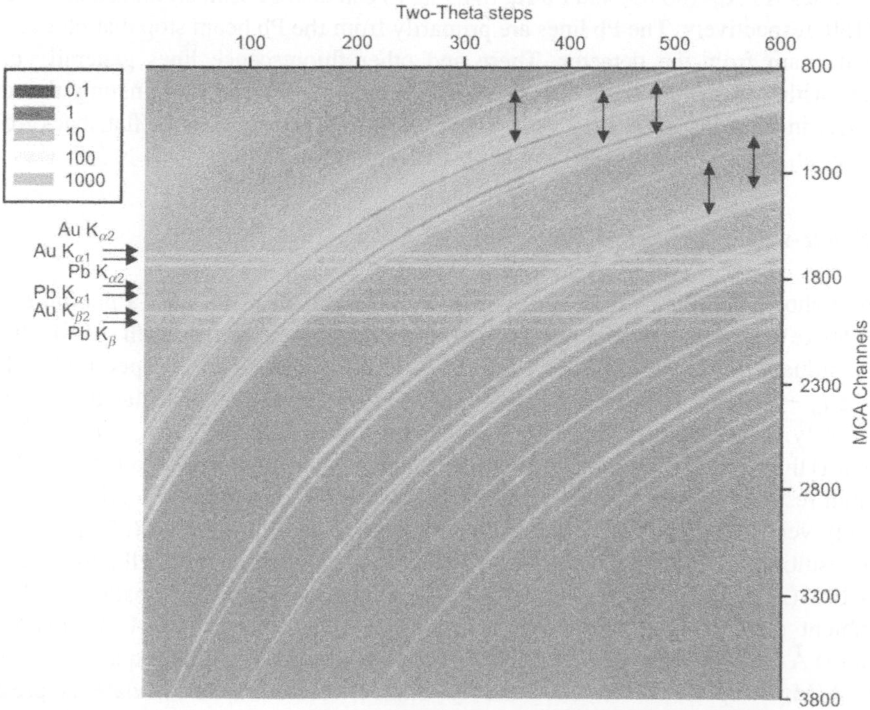

Figure 2. A 2D plot of intensity collected on MgO + Au against MCA channel numbers and 2θ steps (step 0 corresponds to $2\theta = 3°$, step size = 0.01°). Six MCA channel regions exhibit consistently higher intensities near 1694, 1741, 1843, 1898, 2035, and 2150, corresponding to the fluorescence peaks Au $K_{\alpha2}$ (66.9895 keV), $K_{\alpha1}$ (68.8037), Pb $K_{\alpha2}$ (72.794), Pb $K_{\alpha1}$ (74.956), Au $K_{\beta1}$ (77.984), Au $K_{\beta2}$ (80.03), and Pb K_{β} (84.9 keV), respectively.

~12 h intervals. We use the ring current to rescale observed intensities for each EDD file, and the analysis presented here is after this initial intensity correction.

In addition to sample diffraction lines, some extra lines are present in Figure 2, which are more pronounced at low energies and low 2θ angles. According to Bragg's law,

$$\lambda = 2d \sin \theta, \quad (1)$$

where λ is the wavelength of the radiation, any diffraction line that satisfies Eq. (1) should follow the constant $1/(2E\sin\theta)$ (or $\lambda/(2\sin\theta)$) relation. Those additional lines do not follow this requirement; rather, they are shifted from the intense sample diffraction peaks by a constant increment in MCA channels (or energies) and are identified as Ge escape peaks, which occur when X-rays excite a Ge atom in the detector and the resulting Ge K_α fluorescent X-rays ($K_{\alpha1} = 9.886$ keV, $K_{\alpha2} = 9.855$ keV) manage to escape the detector crystal. Some of the relatively intense escape lines are tied to their parent diffraction lines in Figure 2. These escape lines sometimes can cause confusion in EDD data analysis, but they are easily identifiable in 2D plots.

Also note the consistently higher intensities around channels 1694, 1741, 1843, and 1898, which correspond to the fluorescence lines Au $K_{\alpha2}$ (66.9895 keV), $K_{\alpha1}$ (68.8037), Pb $K_{\alpha2}$ (72.794), and Pb $K_{\alpha1}$ (74.956), and Au $K_{\beta1}$ (77.984), respectively. Additional weaker lines Au $K_{\beta2}$ (80.03) and Pb K_β (84.9 keV) can also be seen around channels 2035 and 2150, respectively. The Pb lines are primarily from the Pb beam stop that blocked the incident beam from the detector. These and other fluorescence lines generally cause overlap with diffraction peaks in an EDD spectra, thereby contaminating the data. However, in an ADD plot, they only contribute to a higher, but still flat, background (Wang et al., 2004) in very specific energy (wavelength) channels.

3.2. *Single-wavelength refinement*

Figure 3 shows the results from an analysis by fitting a spectrum with a photon energy of 90.60 keV (MCA channel 2298), corresponding to a wavelength of 0.1368 Å. The residuals Rp and wRp for the fit are 0.22 and 0.26, respectively. Here $\text{Rp} = \Sigma|I_o - I_c|/\Sigma I_o$, in which I_o and I_c are the observed and calculated intensities, respectively. The weighted residual wRp is defined as $[w\text{Rp}]^2 = [\Sigma w(I_o - I_c)^2]/[\Sigma w I_o^2]$, where w is the weighing function for the observed intensity (Larson and Von Dreele, 2001). The high residual is primarily due to the fact that the background of the ADD spectra is generally very low; hence the residuals for the background fit are high (often as high as 50%), resulting in elevated overall fitting residuals. The resulting unit-cell parameters are $a = 4.1493(2)$ Å for MgO and $a = 4.0263(1)$ Å for Au. Using the EDD patterns collected at ambient conditions, the zero-pressure unit-cell edge lengths are 4.2130(4) Å and 4.0780(3) Å for MgO and Au, respectively. Using recently reported equations of state reported (MgO: Speziale et al., 2001 and Au: Shim et al., 2002), we calculate the pressure of the sample to be 8.0 and 7.0 GPa for MgO and Au, respectively. The difference in pressure may be due to discrepancies in the thermoelastic parameters in the two equations of state as well as effects of non-hydrostatic stress (Wang et al., 1998).

In this and all of the following fits, the observed intensities were corrected for the variation in diffraction volume with 2θ as defined by the incident and diffraction beams.

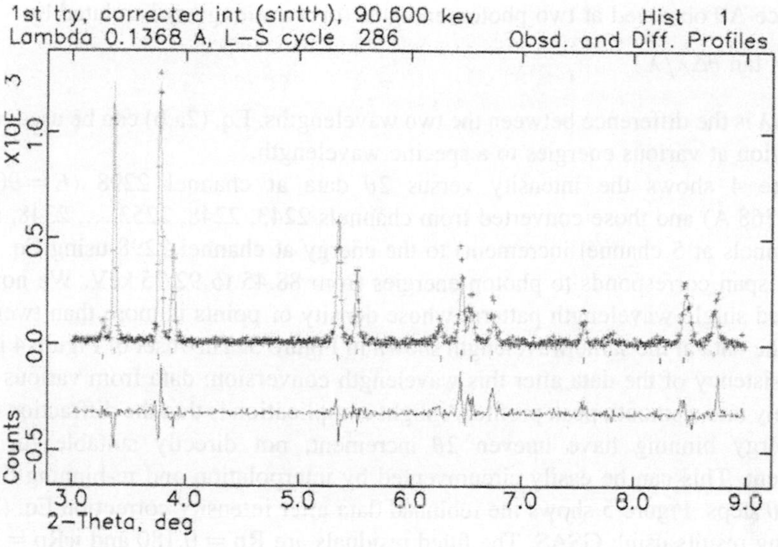

1st try, corrected int (sintth), 90.600 kev
Lambda 0.1368 A, L-S cycle 286
Hist 1
Obsd. and Diff. Profiles
X10E 3
Counts
1.0
0.5
0.0
0.5
3.0
4.0
5.0
6.0
7.0
8.0
9.0
2-Theta, deg

difference $\Delta\theta$ observed at two photon energies (or wavelengths) is related by

$$\Delta\theta \approx \tan\theta\Delta\lambda/\lambda \tag{3}$$

where $\Delta\lambda$ is the difference between the two wavelengths. Eq. (2a,b) can be used to convert observation at various energies to a specific wavelength.

Figure 4 shows the intensity versus 2θ data at channel 2298 ($E = 90.60$ keV, $\lambda = 0.1368$ Å) and those converted from channels 2243, 2248, 2253,..., 2348, and 2353 (22 channels at 5 channel increment) to the energy at channel 2298 using Eq. (3). This channel span corresponds to photon energies from 88.45 to 92.75 keV. We now have a combined single-wavelength pattern, whose density of points is more than twenty times that of the data at the same wavelength shown in Figure 3. The inset of Figure 4 illustrates the consistency of the data after this wavelength conversion: data from various channels fall nicely into a smooth peak profile. A slight complication is that the diffraction data after this energy binning have uneven 2θ increment, not directly suitable for Rietveld refinement. This can be easily circumvented by interpolation and re-binning the data at equal 2θ steps. Figure 5 shows the rebinned data after intensity correction Eq. (2a,b) and the fitting results using GSAS. The fitted residuals are Rp = 0.180 and *w*Rp = 0.226.

3.4. Simultaneous multiple wavelength fit

Because an SSD can record data from a wide range of photon energies simultaneously, it is advantageous to utilize multiple wavelengths for structural refinement for better coverage. For the limited 2θ range of the scans performed in this study (3°–9°), a 40 keV ($\lambda = 0.310$ Å) radiation will cover a *d*-spacing range from 5.92 to 1.98 Å; whereas a 120 keV (0.103 Å) radiation will cover 1.97 to 0.656 Å. Thus, a combined data set will

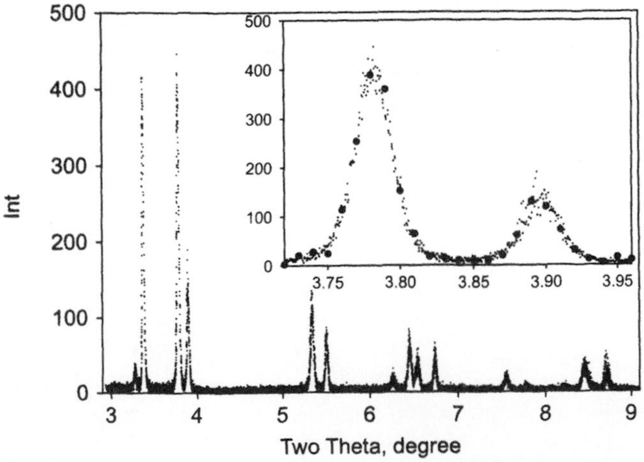

Figure 4. A binned spectra of 23 channels, centered at channel 2298 (90.60 keV), at 5 channel steps. Inset shows two peaks between 2θ values 3.72 and 3.96°. Large filled circles are the data at channel 2298, whereas small dots are data from the neighboring 22 channels after binning using Eq. (3). Consistency among the data points is evident.

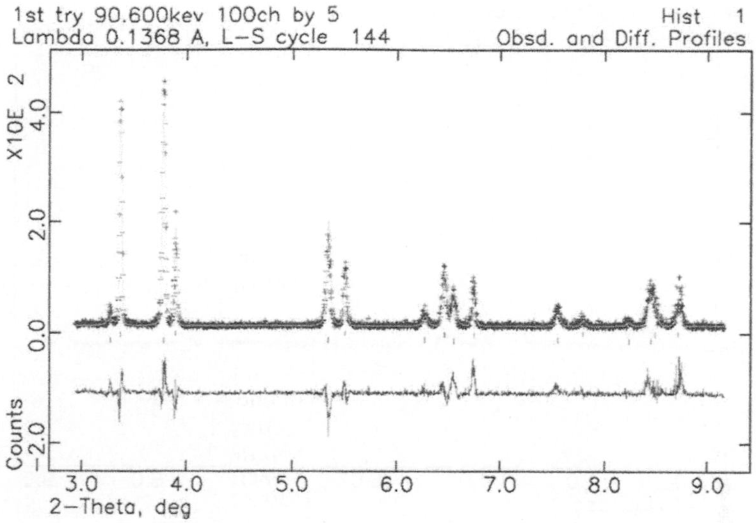

Figure 5. Rietveld refinement results on the binned data shown in Figure 4. The wRp value is 0.226.

cover a much wider d-spacing from 0.656 to 5.92 Å. This increase in d range is particularly beneficial for studies where X-ray access is limited.

We performed a simultaneous Rietveld fit with intensity data centered at three energies, 65.02 (channel 1844), 90.60, and 101.00 keV (channel 2735). The 65.02 keV data are from a single channel, whereas, data centered at 90.60 and 101.00 keV are binned using 23 channels, spanning ± 50 channels, as discussed in Section 3.2. The results are summarized in Table 1 and a fitting example is shown in Figure 6. There are no free atomic positions in the two phases (MgO and Au) to be fit, and the cell parameters obtained are virtually identical to the single-wavelength fit described in Section 3.2. The Au element fractions (0.011–0.015) obtained within 2σ (Table 1). More importantly, the residuals (Rp and wRp) from individual fits indicate that combining intensity data over a $\pm 5\%$ energy band does not compromise data quality.

In fact, the resolved peak width for MgO appears to be slightly narrower with the binned dataset. For the single-wavelength fit (Fig. 4), the full-width-at-half-maximum (FWHM) varies from 0.031° to 0.043° over the 2θ range from 3° to 7°, whereas the

Table 1. Rietveld refinement results of a combined fit to data at three wavelengths (65.02, 90.60, and 110.00 KeV). Elemental fraction of Au (Au frac) is defined by $X_{Au}/(X_{MgO} + X_{Au})$. Variation in the scaling factor represents relative source spectrum intensity variation as a function of energy, as well as SSD sensitivity to photons at higher energies.

Wavelength (Å)	Energy (keV)	Au (frac)	Scaling factor	wRp	Rp
0.190671	65.02	0.0112(14)	0.0297	0.218	0.156
0.136843	90.60	0.0147(12)	0.0343	0.231	0.183
0.112711	110.00	0.0140(8)	0.0008	0.204	0.156
Overall				0.221	0.171

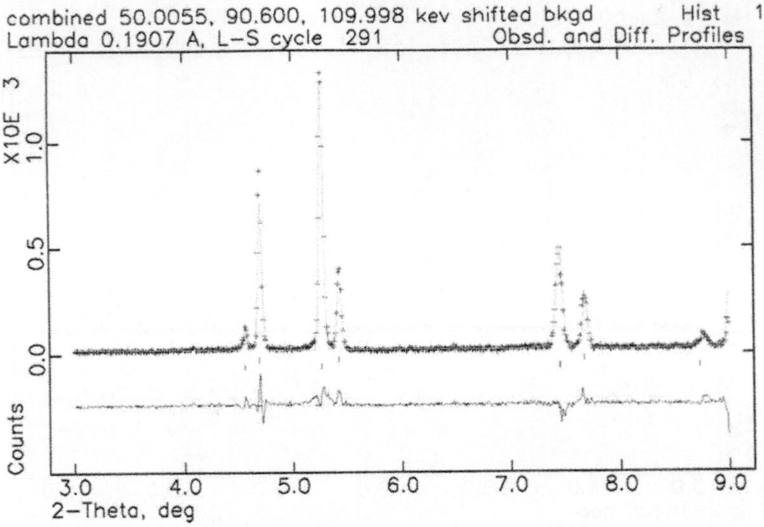
combined 50.0055, 90.600, 109.998 kev shifted bkgd
Hist 1
Lambda 0.1907 A, L-S cycle 291
Obsd. and Diff. Profiles
X10E 3
1.0
0.5
0.0
Counts
3.0
4.0
5.0
6.0
7.0
8.0
9.0
2-Theta, deg

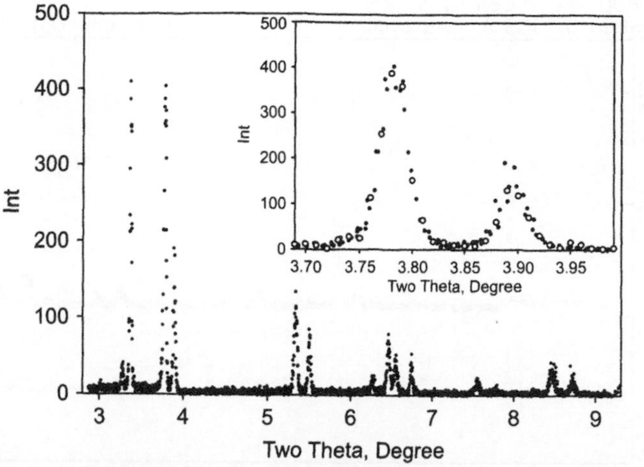

Figure 7. A simulated 0.1° step scan (solid dots) binned with 41 channels 2198, 2203, 2208, 2103,..., 2398, centered at channel 2298 (90.60 keV), compared to a single channel 2298 (90.60 keV) data (large open circles). A total of 41 channels are used, corresponding to MCA channels from 2198 to 2398, at a 5 channel increment.

the 41 channel binned spectrum (step size 0.1°). Compared to the single monochromatic 2θ scan at 0.01° steps, the binned data density is increased by a factor of ~ 10, despite the fact that the step size is 10 times coarser. The actual coverage in 2θ is also increased slightly. Note that we only binned 1/5 of the available energies in the $\Delta E/E$ range; therefore, the density of coverage can be further increased by another factor of 5. No significant increase in peak width is observed, suggesting that we have not compromised the resolution in 2θ. Thus, we conclude that binning over a $\pm 5\%$ energy range does not significantly affect data quality.

The binned data shown in Figure 7 are interpolated to equal 2θ steps before performing the Rietveld refinement. Figure 8 shows the results, with residuals $\mathrm{Rp} = 0.175$ and $w\mathrm{Rp} = 0.225$, essentially identical to those in a single wavelength, 0.01° step scan shown in Section 3.2. Based on this analysis, we conclude that a reasonable structural refinement can be obtained using 0.1° or even 0.2° step sizes. For a 50 s EDD data collection, approximately 50 min is required for a complete 2θ scan from 3° to 9° with a step size of 0.1°. Note that the 50 s collection time was based on <5% dead time and over 1000 counts/channel for the average peak intensity. With properly configured electronics and strong incident beam, it is, therefore, possible to reduce this time by a factor of 5–10, making it feasible for collecting monochromatic data within 10 min.

4. Conclusions

We have demonstrated the feasibility of the CAESAR technique in a MAP, where X-ray access is limited in 2θ to within 10°. Beam collimation in an EDD setup makes it feasible to collect clean intensity information from a sample with diameters of about 1 mm, without being contaminated by scatter from the pressure media, heating elements, and

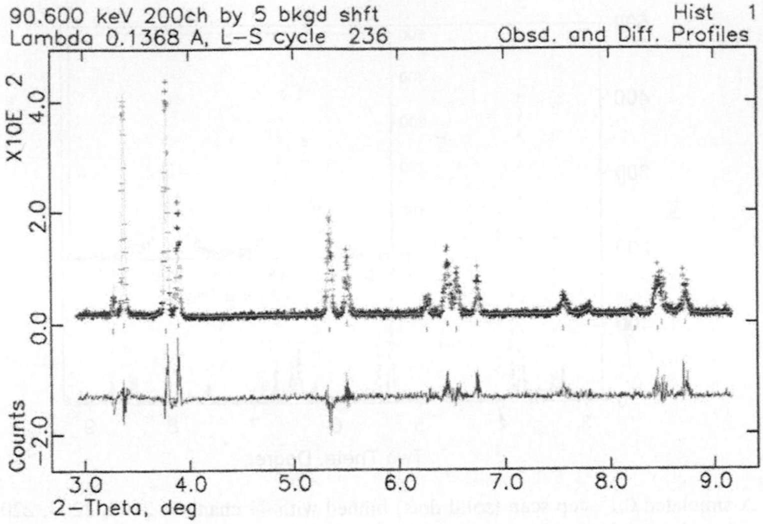
90.600 keV 200ch by 5 bkgd shft
Hist 1
Lambda 0.1368 A, L-S cycle 236
Obsd. and Diff. Profiles
Counts
X10E 2
4.0
2.0
0.0
-2.0
3.0
4.0
5.0
6.0
7.0
8.0
9.0
2-Theta, deg

identification. At a critical point under certain pressure and temperature conditions, one can then conduct a step-scan to collect ADD data. This allows one to take advantages of both energy dispersive and angle-dispersive methods. Lastly, with multiple wavelength data recorded in an SSD, a large range of energies (from 40 to 140 keV) is available, corresponding to wavelengths from 0.31 to 0.089 Å. For the extremely limited 2θ range tested (3°–9°), the Q ($= 2\pi\sin\theta/\lambda$) range is from 0.46 to 5.54 $Å^{-1}$.

It is difficult to evaluate the energy resolution of the CAESAR technique theoretically. Based on Rietveld refinement on NIST standard SRM 674a at ambient conditions and analysis on energy resolution of a SSD, Wang et al. (2004) concluded that an energy resolution of $\Delta E/E \approx 0.006$ can be achieved at about 78 keV. At this energy, the angular resolution $\Delta\theta/\theta$ (in terms of refined FWHM) is about 0.006 for the 2θ range between 3° and 10°.

The SPEED-1500 has demonstrated to be capable of generating up to 35 GPa and 2800 K simultaneously (Ito et al., 2004). With careful alignment and sample preparation, the CAESAR technique makes it possible to perform routine structural refinements under these conditions. The main disadvantage of this technique is that, like conventional 1D step-scan techniques, the scan is limited to a specific direction and does not integrate most or all of the available diffracted signals, as a 2D detector does. In cases where few crystallites are present, or preferred orientation develops in the sample, counting statistics will be significantly affected. Effects of poor counting statistics can be minimized by oscillating the sample. In fact, sample oscillation is frequently practiced in the DAC with EDD. In the MAP, the new system SPEED-Mk.II (Katsura et al., 2004) also has the oscillation capability, although a larger beam path opening will be needed for the first stage in order to apply the CAESAR technique.

Acknowledgements

We thank Mark Rivers (GSECARS) and Wataru Utsumi (JAERI) for their support during the development of this technique and two anonymous reviewers and J. Chen for their constructive comments. GSECARS is supported by the National Science Foundation – Earth Sciences, Department of Energy – Geosciences, W.M. Keck Foundation, and the US Department of Agriculture. Use of the Advanced Photon Source was supported by the US Department of Energy, Office of Science, Basic Energy Sciences, under Contract No. W-31-109-Eng-38. Work supported by the NSF grant EAR001188 (YW).

References

Angel, R.J., Downs, R.T., Finger, L.W., 2000. High-temperature-high-pressure diffractometry. In: Hazen, R.M., Downs, R.T. (Eds), High-Temperature and High-Pressure Crystal Chemistry. Mineralogical Society of America, Washington, DC, pp. 559–596.

Ballirano, P., Caminti, R., 2001. Rietveld refinements on laboratory energy dispersive X-ray diffraction (EDXD) data. J. Appl. Cryst. 34, 757–762.

Chen, J., Kikegawa, T., Shimomura, O., Iwasaki, H., 1997. Application of an imaging plate to large volume press MAX80 at photon factory. J. Synchrotron Radiat. 4, 21–27.

Ito, E., Kubo, A., Katsura, T., Walter, M.J., 2004. Melting experiments of mantle materials under lower mantle conditions with implications for magma ocean differentiation. Phys. Earth Planet. Interiors 143–144, 397–406.

Katsura, T., Funakoshi, K., Kubo, A., Nishiyama, N., Tange, Y., Sueda, Y., Kubo, T., Utsumi, W., 2004. A large-volume high-pressure and high-temperature apparatus for in situ X-ray observation, SPEED-Mk.II. Phys. Earth Planet. Interiors 143–144, 497–506.

Larson, A.C., Von Dreele, R.B., 2001. General Structure Analysis System (GSAS). Los Alamos National Laboratory Report LAUR 86-748.

Shim, S.-H., Duffy, T.S., Takemura, T., 2002. Equation of state of gold and its application to the phase boundaries near 660 km depth in Earth's mantle. Earth Planet. Sci. Lett. 203, 729–739.

Speziale, S., Zha, C.-S., Duffy, T.S., Hemley, R.J., Mao, H.-K., 2001. Quasi-hydrostatic compression of magnesium oxide to 52 GPa: implications for the pressure–volume–temperature equation of state. J. Geophys. Res. 106, 515–528.

Toby, B.H., 2001. EXPGUI, a graphical user interface for GSAS. J. Appl. Crystallogr. 34, 210–213.

Urakawa, S., Igawa, N., Umesaki, N., Igarashi, K., Shimomura, O., Ohno, H., 1996. Pressure-induced structure change of molten KCl. High Pressure Res. 14, 375–382.

Utsumi, W., Funakoshi, K., Urakawa, S., Yamakata, M., Tsuji, K., Konishi, H., Shimomura, O., 1998. SPring-8 beamline for high pressure science with multi-anvil apparatus. Rev. High Pressure Sci. Technol. 7, 1484–1486.

Wang, Y., Weidner, D.J., Meng, Y., 1998. Advances in equation of state measurements in SAM-85. In: Manghnani, M.H., Yagi, T. (Eds), Properties of Earth and Planetary Materials at High Pressure and Temperature. American Geophysical Union, Washington, DC, pp. 365–372.

Wang, Y., Uchida, T., Von Dreele, R.B., Rivers, M.L., Nishiyama, N., Funakoshi, K., Nozawa, A., Kaneko, H., 2004. A new technique for angle-dispersive powder diffraction using an energy-dispersive setup and synchrotron radiation. J. Appl. Cryst. 37, 947–956.

Yamanaka, T., Ogata, K., 1991. Structure refinement of GeO_2 polymorphs at high pressures and temperatures by energy-dispersive spectra of powder diffraction. J. Appl. Cryst. 24, 111–118.

Yamanaka, T., Sugiyama, K., Ogata, K., 1992. Kinetic study of the GeO2 transition under high pressures using synchrotron X-radiation. J. Appl. Cryst. 25, 11–15.

Yaoita, K., Katayama, Y., Tsuiji, K., Kikegawa, T., Shimomura, O., 1997. Angle-dispersive diffraction measurement system for high-pressure experiments using a multichannel collimator. Rev. Sci. Instrum. 68, 2106–2110.

Zhao, Y., Parise, J.B., Wang, Y., Kusaba, K., Vaughan, M.T., Weidner, D.J., Kikegawa, T., Chen, J., Shimomura, O., 1994. High-pressure crystal chemistry of $NaMgF_3$: An angle dispersive diffraction study using monochromatic synchrotron X-radiation. Am. Mineral. 79, 615–621.

Advances in High-Pressure Technology for Geophysical Applications
Jiuhua Chen, Yanbin Wang, T.S. Duffy, Guoyin Shen, L.F. Dobrzhinetskaya, editors

Chapter 17

Methods and application of the Paris–Edinburgh Press to X-ray diffraction structure solution with large-volume samples at high pressures and temperatures

Wilson A. Crichton and Mohamed Mezouar

Abstract

We review the experimental methodology and examples of the potential of the Paris–Edinburgh large-volume apparatus in structure solution via angle-dispersive X-ray powder diffraction methods. While the examples considered are simple in their chemistry and structure, they have been – in most cases first solved from data collected in situ at the ESRF and hopefully demonstrate the utility of this technique and its potential for other, more complicated, systems in the fields of materials science and geology.

1. Introduction

The Paris–Edinburgh (P–E) large-volume apparatus has been initially developed for neutron scattering experiments and is capable of pressures in excess of 25 GPa (Besson et al., 1992; Klotz et al., 1996). For use at an X-ray synchrotron there have been several significant adaptations necessary; these can be divided into those required by the materials employed in construction of the cell assembly and those related to the scattering of X-rays and geometry of detection systems. The system presented here is capable of generating pressures up to 15 GPa with temperatures in excess of 2500 K, at $P < 10$ GPa and to ~1000 K for $P > 10$ GPa, using standard graphite furnaces.

At neutron sources, e.g. in time-of-flight operation at spallation source at ISIS (Marshall and Francis, 2002) the P–E cell is mounted with the incident beam direction travelling along the main axis of the press, with the press lying horizontally and scattering vertically and radially, through the normally toroidal gasket. In order to have a system that will allow scattering through the gasket, users of this device make use of a particular property of neutron scattering lengths that can be both positive and negative in sense. By preparing a suitable alloy, e.g. of TiZr, the effect is that a null scattering gasket can be prepared that has little effect on attenuating the signal that comes from the sample. Recent studies have demonstrated the improved quality of neutron scattering data available using this geometry with included liquids (deuterated methanol:ethanol) as pressure transmitting medium (Marshall and Francis, 2002).

Other geometries are of course available at neutron sources and vary by the type of source of the neutrons. For example, very recently, successful test experiments have been carried out at the ILL reactor source at Grenoble, where the incident beam and scattering vectors were in the same equatorial plane as the gasket (Parise et al., 2004). In that configuration the press was mounted vertically and used the same null scattering gasket material to obtain angle-dispersive data on the magnetic ordering of CoO under pressure and of the behaviour of D_2O ices. Interestingly, this experiment also used, somewhat counter-intuitively, c-BN (rather than sintered diamond or WC) as anvil material. Cubic BN is a strong neutron absorber and had the effect of auto-collimating the incoming neutron beam to the size of the anvil gap (Parise et al., 2004) and significantly improving the signal/background ratio. Thus, the materials that are chosen for anvil and as gasket assembly materials have specific purposes and properties. The same is the case for materials employed in the X-ray setting.

In standard operation the X-ray gasket will be made of a mixture of boron:epoxy resin, hardened and shaped to fit the conical profile of the anvils. These are the first major differences between the neutron and X-ray set-ups, the use of boron and the shape of the anvil. Boron is a low-Z element, as such does not absorb, nor consequently scatter, X-radiation very significantly. The choice of conical anvil profile is made for several reasons; most importantly, so it allows increased access to the solid 2D angle from the sample between the anvil gap and profile. This increase in solid angle warrants the use of 2D detector systems that allow the collection of a greater extent of the Debye–Scherrer rings that, once integrated, become the 1D 2θ-intensity data that we are most familiar with. In doing so, the effects of preferred orientation due to strain or sample preparation and "spottiness" due to recrystallisation could be assessed and treated. This, in addition to the improvement in counting statistics from radial averaging of the 2D data in reduction to the conventional 1D pattern, results in a much more robust dataset, than compared to that obtained, for example, in 1D energy-dispersive mode.

If we consider further the requirements for the X-ray version, all pieces of the assembly must be made of low-scattering materials with desirable physical properties (see, for example, Zhao et al., 1999 and http://multianvil.asu.edu for an extensive list of materials in common laboratory use). In general, those properties required by limitations due to the use of X-rays are low-Z and high symmetry (though this is not strict and deviations from this rule-of-thumb are sometimes very useful, e.g. use of a Re furnace for very high pressure, >10 GPa. Re is high-Z element and it absorbs a high proportion of the incident beam, but it also absorbs scattering from the incident-beam side of the gasket – removing this from the resulting diffraction pattern). Satisfying these simple requirements, graphite is an excellent choice as a furnace material, with its lowest angle peak at $d_{(002)} \approx 3.4$ Å. The hexagonal form of BN, with a very similar structure is also a good choice as sample capsule. Both materials have also favourable mechanical properties in that they are easily machined to small diameter and are soft with bulk moduli between 30 and 40 GPa. Because of this, they transmit pressure effectively and are sensitive to pressure increase. They can, therefore, be used as internal pressure markers through their p–V–T equation of state (Le Godec et al., 2000).

The system in operation at ID30 at the ESRF is not the only to take advantage of the angular-dispersive geometry afforded through using an image plate; similar devices are available, for examples, at the Photon Factory (Kikegawa et al., 1995; Chen et al., 1997; Yaoita et al. 1997) where the system is also coupled to multi-channel collimating slits that have allowed the investigation of liquids and powder samples at high pressure possible, the National Synchrotron Light Source (Chen et al., 1998a,b) and SPring-8 (Ohtaka et al., 2002).

2. ADX geometry at ID30

2.1. Generalities

The angle-dispersive X-ray geometry used at ID30 at the ESRF, Grenoble, (Fig. 1) has been previously described elsewhere (Mezouar et al., 1999) and we will limit ourselves to discussion of more recent features. In the X-ray beam we will have the B:epoxy gasket, graphite furnace, h-BN capsule and sample and all will naturally contribute to the resulting diffraction pattern. If we are to collect "clean" data from the sample, the diffraction peaks from the surrounding materials must be excluded by geometrical methods. This is particularly apparent when little or no residual peaks can be tolerated; for example, in auto-indexing unknown phases, or the collection of data from amorphous, liquid or disordered materials.

2.2. Oscillating slit system

As an analogy, in laboratory tube-sourced powder diffractometers a system of incident- and diffracted-beam parallel-plate collimators can be used to decrease the axial divergence that results from use of a LFF tube in parafocussing, or Bragg–Brentano, geometry. If not eliminated, this axial divergence would lead to peak broadening due to the differences in the source–sample and sample–detector distances over the surface of the flat-plate sample. Thus, unwanted scattering is removed by the use of these collimators in conjunction with diffracted beam slits by illuminating a selected area of the surface of the

Figure 1. The angular dispersive setup at ID30, ESRF, showing (*from left*) the position of the Mar345 IP detector, the mounted oscillating (Soller) slits, *central*, and the P–E cell on the *right*. The beam direction if from right to left.

possible, the National Synchrotron Light Source (Chen et al., 1998b) and Spring-8 (Utsuka et al., 2002).

2. ADX geometry in H<S>

2.1 [illegible]

[illegible] 30 [illegible] been previously described [illegible] all [illegible] the Bragg [illegible] of the dense [illegible] capsule and sample and [illegible] material, which [illegible] the resulting diffraction pattern. If we are to collect 'clean' data from the sample, the diffraction peaks from the surrounding materials must be excluded by geometrical methods. This is particularly important when little or no a priori knowledge can be obtained for the sample, in situ including unknown phases, or the collection of data from amorphous, liquid or disordered materials.

2.2 Defining [illegible]

As an analogy to laboratory tube-sourced powder diffractometers, a system of incident and diffracted beam parallel-plate collimation can be used [illegible] the axial divergence [illegible] in [illegible] the [illegible] but the differences in the source, sample and sample-detector distances [illegible] of the data [illegible] unwanted scattering is removed [illegible] of these [illegible] in conjunction with diffracted beam [illegible]

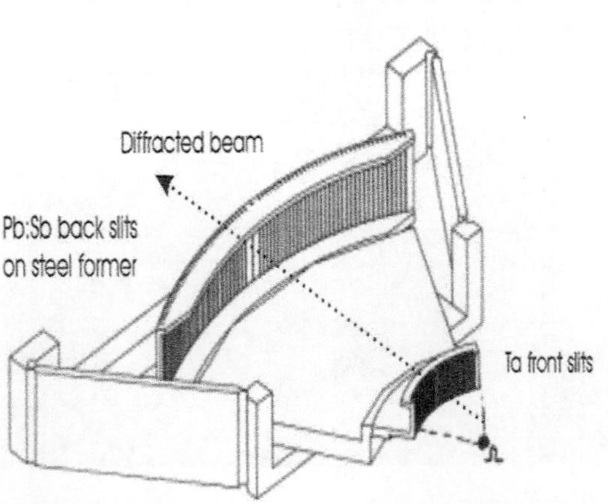

Figure 1. The angular dispersive setup at ID30, ESRF, showing (from left) the position of the MAR345 detector, the mounted oscillating (Soller) slits (Pb:Sb), and the Ta slits of the press. The beam direction is from right to left.

adjustments are made prior to and during data collection:

1. The central slit is aligned, initially manually using a laser that delineates a rough beam path. This is followed-up by alignment with the beam.
2. Comparing the foreshortening of the data range in 2θ will allow the user to correctly place the sample at the centre of rotation of the slits; i.e with the sample too far from the rotation of the slits the detector will not see the sample, so no diffraction from the sample will be evident. This is because increasing 2θ reduces the length of the diffracting volume along the beam direction. As the correct position is reached, intensities at expected peak positions will increase, go through a maximum and then decrease as the sample placement improves, is correct, and then degrades. The sample will be placed at the position that corresponds to a maximum in diffracted sample intensity (which is not necessarily the geometric centre of the sample).
3. Using a glass sample the oscillation range of the slits is verified. Simply, if the range of the slit movement is too large, then there will be overlap between the range in 2θ over which one slit and its neighbour covers – giving a series of thin rectangles (one for each slit) on the detector that are over-exposed with respect to the rest. In the integrated 1D pattern this is expressed as a series of equidistant peaks. If the oscillation range is too small, the opposite case is true, that we shall observe lighter areas, not exposed during the progress of the slits and troughs, below background level on the 1D dataset.
4. Quite naturally, when the slits oscillate they reach the end of the travel, before which the motor driving the rotation will decelerate, stop, change direction and accelerate to full speed to cover the range in the opposite direction. Code is written that controls the opening and closing of a fast-shutter, via measurement of the frequency of the motors so that when the deceleration state is active, the shutter is closed and the motor drives past the target point and stops. It is only when the motor is at top speed and passing the target position again that the shutter is reopened, thus ensuring that the slits are not exposing the plate when the slits are momentarily stationary. This would lead to equidistant darkened regions on the plate, corresponding to the extremes of the oscillation movement.
5. As the sample type changes, we are required to expose for longer or shorter periods. Again, the code driving the slits takes care of this by (a) changing the motor speed automatically and (b) if this is not sufficient, by changing the number of oscillations made.

It is through use of these techniques that we can achieve near clean data for most sample types in the ranges above a few degrees 2θ. However, the system is not without its limits. The slits are not effective along scattering directions sub-parallel to the beam direction, as the diffracting lozenge tends towards infinity in length along this direction as 2θ tends to 0°. Therefore, attempting to limit the illuminated region at low 2θ is ineffective. The angle at which the slits become efficient is largely limited by the slit gap and distance from the sample. In the case of the oscillating slits this is limited by simply what is machinable and in practice this is at angles greater than about 8°, thus time must be taken to choose the most appropriate incident wavelength so as to have peaks from the sample environment shifted, through Bragg's equation, to higher 2θ, preferably above 8°. In these terms, it is clear that at ID30's lowest practicable energy (20 keV or $\lambda \approx 0.62$ Å) would result in the best signal:background ratio, but, for a detector capable of $\pm 30°$ 2θ

this would correspond to a minimum available d-spacing of 1.2 Å, probably not sufficient to prudently refine bond-lengths, nor reliably normalise data from a glass or liquid sample. If we consider the background materials and how much they could contribute to the diffraction pattern, at ambient conditions graphite has its lowest 2θ peak (at 20 keV) at 10.5° 2θ and h-BN at 10.7° 2θ, thus these would not be visible for normal sample geometries. The situation with B:expoxy is different; we cannot eliminate the scattering from this material from the diffraction pattern at this (or any other energy) as there are many peaks below 8°. What is more, as the incident energy decreases, so lower Z materials absorb more and diffract more, thus there will be a stronger signal from B:epoxy at 20 keV than there will be at 80 keV. It is evident that there are trade-offs to be made, principally:

- involving the diameter of the sample relative to its X-ray absorption cross-section (a moderate Z sample cannot be 2.5 mm thick at 20 keV). Both a smaller sample and higher energy data collection will lead to an increased background contribution.
- the desired data analysis procedure. For example, to obtain anisotropic temperature factors we need data to around $d = 0.6$ Å. However, to obtain data to 0.6 Å, say, we are required to work at energies higher than 35 keV. Again increasing the background, at the expense of real-space resolution.
- the detector position and the desired resolution of overlapped peaks (based on the symmetry and cell parameters) relative to the real-space resolution required.

In extreme cases, where reliable intensity and information at high momentum transfer (e.g. in pair-distribution function analysis), $Q = 2\pi/d$, is strictly required, the full range of energies (from 20 to 100 keV) and detector positions (horizontal and sample–detector distance translations) may be used during the course of data collection in any one study, or, even pressure–temperature point!

2.3. The P–E cell for refinement and structure solution

When samples are compressed we observe the various effects of strain and how it is distributed in the sample. These, in the simplest case lead to peak broadening and if the sample is prone or the pressure poorly distributed, will lead to preferred orientation. These effects, in addition to those inherent to the sample, which may be due to production of staking faults or other defects, degrade the quality of the diffraction peaks and our ability to extract useful information from them. In general, until the sample is loaded in a hydrostatic medium, we must be prepared to heat the sample under high-pressure conditions to obtain robust data. In more specific cases, for example, where peak broadening occurs close to a phase transition and is due to kinetic effects, we have no option but to heat to allow the phase to express itself in the high-pressure (or temperature) form. Indeed, several previously reported cases of pressure-induced amorphisation have masked just such a transition, e.g. in covellite, CuS (Crichton et al., 2003a).

Before taking this further, we must first answer the question, what are the advantages and disadvantages to using the P–E cell opposed to using, for example, a diamond anvil cell (DAC) in ADX mode?

It is clear that when we can load a DAC with nothing other than sample, pressure-transmitting medium and a ruby for pressure measurement, that the data will be "cleaner".

It is also clear with the correct choice of DAC a typical user setup will allow for data collections to 40–50 GPa. Yet for high-temperature data collection at pressures in the range of the P–E cell, it is the superior tool. With the P–E cell we have access to a wide range of temperatures; where the typical DAC cannot be heated evenly for the temperature range between 350 and 1000°C, which are respectively, the upper limit for external collar-type heating and an approximate lower limit for laser heating. The P–E cell will heat over a continuous range from room temperature to in excess of 2500°C over its full pressure range, with greater precision and more even temperature gradients. What is more, the temperature can be measured directly over most of this range by thermocouples. If we use the rule-of-thumb that the cell will anneal out strain in the cell at temperatures above 400 °C and recrystallisation will not be seen much below 1/2–2/3 melting point, this is exactly the range in which the vast majority of DACs cannot be heat. Another important point is scattering from the DAC itself. Reflections from diamonds, which can be considerable, are easily taken care of and have a minor effect on the quality of the resulting powder pattern (the same cannot be said about a liquid pattern). However, the Compton scattering from diamonds prove to give rise to considerable structured background and limit the maximum exposure time available. The Compton scattering from the materials in the P–E cell is only more important when considering liquid samples at high-Q. Thus, with the P–E cell, counting times are longer, with a much increased signal:background ratio. Typical DAC experiments are carried out at one energy, with a cell of fixed geometry; therefore, there is not the same degree of versatility in matching large-unit cells with low-energies and variable real-space resolution than there is with the P–E cell. These elements combined with the temperature range and stability makes the P–E cell ideal for high-quality ADX diffraction measurements.

Once pressurised to the desired region of P-space and annealed to decrease peak broadening, the first step in data collection for solution of new structures is made by identifying the source of all peaks in the data. We can alleviate this task by translating the cell across the beam to illuminate the various components of the cell assembly, the h-BN, graphite, pT markers and gasket to observe how their various patterns change with p, T and also to use this information as aids to calibration of the pT conditions, if no thermocouple it used (e.g. Crichton and Mezouar, 2002). We cannot understate how important it is to source all peaks in the diffraction pattern, especially if the structure is new and an analogous phase neither known nor obvious.

3. Data collection and treatment

3.1. Calibration of detector

When using a fixed IP detector we are required to characterise and correct:

i. The sample–detector distance,
ii. The position of the beam centre, i.e. 0° 2θ,
iii. The orientation of the detector about the beam centre, its tilt (about a horizontal axis through the incident beam position) and rotation, around the vertical axis through the incident beam position.

Typically, these measurements will be made with either NBS Si (Standard Reference Material 640c) or NBS LaB_6 (SRM 660a). These are both Line Position standards, i.e. have NBS-calibrated and certified *d*-spacing positions, with which we can use to calculate the sample–detector position for known or estimated wavelengths.

The rings appearing on the IP surface have constant angle from the incident beam direction in 2θ, as given by the Bragg Equation, but the distance from the direct beam position and the ring on the image plate varies, increasing in radius with distance. When the sample–detector distance is correct, the linear distance on the plate from the direct-beam centre to the ring, its radius will correspond to the expected 2θ position for the standard material. This is from the simple expression of the $\tan 2\theta = \text{radius/(sample–detector distance)}$. The radius distance is measured in pixels and each pixel will correspond to either 150 or 100 μm in linear dimension. However, there are second-order effects to consider; the diffractometer will not be sitting perpendicular to the incident beam and as this non-orthogonality increases, so does the quality of the peaks degrade through artificial broadening. During automatic calibration, e.g. with Fit2D (Hammersley et al., 1996) the sample–detector distance estimation is conducted at the same time as beam centre location. Following this, initial estimates on the non-orthogonality of the detector around the beam centre position are made, as these will be expressed through the diffraction ring developing an elliptical shape. Using a list of calibrated *d*-spacings and initial estimates of the sample–detector distance and beam centre the program will search in the expected 2θ region for the next standard peak, conduct 1D fits around this ring to estimate the ellipticity and re-estimate all detector parameters. Once searching and fitting of all available rings is completed the detector is ready for use.

Another advantage of the use of these standard materials is to estimate the contribution of the instrument (i.e. everything except the sample that is in the beam) to the overall peak profile shape. However, in practice, this is of less importance than for example, laboratory sources or neutron diffractometers, for the reason that synchrotron X-ray diffractometers are inherently high resolution due to the high collimation of the incident beam, with small sampling volumes and usually in transmission and the effects of crystallite size, strain broadening and sample defects all contribute to mask the effect of the instrument on the overall peak shape function.

3.2. Peak identification, indexing and refinement

One of the common points in the following examples of simple structure solution is the care that must be taken in the initial peak assignment, for this influences the choice of unit cell and all consequent treatments. It can be very difficult to establish whether or not a structural transition has taken place within a single-phase material, when there can be peaks due to scattering from the sample assembly, reaction with the assembly, reduction of the sample and even phase dissociation are all possible (e.g. in Ag_2HgI_4, Parfitt et al., 2004, or CuS, Crichton et al., 2003a, or, $YNaF_4$, Grzechnik et al., 2002).

The simplest way to help identify these effects is through evidence of texture on the 2D detector image, spottiness due to crystal growth, and any other effects that may distinguish the origin of different maxima through different loading responses, e.g. oscillating zones of strong–weak intensity around a ring (from sample preparation or developed preferred orientation). In Figure 3, it is clear that there are at least two populations of textures in the

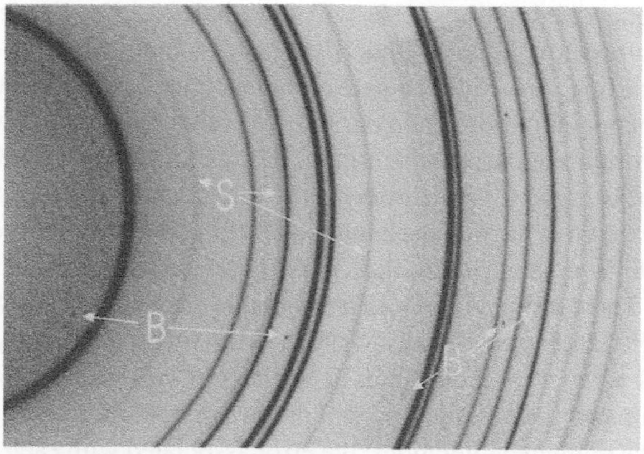

Figure 3. Two-dimensional pattern of the incomplete reaction between $Mg(OH)_2$ (brucite, "B") + MgF_2 (sellaite, "S") to form MgOHF. The difference in texture is obvious, and a good aid to identification. The continuous rings belong to MgOHF whilst the most spotty are those of brucite.

pattern, shown for the reaction of brucite and sellaite to form MgOHF (Crichton et al., 2003b). Identification of the spotty peaks indicates that there is brucite, therefore the reaction is incomplete and we should expect sellaite in addition to MgOHF, though there is no difference in the texture of the rings from sellaite and MgOHF and just removing those that are spotty would not be sufficient for correct indexing. In this example, simply knowing that there are unreacted materials in the sample charge is perhaps sufficient to improve synthesis procedures or to allow us to proceed with more sophisticated fitting as we must account for those intensities that are overlapped – due to the similarities of these three structures.

We must be aware when checking data derived from high pressures and temperatures. Cell parameters, and consequently, *d*-spacings are – quite naturally – most likely not those in a database that we might use for comparison due to the heating and compressing of the sample. In combination, the effects of preferred orientation or strong crystallite growth might also lead to difficulties if we were to rely on limited Hanawalt searches using the first three strongest lines (e.g. see Crichton et al., 2003c for high p, T data on black-P). These problems are somewhat circumvented with modern search–match software algorithms. However, used manually, one is advised to proceed being aware of these effects, estimating likely changes in unit cell dimensions from known compressibilities and thermal expansivities of similar structures.

After identification of all peaks in the pattern, the next stage in solution of an unknown structure is to index the unit cell. In order to obtain accurate peak positions we require some fitting procedure and there are many programs available for this exercise (e.g. XFIT, Cheary and Coelho, 1996; POWDERV2.00, Dragoe, 1999). Normally one might progress by background fitting and removal, finding the peaks by inspection of the second-derivative of the 2θ-intensity data or by being above threshold value and then fitting them with a suitable function, e.g. Lorenzian or psuedo-Voigt (see Howard and Preston, 1989; Finger et al., 1994 and references therein). Using suitable interfaces, such as that in CRYSFIRE (Shirley, 2000) we can control the indexing routines that will attempt to fit a unit cell that accounts for most or all of the peaks.

We point out again that impurity peaks will contribute to a low figure of merit. If there are a sufficient number of peaks at low 2θ angles, indexing can fail completely. The combination of indexing runs with comparison of calculated powder patterns will help in this stage in eliminating impurities from our indexed cells; e.g. using POWDERCELL (Kraus and Nolze, 1999), or through using CHECKCELL (Laugier and Bochu, 2002) which has the added capability of being able to read the results of CRYSFIRE solutions directly.

It is essential to improve the initial estimation of cell parameters before the next stage in the structure solution process; often, individual peaks from the experimental data will multiply indexed. There are many programs available for this task. A. Le Bail's ERACEL and UNITCELL (Holland and Redfern, 1997) will refine symmetry constrained or unconstrained cell parameters from input of a file of *d*-spacings or 2θ values and their corresponding estimated indices. ERACEL will produce an output that displays the variable shifts for each of the refinement iterations; the error associated with each peak position and produces a cell (with associated errors) as well as a pseudo-PDF card for it. UNITCELL goes further with a more sophisticated statistical treatment of the input data that gives information of potentially detrimental indices or reflections and on those that weight the fit the most. The mode of operation is controlled in the input file of ERACEL and through the interface of UNITCELL and the construction of the input files in each is simple to use. Others, such as POWDERV2.00 and CHECKCELL, provide a graphical user interface (GUI) with which to work and accept multiple file-types. Peak positions can be searched within these GUIs, directly in the case of POWDERV2.00 and in conjunction with CELREF in the case of CHECKCELL.

CHECKCELL offers the advantage that the user can input directly the entire summary of indexing trials from CRYSFIRE, investigate the possibility of higher symmetry cells and the number of formula units per cell. We can also do the least-squares fit of cell parameters to indexed reflections and estimate the best space group assignment – manually or automatically. POWDERV2.00 has additional features such as cell-refinement and auto-calibration versus known indexing or standard material. In these respects, the latter two programs are more "user-friendly" providing a link (after initial reformatting of data) from initial data through to refined cell and possible space-group and *Z* information, within one program.

At this point it may be advantageous to illustrate the use of these programs through an example. We draw on the solution of trigonal-S9, an unquenchable phase (Crichton et al., 2001). The new structure was obtained from the α-S eight-ring molecular structure, which is stable at ambient conditions, upon pressurising to 3 GPa, and heating to 400°C. The sample was loaded inside a BN capsule and (unusually) a 10 mm diameter fired pyrophyllite gasket, the furnace was graphite. During pressurising and heat treatment, the gasket, h-BN and furnace peak positions were monitored to ensure that we could identify these if they occurred in the sample pattern one crystallisation of the new phase was reached.

The sample–detector distance and detector offset parameters were calibrated previously by SRM 640 and NaCl placed at the sample position.

The original solution proceeded thus:

- Peak finding and fitting with POWDERV2.00
- Indexing with DICVOL91 (Boultif and Loüer, 1991), run from the POWDERV2.00 interface

- Cell refinement
- Solution with ENDEAVOUR (Putz et al., 1999) using variable Z information (no knowledge of change of volume on transformation) and no symmetry information (no unique space group assignment)
- Symmetry element search and application
- Refinement with GSAS

The diffraction pattern, Figure 4, was clean, thus it was relatively easy to extract peak positions for the first, most intense and not obviously overlapped peaks. From these 19 peaks DIVCOL91 gave a hexagonal cell with an M(19) of 84.4 and F(19) of 636.7, for $a = 7.0809(5)$ Å and $c = 4.3028(6)$ Å. This exceptional figure of merit (see, for example, Wolff, 1972) is testament not only to the high symmetry of obtained cell and unambiguous peak assignments, but also to the stability of the detector and robust peak fitting. For the purpose of demonstration, if we rerun the auto-indexing search through the CRYSFIRE suite, we obtain similar solutions from ITO (Visser, 1969), TAUP (Taupin, 1973) and DICVOL91, namely that of a primitive orthorhombic cell with $a = 4.302$ Å, $b = 6.140$ Å and $c = 3.544$ Å. More importantly, ITO suggests that the cell is probably hexagonal, $a = 7.09$ Å, $c = 4.30$ Å, through $a_{\mathrm{HEX}}^2 = 7.09^2 = b_{\mathrm{ORTH}}^2 + c_{\mathrm{ORTH}}^2 = 6.140^2 + 3.554^2$. The hexagonal solution is then refined. Similar geometrical relationships between sub/supercells are checked in stand-alone programs such as LEPAGE or TRUECELL, both of which are also incorporated in CHECKCELL, and find the hexagonal unit cell when the orthorhombic is used as initial. TAUP and DICVOL91 incidentally, also find the hexagonal lattice as a possible solution.

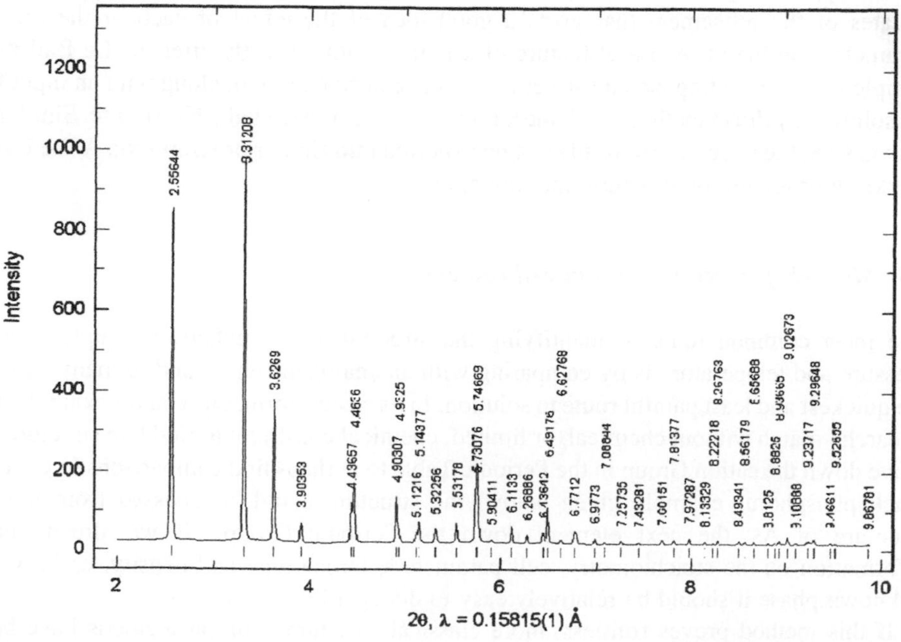

Figure 4. In situ X-ray powder diffraction pattern for S9, trigonal sulfur, obtained at 3 GPa and 400°C.

In cases like this, where we obtain results, such that all different indexing algorithms finding the same hexagonal cell and/or an orthorhombic subcell with high figures of merit, it is safe to assume that we can continue to the next stage of the solution process. Depending on the targeted method this can be least-squares refinement of cell and determination of symmetry, Le Bail fitting if the space-group has been identified with "Best Group" or "Best Solution" in CHECKCELL (or chemical, optical or other methods) or to directly proceed with *ab initio* energy minimisation solution without symmetry constraints. With our experience, and pre-written re-formatting codes, CHECKCELL combined with CRYSFIRE offers the most versatile, intuitive and user-friendly approach to excluding possible and identifying potential solutions.

Further estimation of the correctness of the proposed indexing and space group solutions is available from structureless whole-pattern profile-fitting; e.g. by Pawley (Pawley, 1980) or Le Bail (Le Bail et al., 1988) methods. The profile-pattern fitting procedures do not calculate intensities based on a model structure and fitting will only have background, cell parameters and peak-shape parameters as variables. The quality of the fit will depend largely on the cell parameter and symmetry combination – though does not guarantee if either or the combination is correct (symmetries with the same systematic absences will fit equally). These profile fits are also useful for routine fitting of unit cells, for example for equation of state measurement or where particular problems prevent full structural analysis. These may be various, for example, extreme preferred orientation or large contrast in scattering powers rendering one atomic species hard to refine. We can, however, derive no information on the amount that each phase contributes to the pattern. Programs like RIETICA (Hunter, 1997) and GSAS (Larson and Von Dreele, 2004), with EXPGUI (Toby, 2001) are very easy to interact with to control the progression of the Le Bail-type fitting to obtain cell parameters and volumes, both allow real-time graphical updates of the refinement that gives a good idea of the effect of each of the refined parameters on the fit. A useful feature of RIETICA is that directly after the Le Bail fit is completed, the resulting *hkl* and structure factors can be exported, along with an input file for solution by direct methods with the SIR97/EXPO (Altomare et al., 2003) suite. Similarly, with GSAS, the structure factor file can be exported into similar programs via A. Le Bail's OVERLAP program, for structural interpretation.

3.3. Methods for location of unit-cell contents

The most common route to identifying the structure of an unknown sample at high pressure and temperature is by comparing with an analogous phase and in many cases is the quickest and least painful route to solution. In its easiest form, one can for example, run a search–match without chemical, or limited, chemical constraint to find like structures or move down the cation Group in the Periodic Table to perhaps find comparisons (e.g. in the black phosphorus example given below, the structure could be guessed from the *A*7 structure of As, the next element down the Periodic Group). If we already have information on the stoichiometry, cell parameters, lattice-type and density, or *Z*, of the unknown phase it should be relatively easy to disregard many options.

If this method proves fruitless, more classical structure solution methods have been used successfully to locate atomic positions inside target unit cells. These might include direct methods of phase solution and electronic map interpretation available with EXPO, by global optimisation/simulated annealing (e.g. ENDEAVOUR), or using Monte Carlo minimisation functions; e.g. ESPOIR (Le Bail, 2001). Often these can be run with partially known fragments of the structure (in which case difference Fourier maps can be also used), or by minimising the position of rigid molecules or bodies such as polyhedra or sheets, e.g. with FOX (Favre-Nicolin and Cernỳ, 2002).

In its most simplest form, ENDEAVOUR can be run without assignment of space group as it will search for space groups based on reduced cells coupled with translations and atom positions that it finds whilst calculation proceeds – so we could have skipped the space group assignment stage above, in fact the original S9 solution was carried out in this fashion. The input to ENDEAVOUR is a minimum of a unit cell, not even necessarily the correct one (which is particularly useful as very few space groups are uniquely identifiable from diffraction data) – the subcell from ITO would have sufficed, a suitable bond length, number of formula units per cell and the diffraction profile.

In the case of S9, several runs were carried out, varying Z, but fixing the repulsion potentials and using the DICVOL91 unit cell. The best two solutions were near identical, for $Z = 9$. It was evident that the two solutions were related (were mirror images of each other) and all we required was to test these by Rietveld refinement of the symmetry, atom position combination. However, since these runs were conducted in $P1$, i.e. no symmetry constraints, there were nine unrelated atoms in the unit cell. ENDEAVOUR can use this information to correlate the position one atom to the next via symmetry elements and establish the full symmetry of the unit cell automatically (Hannemann et al., 1998; Hundt et al., 1999). Doing this we obtain the enantiomorphic pairs, $P3_221$ and $P3_112$. This same result is found running the entire CRYSFIRE output through CHECKCELL and the "Best Solution" function. A check to see if the bond lengths are reasonable (remembering that these change with pressure and temperature) is sufficient information to finish with the calculation and try a Rietveld refinement with the final positional parameters and space group assigned by ENDEAVOUR. Similar treatment was demonstrated on the equally simple black-P system (Crichton et al., 2003c) and the much more complex high-pressure post-dickite layer structure (Dera et al., 2003) to minimise the H positions from *in situ* DAC data.

We come back to the MgOHF example to demonstrate another method of solution. The unit cell was previously obtained through use of DICVOL91 (via CRYSFIRE) with $F(20) = 58.7$, Z was estimated at four and the space group was unclear, but estimated at no less symmetrical than $P2_12_12_1$. The cell was checked by profile refinement using RIETICA (Rp = 2.47% wRp = 3.57%) before export to EXPO for location of unit-cell contents. The input to EXPO consisted of the diffraction pattern, wavelength, unit-cell size and symmetry and unit-cell contents. Going through the various stages of single-peak fitting, background refinement and whole-pattern fitting produced, the first extraction stage finished with Rp = 4.45% and the initial phasing run produced the three-atom structure with Mg, O, F unambiguously located. Indeed, there was an additional Q (unassigned) peak at the position associated with the hydrogen site in isostructural diaspore. Refinement of this solution (still in EXPO) produced a moderate improvement in the quality of the fit and Rietveld refinement was made with GSAS before reassessment of the structure was made with Newsym and Addsym routines of Platon (Spek, 2003) (by running DISAGL and GSAS2CIF and reading the cif file (Hall et al., 1991) in PLATON). The correct space group of *Pnma* for this diaspore-type structure was found and later combined refinement with neutron diffraction data confirmed the H position in the same, undeuterated sample (Crichton et al., 2003b). Thus, the more classical solutions are also available to users, and are indeed more versatile; allowing user control over the choice of peak-shape functions, background, partitioning of data ranges, etc in addition to visualisation of the electron density maps (also in RIETICA and GSAS).

However, it is not always as easy as these examples. The application of all the above methods failed even to reproduce the following sample's chemistry (Grzechnik et al., 2001). The sample in this case is ZrW_2O_8, the aristotype negative-thermal-expansion framework structure, which undergoes pressure-induced amorphisation at low pressure. On applying pressure, the amorphous-like diffraction pattern was obtained and with heating it was evidenced that crystalline peaks were appearing from this material. These peaks were picked and fitted from which the initial cell was indexed as hexagonal with $a = 6.4145(3)$ Å and $c = 3.7941(2)$ Å, with high figures of merit of $M(19) = 97.7$ and $F(19) = 244.9$. Examination of the refined and indexed reflections did not reveal any systematic absences. After the normal routes to solution failed (tested by calculating bond lengths and overall chemistry to have the correct 1:2 Zr:W ratio), initial searches in the literature found only one AB_2X_8 phase with similar cell parameters; that of UTa_2O_8. A pattern was calculated for this structure (using POWDERCELL), but by replacing the U with W and Ta with Zr; it was clearly not the same. Thus, even the route through analogy, was not obvious. The next step considered that the Zr and W positions may be disordered, so a literature search for A_3X_8-type chemistries with similar cell parameters was tried. If we inspect the results of this search (e.g., on the ICSD database) we find that α-U_3O_8 has similar cell parameters, has $Z = 1$ and is in space group *P*-62m, which has no systematic absences (Ackermann et al., 1977). Therefore, it ties in with the information we have gleaned from the diffraction pattern ZrW_2O_8. Trying a calculated powder diffraction

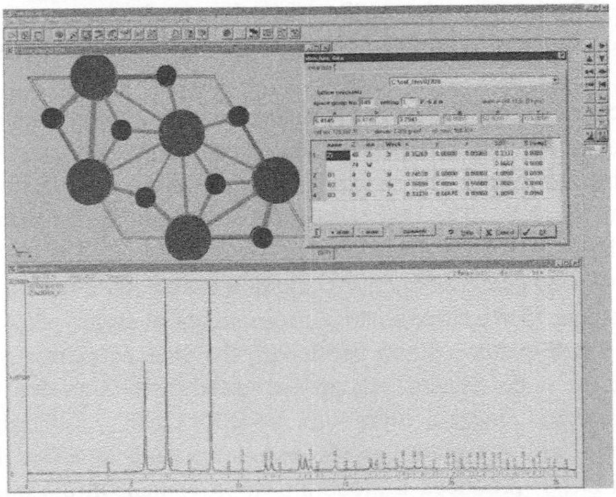

Figure 5. Comparison of the experimental pattern for ZrW_2O_8 and that calculated by using the atomic coordinates of α-U_3O_8 of Ackermann et al. (1977), replacing the cell parameters with those determined in auto-indexing and the U-site occupied by 1/3 Zr + 2/3 W.

pattern for this phase, by replacing U with 1/2 Zr and 2/3 W, for example with POWDERCELL, we obtain a diffraction pattern very similar to the experimental diffraction pattern, Fig. 5. Since there was a close match with our diffraction data, the disordered α-U_3O_8 structure was used as the initial model for fitting. Rietveld refinement of the structure confirmed that ZrW_2O_8 has the α-U_3O_8 structure at high pressure and temperature. With hindsight, it is now evident that the normal structure solution routines were failing because of the disorder of the Zr and W occupancies on the same crystallographic site.

4. Conclusion

We have highlighted the advances made at ID30 in producing high-quality clean diffraction patterns, particularly through the introduction of the oscillating slit system that can remove a large proportion of the background contribution. We have shown several examples of simple structures, being solved by just a few of the, mostly free, available suites of programs available, principally from www.ccp14.ac.uk.

We hope that with these examples that we have demonstrated that classical and more routine structure solution routes are available for treatment of data collected *in situ*, on samples under high pressures and temperatures. This is a situation that is extremely interesting and will prove to be of tremendous value in the interpretation of phase diagrams and stability of geologically relevant systems, particularly in those cases where high *p*, *T* phases prove to be unquenchable.

References

Ackermann, R.J., Chang, A.T., Sorrell, C.A., 1977. Thermal expansion and phase transition of the U_3O_8-Z phase in air. J. Inorg. Nucl. Chem. 39, 75–85.

Altomare, A., Burla, M.C., Carnalli, M., Carrozzini, B., Cascarano, G., Giacovazzo, C., Guagliardi, A., Moliterni, A.G.G., Polodori, G., Rizzi, R., 2003. EXPO and SIR97 Release 1.01.

Besson, J.M., Nelmes, R.J., Hamel, G., Loveday, J.S., Weill, G., Hull, S., 1992. Neutron powder diffraction above 10-GPa. Physica B 180, 907–910.

Boultif, A., Loüer, D., 1991. Indexing of powder diffraction patterns for low symmetry lattices by the successive dichotomy method. J. Appl. Crystallogr. 24, 987–993.

Cheary, R.W., Coelho, A.A., 1996. XFIT deposited, with tutorials at www.ccp14.ac.uk/tutorial/xfit-95/xfit.htm.

Chen, J., Kikegawa, T., Shimomura, O., Iwasaki, H., 1997. Application of an imaging plate to large volume press MAX80 at Photon Factory. J. Synchrotron Radiat. 4, 21–27.

Chen, J., Parise, J.B., Li, R., Weidner, D.J., Vaughan, M.T., 1998a. The imaging plate system interfaced to the large volume press at beamline X17B1 of the National Synchrotron Light Source. In: Manghnani, M.H., Yagi, T. (Eds), High-Pressure Temperature Research: Properties of The Earth and Planetary Materials, AGU, Washington, DC, pp. 129–134.

Chen, J., Weidner, D.J., Vaughan, M.T., Li, R., Parise, J.B., Koleda, C.C., Baldwin, K.J., 1998b. Time resolved diffraction measurement with an imaging plate at high pressure and temperature. Rev. High Pressure Sci. Technol. 7, 272–274.

Crichton, W.A., Mezouar, M., 2002. Non-invasive pressure and temperature estimation in large-volume apparatus by equation-of-state cross-calibration. High Temp.–High Pressure 34, 235–242.

Crichton, W.A., Vaughan, G.M.B., Mezouar, M., 2001. In situ structure solution of helical sulfur at 3 GPa and 400°C. Z. Kristallogr. 216, 417–419.

Crichton, W.A., Aquilanti, G., Gzrechnik, A., Pascarelli, S., 2003a. High Pressure and Temperature Behavior of Cu_XS_Y Phases with Initial CuIIS Stoichiometry, Geological Society of America, Seattle, Abstract.

Crichton, W.A., Parise, J.B., Breger, J., Marshall, W.G., Müller, H., Welch, M.D., 2003b. MgOHF and CuOHF: Synthesis, Structure and Crystal Chemistry of these Diverse Brucite- and Diaspore-Like Phases, Geological Society of America, Seattle, Abstract.

Crichton, W.A., Mezouar, M., Monaco, G., Falconi, S., 2003c. Phosphorus: new in situ powder data from large-volume apparatus. Powder Diffr. 18, 155–158.

Dera, P., Prewitt, C.T., Japel, S., Bish, D.L., Johnson, C.T., 2003. Pressure-controlled polytypism in hydrous layered materials. Am. Mineral. 88, 1428–1435.

Dragoe, N., 1999. Powder V2.00 version E: 1999-03-03, deposited at www.ccp14.ac.uk.

Favre-Nicolin, N., Cernỳ, R., 2002. FOX, free objects for crystallography: a modular approach to ab initio structure determination from powder diffraction. J. Appl. Crystallogr. 35, 734–743.

Finger, L.W., Cox, D.E., Jephcoat, A.P., 1994. A correction for powder diffraction peak asymmetry due to axial divergence. J. Appl. Crystallogr. 27, 892–900.

Grzechnik, A., Crichton, W.A., Syassen, K., Adler, P., Mezouar, M., 2001. A new polymorph of ZrW_2O_8 synthesized at high pressures and temperatures. Chem. Mater. 14, 4255–4259.

Grzechnik, A., Bouvier, P., Crichton, W.A., Farina, L., Köhler, J., 2002. Metastable $NaYF_4$ fluorite at high pressures and high temperatures. Solid State Sci. 4, 895–899.

Hall, S.R., Allen, F.H., Brown, I.D., 1991. The Crystallographic Information File (CIF): a new standard archive file for crystallography. Acta Crystallogr. A47, 655–685.

Hammersley, A.P., Svensson, S.O., Hanfland, M., Fitch, A.N., Haüsermann, D., 1996. Two-dimensional detector systems: from real detector to idealised image of two-theta scan. High Pressure Res. 14, 235–248.

Hannemann, A., Hundt, R., Schön, J.C., Jansen, M., 1998. A new algorithm for space-group determination. J. Appl. Crystallogr. 31, 922–928.

Kikegawa, T., Chen, J., Yaoita, K., Shimomura, O., 1995. DDX diffraction system: a combined diffraction system with EDX and ADX for high pressure structural studies. Rev. Sci. Instrum. 66 (2), 1335–1337.

Holland, T.J.B., Redfern, S.A.T., 1997. Unit cell refinement from powder diffraction data: the use of regression diagnostics. Mineral. Mag. 61, 65–77.

Howard, S.A., Preston, K.D., 1989. Profile fitting of powder diffraction patterns. In: Bish, D.L., Post, J.E. (Eds), Modern Powder Diffaction, Reviews in Mineralogy, Vol. 20. Mineralogical Society of America, Washington, DC, pp. 217–275.

Hundt, R., Schön, J.C., Hannemann, A., Jansen, M., 1999. Determination of symmetries and idealized cell parameters for simulated structures. J. Appl. Crystallogr. 32, 413–416.

Hunter, B.A., 1997. RIETICA version 1.7.7, IUCR Powder Diffraction, 22, 21.

Klotz, S., Besson, J.M., Hamel, G., Nelmes, R.J., Loveday, J.S., Marshall, W.G., 1996. High pressure neutron diffraction using the Paris–Edinburgh cell: experimental possibilities and future prospects. High Pressure Res. 14, 249–255.

Kraus, W., Nolze, G., 1999. POWDERCELL. Federal Institute for Materials Research and Testing (BAM), Berlin, version 2.3.

Larson, A.C., Von Dreele, R.B., 2004. General Structure Analysis System (GSAS) Los Alamos National Laboratory Report LAUR 86-748.

Laugier, J., Bochu, B., 2002. CHECKCELL and CELREF part of the suite of programs written at the Laboratoire des Materiaux et du Génie Physique de l'Ecole Supérieure de Physique de Grenoble by Jean Laugier and Bernard Bochu. Deposited at www.ccp14.ac.uk.

Le Bail, A., 2001. ESPOIR: a program for solving crystal structures by Monte Carlo analysis of powder diffraction data. Mater. Sci. Forum 378, 47–52.

Le Bail, A., Duroy, H., Fourquet, J.L., 1988. Ab-initio structure determination of $LiSbWO_6$ by X-ray powder diffraction. Mater. Res. Bull. 23, 447–452.

Le Godec, Y., Martinez-Garcia, D., Mezouar, M., Syfosse, G., Itié, J.P., Besson, J.M., 2000. Thermoelastic behaviour of hexagonal graphite-like boron nitride. High Pressure Res. 17, 35–46.

Marshall, W.G., Francis, D.J., 2002. Attainment of near-hydrostatic compression conditions using the Paris–Edinburgh cell. J. Appl. Crystallogr. 35, 122–155.

Mezouar, M., Le Bihan, T., Libotte, H., Le Godec, Y., Hausermann, D., 1999. Paris–Edinburgh large-volume cell coupled with a fast imaging-plate system for structural investigation at high pressure and high temperature. J. Synchrotron Radiat. 6, 1115–1119.

Mezouar, M., Faure, P., Crichton, W., Rambert, N., Sitaud, B., Bauchau, S., Blattmann, G., 2002. Multichannel colimator for structural investigation of liquids and amorphous materials at high pressures and temperatures. Rev. Sci. Instrum. 73, 3570–3574.

Ohtaka, O., Takebe, H., Yoshiasa, A., Fukui, H., Katayama, Y., 2002. Phase relations of AgI under high pressure and high temperature. Solid State Commun. 123, 213–216.

Parfitt, D.C., Hull, S., Keen, D.A., Crichton, W., 2004. High pressure dissociation of silver mercury iodide Ag_2HgI_4. J. Solid State Chem. 117, 3715–3720.

Parise, J.B., Klotz, S., Hamel, G., Straessle, T., Crichton, W., Mezouar, M., Ling, C., 2004. Installation of The Paris–Edinburgh Cell at D20, ILL Experimental Report number 5-24-182.

Pawley, G.S., 1980. EDINP, the Edinburgh powder profile refinement program. J. Appl. Crystallogr. 13, 630–633.

Putz, H., Schön, J.C., Jansen, M., 1999. Combined method for ab initio structure solution from powder diffraction data. J. Appl. Crystallogr. 32, 864–870.

Shirley, R., 2000. The CRYSFIRE System for Automatic Powder Indexing: User's Manual, The Lattice Press, 41 Guilford Park Avenue, Guilford, Surrey, England.

Spek, A.L., 2003. PLATON, A Multipurpose Crystallographic Tool, Utrecht University, Utrecht, The Netherlands.

Taupin, D., 1973. A powder-diagram automatic-indexing program. J. Appl. Crystallogr. 6, 380–385.

Toby, B.H., 2001. EXPGUI, a graphical user interface for GSAS. J. Appl. Crystallogr. 34, 210–213.

Visser, J.W., 1969. A fully automatic program for finding the unit cell from powder data. J. Appl. Crystallogr. 2, 89–95.

de Wolff, P.M., 1972. The definition of the indexing figure of merit M(20). J. Appl. Crystallogr. 5, 243–245.

Yaoita, K., Katayama, Y., Tsuji, K., Kikegawa, T., Shimomura, O., 1997. Angle-dispersive diffraction measurement system for high-pressure experiments using a multichannel collimator. Rev. Sci. Instrum. 68, 2106–2110.

Zhao, Y.S., von Dreele, R.B., Morgan, J.G., 1999. A high P–T cell assembly for neutron diffraction up to 10 GPa and 1500 K. High Pressure Res. 16, 161–177.

Advances in High-Pressure Technology for Geophysical Applications
Jiuhua Chen, Yanbin Wang, T.S. Duffy, Guoyin Shen, L.F. Dobrzhinetskaya, editors

Chapter 18
High-pressure structure determination and refinement by X-ray diffraction

Ross J. Angel

Abstract

The determination of the structure of a crystalline material at high pressures is dependent upon the accurate measurement of the intensities of X-ray or neutron beams diffracted by the sample, and the precise correction of the intensities for the various effects of the high-pressure device. In this contribution the methods required to obtain accurate structure refinements from high-pressure single-crystal X-ray diffraction measurements are reviewed and the potential of new measurement methods with single crystals and synchrotron X-ray sources is assessed.

1. Introduction

The determination of the arrangement of atoms within a crystalline material relies upon the accurate measurement of the intensity of radiation diffracted by a sample of that material. Both neutrons and X-rays can be used with both powder and single-crystal samples. The choice between using X-rays or neutrons for high-pressure studies is essentially the same as for studies at ambient pressure; neutrons are useful for samples containing both heavy and light atoms, those in which it is necessary to distinguish between elements of similar atomic number, and those experiments in which magnetic scattering is of interest. The lower flux from neutron sources, however, has limited the maximum pressures to less than those achieved by X-ray diffraction because of the need for larger samples. Nonetheless, impressive progress has been made in recent years in high-pressure neutron diffraction both with "large-volume cells" that can achieve pressures in excess of 10 GPa (Besson et al., 1992; Marshall and Francis, 2002) and large-anvil cells have been used for experiments as high as 50 GPa (Goncharenko et al., 2000). The coupling of the latter to more intense neutron sources such as the SNS, currently under construction at Oak Ridge, opens the opportunity to pursue such measurements to much higher pressures than is currently achievable on a routine basis. The issues concerning high-pressure measurements by neutron diffraction are dealt with elsewhere in this volume (Zhao et al., 2005).

High-pressure X-ray diffraction measurements have been traditionally made by single-crystal diffraction in the laboratory, using methods adopted from traditional crystallography. The measurements are made with radiation of a specific wavelength in an angle-dispersive mode. The diffracted beams from a powder specimen are much weaker than those from the same volume of a single crystal, and thus synchrotrons have become the X-ray source of choice for high-pressure powder X-ray diffraction. In contrast to a conventional laboratory X-ray source, a synchrotron produces an intense white radiation spectrum containing a range of wavelengths that can be employed in two different experimental configurations. If the entire white-beam spectrum is used then "energy-dispersive" powder diffraction can be performed, although it is difficult to obtain reproducible and quantitative diffracted intensities by the method. Alternatively, a narrow range of wavelength can be selected from the incident beam and be used to perform more conventional angle-dispersive powder diffraction. The same choice of techniques is available for single-crystal diffraction at the synchrotron. The entire white-beam spectrum has been used in the "Laue" technique for some structural studies of proteins (e.g. Nieh et al., 1999; Ren et al., 1999), especially for studies in which data collection time is limited. Conventional angle-dispersive single-crystal diffraction with a monochromatized beam is also widely used at synchrotrons, both for small-molecule and protein crystallography.

Crystallographers have spent more than two decades developing the measurement techniques and methods for data reduction for single-crystal diffraction in the diamond anvil pressure cell, mostly through painstaking development in individual laboratories with conventional X-ray sources. The resulting structural data are superior to that obtained by powder diffraction because the intensity data is intrinsically three-dimensional and, therefore, suffers from none of the resolution and overlap problems intrinsic to powder data that often precludes structure solution from all but the simplest of compounds. For example, single-crystal data can be used to determine acentric crystal structures that cannot be refined from powder data (e.g. Lager et al., 2002) or to resolve discrepancies between powder diffraction datasets (e.g. Dera et al., 2002). Powder data, especially in the tiny samples used in the DAC, also has the potential to be affected by the absence of a true powder average, and by the effects of preferred orientation, neither of which can be unambiguously determined. By contrast, the intensities diffracted from a single crystal are unaffected by such issues. Nonetheless, powder diffraction is in widespread and routine use at synchrotron sources for high-pressure studies that employ both diamond anvil cells (e.g. Fei and Wang, 2000) and large-volume presses (e.g. Chen et al., 1998). Because the requirements for access to the sample under high pressure are less stringent for powder diffraction than for single-crystal diffraction, powder diffraction has also been used to study structures at simultaneous high pressures and temperatures (e.g. Chen et al., 1997).

Single-crystal diffraction methods have, therefore, fallen behind powder diffraction in terms of usage and in terms of pressure range over which it is applied. The reasons for this are fourfold. First, it is often more difficult to preserve single crystals to high pressures because of either phase transitions or the small volume available for the sample in the high-pressure device. The advent of synchrotron radiation as an X-ray source, however, allows micron-sized crystals to be used. It is believed that such smaller samples may stand a better chance of surviving phase transitions without loss of crystallinity, although there are plenty of cases of larger crystals being pressurized through first-order transitions (e.g. Angel et al., 1992; Yamaguchi et al., 1992; Arlt and Angel, 2000; Dera et al., 2002). Second, the quality of single-crystal diffraction patterns is degraded by the presence of non-hydrostatic stresses, but laser annealing of sample assemblies is now used routinely to relieve such stresses and maintain a hydrostatic environment, and the same result can also be achieved through external heating. Third, monochromatic single-crystal measurements require customized four-circle goniometers and sophisticated control software that is not available at most synchrotron beamlines dedicated to high-pressure research. Lastly, single-crystal diffraction experiments require greater knowledge and expertise in crystallography on behalf of the experimentalist in designing and setting up the experiment than is required for a powder diffraction experiment. This contribution is intended to try and address this last difficulty by providing some guidelines to the issues that have to be addressed in single-crystal measurements at high pressure intended to produce data for accurate structure determination and refinement.

2. Theory

Diffraction experiments produce two basic kinds of information about the structure of a crystalline sample. The positions of the diffracted beams are determined by the geometry of the instrument, the wavelength of the radiation and, most importantly, by the size and shape of the unit cell of the sample. The cell parameters of a crystal held at high pressure are, therefore, determined entirely from the diffracted beam *positions*. The determination is direct and unambiguous, provided the data are of sufficient quality, and the reader is advised to consult recent reviews for details of the experimental methods (e.g. Hazen and Downs, 2000).

The diffracted beam positions and the unit-cell parameters contain essentially no information about the arrangement of the atoms within the unit cell of the crystal except in those few special cases of crystal structures in which all of the atoms occupy positions within the unit cell that are fully constrained by symmetry. Cubic perovskites and MgO are examples of such structures. However, in the vast majority of crystal structures the atoms occupy positions within the unit cell that are not fixed by symmetry and these atom positions, and thus the bond-lengths and angles within the crystal structure, can only be determined from the *intensities* of the diffracted beams.

The relationship between the fractional coordinates of the atoms and the diffracted intensity of a reflection indexed as $\mathbf{H} = hkl$ is a represented by the structure-factor equation:

$$F(hkl) = \sum_j f_j \exp(2\pi i\mathbf{H}\cdot\mathbf{X}_j)\exp(-B_j \sin^2\theta/\lambda^2)$$

$$= \sum_j f_j(\cos(2\pi\mathbf{H}\cdot\mathbf{X}_j) + i\sin(2\pi\mathbf{H}\cdot\mathbf{X}_j))\exp(-B_j \sin^2\theta/\lambda^2) \quad (1)$$

The summation is over all of the atoms j in the unit cell with f_j being the scattering power of the jth atom at fractional coordinates x_j, y_j, z_j represented in Eq. (1) by the vector $\mathbf{X}_j$. Note that the f_j vary with both the diffraction angle and the wavelength of radiation. The atomic displacement parameters, which reflect the apparent deviation of each atom j from an ideal point scatterer, are represented by the parameter B_j. Note that $2\pi\mathbf{X}_j\cdot\mathbf{H} = 2\pi(hx_j + ky_j + lz_j)$ for each atom. The structure-factor $F(hkl) = F(\mathbf{H})$ is a complex quantity, and is related to the diffracted beam intensity as $I(hkl) = F(hkl)F^*(hkl)$. The structure-factor equation clearly allows us to calculate $F(hkl)$ and the corresponding intensity for any reflection once we know the atomic coordinates of all of the atoms in the unit cell.

The structure-factor equation in Eq. (1) is a Fourier transform, so one can invert the equation and write an equation for the atomic coordinates in terms of the *F*(*hkl*). However, because the structure factors are complex quantities they can also be described in terms of an amplitude and a phase. But in a normal diffraction experiment it is the *Intensity* of the reflection *hkl* that is measured, *not the amplitude and the phase*. Therefore, the *phase* information is not obtained in an experiment, only the square of the amplitude. Therefore, the inversion of the structure-factor equation cannot be performed, and one cannot obtain the positions of the atoms directly from the intensities. This is known as *the phase problem*, and is the reason why crystallographers have jobs. Instead, in order to determine a structure, one has to make a guess (intelligent or otherwise) at the positions of the atoms in the unit cell, and then use this "trial structure" to calculate the *F*(*hkl*) and thus *I*(*hkl*) through Eq. (1). This first step is known as *structure solution*. It is usually not necessary in a high-pressure diffraction study unless the crystal has undergone a phase transition at high pressure, or has been synthesized *in situ*. As we will see, the quality of data collected in a high-pressure experiment is degraded compared to that collected from the sample under ambient conditions, so the process of structure solution even of relatively simple structures at high pressures can become a significant challenge. In this respect the three-dimensional nature of single-crystal data can be of considerable advantage over powder data (e.g. Dera et al., 2002; Nelmes et al., 2002; McMahon et al., 2003; McMahon, 2004).

The more common case is that one already knows the structure of the crystal, and one is interested in determining the relatively small change in bond lengths and angles as pressure or temperature are changed. This requires only the process of *structure refinement*. In this process the intensities calculated from a trial structure via the structure-factor equation are compared to the observed intensities, and the parameters of the trial structure are then adjusted so as to reduce the discrepancy between the calculated and observed intensity data.

The reason for presenting this revision of the very basic principles of diffraction is to reinforce the point that the measurements required to determine unit-cell parameters (and Equations of State) are fundamentally different from those required to determine the arrangements of the atoms within the structure, and thus bond lengths and angles. It should, therefore, not be surprising that the optimal experimental configuration for making these two types of experimental measurement can be fundamentally different (see review by Angel et al., 2000). Furthermore, in order to obtain accurate and precise structures, one needs to have accurate and precise measurements of the intensities of diffracted beams from the sample. In this respect, single-crystal diffraction offers significant advantages over powder diffraction especially because the diffracted intensities are higher for a given sample volume, and there are no issues of peak overlap that can occur in powder diffraction for either complex crystal structures, those structures with pseudo-symmetry, or acentric structures in which the *hkl* and $\bar{h}\bar{k}\bar{l}$ reflections overlap exactly (e.g. Lager et al., 2002). Lastly, one expects the structures determined from single-crystal data to be more precise because there is no need to simultaneously refine profile and background parameters. The majority of the following discussion will, therefore, be restricted to single-crystal X-ray diffraction studies, with explicit note made when the issues need also to be addressed when collecting or processing powder diffraction data. The issues discussed here will apply equally to single-crystal data collected in the laboratory and at synchrotron beam lines.

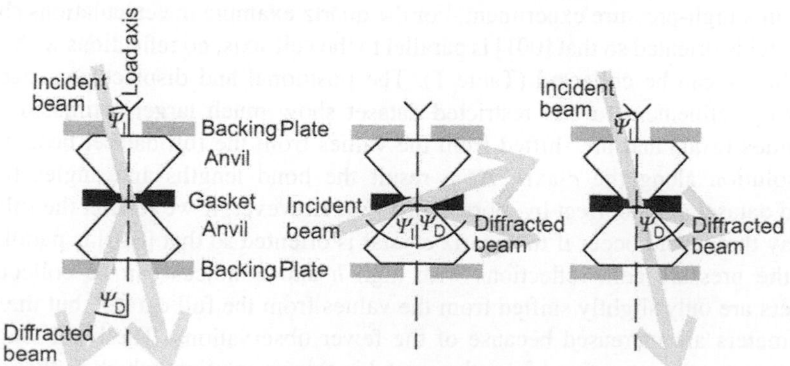

Load axis
Incident beam
Ψ_I
Backing Plate
Anvil
Gasket
Anvil
Backing Plate
Ψ_D
Diffracted beam
Incident beam
Ψ_I Ψ_D
Diffracted beam
Incident beam
Ψ_I
Ψ_D
Diffracted beam

review of DACs for single-crystal diffraction and their relative merits is provided by Miletich et al. (2000).

For most transmission DACs the body of the diamond cell restricts the maximum values of Ψ_I and Ψ_D to 40 degrees or less (Fig. 1). These restrictions apply equally whatever type of detector, point or area detector, is being used. On an area detector the limits to the ψ angles appear on the data images as shadowed areas of the detector. For a point detector the diffractometer control software is set up so that reflections obscured by the pressure cell are not collected. Maximum access to reflections is achieved by operating the goniometer in "fixed-phi" mode (Finger and King, 1978; Angel et al., 2000). Nonetheless, for a transmission cell with ψ restricted to 40° or less, only one-third of the total number of reflections out to 60° in 2θ are accessible. Furthermore, these accessible reflections are not distributed equally over reciprocal space, but are restricted to those reciprocal lattice vectors that are close to being perpendicular to the axis of the DAC. In order to achieve high precision in all three fractional coordinates of all of the atoms in the structure, one must, therefore, take care to ensure that the crystal is oriented in such a way so as to allow access to as large a portion as possible of an asymmetric unit of reciprocal space.

For crystals that are loaded at room pressure, as opposed to being grown *in situ* in the DAC, the effect of the restricted access can be simulated prior to preparation of the crystal and loading the diamond anvil cell. This allows the optimal orientation of the crystal to be calculated in advance and the sample to be prepared appropriately for the high-pressure experiment. This process is illustrated here with quartz in a transmission-geometry DAC as the example. The first column of Table 1 lists the various statistical parameters as well as the structural parameters obtained by refinement to the complete dataset collected from the quartz crystal mounted in air. These parameters represent what we are then trying to achieve from the same crystal mounted in the pressure cell. We now use a computer program (Angel, 2004c) to select from this dataset only those reflections that would be accessible for a given orientation in the DAC. Structure refinements are then performed with these restricted portions of the entire dataset to simulate the results that would be obtained in a high-pressure experiment. For the quartz example the calculations show that if the crystal is oriented so that [001] is parallel to the cell-axis, no reflections with l indices greater than 1 can be collected (Table 1). The positional and displacement parameters obtained by refinement to this restricted dataset show much larger estimated standard uncertainties (*esu*s) and are shifted from the values from the full dataset because of the poor resolution along the c-axis. As a result the bond lengths and angles from this restricted dataset are incorrect by more than 1 *esu*. However, if we restrict the full dataset in the way that would occur if the quartz crystal is oriented so that [010] is parallel to the axis of the pressure cell, reflections with high h and l indices can be collected. The parameters are only slightly shifted from the values from the full dataset, but the *esu*'s of the parameters are increased because of the fewer observations. The bond lengths and angles are correctly determined from the crystal in this orientation, which is, therefore, the one that would be selected for a high-pressure experiment. As a test, the last column of Table 1 shows the actual results for data collected from the same crystal in a DAC, oriented with [010] parallel to the cell-axis. Note that the parameters show similar shifts to the simulation, but the *esu*s are increased because of the lower signal-to-noise ratio of the data that arises from the absorption and additional background scattering from the DAC.

Table 1. Data simulation and structure refinements of quartz for a transmission DAC.

Sample environment	Air	Air	Air	DAC
Access restrictions	None	[001]//DAC axis	[010]//DAC axis	[010]//DAC axis
h,k max	8	7	6	6
l max	9	1	9	9
R_{int}	0.023	0.019	0.016	0.034
R_{sig}	0.028	0.028	0.026	0.025
Refinement				
N(observed reflections)	270	50	153	149
N(obs. reflections)/*N*(parameters)	15.9	2.9	9.0	8.8
R_u	0.015	0.013	0.023	0.028
R_w	0.020	0.017	0.017	0.028
Si–O: Å	1.6077(8)	1.6029(37)	1.6074(11)	1.6059(19)
Si–O′: Å	1.6115(7)	1.6187(11)	1.6115(11)	1.6102(19)
O–Si–O: deg	143.56(5)	143.30(33)	143.59(8)	143.89(15)

Notes: The statistical measures reported in the Table are defined as follows: R_{int} is the internal agreement index of symmetry-equivalent reflections defined as $\sum|F^2 - \bar{F}^2|/\sum F^2$, in which F^2 is the square of an individual structure factor and $\bar{F}^2$ is the corresponding averaged structure factor. The summations are performed over all observed reflections. R_{sig} is a measure of the signal-to-noise of the dataset, calculated as $\sum\sigma(I)/\sum I$ where I are the measured intensities and $\sigma(I)$ are their estimated standard uncertainties. R_u is the conventional discrepancy index for the structure refinement, defined as $\sum||F_0| - |F_c||/\sum|F_0|$, in which F_0 is the measured value of the structure factor and F_c is the value of the structure factor calculated from the refined structure model. The summation is over all observed averaged structure factors. R_w is the weighted residual for the structure refinement, defined as $R_w^2 = \sum w(F_0 - F_c)^2/\sum wF_0^2$. The summation is over all observed averaged structure factors.

This example illustrates the general principle that, for measurements in transmission-geometry DACs, crystals belonging to uni-axial crystal systems should whenever possible be prepared as *hk*0 crystal plates, so that the *c*-axis lies parallel to the culet surface of the diamonds. This ensures that reflections with both high values of *l* and high values of *h* and/or *k* are accessible. For lower symmetry crystal systems it is often possible to choose an orientation that still provides reasonable resolution along all three reciprocal lattice vectors (e.g. Downs et al., 1996). When one data collection is not sufficient to provide the necessary data coverage, it is then necessary to collect two separate datasets. This can be achieved by reloading the same crystal twice into a cell in different orientations and collecting data at the same pressures for each orientation (e.g. Angel, 1988). An alternative is to load together two (or more) crystals of the same sample in different orientations in the diamond anvil cell and to collect the data from each crystal.

The major restriction on access to reflections in transverse-geometry DACs is the body of the two halves of the cell and the pillars that connect them. Maximum access to the reflections from a crystal in such cells is obtained by generally operating the diffractometer in bisecting mode (Ashbahs, 1984, 1987). In some positions the pillars that connect the two halves of the cell will obscure reflections and for these a small rotation of the cell about the diffraction vector will normally bring the reflection into a measurable position (Koepke et al., 1985). This procedure allows up to 95% of the reflections in one hemisphere of reciprocal space to be measured provided that the gasket is made of an X-ray transparent material. Following the initial development of beryllium gaskets (Koepke et al., 1985) other materials including boron (Lin et al., 2003) and diamond (Zou et al., 2001) have been developed for use as X-ray transparent gaskets. If, however, the gasket is highly absorbing then the proportion of accessible reflections drops considerably, and the corrections for absorption become very difficult to perform reliably (e.g. D'Amour et al., 1978).

The region of accessible reflections in a mixed-mode cell depends critically upon the exact details of the cell design, and cannot, therefore, be generalized. As an example, the design of Malinowski (1987) has most of the cell body on the diffracted-beam side of the DAC removed. This allows reflections to be collected at 2θ angles as high as 160° and provides access to 85% of all reflections (Malinowski, 1987).

4. Data collection and integration

There are two types of detector geometry available for X-ray diffraction, area detectors such as film, CCD detectors and image plates, and point detectors. The differences in data collection procedures for the two types of detector are summarized in Figure 2. It is clear that more expertise is required in the preparation for a measurement with a point detector, as the orientation of the crystal has to be first determined by a search for diffraction maxima followed by indexing and accurate centering of these positions so as to obtain an orientation (UB) matrix. This matrix is then used by the controlling computer to position the diffractometer so as to collect each diffraction intensity in sequence. By contrast, data can be collected with an area detector without specifying an orientation matrix and thus the preparation time for a measurement is less. Why then bother with a point detector that in principle takes longer to set up, and longer to collect data? The reason is that the point

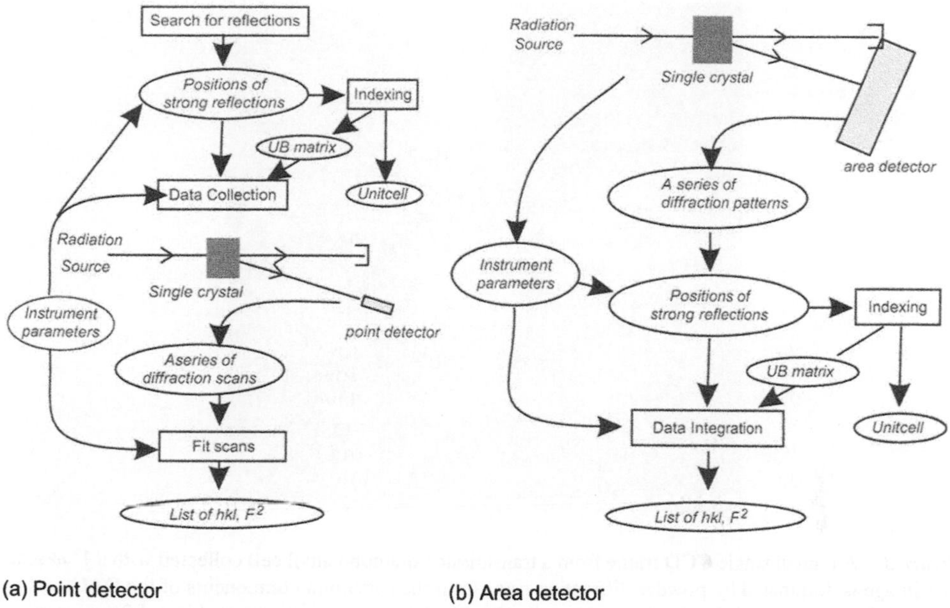

Figure 2. Flow charts outlining the basic steps required for data collection and integration for a single-crystal diffraction experiment. For a point detector (a) the crystal has to be oriented prior to the data collection, whereas for an area detector (b) the indexing and orientation is performed *off-line* after the data collection is complete.

detector can be collimated so as to remove all of the scattered background from the DAC components that dominates images from an area detector (Fig. 3). And the point detector system can be programmed to spend a larger proportion of the data collection time on the weaker intensities so as to optimize the information content of the final dataset.

Both aspects mean that the point detector often delivers more precise data, as illustrated in Table 2 for a crystal of $YAlO_3$ perovskite (*Pbnm*, unit-cell volume 204 A^3) measured on two laboratory diffractometer systems. While the CCD data shows better apparent signal-to-noise (R_{sig}) this statistic is dominated by the strongest reflections and the true quality of the datasets is better indicated by the measure of internal consistency (R_{int}) The point detector data has smaller R_{int} and, as a result, the final bond lengths and angles obtained from the refined structure are both more precise and more accurate. The point detector is, therefore, the detector of choice when the primitive unit cell is relatively small, perhaps up to 500–1000 A^3 because there are relatively few reflections to be collected and each CCD image, therefore, contains only one or two reflections (e.g. Fig. 3). By contrast, the CCD is much more efficient for searching for reflections, and for identifying twinning, diffuse scattering, and crystal quality, even for small unit-cell crystals. And as the unit-cell size of the sample becomes bigger, the number of reflections per CCD frame becomes larger, so in contrast to the point detector the data collection time does not increase. This often outweighs the disadvantage of not being able to collimate out the background scattering or to perform constant-precision scans. So, for structures with larger unit cells the CCD coupled with appropriate integration algorithms becomes the detector of choice for intensity

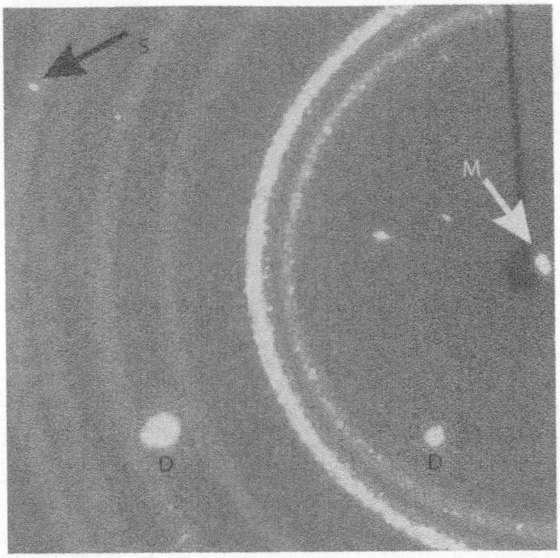

Figure 3. A typical single CCD frame from a transmission diamond anvil cell collected with a 1° ω-scan. The image is dominated by powder diffraction rings from the beryllium components of the DAC, and by strong diffraction peaks from the diamond anvils (D). The latter give rise to the multiple diffraction peaks (M) near the shadow of the beam stop. The only sample reflection in this frame is arrowed at top left (S).

measurement. Indeed, there is much to be said for using both types of detector in tandem on a single diffractometer as to be able to use each for the tasks to which it is best suited.

4.1. Point detector

The collection of intensity data from a single crystal mounted in a DAC with a diffractometer equipped with a point detector has been covered in considerable detail in the literature (e.g. Denner et al., 1978; Finger and King, 1978; Hazen and Finger, 1982; Ashbahs, 1987; Angel et al., 2000) so only a brief review of the most salient points will be given here. The first stage of the experiment is to optically align the DAC to the center of the diffractometer. Then a search procedure is initiated to find the setting angles of strong reflections from which a UB, or orientation matrix, and unit-cell parameters can be determined. Alternatively, the UB matrix can be initially determined from measurements with an area detector system or even film (e.g. Hazen and Finger, 1982). Once an approximate UB matrix is determined, it can be used by the diffractometer control system to find and accurately center a small number of strong reflections. The use of the "8-position centering technique" of Hamilton (1974a) as adapted for DAC measurements by King and Finger (1979) allows the crystal offsets from the diffractometer center to be determined and the DAC to be moved to reduce them. It is absolutely critical to reduce these offsets to a few microns in order to obtain accurate intensity data.

It is not always possible to perform full 8-position centering on many single-crystal diffractometers either because of restrictions on the motion of the goniometer circles or because the goniometer geometry does not allow it for a wide range of reflections.

Table 2. Comparison of point and CCD detector systems.

Experiment	CCD, crystal in DAC	Point detector, crystal in DAC	Point detector, crystal in air	CCD, crystal in offset DAC
Goniometer	Xcalibur-2	Xcalibur-1	Xcalibur-1	Xcalibur-2
X-ray optics	0.3 mm enhance optic	0.5 mm collimator	0.5 mm collimator	0.3 mm enhance optic
Detector	Sapphire-II	Point detector	Point detector	Sapphire-II
Data collection	20 h, 680 frames of 1° and 40 s each	20 h, constant precision scans	55 h, constant precision scans	20 h, 680 frames of 1° and 40 s each
Integration	Crysalis (Oxford Diffraction, 2003)	WinIntegrStp profile integration (Angel, 2003)	WinIntegrStp	Crysalis (Oxford Diffraction, 2003)
Nrefln	930	543	688	833
Data reduction				
R_{sig}	0.020	0.027	0.032	0.020
Navg	179	188	209	152
R_{int}	0.042	0.029	0.042	0.101
Refinement				
$N(F > 4\sigma(F))$	140	165	180	98
R_u	0.020	0.021	0.019	0.042
R_w	0.025	0.018	0.023	0.030
Al–O1 (Å)	1.8996(20)	1.8960(10)	1.8947(9)	1.8956(36)
Al–O2 (Å)	1.9128(44)	1.9070(22)	1.9049(26)	1.9003(88)
Al–O2′ (Å)	1.9167(43)	1.9221(22)	1.9200(28)	1.9297(93)

The kappa-geometry goniometers that are now widely used for laboratory measurements are one example, for which the 8-positioning centering is restricted to a small number of reflections with setting angles close to Eulerian $\chi = 90°$. Rather than rely on the measurement of one or two reflections we have obtained more precise centering results from performing a small data collection of 30–40 low-angle strong reflections. The omega scans are then stored and the exact positions of the maxima are then obtained by profile fitting (e.g. Angel, 2003) and stored along with the other setting angles of each reflection. The crystal offsets and the unconstrained unit-cell parameters are then obtained from these setting angles by the iterative method of Dera and Katrusiak (1999) and the position of the DAC on the diffractometer is adjusted accordingly. This entire procedure of data collection and refinement is then repeated until the offsets are effectively zero, although with care the offsets can be reduced to zero at the first adjustment.

Once the UB matrix is determined and the crystal offsets have been reduced to zero the data collection can be initiated. Extra collimation is often added to the detector in order to prevent radiation scattered from the DAC components from entering the detector (e.g. D'Amour et al., 1978; Wittlinger, 1997; Li and Ashbahs, 1998; Wittlinger et al., 1998) as illustrated in Figure 4. Such collimation must be carefully aligned in order to avoid the measured diffraction intensities being inadvertently reduced (e.g. Ashbahs, 1987). Once the cell and collimation is aligned the data collection proceeds in a normal fashion, with the diffractometer control software only scanning those diffracted beam positions known to be accessible given the restrictions of the DAC, and the step scan data are stored.

Integration of the data plays a critical role in alleviating the effects on data quality caused by lower signal-to-noise levels in high-pressure data. For data step-scanned with a point detector one can use the strong reflections to determine both the peak shapes and the

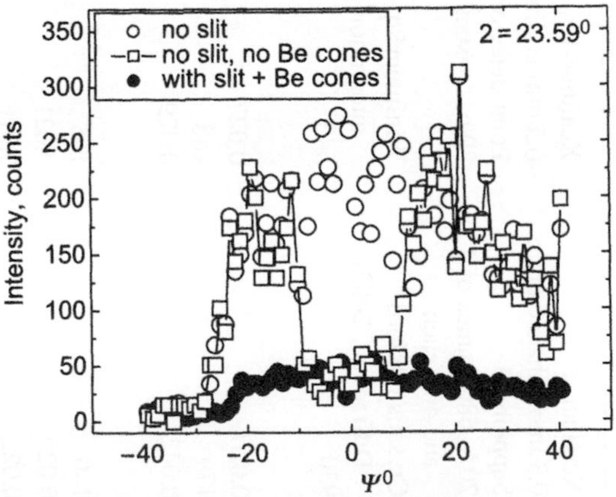

Figure 4. The background intensity from a transmission DAC with Be seats at low diffraction angles plotted as a function of ψ. Most of the background scattering comes from the Be seats being illuminated by the direct beam (open circles). This can be reduced by removing the Be plug that normally fills the optical access holes in the seats (open squares), but at the expense of making absorption corrections more difficult and less reliable (Angel et al., 2000). The introduction of an additional collimation slit in front of the detector almost completely removes the background from the measured data (filled symbols). See also Li and Ashbahs (1998).

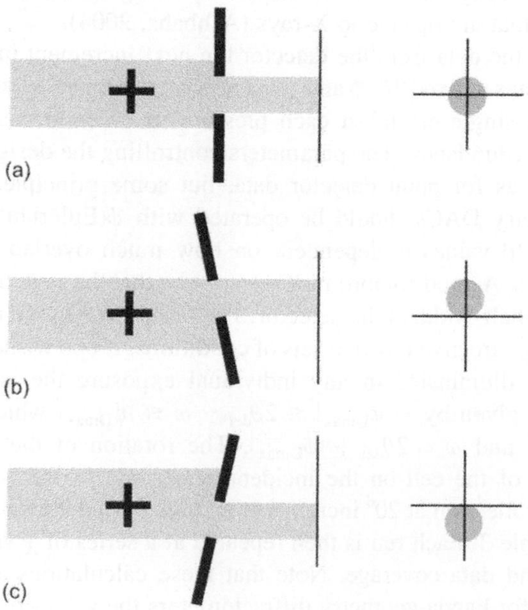
(a)
(b)
(c)

position of the transmitted beam moves with cell rotation then the DAC can be moved along the beam until motion of the transmitted beam is eliminated. On a laboratory diffractometer this method allows the cell to be centered to within 30 μm of the diffractometer center. More reliable centering can only be achieved by diffracted-beam centering with a point detector so there is a case for having both detector systems available on a single diffractometer. The effect of a crystal offset of only 80 μm which is typical of the result of optical centering is illustrated in the last column of Table 2. The data from the offset crystal are of significantly lower quality and result in bond lengths with both significantly higher *esu*'s and incorrect values.

Data collection procedures are also different with an area detector (Fig. 2b) because the orientation of the crystal is not required prior to the data collection. Instead the data collection is designed to cover all accessible positions of diffraction maxima, which are subsequently processed off-line; a "shoot first and ask questions later" strategy! The usual data collection strategy is to set the center of the detector at some fixed 2θ value, and the goniometer carrying the DAC at some fixed ϕ and χ (or κ for kappa goniometers). The DAC is then rotated through a small angle in omega while the detector collects diffracted intensities. The optimal width of this omega scan depends on both the experiment and on the details of the detector technology and integration software but is typically in the range of 0.1 to 1.0 degrees for CCD detectors but greater for detectors such as image plates with longer read-out times. Because the source of the background scattering in the data images is mostly beryllium and diamond that does not lie at the center of the diffractometer, wider omega scans will smear these rings and thus obscure more data than would be affected in a narrower scan. This background can be reduced by using transmission-geometry DACs with diamond seats that are opaque to X-rays (Ashbahs, 2004).

After read-out of the data from the detector the next increment in omega is scanned; each set of omega scans at fixed 2θ, ϕ and χ (or κ) is often termed a "run". Many such runs are performed for a single crystal at each pressure in order to ensure maximum data coverage and some redundancy. The parameters controlling the design of the run list are not as well defined as for point detector data, but some principles remain the same. Transmission-geometry DACs should be operated with ϕ(Eulerian) fixed at zero. The choice of detector 2θ values is dependent on how much overlap or redundancy one requires between runs. A good compromise seems to be that the step in 2θ should be a little less than the angular half-width of the detector. The limits on the total range of ω at a given detector position $2\theta_{det}$ are given by two sets of conditions. If one wishes at least half of the detector area to be illuminated in any individual exposure the access limits for the diffracted beam are given by $-|\psi_{Dmax}| \leq 2\theta_{det} - \omega \leq |\psi_{Dmax}|$ which are equivalent to $\omega \geq 2\theta_{det} - |\psi_{Dmax}|$ and $\omega \leq 2\theta_{det} + |\psi_{Dmax}|$. The rotation of the cell is then further limited by aperture of the cell on the incident-beam side of the cell, $|\omega| \leq \psi_{Imax}$. An example for a data collection at 20° increments in $2\theta_{det}$ for a detector with a half-width of 22.5° is listed in Table 3. Each run is then repeated at a series of χ values so as to obtain good redundancy and data coverage. Note that these calculations are all performed in Eulerian geometry; for kappa-geometry diffractometers the values of ϕ(Kappa) and κ for each run must be obtained by conversion from Eulerian geometry.

Once the run list has been designed the exposure time per data frame must be chosen. Many CCD control systems automatically repeat the data collection exposure if the CCD is saturated by a strong reflection. With most DAC measurements the sample reflections

Table 3. Example goniometer settings for a transmission DAC and area detector for $\Psi_{\text{Imax}} = \Psi_{\text{Dmax}} = 40°$.

2θ	ω_{abs} allowed by Ψ_{Dmax}	ω_{abs} allowed by Ψ_{Dmax} and Ψ_{Imax}
−80	−120 to −40	−40 to −40
−60	−100 to −20	−40 to −20
−40	−80 to 0	−40 to 0
−20	−60 to +20	−40 to +20
0	−40 to +40	−40 to +40
20	−20 to +60	−20 to +40
40	0 to +80	0 to +40
60	+20 to +100	+20 to +40
80	+40 to +120	+40 to +40

Note: The restrictions at $2\theta = \pm 80°$ mean that no scan is performed at these detector settings.

are not usually strong enough to saturate the detector, but the diffraction maxima from the diamond anvils will certainly do so. Data collection times can, therefore, be safely reduced by suppressing any such re-measurement procedures.

On a conventional laboratory diffractometer the incident beam is larger than the gasket hole and both the gasket and the beryllium seats are illuminated by the direct beam. Scattering from these DAC components, therefore, dominates the CCD images (Fig. 3) because the detector cannot be collimated like a point detector. It is, therefore, essential to have sophisticated peak hunting and integration algorithms in order to extract meaningful reflection intensities from the raw images. The first step in data analysis is to identify the diffraction maxima from the crystal in the presence of the strong and structured background scattering from the cell components. If the data have been collected with a step size in omega that is smaller than the rocking curve of the crystal, the crystal peaks can be identified by a primitive profile analysis. An alternative is simply to mask and not search those regions of the data images that are affected by powder rings and diamond reflections. Hand-picking reflections provides a third, but tedious, alternative that is sometimes necessary. Once sufficient crystal reflections are identified then normal indexing procedures can be used to attempt to find a UB matrix and cell parameters. Integration methods are very dependent upon the details of detector design and, to some extent, the frame width in omega of the individual data images should be chosen to best match the integration method.

Once a UB matrix has been determined the data images from the detector are processed again to extract intensity data from the positions where diffraction maxima are calculated to occur. For the integration process to provide reliable integrated intensities in the presence of strongly varying background from the DAC it must provide an algorithm that reliably estimates and subtracts the background in the immediate neighborhood of the peak. It appears that the combination of a "regional" method of background evaluation coupled with a local evaluation provides more reliable integrated intensities than a purely global fit of the background that is used for ambient pressure data. This is not really surprising given the highly structured nature of the scattering from the diamond cell.

The peak search and background evaluation must, of course, be limited to those areas of the data frames that are not obscured by the body of the diamond anvil cell. This is a trivial calculation for transmission geometry cells but is considerably more complex for transverse and mixed-mode DACs.

5. Data corrections

The output from the data integration and reduction procedures applied to single-crystal diffraction data is the same for all detector types and radiation sources. The data consists of a list of unmerged but indexed structure factors (or the squares of structure factors, depending upon the file format). Each structure factor corresponds to a single observation of a reflection from the crystal that has been corrected for instrumental effects such as intensity variations of the incident beam (at synchrotron sources), detector sensitivity and the Lorentz-polarization factor. The structure factors at this stage are not yet corrected for the effects of absorption or other diffraction events in the pressure cell. In order to make these corrections, it is necessary for the structure factors to be accompanied by geometric information that is sufficient to specify the directions of the incident and diffracted beams with respect to both the crystal and the pressure cell. The most common form for this beam path information from area detector systems is as the direction cosines with respect to the crystal axes. These can be converted to the more useful ϕ-axis system of Busing and Levy (1967) by the method of Allan et al. (2000). Data from point detectors is often accompanied by the setting angles of the diffractometer that are sufficient to completely specify the necessary beam path information. Full details of these issues can be found in Allan et al. (2000); Angel et al. (2000); Angel (2004a); Dawson et al. (2004) and references therein. Figure 6 is offered as an overall guide to the entire process of data correction, structure solution and refinement.

5.1. Absorption

The absorption of X-rays is characterized by the transmission factor, T, often called the "transmission coefficient". It is used to correct the measured diffracted beam intensities for absorption as $I_{corr} = I_{obs}T^{-1}$. The transmission coefficient for radiation passing through a material of thickness t and linear absorption coefficient μ_l is $T = \exp(-\mu_l t)$. Values for μ_l for any material can be calculated from the values tabulated in International Tables for Crystallography (Creagh and Hubbell, 1992). For a crystal in an ambient-pressure diffraction experiment, the path lengths of the incident and diffracted beams vary across the crystal and the expression for the transmission coefficient for a reflection intensity becomes an integral over the crystal volume, V, of the path lengths of the incident (t_I) and diffracted beams (t_D) :

$$T = V^{-1}\int_V \exp(-\mu(t_I + t_D))\mathrm{d}V \tag{2}$$

For a crystal mounted in a pressure cell, corrections also have to be made for the absorption by the components of the pressure cell through which the X-rays pass.

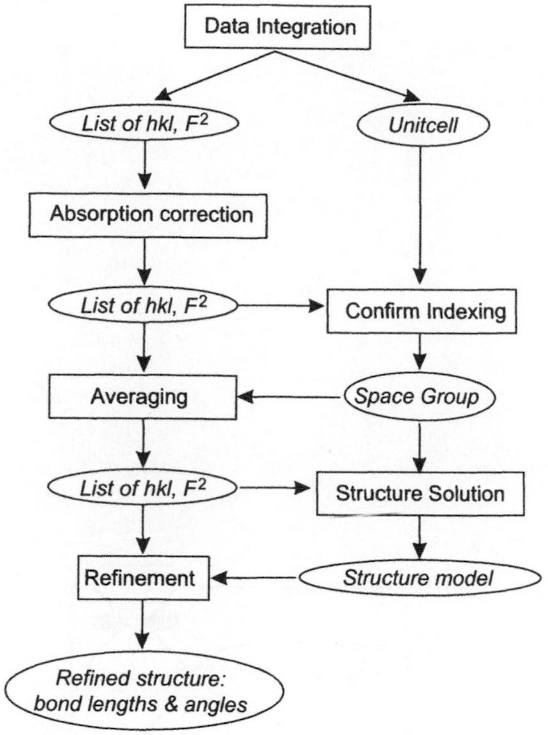

Figure 6. A flow chart outlining the various procedures (boxes) and information (ellipses) necessary for the correction of intensity data collected from a single crystal in a DAC. If the structure is already known, then only the series of steps shown on the left is necessary to obtain a refined structure. If the structure is unknown, both series of procedures needs to be followed in parallel.

The general expression for the transmission coefficient (e.g. Santoro et al., 1968) becomes:

$$T = V^{-1} \int_V \exp\left(- \sum_i \mu_i (t_{\mathrm{I}i} + t_{\mathrm{D}i}) \right) \mathrm{d}V \tag{3}$$

The factor

$$\exp\left(- \sum_i \mu_i (t_{\mathrm{I}i} + t_{\mathrm{D}i}) \right)$$

is the total transmission factor associated with a volume element $\mathrm{d}V$ of the crystal that is illuminated by the incident beam. Note that for experiments in which the beam is *smaller* than the crystal, an additional normalization for the illuminated volume of the crystal must also be applied. Each material traversed by the beam has a linear absorption coefficient μ_i. The $t_{\mathrm{I}i}$ and $t_{\mathrm{D}i}$ are the path lengths for the incident and diffracted X-ray beams in each different material i, including the sample crystal itself. For transmission-geometry pressure cells the materials will generally be the anvils, the pressure medium and the crystal for all reflections. Beams for reflections collected with higher values of ψ will also

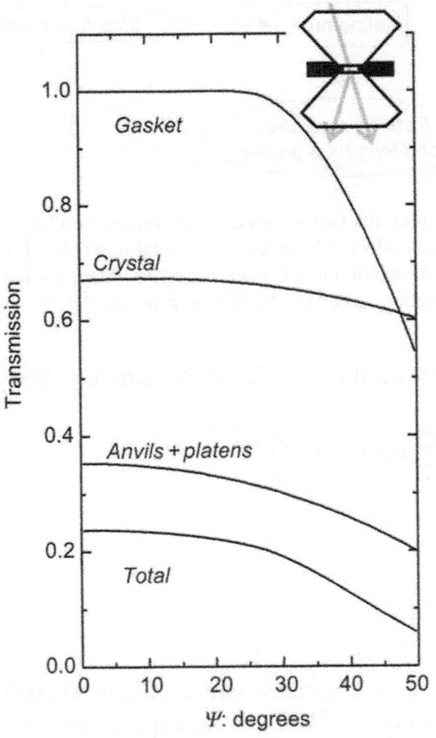
Gasket
Crystal
Anvils + platens
Total
Transmission
Ψ: degrees

multiplier to the separately calculated absorption correction due to the crystal and other materials (Santoro et al., 1968). Thus:

$$T = V^{-1}T(\Psi_\mathrm{I})T(\Psi_\mathrm{D})\int_V \exp\left(-\sum_i \mu_i(t_{\mathrm{I}i} + t_{\mathrm{D}i})\right)\mathrm{d}V \tag{4}$$

in which $T(\psi)$ denotes the transmission function for an incident or diffracted beam in a single backing plate and anvil. However, the path lengths in the gasket and the pressure medium will be different for different points in the crystal and cannot be removed from inside the integration in Eq. (4) and must be evaluated point-by-point.

Evaluation of the integral in Eq. (4) must be performed by a numerical integration technique. The normal method (*cf.* Busing and Levy, 1957; Wuensch and Prewitt, 1965; Burnham, 1966) is to describe the crystal size and shape on a Cartesian coordinate system and to set up a Gaussian grid of unequally-spaced points that spans the crystal volume. The directions of the incident and diffracted beams relative to the Cartesian coordinate system for each reflection are calculated from the information associated with the structure-factor data, as discussed above. The path lengths of the incident and diffracted beams are then calculated for each point in the crystal, and the transmission function for a reflection is calculated as:

$$T = V^{-1}T(\Psi_\mathrm{I})T(\Psi_\mathrm{D})\sum_{i,j,k} w_i w_j w_k \exp\left(-\sum_i \mu_i(t_{\mathrm{I}i} + t_{\mathrm{D}i})\right) \tag{5}$$

The w_i, w_j, w_k are weights pre-assigned to each grid point to compensate for the unequal spacing of the points. Further details about the absorption corrections made by this method of calculation and a program to perform them can be found in Angel (2004a). Further simplifications to the integral for the transmission function can be made for certain experimental configurations. When the sample crystal is formed by condensing gases or fluids *in situ* in the DAC the crystal fills the gasket hole (e.g. Miletich et al., 2000). This allows the effect of gasket shadowing to be calculated analytically (e.g. Santoro et al., 1968), especially if the crystal is non-absorbing and the gasket is considered opaque (Von Dreele and Hanson, 1984).

The common geometry of high-pressure powder diffraction measurements makes the corrections for absorption much easier than for single-crystal data. The incident beam is usually kept parallel to the load-axis of the cell, but even if it is not, the cell is not normally rotated during the measurement except perhaps for rocking motions to improve the powder average. Therefore, unlike a single-crystal experiment, the incident beam therefore suffers the same amount of absorption by the anvil and its backing plate for the entire powder pattern, and it is therefore not necessary to apply a correction for this absorption. In transmission measurements the diffracted beams leave the cell at an angle 2θ to the load axis, so the transmission function of the cell varies across the diffraction pattern as simply:

$$T(2\theta) = \exp(-(\mu_\mathrm{Dia}T_\mathrm{Dia} + \mu_\mathrm{P}T_\mathrm{P})/\cos 2\theta) \tag{6}$$

in which T_Dia, and T_P are the thickness of the anvil and platen, (*not* the path lengths) respectively. If, as often the case, X-ray opaque seats are used to support the diamond anvils then only correction for absorption by the anvil is necessary. Provided the beam size is significantly smaller than the gasket hole no correction for gasket shadowing is necessary. The transmission function of the beam through the powder sample for this geometry can be readily calculated by integration of Eq. (4) to obtain:

$$T(2\theta) = \frac{1}{\mu_X T_X(1 - \sec 2\theta)}[\exp(-\mu_X T_X \sec 2\theta) - \exp(-\mu_X T_X)] \tag{7}$$

in which T_X is the thickness of the sample. Further details of the corrections for powder data are provided by Katrusiak (2004). While the variation in the transmission function is small for most powder measurements, failure to make the correction will certainly bias the values of refined atomic displacement parameters to incorrect values. Once the absorption correction has been made to the raw data, the powder pattern is ready for refinement.

5.2. *Averaging*

The absorption correction does not correct the data for the effects of the variety of multiple diffraction events that occur within the diamonds (Loveday et al., 1990). Those events that result in a diffracted beam from the anvil entering the detector are immediately recognizable by their intensity and peak shape and are usually eliminated from the dataset at the earlier stage of data integration. On the other hand, the dips in the scan shown in Figure 8 are caused by two other types of event. Either the anvil on the incident-beam side of the cell diffracts away some of the incident intensity. Or the anvil on the detector-side of the cell diffracts away part of the diffracted beam from the crystal. These affected reflections could, in principle, be determined by calculations of the form proposed by Rossmanith et al. (1990) if the orientation matrices of the anvils are known in addition to that of the crystal sample. However, they may be readily be detected after data collection by comparing the intensities of symmetry-equivalent reflections and rejecting those reflections (so-called "outliers") whose intensities differ significantly from the average of the set of equivalents. Such a procedure with a sound statistical basis has been available for many years (Blessing, 1987) and can be modified for DAC data by making use of the fact that the reflections affected are always reduced in intensity. These are then rejected in a first-pass through the dataset by applying the Blessing criteria to only those reflections with intensities less than the mean intensity of the set of symmetry-equivalent reflections (Angel, 2004b). A second pass through the unrejected data is then performed to average the remaining data according to the normal statistical criteria (Blessing, 1987). This combined procedure also serves to remove individual outliers that may arise from a variety of other causes.

6. Structure solution and refinement

Once one has obtained an intensity dataset corrected for the effects of the pressure cell and from which aberrant reflections have been removed, the process of structure solution and refinement can proceed much as with a dataset collected from a crystal at ambient pressure. As for ambient pressure data, structure solution of single-crystal data is relatively

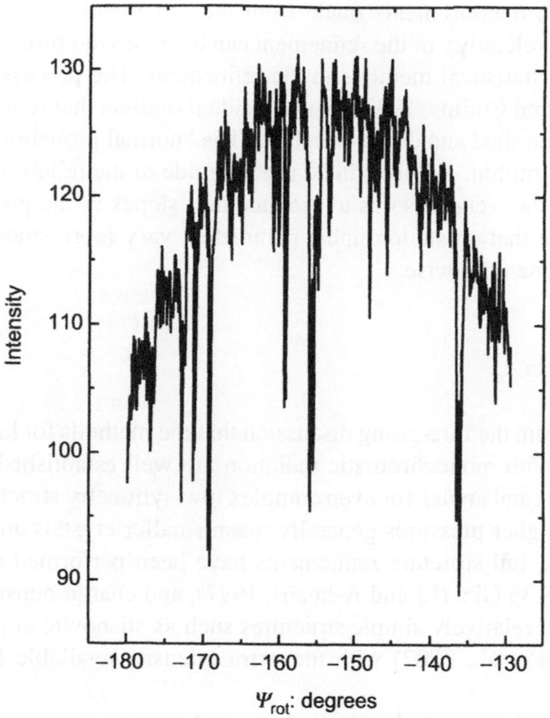

Figure 8. The measured variation in intensity of a reflection from a crystal in a transmission DAC as a function of rotation of the crystal around the scattering vector. The overall curvature of the measured intensity is due to the combined effects of absorption by the cell components, gasket and crystal and can be corrected by absorption programs. The sharp dips in intensity are due to simultaneous diffraction events in the diamond anvils that reduce the intensity diffracted from the crystal that actually reaches the detector (see Loveday et al., 1990). These so-called "diamond dips" can only be identified at the averaging stage of the data correction process.

trivial compared to using powder diffraction data. Indeed there are several examples when poor-quality single-crystal data has proved essential for solving structures (e.g. McMahon et al., 2003) which had proved impossible to determine by powder diffraction. The only real drawback with single-crystal data is the issue of data coverage. It is possible that with certain orientations of crystals in certain geometries of diamond anvil cells, the data will be insufficient to allow a direct structure solution to be obtained. Even in such cases the use of Harker sections and projections calculated from the data should yield partial structure solutions that could be used to guide the full solution through phasing.

In principle, structure refinement to diffraction data collected at high pressures should proceed exactly as a refinement to data collected at room conditions. In reality, however, limited resolution and lower signal-to-noise will degrade the quality of the data. Furthermore, however well the corrections to intensity data are made, they are never perfect. Taken together these effects mean that there are often systematic errors in the structural parameters determined for crystals held in diamond cells. These systematic errors can be identified by making measurements of the crystal at room conditions *in the device* both before and after the high-*P* or high-*T* experiment, and comparing the results to those obtained from measurements made from the same crystal in air, as illustrated in Tables 1 and 2. The reliability of the refinement can be improved further by the application of two well-known statistical methods in the refinement. The process of robust-resistant refinement (Prince and Collins, 1992) allows residual outliers that remain in the dataset to be automatically identified and down-weighted. The "normal probability plot" (Abrahams and Keve, 1971; Hamilton, 1974b) can be used a guide to the reliability of the weighting scheme. Adjusting the weights so as to produce unit slopes in the probability plot result also helps to ensure that crystallographic parameters vary more smoothly with pressure than refinements done otherwise.

7. Future prospects

It should be clear from the fore-going discussion that the methods for high-pressure single-crystal diffraction with monochromatic radiation are well established and are providing precise bond lengths and angles for even complex low-symmetry structures at pressures up to 10 or 15 GPa. Higher pressures generally mean smaller crystals and thus lower signal levels. Nonetheless, full structure refinements have been performed in the laboratory at pressures as high as 33 GPa (Li and Ashbahs, 1997), and charge-density distributions can be determined from relatively simple structures such as stishovite at pressures as high as 50 GPa (Yamanaka et al., 2002) with the extra intensity available from a synchrotron source.

Synchrotron radiation coupled with single-crystal diffraction does offer, in tandem with laboratory-based studies, several solutions to the limitations on single-crystal data quality that we have discussed above. One approach is to only exploit the intensity gain of a synchrotron over a sealed-tube source (at least 10^4 for a monochromatic beam) and perform conventional angle-dispersive single-crystal diffraction measurements. The higher incident intensity immediately allows the use of smaller single-crystal samples that in turn brings further advantages. For example, one can use several small crystals mounted simultaneously in a transmission-geometry DAC to overcome the problems of access to reflections. This would be preferable to trying to correct diffracted beam intensities for absorption by the gasket in mixed-mode geometry cells which would otherwise provide far greater data access. The data from such multiple-crystal mounts can be collected with an area detector and processed off-line (thus saving valuable beam time compared to a point detector experiment) because the two big drawbacks of area detectors discussed above can also be overcome. First, the high intensity of the incident beam means that there is nothing to be gained in performing constant-precision scans. And the high backgrounds on area detectors (Fig. 3) can be avoided by either collimating the incident beam down to a radius significantly smaller than the gasket hole or constructing cells with X-ray opaque seats (e.g. Ashbahs, 2004). Note that this cannot be achieved with the transverse geometry cells (Fig. 1) because the gasket always remains illuminated by the incident beam. Further, the use of smaller crystals immediately means that higher pressures will be attainable.

An alternative approach may be to exploit the white beam available from synchrotron sources and to employ the Laue diffraction technique that has been used for protein crystallography at ambient pressure (e.g. Nieh et al., 1999; Ren et al., 1999).

Laue diffraction relies on the fact that the X-ray beam from a synchrotron contains a continuous range of wavelengths. This range of wavelengths means that the Bragg equation is satisfied for many "reflections" simultaneously even for a stationary crystal and many diffracted beams are produced simultaneously. The Laue method also shares the advantages of fixed-wavelength, angle-dispersive diffraction at the synchrotron in that the beam can be collimated down to eliminate scattering from the components of a DAC used in transmission mode. The background from such polycrystalline components would in any case be less problematic because in the Laue method it appears as an extended pseudo-continuous variation rather than the discrete diffraction rings observed with monochromatic radiation (Fig. 3). There are some further advantages for Laue diffraction. The data are collected "blind" with an area detector so that beam time is not wasted in orienting samples prior to data collection. In principle an entire dataset could be collected with a single exposure, with the diamond anvil cell fixed in position. In practice a small number of exposures is probably to be preferred in order to obtain the data redundancy necessary to verify the more complex data reduction procedures. Data collection times will certainly be shorter than for measurements with monochromatic radiation because the beam intensity will be higher and only a few exposures will be necessary. Data access is alleviated for Laue diffraction with a transmission-geometry DAC even though the same restriction applies to ψ angles as for monochromatic measurements. But the total number of accessible reflections will be increased compared to the same crystal measured in fixed-wavelength angle-dispersive mode, because higher order reflections will be shifted to lower 2θ values and hence lower ψ angles by diffraction by the shorter wavelength components of the incident X-ray beam. Reflection access for Laue diffraction is, therefore, determined by both the crystal orientation in the DAC and by the usable wavelength range of the incident X-ray beam. The greater challenge is in data integration and subsequent correction of Laue diffraction data. Each diffraction maxima in a Laue pattern consists of several harmonics of different energies. That is multiple orders of a reflection *hkl* will all satisfy the Bragg equation $\lambda/n = 2d(hkl)\sin\theta/n$ contribute to one measured diffraction intensity. These contributions have to be separated prior to performance of the absorption correction, as the absorption coefficients vary with wavelength. Similarly, the refinement programs also require modification because the scattering power of the atoms represented by the scattering factors in Eq. (1) also varies with wavelength. Addressing these issues will require the investment of some time in development. Nonetheless the potential return in the form of high-quality structure refinements of crystals at high pressures makes the investment in both monochromatic and white-beam single-crystal diffraction measurements at synchrotron sources worthwhile and the future of high-pressure crystallography an exciting one.

Acknowledgements

Recent developments of single-crystal X-ray diffraction methods at Virginia Tech have been supported by Oxford Diffraction and NSF grants EAR-0105864 and EAR-0408460 to NL Ross and RJ Angel. Mathias Meyer and Jing Zhao are thanked for collaborations on technical developments. Jing Zhao also kindly provided the data shown in Figures 4 and 8.

References

Abrahams, S., Keve, E., 1971. Normal probability plot analysis of error in measured and derived quantities and standard deviations. Acta Crystallogr. A27, 157–165.

Allan, D.R., Clark, S.J., Parsons, S., Ruf, M., 2000. A high-pressure structural study of propionic acid and the application of CCD detectors in high-pressure single-crystal X-ray diffraction. J. Phys. C: Condens. Matter 12, L613–L618.

Angel, R.J., 1988. High-pressure structure of anorthite. Am. Mineral. 73, 1114–1119.

Angel, R.J., 2003. Automated profile analysis for single-crystal diffraction data. J. Appl. Crystallogr. 36, 295–300.

Angel, R.J., 2004a. Absorption corrections for diamond-anvil pressure cells implemented in a software package Absorb-6.0. J. Appl. Crystallogr. 37, 486–492.

Angel, R.J., 2004b. Average www.crystal.vt.edu.

Angel, R.J., 2004c. DAC_restrict www.crystal.vt.edu.

Angel, R.J., Chopelas, A., Ross, N.L., 1992. Stability of high-density clinoenstatite at upper-mantle pressures. Nature 358, 322–324.

Angel, R.J., Downs, R.T., Finger, L.W., 2000. High-pressure, high-temperature diffractometry. In: Hazen, R.M., Downs, R.T. (Eds), High-Pressure and High-Temperature Crystal Chemistry. Mineralogical Society of America, Washington, DC, pp. 559–596.

Arlt, T., Angel, R.J., 2000. Pressure buffering in a diamond anvil cell. Mineral. Mag. 64, 241–245.

Ashbahs, H., 1984. Diamond-anvil high-pressure cell for improved single-crystal X-ray diffraction measurements. Rev. Sci. Instrum. 55, 99–102.

Ashbahs, H., 1987. X-ray diffraction on single crystals at high pressure. Prog. Cryst. Growth Charact. 14, 263–302.

Ashbahs, H., 2004. New pressure cell for single-crystal X-ray investigations on diffractometers with area detectors. Z. Kristallogr. 219, 305–308.

Besson, J.M., Nelmes, R.J., Hamel, G., Loveday, J.S., Weill, G., Hull, S., 1992. Neutron powder diffraction above 10-GPa. Physica B 180, 907–910.

Blessing, 1987. Data reduction and error analysis for accurate single crystal diffraction intensities. Crystallogr. Rev. 1, 3–58.

Burnham, C.W., 1966. Computation of absorption corrections and the significance of end effects. Am. Mineral. 51, 159–167.

Busing, W.R., Levy, H.A., 1957. High-speed computation of the absorption correction for single crystal diffraction measurements. Acta Crystallogr. 10, 180–182.

Busing, W., Levy, H., 1967. Angle calculations for 3- and 4-circle X-ray and neutron diffractometers. J. Appl. Crystallogr. 22, 457–464.

Chen, J., Iwasaki, H., Kikegawa, T., 1997. Crystal structure of the high-pressure–high-temperature phase of bismuth using high-energy synchrotron radiation. J. Phys. Chem. Solids 58, 247–255.

Chen, J., Parise, J.B., Li, R., Weidner, D.J., Vaughan, M.T., 1998. The imaging plate system interfaced to the large volume press at beamline X17B1 of the National Synchrotron Light Source. In: Manghnani, M.H., Yagi, T. (Eds), High-Pressure Temperature Research: Properties of The Earth and Planetary Materials. American Geophysical Union, Washington, DC, pp. 129–134.

Creagh, D., Hubbell, J.H., 1992. X-ray absorption (or attenuation) coefficients. In: Wilson, A. (Ed.), International Tables for Crystallography. Kluwer Academic Publishers, Dordrecht, pp. 189–205.

D'Amour, H., Schiferl, D., Denner, W., Schulz, H., Holzapfel, W.B., 1978. High-pressure single-crystal structure determinations for ruby up to 90 kbar using an automatic diffractometer. J. Appl. Crystallogr. 49, 4411–4417.

Dawson, A., Allan, D.R., Parsons, S., Ruf, M., 2004. The use of CCD detectors in crystal structure determinations at high pressure. J. Appl. Crystallogr. 37, 410–416.

Denner, W., Schulz, H., D'Amour, H., 1978. A new measuring procedure for data collection with a high-pressure cell on an X-ray four-circle diffractometer. J. Appl. Crystallogr. 11, 260–264.

Dera, P., Katrusiak, A., 1999. Diffractometric crystal centering. J. Appl. Crystallogr. 32, 510–515.

Dera, P., Jayaraman, A., Prewitt, C.T., Gramsch, S.A., 2002. Structural basis for high-pressure polymorphism in $CuGeO_3$. Phys. Rev. B 65, 134105-1-10.

Downs, R.T., Zha, C.S., Duffy, T.S., Finger, L.W., 1996. The equation of state of forsterite to 17.2 GPa and effects of pressure media. Am. Mineral. 81, 51–55.

Fei, Y.W., Wang, Y.B., 2000. High-pressure and high-temperature powder diffraction. In: Hazen, R.M., Downs, R.T. (Eds), High-Temperature and High-Pressure Crystal Chemistry. Mineralogical Society of America, Washington, DC, pp. 521–557.

Finger, L.W., King, H., 1978. A revised method of operation of the single-crystal diamond cell and refinement of the structure of NaCl at 32 kbar. Am. Mineral. 63, 337–342.

Goncharenko, I.N., Mirebeau, I., Ochiai, A., 2000. Magnetic neutron diffraction under pressures up to 43 GPa. Study of the EuX and GdX compounds. Hyperfine Interact. 128, 225–244.

Hamilton, W.C., 1974a. Angle Settings for Four-Circle Diffractometers, International Tables for X-ray Crystallography. Kynoch Press, Birmingham, pp. 273–284.

Hamilton, W.C., 1974b. Normal probability plots. In: International Tables for X-ray Crystallography. Kynoch Press, Birmingham, pp. 293–294.

Hazen, R.M., 1976. Effects of temperature and pressure on the cell dimension and X-ray temperature factors of periclase. Am. Mineral. 61, 266–271.

Hazen, R.M., Downs, R.T., 2000. High-Pressure and High-Temperature Crystal Chemistry, Reviews in Mineralogy and Geochemistry. Mineralogical Society of America, Washington, DC.

Hazen, R.M., Finger, L.W., 1982. Comparative Crystal Chemistry. Wiley, New York.

Katrusiak, A., 2004. Shadowing and absorption corrections of high-pressure powder diffraction data: toward accurate electron density determinations. Acta Crystallogr. A 60, 409–417.

King, H., Finger, L.W., 1979. Diffracted beam crystal centering and its application to high-pressure crystallography. J. Appl. Crystallogr. 12, 374–378.

Koepke, J., Dieterich, W., Glinnemann, J., Schulz, H., 1985. Improved diamond anvil high-pressure cell for single-crystal work. Rev. Sci. Instrum. 56, 2119–2122.

Lager, G.A., Downs, R.T., Origlieri, M., Garoutte, R., 2002. High-pressure single-crystal X-ray diffraction study of katoite hydrogarnet: evidence for a phase transition from $Ia3d \rightarrow I\bar{4}3d$ symmetry at 5 GPa. Am. Mineral. 87, 642–647.

Li, Z., Ashbahs, H., 1997. Hydrostatic compression and crystal structure of pyrope to 33 GPa. Phys. Chem. Miner. 25, 301–307.

Li, Z., Ashbahs, H., 1998. New pressure domain in single-crystal X-ray diffraction using a sealed source. Rev. High Pressure Sci. Technol. 7, 145–147.

Lin, J.F., Shu, J., Mao, H.K., Hemley, R., Shen, G., 2003. Amorphous boron gasket in diamond anvil cell research. Rev. Sci. Instrum. 74, 4732–4736.

Loveday, J.S., McMahon, M.I., Nelmes, R.J., 1990. The effect of diffraction by the diamonds of a diamond-anvil cell on single-crystal sample intensities. J. Appl. Crystallogr. 23, 392–396.

Malinowski, M., 1987. A diamond-anvil high-pressure cell for X-ray diffraction on a single crystal. J. Appl. Crystallogr. 20, 379–382.

Marshall, W.G., Francis, D.J., 2002. Attainment of near-hydrostatic compression conditions using the Paris–Edinburgh cell. J. Appl. Crystallogr. 35, 122–125.

McMahon, M.I., 2004. High-pressure diffraction from good powders, poor powders and poor single crystals. In: Katrusiak, A., McMillan, P.F. (Eds), High-Pressure Crystallography. Kluwer, Dordrecht, pp. 1–20.

McMahon, M.I., Degtyareva, O., Hejny, C., Nelmes, R.J., 2003. New results on old problems: the use of single-crystals in high-pressure structural studies. High Pressure Res. 23, 289–299.

Miletich, R., Allan, D.R., Kuhs, W.F., 2000. High-pressure single-crystal techniques. In: Hazen, R.M., Downs, R.T. (Eds), High-Temperature and High-Pressure Crystal Chemistry. Mineralogical Society of America, Washington, DC, pp. 445–519.

Nelmes, R.J., McMahon, M.I., Loveday, J.S., Rekhi, S., 2002. Structure of Rb-III: novel modulated stacking structures in alkali metals. Phys. Rev. Lett. 88, 155503-1-4.

Nieh, Y.P., Raftery, J., Weisgerber, S., Habash, J., Schotte, F., Ursby, T., Wulff, M., Hädener, A., Campbell, J.W., Hao, Q., Helliwell, J.R., 1999. Accurate and highly complete synchrotron protein crystal Laue diffraction data using the ESRF CCD and the Daresbury Laue software. J. Synchrotron Radiat. 6, 995–1006.

Oxford Diffraction, 2003. Crysalis software. Wroclaw, Poland.

Prince, E., Collins, D.M., 1992. Refinement of structural parameters, 8.2: other refinement methods. In: Wilson, A. (Ed.), International Tables for X-ray Crystallography Volume C, International Union of Crystallography. Kluwer Academic Publishers, Dordrecht, pp. 605–608.

Reichmann, H.J., Angel, R.J., Spetzler, H., Bassett, W.A., 1998. Ultrasonic interferometry and X-ray measurements on MgO in a new diamond anvil cell. Am. Mineral. 83, 1357–1360.

Ren, Z., Bourgeois, D., Helliwell, J.R., Moffat, K., Srajer, V., Stoddard, B.L., 1999. Laue crystallography: coming of age. J. Synchrotron Radiat. 6, 891–917.

Rossmanith, E., Kumpat, G., Schulz, A., 1990. N-beam interactions examined with the help of the computer programs PSIINT and PSILAM. J. Appl. Crystallogr. 23, 99–104.

Santoro, A., Weir, C.E., Block, S., Piermarini, G.J., 1968. Absorption corrections in complex cases. Application to single crystal diffraction studies at high pressure. J. Appl. Crystallogr. 1, 101–107.

Von Dreele, R.B., Hanson, R.C., 1984. Structure of NH_3-III at 1.28 GPa and room temperature. Acta Crystallogr. C40, 1635–1638.

Wittlinger, J., 1997. Vergleichende Strukturelle Untersuchungen an $MgAl_2O_4$- und $ZnCr_2S_4$-Spinelleinkristallen unter Hochdruck. Ph.D. thesis, Ludwigs-Maximillian-Universität, Muenchen.

Wittlinger, J., Werner, S., Schulz, H., 1998. Pressure-induced order–disorder phase transition of spinel single crystals. Acta Crystallogr. B54, 714–721.

Wuensch, B.J., Prewitt, C.T., 1965. Corrections for X-ray absorption by a crystal of arbitrary shape. Z. Kristalogr. 122, 24–59.

Yamaguchi, M., Yagi, T., Hamaya, N., 1992. Brillouin-scattering study of pressure-induced phase-transition in MnF_2. J. Phys. Soc. Jpn 61, 3883–3886.

Yamanaka, T., Fukuda, T., Komatsu, Y., Sumiya, H., 2002. Charge density analysis of SiO_2 under pressures over 50 GPa using a new diamond anvil cell for single-crystal structure analysis. J. Phys. Condens. Matter 14, 10545–10551.

Zhao, Y., He, D.W., Qian, J., Pantea, C., Lokshin, K.A., Zhang, J., Daemen, L.L., 2005. Development of High *P–T* Neutron Diffraction at LANSCE. In: Chen, J., Wang, Y., Duffy, T.S., Shen, G., Dobrzhinetskaya, L.F.(Eds), Advances in High-Pressure Technology for Geophysical Applications, Elsevier, Amsterdam, pp. 461–474.

Zou, G.T., Ma, Y.Z., Mao, H.K., Hemley, R.J., Gramsch, S.A., 2001. A diamond gasket for the laser-heated diamond anvil cell. Rev. Sci. Instrum. 72, 1298–1301.

Advances in High-Pressure Technology for Geophysical Applications
Jiuhua Chen, Yanbin Wang, T.S. Duffy, Guoyin Shen, L.F. Dobrzhinetskaya, editors

Chapter 19

Nuclear resonant inelastic X-ray scattering and synchrotron Mössbauer spectroscopy with laser-heated diamond anvil cells

Jung-Fu Lin, Wolfgang Sturhahn, Jiyong Zhao, Guoyin Shen, Ho-kwang Mao and Russell J. Hemley

Abstract

The laser-heated diamond anvil cell (LHDAC) technique is a uniquely powerful method for generating the ultrahigh static pressures and temperatures ($P > 100$ GPa and $T > 3000$ K) found deep within planetary interiors. Here we show that the LHDAC technique can be used in conjunction with nuclear resonant inelastic X-ray scattering and synchrotron Mössbauer spectroscopy for studying magnetic, elastic, thermodynamic, and vibrational properties of ^{57}Fe-containing materials under high pressures and temperatures. A Nd:YLF laser, operating in continuous donut mode (TEM_{01}), has been used to heat a sample of ^{57}Fe-enriched hematite (Fe_2O_3) and iron from both sides of a diamond cell. Temperatures of the laser-heated sample are measured by means of spectroradiometry and the detailed balance principle of the energy spectra. The detailed balance principle applied to the inelastic X-ray scattering spectra provides absolute temperatures of the laser-heated sample. When the sample was heated evenly on both sides, these temperatures were in very good agreement with values determined from the thermal radiation spectra fitted to the Planck radiation function. Synchrotron Mössbauer spectra and partial phonon density of states (PDOS) of Fe_2O_3 have been obtained with these techniques up to 24 GPa and 1400 K, providing rich information for understanding physical properties of the sample under high pressures and temperatures. Time spectra of the synchrotron Mössbauer spectroscopy at 10 and 24 GPa upon laser heating reveal that Fe_2O_3 undergoes a magnetic-to-nonmagnetic transition at 900 ($\pm$100) K and 1000 ($\pm$100) K, respectively, from a room-temperature magnetic state to a high-temperature nonmagnetic state. The PDOS of Fe_2O_3 are shifted to lower energies at 1400 K, indicating the softening of the lattice vibrations at high temperatures. This study demonstrates a new arsenal of in situ probes to study magnetic, vibrational, elastic, and thermodynamic properties of ^{57}Fe-containing materials, such as metallic Fe, Fe alloys, and iron-bearing oxides and silicates [(Mg,Fe)O and (Mg,Fe)SiO_3], in the Earth's interior.

1. Introduction

Since the birth of the laser-heated diamond anvil cell (LHDAC) in the late 1960s (Ming and Bassett, 1974; Bassett, 2001), the LHDAC technique has been widely used with *in situ* X-ray diffraction, melting point studies, phase equilibrium studies, and chemical analyses of quenched samples (Boehler, 1986; Shen et al., 1998; Boehler and Chopelas, 1991; Lazor et al., 1993; Dewaele et al., 2000; Shen et al., 2001). *In situ* X-ray diffraction studies of materials deep in the Earth's mantle and core have advanced our understanding of the possible mineralogy and composition of the Earth's interior (Hemley et al., 2000).

Recently, the LHDAC technique has also been used with *in situ* high-pressure and high-temperature Raman spectroscopy to study vibrational properties of materials under these conditions (Lin et al., 2004a; Santoro et al., 2004). These studies have yielded data on the phase diagram, equation of state (EOS), elasticity, composition, and melting curve of planetary materials. However, important physical properties of planetary materials, such as magnetic, elastic, and thermodynamic properties under simultaneous high pressure and temperature conditions, remain to be explored with other techniques.

The nuclear resonant inelastic X-ray scattering (NRIXS) technique provides a direct probe of the phonon density of states (DOS) of a resonant isotope (Seto et al., 1995; Sturhahn et al, 1995, 1998; Lübbers et al., 1999a; Rüffer and Chumakov, 2000; Sturhahn, 2000; Alp et al., 2001; Sturhahn, 2004), e.g. using the 14.4125 keV transition of ^{57}Fe, whereas the synchrotron Mössbauer spectroscopy (SMS) technique probes the magnetic properties of a sample (Hastings et al., 1991; Lübbers et al., 1999b). The most suitable nuclear resonant isotope is ^{57}Fe, and iron-containing materials, such as hcp-Fe, magnesiowüstite [(Mg,Fe)O], and silicate perovskite [$(Mg,Fe)SiO_3$], are of tremendous interest in mineral physics. The NRIXS technique is truly unique for the study of lattice vibrations because the NRIXS signals only originate from particular nuclei with a complete isotope selectivity, and materials surrounding the sample that do not contain resonant nuclei produce no unwanted background, permitting detailed studies of iron-containing materials in a diamond anvil cell (DAC). For pure ^{57}Fe sample, one can obtain full phonon DOS and derive elastic, thermodynamic, and vibrational parameters fully contributed from the ^{57}Fe sample. For ^{57}Fe-containing materials, such as magnesiowüstite [(Mg,Fe)O] and silicate perovskite [$(Mg,Fe)SiO_3$], a partial DOS is obtained from the contribution of the Fe sublattice. Physical parameters derived from the partial DOS only represent the contribution of the Fe sublattice vibrations, and theoretical calculations may be needed to complement experimental results.

The NRIXS and SMS techniques have been used to study elastic, magnetic, thermodynamic, and vibrational properties of hexagonal close-packed (hcp) Fe at high pressures (Lübbers et al., 2000; Mao et al., 2001; Gieffers et al., 2002; Struzhkin et al., 2004), hcp-Fe to 29 GPa and 920 K by wire heating (Shen et al., 2004), Fe–Ni, Fe–Si, Fe–S, and Fe–H alloys (Lin et al., 2003; Lin et al., 2004b; Mao et al., 2004), iron valencies in $(Mg,Fe)SiO_3$ perovskite to 120 GPa (Jackson et al., 2005), and high-pressure and/or high-temperature study of magnetism (Lübbers et al., 1999b; Rupprecht et al., 2002; Pasternak et al., 2004), providing important information on the physical properties of Fe and its alloys, such as compressional wave velocity (V_P), shear wave velocity (V_S), and shear modulus (G) under high pressures. However, NRIXS and SMS studies under the pressure and temperature conditions of the Earth's interior remain to be explored, mainly due to the weak NRIXS signals, long data collecting time, and long-term laser heating stability. The typical collecting time for a meaningful NRIXS spectrum of a metallic ^{57}Fe in a DAC at the 3ID beamline of the Advanced Photon Source (APS), Argonne National Laboratory (ANL) is at least 4–6 h, suggesting that a LHDAC system that provides stable heating conditions on the order of hours needs to be built for such studies to be successful. Moreover, the use of a high-resolution monochromator with meV bandwidth and avalanche photodiode detectors (APD) requires extremely sensitive hutch-temperature control in the mK range, complicating the combination of the LHDAC and NRIXS techniques.

Here we describe the application of the LHDAC technique in NRIXS and SMS studies using hematite (Fe_2O_3) as an example. The SMS spectra and partial phonon density of states (PDOS) of hematite have been obtained with these techniques up to 24 GPa and 1400 K. Time spectra were evaluated with the CONUSS (coherent nuclear resonant scattering by single crystals) programs to permit deviation of magnetic hyperfine parameters, whereas elastic, thermodynamic, and vibrational parameters were derived by evaluating the energy spectra using the PHOENIX (phonon excitation by nuclear inelastic scattering of X-rays) programs (Sturhahn, 2000).

2. Laser-heating system set-up and sample configuration

We have built a double-sided laser-heating system at sector 3 of the APS, ANL for NRIXS and SMS studies of materials under high pressures and high temperatures (Figs 1 and 2). A Nd:YLF laser, operating in continuous donut mode (TEM_{01}), was used to heat a sample of ^{57}Fe-enriched hematite (Fe_2O_3, 96% enrichment or better) from both sides of a DAC (Shen et al., 2001). Flat diamonds with a culet of 400 μm were used to pre-indent a beryllium gasket to a thickness of 30 μm. A hole with a 150 μm diameter was drilled in the indented area and filled with a sandwich configuration of hematite sample and dried NaCl, for the thermal insulator and pressure medium, on both sides of the sample (Fig. 2). During the experiments, the diameter of the laser beam at the sample position was about 40 μm. Greybody temperatures were determined by fitting the thermal radiation spectrum between 670 and 830 nm to the Planck radiation function (Shen et al., 2001). We also measured temperatures of the laser-heated sample by the temperature-dependent intensity asymmetry of the energy spectra based on the detailed balance principle (Lin et al., 2004c; Shen et al., 2004). The detailed designs and experimental procedures of the laser-heating system are similar to a system installed at GSECARS of the APS (Shen et al., 2001; Lin et al., 2004c). A back-illuminated CCD enabled us to measure temperatures above 1000 K. Pressures were measured before and after laser heating from rubies placed in the NaCl medium using the ruby pressure scale (Mao et al., 1978).

3. NRIXS and SMS experiments

The NRIXS experiments were conducted using a high-resolution monochromator with 1 meV energy bandwidth. The high-energy resolution is essential for deriving accurate Debye sound velocities (V_D) from the DOS (the Debye parabola is best constrained at the lower energy limit) (Hu et al., 2003; Sturhahn, 2004). Energy spectra were obtained by tuning the X-ray energy (±70 meV) around the nuclear transition energy of 14.4125 keV and collecting the Fe K-fluorescence radiation that was emitted with time delay relative to the incident X-ray pulses. The fluorescence radiation was collected by three APDs, and the fourth APD along the beam was used to record synchrotron Mössbauer spectra in which the incident X-ray energy was fixed at 14.4125 keV (Figs 1 and 2) (Sturhahn, 2004). The diameter of the focused X-ray beam was 6–7 μm (FWHM) in both vertical and horizontal directions; the small beam size insured that the signal from the sample was only measured within the laser-heated spot of 40 μm (Fig. 2). The beam size was measured by scanning a

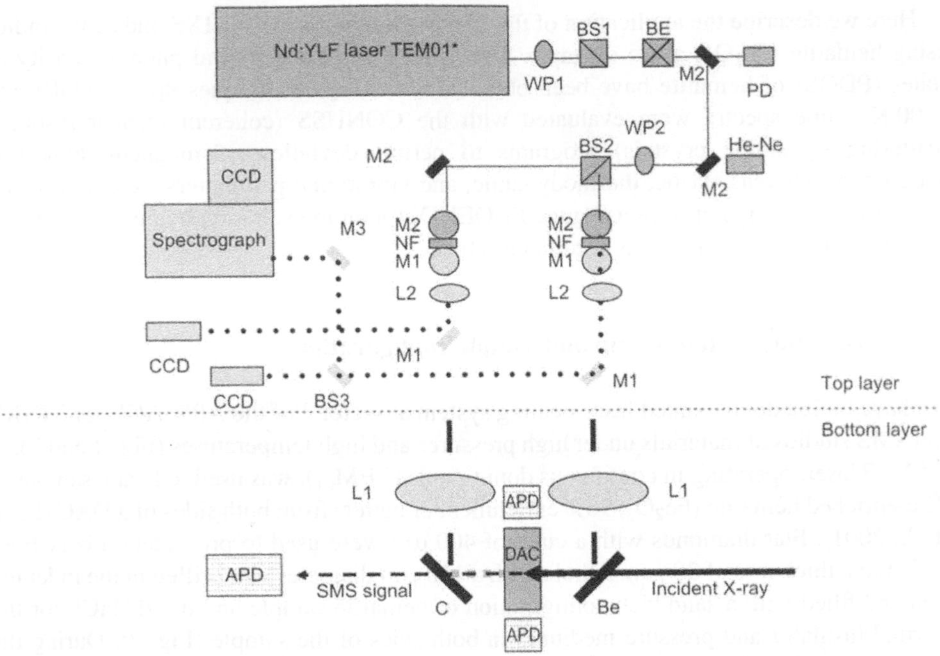

Figure 1. Schematic of the double-sided laser heating system combined with the NRIXS technique at sector 3 of the Advanced Photon Source. A continuous wave Nd:YLF laser beam in TEM_{01} mode (donut mode) with a maximum output power of 80 W was regulated by the wave plate (WP1) and polarized beam splitter (BS1). The laser beam was split into two beams by the second wave plate and polarized beam splitter (WP2 and BS2), which were also used to balance the laser power between the upstream and downstream sides. The laser beams were focused onto ~40 μm at the sample. An upstream beryllium mirror coated with gold and a downstream carbon mirror coated with silver were used to guide the laser beam and to reflect the thermal radiation. The wide-angle DAC was constantly cooled by the attached cooling plates insuring the stability of the cell temperature and the sample position during laser heating. The temperature of the DAC stays below than 310 K during data collection, which ensures mechanical stability. The front entrance of each APD detector was covered with a thin Be window to protect the device from light. The Fe K-fluorescence radiation was collected by three APDs surrounding the DAC. The fourth APD in transmission was used to record the SMS signal. Thin black line: YLF laser; thick black line: incident X-rays; dotted grey line: white light/thermal emission; dash/dotted line: laser and white light path. WP and BS: wave plate and polarized beamsplitter to regulate the Nd:YLF laser; BE: beam expander; M2: dichroic laser mirror reflects at least 99.5% of the YLF laser and transmits more than 90% of the visible light; PD: photodiode; He–Ne: He–Ne laser for alignment; M1: aluminum-coated mirror; NF: notch filter; L1: laser lens; L2: achromatic lens; BS3: 50/50 beam splitter.

Si edge (coated with Cr) across the X-ray beam in both the horizontal and vertical directions while collecting Cr K florescence with a Si(Li) detector. The counting time for each spectrum was 45 min, and between 10–15 spectra were collected at the same pressure and temperature conditions and added. At present a total collection time of up to 10 h is required to have good statistical accuracy of the energy spectra. In fact, continuously stable laser heating of the sample in a DAC for more than half a day has been achieved in this study. Elastic, thermodynamic, and vibrational parameters of the measured sample under high pressures were derived by evaluating the energy spectra

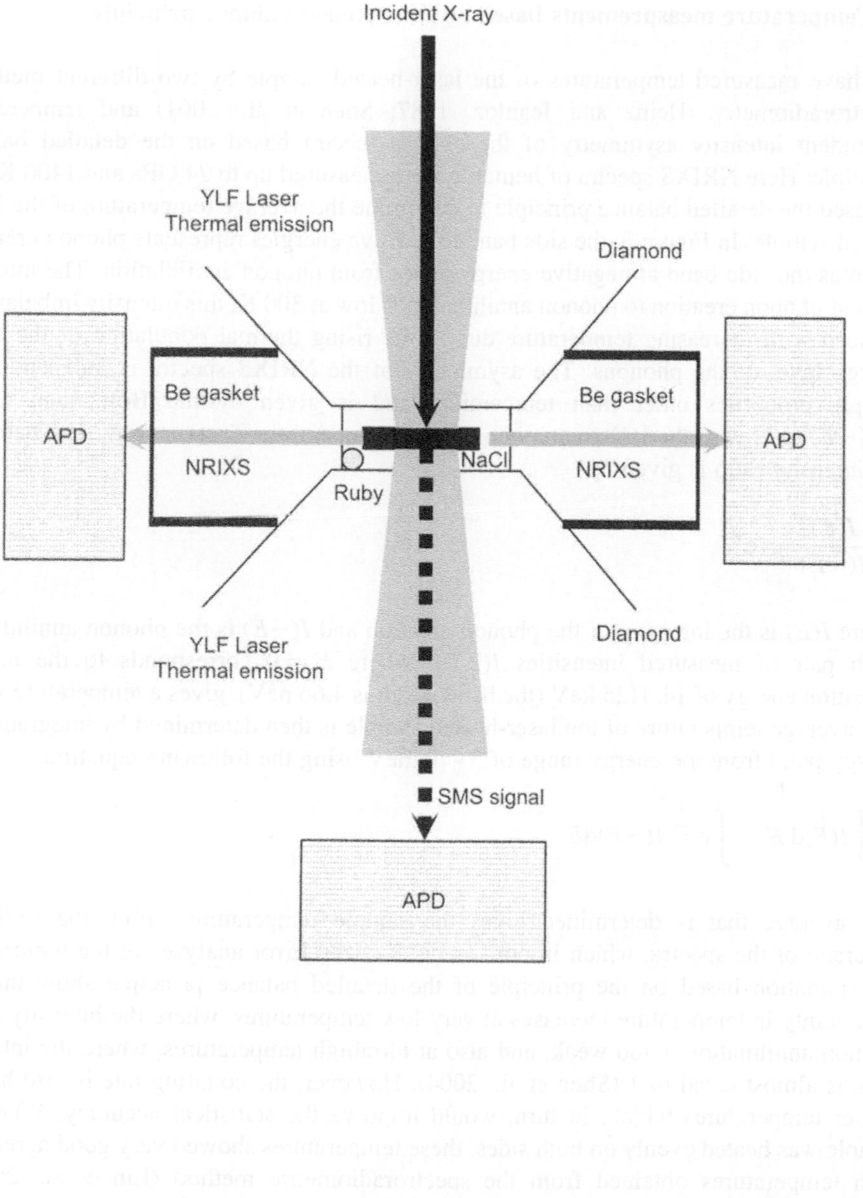

Incident X-ray
YLF Laser
Thermal emission
Diamond
Be gasket
Be gasket
APD
APD
NRIXS
NRIXS
NaCl
Ruby
Diamond
YLF Laser
Thermal emission
SMS signal
APD

4. Temperature measurements based on the detailed balance principle

We have measured temperatures of the laser-heated sample by two different methods: spectroradiometry (Heinz and Jeanloz, 1987; Shen et al., 2001) and temperature-dependent intensity asymmetry of the energy spectra based on the detailed balance principle. Here NRIXS spectra of hematite were measured up to 24 GPa and 1400 K, and we used the detailed balance principle to determine the average temperature of the laser-heated sample. In Figure 3, the side band at positive energies represents phonon creation, whereas the side band at negative energy arises from phonon annihilation. The intensity ratio of phonon creation to phonon annihilation is low at 300 K; this intensity imbalance is reduced with increasing temperature due to the rising thermal population of the upper energy level of the phonons. The asymmetry of the NRIXS spectra is independent of sample properties other than temperature and is given by the Boltzmann factor, $\exp[-E/k_B T]$ with the Boltzmann constant k_B, temperature T, and energy E. Therefore, the intensity ratio is given by

$$\frac{I(E)}{I(-E)} = e^{\frac{E}{k_B T}} \tag{1}$$

where $I(E)$ is the intensity of the phonon creation and $I(-E)$ is the phonon annihilation. Each pair of measured intensities $I(\pm E)$, where $E = 0$ corresponds to the nuclear transition energy of 14.4125 keV (the band width is 4.66 neV), gives a temperature value. The average temperature of the laser-heated sample is then determined by integrating all energy pairs from the energy range of 5–70 meV using the following equation:

$$\int I(E)\mathrm{d}E = \int e^{\frac{E}{k_B T}} I(-E)\mathrm{d}E \tag{2}$$

The average that is determined gives the sample temperature within the statistical accuracy of the spectra, which in our case is 5–10%. Error analyses of the temperature determination based on the principle of the detailed balance principle show that the uncertainty in temperature increases at very low temperatures, where the intensity of the phonon annihilation is too weak, and also at ultrahigh temperatures, where the intensity ratio is almost equal to 1 (Shen et al., 2004). However, the counting rate is also high at higher temperatures which, in turn, would improve the statistical accuracy. When the sample was heated evenly on both sides, these temperatures showed very good agreement with temperatures obtained from the spectroradiometric method (Lin et al., 2004c). The spectroradiometric method for temperature measurement in the LHDAC has been widely used to study planetary materials, but the accuracy of obtained temperature values has been debated (Heinz and Jeanloz, 1987). The absolute temperature determination using the energy spectra and the detailed balance principle independently confirm the validity of temperatures determined from the Planck radiation law in LHDAC experiments. However, temperature-dependent spectral emissivity should be taken into account in the spectroradiometric method at higher temperatures where the spectral emissivity may be affected by high temperature (Heinz and Jeanloz, 1987; Shen et al., 2001; Lin et al., 2004c).

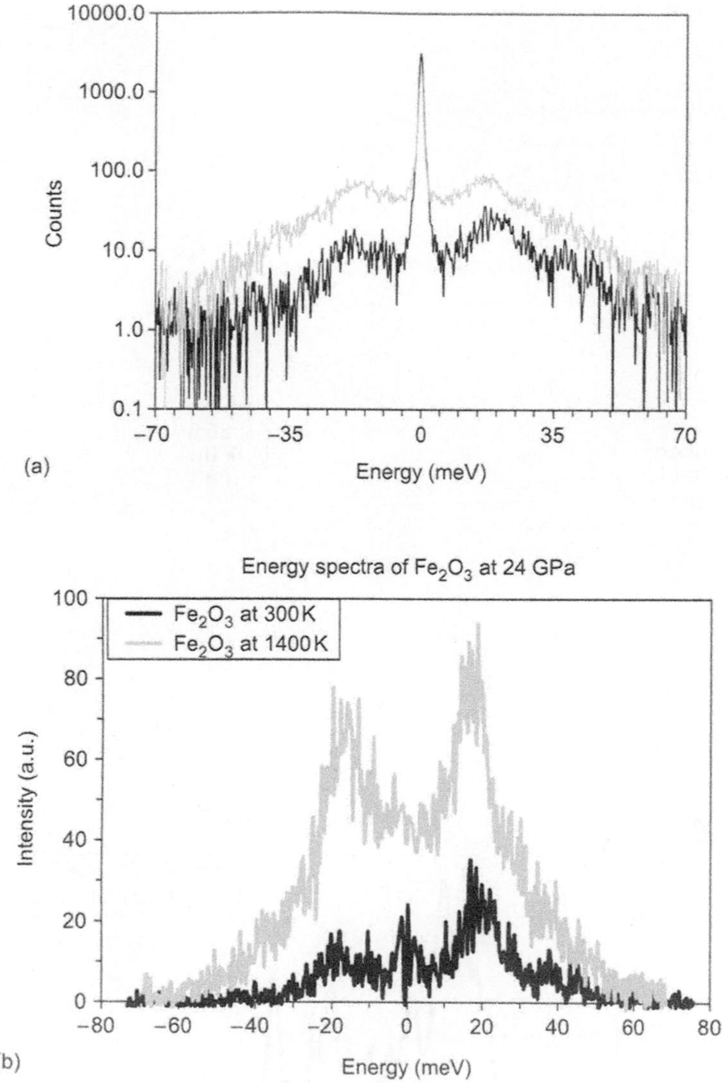
10000.0
1000.0
100.0
10.0
1.0
0.1
Counts
–70
–35
0
35
70
(a)
Energy (meV)
Energy spectra of Fe2O3 at 24 GPa
Fe2O3 at 300K
Fe2O3 at 1400K
100
80
60
40
20
0
Intensity (a.u.)
–80
–60
–40
–20
0
20
40
60
80
(b)
Energy (meV)

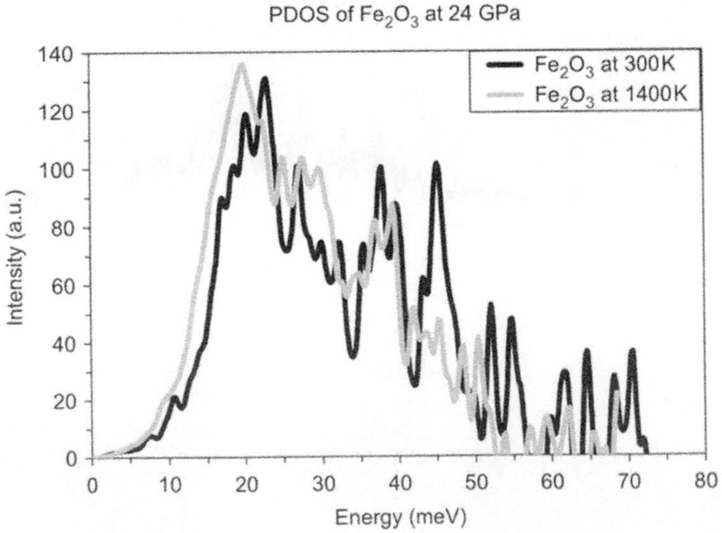

PDOS of Fe2O3 at 24 GPa
Fe2O3 at 300K
Fe2O3 at 1400K
Intensity (a.u.)
Energy (meV)
0
20
40
60
80
100
120
140
0
10
20
30
40
50
60
70
80

Since the thermal EOS of Fe_2O_3 is still unknown, we discuss here the general procedure of deriving the aggregate compressional wave velocity (V_P), the shear wave velocity (V_S), and the average shear modulus (G) from the Debye sound velocity (V_D) and the EOS. The value of V_D was derived from parabolic fitting of the low-energy slope of the PDOS in the range of 3.5–14 meV after applying a correction factor, the cube root of the ratio of the mass of the nuclear resonant isotope (^{57}Fe) to the average atomic mass of the sample (Hu et al., 2003). While the obtained partial DOS gives only part of the lattice dynamics of the material, the low-energy portion of the partial DOS provides the bulk V_D. The method of extracting the V_D from partial DOS, based on the quasi-harmonic approximation, has been tested for different ^{57}Fe-bearing materials with Debye-like low-frequency dynamics (Hu et al., 2003) and used to derive the sound velocities of Fe–Ni, Fe–Si, Fe–H, and Fe–S alloys under high pressures (Lin et al., 2003; Lin et al., 2004b; Mao et al., 2004). The vibrational kinetic energy (E_k), zero-point energy (E_z), vibrational heat capacity (C_{vib}), vibrational entropy (S_{vib}), and other thermodynamic and vibrational parameters were also calculated by integrating the partial DOS, which represent the contribution of the Fe sublattice (Table 1) (Sturhahn, 2000; Mao et al., 2001). These parameters are useful for testing theoretical calculations on Fe-containing materials under high pressures and high temperatures. In most applications, the resonant isotope (^{57}Fe) is only one of the several constituents of the materials under investigation, and the quantities derived from the integration of the partial DOS only represent the contribution of the resonant isotope, ^{57}Fe. These quantities may not be representative for the full thermodynamic behaviour of the sample, and theoretical calculations may be needed to complement experimental results.

The procedure for deriving aggregate V_P, V_S, and G from the V_D and EOS parameters, namely the adiabatic bulk modulus (K_S) and density (ρ), has been described previously (Mao et al., 2001; Lin et al., 2003; Lin et al., 2004b; Struzhkin et al., 2004). The K_S and ρ

Table 1. Elastic, vibrational, and thermodynamic parameters of Fe_2O_3 at 24 GPa.

T (K)	300	1400 (150)
T_{D1} (K)	520 (60)	440 (440)
T_{D2} (K)	453 (10)	404 (40)
V_D (THz × (vol)$^{1/3}$)	15.68 (0.64)	14.11 (0.03)
E_k (meV/atom)	14.36 (0.67)	60.6 (1.1)
D_{av} (N/m)	256 (28)	190 (13)
f_{LM}	0.8074 (0.0058)	0.3132 (0.0023)
$f_{TM,T=0}$	0.9275 (0.0016)	0.9191 (0.0006)
E_Z (meV/atom)	7.92 (0.52)	6.89 (0.21)
C_{vib} (k_B/atom)	2.593 (0.082)	2.985 (0.045)
S_{vib} (k_B/atom)	2.777 (0.064)	7.672 (0.085)

Lamb–Mössbauer factor, f_{LM}; kinetic energy, E_k; mean force constant, D_{av}; Lamb–Mössbauer factor at $T=0$, $f_{TM,T=0}$; kinetic energy at $T=0$, E_Z; vibrational specific heat, C_{vib}; vibrational entropy, S_{vib}; Boltzmann constant, k_B. T_{D1} is the Debye temperature calculated from the specific heat. The large error for the Debye temperature at 1400 K is typical because the specific heat approaches $3k_B$/atom and is very insensitive to the details of the PDOS. T_{D2} is the Debye temperature calculated from the Lamb–Mössbauer factor at 0 K. We note that these values only represent the contribution of the Fe sublattice. The Debye temperatures slightly decrease with increasing temperature.

of the material can be calculated using the Birch–Murnaghan EOS. The adiabatic bulk modulus at zero pressure (K_{0S}) is calculated as:

$$K_{0S} = K_{0T}(1 + \alpha\gamma T), \tag{3}$$

where K_{0T} is the isothermal bulk modulus at zero pressure, α is the thermal expansion coefficient, γ is the Grüneisen parameter, and T is temperature (300 K). The K_S, ρ, and V_D are used to solve for the aggregate V_P, V_S, and G by the following equations:

$$\frac{K_S}{\rho} = V_P^2 - \frac{4}{3}V_S^2 \tag{4}$$

$$\frac{G}{\rho} = V_S^2 \tag{5}$$

$$\frac{3}{V_D^3} = \frac{1}{V_P^3} + \frac{2}{V_S^3}. \tag{6}$$

As shown in the above equations, V_S is relatively insensitive to the differences in the EOS data. Therefore, the NRIXS technique is particular good at constraining V_S with a precise measurement of V_D. However, the use of the EOS parameters (K_S, ρ) introduces relatively high uncertainties in the aggregate V_P. Although previous NRIXS studies on the temperature dependence of the DOS of iron to 920 K did not show any evident deviation from the harmonic approximation (Chumakov et al., 1997; Shen et al., 2004), the potential deviation of the Debye-like behaviour of the partial DOS at higher temperatures would also affect the accuracy of deriving the sound velocities from the low-energy region of the partial DOS.

6. SMS spectra under high pressures and temperatures

In the SMS experiments, the time spectrum is a collection of events that reveal the time span between the arrival of the synchrotron radiation pulses and the arrival of scattered photons. The time spectra were recorded by an APD in the forward direction (Figs 1 and 2). Time spectra of Fe_2O_3 have been collected up to 24 GPa and 1400 K in steps of ~100 K (Fig. 5a–b). Time spectra at 10 and 24 GPa upon laser heating reveal that Fe_2O_3 undergoes a magnetic to nonmagnetic transition at 900 (±100) K and 1000 (±100) K, respectively, from a room-temperature magnetic state to a high-temperature nonmagnetic state (Fig. 5a, b). This magnetic transition is found to be reversible with temperature. Since the Néel temperature of hematite is 956 K at ambient pressure, our results show that the Néel temperature remains unchanged with the present accuracy up to 24 GPa. A magnetic-to-nonmagnetic transition in hematite has also been observed under high pressures at ~50 GPa and 300 K; although, the magnetic breakdown is reported to be connected with a first-order structural transition (Pasternak et al., 1999; Badro et al., 2002). The time spectra are evaluated with the CONUSS programs to permit derivation of magnetic hyperfine parameters (Sturhahn, 2000) (Fig. 6). The magnetic hyperfine field is 52.20 T (±0.02) at 24 GPa and 300 K, a typical value of the hyperfine field for ionic ferric oxide bonding, consistent with previous Mössbauer studies (Pasternak et al., 1999).

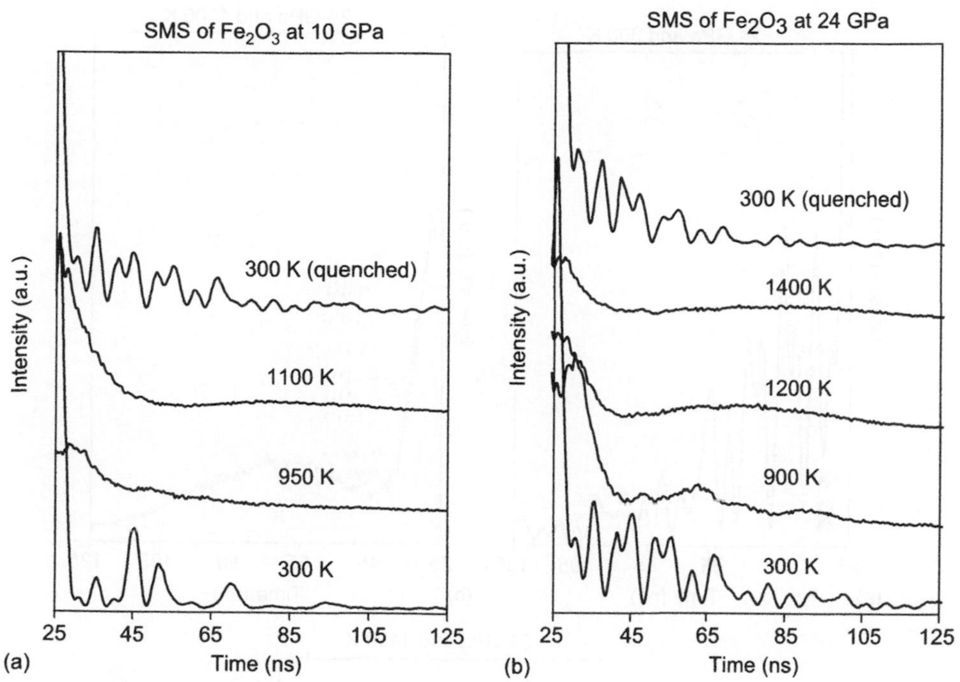

Figure 5. Representative time spectra of the SMS of Fe_2O_3 at 10 GPa (a) and 24 GPa (b) upon laser heating. Temperatures were determined from the energy spectra based on the detailed balance principle. The decay time is dictated by the nuclear lifetime of the ^{57}Fe nuclei in Fe_2O_3. Oscillations in the time spectra are observed that originate from the nuclear-level splitting (Sturhahn, 2004), whereas the flat feature indicates a nonmagnetic state. The spectra were collected at a temperature step of ~100 K from 300 to 1400 K.

At approximately 1100 K, a magnetic-to-nonmagnetic transition occurs; the magnetic component is highly reduced to approximately 5% at 1100 K. A pure quadrupole spectrum is observed at 1400 K. We note that the absolute temperature measurement is also very useful in the SMS study at temperatures below 1000 K, where the spectroradiometric method is limited due to the weak thermal emission. The sample temperature can be directly determined from the energy spectra, whereas the time spectra reveal magnetic ordering within the sample.

7. Conclusions

Here we have interfaced the LHDAC system with the NRIXS and SMS techniques to explore particular physical properties of ^{57}Fe-containing materials as they exist deep in the Earth's interior. The SMS spectra and the partial DOS of Fe_2O_3 have been measured with these techniques up to 24 GPa and 1400 K, providing rich information on the physical properties of the sample under high pressures and temperatures. The detailed balance principle applied to the NRIXS energy spectra provides absolute temperatures

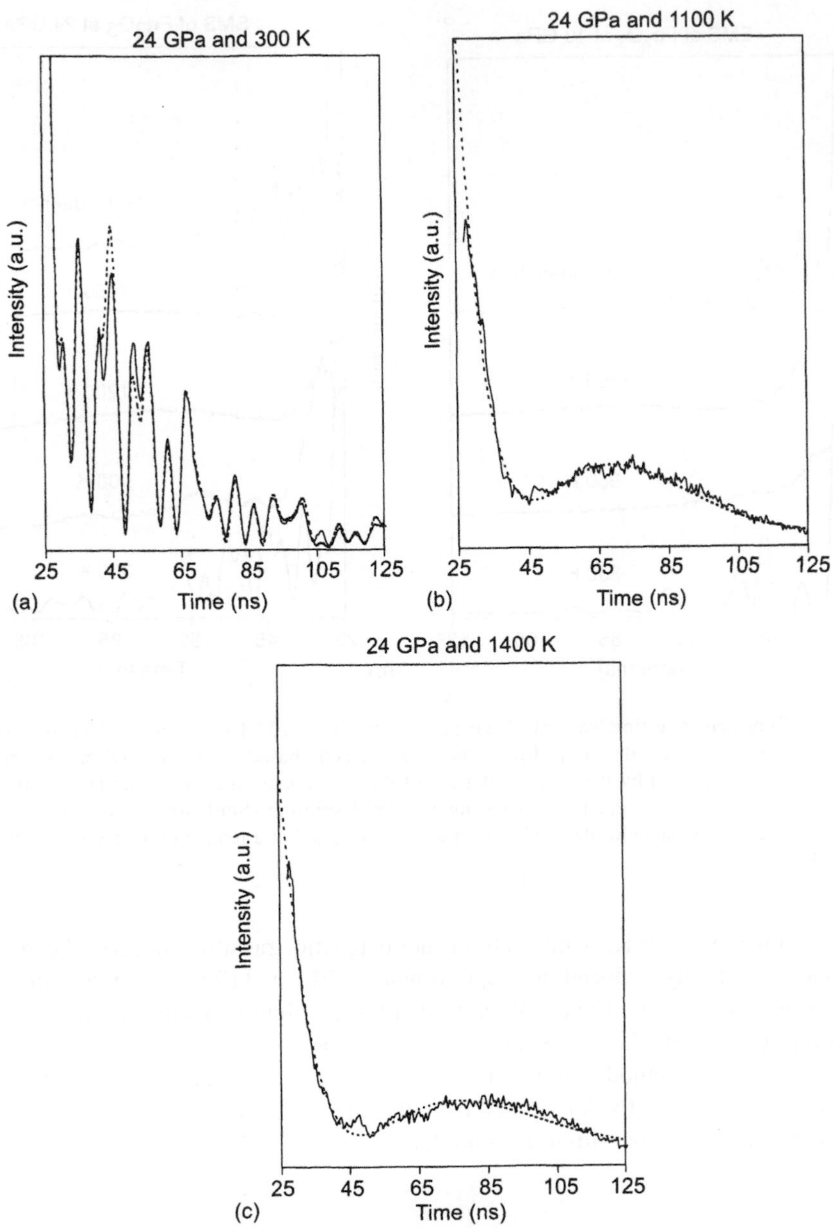

Figure 6. The SMS spectra at 300 K (a), 1100 K (b), and 1400 K (c) are evaluated by CONUSS programs to permit derivation of magnetic hyperfine parameters (Sturhahn, 2000). The magnetic hyperfine field is 52.20 T (±0.02) at 24 GPa and 300 K, a typical value of the hyperfine field for ionic ferric oxide bonding, which is also consistent with previous Mössbauer studies (Pasternak et al., 1999). At approximately 1100 K, a magnetic-to-nonmagnetic transition occurs; the magnetic component is highly reduced to approximately 5% at 1100 K. A pure quadrupole spectrum is observed at 1400 K. Solid curves: experimental time spectra; dashed curves: time spectra calculated from the CONUSS programs.

of the laser-heated sample. Time spectra of the SMS can be used to determine magnetic phase diagrams under high pressures and temperatures. The application of the LHDAC technique with NRIXS and SMS provides a new arsenal of probes to study the magnetic, vibrational, elastic, and thermodynamic properties of ^{57}Fe-containing materials, such as Fe, Fe alloys, (Mg,Fe)O, and (Mg,Fe)SiO_3 as they exist in the Earth's interior. The SMS technique is based on coherent elastic scattering from nuclear resonances in the forward direction. For a given experimental situation, the scattering intensity will decrease if the probability for recoilless absorption of the X-rays by the nuclei is reduced. The probability for recoilless absorption is also known as the Lamb–Mössbauer factor, and, more importantly, it vanishes for liquids and gases. In contrast to diffraction technique, the SMS signal only depends on the solid state of the sample but not on spatial order. Therefore, our techniques can also be used to measure the melting curves of iron-containing materials. These studies also provide useful parameters for testing theoretical calculations on Fe-containing materials under high pressures and temperatures.

Acknowledgements

This work and use of the APS are supported by US Department of Energy, Basic Energy Sciences, Office of Science under contract No. W-31-109-ENG-38, and the State of Illinois under HECA. We thank GSECARS, APS for the use of the ruby fluorescence system. We also thank E.E. Alp, D. Errandonea, S.-K. Lee, V. Struzhkin, M. Hu, D.L. Heinz, G. Steinle-Neumann, R. Cohen, J. Burke, V. Prakapenka, M. Rivers, S. Hardy, S. Jacobsen, and J.M. Jackson for their help and discussions. Work at Carnegie was supported by DOE/BES, DOE/NNSA (CDAC grant DE-FC03-03NA00144), NASA, NSF, and the W.M. Keck Foundation.

References

Alp, E.E., Sturhahn, W., Toellner, T.S., 2001. Lattice dynamics and inelastic nuclear resonant x-ray scattering. J. Phys.: Condens. Matter 13, 7645–7658.

Badro, J., Fiquet, G., Struzhkin, V.V., Somayazulu, M., Mao, H.K., Shen, G., Le Bihan, T., 2002. Nature of the high-pressure transition in Fe_2O_3 hematite. Phys. Rev. Lett. 89, 205504.

Bassett, W.A., 2001. The birth and development of laser heating in diamond anvil cells. Rev. Sci. Instrum. 72, 1270–1272.

Boehler, R., 1986. The phase diagram of iron to 430 kbar. Geophys. Res. Lett. 13, 1153–1156.

Boehler, R., Chopelas, A., 1991. A new approach to laser heating in high pressure mineral physics. Geophys. Res. Lett. 18, 1147–1150.

Chumakov, A.I., Rüffer, R., Baron, A.Q.R., Grünsteudel, H., Grünsteudel, H.F., 1997. Temperature dependence of nuclear inelastic absorption of synchrotron radiation in α-^{57}Fe. Phys. Rev. B 54, R9596–R9599.

Dewaele, A., Figuet, G., Andrault, D., Hausermann, D., 2000. P–V–T equation of state of periclase from synchrotron radiation measurements. J. Geophys. Res. 105, 2869–2877.

Gieffers, H., Lübbers, R., Rupprecht, K., Wortmann, G., Alfé, D., Chumakov, A.I., 2002. Phonon spectroscopy of oriented hcp iron. High Pressure Res. 22, 501–506.

Hastings, J.B., Siddons, D.P., van Bürck, U., Hollatz, R., Bergmann, U., 1991. Mössbauer spectroscopy using synchrotron radiation. Phys. Rev. Lett. 66, 770–773.

Heinz, D.L., Jeanloz, R., 1987. Temperature measurements in the laser-heated diamond cell. In: Manghnani, M.H., Syono, Y. (Eds), High-Pressure Research in Mineral Physics. Terra Scientific Publishing Company (TERRAPUB), Tokyo/American Geophysical Union, Washington, DC, pp. 113–127.

Hemley, R.J., Mao, H.K., Gramsch, S.A., 2000. Pressure-induced transformations in deep mantle and core minerals. Mineral. Mag. 64, 157–184.

Hu, M., Sturhahn, W., Toellner, T.S., Mannheim, P.D., Brown, D.E., Zhao, J., Alp, E.E., 2003. Measuring velocity of sound with nuclear resonant inelastic x-ray scattering. Phys. Rev. B 67, 094304.

Jackson, J.M., Sturhahn, W., Shen, G., Zhao, J., Hu, M.Y., Errandonea, D., Bass, J.D., Fei, Y., 2005. A synchrotron Mössbauer spectroscopy study of (Mg,Fe)SiO_3 perovskite up to 120 GPa. Am. Miner., 90, 199–205.

Lazor, P., Shen, G., Saxena, S.K., 1993. Laser-heated diamond anvil cell experiments at high pressure: melting curve of nickel up to 700 kbar. Phys. Chem. Miner. 20, 86–90.

Lin, J.F., Struzhkin, V.V., Sturhahn, W., Huang, E., Zhao, J., Hu, M.Y., Alp, E., Mao, H.K., Boctor, N., Hemley, R.J., 2003. Sound velocities of iron–nickel and iron–silicon alloys in the Earth's core. Geophys. Res. Lett. 30, 2112.

Lin, J.F., Santoro, M., Struzhkin, V.V., Mao, H.K., Hemley, R.J., 2004a. *In situ* high pressure-temperature Raman spectroscopy technique with laser-heated diamond anvil cells. Rev. Sci. Instrum. 75, 3302–3306.

Lin, J.F., Fei, Y., Sturhahn, W., Zhao, J., Mao, H.K., Hemley, R.J., 2004b. Magnetic transition and sound velocities of Fe_3S at high pressure. Earth Planet. Sci. Lett. 226, 33–40.

Lin, J.F., Sturhahn, W., Zhao, J., Shen, G., Mao, H.K., Hemley, R.J., 2004c. Absolute temperature measurement in a laser-heated diamond anvil cell. Geophys. Res. Lett. 31, L14611.

Lübbers, R., Wortmann, G., Grünsteudel, H.F., 1999a. High-pressure studies with nuclear scattering of synchrotron radiation. Hyperfine Interact. 123/124, 529–559.

Lübbers, R., Pleines, M., Hesse, H.J., Wortmann, G., Grünsteudel, H.F., Rüffer, R., Leupold, O., Zukrowski, J., 1999b. Magnetism under high pressure studied by ^{57}Fe and ^{151}Eu nuclear scattering of synchrotron radiation. Hyperfine Interact. 120/121, 49–58.

Lübbers, R., Grünsteudel, H.F., Chumakov, A.I., Wortmann, G., 2000. Density of phonon states in iron at high pressure. Science 287, 1250–1253.

Mao, H.K., Bell, P.M., Shaner, J.W., Steinberg, D.J., 1978. Specific volume measurements of Cu, Mo, Pd, and Ag and calibration of the ruby R_1 fluorescence pressure gauge from 0.06 to 1 Mbar. J. Appl. Phys. 49, 3276–3283.

Mao, H.K., Xu, J., Struzhkin, V.V., Shu, J., Hemley, R.J., Sturhahn, W., Hu, M.Y., Alp, E.E., Vocadlo, L., Alfé, D., Price, G.D., Gillan, M.J., Schwoerer-Böhning, M., Häusermann, D., End, P., Shen, G., Giefers, H., Lübbers, R., Wortmann, G., 2001. Phonon density of states of iron up to 153 gigapascals. Science 292, 914–916.

Mao, W.L., Sturhahn, W., Heinz, D.L., Mao, H.K., Shu, J., Hemley, R.J., 2004. Nuclear resonant x-ray scattering of iron hydride at high pressure. Geophys. Res. Lett. 31, L15618.

Ming, L.C., Bassett, W.A., 1974. Laser heating in the diamond anvil press up to 2000°C sustained and 3000°C pulsed at pressures up to 260 kilobars. Rev. Sci. Instrum. 45, 1115–1118.

Pasternak, M.P., Rozenberg, G.Kh., Machavariani, G.Yu., Naaman, O., Taylor, R.D., Jeanloz, R., 1999. Breakdown of the Mott–Hubbard state in Fe_2O_3: a first-order insulator-metal transition with collapse of magnetism at 50 GPa. Phys. Rev. Lett. 82, 4663–4666.

Pasternak, M.P., Rozenberg, G.Kh., Xu, W.M., Taylor, R.D., 2004. Effect of very high pressure on the magnetic state of transition metal compounds. High Pressure Res. 24, 33–43.

Rüffer, R., Chumakov, A.I., 2000. Nuclear inelastic scattering. Hyperfine Interact. 128, 255–272.

Rupprecht, K., Friedmann, T., Gieffers, H., Wortmann, G., Doyle, B., Zukrowski, J., 2002. High-pressure/high-temperature NFS study of magnetism in $LuFe_2$ and $ScFe_2$. High Pressure Res. 22, 189–194.

Santoro, M., Lin, J.F., Mao, H.K., Hemley, R.J., 2004. *In situ* high P–T Raman spectroscopy and laser heating of carbon dioxide. J. Chem. Phys. 121, 2780–2787.

Seto, M., Yoda, Y., Kikuta, S., Zhang, X.W., Ando, M., 1995. Observation of nuclear resonant scattering accompanied by phonon excitation using synchrotron radiation. Phys. Rev. Lett. 74, 3828–3831.

Shen, G., Mao, H.K., Hemley, R.J., Duffy, T.S., Rivers, M.L., 1998. Melting and crystal structure of iron at high pressures and temperatures. Geophys. Res. Lett. 25, 373–376.

Shen, G., Rivers, M.L., Wang, Y., Sutton, S.R., 2001. Laser heated diamond cell system at the Advanced Photon Source for *in situ* x-ray measurements at high pressure and temperature. Rev. Sci. Instrum. 72, 1273–1282.

Shen, G., Sturhahn, W., Alp, E.E., Zhao, J., Toellner, T.S., Prakapenka, V.B., Meng, Y., Mao, H.K., 2004. Phonon density of states in iron at high pressures and high temperatures. Phys. Chem. Miner. 31, 353–359.

Struzhkin, V.V., Mao, H.K., Mao, W.L., Hemley, R.J., Sturhahn, W., Alp, E.E., L'Abbe, C., Hu, M.Y., Errandonea, D., 2004. Phonon density of states and elastic properties of Fe-based materials under compression. Hyperfine Interact. 153, 3–15.

Sturhahn, W., Toellner, T.S., Alp, E.E., Zhang, X., Ando, M., Yoda, Y., Kikuta, S., Seto, M., Kimball, C.W., Dabrowski, B., 1995. Phonon density of states measured by inelastic nuclear resonant scattering. Phys. Rev. Lett. 74, 3832–3835.

Sturhahn, W., Alp, E.E., Toellner, T.S., Hession, P., Hu, M., Sutter, J., 1998. Introduction to nuclear resonant scattering with synchrotron radiation. Hyperfine Interact. 113, 47–58.

Sturhahn, W., 2000. CONUSS and PHOENIX: Evaluation of nuclear resonant scattering data. Hyperfine Interact. 125, 149–172.

Sturhahn, W., 2004. Nuclear resonant spectroscopy. J. Phys.: Condens. Matter 16, S497–S530.

Advances in High-Pressure Technology for Geophysical Applications
Jiuhua Chen, Yanbin Wang, T.S. Duffy, Guoyin Shen, L.F. Dobrzhinetskaya, editors

Chapter 20

In situ Raman spectroscopy with laser-heated diamond anvil cells

Mario Santoro, Jung-Fu Lin, Viktor V. Struzhkin, Ho-kwang Mao and Russell J. Hemley

Abstract

We have built a micro-optical spectroscopy system coupled with a Nd:YLF laser heating system for performing high pressure–temperature in situ Raman measurements in diamond anvil cells (DAC). A variety of materials can be investigated, providing information about structural and dynamical properties of condensed matter under extreme conditions. We report on a method for laser heating transparent samples using a metallic foil (Pt, Re, Mo, or W) as the infrared laser absorber (internal heating furnace) in the DAC. Metal foils of 5–15 μm in thickness with a small hole of 10–20 μm at the center are irradiated by the Nd:YLF laser beam directed into one side of the cell; the transparent sample in the small hole is uniformly heated and the Raman signals excited by an Ar^+ or Kr^+ laser are measured from the opposite side of the cell. The temperature of foil is measured by means of spectroradiometry, whereas the average temperature of sample is determined from the intensity ratios of Stokes/anti-Stokes pairs according to the principle of detailed balance. The average overall pairs give the sample temperature with the statistical accuracy of the Raman spectra, which is about ±50–100 K. Transparent samples such as CO_2 have been heated up to 1600 K and 65 GPa, indicating the high efficiency of the internal metal furnace method. In situ Raman spectroscopy in the laser-heated DAC represents a powerful technique to characterize high P–T properties of materials including dense planetary gases and ices.

1. Introduction

High-pressure conditions strongly modify atomic and molecular interactions in condensed matter. Moreover, the combination of high pressures and temperatures is of paramount importance for discovering possible new phases and for overcoming kinetic barriers that often hinder the high-pressure transformations. The laser-heated diamond anvil cell (LHDAC) technique has been a unique method to reach ultrahigh static pressure and temperature conditions, including those that prevail in deep planetary interiors, within $P > 100$ GPa and $T > 3000$ K (Heinz and Jeanloz, 1987; Gillet et al., 1998; Bassett, 2001; Shen et al., 2001). Since the introduction of LHDAC methods in late 1960s (see Bassett, 2001), the LHDAC technique has been widely used with *in situ* X-ray diffraction, melting studies, and chemical analyses of the quenched samples (Gillet et al., 1998; Shen et al., 2001). Equations of state and crystal structures of a variety of materials such as silicates, oxides, metals, and metal alloys have been examined. The technique has also been widely used to synthesize new materials. Recently, this method has been coupled with nuclear resonant inelastic X-ray scattering (NRIXS) and synchrotron Mössbauer spectroscopy (SMS) to study the sound velocities and magnetic hyperfine parameters of ^{57}Fe-enriched materials (Lin et al., 2004b, p. L14611). These studies are providing rich information essential for understanding the Earth's mantle and core.

The LHDAC technique has also been applied to study planetary molecular materials such as CH_4 and CO_2. These studies are often limited to examining room temperature quenched samples after having performed the laser heating runs (Benedetti et al., 1999; Iota et al., 1999; Tschauner et al., 2001), which makes the interpretation of the data difficult. Simple molecular systems such as H_2, CH_4, H_2O, NH_3, and CO_2 are important components of planetary interiors such as those of Jupiter, Saturn, and icy satellites. These species are also model systems for high-pressure physics and chemistry, and their *in situ* investigation under simultaneous high $P-T$ conditions is an important aim. However, *in situ* X-ray diffraction studies in a LHDAC are technically challenging for simple molecular systems, for they consist of light elements (H, C, O, N) which are weak X-ray scatters. On the other hand, Raman spectroscopy in a LHDAC is a powerful method for performing *in situ* HP–HT studies of these systems (Santoro et al., 2004), because they have high Raman cross-sections.

Dynamic compression methods (e.g. shock-wave) represent another approach for characterizing properties of planetary materials at ultrahigh pressures and temperatures (Holmes et al., 1985; Ahrens, 1987). However, there can be drawbacks with these techniques: temperature and pressure are coupled in a transient experiment and the employment of *in situ* techniques is often limited because of a short time scale. Nevertheless, shock-waves have been employed to derive a variety of different equation of state models and investigate molecular systems both in the fluid and solid state. For example, Raman spectra of fluid water have been measured up to 26 GPa and 1700 K in shock-wave experiments (Holmes et al., 1985). On the other hand, the externally heated DAC (i.e. with resistive heating) has been commonly used as the main static tool to study molecular systems *in situ* (Bassett et al., 1993; Bassett et al., 2000; Gregoryanz et al., 2003; Lin et al., 2004a), but the temperature is often limited to ~1300 K due to the lack of strength of diamonds and of the DAC itself. Recently, an internal resistive heater in a DAC has been used for performing experiments up to 10 GPa and 3000 K and *in situ* Raman spectra of BN and SiO_2 have been collected up to 1700 K (Zha and Bassett, 2003). Moreover, *in situ* Raman spectroscopy in a LHDAC using type IIa diamonds and CO_2 laser as the heating source has also been employed to study carbonates and calcium perovskite ($CaTiO_3$) (Boehler and Chopelas, 1991; Gillet et al., 1993; Gillet, 1996). The use of the CO_2 laser (10.64 μm) generally requires expensive type IIa diamonds (transparent to the laser wavelength) and special optics, which poses some practical limitations on the technique.

We have built an *in situ* Raman system to study materials in a LHDAC, using a Nd:YLF laser (1053 nm) as the heating source and an ion laser for excitation of the Raman signals. In fact, this system has a wide range of applications in studying materials, but here we emphasize the use of a small piece of metal foil (Re, Pt, W, or Mo of 5–20 μm in thickness) inserted in the sample chamber to effectively absorb the laser energy and transfer it to transparent samples (Chudinovskikh and Boehler, 2001). Temperatures of the laser-heated foil are determined from the spectrum of the thermal radiation fitted to the Planck's function (Heinz and Jeanloz, 1987; Shen et al., 2001) whereas the average temperature of the heated sample is determined from the temperature-dependent intensity asymmetry of the Stokes/anti-Stokes Raman spectra based on the principle of detailed balance (Long, 1977; Nemanich et al., 1984; Herchen and Cappelli, 1991; Cui et al., 1998; Lin et al., 2004b, p. L14611; Santoro et al., 2004). Application of the LHDAC technique in Raman spectroscopy provides a new arsenal to unveil vibrational properties, structures, chemical reactivity, and phase transformations of planetary materials *in situ* at high $P-T$ conditions.

2. System setup and sample preparation

Figure 1 shows a schematic diagram of the setup consisting of optics for Nd:YLF laser heating, for temperature measurement and imaging, and for Raman spectroscopy. A Nd:YLF laser, operating in continuous donut mode (TEM_{01}^*), is used to heat a metal foil from one side of the DAC (Fig. 2). The employment of the TEM_{01}^* mode profile leads to reducing the radial temperature gradients with respect to the sharp Gaussian TEM_{00} mode (Shen et al., 2001; Mao et al., 1998). The diameter of the focused laser beam on the metal foil is approximately 40 μm. For improved beam stability, we operate the YLF laser close to its maximum power (35 A, 80 W). The horizontally polarized laser beam passes thorough a wave plate (WP) and a polarized beamsplitter (BS1), and the actual incident power to the sample was regulated by rotating the wave plate (Shen et al., 2001). Dichroic laser mirrors (M1) reflect the laser beam and transmit the visible light; therefore, the thermal radiation from the laser-heated spot passes through one of the dichroic mirrors, get dispersed by a monochromator, and measured by a charge coupled device (CCD). The system spectral response is calibrated using a standard tungsten ribbon lamp with known radiance, and grey-body temperatures of the heated metal foil are determined by fitting the thermal radiation spectrum between 670 and 830 nm to the Planck radiation function (Heinz and Jeanloz, 1987; Shen et al., 2001). The system can be easily modified to a double-sided laser heating system so that one can heat a sample from both sides.

Raman spectra are measured from the other side of the DAC in a back-scattering geometry. An Ar^+ laser with a maximum output power of a few hundred mW (e.g. 457, 488, or 514 nm lines) was used as the Raman excitation source, and Raman signals were dispersed by the high throughput-single grating Jobin-Yvon HR460 monochromator and collected by a CCD detector (Santoro et al., 2004). The diameter of the focused laser beam was as small as a few microns; the small beam spot insured that the signal from the sample was only measured within the small holes of the metal foil (Figs 2 and 3) and unwanted thermal radiation background was significantly reduced by a spatial filter (SF). Blue and UV Ar^+ or Kr^+ laser lines are preferred in this type of experiment, in order to minimize the thermal radiation background. When the 514 nm green line of the Ar^+ laser was used, strong thermal radiation signals overwhelmed the Raman signals above ~1600 K, as the thermal emission intensity increases proportionally to T^4.

A rhenium gasket with an initial thickness of 250 μm was pre-indented to a thickness of 25–30 μm between two diamonds with a culet of 200–400 μm. A hole of 100–150 μm in diameter was drilled in the indented area. A metal foil (Pt, Re, Mo, or W) with size of 50–100 μm and thickness of 5–15 μm, having a small hole(s) of 10–20 μm at the center, was prepared by a pulsed Nd:YAG laser (Figs 2 and 3) and inserted into the sample

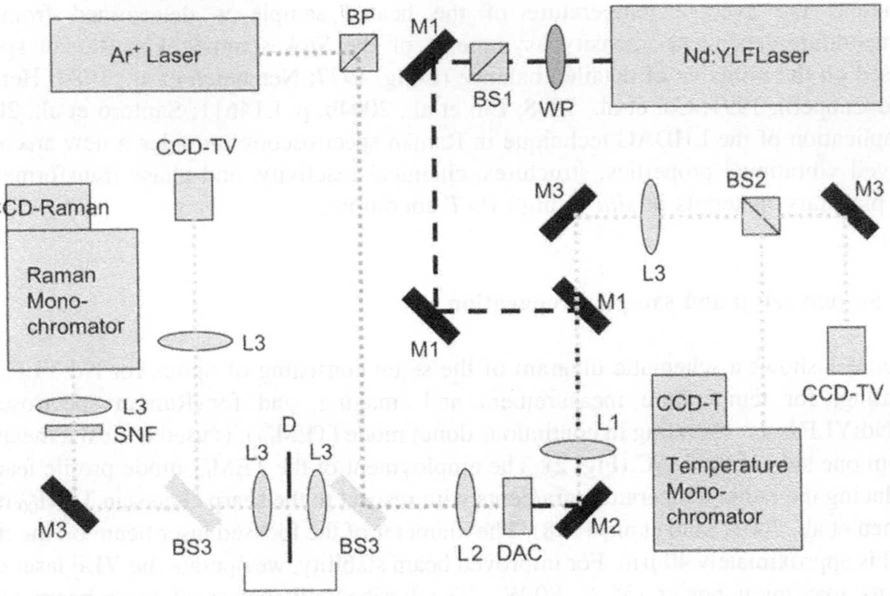

Ar+ Laser
BP
M1
BS1
WP
Nd:YLFLaser
CCD-TV
CCD-Raman
Raman Mono-chromator
M3
BS2
M3
L3
M1
M1
L3
CCD-TV
CCD-T
L3
SNF
D
L3
L3
L1
Temperature Mono-chromator
M3
BS3
BS3
L2
DAC
M2
SF

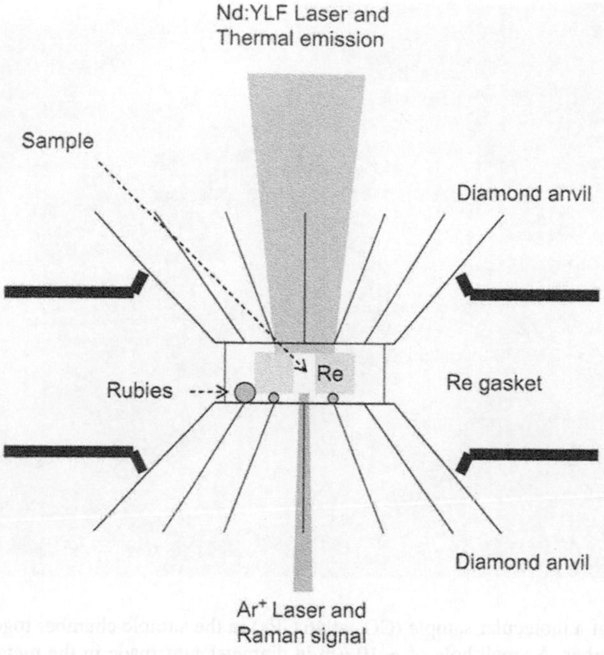
Nd:YLF Laser and
Thermal emission
Sample
Diamond anvil
Re
Rubies
Re gasket
Diamond anvil
Ar+ Laser and
Raman signal

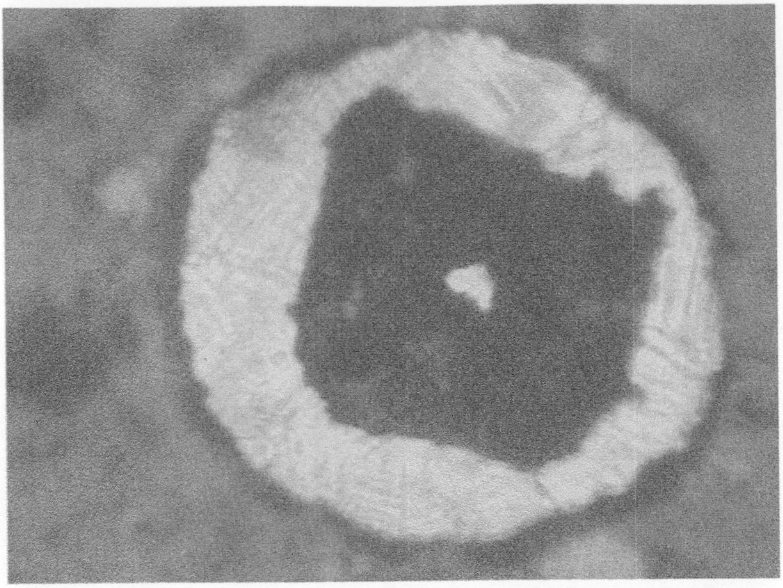

Figure 3. Photo of a molecular sample (CO_2 at 46 GPa) in the sample chamber together with a Pt metal foil as a laser absorber. A small hole of $\sim 10\ \mu m$ in diameter was made in the metal foil using a pulsed YAG laser (see Fig. 2). Ruby chips were used for the pressure measurement.

The temperature-dependent intensity asymmetry of the Raman spectra is thus used to independently determine the average temperature of the heated sample (T_R). This method relies on the principle of detailed balance and has been applied to determine temperatures of laser heated iron under pressure from the energy spectra of the NRIXS experiments (Lin et al., 2004b, p. L14611). The intensity asymmetry is independent of sample properties other than temperature and is driven by the Boltzmann factor: $\exp(-hc\Delta\nu/k_B T)$ with the Boltzmann constant k_B, temperature T, and transition energy $hc\Delta\nu$ (h: Plack constant; c: speed of light; $\Delta\nu$: transition energy wave number (absolute value)). Each Stokes/anti-Stokes excitation pair of measured intensities $I(\nu_0 \pm \Delta\nu)$ (ν_0: wave number of the exciting laser energy) gives a temperature value through the following equation:

$$\frac{I(\nu_0 + \Delta\nu)}{I(\nu_0 - \Delta\nu)} = \gamma\left(\frac{\nu_0 + \Delta\nu}{\nu_0 - \Delta\nu}\right)^4 \exp\left(\frac{-hc\Delta\nu}{k_B T}\right) = \gamma\beta \exp\left(\frac{-hc\Delta\nu}{k_B T}\right),$$

where γ is related to the spectral response of the system such as quantum efficiency of the CCD, optics, and monochromator. The factor $\gamma\beta$ was determined here for each Raman pair by measuring the intensity ratio of that pair at room temperature (298 K); the small temperature dependence the $\Delta\nu$ quantity can be neglected in such calibration procedure, because of the smooth frequency dependence of the $\gamma\beta$ factor. The average overall data pairs give the sample temperature within the statistical accuracy of the spectra, which is approximately $\pm 50-100$ K.

In Figure 4 we show representative high-temperature Raman spectra of carbon dioxide at two different pressures; the corresponding spectra at 300 K, collected after having

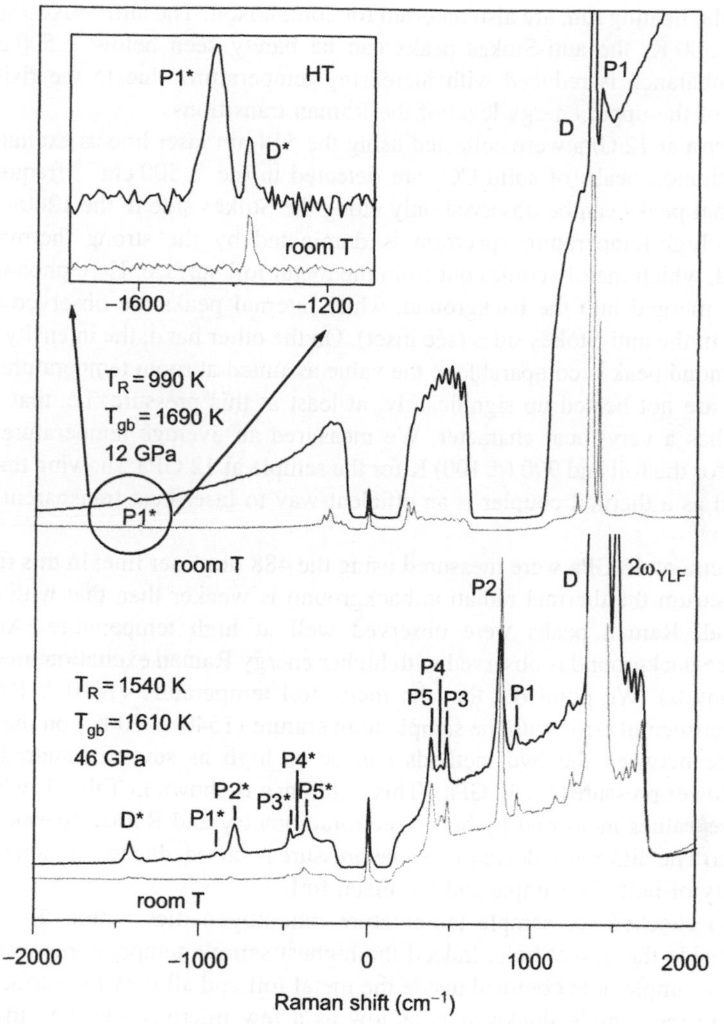

Figure 4. Representative Raman spectra of a CO_2 sample measured in a LHDAC at 12 and 46 GPa using the green (514 nm) and the blue (488 nm) laser lines, respectively; the increased intensities of the anti-Stokes peaks in the heated sample prove that it has been subjected to high temperatures. The intensity asymmetry of these spectra was used to determine the average sample temperature (see text). Thermal radiation background increases as the temperature is raised, and the background is higher at longer wavelengths at these temperatures. Black line: spectra at high temperatures; grey line: corresponding Raman spectra at 300 K; T_{gb} : grey body temperatures determined by means of spectral radiometry; T_R: temperatures determined from Raman peaks and the principle of detailed balance; $P_\#$ and $P^*_\#$: Stokes and anti-Stokes Raman peaks of CO_2, respectively (the labels of the peaks are for the purpose of excitation pairs only); D and D* Stokes and anti-Stokes diamond peaks respectively; and $2\omega_{YLF}$: second harmonic signal of the Nd:YLF laser, generated by the sample. Suppression of the optical signals are produced by the super notch laser filters. Inset: detail of the anti-Stokes side frequency region lying around the diamond peak for spectra recorded at 12 GPa.

performed the heating run, are also reported for comparison. The anti-Stokes/Stokes ratios are low at 300 K; the anti-Stokes peaks can be barely seen below $-500\ cm^{-1}$. This intensity imbalance is reduced with increasing temperature, due to the rising thermal population of the upper energy level of the Raman transitions.

The spectra at 12 GPa were collected using the 514 nm laser line as excitation source. At 300 K phonon peaks of solid CO_2 are detected in the $\pm 500\ cm^{-1}$ frequency range, while internal peaks can be observed only along the Stokes side in the $1200-1500\ cm^{-1}$ range. The high-temperature spectrum is dominated by the strong thermal radiation background, which mainly comes out from the metal foil surface. Here phonon peaks are completely merged into the background, while internal peaks are observed both in the Stokes and in the anti-Stokes sides (see inset). On the other hand, the intensity of the anti-Stokes diamond peak is comparable to the value assumed at room temperature, indicating that anvils are not heated up significantly, at least at this pressure; i.e. that the heating procedure has a very local character. We measured an average temperature of 1690 K ($\pm$ 100 K) for the foil and 990 ($\pm$ 100) K for the sample at 12 GPa, showing that the use of a metal foil as a thermal coupler is an efficient way to laser-heat transparent samples in the DAC.

The spectra at 46 GPa were measured using the 488 nm laser line; in this region of the visible spectrum the thermal radiation background is weaker than that with the 514 nm line, and all Raman peaks were observed well at high temperature. An increased fluorescence background is observed with higher energy Raman excitation lines (e.g. from diamond anvils). We point out that the metal foil temperature (1610 $\pm$ 100 K) agrees within experimental error with the sample temperature (1540 $\pm$ 70 K); on the other hand, a difference between the two methods can be as high as several hundred degrees at relatively lower pressures (<40 GPa). This difference is shown in Table 1, which lists the temperature values measured by both spectroradiometry and Raman thermometry at 12 and 46 GPa. The difference decreases as the pressure is raised, due to the increased thermal conductivity of both the sample and the metal foil.

We also checked for sample temperature inhomogeneities within the entire space enclosed inside the gasket hole. Indeed the highest sample temperatures were measured in the small sample hole confined inside the metal foil and all over the surface of the foil itself, where the sample thickness is as low as a few microns. On the other hand, the temperature drops by several hundreds of degree beyond the external edge of the foil. The temperature field was therefore relatively flat and well confined inside the metal foil area.

To the best of our knowledge, our study represents the first *in situ* Raman measurements in a LHDAC using the principle of detailed balance as an independent determination of the sample temperature, providing an accurate technique to quantitatively investigate vibrational properties, structures, chemical reactivity, and phase transformations of planetary materials. We have independently tested the validity of the method of temperature determination by collecting Raman spectra in an externally heated DAC using two thermocouples attached to the diamond surface as the temperature sensors up to ~1000 K (Bassett et al., 1993, 2000). The temperature difference between thermocouples and detailed balance method is a statistical variation ranging within about $\pm 50-100$ K, i.e. within the statistical errors affecting the determination of temperatures by Raman thermometry. The sensitivity of the intensity imbalance for low-frequency phonon peaks

Table 1. Comparison between temperature values measured by spectral radiometry, T_{gb}, and temperature-dependent intensity asymmetry of the Raman spectra of a CO_2 sample (see text and Fig. 4), T_R, at two different pressures.

$P = 12(\pm 1)$ GPa		$P = 46(\pm 2)$ GPa	
T_{gb} (K) ($\pm$ 100 K)	T_R (K) ($\pm$ 100 K)	T_{gb} (K) ($\pm$ 100 K)	T_R (K) ($\pm$ 70 K)
1200	760	1150	1000
1400	840	1320	1180
1540	960	1400	1300
1690	990	1430	1350
1790	1100	1500	1370
		1560	1530
		1600	1510
		1610	1540

A Pt foil was used in both cases as a laser absorber. The integration times employed to acquire both the Raman and the thermal emission spectra were typically around 5–10 s. Error bars on T_R were estimated combining the inaccuracies related to the actual spectral intensities and the intensity calibration procedure. T_{gb} was determined just before and after having collected each Raman spectrum, and related error bars are mainly due to the difference between the two measurements. At 12 GPa T_R was evaluated from the pair of peaks assigned to the molecular symmetric stretching mode of CO_2, positioned around ± 1416 cm^{-1}, which by far is the most intense spectral feature at this pressure. At 46 GPa several pairs of phonon lines were available to determine T_R. The results obtained from different pairs agree within the reported error bars. The agreement between T_R and T_{gb} increases as the pressure is stepped up, due to the increased thermal conductivity of both the sample and the metal absorber.

(<1000 cm^{-1}) is expected to be strongly reduced at very high temperatures ($\gtrsim 2000$ K); on the other hand, vibron peaks in molecular systems, which are located at higher frequencies, can be used to obtain the absolute temperature in this regime. However, the high frequency Stokes peaks are more susceptible to interference from thermal emission. The use of either UV Raman spectroscopy (Yashima et al., 1997; Fujimori et al., 2001) or pulsed Raman spectroscopy with gated detectors (Exarhos and Schaaf, 1991; Simon et al., 2003) in the LHDAC would then reduce the unwanted thermal emission background.

4. Conclusions

An *in situ* Raman system for LHDACs employing a Nd:YLF laser as the heating source has been developed to study physical and chemical properties of materials under high pressures and temperatures. The system has been tested in studies of molecular systems such as CO_2 up to 65 GPa and 1600 K. A metal foil is used as an internal heating furnace to effectively heat transparent samples to thousands of degrees. The temperature-dependent asymmetry of the Raman spectra is used to determine the average temperature of the sample. The *in situ* high *P–T* Raman technique in the LHDAC represents an important new method to characterize the structures, chemical reactions, and phase diagrams of materials under extremely high pressures and temperatures. The use of other wavelengths such as those from CO_2 lasers and UV sources for heating and Raman excitation, respectively, can be used to increase the temperature range. Moreover, a gated-CCD coupled with a pulsed laser as Raman apparatus could be employed to suppress the thermal radiation background.

Acknowledgements

We thank E. Gregoryanz and M. Wolf for their help in the study. We also thank G. Shen for the use of his software. M.S. acknowledges the European Union for supporting LENS under contract RII3-CT-2003-506350. V.V.S. acknowledges financial support from the Department of Energy under grant #DE-FG02-02ER45955. Work at Carnegie was supported by DOE/BES, DOE/NNSA (CDAC), NASA, NSF, and the W.M. Keck Foundation.

References

Ahrens, T.J., 1987. Shock wave techniques for geophysics and planetary physics. In: Sammis, C.G., Henyey, T.L. (Eds), Methods of Experimental Physics, Vol. 24, Part A. Academic Press, Orlando, FL, pp. 185–235.

Bassett, W.A., 2001. The birth and development of laser heating in diamond anvil cells. Rev. Sci. Instrum. 72, 1270–1272.

Bassett, W.A., Shen, A.H., Bucknum, M., Chou, I.M., 1993. A new diamond anvil cell for hydrothermal studies to 2.5 GPa and from − 190 to 1200°C. Rev. Sci. Instrum. 64, 2340–2345.

Bassett, W.A., Anderson, A.J., Mayanovic, R.A., Chou, I.M., 2000. Hydrothermal diamond anvil cell for XAFS studies of first-row transition elements in aqueous solution up to supercritical conditions. Chem. Geol. 167, 3–10.

Benedetti, L.R., Nguyen, J.H., Caldwell, W.A., Liu, H.J., Kruger, M., Jeanloz, R., 1999. Dissociation of CH_4 at high pressures and temperatures: diamond formation in giant planet interiors? Science 286, 100–102.

Boehler, R., Chopelas, A., 1991. A new approach to laser heating in high pressure mineral physics. Geophys. Res. Lett. 18, 1147–1150.

Chudinovskikh, L., Boehler, R., 2001. High-pressure polymorphs of olivine and the 660-km seismic discontinuity. Nature 411, 574–577.

Cui, J.B., Amtmann, K., Ristein, J., Ley, L., 1998. Noncontact temperature measurements of diamond by Raman scattering spectroscopy. Appl. Phys. 83, 7929–7933.

Exarhos, G.J., Schaaf, J.W., 1991. Raman scattering from boron nitride coatings at high temperatures. J. Appl. Phys. 69, 2543–2548.

Fujimori, H., Kakihana, M., Ioku, K., Goto, S., Yoshimura, M., 2001. Advantages of anti-Stokes Raman scattering for high-temperature measurements. Appl. Phys. Lett. 79, 937–939.

Gillet, P., 1996. Raman spectroscopy at pressure and high temperature. Phase transitions and thermodynamic properties of minerals. Phys. Chem. Miner. 23, 263–275.

Gillet, P., Fiquet, G., Daniel, I., Reynard, B., 1993. Raman spectroscopy at mantle pressure and temperature conditions: experimental set-up and the example of $CaTiO_3$ perovskite. Geophys. Res. Lett. 20, 1931–1934.

Gillet, P., Hemley, R.J., McMillan, P.F., 1998. Vibrational properties at high pressures and temperatures. In: Hemley, R.J. (Ed.), Ultrahigh-Pressure Mineralogy: Physics and Chemistry of the Earth's Deep Interior, Reviews of Mineralogy, Vol. 37. Mineralogical Society of America, Washington, DC, pp. 525–581.

Gregoryanz, E., Goncharov, A.F., Matsuishi, K., Mao, H.K., Hemley, R.J., 2003. Raman spectroscopy of hot dense hydrogen. Phys. Rev. Lett. 90, 175701.

Heinz, D.L., Jeanloz, R., 1987. Temperature measurements in the laser-heated diamond anvil cell. In: Manghnani, M.H., Syono, Y. (Eds), High-Pressure Research in Mineral Physics. TERRAPUB/American Geophysical Union, Tokyo/Washington, DC, pp. 113–127.

Herchen, H., Cappelli, M.A., 1991. First-order Raman spectrum of diamond at high temperatures. Phys. Rev. B 43, 11740–11744.

Holmes, N.C., Nellis, W.J., Graham, W.B., Walrafen, G.E., 1985. Spontaneous Raman scattering from shocked water. Phys. Rev. Lett. 55, 2433–2436.

Iota, V., Yoo, C.S., Cynn, H., 1999. Quartz like carbon dioxide: an optically nonlinear extended solid at high pressures and temperatures. Science 283, 1510–1513.

Lin, J.F., Militzer, B., Struzhkin, V.V., Gregoryanz, E., Mao, H.K., Hemley, R.J., 2004a. High pressure–temperature Raman measurements of H_2O melting to 22 GPa and 900 K. J. Chem. Phys. 121, 8423–8427.

Lin, J.F., Sturhahn, W., Zhao, J., Shen, G., Mao, H.K., Hemley, R.J., 2004b. Absolute temperature measurement in a laser-heated diamond anvil cell. Geophys. Res. Lett. 31, L14611.

Long, D.A., 1977. Raman Spectroscopy. McGraw-Hill, New York.

Mao, H.K., Shen, G., Hemley, R.J., Duffy, T.S., 1998. X-ray diffraction with a double hot-plate laser-heated diamond cell. In: Manghnani, M.H., Yagi, T. (Eds), Properties of Earth and Planetary Materials at High Pressure and Temperature, Geophysical Monograph, Vol. 101. American Geophysical Union, Washington, DC, pp. 27–34.

Nemanich, R.J., Biegelsen, D.K., Street, R.A., Fennel, L.E., 1984. Raman scattering from solid silicon at melting temperature. Phys. Rev. B 29, 6005–6007.

Santoro, M., Lin, J.F., Mao, H.K., Hemley, R.J., 2004. High pressure and temperature carbon dioxide transformations studied by in situ Raman spectroscopy and laser heating. J. Chem. Phys. 121, 2780–2787.

Shen, G., Rivers, M.L., Wang, Y., Sutton, S.R., 2001. Laser heated diamond cell system at the advanced photon source for in situ X-ray measurements at high pressure and temperature. Rev. Sci. Instrum. 72, 1273–1282.

Simon, P., Moulin, B., Buixaderas, E., Raimboux, N., Herault, E., Chazallon, B., Cattey, H., Magneron, N., Oswalt, J., Hocrelle, D., 2003. High temperatures and Raman scattering through pulsed spectroscopy and CCD detection. J. Raman Spectrosc. 34, 497–504.

Tschauner, O., Mao, H.K., Hemley, R.J., 2001. New transformations of CO_2 at high pressures and temperatures. Phys. Rev. Lett. 87, 075701.

Yashima, M., Kakihana, M., Shimidzu, R., Fujimori, H., Yoshimura, M., 1997. Ultraviolet 363.8-nm Raman spectroscopic system for in situ measurements at high temperatures. Appl. Spectrosc. 51, 1224–1228.

Zha, C.S., Bassett, W.A., 2003. Internal resistive heating in diamond anvil cell for in situ X-ray diffraction and Raman scattering. Rev. Sci. Instrum. 74, 1255–1262.

VI
Pressure calibration and generation

Advances in High-Pressure Technology for Geophysical Applications
Jiuhua Chen, Yanbin Wang, T.S. Duffy, Guoyin Shen, L.F. Dobrzhinetskaya, editors
Published by Elsevier B.V. (2005)

Chapter 21
Calibration based on a primary pressure scale in a multi-anvil device

Hans J. Mueller, Frank R. Schilling, Christian Lathe and Joern Lauterjung

Abstract

A key question to all high-pressure research arises from the reliability of pressure standards. There is some indication and discussion of an uncertainty of 10–20% for higher pressures in all standards. Independent and simultaneous investigation of the dynamical (ultrasonic interferometry of elastic wave velocities) and static (XRD-measurement of the pressure-induced volume decline) compressibility on a sample reveal the possibility of a standard-free pressure calibration and, consequently an absolute pressure measurement, because all required parameter are collected directly; no additional data, e.g. the volume dependence of the Grüneisen parameter etc. are needed. Ultrasonic interferometry is used to measure velocities of elastic compressional and shear waves in the multi-anvil high-pressure device MAX80 at HASYLAB Hamburg enables XRD, X-radiography, and ultrasonic experiments. Two of the six anvils were equipped with lithium niobate transducers of 33.3 MHz natural frequency. NaCl was used as pressure calibrant, using the equation of state (EoS) of [J. Appl. Phys. 42 (1971) 3239], and sample for ultrasonic interferometry at the same time. From the ultrasonic wave velocity data, v_p and v_s, we calculated the compressibility of NaCl as a function of pressure independent from NaCl-pressure calibrant. To derive the ultrasonic wave velocities from the interferometric frequencies of constructive and destructive interference requires precise in situ sample length measurements. For a NaCl-sample this is of particular importance, because the sample is the most ductile part of the whole set-up. We measured the sample length by XRD-scanning and by X-radiography. The compressibility results, derived from the ultrasonic data, were compared with data of static compression experiments up to 5 GPa [Phys. Rev. 57 (1940) 237] and up to 30 GPa [J. Geophys. Res. 91 (1986) 4949] using experimental data from [J. Phys. Chem. Solids 41 (1980) 517] and [Accurate Characterization of the High Pressure Environment]. At 1.2 and 5.3 GPa our velocity-derived compressibility data agree with the results of static compression. In the range between 2 and 4 GPa our dynamical data have 1.5–3% higher values. In general, the pressure revealed according to [J. Appl. Phys. 42 (1971) 3239] is in accordance to our standard-free pressure calibration. Consequently, up to 8 GPa the NaCl pressure standard has a reliability of at least 1%. However, there is some evidence that at higher pressures the inaccuracy of the NaCl standard seems to exceed 1%. Extrapolation of the compressibility data to higher pressures would also result in an increasing deviation, for EoS-fit and numerical fit of the density more than for the deformation fit.

1. Introduction

Multi-anvil devices are a very successful tool for experimental simulation of mantle conditions with relatively large samples. Accurate pressure determinations are critical to most high-pressure measurements. However, pressure calibration and the reliability of pressure standards are discussed controversially.

The formation and the development of gaskets between the anvils causes a deviation between load per anvil surface and pressure inside the set-up because of friction, material variation of the pressure transmitting medium, minor fit variation in the set-up, minor adjustment variation of the set-up and the anvils to each other, and different compressibility of the samples. Recent pressure determinations in a gas piston–cylinder apparatus successfully reduced the uncertainty to 0.2%, which is as low as that of free piston gages at 2.5 GPa (Getting, 1998). Therefore, *in situ* pressure measurements and precise standards are very important for this type of experiment.

Different options for pressure calibration exist

- using the known pressure of mineral reactions due to phase transitions, e.g. by measuring the change of electrical conductivity or using petrological experiments to determine mineral reactions. Several discrete measurements result in a pressure calibration curve (Luth, 1993),
- spectroscopic observation of a pressure-dependent absorption band or peak, e.g. ruby chip (Piermarini et al., 1975; Mao et al., 1986) (standard method for diamond anvil cells, not suitable for multi-anvil cells),
- continuous determination of the pressure-dependent unit cell size of a standard by X-ray diffraction, using the pressure marker's equation of state (EoS) (Decker, 1971; Chen et al., 2000).

The most common material to calibrate for conditions simulating the upper mantle is NaCl, following the EoS published by Decker (1971), recently revised by Brown (1999). At the time that Decker made his calculations, the EoS was based on first principles and therefore as independent as possible. Ruby fluorescence is a secondary pressure scale and is usually calibrated against NaCl at less high pressures. Progress in indirect pressure scale measurements has led to precision, which exceeds the accepted uncertainty of the practical pressure scale by a factor of as much as five. A new indirect pressure scale would become available from the over-determination of the EoS of a reference material by simultaneous X-ray and ultrasonic measurements (Ruoff et al., 1973; Yoneda et al., 1994; Getting, 1998; Zha et al., 1998, 2000; Bassett et al., 2000).

MAX80 is a single-stage multi-anvil apparatus (Yagi, 1988) equipped for ultrasonic interferometry (Mueller et al., 2002, 2003) and permanently located at HASYLAB, Hamburg for having access to synchrotron radiation for *in situ* XRD measurements. We present simultaneous XRD- and high-pressure ultrasonic interferometry measurements of compressional and shear wave velocities of polycrystalline NaCl to determine a standard-free pressure scale and to test the existing EoS by Decker (1971) and Brown (1999). *In situ* sample length measurement, necessary for high-precision ultrasonic interferometry, were performed by scanning both sample interfaces to the adjacent buffer and reflector and evaluating the XRD-spectra, as well as by X-radiography, i.e. taking X-ray shadow graphs of the set-up, recently installed at MAX80.

2. Techniques, methods, and materials description

2.1. Multi-anvil high-pressure apparatus MAX80

MAX80 (Fig. 1) is a single-stage multi-anvil high-pressure apparatus with six tungsten carbide anvils to compress a cubic sample volume of maximum $8 \times 8 \times 8$ mm^3. The anvils are driven by a 2500 *N* uniaxial hydraulic ram, the top and bottom anvil directly, the lateral anvils by two load frames and four reaction bolsters, see Figure 2. Three anvil-sets with different truncations exist – 6, 5, 3.5 mm. The maximum attainable pressures using 3.5 mm tungsten carbide anvils reach 12 GPa at 2000 K produced by an internal graphite heater. The 6 mm truncation limits the maximum pressure to approximately 7 GPa.

Diffraction patterns are recorded in an energy-dispersive mode (XRD) using white X-rays from the storage ring DORIS III at HASYLAB. MAX80 is equipped with a germanium solid-state detector analyzing the diffracted white beam at a fixed angle with a resolution of 135 eV for 6.3 keV and 450 eV for 122 keV. Using a double-crystal, fixed-offset monochromator with silicon (311) single crystals, calibrated in the wavelength range of 0.4–0.6 Å, and a 2048×2048 pixels CCD-camera angle-dispersive X-ray diffraction (not used in this study) is also available.

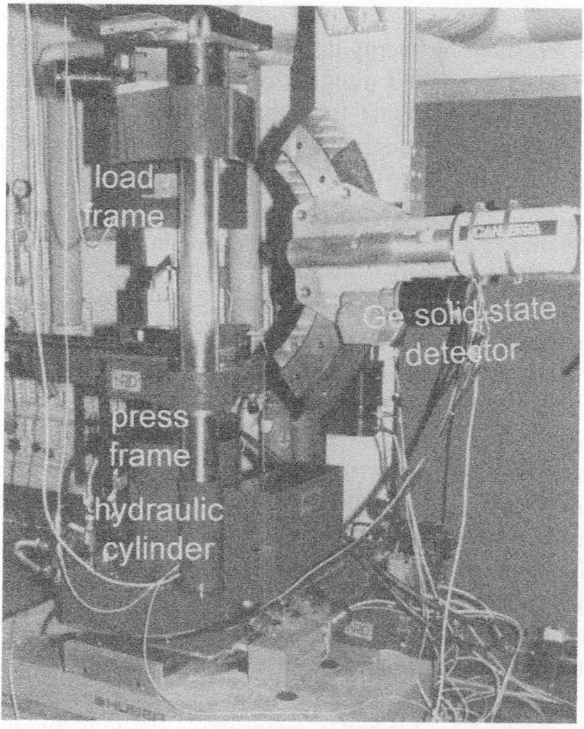

Figure 1. DIA-type multi-anvil apparatus MAX80 with Ge solid-state detector. The load frames are assembled at a 250 tons hydraulic ram. The Ge solid-state detector is also assembled at the press frame and follows the adjustment of the whole apparatus in relation to the X-ray beam.

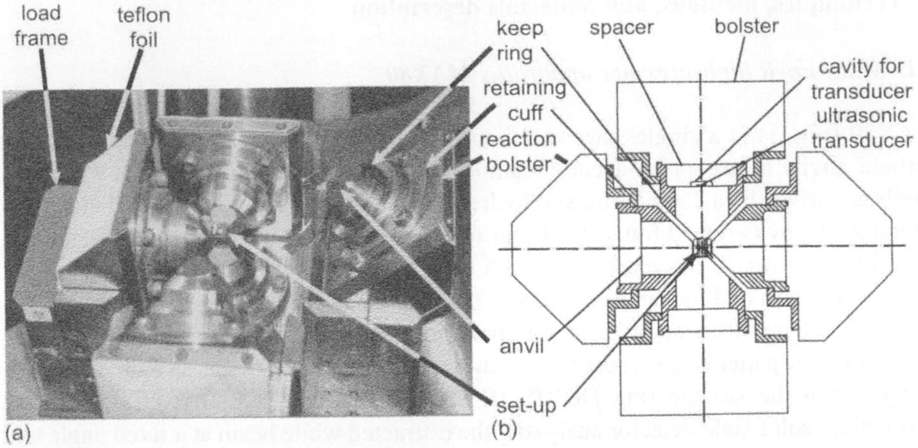

Figure 2. Load frame, anvils and sample arrangement of MAX80. (a) Apparatus opened for sample exchange and (b) vertical cross-section with transducer installation at the top anvil. The tungsten carbide anvils with a steel made keep ring are assembled to bolsters and reaction bolsters, respectively. Driven by the vertical movement of the hydraulic cylinder the load frame and the reaction bolsters generate the movement of the lateral anvils.

The pressure is measured by energy-dispersive XRD using the high-pressure EoS for NaCl (Decker, 1971). The method uses the observation of the elementary lattice cell compression of cubic NaCl crystals to derive the pressure *in situ*. These data are implemented in an in-house PC-program to calculate the resulting pressure at normal or given temperature. For details see Shimomura et al. (1985), Vaughan (1993), and Zinn et al. (1997).

In general, differential stress in the sample has the potential to affect the Decker scale. The first effect is that the volume change of NaCl might be overestimated. If the differential stress is greatest along the axis of their sample, then the added stress along this axis will also elastically shorten the sample resulting in a volume error that becomes interpreted as higher pressure. To estimate the value of differential stress, we performed a simple stress test by calculating the volume of the unit cell from 111 to 200 under high-pressure conditions. Generally, if the 111 suggests a smaller unit cell volume than the 200, this would indicate a tendency to underestimate the sample volume. For run 3.27 we found a quotient of the unit cell volumes V_{111}/V_{200} between +0.03 and +0.25%, i.e. any significant differential stress resulting in negative quotients was not found. The only indication for minor differential stress we noticed is the decrease of the 111 intensity compared to normal conditions. Because of the very low strength of NaCl differential stress seems to be much less important than for mineral samples. Therefore, NaCl is widely used as pressure transmitting medium, e.g. in piston–cylinder apparatus.

The high-pressure cell consists of a cube made of epoxy resin mixed with amorphous boron with the weight ratio 1:4 for better compressive strength containing the ultrasonic configuration, the heater, the pressure standard, and the thermocouple. Although the graphite heater was not necessary for the experiments presented here it was not removed from the set-up for 6 mm anvil truncation to use the standard ultrasonic configuration of

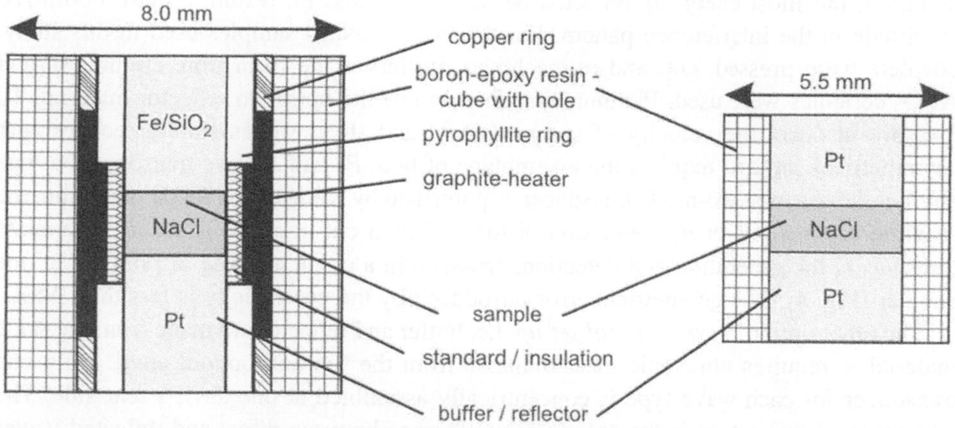

Figure 3. Ultrasonic high-pressure set-ups for anvil truncations of 6 and 3.5 mm. The smaller set-up was not equipped with a heater to keep the sample cross section as big as possible, because the strength of reflected ultrasonic waves is a function of the sample diameter.

MAX80 (Mueller et al., 2002). Removing the heater would result in pressure data not representative for the standard configuration. On the other hand, the 5 mm cube set-up was especially designed without heater to keep the sample surface bigger for stronger ultrasonic reflections, and to enhance the signal-to-noise-ratio. All interfaces between the sample and the close-fitting buffer rods/reflector bars are polished for optimal ultrasonic coupling (Fig. 3). Additional coupling media were not used. Copper rings contact the heater at the top and bottom anvils. The sample is surrounded by rings made from boron nitride or glass ceramics for electrical insulation and as a quasi-hydrostatic pressure transmitting medium. Further details of the apparatus are described by Mueller et al. (2005) on pages 67–94, this volume and also by Mueller et al. (2002).

2.2. *Ultrasonic interferometry*

Ultrasonic interferometry, using the interference between the incident and reflected waves inside the sample, was first described by McSkimin (1950). Piezoelectric transducers for the generation and detection of ultrasonic waves are cemented at the polished rear anvil's side outside the true pressure cell. One or two of the original MAX80 anvil spacers (see Figure 2) were replaced by redesigned parts for ultrasonic experiments. The new spacers have a cavity in their center to keep the ultrasonic transducer free of any stress. In principle, two types of ultrasonic set-ups were used in the presented experiments.

Asymmetrical set-ups are characterized by the optimization of buffers and reflectors, i.e. the buffer is made of a material resulting in intermediate acoustic impedance contrasts at both interfaces (anvil–buffer, buffer–sample) and the reflector material is selected for maximum reflection at the rear side of the sample. At ambient pressure, the reflection coefficient for the NaCl–Pt interface is 80%, the reflection coefficient for the Pt–TC interface only 20%. This means, that only a minor amount is reflected between anvil and platinum, but most energy is reflected between NaCl and Pt, resulting in an optimized amplitude in the interference pattern. For the massive NaCl samples used in this study, powders were pressed, cut, and re-machined. Buffers made from iron, aluminum, and Al_2O_3 ceramics were used. Platinum was found to be the optimum reflector material. To measure at once, the velocity of compressional and shear waves simultaneously with asymmetrical set-ups, requires the assemblage of both *P*- and *S*-wave transducers at one anvil or the use of a two-mode transducer as published by Kung et al. (2000). To ensure the maximum ultrasonic energy emission of the optimum cut transducers we used separate transducers for generation and detection, arranged in a circle as close as possible to one another (Fig. 4). The geometrical error introduced by the eccentricity is less than 0.5%.

The other option – *symmetrical set-up*, i.e. buffer and reflector are made from the same material – requires ultrasonic measurements from the top and bottom anvil. Only one transducer for each wave type is concentrically assembled at one anvil's rear side. The advantage of this set-up is the optimum interference between direct and reflected waves because the transducer receives the reflected and interfered waves without any angular loss. On the other hand, the symmetrical set-up results in additional energy losses due to non-optimum impedance contrasts between sample and buffer/reflector. For symmetrical set-ups we used platinum at both sides, which is an optimum reflector, but a poor buffer resulting in additional reflection losses, especially at the platinum–NaCl interface. In case of measurements at elevated temperatures, not performed in this study, only one of the "ultrasonic" anvils can be grounded. Even by using a dc-power supply small fluctuations of the current result in interference with the ultrasonic signals.

For generation and detection of the ultrasonic waves we used lithium niobate transducers, cold covered, overtone polished with a natural frequency of 33.3 MHz and a diameter of 5 mm. They were cemented at the polished rear anvil side using epoxy resin diluted by acetone to reduce its viscosity for minimizing the thickness of the glue film. This is of fundamental importance for the interferometric method to ensure rigid coupling to the anvil, because it requires a broadband characteristics of the transducer as a result of

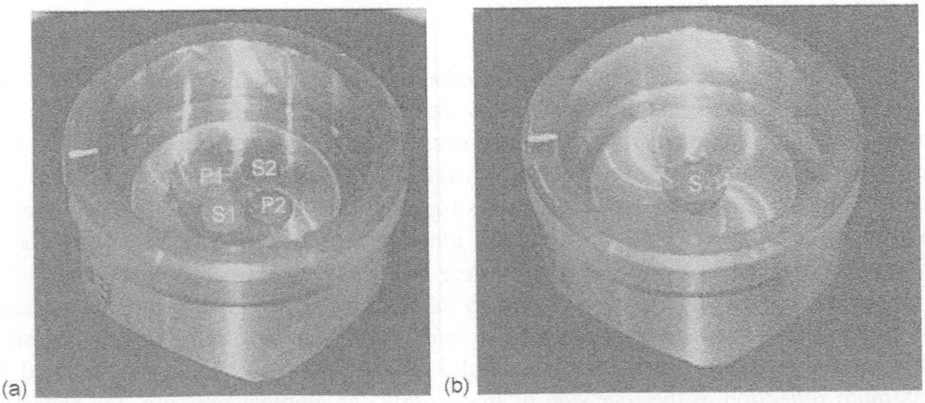

Figure 4. Transducer arrangements on the rear side of MAX80 anvils: (a) two transducer couples for asymmetrical set-ups; (b) single transducer for symmetrical set-up.

strong attenuation. The ultrasonic anvils are equipped with one *P*-wave or *S*-wave transducer, or with two couples of *P*-wave and *S*-wave transducers, respectively, depending on whether an asymmetrical set-up and two-transducer method, or a symmetrical set-up and single transducer method was used. First tests were performed with 3.5 mm truncation anvils made from cubic boron nitride (cBN).

Figure 5 shows the electronic equipment for ultrasonic interferometry at MAX80. A PC-program controls the frequency sweep of the rf-generator by a frequency step of 100 kHz. An arbitrary waveform generator cuts wavelets (Shen et al., 1998) or double wavelets (Li et al., 1998) with a duration of 20 ns to 4 μs from the continuous sinusoidal signal of the rf-generator. The ultrasonic generator delivers the master trigger pulse and amplifies the received signal. For single transducer configurations, i.e. the transducer acts sequentially as generator and receiver of ultrasonic waves, a directional bridge is used to prevent the strong excitation wavelet from hitting the sensitive input of the receiving amplifiers. A power amplifier and pre-amplifier are used for samples with high damping or strong reflection losses at the interfaces. The multi-channel oscilloscope displays and digitizes the interference signals, finally stored on the PC's hard drive. The evaluation using an in-house computer program includes the selection and copying of the critical signal ranges, i.e. the buffer and sample reflections, their subtraction to isolate the interference between the signals, digital filtering, displaying the resulting periodic energy levels (constructive and destructive interferences) as a function of frequency, and finally displaying the resulting travel-time curves as a function of frequency as well (Mueller et al., 2003). The determined two-way travel time or its multiple inside the sample is represented by the bold straight line between the curves of opposite curvature in Figure 6.

2.3. *Determination of sample length*

The result of ultrasonic interferometric measurements is an equidistant sequence of critical frequencies for constructive and destructive interference of the reflected waves from the plane-parallel surfaces of the sample rod. Unfortunately the interference pattern does not only depend on the material properties of the sample, but also on sample length. Due to the sharp interference pattern, the travel time is determined with high precision – better than 0.4% – and the accuracy of the velocity determination mainly depends on the precision of the length measurement (Li et al., 2001). *In situ* sample length measurement in multi-anvil devices is not trivial, as it cannot simply be derived from measurements of the advance of anvils. Therefore, sample deformation models, derived from direct length measurements prior and after the experiment (Knoche et al., 1997, 1998) are common usage, or it is assumed that the sample deforms purely elastically. The so-called Cook's method (Cook, 1957) calculates the *in situ* sample length from the compressibility derived from measured elastic wave velocities. Our measurements with different samples in a variety of configurations showed that this assumption is only valid, if the sample is the strongest part of the buffer–sample–reflector combination. Knoche et al. (1997, 1998) had a hot isostatically pressed forsterite sample between two platinum buffer rods. Consequently, the condition mentioned can be a good approximation for high-strength samples, as it was also the case in our experiments with San Carlos olivine, anorthite, clinoenstatite, and

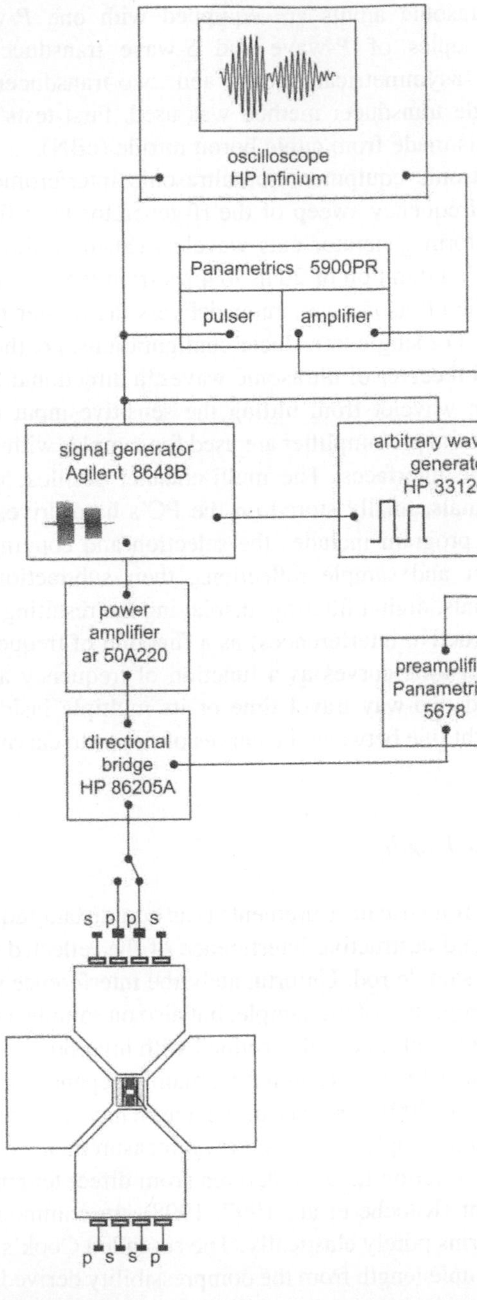

Figure 5. **Electronic equipment for ultrasonic interferometry at MAX80. Rectangular pulses made by an arbitrary waveform generator gate a signal generator resulting in rf pulses or double-pulses. A directional bridge prevents the power burst from hitting the sensitive pre-amplifier. The oscilloscope displays and digitizes the received ultrasonic signals. Amplifiers and an integrated trigger source (5900PR) were used. Transducers were installed on modified top and bottom anvils of MAX80. A computer-controlled switch selects the active transducer or transducer pair.**

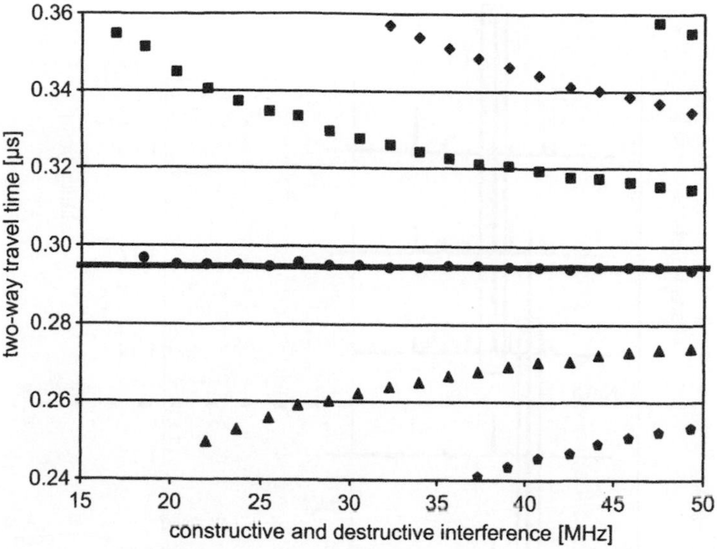

Figure 6. Typical resultant travel-time determination using interference pattern. Travel-time curves are plotted as a function of frequency at 7.71 GPa. Each point represents a frequency for constructive or destructive interference, and hence can be considered as an independent travel-time determination. The symbols fitted by the horizontal line represent the revealed travel time. The upper and lower curves represent neighboring fringes of interference pattern.

quartz (see Mueller et al., 2005, pages 67–94, this volume), but not if a ductile sample is taken into account.

2.3.1. XRD-scanning

Contrary to experiments with high-strength samples, a NaCl-specimen between Al_2O_3 ceramics or iron and platinum buffers is the most ductile part and will accommodate large parts of the total deformation (Mueller et al., 2003). Because sodium chloride deforms as a combination of ductile and elastic behavior, simple deformation models are not useful and measurements under *in situ* conditions are necessary. An advantage of ultrasonic measurements at a radiation source is, that the sample length can be determined independently from the ultrasonic experiments. The first option is to scan the buffer–sample–reflector combination stepwise, crossing both the interfaces and determine the sample length by evaluating the *in situ* XRD-spectra (Fig. 7). The circles represent the X-ray beam radius of about 50 μm. The XRD-spectra close to the interface are a superposition of two spectra, because the X-ray beam penetrates both materials, i.e. Pt and NaCl, to some degree. The whole press (including the multi-anvil device) can be lifted by stepper motors with an accuracy of 1 μm. By calculating the interface from the last and first pure spectrum the sample length can be determined much more precisely than the beam diameter is, that is, an accuracy of 5–10 μm. An advantage of this method is, it requires no additional equipment and results in sufficient accuracy. The drawback is, it is highly time-consuming, about 20 min using the lowest step rate.

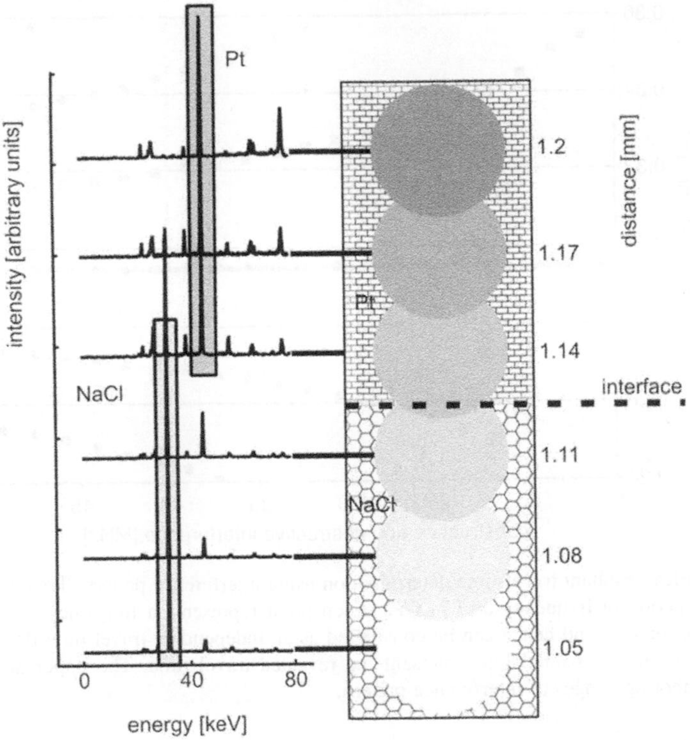

Figure 7. Determination of sample length under *in situ* conditions by XRD-scanning. The position of the interface is calculated as half the distance between the last and first appearance of pure XRD-spectra.

2.3.2. X-radiography

Li et al. (2001) used X-radiography to measure the sample length in multi-anvil devices under *in situ* conditions, after the method was established in the 1990s by other authors to observe under high-pressure conditions falling spheres in melts for viscosity measurements. At the latest when we introduced an ultrasonic data transfer function technique at MAX80 (see Mueller et al., 2005, pages 67–94, this volume) the XRD-scanning method was no longer adequate as the only available length measurement technique. A digital ultrasonic sweep for v_p and v_s lasts about 90 min. Consequently, a duration of about 20 min for a XRD-scan of both interfaces was acceptable. But if the recording of two data transfer functions representing the whole v_p- and v_s-data requires only some seconds, the length measurement becomes the limiting factor.

As the first step to establish a X-radiography system the fixed double-slits unit of MAX80 was exchanged by an adjustable slits system. We used a four-blade high-precision slits system of ADC (Fig. 8) equipped with four independent stepper motors including all the control electronics onboard. The maximum slits opening is 1 in. The motion repeatability is 1 μm with a motion resolution of 0.4 μm. The MS Windows compatible IMS terminal software allows to control the slits system simply by the PC, already

Figure 8. High-precision four-blade slits system.

installed for the ultrasonic measurements. Because the four blades can be moved independently from each other the slits system is able to define the X-ray beam position and size. Differently from the original state the X-ray beam position can be controlled by the slits and the positioning table of MAX80 now. For X-radiography the blades are opened so far, that the X-ray beam covers the whole sample length including a part of the adjacent buffer and reflector rods. Using tungsten carbide anvils absorbing the synchrotron radiation (intense X-rays) the maximum vertical opening of the beam is adapted to the maximum available gap between the lateral anvils, of about 1.5 mm at normal pressure and less than 0.5 mm at maximum conditions. To limit the scattered radiation inside the hutch, the slits is only opened to the size necessary for the sample length measurement.

First of all the X-radiography system (Fig. 9) consists of an 0.1 mm thick Ce:YAG-crystal (by courtesy of IKZ) of 15 mm diameter in an adjustable aluminum mounting. It partially converts the X-ray shadow graph after passing through the set-up by

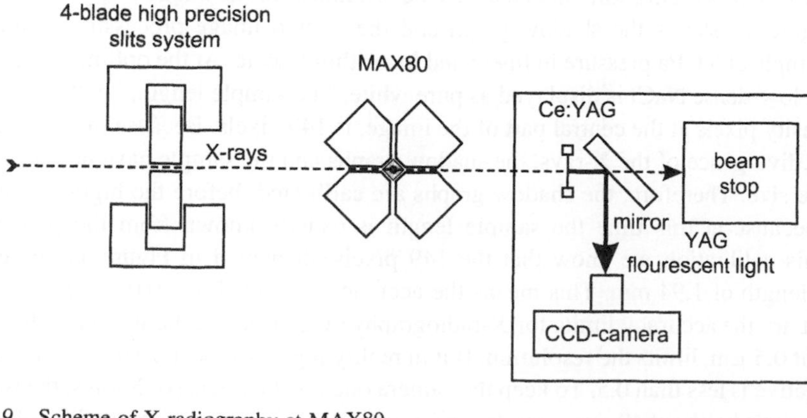

Figure 9. Scheme of X-radiography at MAX80.

Figure 10. X-radiography system without camera inside PB-shielding below the XRD-detector.

fluorescence to an optical image of about 540 nm wavelength (light green), which is redirected to a CCD-camera by an aluminum-coated mirror. A beam-stop behind the mirror absorbs the non-converted X-rays. The Ce:YAG-crystal should be made as thin as possible to limit the warming up by X-ray absorption and to keep the optical image as sharp as possible, because the fluorescence creates optical images at all atomic planes inside the crystal. Extensive use of aluminum for X-ray exposed components is recommended to limit the warming of the parts by absorption. The decoupling of the optical image from the X-ray shadow graph by the mirror is necessary to prevent the CCD-camera from direct X-ray flux. The whole system is covered by a 2.5 mm thick Pb-casing for shielding from scattered radiation inside the hutch (Fig. 10).

For taking images optimum for the following evaluation each shadow graph was recorded with three different exposure times, differing from each other by the fourfold exposure. The automatic exposure control failed because of the high-intensity contrasts of the images. The evaluation of the shadow graphs is performed by densitometry profiling, i.e. the image processing software analyzes the brightness of the image along a pre-defined line. Figure 11 shows the shadow graph and the related image processing result for a NaCl-sample at 5 GPa pressure in linear and logarithmic scale. At the optimum exposition time the low-dense NaCl is displayed as pure white. The sample length, i.e. the number of zero density pixels at the central part of the image, is 149 pixels. Because of the small, but existing divergence of the X-rays, the shadow graphs and the sample have not necessarily the same size. Therefore, the shadow graphs are calibrated, before the high-pressure run starts, because at this time the sample length is exactly known from the preparation. From this calibration we know that the 149 pixels, displayed in Figure 11 represent a sample length of 1.94 mm. This means the accuracy is 1 pixel, i.e. 0.013 mm.

What are the accuracy limits for X-radiography? On principle the wavelength of light, i.e. about 0.5 μm, limits the resolution. But in reality it gets worse, because the aperture of the objective is less than 0.5. To keep the camera outside the intensive X-rays, the working distance must be about 40 mm, very large, i.e. very disadvantageous for a micro objective.

(a)

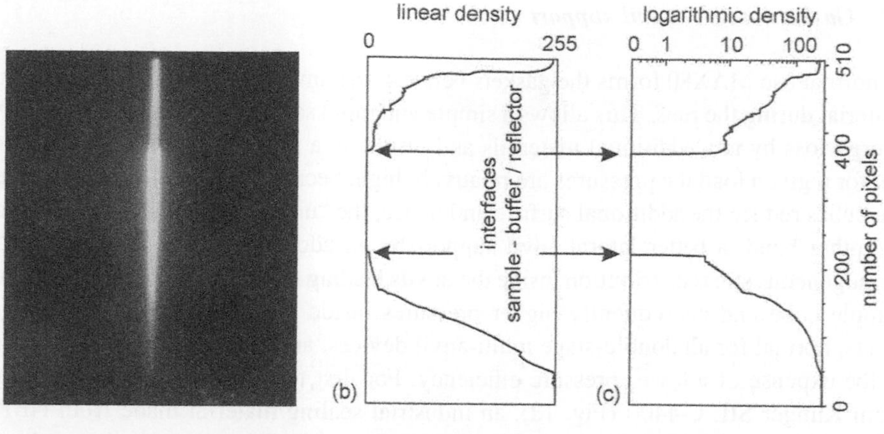

Figure 11. X-ray shadow graph (a) and its evaluation by density analysis in linear (b) and logarithmic scale (c).

This limits the practical accuracy to about 1 μm, which is a half order of magnitude better than X-ray scanning at the minimum. First results with a conventional consumer 5 megapixels color camera with a minimum working distance of 70 mm demonstrate the potential of the used set-up and confirm the results of XRD-scanning. Because the image processing only uses the density of the image, first the color image is converted to a gray scale one. Therefore, in the next experiments a 6 megapixels black and white CCD-camera will be used at a working distance of 40 mm to guarantee a 1 μm resolution.

2.4. Experimental procedures

Three experiments were included in the evaluation. Run 2.2 is one of the first ultrasonic experiments at all performed simultaneously with synchrotron XRD-maintenance in MAX80. Due to the similarity between the set-up of run 3.10 and 2.2, the sample deformation measured by synchrotron radiation during run 3.10 was used for both experiments. Set-up 2.2 had a buffer made of glass ceramics; set-up 3.10 had an iron buffer. Both asymmetrical set-ups had platinum reflectors. To make the pressure per load and deformation results comparable to other experiments the set-ups had a stepped graphite heater which was not in use during these experiments.

Run 3.27 used six cBN anvils with 3.5 mm truncation to increase the maximum pressure. Because cBN is an electrical insulator the rear side of the top and bottom anvil got a gold–platinum electrode for the transducers by sputtering. The top anvil was equipped with pairs of *p*- and *s*-wave transducers. In addition to that, the bottom anvil was equipped with a single *p*-wave transducer to compare the results of both configurations (see Figure 4). Due to electrical contact failure at the bottom piston only the symmetrical set-up with two platinum buffers could be used. The much smaller anvil truncation require boron-epoxy cubes of 5.5 mm length. To enlarge the reflection surface, i.e. to have a higher sample diameter a special set-up was designed without heater and insulator tube (Fig. 3). The experiment showed that the friction between anvil's surface and gaskets was much higher than using tungsten carbide anvils resulting in a maximum pressure of 7.71 GPa.

2.5. Gasket insets – anvil support

In normal use MAX80 forms the gaskets between the anvils from the boron epoxy cube's material during the runs. This allows a simple and rapid sample change, ensures low X-ray energy loss by any additional materials and results in a good "high-pressure efficiency", i.e. for a given load the pressures are relatively high, because the small gaskets formed by the cubes reduce the additional surface and hence, the "unproductive" part of the load. On the other hand, a better lateral anvil support by an additional gasket results in a more homogeneous stress distribution inside the anvils leading to a higher maximum force to the sample cube and consequently higher pressures inside the set-up. Prefabricated gasket insets, normal for all double-stage multi-anvil devices, are a way for lateral anvil support at the expense of a lower pressure efficiency. For first tests we used gasket strips made from Klinger SIL C-4400 (Fig. 12), an industrial sealing material made from NBR tied *p*-aramide fibers for tungsten carbide and cBN anvils. The post-experimental optical inspection of the tungsten carbide anvils showed that the material starts to flow at the corners of the front face without any failure of the anvil. Because the gap width between the anvils was larger at elevated pressures the X-ray intensity was higher and the adjustment of the ray was easier. In other words, due to the reduced pressure efficiency the maximum pressure could be enhanced and the XRD measurements could be improved. For cBN anvils the friction between the gasket material and the anvils was too high, resulting in a stick-slip behavior and material failure. Material and shape of the gasket insets will further be optimized for future experiments.

The first experiments with tungsten carbide anvils showed ≈25% higher maximum pressures compared to the standard MAX80 configuration because of increased lateral

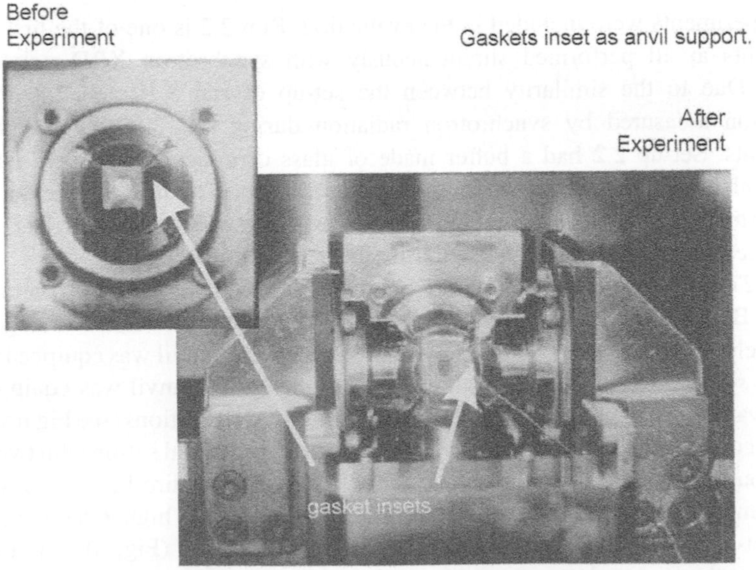

Figure 12. Prefabricated gasket insets prior and after the high-pressure run.

anvil support, reduced number of blow-outs, higher X-ray intensity, and a reduced probability of thermocouple cut-off during the experiments.

2.6. Samples

Sodium chloride powder of 99.5% purity (analytical grade by Merck) was used as starting material. The mean grain size was 50 μm. The powder was pressed to a sample cylinder of 10 mm diameter and a length of 20 mm using a load of 6 tons resulting in an effective pressure of 0.25–0.3 GPa. The millimeter sized samples (diameter 2.4 and 1.6 mm length for 6 mm anvil truncation, and diameter 3.1 and 1.1 mm length for 3.5 mm anvil truncation) for the high-pressure experiments were shaped with a high-precision (±0.5 μm) cylindrical grinding machine and polished at the plane-parallel faces of the sample rod.

3. Results and discussion

The digitized interferometric signals stored on a PC's hard drive were processed using an in-house program. The resulting sequence of maxima and minima represents the frequencies for constructive and destructive interference. Picking all available maxima and minima as a function of frequency ν allows the determination of the travel time τ inside the sample as the regression result for the horizontal point sequence between the curves of opposite curvature (Fig. 6). The curvature is the result of an inappropriate use of the order of interference n according to $\tau = n1/\nu$.

The calculation of wave velocities requires the sample length as a function of pressure. Consequently, the precise measurement of sample deformation during the experiment is essential for the accuracy of the whole method, because for higher degrees of deformation this contribution to the critical frequency interval can be higher than that of the variation of sample's elastic properties. Figure 13 is the plot of v_p and v_s for the three experimental runs. The results for our runs are in agreement with previous results published by Frankel et al. (1976) within the limit of experimental errors (~1.5%).

The velocities of run 3.27 are located between the values of the other two experiments and are used as average value. This run reaches the highest pressure and was used for further modeling.

The measured elastic wave velocities v_p (compressional wave) and v_s (shear wave) were used to calculate the adiabatic bulk modulus K_S and the corresponding compressibility κ_S.

$$K_S = \rho\left(v_p^2 - \frac{4}{3}v_s^2\right) \tag{1}$$

and

$$\kappa = \frac{1}{K_S} \tag{2}$$

This calculation requires the density ρ of the sample as a function of pressure which is directly obtained by XRD measurements. *In situ* sample length measurement is the

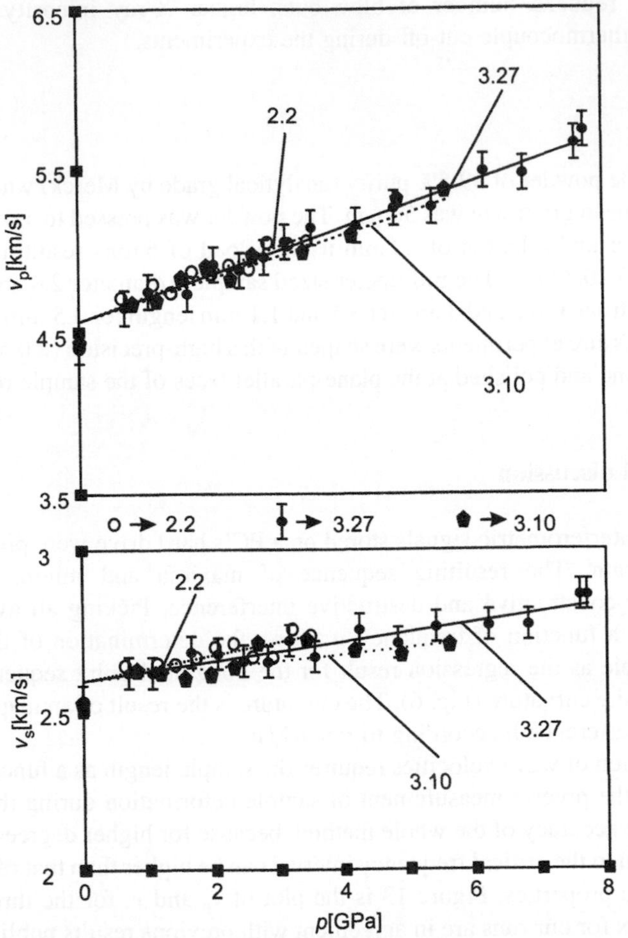

Figure 13. Elastic wave velocities v_p and v_s of polycrystalline NaCl at high pressure. Runs 2.2 and 3.10 use 8 mm set-ups for 6 mm anvil truncation; run 3.27 uses a 5.5 mm set-up for 3.5 mm anvil truncation.

important basis, but in addition to that the study of the lateral deformation is also necessary. That becomes the more important, the less hydrostatic the pressure and the more ductile the sample is. Therefore, a user of any multi-anvil device has to take care of set-up deformation. Different methods exist to meet the demands.

The general form of the EoS is:

$$P(V,T) = P_1(V) + P_{TH}(V,T) \tag{3}$$

where P_1 refers to the isothermal EoS, and P_{TH} refers to the thermal pressure. For small compressions, the isothermal bulk modulus K_T can be approximated by:

$$K_T = -V\left(\frac{\partial P}{\partial V}\right)_T = K_0 + PK_0' + \frac{P^2 K_0''}{2} + \cdots \tag{4}$$

Here K_0, K_0', and K_0'' are zero-pressure values of K and its first and second pressure P derivatives, at constant temperature. The first two terms usually suffice to represent ultrasonic measurements, but K_0'' appears to be negative and of a magnitude such that a quadratic in P leads to $K = 0$ (Birch, 1978). Therefore, only K_0 and K_0' are used. Using published data for K, K' and density at normal pressure ρ_0 (Birch, 1978, 1986; Holland and Ahrens, 1998) the density at given pressure can be calculated:

$$\frac{V}{V_0} = 1 - \kappa P \tag{5}$$

and

$$\rho_P = \frac{V_0}{V}\rho_0 \tag{6}$$

Another widely used approach is measuring and deriving the deformation of the sample from the ultrasonic experiment itself, called Cook's method (Cook, 1957; Kung et al., 2001a,b).

$$S = 1 + \frac{1 + \alpha\gamma T}{3h_0}\int_0^P \frac{\mathrm{d}P}{\left(\frac{1}{t_p^2} - \frac{4}{3}\frac{1}{t_s^2}\right)} \tag{7}$$

$$h_0 = 4\rho_0 l_0^2 \tag{8}$$

where S is linear compression, α linear thermal expansion coefficient, γ thermodynamical Grüneisen parameter, T absolute temperature, P pressure, ρ_0 density at zero-pressure, l_0 sample length at zero-pressure, t_p travel time of compressional waves along the sample, and t_s is travel time of shear waves along the sample.

But this is only valid for

$$\frac{\rho}{\rho_0} = \left(\frac{l_0}{l}\right)^3 = S^3 \tag{9}$$

which means the deformation is purely hydrostatic, i.e. uniform in all directions of space. However, our post-experimental examination of the set-up showed that this boundary condition is not achieved for our set-up and non-encapsulated NaCl-samples, because the sample is the most ductile part of the set-up. As a consequence of the gasket formation there is a reel-shaped deformation of the sample, i.e. the length decreases, the diameter at half the sample length slightly decreases or keeps constant, but the diameter at the front faces increases. Some minor parts of the material can be even squeezed out there. Therefore, we used a more generalized equation published by Frankel et al. (1976). For a material whose EoS is unknown, Katz and Ahrens (1963) showed that an EoS can be solved for by assuming that the geometry of the specimen changes under pressure such that

$$\rho = \rho_0 X^n \tag{10}$$

where

$$X = \frac{l_0}{l} \tag{11}$$

where X is geometry characteristics.

The parameter n is any positive number and is assumed to be independent of pressure. The change in the specimen density and thickness can be determined from the data as follows:

$$X^{n-2} = 1 + \left(\frac{n-2}{n}\right)\frac{1}{4l_0^2\rho_0}\int_0^P Y\,\mathrm{d}P \tag{12}$$

For $n \neq 2$, and

$$X = \exp\left[\frac{1}{8l_0^2\rho_0}\int_0^P Y\,\mathrm{d}P\right] \tag{13}$$

For $n = 2$, where

$$Y = \frac{1+\Delta}{\Delta f_\mathrm{p}^2 - \frac{4}{3}\Delta f_\mathrm{s}^2} \tag{14}$$

$$\Delta f_\mathrm{p} = \frac{v_\mathrm{p}}{2l} \tag{15}$$

$$\Delta f_\mathrm{s} = \frac{v_\mathrm{s}}{2l} \tag{16}$$

$$\Delta = \frac{9\alpha^2 T K_\mathrm{S}}{\rho C_\mathrm{p}} \tag{17}$$

where Δf_p is frequency interval between two critical frequencies for compressional waves, Δf_s frequency interval between two critical frequencies for shear waves, and C_p is specific heat at constant pressure.

If the forces acting upon a specimen are perfectly balanced, such as they are in a liquid pressure transmitting medium, the parameter n in Eq. (10) is equal to 3.0. All strains are due to hydrostatic stresses. The assumption of hydrostatic compression led Ahrens and Katz (1962) to use an expression as Cook's method identical to Eq. (12) with $n = 3$. If the deformation of the specimen is piston-like, i.e. the side walls are rigid and only the thickness changes, then the value of n is 1.0. If the sidewalls – as in our experiments – are rigid or more easily deformable than the buffers in axial direction, $n \geq 1.0$. We found $n = 0.622$ for run 3.27.

A third possibility to determine density as a function of pressure is an iterative numerical approach. The calculation of the adiabatic bulk modulus at the first pressure step starts with the assumption $\rho = \rho_0$, i.e. the density do not change within this small pressure-interval. The resulting compressibility is used to calculate the increased density at this pressure, which is used for the next calculation cycle. The result is very close to the data determined by using the EoS published by Ahrens and Katz (1962) and Birch (1986).

The *in situ* density evaluation was performed while using unit cell parameters of NaCl derived by XRD.

K_0 – and K_T' – values published by Birch (1986) were used to calculate the isothermal bulk modulus K_T and the corresponding compressibility κ_T using Eqs. (2) and (4). The V/V_0 values published by Bridgman (1940) (see also Fritz et al., 1971; Boehler and Kennedy, 1980) were also used to calculate the isothermal compressibility. Both values agree very well (Fig. 14). The difference between the adiabatic (K_S) and isothermal (K_T) bulk moduli is

$$K_S = K_T(1 + \alpha\gamma T) \tag{18}$$

and $\alpha\gamma T \approx 0.01$ at room temperature (Kung and Rigden, 1999) was taken into account.

The detailed comparison of the data showed minor differences. The ultrasonic curves cross the static compression graphs twice. At ≈ 1.2 GPa the compressibility graph derived from the EoS-fitted ultrasonic data intersects the static compression graph first. At ambient conditions the static compressibility is 7% higher than the dynamical compressibility derived from ultrasonic measurements. This seems to be the result of non-intrinsic compression, e.g. due to a closure of micro-cracks at the early compression stage in static compression experiments.

Between 2 and 4 GPa the graphs are nearly parallel with up to 3% higher compressibility derived from ultrasonic measurements. The high-pressure intersection is located at 5.3 GPa. At higher pressures the difference seems to increase. At our maximum pressure of 7.71 GPa the static compressibility is again 6.6% higher than the presented value. This may lead to significant errors for the pressure standard at higher pressure.

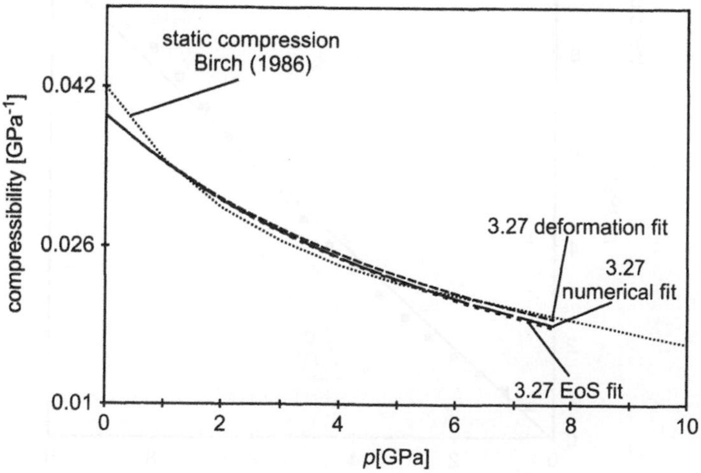

Figure 14. Compressibility of NaCl measured by ultrasonic interferometry and static compression: The calculation of compressibility from elastic wave velocities require the density as a function of pressure. The *in situ* density was determined by analyzing the sample deformation (deformation fit), using published EoS (EoS-fit) and successive approximation. The *X*-axis is related to the Decker (1971) pressure scale.

Table 1. Polynomial fit coefficients for the compressibility of NaCl measured by ultrasonic interferometry and static compression.

Polynomial fit coefficients (Eq. (19))	Static compression[a]	Ultrasonics, ρ from EoS[b]	Ultrasonics, ρ from deformation[c]	Ultrasonics, ρ from num. approach[d]
A	0.04191	0.03907	0.03907	0.03907
B_1	-0.0088	-0.00523	-0.00514	-0.00514
B_2	0.0018	6.51178×10^{-4}	6.50735×10^{-4}	6.00421×10^{-4}
B_3	-2.43008×10^{-4}	-1.09931×10^{-4}	-1.02209×10^{-4}	-9.2428×10^{-5}
B_4	1.93729×10^{-5}	2.19162×10^{-5}	1.86429×10^{-5}	1.83552×10^{-5}
B_5	-8.72196×10^{-7}	-3.17726×10^{-6}	-2.56361×10^{-6}	-2.71682×10^{-6}

[a] Compressibility measured by static compression (Birch, 1986).
[b] Compressibility measured by ultrasonic interferometry (this work), density ρ was derived from EoS (Birch, 1986).
[c] Compressibility measured by ultrasonic interferometry (this work), density ρ was derived from sample deformation (Ahrens and Katz, 1962).
[d] Compressibility measured by ultrasonic interferometry (this work), density ρ was calculated by an iterative numerical approach.

In terms of pressure measurement the compressibility calculated from ultrasonic data indicate at 3 GPa about 0.25 GPa higher pressures than derived from static compression data by Bridgman (1940). The ultrasonic data are related to Decker (1971) pressure scale.

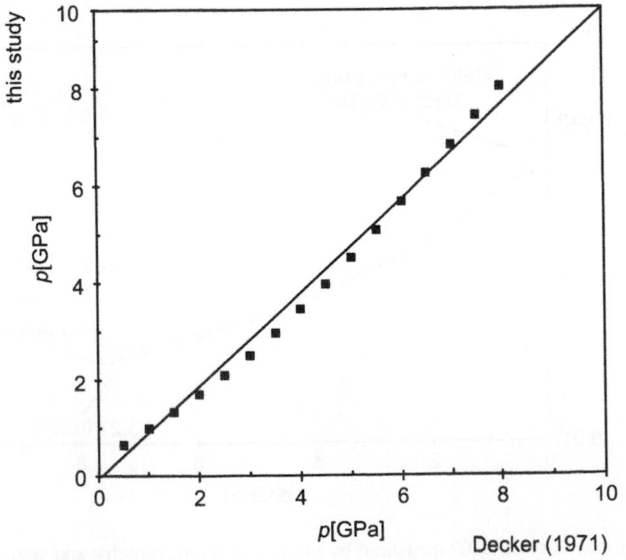

Figure 15. Pressure measured by ultrasonic interferometry in this study vs. Decker (1971) pressure scale related to the EoS by Birch (1986).

The graphs for all calculated NaCl compressibilities, i.e. derived from static compression and from ultrasonic measurements using the EoS, the empirical deformation model, and the numerical approach, were polynomial fitted up to the power of 5, which is required for the range between 1 and 5 GPa. Table 1 presents the coefficients of this fits according to Eq. (19).

$$K_T = A + B_1P + B_2P^2 + B_3P^3 + B_4P^4 + B_5P^5 \tag{19}$$

Figure 15 shows the relation between the Decker (1971) pressure scale and the pressure derived from the ultrasonic measurements of this study using the Eos by Birch (1986). The data were also polynomial fitted up to the power of 5:

$$p_{us} = 0.34611 + 0.6807\, p_{De} + 0.01921\, p_{De}^2 + 0.00246\, p_{De}^3 + 8.4777 \times 10^{-4} p_{De}^4 + 5.75971 \times 10^{-5} p_{De}^5 \tag{20}$$

where p_{us} is pressure derived from ultrasonic measurements of this study and p_{De} is pressure according to Decker (1971).

4. Conclusions

The results demonstrate the ability to measure the pressure inside of multi-anvil pressure cells standard-free by ultrasonic interferometry. The synchrotron radiation is used to measure the pressure by XRD-techniques using EoS after Decker (1971). The synchrotron radiation is also used for precise *in situ* sample length and density determination required for the ultrasonic method. Different ways of density determinations were used (using the EoS for NaCl, published by Birch (1986), analyzing the deformation (Ahrens and Katz, 1962), and using an iterative numerical approach) and agreed within <0.1%. Ultrasonic pressure measurement will probably not substitute the XRD-determination completely, because of its higher technical expense, but might be important for a calibrant-free pressure scale determined at very high pressures. However, it seems to become a standard high-pressure method to determine elastic properties of polycrystalline samples parallel to the growing amount and quality of ultrasonic measurements on single crystals under experimental simulated Earth's mantle conditions.

Acknowledgements

We would like to express our special thanks to the editors J. Chen, Y. Wang, T. Duffy, G. Shen, and L. Dobrzhinetskaya for their initiative, patience, and guidance, as well as two unknown referees for their very constructive reviews. We are especially grateful to S. Ganschow for the Ce:YAG crystal, manufactured and put at our disposal by IKZ and J. Kulesza (ADC) for his untiring support. The authors thank M. Kreplin and G. Berger for fabricating tools and cell parts, C. Karger for her software contribution, and H. Witzki and W. Steiner for many initiatives and important technical contributions as well as all colleagues of the high-pressure mechanical workshop for their dedicated support.

References

Ahrens, T.J., Katz, S., 1962. An ultrasonic interferometer for high-pressure research. J. Geophys. Res. 67, 2935–2944.

Bassett, W.A., Reichmann, H.-J., Angel, R.J., Spetzler, H., Smyth, J.R., 2000. New diamond anvil cells for gigahertz ultrasonic interferometry and X-ray diffraction. Am. Mineral. 85, 283–287.

Birch, F., 1978. Finite strain isotherm and velocities for single-crystal and polycrystalline NaCl at high pressures and 300 K. J. Geophys. Res. 83, 1257–1268.

Birch, F., 1986. Equation of state and thermodynamic parameters of NaCl to 300 kbar in the high-temperature domain. J. Geophys. Res. 91, 4949–4954.

Boehler, R., Kennedy, G.C., 1980. Equation of state of sodium chloride up to 32 kbar and 500°C. J. Phys. Chem. Solids 41, 517–523.

Bridgman, P.W., 1940. Compressions to 50,000 kg/cm^2. Phys. Rev. 57, 237–239.

Brown, J.M., 1999. NaCl pressure standard. J. Appl. Phys. 86, 5801–5808.

Chen, J., Parise, J.B., Li, R., Weidner, D.J., Vaughan, M., 2000. The imaging plate system interfaced to the large-volume press at beamline X17B1 of the national synchrotron light source. In: Manghnani, M.H., Yagi, T. (Eds), Properties of Earth and Planetary Materials at High Pressure and Temperature, Geophysical Monograph 101. AGU, Washington, DC, pp. 139–144.

Cook, R.K., 1957. Variation of elastic constants and static strains with hydrostatic pressure: a method for calculation from ultrasonic measurements. J. Acoust. Soc. Am. 29, 445–449.

Decker, D.L., 1971. High-pressure equation of state for NaCl, KCl, and CsCl. J. Appl. Phys. 42, 3239–3244.

Frankel, J., Rich, F.J., Homan, C.G., 1976. Acoustic velocities in polycrystalline NaCl at 300 K measured at static pressures from 25 to 270 kbar. J. Geophys. Res. 81, 6357–6363.

Fritz, J.N., Marsh, S.P., Carter, W.J., Mcqueen, R.G., 1971. The Hugonoit equation of state of sodium chloride in the sodium chloride structure. In: Lloyd, E.C. (Ed.), Accurate Characterization of the High Pressure Environment, NBS Special Publication, Vol. 326, pp. 201–208.

Getting, I.C., 1998. The practical pressure scale: fixing fixed points and future prospects. Eos Trans., AGU 79, F830.

Holland, K.G., Ahrens, T.J., 1998. Properties of LiF and Al_2O_3 to 240 GPa for metal shock temperature measurements. In: Manghnani, M.H., Yagi, T. (Eds), Properties of Earth and Planetary Materials at High Pressure and Temperature, Geophysical Monograph 101. AGU, Washington, DC, pp. 335–343.

Katz, S., Ahrens, T.J., 1963. Ultrasonic measurements of elastic properties of small specimens at high pressure. High Pressure Meas. Pap. 1962, 246–261.

Knoche, R., Webb, S.L., Rubie, D.C., 1997. Experimental determination of acoustic wave velocities at Earth mantle conditions using a multianvil press. Phys. Chem. Earth 22, 125–130.

Knoche, R., Webb, S.L., Rubie, D.C., 1998. Measurements of acoustic wave velocities at *P–T* conditions of the Earth's mantle. In: Manghnani, M.H., Yagi, T. (Eds), Properties of Earth and Planetary Materials at High Pressure and Temperature, Geophysical Monograph 101. AGU, Washington, DC, pp. 119–128.

Kung, J., Rigden, S., 1999. Oxide perovskites: pressure derivatives of the bulk and shear moduli. Phys. Chem. Miner. 26, 234–241.

Kung, J., Gwanmesia, G.D., Liu, J., Li, B., Liebermann, R.C., 2000. PV3T experiments: simultaneous measurement of sound velocities (V_p and V_s) and sample volume (V) of polycrystalline specimens of mantle minerals at high pressures (P) and temperatures (T). Eos Trans., AGU 81, 48, F1151.

Kung, J., Angel, R.J., Ross, N.L., 2001a. Elasticity of $CaSnO_3$ perovskite. Phys. Chem. Miner. 28, 35–43.

Kung, J., Weidner, D.J., Li, B., Liebermann, R.C., 2001b. Determination of the elastic properties at high pressure without pressure scale. Eos Trans., AGU 82, F1383.

Li, B., Chen, G., Gwanmesia, G.D., Liebermann, R.C., 1998. Sound velocity measurements at mantle transition zone conditions of pressure and temperature using ultrasonic interferometry in a multianvil apparatus. In: Manghnani, M.H., Yagi, T. (Eds), Properties of Earth and Planetary Materials at High Pressure and Temperature, Geophysical Monograph 101. AGU, Washington, DC, pp. 41–61.

Li, B., Vaughan, M.T., Kung, J., Weidner, D.J., 2001. Direct Length Measurement Using X-Radiography for the Determination of Acoustic Velocities at High Pressure and Temperature. NSLS Activity Report, 2, pp. 103–106.

Luth, W.L., 1993. Measurement and control of intensive parameters in experiments at high pressure in solid-media apparatus. In: Luth, R.W. (Ed.), Experiments at High Pressure and Applications to the Earth's Mantle, Short Course Handbook 21. Mineral Association Canada, Edmonton, Alta., pp. 15–37.

Mao, H.K., Xu, J., Bell, P.M., 1986. Calibration of the ruby pressure gauge to 800 kbar under quasi-hydrostatic conditions. J. Geophys. Res. 91 (B5), 4673–4676.

McSkimin, H.J., 1950. Ultrasonic measurement techniques applicable to small solid specimens. J. Acoust. Soc. Am. 22, 413–418.

Mueller, H.J., Lauterjung, J., Schilling, F.R., Lathe, C., Nover, G., 2002. Symmetric and asymmetric interferometric method for ultrasonic compressional and shear wave velocity measurements in piston–cylinder and multi-anvil high-pressure apparatus. Eur. J. Mineral. 14, 581–589.

Mueller, H.J., Schilling, F.R., Lauterjung, J., Lathe, C., 2003. A standard free pressure calibration using simultaneous XRD and elastic property measurements in a multi-anvil device. Eur. J. Mineral. 15, 865–873.

Mueller, H.J., Lathe, C., Schilling, F.R., 2005. Simultaneous determination of elastic and structural properties under simulated mantle conditions using multi-anvil device MAX80. In: Chen, J., Wang, Y., Duffy, T., Shen, G., Dobrzhinetskaya, L. (Eds), Frontiers in High Pressure Research: Geophysical Applications. Elsevier Science, Amsterdam, pp. 67–94.

Piermarini, G.J., Block, S., Barnet, J.D., Forman, R.A., 1975. Calibration of the pressure dependence of the R_1 ruby fluorescence line to 195 kbar. J. Appl. Phys. 46, 2774–2780.

Ruoff, A.L., Lincoln, R.C., Chen, Y.C., 1973. A new method of absolute high pressure determination. J. Phys. D: Appl. Phys. 6, 1295–1306.

Shen, A.H., Reichmann, H.-J., Chen, G., Angel, R.J., Bassett, W.A., Spetzler, H., 1998. GHz ultrasonic interferometry in a diamond anvil cell: P-wave velocities in periclase to 4.4 GPa and 207°C. In: Manghnani, M.H., Yagi, T. (Eds), Properties of Earth and Planetary Materials at High Pressure and Temperature, Geophysical Monograph 101. AGU, Washington, DC, pp. 71–77.

Shimomura, O., Yamaoka, S., Yagi, T., Wakutsuki, M., Tsuji, K., Kawamura, H., Hamaya, N., Fukunaga, O., Aoki, K., Akimoto, S., 1985. Multi-anvil type X-ray system for synchrotron radiation. In: Minomura, S. (Ed.), Solid State Physics Under Pressure. Terra Sci. Publ. Co., Tokyo, pp. 351–356.

Vaughan, M.T., 1993. In situ X-ray diffraction using synchrotron radiation at high *P* and *T* in a multi-anvil device. In: Luth, R.W. (Ed.), Experiments at High Pressure and Applications to the Earth's Mantle, Short Course Handbook 21. Mineral Association Canada, Edmonton, Alta., pp. 95–130.

Yagi, T., 1988. MAX80: large-volume high-pressure apparatus combined with synchrotron radiation. Eos Trans., AGU 69 12, 18–27.

Yoneda, A., Spetzler, H., Getting, I., 1994. Implication of the complete travel time equation of state for a new pressure scale. In: Schmidt, J.W., Shaner, G.A., Samara, G., Ross, M. (Eds), High-Pressure Science and Technology – 1993. Proceedings of International Association of Research Advancement of High Pressure Science and Technology and the American Physical Society. American Institute of Physics, New York, pp. 1609–1612.

Zha, C.-S., Duffy, T.S., Downs, R.T., Mao, H.K., Hemley, R.J., 1998. Brillouin scattering and X-ray diffraction of San Carlos olivine: direct pressure determination to 32 GPa. Earth Planet. Sci. Lett. 159, 25–33.

Zha, C.-S., Mao, H.K., Hemley, R.J., 2000. Elasiticity of MgO and a primary pressure scale to 55 GPa. Proc. Natl Acad. Sci. 97, 13494–13499.

Zinn, P., Hinze, E., Lauterjung, J., Wirth, R., 1997. Kinetic and microstructural studies of the quartz–coesite phase transition. Phys. Chem. Earth 22, 105–111.

Advances in High-Pressure Technology for Geophysical Applications
Jiuhua Chen, Yanbin Wang, T.S. Duffy, Guoyin Shen, L.F. Dobrzhinetskaya, editors

Chapter 22

High-pressure generation in the Kawai-type apparatus equipped with sintered diamond anvils: application to the wurtzite–rocksalt transformation in GaN

Eiji Ito, Tomoo Katsura, Yoshitaka Aizawa,
Kazuyuki Kawabe, Sho Yokoshi, Atsushi Kubo,
Akifumi Nozawa and Ken-ichi Funakoshi

Abstract

Technical development of the Kawai-type apparatus has briefly been described from the original single-staged split-sphere to the recent 6–8 double-staged devise in which sintered diamond (SD) is equipped as the anvil material. In this article, high-pressure generation using sintered diamond cubes with 14.0 mm edge length and 1.5 mm corner truncation has been reported together with results for the wurtzite–rocksalt transformation in GaN at high pressure. Octahedral specimen assembly with 5.0 mm edge length containing GaN powder in an MgO capsule (pressure marker) was compressed in a DIA-type press SPEED MkII recently installed at SPring-8, Japan. Pressure value was determined from compression of the MgO marker and the transformation of GaN was monitored by means of in situ X-ray diffraction. The maximum pressure of 63.3 ± 0.4 GPa was reached based on the MgO scale by Jamieson et al. at 300 K. Onset of the wurtzite–rocksalt transformation in GaN was observed at 54 GPa and 300 K and at 51.4 GPa and 750 K, suggesting a negative Clapeyron slope of −170 K/GPa. The rocksalt type of GaN is judged to be quenchable to the ambient conditions because no reverse transformation was observed. Noticeable change in electric resistance may not be accompanied with the transformation. Moreover, the sluggishness also prevents the phase transformation in GaN from being used as a pressure fixed point.

1. Introduction

A brief history of development of the Kawai-type high-pressure apparatus. In 1965, the late Prof. Naoto Kawai and his colleagues began to develop a novel high-pressure apparatus at Osaka University, Japan. Traces of the technological innovation are schematically illustrated in Figure 1. Basic idea was as follows (Kawai, 1966). A sphere of tungsten carbide (WC) is split into eight identical blocks by three planes mutually perpendicular and crossing at the center; each of which is surrounded by one curved bottom and three inclining side surfaces. The top of each block is truncated in such a way that, when all the parts are put together, they form an octahedral space at the center. Into this hollow space is put an octahedral pressure medium that is somewhat larger than the void. Compressive and insulating gaskets are sandwiched in-between all neighboring blocks just behind the pressure medium, so that all blocks are separated from each other. The whole sphere comprising these blocks is covered by a thick spherical rubber capsule

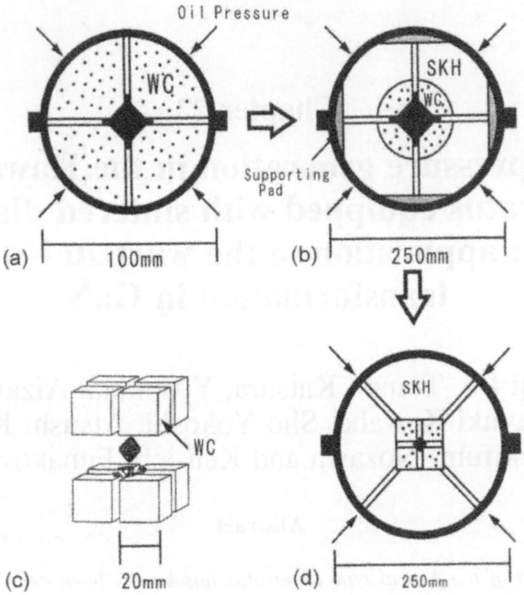
Oil Pressure
WC
SKH
WC
Supporting Pad
(a)
100mm
(b)
250mm
SKH
WC
(c)
20mm
(d)
250mm

Figure 1c. Through these procedures, the original single-staged split sphere apparatus evolved into the double-staged 6–8 type one (Kawai and Endo, 1970). It should be noted that versatility of the apparatus was remarkably increased by adopting a disposable cubic WC anvil that is much easier in machining than a split sphere. Three of six outer anvils were later glued together in a semispherical cavity of a steel vessel to form a guide block, and the two guide blocks were aligned in a hydraulic press to compress the cubic inner anvil assembly by the uniaxial force (Kawai et al., 1973).

Yagi (2001) strongly recommended calling the double-staged 6–8 type apparatus as Kawai-type after the late Prof. Kawai because of his leadership in the course of development of the apparatus. Hereafter, I name the assembly of eight cubic anvils and the central octahedral specimen (Fig. 1d) "Kawai-cell".

One of the most remarkable advantages of the Kawai-type apparatus is the capability to compress specimens of 10^3 to 1 mm^3 by selecting various sizes of truncation made at a corner of cubic anvils, which makes it possible to keep a sample of at least several milligrams under the controlled pressure and temperature conditions for desired duration. Another merit of the apparatus is that the sample in the Kawai-cell is kept in quasi-hydrostatic state despite usage of the solid pressure medium due to compression from multi-directions. The Kawai-type apparatus has been widely used in the mineral physics field; especially for determination of phase equilibria of a system representative of the mantle chemistry under precisely controlled pressure and temperature conditions (e.g. Ito and Takahashi, 1989), crystal growth of important mantle material (e.g. Ito and Weidner, 1986), and synthesis of a large quantity of high-pressure phase for thermochemical measurement (e.g. Ito et al., 1990) or a bulk sintered aggregate for physical property measurement (e.g. Osako and Ito, 1991). These performances are in some sense complementary to those of diamond anvil cell (DAC), which is a tool of extraordinary versatility to observe the state of a very small amount of sample under pressures up to few hundred GPa. In recent *in situ* X-ray measurement using synchrotron radiation, squeezing of the Kawai-cell using a cubic press (DIA type apparatus) (Inoue and Asada, 1973) has become a popular technique, because the combination of both devices is the most feasible for X-ray optics (Utsumi et al., 1998). In spite of the versatile abilities of the Kawai-type apparatus, it has been realized in last two decades that the maximum attainable pressure is limited to about 28 GPa (Kubo and Akaogi, 2000), almost one order of magnitude lower than that in DAC, so far as WC is used as the anvil material.

Sintered diamond has become available as the anvil of Kawai-cell in the late 1980s. As the hardness of SD is at least twice higher than that of WC (Sung and Sung, 1996), attainable pressure range of the Kawai apparatus has been extended. Ohtani et al. (1989) first squeezed Kawai-cell using SD cubes of 4.85 mm edge in DIA type press at KEK, Japan and generated 41 GPa based on the Au scale of Jamieson et al. (1982). Kondo et al. (1993) successfully generated the conditions of 30 GPa and 2100 K employing 10 mm cubes and an internal heater in the pressure medium. Funamori et al. (1996) determined an equation of state for $MgSiO_3$ perovskite up to 30 GPa and 2000 K by means of *in situ* X-ray diffraction methods.

Purpose of the present study. Under these circumstances, we have also adopted SD anvils in Kawai-type apparatus, and carried out quenching phase equilibrium studies (Ito et al., 1998; Ono et al., 2001; Ito et al., 2004) and *in situ* X-ray observations (Ono et al., 2000; Kubo et al., 2003). The pressure range of these studies exceeded those of similar studies previously conducted using WC anvils by 10–15 GPa. In this chapter, we report experimental results to evaluate the maximum attainable pressure by the equipment of SD anvil and to examine the wurtzite–rocksalt transformation in GaN (Uehara et al., 1997) up to 63 GPa. The possibility of this transformation serving as a high-pressure calibrant above Zr point (33 GPa) (Xia et. al., 1990; Akahama et al., 1991) is discussed.

2. Experimental

We adopted SD cubes edge length of 14 mm and a truncated corner of 1.5 mm, which were manufactured by Sumitomo Electric Industries, Ltd. A small extra slope of $\tan^{-1}(1/400)$ was made from the truncated center to the opposite corner for three surfaces of each cube to improve the efficiency of pressure generation. Octahedron of MgO + 5% Cr_2O_3 with an edge length of 5.0 mm was used as pressure medium with small pyrophyllite gasket, which was heated at 900°C for 20 min to increase its hardness. The Kawai-cell thus constructed is shown in Figure 2.

The Kawai-cell was compressed using a new DIA type press *SPEED MkII* installed at SPring-8, Japan (Katsura et al., 2004). Two experimental runs, M139 and M148, were performed to examine the phase transformation of GaN, employing specimen assemblies schematically shown in Figure 3a and b, respectively. Run M139 was conducted at 300 K, in which powdered GaN sample was packed in the central 0.5 mm length between lumps of fine diamond powder in the MgO capsule whose sizes are 1.7 mm in length, 0.4 and 1.2 mm in inner and outer diameters. In run M148, however, the sample was heated up to 850 K. The heating system shown in Figure 3b is similar to that used by Kubo et al. (2003), with minor modification in sizes of the parts. For example, outer and inner diameters of $LaCrO_3$ were 1.7 and 0.8 mm, respectively. Nichrome foil and small rods of sintered TiC were used as the heater and electrodes, respectively. The cylindrical assemblies were set in the center of the octahedral pressure media so that the longitudinal axes were parallel to the incident X-ray. In both the runs, the MgO capsules served as the pressure marker.

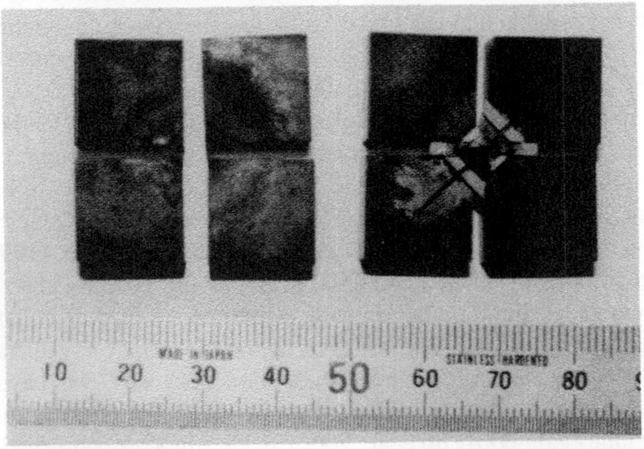

Figure 2. Photos of the Kawai-cell composed of eight SD cubic anvils with an edge length of 14 mm. An octahedral pressure medium of MgO + 5% Cr_2O_3 and pyrophyllite gasket pieces are placed at the center.

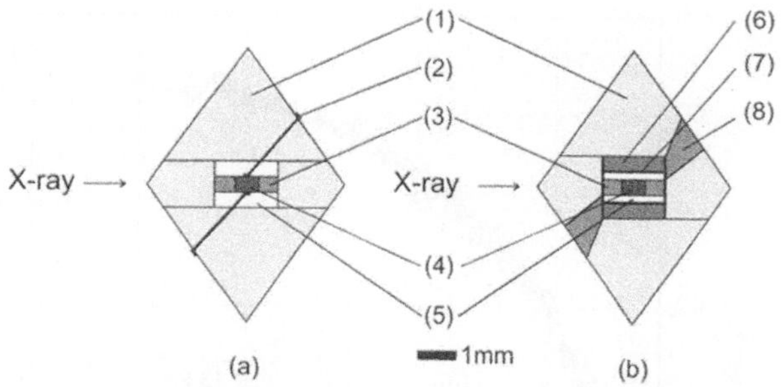

Figure 3. Cross-sections of the specimen assemblies employed in GaN runs of M139 (a) and M148 (b). (1): Pressure medium (MgO + 5wt%Cr_2O_3), (2): Electrode (Pt), (3): Diamond powder, (4): Sample, (5): Sample capsule (pressure standard) (MgO), (6): Thermal insulation sleeve ($LaCrO_3$), (7): Cylindrical heater (nichrome), (8): Electrode (sintered TiC). Thermocouple (not shown) is perpendicular to the paper and in contact with outer surface of the heater.

As reagent grade GaN (99.99%) powder was not well-crystallized, showing very broad X-ray diffraction pattern, it was treated at 6 GPa and 1800 K for 1 h in a BN capsule. Unfortunately, X-ray powder diffraction of the recovered sample showed that it contained substantial amount of Ga_2O_3 with well-crystallized GaN. The mixture of GaN and Ga_2O_3 thus obtained and the original poorly crystalline GaN powder were both used as starting materials for runs M139 and M148, respectively.

Phase identification of GaN sample and pressure determination were performed by *in situ* energy dispersive X-ray diffraction methods using polychromatic synchrotron radiation on the beam port BL04B1, SPring-8. An X-ray beam collimated to 100 μm vertically and 50 μm horizontally was introduced to the sample and the MgO standard (capsule), independently. The diffracted beam was detected by a Ge solid-state detector (SSD) through a collimator of 50 μm width to ensure a diffraction angle 2θ of ca. 6.0°, which was calibrated using *d*-spacings of MgO (111), (200), (220), (311), (222), (400), and (331) at ambient conditions with accuracy of $\pm 0.001°$. The pressure was determined from the measured unit cell volume of the MgO next to the sample using the equation of state for MgO by Jamieson et al. (1982). Six diffraction peaks of MgO, (111), (200), (220), (311), (222), and (400), were mostly used for the unit cell volume calculation. Uncertainties of the pressure determination were mostly within ± 0.5 GPa. Acquisition time for diffraction pattern at each P/T condition was typically 600 s.

In run M139, reconnaissance measurement of electrical resistance of GaN was also carried out by inserting two Pt wires to opposite positions on the cylindrical outer surface of the sample. The terminal was taken out to the top surfaces of the SD anvils whose electrical resistance are several tens of mΩ and served as electrodes.

3. Results and discussion

Pressure generation. Generated pressures at 300 K are plotted against the press load in Figure 4 for runs M139, M148 and independent runs M162 and M181 carried out using same assembly to Figure 3a. It is remarkable that pressure is generated in a repeatable accuracy of 2–3% up to ca. 60 GPa over the four runs. In run M139, after the maximum

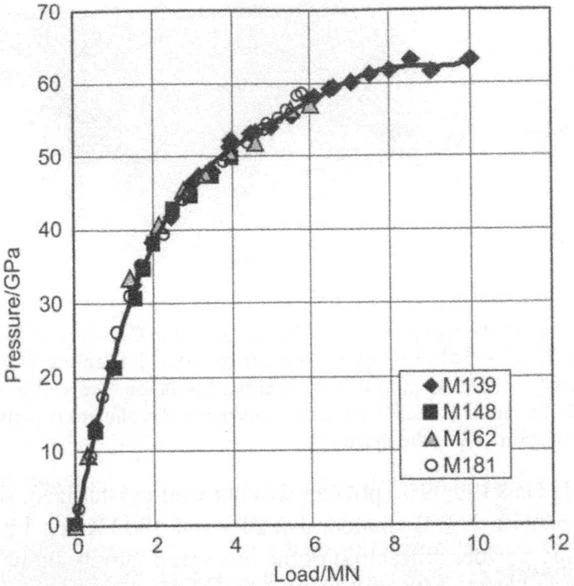

Figure 4. Results of pressure determination at 300 K for the Kawai-type high-pressure apparatus equipped with SD cubic anvils with truncated edge length of 1.5 mm and an octahedral pressure medium of MgO + 5% Cr_2O_3 with 5.0 mm edge length. Generated pressures determined from the MgO scale (Jamieson et al., 1982) in four runs (M139, M148, M162, and M181) are plotted against the press load.

pressure of 63.3 ± 0.4 GPa was achieved at load of 8.6 mega Newton (MN), it slightly dropped to 61.7 ± 0.8 GPa at 9.0 MN. Then the pressure increases to 63.0 ± 0.2 GPa by further compression to 9.9 MN. However, the run was terminated at 10.0 MN by a big explosion presumably extending to the overall Kawai-cell, and all SD cubes were damaged. It is possible that deformation or fracture of SD anvils would have started at around 8.6 MN (ca. 63 GPa) and progressed catastrophically at 10.0 MN. In order to achieve higher pressure, therefore, improvement in quality of SD is the most urgent requirement.

Present study extended the pressure range of the Kawai apparatus to 63 GPa, which is almost twice as high as those in early works using SD anvils by Kondo et al. (1993) and Funamori et al. (1996). An important factor for the extension was that we used larger SD cubic anvils than them; i.e. 14 versus 10 mm edge lengths. Accordingly we could compress the Kawai-cell to 10 MN, whereas the loads in the previous works were limited to 5 MN. The larger size and higher brittleness are the most decisive factors for SD cubes to generate high pressure in the Kawai-type apparatus. Both the requirements are somewhat contradictory, because SD is produced only under conditions of several GPa and higher than 1800 K. Nevertheless, we recently have opportunity to use SD cubes with 20 mm edge length, and the improvement in quality of SD has actively been challenged in industry.

Phase transformation of GaN. Some characteristic diffraction patterns are reproduced with peak assignment in Figure 5. At pressures higher than 50 GPa, contraction and deformation of the specimen made it difficult to acquire diffraction profile of the sample

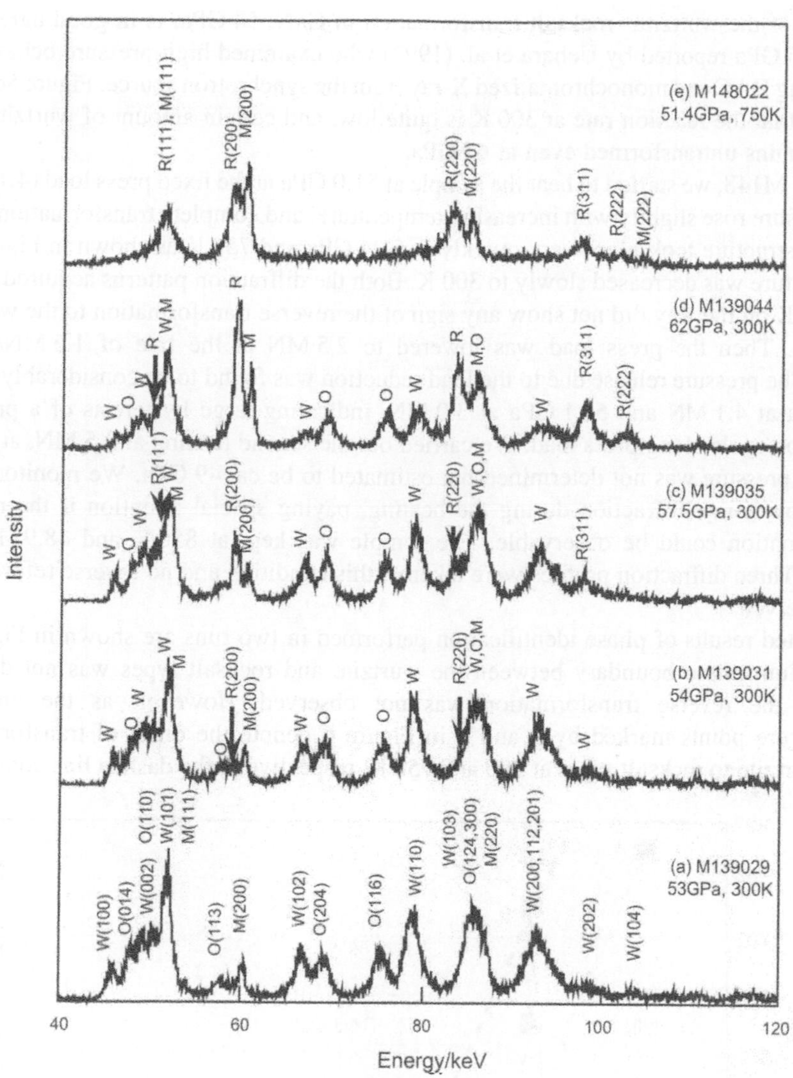

Figure 5. Selected diffraction profiles of GaN sample. (a), (b), (c), and (d): sequence of diffraction profiles collected at 300 K in run M139. Diffraction peaks of the rocksalt type of GaN are first present at 54 GPa and become pronounced with increasing pressure. (E): collected at 51.4 GPa and 750 K, which shows complete conversion to the rocksalt structure. Abbreviations attached to diffraction peaks; W: wurtzite type GaN, R: rocksalt type GaN, M: MgO, O: Ga_2O_3.

separately from that of MgO capsule. In run M139, diffraction profiles were further complicated by existence of Ga_2O_3 as shown in Figure 5a to d. Fortunately, the complexity in the diffraction pattern due to the additional phases did not cause any difficulty in identifying the phases of GaN. Wurtzite-type of GaN is stable up to 53 GPa as seen from pattern (a) in Figure 5. However, the diffraction pattern at 54 GPa in (b) shows the presence of small amount of GaN possessing the rocksalt type of structure. The onset pressure of the wurtzite–rocksalt transformation in GaN, 54 GPa, is in good agreement with 53.6 GPa reported by Uehara et al. (1997) who examined high-pressure behavior of GaN using DAC and monochromatized X-ray from the synchrotron source. Figure 5c and d indicate that the reaction rate at 300 K is quite low, and certain amount of wurtzite-type GaN remains untransformed even at 62 GPa.

In run M148, we started to heat the sample at 51.0 GPa at the fixed press load (4.1 MN). The pressure rose slightly with increasing temperature, and complete transformation to the rocksalt structure took place very quickly at 51.4 GPa and 750 K as shown in Figure 5e. Temperature was decreased slowly to 300 K. Both the diffraction patterns acquired at 550 and 300 K on the way did not show any sign of the reverse transformation to the wurtzite structure. Then the press load was lowered to 2.5 MN at the rate of 1.6 MN/7 h at 300 K. The pressure release due to the load reduction was found to be considerably small, 52.0 GPa at 4.1 MN and 50.1 GPa at 3.0 MN, indicating large hysteresis of a pressure generation cycle along press load. We carried out the second heating at 2.5 MN, at which load the pressure was not determined but estimated to be ca. 49 GPa. We monitored the sample by X-ray diffraction during the heating, paying special attention if the reverse transformation could be observable. The sample was kept at 850 K and 48.9 GPa for 90 min. Three diffraction profiles were taken at this condition and no reverse reformation was observed.

Selected results of phase identification performed in two runs are shown in Figure 6. Equilibrium phase boundary between the wurtzite and rocksalt types was not defined, because the reverse transformation was not observed. However, as the pressure/temperature points marked by *a* and *b* in Figure 6 denote the onset of transformation from wurtzite to rocksalt types at 300 and 750 K, respectively, the dashed line connecting

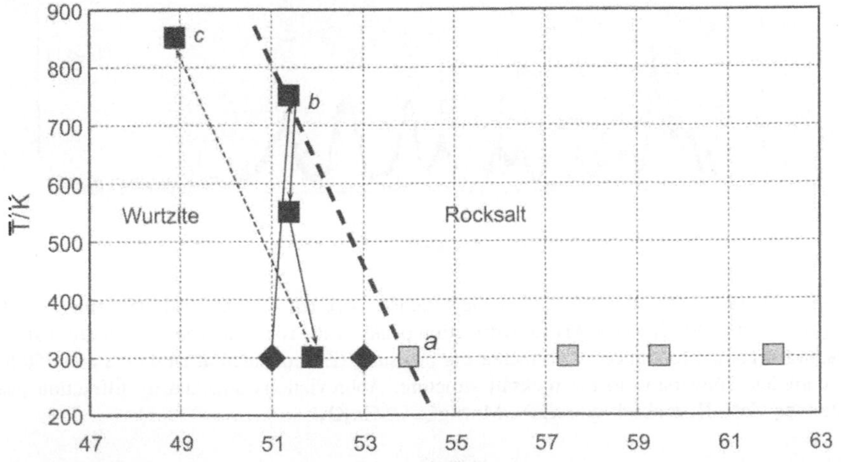

Figure 6. Results of synthesis experiments on GaN. *a* and *b* denote the conditions of onset of the wurtzite–rocksalt transformation at 300 and 750 K, respectively. The dashed line is drawn as a tentative phase boundary. Light arrow lines show sequence of the experimental conditions in run M148. The light broken arrow line does not show a real trajectory. Sample pressure was first decreased to ca. 49 GPa and then the temperature was increased to 850 K (see text). Solid diamond: wurtzite type; solid square: rocksalt type; gray square: composite of wurtzite and rocksalt types.

a and *b* points would be close to the equilibrium phase boundary. Therefore, Clapeyron slope for the phase boundary is suggested to be negative around $dT/dP = -170$ K/GPa. On the other hand, presence of the rocksalt type at *c* at which pressure was about 2 GPa lower than the dashed line at 850 K, indicates that the activation energy for the reverse transformation is relatively high compared to the thermal energy of the rocksalt phase at this condition. Therefore, the rocksalt type persists as a metastable phase over a wide range of P/T conditions, and presumably quenchable to the ambient conditions. Unfortunately, this run also ended up at 1.28 MN; we could not recover the sample.

Electrical resistance measured by a conventional multiple meter in run M139 was tens of MΩ up to about 62 GPa, and no remarkable change in resistance was observed on the wurtzite–rocksalt transformation. As the sample contained substantial amount of Ga_2O_3, a conclusion on the electrical resistance change should be postponed. However, the sluggishness shown in the present study does not allow this transformation to be used as a pressure fixed point at room temperature.

Acknowledgements

We are grateful to A. Yoneda for his discussion and comments. We also thank C. Oka and Y. Shimizu for their technical assistance. Critical review and comments by two anonymous reviewers and J. Chen are much appreciated. This work was financially supported by grants from Ministry of Education, Culture, Sports, Science and Technology, Japan (proposal number, 13852005). *In situ* X-ray experiments were carried out at SPring-8 (proposal number, 2004A0432-CD2b-np).

References

Akahama, Y., Kobayashi, M., Kawamura, H., 1991. High-pressure X-ray diffraction study in electronic *s–d* transition in zirconium. J. Phys. Soc. Jpn 60, 3211–3214.

Funamori, N., Yagi, T., Utsumi, W., Kondo, T., Uchida, Y., Funamori, M., 1996. Thermoelastic properties of $MgSiO_3$ perovskite determined by in situ X-ray observations up to 30 GPa and 2000 K. J. Geophys. Res. 101, 825–826.

Inoue, K., Asada, T., 1973. Cubic anvil X-ray diffraction press up to 100 kbar and 1000°C. Jpn. J. Appl. Phys. 12, 1786–1793.

Ito, E., Takahashi, E., 1989. Post-spinel transformations in the system Mg_2SiO_4–Fe_2SiO_4 and some geophysical implications. J. Geophys. Res. 94, 10637–10646.

Ito, E., Weidner, D.J., 1986. Crystal growth of $MgSiO_3$ perovskite. Geophys. Res. Lett. 13, 464–466.

Ito, E., Akaogi, M., Topor, L., Navrotsky, A., 1990. Negative pressure–temperature slopes for reactions forming $MgSiO_3$ perovskite. Science 249, 1274–1278.

Ito, E., Kubo, A., Katsura, T., Akaogi, M., Fujita, T., 1998. High-pressure transformation of pyrope ($Mg_3Al_3Si_3O_{12}$) in a sintered diamond cubic anvil assembly. Geophys. Res. Lett. 25, 821–824.

Ito, E., Kubo, A., Katsura, T., Walter, M.J., 2004. Melting experiments of mantle materials under lower mantle conditions with implications of magma ocean differentiation. Phys. Earth Planet. Interiors 143–144, 397–406.

Jamieson, J.C., Fritz, J.N., Manghnani, M.N., 1982. Pressure measurement at high temperature in X-ray diffraction studies: gold as a primary standard. In: Akimoto, S., Manghnani, M.H. (Eds), High Pressure Research in Geophysics, AEPS, 12, CAPJ, Tokyo, pp. 27–48.

Katsura, T., Funakoshi, K., Kubo, A., Nishiyama, N., Tange, Y., Sueda, Y., Kubo, T., Utsumi, W., 2004. A large-volume high P–T apparatus for in situ X-ray observation 'SPEED-mkII'. Phys. Earth Planet. Interiors 143–144, 497–506.

Kawai, N., 1966. A static high pressure apparatus with tapered multi-piston formed a sphere I. Proc. Jpn. Acad. 42, 385–388.
Kawai, N., Endo, S., 1970. The generation of ultrahigh hydrostatic pressure by a split sphere apparatus. Rev. Sci. Instrum. 4, 425–428.
Kawai, N., Togaya, M., Onodera, A., 1973. A new device for high pressure vessels. Proc. Jpn. Acad. 49, 623–626.
Kondo, T., Sawamoto, H., Yoneda, A., Kato, M., Motsumuro, A., Yagi, T., 1993. Ultrahigh-pressure and high temperature generation by use of the MA8 system with sintered diamond anvils. High Temp.–High Pressure 25, 105–112.
Kubo, A., Akaogi, M., 2000. Post-garnet transitions in the system $Mg_4Si_4O_{12}$–$Mg_3Al_2Si_3O_{12}$ up to 28 GPa: phase relations of garnet, ilmenite and perovskite. Phys. Earth. Planet. Interiors 121, 85–102.
Kubo, A., Ito, E., Katsura, T., Shinmei, T., Yamada, H., Nishikawa, O., Song, M., Funakoshi, K., 2003. *In situ* X-ray observation of iron using Kawai-type apparatus equipped with sintered diamond: absence of β phase up to 44 GPa and 2100 K. Geophys. Res. Lett. 30, 1126, doi: 10.1029/2002GL01394.
Ohtani, E., Kagawa, K., Shimomura, O., Togaya, M., Suito, K., Onodera, A., Sawamoto, H., Yoneda, A., Utsumi, W., Ito, E., Matsumuro, A., Kikegawa, T., 1989. High-pressure generation by a multiple anvil system with sintered diamond anvils. Rev. Sci. Instrum. 60, 922–925.
Ono, S., Ito, E., Katsura, T., 2001. Mineralogy of subducted basaltic crust (MORBO) from 25 to 37 GPa, and chemical heterogeneity of the lower mantle. Earth Planet. Sci. Lett. 190, 57–63.
Ono, S., Ito, E., Katsura, T., Yoneda, A., Walter, M.J., Urakawa, S., Utsumi, W., Funakoshi, K., 2000. Thermoelastic properties of the high-pressure phase of SnO_2 determined by in situ X-ray observations up to 30 GPa and 1400 K. Phys. Chem. Miner. 27, 1618–1622.
Osako, M., Ito, E., 1991. Thermal diffusibility of $MgSiO_3$ perovskite. Geophys. Res. Lett. 18, 239–242.
Sung, C.-M., Sung, M., 1996. Carbon nitrite and other speculative super hard materials. Mater. Chem. Phys. 43, 1018.
Uehara, S., Masamoto, T., Onodera, A., Ueno, M., Shimomura, O., Takemura, K., 1997. Equation of state of the rocksalt phase of III–V nitrides to 72 GPa or higher. J. Phys. Chem. Solids 58, 2093–2099.
Utsumi, W., Funakoshi, K., Urakawa, S., Yamakata, M., Tsuji, K., Konishi, H., Shimomura, O., 1998. Spring-8 beamlines for high pressure science with multi-anvil apparatus. Rev. High Pressure Sci. Tech. 7, 1484–1486.
Xia, H., Parthasarathy, G., Luo, H., Vorha, Y.K., Ruoff, A.L., 1990. Crystal structures of group IVa metals at ultrahigh pressures. Phys. Rev. B 42, 6736–6738.
Yagi, T., 2001. KAWAI-type apparatus. Rev. High Pressure Sci. Tech. 11, 171, in Japanese.

Advances in High-Pressure Technology for Geophysical Applications
Jiuhua Chen, Yanbin Wang, T.S. Duffy, Guoyin Shen, L.F. Dobrzhinetskaya, editors

Chapter 23
Development of high *P–T* neutron diffraction at LANSCE – toroidal anvil press, TAP-98, in the HiPPO diffractometer

Yusheng Zhao, Duanwei He, Jiang Qian, Cristian Pantea, Konstantin A. Lokshin, Jianzhong Zhang and Luke L. Daemen

Abstract

The development of neutron diffraction under extreme pressure (P) and temperature (T) conditions is highly valuable to condensed matter physics, crystal chemistry, materials sciences, as well as the Earth and planetary sciences. We have incorporated a 500-ton press, TAP-98 (Toroidal Anvil Press, designed in 1998) into the HiPPO diffractometer at LANSCE to conduct in situ high P–T neutron diffraction experiments. The technical issues related to press design, P–T control, translation stage, diffraction optics, anvil package, and cell assembly are discussed in this chapter. Pressure and temperature calibrations and preliminary experimental data are also presented in this chapter.

1. Introduction

Neutron scattering continues to play an increasingly important role in the determination of crystal structure, lattice dynamics, texture development, and magnetic excitations. It has been used advantageously to detect atom position and to study atom thermal vibrations (Larsen et al., 1982; Rupp et al., 1988; Kampermann et al., 1995). Neutrons interact directly with nuclei so that, unlike X-ray, scattering is independent of atomic number (Z), but can be used to differentiate isotopes of the same element. Neutron scattering factors, again unlike X-rays, do not decrease for high index (*hkl*) reflections, with the consequence that much more crystallographic information can be obtained. The large Q ($= 2\pi/\lambda$) coverage of neutron diffraction allows for the detailed study of crystal structure, hydrogen bonding, magnetism, and thermal parameters of light elements (e.g. H, Li, B) and heavy elements (e.g. Ta, U, Pu,) compounds, that are difficult to work with using X-ray diffraction techniques (Ptasiewiczbak et al., 1981; Lawson et al., 1991; Rodriguez-Carvajal et al., 1991; Steiner and Saenger, 1992; Krimmel et al., 1992). Furthermore, neutrons are also sensitive to the magnetic moments of atoms and can therefore be used for studies of magnetic order and disorder, colossal magneto-resistance, and spin glasses (DeTeresa et al., 1996; Niraimathi and Hofmann, 2000).

In situ high *P–T* neutron diffraction experiments provide unique opportunities to study materials under extreme conditions. The high penetrating power of neutrons presents great advantages in designing a pressure cell with various kinds of high strength metals, alloys, and ceramics. Some extremely difficult studies, e.g. Debye–Waller factor as a function of pressure and temperature can only be derived using *in situ* high $P-T$ neutron diffraction techniques. High $P-T$ neutron diffraction realizes its full potential in a broad spectrum of scientific problems. For instance, puzzles in Earth science such as the global carbon cycle and the role of hydrous minerals for water exchange between lithosphere and biosphere can be directly addressed. The pressure and temperature effects on hydrogen bonding in ice, hydrates, and hydrocarbon phases can only be refined using high-resolution neutron diffraction. The uniqueness of neutron diffraction in resolving pair-distribution-functions can advance the study of liquid, melts, and amorphous phases under high $P-T$ conditions (Proffen and Billinge, 2002; Palosz et al., 2002). Moreover, by introducing *in situ* shear and/or differential stress, the yield strength and texture development of metals and minerals accompanied with phase transitions at high $P-T$ conditions can also be studied by high $P-T$ neutron diffraction (Nikitin et al., 1995; He et al., 2004).

With the development of the Paris–Edinburgh (PE) toroidal anvil press, significant progress has been made regarding neutron diffraction experiments at high pressure (Besson et al., 1992a,b, 1995; Loveday et al., 1995; Von Dreele, 1995; Klotz et al., 1996). The new capacity to conduct neutron diffraction experiments simultaneously at high pressures and high temperatures has extended the scope of possible investigations. Zhao et al. (1999a); Le Godec et al. (2001) have overcome severe difficulties and made the cell assemblies work under simultaneous high $P-T$ conditions. Large sample volumes, long data collection times, wide detector areas, and restricted diffraction geometry are inevitable in high $P-T$ neutron diffraction experiments. We have successfully conducted high $P-T$ neutron diffraction experiments at Los Alamos Neutron Science Center (LANSCE) and achieved simultaneous high pressures and temperatures of 10 GPa and 1500 K using a PE cell with High Intensity Powder Diffractometer (HIPD). With an average of $6 \sim 12$ h of data collection, the diffraction data are of sufficiently high quality for the determination of structural parameters and thermal vibrations. We have studied ceramics, minerals, and metals using high $P-T$ neutron diffraction techniques (Parise et al., 1994; Zhao et al., 2000; He et al., 2004). The present study describes recent developments at LANSCE with the implementation of TAP-98 and *H*igh-*P*ressure and *P*referred-*O*rientation (HiPPO) instrument from 2000 to 2003.

2. Press design for HiPPO

A new time-of-flight (TOF) neutron diffractometer, HiPPO, for materials studies has been designed and constructed, and is now operating at LANSCE (Bennett et al., 2000; Wenk et al., 2003; Vogel et al., 2004). The HiPPO diffractometer takes advantage of the improved spallation neutron source at LANSCE and a short flight path to achieve a neutron flux at the sample of 10^8 n/cm^2/s, nominally. The 3D detector banks consist of about 1400 ^{3}He gas-filled detector tubes and allow the collection of diffracted neutrons in virtually all directions in space. The new HiPPO beam-line increases more than five times the detector area for the 90° diffraction banks (*used mostly for high-pressure experiments*) compared to the old HIPD beam-line. However, the existing Paris–Edinburgh cell, when used with HiPPO, projects a rather large shadow because of its four loading posts, which block the diffracted beam azimuthally by as much as 156° (shadow cast ~43% in equatorial plane). Furthermore, the PE anvil with a slope of only 7° (axial aperture of 14°) cuts off 30% of the diffracted beam illuminating the $2\theta = 90°$ diffraction banks (of 20° axial

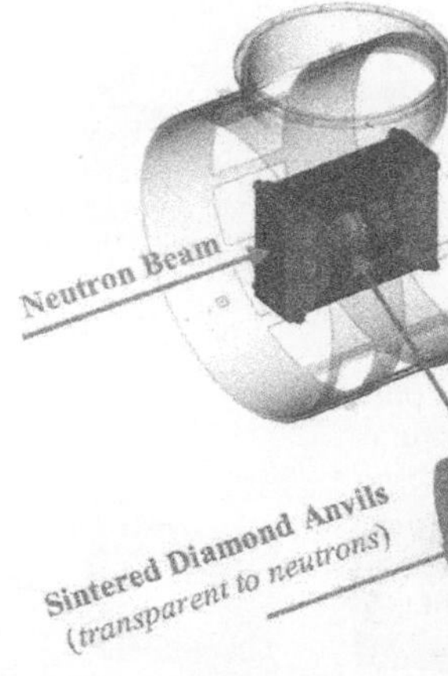

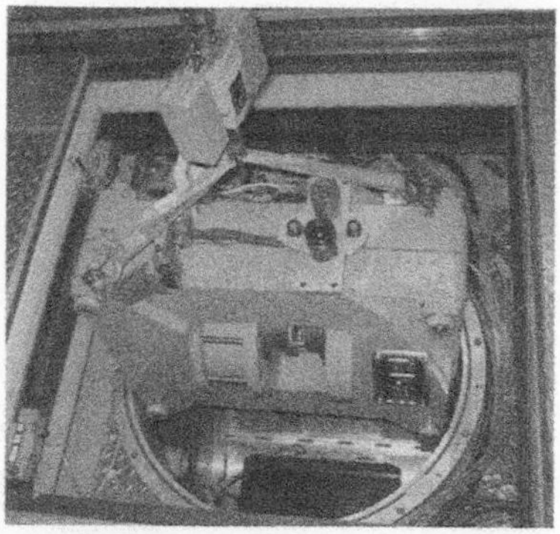

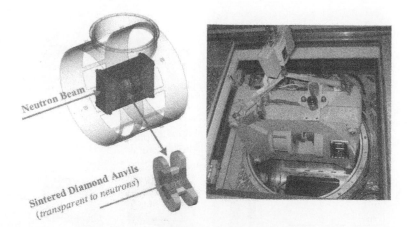

Figure 1. (Left) Schematic drawing of TAP-98 inside the HiPPO sample chamber. The composite curvature opening of the press frame and toroidal cell assembly design allow the maximum use of the $2\theta = 90°$ diffraction windows available to HiPPO. (Right) The custom-tailored TAP-98 fits into the HiPPO sample chamber with a rim opening of 73.66 cm (29 in.).

aperture for the HiPPO detectors). The available tooling space is less than 819 cm^3 (50 $in.^3$) around the anvil package, which significantly limits encapsulation and the use of attachments around the cell assembly.

To fully utilize the capabilities of the HiPPO diffractometer, we have designed a new toroidal anvil press, TAP-98, better adapted to HiPPO (Fig. 1) for neutron diffraction at high pressures and high temperatures (Zhao et al., 1999b). The new press design employs a dual-plate steel frame with composite curvature window opening to achieve a loading capacity of 5.0 MN (500 metric ton). The compact size of the TAP-98 loading frame, 71.12 cm × 53.34 cm × 17.78 cm (28 in. × 21 in. × 7 in.), is custom-tailored/trimmed to fit into the HiPPO sample chamber, which has a rim opening of 73.66 cm (29 in.). The dual-plate frame in a split arrangement allows incident neutrons to have multiple access beam ports. The composite curvature window opening minimizes the stress concentration and maximizes the detection of the diffracted neutrons. The TAP-98 design has enlarged the tooling space so that we can encapsulate the anvil package for environmental, safety, and health (ESH) considerations. This implementation is critical for high *P–T* neutron diffraction experiments with radioactive, toxic, and energetic sample materials. Specifically, the new design of TAP-98 with the open rectangular window reduces the shadow angles to 110° (a 29% reduction) and, at the same time, the new anvils with a slope of 10° fully use the $2\theta = 90°$ diffraction bank. The new design of the TAP-98 press and anvil has an increased diffraction detecting area about 158% compared to the PE cell when used on HiPPO. The enlarged tooling space of about 4916 cm^3 (~300 $in.^3$) allows easy installation of high temperature and high magnetic field attachments for the cell assemblies.

3. The mechanics of TAP-98

TAP-98 uses a steel multi-plate frame with a composite curvature window opening to contain the hydraulic jack, loading plates, and the toroidal anvil assemblies. The frame

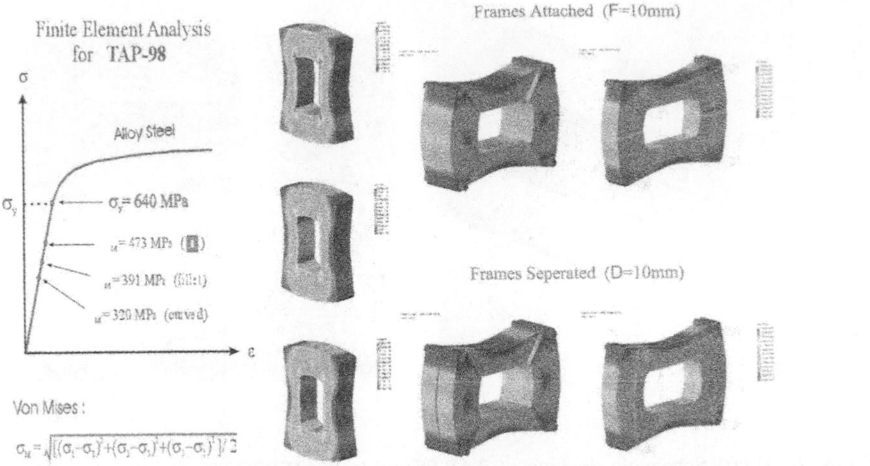

Figure 2. 3D finite element analysis (FEA) of the TAP-98 press frame. (Left) The maximum Von Mises stress occurs at the four corners of the press opening and at the center parts of the press vertical axis. A combined feature of curved window opening and small radius fillets leads to a 32% reduction in peak stress over the rectangular window designs (i.e. a decrease from 473 to 320 MPa, to be compared to a yield strength of 650 MPa of the frame material). (Right) Von Mises stress distribution of two-frame bundling scenarios. The 3D FEA shows that the attached and separated cases have approximately the same stress levels.

loading plates, made of alloy steel (T1, *ASTM A514*) with a high yield strength of 650 MPa, contribute large mechanical safety factors to the toroidal anvil press. The curved shape of the window opening blends small radius fillets in the corners with a large radius side shape. This compound radius shape reduces peak stresses significantly compared to straight-sided window designs, (Fig. 2). A detailed 3D finite element analysis (FEA) indicates a 32% reduction in stress concentration and a relatively even distribution of the maximum effective stresses (Von Mises) along the sides of the opening. The press frame material and design stress levels indicate a fatigue life expectancy well in excess of 10^4 cycles at full design load.

We have also analyzed the stress distribution for two-frame bundling scenarios, namely the attached case and the separated case. The 3D FEA study shows that these two cases have about the same stress level. The high $P-T$ neutron diffraction experiments need to take advantage of the separated frame arrangement so that the incident neutron beam can have multiple access (top, side, and front) to the sample volume. Such frame arrangement also leaves room for high $P-T$ texture (preferred orientation) studies using neutron diffraction techniques.

4. Pressure and temperature control

The pressure control system is constructed with a servo motor driven syringe-type positive-displacement pump for fine control of the pressurization and depressurization cycle, (Fig. 3). It also consists of a hand pump for initial anvil engagement and for low-pressure compression. The range of the hydraulic pressure control system is from 1 to

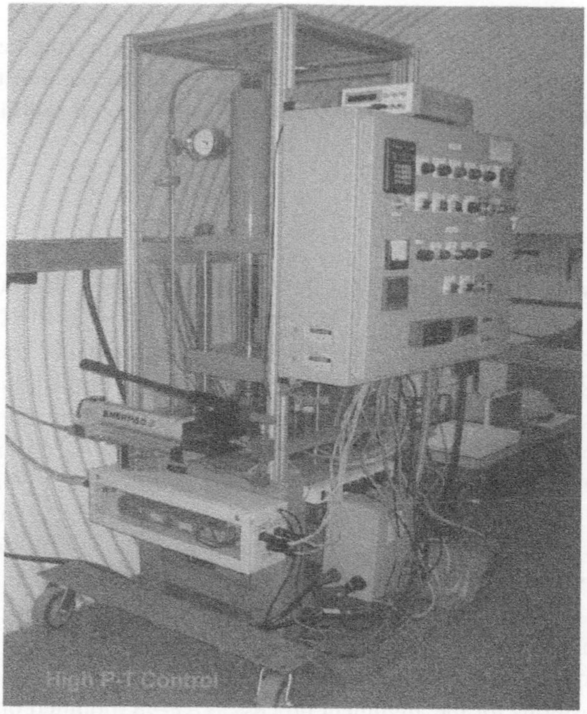

Figure 3. A photograph of the pressure–temperature control system for TAP-98 experiments. This system also includes the displacement (LVDT) meters, the translational stage control, and the remote communication box. This unit is compact and portable and can be used close to and away from neutron flight paths.

200 MPa (10 to 2000 bar) and produces a loading force up to 500 tons acting on the 17.78 cm (7-inch) diameter TAP-98 piston. The combined volume capacity of the hand pump and syringe pump to contain hydraulic oil is more than 600 cm^3, more than enough to feed the maximum piston advancement of 5 cm. The system includes displacement transducers, a digital readout, and an analog display gauge. The computer control of the pressure system is able to ramp up/down to a specified pressure at a pumping rate of 1.6 ~ 40.0 MPa/min and to hold hydraulic pressure accurately within a ±1% drift in 24 h. The pressure control system is compact and portable (on wheels) so that it can be placed readily near the neutron flight path during experiments.

The temperature control system is composed of three different but correlated components: AC electrical heating with power up to 5000 W ($V = 10 \sim 20$ V; $I = 200 \sim 500$ A), automatic microprocessor control, and Labview monitoring. A graphite furnace serves as a resistance heater; its lifetime at high P–T is between a few hours and several days and is discarded upon disassembly of the experiment. With the input from the C-type thermocouple placed inside the graphite furnace and specific settings such as temperature set point, ramp rate and dwell time, a programmable processor (Eurotherm-2404) regulates the heating power through the silicon controlled rectifier. In case of thermocouple failure, heating control can still be performed manually.

In order to avoid a possible catastrophic failure due to overheating and because of other safety concerns, we also built a Labview application monitoring the heating power and furnace resistance during the experiment. The temperature has been tested up to 2000 K for the large volume ($V = 150\ mm^3$) cell assembly, but we keep routine experiments at $T < 1500$ K with auto-control.

5. Translational stage

A translator was specifically designed for the newly established "HiPPO" beam-line for use with the new toroidal anvil press TAP-98 at LANSCE, (Fig. 4). The design of the translator for HiPPO/TAP-98 employs a multi-driver assembly with a translation/rotation/tilting five degrees of freedom, x-y-z_ω_α. Our specific design of the translator is based on the following considerations: (1) to align TAP-98 inside of HiPPO to a precision of 10 μm and 0.01°; (2) to be able to move a 680 kg (1500 lbs) weight by 6.35 mm (1/4 in.) for the y-z_ω_α adjustment and by as much as 5 cm along the neutron beam x-direction; (3) to be able to separate to two stages so that the standing position of TAP-98 can be horizontal for a double-deck or vertical for a single-deck configuration; (4) to have a compact size so that it can fit inside the HiPPO sample chamber and be easily put in and taken out; and (5) the translator should be hidden within the diffraction shadow of the TAP-98 frame to maximize neutron detection.

Three laser beams are mounted on the body of TAP-98 to define front–back, left–right, and height–tilt positions for the alignment process. TAP-98 mounted on the translational stage will therefore have a calibrated initial position repeatable for each installation. We have mounted two LVDTs to monitor the piston movement and another two LVDTs to

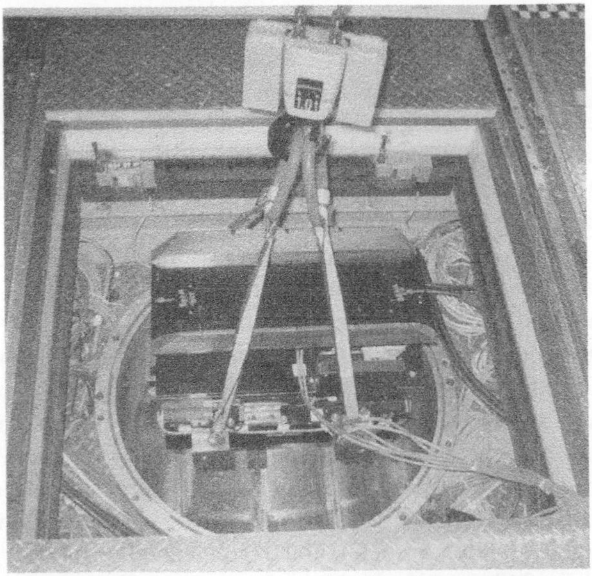

Figure 4. Placement of the translational stage (double-deck) into the HiPPO sample chamber.

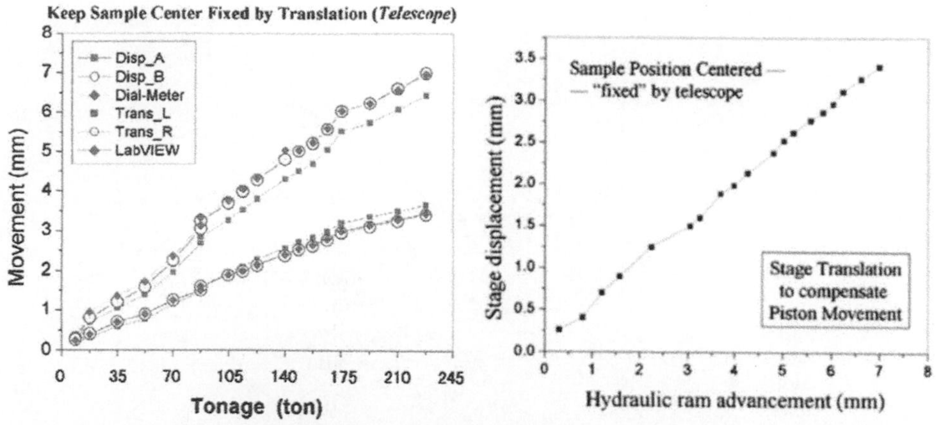

Figure 5. (Left) Hydraulic ram movement under pressure and the corresponding adjustment needed for the translation stage to maintain the sample center at its original position (i.e. the calibrated position at zero pressure). Two LVDTs (Disp_A and Disp_B) and a dial meter were used to measure the hydraulic ram displacement. Trans_L/Trans_R and LabView denote the translation-stage adjustments and the corresponding computer inputs. A high-resolution telescope was used during the experiment to track the diffraction center. (Right) Correlation between translation-stage displacement and hydraulic ram advancement, obtained based on the data shown in the left diagram.

follow x-translations (along the beam direction). A calibration run was needed to derive the correlation between piston advancement and translation adjustment at high pressures, (Fig. 5). These adjustments are necessary to keep the center of diffraction at the geometric center of the detector. We used a high-resolution telescope to monitor the movement of the sample center as pressure increases and always pull the system back to the original (calibrated) position using the translation stage. Two dial-meters were used to calibrate the LVDT readings of the hydraulic piston and the translational stage to ensure the consistency of mechanical and electronic displacement measurements. A rough two-to-one ratio in piston advancement and translation adjustment is established and we apply it in high $P-T$ experiment to maintain the origin of the sample position. This is critical to the precision of the diffraction optics.

6. Anvil and cell assembly

We used electric-discharge machining (EDM) to carve the toroidal profile on the "front-side" of the polycrystalline diamond (PCD) die-blank sintered in a tungsten carbide cylinder (De Beer, SYNDIE). The toroidal profile, first proposed by Khovstantsev et al. (1977), is designed to have all revolving circles tangential to each other so that the concave and convex portion of the toroids will have no stress concentration spots (Fig. 6). The "back-side" of the top-anvil and bottom-anvil are carved differently: the top-anvil has a hemispherical dome-shaped dip for less attenuation of the incident neutron beam, whereas the bottom-anvil has a through-hole to allow a ceramic tube for the thermocouple shielding to enter the sample. A cone-shape dent is carved at the center of the bottom-anvil so that a deformed stainless steel cone under pressures can grasp the thermocouple tube to

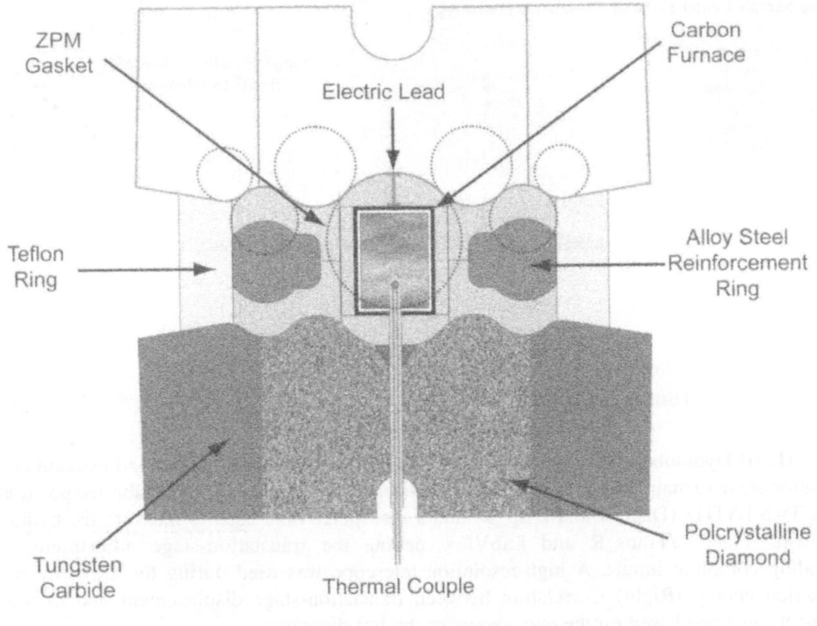

Figure 6. Diagram illustrating key elements of the anvil-cell assembly package. The toroidal profiles formed by revolving circles tangential to each other result in no stress concentration at the concave and convex portions of the anvils. This drawing shows an early design of the cell assembly with super-alloy reinforcement rings being sandwiched between two ZPM ceramic gaskets.

prevent its extrusion. A tapered angle of 1.5° (enclosed angle of 3°) is EDMed on the side of the tungsten carbide (WC) cylinder with a two-thousandth diameter-interference to the center hole of the hardened steel (50 HRc, VascoMax C-300) reinforcement ring (Fig. 7). The PCD anvil is pushed into the steel reinforcement ring with a load of about 10 ~ 15 tons; the corresponding confining pressure acting upon the PCD piece is approximately 400 MPa. A 10° slope is EDMed on the WC portion of the anvil and it extends all the way to the hardened steel reinforcement ring to allow diffracted neutrons access to an enclosed 20° axial aperture window.

Our gasket development is built upon the success of our high P–T cell assembly for the PE cell. Machinable ZPM (zirconium phosphate) ceramic material with very low thermal expansivity ($0.9 \times 10^{-6}K^{-1}$, ~1/8 of alumina and ~1/15 of magnesia) and very low

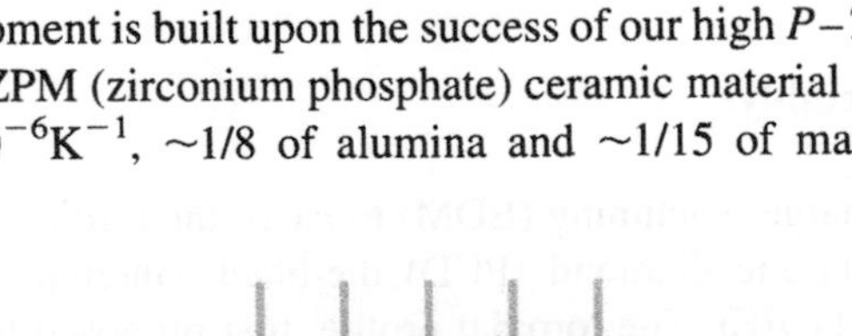

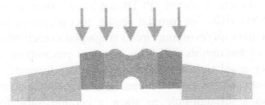

Figure 7. The pushed-in ($F = 10 \sim 15$ ton) assembly of the toroidal anvil (with a tapered angle of 1.5°, i.e. enclosed 3°) into a hardened steel ring.

thermal conductivity (0.8 ~0.9 W/m K, ~1/5 of alumina and ~1/60 of magnesia) is used in the new high $P-T$ cell assembly. The thermomechanical and neutron "transparency" properties of the ZPM ceramic make it an ideal gasket material for high $P-T$ neutron diffraction experiments. We have tested three gasket setups: (1) whole ZPM ceramic gasket; (2) ZPM gasket sandwich with a steel reinforcement ring, (Fig. 6); (3) alloy steel gasket sandwich with a lithographic disk, (Fig. 8). The first one exhibits the best performance at high pressure, however, the sample deforms significantly from a cylinder into a disk shape after pressurization and the remaining diffraction window between the top and bottom-anvils is rather small (<1.0 mm). The second one can keep a better sample shape after the application of pressure, however, it occasionally shorts out the electric heating circuit. The third one performs better than the others in terms of sample shape, insulation, as well as access for the diffracted neutron beam.

We use a Teflon ring to confine the ceramic/lithographic gasket and to avoid electrical shorting-out at high $P-T$ conditions. It stretches/elongates, survives at reasonably high temperatures, and is "transparent" to neutrons. We added an Al-6061 confining ring, which is strong, tough, and expandable, to reinforce the gasket assembly for high $P-T$ neutron diffraction experiments. It also enhances pressure efficiency quite notably. We kept the tubular furnace design, consisting of a cylindrical carbon sleeve and two flat carbon disks, for its advantage in homogeneous temperature distribution. Amorphous carbon is used as furnace material to reduce the contamination of the diffraction pattern with graphite Bragg peaks and to attenuate the mechanical instability of the furnace caused by the slippery nature of graphite. The cylinder/disk carbon furnace does not have direct contact with the anvils and the electric circuit is completed by thin platinum wires and foils (cross sectional area: $\sigma \sim 0.3$ mm^2). The cylinder/disk carbon furnace is encapsulated in a low thermal conducting ceramic, ZPM sleeve and ZPM caps, for electrical and thermal

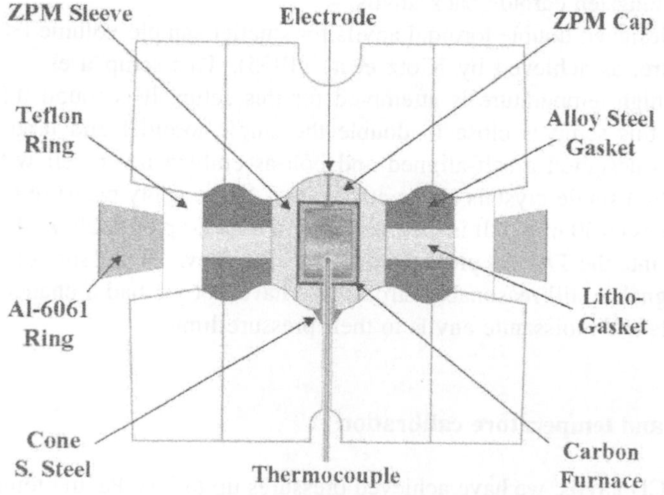

Figure 8. Cell assembly used in high $P-T$ neutron diffraction experiments. The total sample length is typically 6–8 mm. The sample is located at the center of the carbon furnace. The sample diameter is 4 mm for a metallic sample for which an electrical insulation sleeve made of NaCl is needed; the sample diameter is 5.5 mm for ceramic and mineral samples. See text for a detailed discussion of the cell parts.

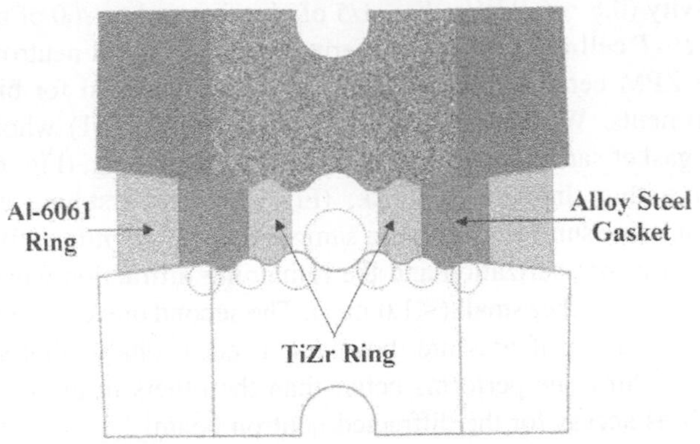

Figure 9. The double toroidal anvil and cell assembly designs for performing room-temperature diffraction experiments. The starting sample is typically 5 mm in length and diameter.

insulation. Thin platinum foils are placed on the PCD anvils to make the electrical connections extend to the WC portion of anvils.

A straight-through thermocouple with ceramic shielding tube is mounted through the central hole of the bottom-anvil. This configuration has the minimum deviatoric stress act upon the thermocouple, thus increasing its survivability. We often use thermocouples made of stronger W/W-Re wires and have achieved high success rate in temperature measurements at high pressure in this fashion. The front side of the top and bottom-anvils is painted with a Gd_2O_3 paste to block unwanted neutron scattering from anvil materials; a hole lets the incident neutron beam pass through. The front anvils and the cell assembly are packaged together after sample loading. The package is then placed inside TAP-98 between two tungsten carbide back anvils.

We have designed double toroidal anvils for smaller sample volume (<100 mm^3) and higher pressure, as achieved by Klotz et al. (1995). This setup uses a metallic gasket (Fig. 9). No high temperature is attempted for this setup. It is found that the pressure efficiency of this setup is close to double the single toroidal and large volume case. We have also designed a self-aligned and bolt-assembled inner cell with gem-quality moissanite (SiC) single crystals as the anvils (Fig. 10) to apply pressure to much smaller sample volumes (<30 mm^3). It is the inner part of a lock-up portable ZAP cell. The inner cell is loaded into the TAP-98 press. During tests it achieved a pressure of 18 GPa and the diffraction signal is still reasonably strong. We have not yet had a chance to test double toroidal anvils and moissanite anvils to their pressure limit.

7. Pressure and temperature calibration

With large PCD anvils, we have achieved pressures up to 10 GPa and temperatures up to 1800 K with a large sample volume of 150 mm^3 (Figs 11, 12) and collected high quality neutron diffraction data for the refinement of crystal structures and Debye–Waller factors. The highest pressure achieved to date using the gem SiC anvils is 18 GPa with a much smaller sample size of 20 mm^3.

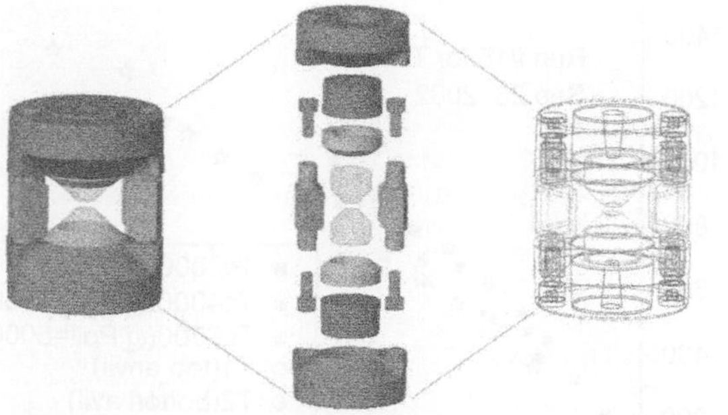

Figure 10. Exploded view of a self-aligned, bolt-assembled room-temperature cell with gem-quality moissanite single crystals as anvil material. This is an inner cell of the ZAP cell and it is designed for achieving pressures up to 20 GPa.

The temperature variation over a time period of several hours and as measured by the thermocouple is mostly within ±10 K at temperatures of 1000 to 1500 K. The sample pressure can be determined using the diffraction lines of pressure calibrants (such as NaCl, CsCl, MgO) and their well-determined thermoelastic equation of state, with the known sample temperature.

The center of the ZPM caps do not experience strong shear stress at high pressure so that both electrical conduction and thermal insulation are ensured. The electric current from an AC power supply runs through the platinum wires and the cylinder/disk carbon furnace and heats up the sample quite efficiently. At a power of 160 W, a sample volume of about 120 mm^3 was heated up to 1670 K. The anvil temperature remained below 450 K and the TAP-98 press body warmed up to 345 K (measured at the hydraulic ram).

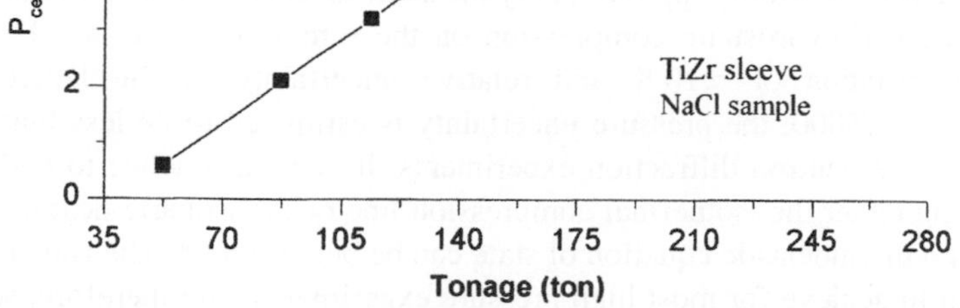

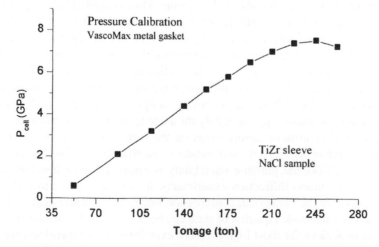

Figure 11. Pressure calibration using NaCl diffraction in a room-temperature cell made of a TiZr alloy as sample container and VascoMax steel as the gasket.

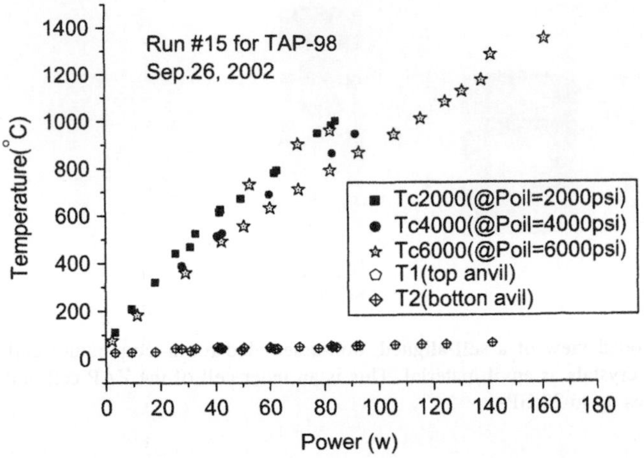

Figure 12. Temperature versus electrical-power relationship showing the heater performance at different loading forces. The calibration experiments were carried out using the cell assembly shown in Figure 8. Also shown are the temperatures measured at the back of the top and bottom-anvils.

Significant deviatoric stress is applied to the polycrystalline sample and pressure calibrant during the initial "cold" compression stage when using a solid pressure medium. The diffraction peaks are drastically broadened when the specimen is under such deviatoric stress, and the quality of the refinement will be affected correspondingly. Upon heating, deviatoric stress in the sample decreases drastically and eventually diminishes as the temperature rises (Weidner et al., 1992; Zhao et al., 1994). A quasi-hydrostatic condition can be attained because the yield strength of the material drops substantially at high temperatures and plastic flow occurs in the sample and surrounding cell assembly. In our high $P-T$ neutron diffraction experiments, we always heat the sample to 800 K at each increment in pressure for about $3 \sim 5$ min. This is applicable only for samples without phase transitions below 800 K. At this temperature, NaCl or CsCl (the pressure calibrants mixed with the powder sample) becomes very soft and the stress field applied to the sample reaches a quasi-hydrostatic pressure condition. The quasi-hydrostatic condition remains for the lower temperatures following cooling and it can be verified by comparing the peak width of Bragg lines in the diffraction patterns at high $P-T$ conditions and at ambient conditions. The diffraction pattern of samples observed at high pressure and temperature has peak widths approximately the same as at ambient conditions, which is a good indication of hydrostatic compression on the sample (Zhao et al., 1994). With a temperature variation of ± 10 K and relative uncertainty of the lattice parameter refinement of $\pm 1/5000$, the pressure uncertainty is estimated to be less than ± 0.1 GPa for our high $P-T$ neutron diffraction experiments. It is good practice to collect $P-V-T$ data following either the isothermal compression line or the isobaric heating line so that the fitting of a thermoelastic equation of state can be better traced. The isobaric heating is very difficult to achieve for most high pressure experiments, and therefore, we adopt the isothermal compression in our experiments. The accuracy of the $P-V-T$ determination for high $P-T$ neutron diffraction experiments has been discussed by Zhao et al. (1998).

8. Concluding remarks

The aim of our research is to accurately map lattice parameters, thermal parameters, bond lengths, bond angles, neighboring atomic environments, and phase stability in *P–T–X* space. Studies based on high-pressure neutron diffraction are important for multidisciplinary science.

Additional equipment for *in situ* high *P–T* neutron diffraction is being designed and tested. An aluminum pressure cell capable of high-*P* (1 GPa) and low-*T* (10 K) has been developed to study hydrostatic pressure effects at low temperature. We have also developed a large gem-crystal anvil cell, ZAP, to conduct neutron diffraction experiments at high-*P* and low-*T*. The ZAP cell can be used with various experimental techniques such as neutron diffraction, laser spectroscopy, and ultrasonic interferometry.

We are developing further high *P–T* technology with a new 2000-ton press, TAPLUS-2000, and a ZIA (D-DIA type, Wang et al., 2003) cubic anvil package to achieve pressures up to 20 GPa and temperatures up to 2000 K in routine experiments. The design of a dedicated high-pressure neutron beam-line, LAPTRON, is also underway for simultaneous high *P–T* neutron diffraction, ultrasonic, calorimetry, radiography, and tomography studies.

References

Bennett, K., Von Dreele, R.B., Wenk, H.R., 2000. HiPPO, the high-pressure preferred orientation diffractometer at LANSCE for characterization of bulk materials. Fifteenth Meeting of the International Collaboration on Advanced Neutron Sources. ICANS-XV, Vol. 1, Tsukuba, Japan, 6–9 November, pp. 473–482.

Besson, J.M., Hamel, G., Grima, T., Nelmes, R.J., Loveday, J.S., Hull, S., Häusermann, D., 1992a. A large volume pressure cell for high temperatures. High Pressure Res. 8, 625–630.

Besson, J.M., Nelmes, R.J., Hamel, G., Loveday, J.S., Weill, G., Hull, S., 1992b. Neutron powder diffractive above 10 GPa. Physica B 180–181, 907–910.

Besson, J.M., Klotz, S., Hamel, G., Grima, T., Makarenko, I., Nelmes, R.J., Loveday, J.S., Wilson, R.M., Marshall, W.G., 1995. High pressure neutron diffraction, present and future possibilities using the Paris–Edinburgh cell. High Pressure Res. 14, 1–6.

DeTeresa, J.M., Ibarra, M.R., Garcia, J., Blasco, J., Ritter, C., Algarabel, P.A., Marquina, C., delMoral, A., 1996. Spin-glass insulator state in (Tb–La)(2/3)Ca1/3MnO3 perovskite. Phys. Rev. Lett. 76, 3392–3395.

He, D., Zhao, Y., Daemen, L.L., Qian, J., Lokshin, K., Shen, T.D., Zhang, J., Lawson, A.C., 2004. Thermoelastic and texture behavior of aluminum at high pressure and high temperature investigated by in situ neutron diffraction. J. Appl. Phys. 95, 4645–4650.

Kampermann, S.P., Sabine, T.M., Craven, B.M., Mcmullan, R.K., 1995. Hexamethylenetetramine-Extinction and thermal vibrations from neutron-diffraction at 6 temperatures. Acta Crystallogr. A 51, 489–497.

Khovstantsev, L.G., Vereshchagin, L.F., Novikov, A.P., 1977. Device of toroid type for high pressure generation. High Temp. – High Pressures 9, 637–639.

Klotz, S., Besson, J.M., Hamel, G., Nelmes, R.J., Loveday, J.S., Marshall, W.G., Wilson, R.M., 1995. Neutron powder diffraction at pressures beyond 25 GPa. Appl. Phys. Lett. 66, 1735–1737.

Klotz, S., Besson, J.M., Hamel, G., Nelmes, R.J., Loveday, J.S., Marshall, W.G., 1996. High pressure neutron diffraction using the Paris–Edinburgh cell: experimental possibilities and future prospects. High Pressure Res. 14, 249–255.

Krimmel, A., Fischer, P., Roessli, B., Maletta, H., Geibel, C., Schank, C., Grauel, A., Loidl, A., Steglich, F., 1992. Neutron diffraction study of the heavy fermion superconductors UM_2Al_3 (M = Pd,Ni). Z. Phys. B 86, 161–162.

Larsen, F.K., Brown, P.J., Lehmann, M.S., Merisalo, M., 1982. Temperature dependence of thermal vibrations in beryllium as determined from short-wavelength neutron diffraction data. Philos. Mag. B 45, 31–50.

Lawson, A.C., Goldstone, J.A., Huber, J.G., Giorgi, A.L., Conant, J.W., Severing, A., Cort, B., Robinson, R.A., 1991. Magnetic structures of actinide materials by pulsed neutron diffraction. J. Appl. Phys. 69, 5112–5116.

Le Godec, Y., Dove, M.T., Francis, D.J., Kohn, S.C., Marshall, W.G., Pawley, A.R., Price, G.D., Redfern, S.A.T., Rhodes, N., Ross, N.L., Schofield, P.F., Schooneveld, E., Syfosse, G., Tucker, M.G., Welch, M.D., 2001. Neutron diffraction at simultaneous high temperatures and pressures, with measurement of temperature by neutron radiography. Mineral. Mag. 65, 737–738.

Loveday, J.S., Nelmes, R.J., Marshall, W.G., Wilson, R.M., Besson, J.M., Klotz, S., Hamel, G., Hull, S., 1995. High pressure neutron powder diffraction: the neutron scattering aspects of work with the Paris–Edinburgh cell. High Pressure Res. 14, 7–12.

Nikitin, A.N., Sukhoparov, W.A., Heinitz, J., Walther, K., 1995. Investigations of texture formation in geomaterials by neutron diffraction with high pressure chambers. High Pressure Res. 14, 155–162.

Niraimathi, A.M., Hofmann, M., 2000. Magnetic properties and ordering of $La_{1-x}Pb_xMnO_3$ with $x = 0.5$. Physica B 276–278, 722–723.

Palosz, B., Grzanka, E., Stel'makh, S., Gierlotka, S., Pielaszek, R., Bismayer, U., Weber, H.P., Proffen, T., Palosz, W., 2002. Application of powder diffraction methods to the analysis of short- and long-range atomic order in nanocrystalline diamond and SiC: the concept of the apparent lattice parameter (alp). Diffus. Defect Data Part B 94, 203–216.

Parise, J.B., Leinenweber, K., Weidner, D.J., Tan, K., Vondreele, R.B., 1994. Pressure-induced H-bonding – neutron-diffraction study of brucite, $Mg(OD)_2$, to 9.3 GPa. Am. Mineral. 79, 193–196.

Proffen, T., Billinge, S.J.L., 2002. Probing the local structure of doped manganites using the atomic pair distribution function. Appl. Phys. A A74, S1770–S1772.

Ptasiewiczbak, H., Leciejewicz, J., Zygmunt, A., 1981. Neutron diffraction study of magnetic ordering in UPd_2Si_2, UPd_2Ge_2, URh_2Si_2 and URh_2Ge_2. J. Phys. F 11, 1225–1235.

Rodriguez-Carvajal, J., Fernandez-Diaz, M.T., Martinez, J.L., 1991. Neutron diffraction study on structural and magnetic properties of La_2NiO_4. J. Phys.: Condens. Matter 3, 3215–3234.

Rupp, B., Fischer, P., Porschke, E., Arons, R.R., Meuffels, P., 1988. Neutron diffraction study of highly oxygen deficient superconducting $YBa_2Cu_3O_{6.39}$. Physica C 156, 559–565.

Steiner, T., Saenger, W., 1992. Geometry of C–H···O hydrogen bonds in carbohydrate crystal structures. Analysis of neutron diffraction data. J. Am. Chem. Soc. 114, 10146–10154.

Vogel, S.C., Hartig, C., Lutterotti, L., Von Dreele, R.B., Wenk, H.R., Williams, D.J., 2004. Texture measurements using the new neutron diffractometer HiPPO and their analysis using the Rietveld method. Powder Diffr. 19, 65–68.

Von Dreele, R.B., 1995. High pressure neutron powder diffraction at LANSCE. High Pressure Res. 14, 13–19.

Wang, Y., Durham, W.B., Getting, I.C., Weidner, D.J., 2003. The deformation-DIA: a new apparatus for high temperature triaxial deformation to pressures up to 15 GPa. Rev. Sci. Instrum. 74 (6), 3002–3011.

Weidner, D.J., Vaughan, M.T., Ko, J., Wang, Y., Leinenweber, K., Liu, X., Yeganeh-Haeri, A., Pacalo, R.E., Zhao, Y., 1992. Large volume high pressure research using the wiggler port at NSLS. High Pressure Res. 8, 617–623.

Wenk, H.R., Lutterotti, L., Vogel, S., 2003. Texture analysis with the new HiPPO TOF diffractometer. Nucl. Instrum. Methods Phys. Res. A 515, 575–588.

Zhao, Y., Weidner, D.J., SAM-85 team, 1994. Perovskite at high $P-T$ conditions: an in-situ synchrotron diffraction study on $NaMgF_3$ perovskite. J. Geophys. Res. 99, 2871–2885.

Zhao, Y., Von Dreele, R.B., Weidner, D.J., 1998. Correction of diffraction optics and $P-V-T$ determination using thermoelastic equations of state of multiple phases. J. Appl. Crystallogr. 32, 218–225.

Zhao, Y., Von Dreele, R.B., Morgan, J.G., 1999a. A high $P-T$ cell assembly for neutron diffraction up to 10 GPa and 1500 K. High Pressure Res. 16, 161–177.

Zhao, Y., Getting, I.C., Von Dreele, R.B., 1999. TAP-98: a new design of toroidal anvil press for high $P-T$ neutron diffraction. International Conference on High Pressure Science and Technology (AIRAPT-17), Hawaii, Abstract, Tu T1 H-13, p. 309.

Zhao, Y.S., Lawson, A.C., Zhang, J.Z., Bennett, B.I., Von Dreele, R.B., 2000. Thermoelastic equation of state of molybdenum. Phys. Rev. B 62, 8766–8776.

Advances in High-Pressure Technology for Geophysical Applications
Jiuhua Chen, Yanbin Wang, T.S. Duffy, Guoyin Shen, L.F. Dobrzhinetskaya, editors

Chapter 24

A new optical capillary cell for spectroscopic studies of geologic fluids at pressures up to 100 MPa

I-Ming Chou, Robert C. Burruss and Wanjun Lu

Abstract

A new optical capillary cell was constructed from square cross-sectioned fused silica-capillary tubing (300 μm × 300 μm with cavities of 100 μm × 100 μm or 50 μm × 50 μm) and a high-pressure valve that allows room-temperature studies of fluids at pressures up to 100 MPa. This capillary cell has the advantage of very precise study of fluid systems under 100 MPa pressures, not easily done in diamond anvil cells. Several key features of this cell include: (1) The ability to directly load sample fluids and monitor pressure during investigation, (2) The lack of optical distortion, (3) The small cell volume suitable for samples of limited supply (e.g. commercially available gas mixtures), (4) The high pressures that can be achieved, (5) The high-magnification, high-numerical aperture objective lens (e.g. 100×) with a short working distance that can be used due to the thin wall of the capillary tube, and (6) The heating–cooling stage or a circulating fluid bath that can be added, allowing for investigations at temperatures other than room temperature, particularly suitable for studies of gas hydrates.

Raman spectra have been collected from the cell at room temperature for methane, ethane, propane, n-butane, and for two gas mixtures containing up to nine components as a function of pressure up to 70 MPa. The spectra document the shift in Raman bands with pressure as well as constrain the detection limits for various gas species in the mixtures. Preliminary experiments on the diffusion of methane in water were conducted by monitoring the concentration of dissolved methane in water as a function of time and distance from the vapor–water boundary, immediately after perturbation of an equilibrium state induced by a sudden incremental change in methane pressure.

1. Introduction

Interpretation of Raman and fluorescence spectra of hydrocarbon fluid inclusions and spectroscopic observations of reactions in hydrocarbon–water systems require high-quality reference spectra of individual gases, gas mixtures, and hydrocarbon–water systems. Several types of high-pressure optical cells (HPOC) have been used in investigations of geologic fluids, particularly those involving studies of gas hydrates, and include the diamond anvil cell (e.g. Chou et al., 2000; Kumazaki et al., 2004), HPOC with glass or sapphire window(s) (e.g. Smelik and King, 1997; Nakano et al., 1999; Schicks and Erzinger, 2003; Schicks and Ripmeester, 2004), and silica or sapphire tubes (e.g. Chou et al., 1990; Sum et al., 1997; Thieu et al., 2000; Hansen et al., 2001a). Each type of HPOC has its own advantages and disadvantages, especially in terms of construction, sample loading, control and measurement of pressure and temperature, accessibility and efficiency for spectroscopic analysis, and safety. With this in mind, we constructed a new optical capillary cell from commercially available square cross-sectioned fused silica-capillary tube (300 μm × 300 μm with cavities of 100 μm × 100 μm or 50 μm × 50 μm) and a high-pressure valve that allows studies of fluids at room temperature and pressures up to 100 MPa. This cell is a modification of the type used by Chou et al. (1990) to calibrate a Raman spectroscopic system for *in-situ* non-destructive analysis of fluid inclusions hosted by minerals as well as to observe morphological changes of methane hydrate crystals as a function of pressure, temperature, and time (Stern et al., 1998). This chapter describes the construction of a new HPOC and its current and potential applications.

2. The new high-pressure optical capillary cell

2.1. The construction of the cell

A square flexible fused silica-capillary tube (300 μm × 300 μm with cavities of 100 μm × 100 μm or 50 μm × 50 μm), commercially available with Polymicro Technologies, LLC (www.polymicro.com), forms the optical window in this cell. This, or other available size (such as 75 μm × 75 μm and 12 μm × 12 μm cavities) was epoxied inside a stainless steel high-pressure capillary tubing (1.59 mm OD, 0.76 mm ID, and 40 mm long), which was connected to a high-pressure valve by a sleeve–gland assembly (Fig. 1). The silica tubes were supplied with a polyimide coating (363 ± 15 μm OD). To provide clear windows for optical observation of samples in the tube, a section of this brown polyimide protection cover was removed using a hydrogen flame. However, to add extra strength to the silica tube, the polyimide layer over the section of the tube to be inserted into the stainless steel tube should not be removed (Fig. 1). To further protect the silica tube, especially under high internal pressure, a slightly longer stainless steel high-pressure capillary tube can be used to enclose the entire silica tube; however, in this case, it is necessary to grind off parts of the stainless steel tube immediately above and below the windows (Fig. 1c). The stainless high-pressure capillary tubing, sleeve, gland, and high-pressure valve are available from High Pressure Equipment Company (www.highpressure.com).

2.2. Sample loading and pressurizing procedures

Figure 2 is a schematic diagram showing the sample loading and pressurizing system for Raman spectroscopic analysis of non-aqueous fluid samples. The procedures are:

(a) evacuate and flush the capillary cell and pressure line with sample fluid before loading: (1) close all valves; (2) open v-1, v-2, v-3, v-4, and v-5, and use the syringe (S-1) to evacuate the line; (3) close v-1 and open v-6 to load the line with the sample fluid from tank 1 (T-1), and use the pressure gauge (G-1) to read the pressure; (4) close v-6 and open v-1 to flush the line with the sample fluid; (5) after repeating the above steps several times, close v-1 and open v-6 to finally load the line with the sample fluid;

(b) load a section of water to separate sample fluid from the pressurizing fluid: (1) close v-2, v-5 and v-6; (2) open v-10 and v-11, and use the syringe (S-2) to fill the line between v-10 and v-11 with water, and then close v-10 and v-11;

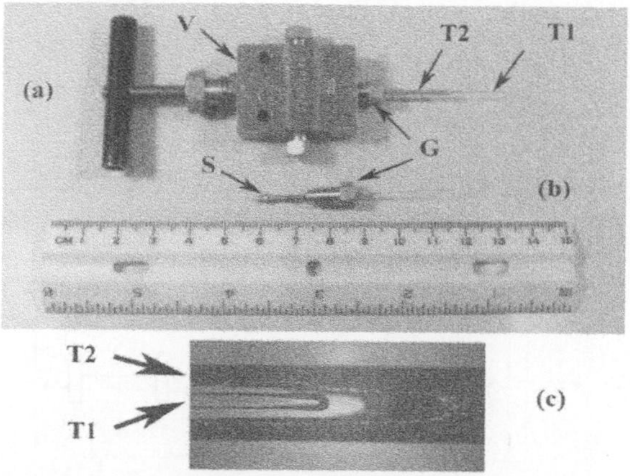

Figure 1. Photograph of the new high-pressure optical cell. The following procedures were used to construct the HPOC as shown in (a): (1) remove the protection cover of polyimide at the window section of the square capillary tube (T-1) by a hydrogen flame; (2) seal this end of the tube by the hydrogen flame; (3) insert the other end of the tube through the high-pressure stainless steel tube (T-2), which already had a sleeve (S) and a gland (G) in place and filled with freshly mixed epoxy; (4) after the epoxy was set, cut off the end of the silica tube near the sleeve, and verify that the cavity was not blocked by epoxy; and (5) connect the assembly (b) to the high-pressure valve (V) as shown in (a). To protect the silica tube, a slightly longer stainless high-pressure capillary tube was used to enclose the entire silica tube, and parts of the stainless steel tube were ground away to expose the window, as shown in (c). Other kinds of protective tube with windows can be constructed and enclosed the silica tube by sliding it along the stainless tube. We have also constructed an HPOC with two open ends, and each end was connected to a high-pressure valve. This cell is slightly bulky, but it was much easier for cleaning and loading samples. However, make sure that the two valves are fixed in position, and that the distance between the two valves is slightly shorter than the silica tube in between, taking the advantage of the flexibility of the silica tube. The ruler shows scales in both centimeters and inches for (a) and (b). For scale in (c), the stainless tubing (T-2) is about 1.6 mm in OD. Part numbers: T-1 (Polymicro Technologies; 2001513, and 2001515), T2 (High Pressure Comp. (HiP); 15-9A1-030, 0.762 mm ID × 1.588 mm OD), V (HiP, 15-14AF1), G (HiP, 15-2AM1), and S (HiP, 15-2A1).

(c) pressurize the sample: (1) open v-9, v-12, v-13, and v-14, and then open v-15 to load the line and pressure generator (PG) with a high-pressure fluid (CH_4, N_2, or CO_2) from T-2; (2) close v-15 and open v-8 and v-7 to pressurize the sample fluid by pushing the water section between v-10 and v-11 into the loop between v-5 and v-7; (3) open v-5 (Fig. 3a), and read the pressure of the sample fluid in the capillary tubing (C) at G-1 or G-2; (4) use the pressure generator to reach higher pressure, but close v-3 before reaching the maximum pressure rating for G-1; (5) close v-3, v-4, and v-5 during Raman analysis, and check the sample pressure after analysis by opening v-3 (for low-pressure reading at G-1) or v-5 and v-4 (for high-pressure reading at G-2); (6) use the pressure generator or the exhaust valve (v-16) to reduce the sample pressure.

To load a sample of water or aqueous fluid at the enclosed end of the tube, use the following procedures: (1) remove the tube from the high-pressure valve; (2) heat the window section of the tube before dipping the open end of the tube in the sample solution;

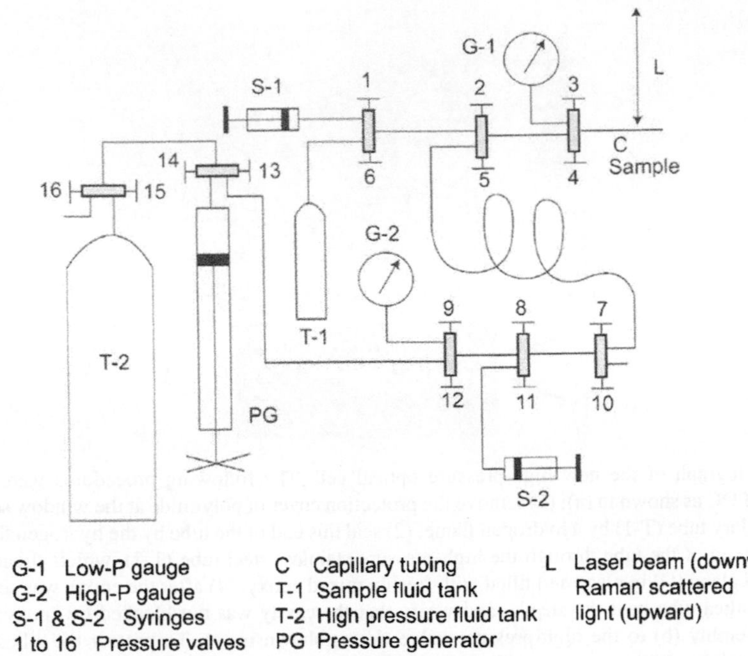

Figure 2. A schematic diagram showing the sample loading and pressurization system for Raman spectroscopic analysis of fluid samples. For details, see text.

(3) after a sufficient amount of time allowing sample fluid to be sucked into the tube during cooling, force the sample solution to the enclosed end using a centrifuge; (4) reconnect the tube to the high-pressure valve and then it is ready for the evacuation and flushing procedures described above before pressurization with other fluids (Fig. 3b).

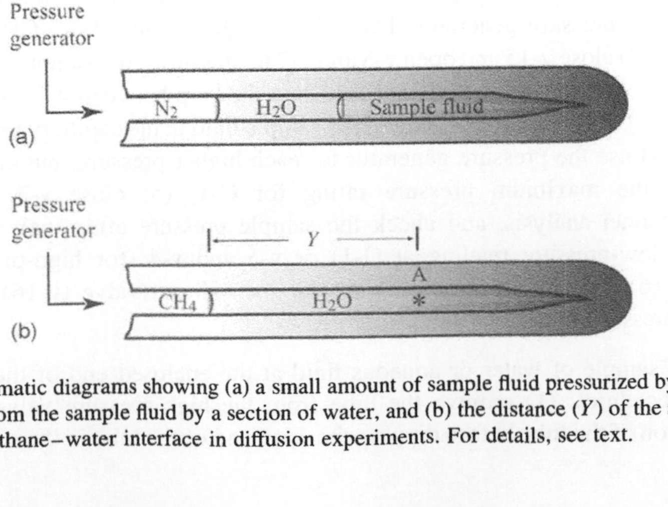

Figure 3. Schematic diagrams showing (a) a small amount of sample fluid pressurized by N_2 gas, which was separated from the sample fluid by a section of water, and (b) the distance (Y) of the sample position (*) from the methane–water interface in diffusion experiments. For details, see text.

3. Applications

3.1. Raman spectra of fluids at various pressures and room temperature

A JY/Horiba LabRam HR Raman system was used in this study, with 532.06 nm (frequency doubled Nd:YAG) laser excitation, a 100× Olympus objective with 0.9 numerical aperture (unless specified otherwise), and ~20 mW laser power at sample surface, with various analysis times and accumulations per spectrum. Spectra were acquired with a 600-groove/mm grating with a spectral resolution of about 2 cm^{-1}. Raman spectra have been collected from the cell for methane, ethane, propane, *n*-butane, and also for two gas mixtures containing up to nine components as a function of pressure up to 70 MPa. For example, Figure 4 shows Raman spectra of methane at room temperature and pressures ranging from 0.1 to 12.1 MPa. Note the strong signals of rotational–vibrational transitions for the asymmetric C–H stretching band (some are marked by $*$) at lower pressures. Significant shifts of up to 7 cm^{-1} of the methane ν_1 peak position near 2918 cm^{-1} were observed as pressure was increased to 70 MPa, in agreement with previous observations (Seitz et al., 1993 and references therein; Thieu et al., 2000). The details will be presented elsewhere (Chou et al., in preparation). Figure 5 shows a resulting Raman spectrum for propane at 0.4 MPa and the background signals of the fused silica tube container, which do not interfere the signals of most of hydrocarbon fluids. Figure 6 shows Raman spectra of a commercial gas mixture (Scotty 14 custom mix, Scotty Specialty Gases, SUPELCO, www.sigma-aldrich.com) containing minor amount of CO_2; the mixture contains 0.506 mole% CO_2, 7.00 mole% propane, and seven other components. Note that the CO_2 Raman signals (indicated by $*$) can be detected at

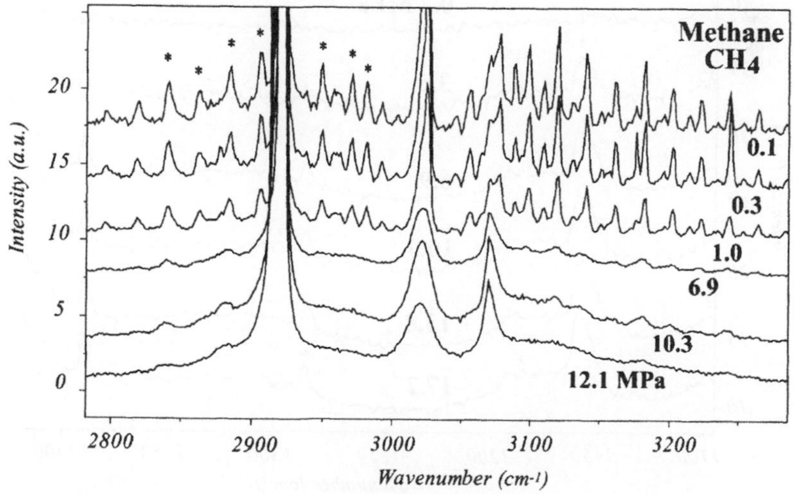

Figure 4. Raman spectra of methane at various pressures, showing the C–H symmetric stretching band (ν_1) near 2918 cm^{-1}, antisymmetric C–H stretching band near 3020 cm^{-1}, and the overtone of antisymmetric bending mode near $2 \times 1534 = 3068$ cm^{-1}. Note the strong signals of rotational–vibrational transitions for the asymmetric C–H stretching band (some are marked by $*$) at lower pressures. Spectra were collected at 0.1 MPa (360 seconds (s) with six accumulations (a)), 0.3 MPa (320 s6a), 1.0 MPa (100 s6a), 6.9 MPa (15 s6a), 10.3 MPa (15 s6a), and 12.1 MPa (10 s6a).

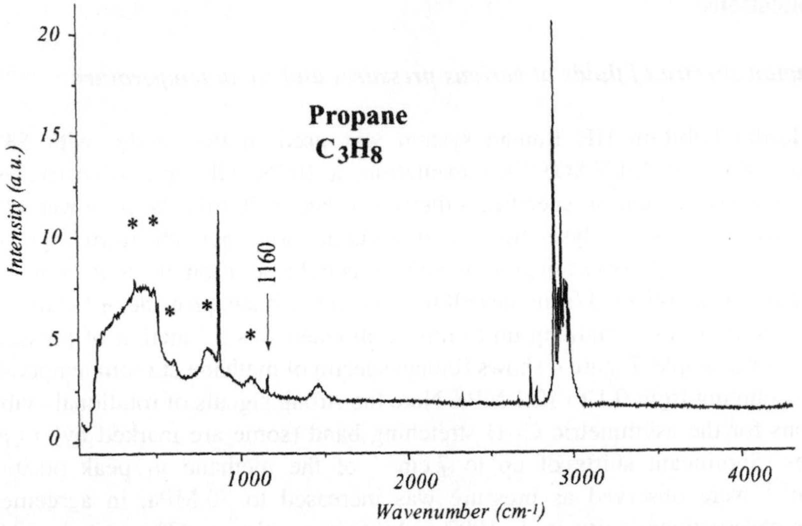

Figure 5. Raman spectrum of propane at 0.4 MPa. Signals from fused silica are indicated by *. The arrow indicates the signal near 1160 cm^{-1} shown in Figure 6. The spectrum was collected for 600 s with two accumulations.

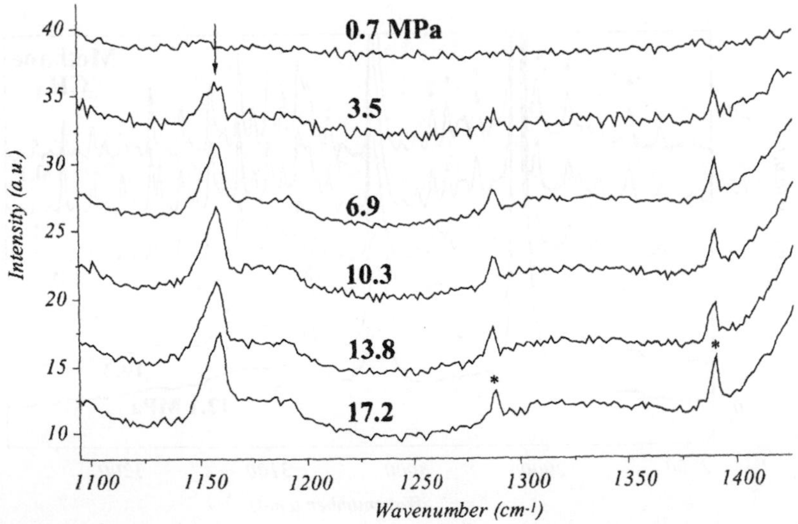

Figure 6. Raman spectra for a commercial gas mixture (Scotty 14 custom mix; Cat. No. 303102, project # 0191116-002; 100 PSIG) containing minor amount of CO_2; the mixture contains 0.506 mole% CO_2, 7.00 mole% propane, and seven other components. The CO_2 Raman signals (*) can be detected at 3.5 MPa total pressure. The Raman signal near 1160 cm^{-1} (arrow) is assigned to propane (see Figure 5). The spectra were collected at 0.7 and 3.5 MPa for 240 s with three accumulations, and the rest were collected for 200 s with three accumulations.

3.5 MPa total pressure, and the Raman signal near 1160 cm^{-1} (arrow) is assigned to propane.

It should be pointed out that solubility of gases in water, which was used to pressurize the sample and in direct contact with the sample, may change the composition of the sample slightly. However, there are two ways to minimize these changes: (1) separate the sample fluids by several small sections of water in the tube, such that maximize the sample fluid/water ratio and saturate the water with gases with only minor change in composition; (2) perform Raman analyses quickly after the sample is pressurized by water taking the advantage of slow diffusion rate of dissolved gas in water, as demonstrated later for the methane–water system.

3.2. *Diffusion of methane in water*

Experiments on the diffusion of methane in water were conducted by monitoring the concentration of dissolved methane in water, as a function of time and distance from the vapor–water boundary, immediately after the perturbation of an equilibrium state induced by a sudden change in methane pressure. Figure 7 shows Raman spectra of methane dissolved in water at room temperature (21.7°C) about 115 min after a 12.2 mm section of water in the sealed end of the tube was pressurized by methane at 24.47 MPa at the positions $Y = 0.70$ mm (a) and 4.12 mm (b) away from the methane vapor–water interface (Fig. 3b). The concentration of dissolved methane in water (in mole fraction, $[X(CH_4)]$) can be calculated from the peak area ratio ($[A(CH_4)_{aq}/A(H_2O)]$) (see Chou and Burruss, 2004) through a newly calibrated linear relation:

$$[X(CH_4)] = 1.0563\,[A(CH_4)_{aq}/A(H_2O)] \quad (1)$$

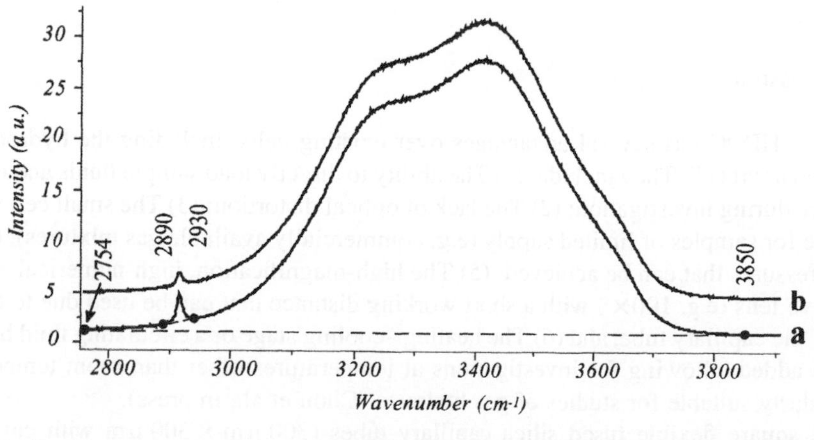

Figure 7. Raman spectra of methane dissolved in water at room temperature about 115 min after the water was pressurized by methane at 24.47 MPa at the positions $Y = 0.70$ mm (a) and 4.12 mm (b) away from the methane vapor–water interface (Fig. 3b). The spectra were collected with a 40× long-working distance objective lens for 40 s with three accumulations. The peak area for methane is integrated between 2890 and 2930 cm^{-1}, and that for water is between 2754 and 3850 cm^{-1}.

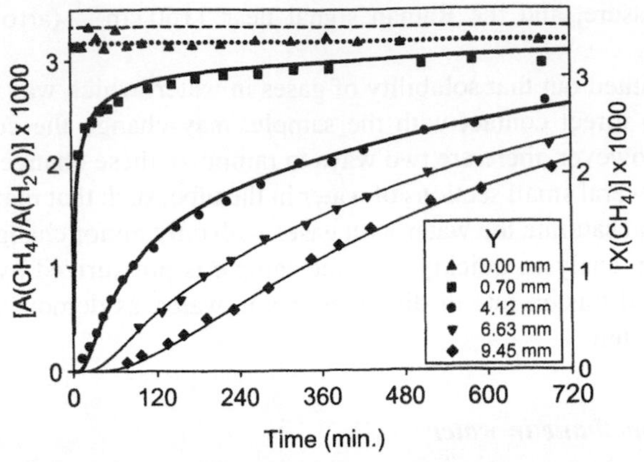

Figure 8. Raman peak area ratios for CH_4/H_2O and concentration of dissolved methane in water (in mole fraction, X, as calculated from Eq. (1)) as a function of distance between the sample spot and methane–water interface (Y, see Figure 3b) and time after the water was pressurized by methane at 24.47 MPa. The dotted curve represents least squares fit of the data at methane–water interface ($Y = 0$ mm). The solid curves represent calculated values for $Y = 0.70$, 4.12, 6.63, and 9.45 mm, assuming the left boundary condition is defined by the dotted curve and a diffusion coefficient of $1.66 \times 10^{-9}\ m^2\ s^{-1}$ for methane in water. This diffusion coefficient gives the best fit of the observed concentration profiles. The horizontal dashed line represents the equilibrium concentration of $X_{CH_4} = 0.00352$ (Sun et al., 2003), which is slightly higher than the dotted line. This difference resulted from the small difference in temperature between current experiment and those experiments for defining the coefficient in Eq. (1). Each spectrum was collected for 40 s with three accumulations. For calculation of peak area, see Figure 7.

Figure 8 shows the experimental results as well as the predicted methane concentration profiles based on a diffusion coefficient of $1.66 \times 10^{-9}\ m^2\ s^{-1}$. Details as well as the effect of pressure on the diffusion coefficient of methane in water will be presented elsewhere (Lu et al., in preparation).

4. Discussion

The new HPOC has several advantages over existing cells, including the hydrothermal diamond anvil cell. They include: (1) The ability to directly load sample fluids and monitor pressure during investigation; (2) The lack of optical distortion; (3) The small cell volume suitable for samples of limited supply (e.g. commercially available gas mixtures); (4) The high pressures that can be achieved; (5) The high-magnification, high-numerical aperture objective lens (e.g. 100×) with a short working distance that can be used due to the thin wall of the capillary tube, and (6) The heating–cooling stage or a circulating fluid bath that can be added, allowing for investigations at temperatures other than room temperature, particularly suitable for studies of gas hydrates (Chou et al., in press).

The square flexible fused silica capillary tubes (300 μm × 300 μm with cavities of 100 μm × 100 μm) can hold pressures up to 100 MPa, and those with 50 μm × 50 μm cavity should be able to hold much higher pressures, but were not tested. Similar tubes from the same source have been used routinely to pressures up to 400 MPa (Yonker et al., 1995). The new HPOC pressurized to 30 MPa has been cooled to liquid-nitrogen temperature, but has not been tested at temperatures greater than 50°C. The manufacturer claims that the tube can be operated at temperatures up to 350°C, and intermittently up to 400°C.

Calibration of the relationship between peak area ratio and the concentration for quantitative Raman spectroscopic method analysis of geologic fluids could be easily performed using this HPOC (Chou and Burruss, 2004). The advantages of this method when compared with the fluid inclusion method of Guillaume et al. (2003) are: (1) elimination of the complicated and time consuming processes for the synthesis and preparation of fluid inclusions; (2) both pressure and temperature of the fluid samples can be controlled and measured directly; and (3) uniform sample geometry, which is not available in synthetic fluid inclusions. Note that the coefficient for Eq. (1) given by Chou and Burruss (2004) is 0.9521 instead of 1.0563, because a different type of HPOC was used, even though both sets of spectra were collected by the same Raman spectrometer.

The use of a high-magnification, high-numerical aperture objective lens improves the efficiency for collecting Raman spectra of samples. This is particularly important for kinetic studies, such as determining the diffusion rate of methane in water described above. The high efficiency increases the number of high-quality spectra that can be collected in a certain period of time. It should also be pointed out that the geometry of the sample in this HPOC is not available in many other types of optical cells (e.g. Smelik and King, 1997; Nakano et al., 1999; Chou et al., 2000; Schicks and Erzinger, 2003; Schicks and Ripmeester, 2004; Kumazaki et al., 2004). The main differences are (1) the orientation of the liquid/vapor interface, which is perpendicular to, instead of parallel with, the windows, and (2) samples far away from this interface are available for observations. For example, when water at the end of the enclosed tube was pressurized by methane gas (Fig. 3b), it allows the concentrations of methane in water being determined as a function of time and distance from the methane vapor–water boundary. Under this configuration, it is also possible to observe metastable phases when methane hydrate crystals were formed at lower temperatures; sI methane hydrate formed in the water-rich region, and the other type of methane hydrate crystals, possibly sII having Raman signatures similar to those reported by Schicks and Erzinger (2003) and Schicks and Ripmeester (2004), were observed to form in the vapor-rich region (Chou et al., in review). On the other hand, when a small volume of methane gas at the end of the tube was pressurized by water, it was possible to observe the growth of methane hydrate crystals at lower temperatures until all methane gas was consumed, and then determine the methane concentration in water in equilibrium with hydrate in the absence of a methane vapor phase.

We have used hydrothermal diamond anvil cell extensively (e.g. Chou et al., 1998, 2000), and it is an excellent optical cell covering a wide pressure–temperature range with many applications (for reviews, see Chou, 2003, Bassett et al., in press). However, when compared to the current HPOC, it has two major drawbacks: (1) it is difficult to load samples quantitatively, and (2) it is also difficult to determine the sample pressure accurately. The current HPOC is similar to our previous optical cell that we used to calibrate a Raman spectrometer (Chou et al., 1990). However, the previous optical cell used an O-ring seal, which was a weak point causing unexpected explosion, and the routine measurements were limited to about 70 MPa (Seitz et al., 1993). Also, the round and thick-walled capillary tubing in that design caused optical distortion (Stern et al., 1998) and low efficiency in collecting Raman spectra. Similarly, the optical cell designed by Hansen et al. (2001a) can not be used for pressures more than 15 MPa (Hansen et al., 2001a,b, 2002), and when compared with our HPOC, the wall thickness of their sapphire tube is quite large (6.35 mm OD × 4.43 mm ID), and the sample volume in their 39.5-mm-long tube is also several orders of magnitude larger.

5. Conclusions

A new high-pressure optical capillary cell was constructed from square cross-sectioned fused silica tubing for spectroscopic studies of geologic fluids at pressures up to 100 MPa. The HPOC has been tested between −190 and 50°C, and the cell may possibly be operational at temperatures up to 400°C, although the maximum pressure may be lower at high temperature. The cell is particularly suitable for investigation of gas hydrates. The construction of the cell is relatively simple. It is portable and easy to use with Raman spectrometers or at synchrotron X-ray facilities for analysis of reaction products. Other advantages of the new HPOC include: (1) The ability to load directly sample fluids and monitor pressure during investigation; (2) The lack of optical distortion; and (3) The use of a high-magnification, high-numerical aperture objective lens with a short working distance due to the thin wall of the capillary tube, such that detailed morphological features of solid samples can be observed and high-quality Raman spectra can be obtained efficiently in kinetic studies.

Using this new HPOC, we have collected high-quality reference Raman spectra of individual gases, gas mixtures, and hydrocarbon–water systems. They can be readily used for interpretation of Raman spectra of hydrocarbon fluid inclusions and spectroscopic observations of reactions in hydrocarbon–water systems. However, it should be emphasized that the strength of the new HPOC has not been fully evaluated. The pressure and temperature limits may depend on some unknown factors. Therefore, sudden pressure and temperature changes should be avoided, and protection devices should be used to minimize possible damages to the optical system as well as to the observers.

Acknowledgements

We would like to thank Anurag Sharma of Rensselaer Polytechnic Institute, Harvey Belkin, and Robert Seal, Jr. of USGS, and an anonymous reviewer for their critical reviews, and Kristin Dennen of USGS for drafting Figure 2. The use of trade, product, industry, or firm names in this report is for descriptive purposes only and does not constitute endorsement by the US Government.

References

Bassett, W.A., Chou, I.M., Anderson, A.J., Mayanovic, R., 2005. Aqueous chemistry in the diamond anvil cell up to and beyond the critical point of water. In: Manaa, M.R. (Ed.), Chemistry at Extreme Conditions. Elsevier, Amsterdam, The Netherlands, pp. 223–240.

Chou, I.M., 2003. Hydrothermal diamond-anvil cell, application to studies of geologic fluids. Acta Petrol. Sin. 19 (2), 213–220.

Chou, I.M., Pasteris, J.D., Seitz, J.C., 1990. High-density volatiles in the system C–O–H–N for the calibration of a laser Raman microprobe. Geochim. Cosmochim. Acta 54, 535–543.

Chou, I.M., Blank, J.G., Goncharov, A.F., Mao, H.K., Hemley, R.J., 1998. In situ observations of a high-pressure phase of H_2O ice. Science 281, 809–812.

Chou, I.M., Sharma, A., Burruss, R.C., Shu, Jinfu, Mao, Ho-kwang, Hemley, R.J., Goncharov, A.F., Stern, L.A., Kirby, S.H., 2000. Transformations in methane hydrates. Proc. Natl Acad. Sci. 97 (25), 13484–13487.

Chou, I.M., Burruss, R.C., 2004. Raman spectroscopic method for quantitative analysis of binary aqueous organic solutions. Proceedings of the 11th International Symposium on Water–Rock Interactions, WRI-11, Vol. 1. 27 June–2 July, Saratoga Springs, New York, USA, pp. 621–624.

Chou, I.M., Wanjun Lu, Burruss, R.C., in press. Simultaneous synthesis of structure I and structure II methane hydrates: a new interpretation. Fifth International Conference on Gas Hydrates, June 13–16, Trondheim, Norway.

Guillaume, D., Teinturier, S., Dubessy, J., Pironon, J., 2003. Calibration of methane analysis by Raman spectroscopy in H_2O–NaCl–CH_4 fluid inclusions. Chem. Geol. 194, 41–49.

Hansen, S.B., Berg, R.W., Stenby, E.H., 2001a. High-pressure measuring cell for Raman spectroscopic studies of natural gas. Appl. Spectrosc. 55, 55–60.

Hansen, S.B., Berg, R.W., Stenby, E.H., 2001b. Raman spectroscopic studies of methane–ethane mixtures as a function of pressure. Appl. Spectrosc. 55, 745–749.

Hansen, S.B., Berg, R.W., Stenby, E.H., 2002. How to determine the pressure of a methane-containing gas mixture by means of two weak Raman bands, ν_3 and $2\nu_2$. J. Raman Spectrosc. 33, 160–164.

Kumazaki, T., Kito, Y., Sasaki, S., Kume, T., Shimizu, H., 2004. Single-crystal growth of the high-pressure phase II of methane hydrate and its Raman scattering study. Chem. Phys. Lett. 388, 18–22.

Lu, W.J., Chou, I.M., Burruss, R.C., Yang, M.Z., in preparation. *In-situ* study of mass transfer in aqueous solutions under high pressures via Raman spectroscopy: I. A new method for the determination of diffusion coefficients of methane in water near hydrate formation conditions (to be submitted to Applied Spectroscopy).

Nakano, S., Moritoki, M., Ohgaki, K., 1999. High-pressure phase equilibrium and Raman microprobe spectroscopic studies on the methane hydrate system. J. Chem. Eng. Data 44, 254–257.

Schicks, J.M., Erzinger, J., 2003. Coexistance of two different methane hydrate phases at moderate pressure and temperature conditions. Eos 84, F843.

Schicks, J.M., Ripmeester, J.A., 2004. The coexistence of two different methane hydrate phases under moderate pressure and temperature conditions: kinetic versus thermodynamic products. Angew. Chem. Int. Ed. 43, 3310–3313.

Seitz, J.C., Pasteris, J.D., Chou, I.M., 1993. Raman spectroscopic characterization of gas mixtures. I. Quantitative composition and density determination of CH_4, N_2, and their mixtures. Am. J. Sci. 293, 297–321.

Smelik, E., King, H.E. Jr., 1997. Crystal-growth studies of natural clathrate hydrates using a pressurized optical cell. Am. Mineral. 82, 88–98.

Stern, L.A., Hogenboom, D.L., Durham, W.B., Kirby, S.H., Chou, I.M., 1998. Optical-cell evidence for superheated ice under gas-hydrate-forming conditions. J. Phys. Chem. B 102, 2627–2632.

Sum, A.K., Burruss, R.C., Sloan, D.S. Jr., 1997. Measurement of clathrate hydrates via Raman spectroscopy. J. Phys. Chem. B 101, 7371–7377.

Sun, R., Huang, Z., Duan, Z., 2003. A new equation of state and Fortran 77 program to calculate vapor–liquid phase equilibria of CH_4–H_2O system at low temperatures. Comput. Geosci. 29, 1291–1299.

Thieu, V., Subramanian, S., Colgate, S.O., Sloan, E.D., Jr. 2000. High-pressure optical cell for hydrate measurements using Raman spectroscopy. In: Gas Hydrates, Challenges for the Future. Ann. NY Acad. Sci. 912, 983–992.

Yonker, C.R., Zemanian, T.S., Wallen, S.L., Linehan, J.C., Franz, J.A., 1995. A new apparatus for the convenient measurement of NMR spectra in high-pressure liquids. J. Magn. Reson. Ser. A 113, 102–107.

Advances in High-Pressure Technology for Geophysical Applications
Jiuhua Chen, Yanbin Wang, T.S. Duffy, Guoyin Shen, L.F. Dobrzhinetskaya, editors

Chapter 25
Internal and external electrical heating in diamond anvil cells

Natalia Dubrovinskaia and Leonid Dubrovinsky

Abstract

Main principles of internal and external electrical heating in diamond anvil cells (DACs) are described. A double-gasket assemblage and its modifications as well as a simple gasket assemblage are discussed for internal heating. Pressures up to ~90 GPa and temperatures up to 3100 K can be achieved using DAC assemblages with internal heating. Simple external heaters, graphite and whole-cell heaters for external heating are described. Pressures up to ~300 GPa and temperatures up to 1500°C have been achieved in experiments using DACs with external heating. A comparison of some advantages and disadvantages of external electrical and laser heating techniques for DACs is presented. Examples of successful high-pressure–high-temperature experiments on electrically conducting and insulating materials (Fe–Mg alloy, $Fe_{0.95}Ni_{0.05}$, *FeO) are given.*

1. Introduction

Understanding the behaviour of crystalline matter at extreme conditions is an integral part of natural sciences. Geoscientists, solid-state physicists and chemists and materials scientists all regularly face problems involving crystalline and amorphous materials at high pressures and temperatures. As a response to growing experimental challenges, numerous high-pressure and high-temperature crystallographic techniques have been developed during the past several decades. Available technology often dictates goals of high-pressure research instead of being guided by specific key problems. For example, progress in mineralogy, crystal chemistry and geochemistry of Earth's upper mantle and transition zone is related to advances in the generation of corresponding pressures and temperatures (to 25–30 GPa and 2300 K) in multi-anvil and diamond anvil apparatus. Development of new *in situ* microanalytical techniques and X-ray sources (synchrotrons and "in-house" facilities) has opened possibilities for modelling processes in the deep Earth and planetary interiors in the megabar pressure range. During the last few decades, diamond anvil cell (DAC) techniques have become the most successful method of pressure generation capable of working in the multi-megabar pressure range. However, there are still problems related to carrying out high-temperature experiments with DACs. Different types of DACs and methodology of their applications, the advantages and problems of DACs with different designs for crystallographic and spectroscopic studies have been widely discussed in the literature (Hazen and Finger, 1982; Eremets, 1996; Hemley, 1998). Here we summarize some methodological aspects of high-temperature experiments in electrically heated DACs, and give examples of successful studies of geophysically important materials.

Methods of the electrical heating can be divided into the internal (heating element is placed inside a pressure chamber) and external (the whole pressure chamber is heated with external high-temperature elements) ones. We consider both methods below.

2. Internal electrical heating

Due to a combination of unique physical and chemical properties, such as hardness, wide band gap, high electron and hole mobility and relative chemical inertness, diamond has found wide applications in modern science and technology (Harlow, 1998). As the least compressible and stiffest substance with exceptional thermal conductivity and extremely low thermal expansion, transparent from the far infrared through the deep ultraviolet, diamond is irreplaceable for the high-pressure technique.

The thermal conductivity of diamond, the highest among solid materials (Harlow, 1998), is important and useful for technological applications, but it creates significant problems for designing high-temperature experiments in DACs. In order to demonstrate the main difficulty associated with internal electrical heating in DACs, let us consider a very simplified one-dimensional task of the heat flow from the heated foil or wire towards diamond anvils separated from the heating element by a pressure medium (for example, periclase, MgO) (Fig. 1). Temperature distribution in such a simple case is described by the equation:

$$T_{\text{sample}} = T_{\text{diamond}} + q\left(\frac{\delta_{\text{MgO}}}{\lambda_{\text{MgO}}} + \frac{\delta_{\text{diamond}}}{\lambda_{\text{diamond}}}\right)$$

where T_{sample} is the temperature at the surface of the heated metal, T_{diamond} is the temperature at the outer diamond surface, λ is the thermal conductivity, δ is the thickness of the thermal isolation layer or diamond, q is the power of the heat source. It is easy to estimate that about 100 W of power is necessary to maintain the temperature of the sample

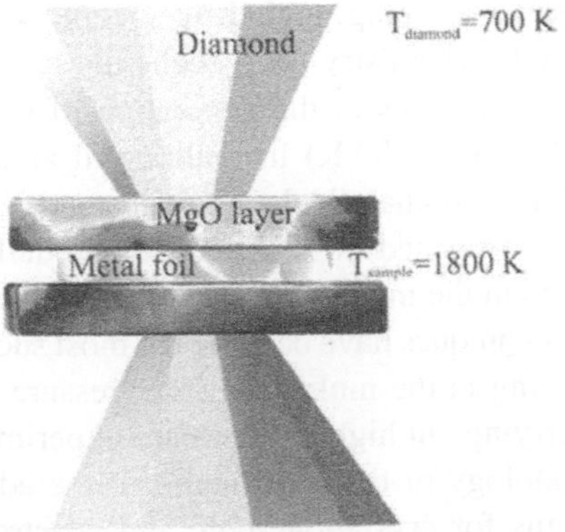

Figure 1. Schematic diagram of the internally heated DAC. A layer of metal (foil or wire) is surrounded by the layer of a thermal insulator (MgO, for example).

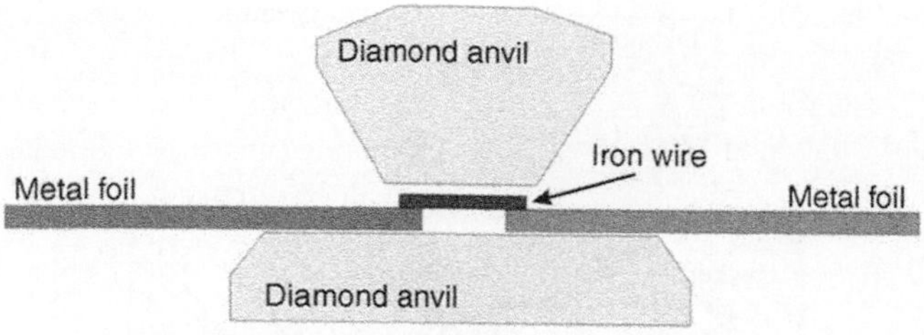

Figure 4. Diagrammatic view of simple gasket assemblies for internal wire (after Boehler 1986) and the foil (after Dubrovinsky and Saxena 1998, Dubrovinskaia et al. 1997). Internal electrical heating in DAC is based on electrical contacts between two metallic gaskets or, as shown here, electrical leads are attached to the foil.

It is convenient to use DAC with large culets (for example, of 0.4–0.8 mm in diameter) (Dubrovinsky et al., 1998). A hole in the center of the flattened part is drilled using an erosion unit (BETSA). The samples are then placed into the holes. The sample diameter is always 25–100 μm less than the diameter of corresponding holes in the gasket. Such an arrangement allows one to heat both conducting materials and insulators electrically (Dubrovinsky et al., 1998).

The wires or foils are heated with DC power (0–18 V/0–20 A power supply, TSX1820P, Thurlby Thandar, Inc., maximum power required 20–40 W). At temperatures higher than 1200 K radiation of visible light from the whole sample is observed. It is known (Dubrovinskaia et al., 1997) that in the wire heating technique at room pressure, the temperature gradient across the gasket hole (100–200 μm in diameter) is not large. Due to the very high thermal conductivity of DAC diamonds, thermal gradients within the hole are in the order of 50 K within a 50 μm diameter. Note that the externally and internally heated experiments are carried out in flowing argon atmosphere which protects the cell from oxidation and cools it simultaneously.

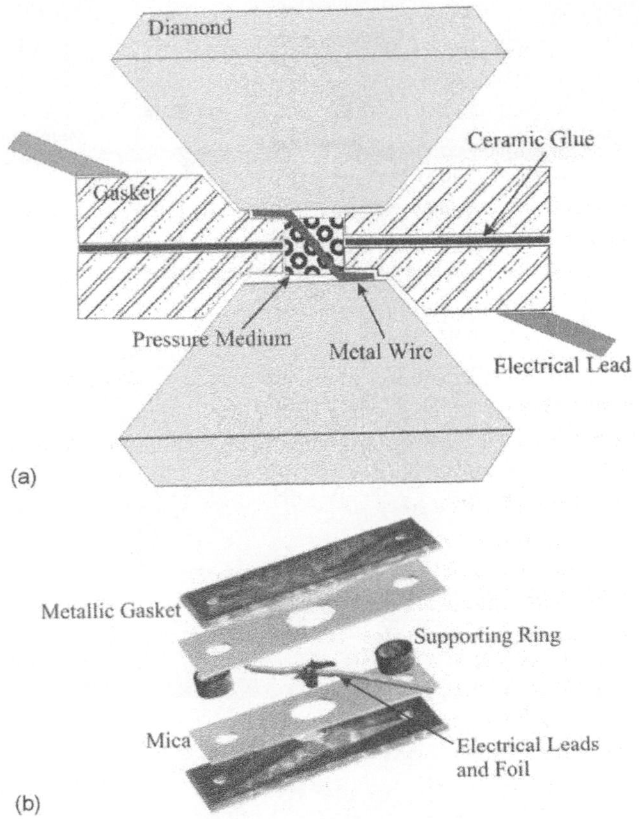

Figure 3. Diagrammatic view of double-gasket assemblages for (a) metallic wire (after Boehler, 1986) and (b) foil (after Dubrovinsky and Saxena, 1998; Dubrovinsky et al., 1998) internal electrical heating in DAC. (A layer of electrical insulator between two metallic gaskets in (b) is not shown, electrical leads are attached to the foil).

its center using the DAC with large culets (for example, of 0.6–0.8 mm in diameter) (Dubrovinsky et al., 1998). A hole in the center of the flattened part is drilled using an erosion drill (BETSA). The sample is then placed into the hole. The sample diameter is always 25–100 μm less than the diameter of corresponding holes in the gasket. Such an arrangement allows one to heat both conducting materials and insulators electrically (Dubrovinsky et al., 1998).

The wires or foils are heated with DC power (0–18 V/0–20 A power supply, TSX1820P, Thurbly Thandar, Inc., maximum power required 20–40 W). At temperatures higher than 1200 K, radiation of visible light from the whole sample is observed. It is known (Dubrovinskaia et al., 1997) that in the wire heating technique at room pressure, the temperature gradient across the gasket hole (100–200 μm in diameter) is not large. Due to the very high thermal conductivity of DAC diamonds, thermal gradients within the hole are in the order of 50 K within a 50 μm diameter. Note that it is recommended experiments are carried out in flowing argon atmosphere, which preserves the cell from oxidation and cools it simultaneously.

One more modified double-gasket assemblage for internal resistive heating method was offered by Zha and Bassett (2003). While Boehler et al. built the high PT chamber by making a gasket hole smaller than the diamond anvil culet in the metallic sandwiched gasket and placing the electrical isolation layer under high pressure, Zha and Bassett employed a nonmetallic gasket for holding the heater-sample assemblage. This design allowed them to avoid a problem of failure of the electrical isolation layer under high pressure. The authors reported pressures up to 10 GPa and temperatures up to 3000 K. They suggested that pressures up to 50 GPa are possible.

Internal electrical heating assemblages based on double-gasket techniques allow an increase in the operational pressure range over 70 GPa and provide stable homogeneous heating of a relatively large area (50–70 μm diameter) at temperatures over 1800 K. They can easily be coupled with synchrotron, or in-house X-ray facilities (Dubrovinsky and Saxena, 1998; Dubrovinsky et al., 1998) as well as with Raman spectrometers. Melting experiments using such a design, however, proved to be difficult, due to significant power loss, and the reduction in strength of the materials at high temperatures results in instability of the complex heating assemblage (Fig. 3b).

2.2. *Simple gasket assemblage*

Electrical leads can be brought into the gasketed DAC in a relatively simple and straightforward manner by covering the metal gasket with electrically isolating material (corundum or periclase, for example) and placing the ends of the leads between the diamonds (Mao et al., 1987). However, due to risk of the diamonds cutting the wire, and/or short-circuits between the gasket and leads, such a simple system works efficiently only up to 35–40 GPa and to about 1800 K (Mao et al., 1987). Significantly higher pressures and temperatures were achieved (Dubrovinsky et al., 2001, 2003) by (i) employing a gasket without a hole, (ii) placing a heating element (wire or foil) *over* the ends of electrodes, and (iii) replacing one diamond with a hard material plate (for example, a 2 mm thick SUMITOMO INC. synthetic double-side polished diamond plate or a 4 mm thick sintered cBN plate produced by LINATEC (Ukraine)) (Fig. 4). Although X-ray diffraction studies using the assemblage without a hole are difficult, such an assemblage can be successfully applied, for example, to investigating chemical reactions between molten metals and/or oxides at megabar pressure range. In particular, Figure 4 shows a series of photographs collected during an experiment to alloy magnesium and iron at a pressure of 89 GPa and a temperature of 3400°C. A backscattering electron image of the melted metal droplet in the recovered sample may be seen. Note that although magnesium melts at a lower temperature than iron, it is possible to reach the conditions at which both iron and magnesium will melt simultaneously at their contact, if one slowly increases the electrical current passing through the magnesium foil.

The internal electrical heating in DAC is attractively simple from a technical point of view (although sample preparation is time consuming), but it has some limitations. Originally it was designed to study samples of electrically conductive materials. Investigation of insulating materials requires some constructive modifications to the experimental set up. Temperature measurements using spectroradiometric methods have the same difficulties as those associated with laser-heated DACs. As with laser heating,

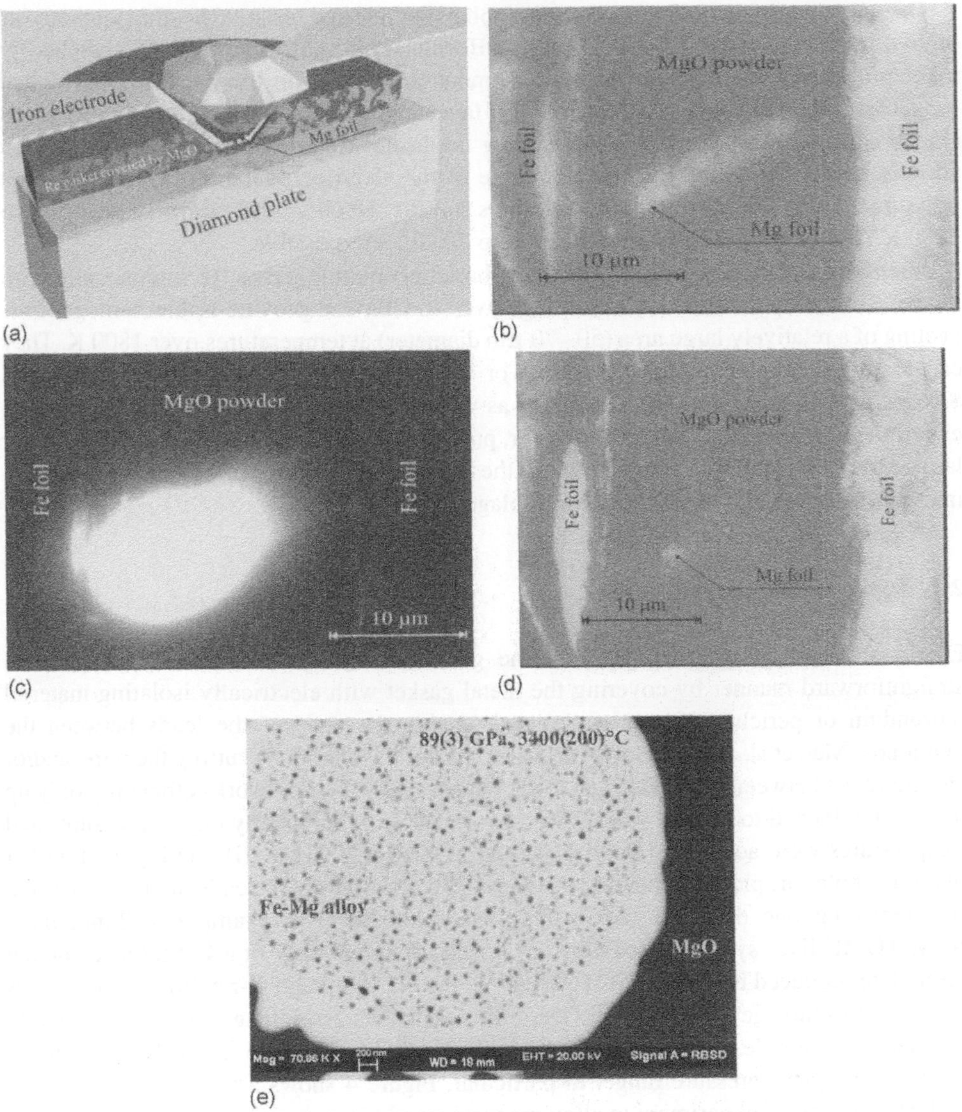

Figure 4. (Supplementary information). (a) Schematic diagram of internal electrical heating assemblage and (b) photographs of the experimental set up before heating, (c) and at 89(3) GPa and 3400(200)°C, (d) and after quenching. A rhenium gasket of 250 μm thickness was indented to 25–30 μm between the diamond anvils with culets of 250 or 300 μm size. The gasket was covered with periclase-based cement and pure periclase was placed into the indentation and around the Mg foil. Iron foil of 0.2 mm thickness flattened to a thickness less than 10 μm at the end was used as electrical leads. The magnesium foil was heated by a DC current with a stabilized power-supply operating at 25 V/25 A range. (e) Back-scattering electron image of the metal droplet in the sample recovered after an experiment in the internally electrically heated diamond anvil cell at 89(3) GPa and 3400(200)°C. On average, the droplet contains 7.8(5) at.% Mg.

large temperature gradients and thermal stresses are created inside the pressure chamber. It should also be noted that there are temperature gradients inside the heated wire or foil, therefore the spectroradiometrically measured surface temperature must be corrected, as proposed, for example, by Williams and Knittle (1991).

3. External electrical heating

External electrical heating allows one to generate temperatures over a wide range (from 300 K to over 1500 K), which are easy to measure (with thermocouples at $T < 2800$ K), and which can be maintained at a constant value (± 5 K at 1500 K) for several hours. Moreover, in an externally heated DAC at $T > 800$ K stresses are practically absent and the heating is largely homogeneous (Eremets, 1996; Dubrovinsky et al., 1999). For these reasons, the external electrical heating is the perfect complementary technique to the laser heating method for studying materials at extreme conditions. However, with external heating many problems arise (Schiferl, 1987; Adams and Christy, 1992; Eremets, 1996): oxidation and graphitisation of diamond, rapid creep deformation of materials, welding of the gasket to the anvils, and so on. Although conducting experiments at constant temperature in resistively heated DACs is relatively easy, maintaining constant pressure during heating is a problem. Differences in the thermal expansion of diamonds, seats, and various mechanical parts of the cell often lead to significant variations in pressure. Externally heated DACs are usually bigger in size than their room- or low-temperature counterparts and hence fitting them to standard analytical equipment (X-ray stages, spectrometers, etc.) may be problematic.

There are large number of designs for externally electrically heated DACs, so only the three most common types are considered below with a resistive wire heater placed around the pressure chamber, with a graphite heating element surrounding the diamonds, and the whole-cell heating assemblage.

3.1. Simple external heaters

The majority of currently available high-temperature cells (including products commercially produced by BETSA and DIACELL) are operated to 900 K and employ resistive wire heaters placed in a cavity just around the diamond anvils (Moore et al., 1970; Kikegawa, 1987; Schiferl, 1987; Adams and Christy, 1992; Bassett et al., 1993; Fei and Mao, 1994; Fei, 1996). Heaters placed around the diamond seats, as in so-called hydrothermal DACs (Bassett et al., 1993), allow temperatures over 1400 K, but only at relatively low pressures (a few GPa). A double-stage heating assemblage for the Mao-Bell type DAC allows one to reach temperatures over 1000 K at megabar pressure ranges (Fei and Mao, 1994; Fei, 1996). In this assemblage one heater is placed around the diamonds inside the cylinder of the Mao-Bell DAC, and another one around the cylinder.

The simple system for external resistive wire heating for a 4-pin DAC made from an inconel alloy is shown in Figure 5. A ceramic cylindrical heater with the outer and inner diameters of 24 and 7 mm, respectively, and a height of 2.5 mm is supplied with 30 holes

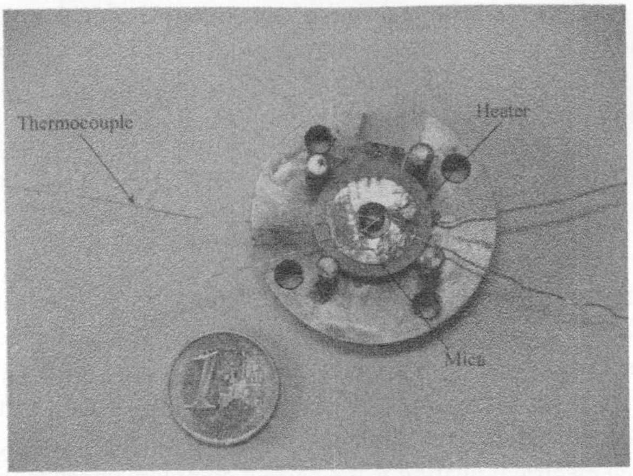

Figure 5. Photographs of the external resistive wire heating assemblage mounted on 4-pin DAC.

(or trenches). It is thermally and electrically isolated from the body of the cell by mica sheets of about 0.1 mm thickness. A platinum wire of 0.5 mm diameter and length of about 50 cm is used as a heating element. An S or K type thermocouple is placed near the hole in the gasket (Fig. 5). The gasket, preferably made from Re or Ir, is mounted on one of the diamonds using graphite bond (AREMCO INC.). Without any special precautions (external cooling, inert atmosphere for diamonds, etc.) such a simple system allows one to reach temperatures up to 900 K and to accomplish any compression experiment in a DAC with high-temperature studies. Another important advantage of the external resistive wire heating assemblage is the high compatibility with different facilities for *in situ* studies – X-ray, Raman and Mössbauer spectroscopy, X-ray absorption measurements, etc. (Fig. 6).

3.2. Graphite heater

A graphite heating assemblage allows one to reach temperatures over 1500 K and operates at pressures over 300 GPa (Dubrovinsky et al., 1999, 2000; Rekhi et al., 1999). Figure 7 shows a typical example of an external electrical heating assemblage (Dubrovinsky et al., 2000; Rekhi et al., 1999) employing flexible graphite foils (GOODFELLOW INC). Two rectangular slabs (approx. 19.5 × 8.5 mm) were cut from a flexible graphite foil of 1 mm thickness. The slabs were indented simultaneously with the Re-gasket sandwiched between them before being put between two thin molybdenum sheets of about 250 μm thickness. The Mo sheets contain holes, so allowing contact between the diamonds and the gasket. The whole arrangement was isolated from the cell body using thin sheets of mica

Figure 6. Photographs of the externally resistively wire-heated DAC mounted at the stage at ID24 beam line at ESRF (X-ray absorption measurements) (a) and on the Mössbauer spectrometer at the BGI (b). (c) Examples of Mössbauer spectra of FeO collected at elevated pressures and temperatures.

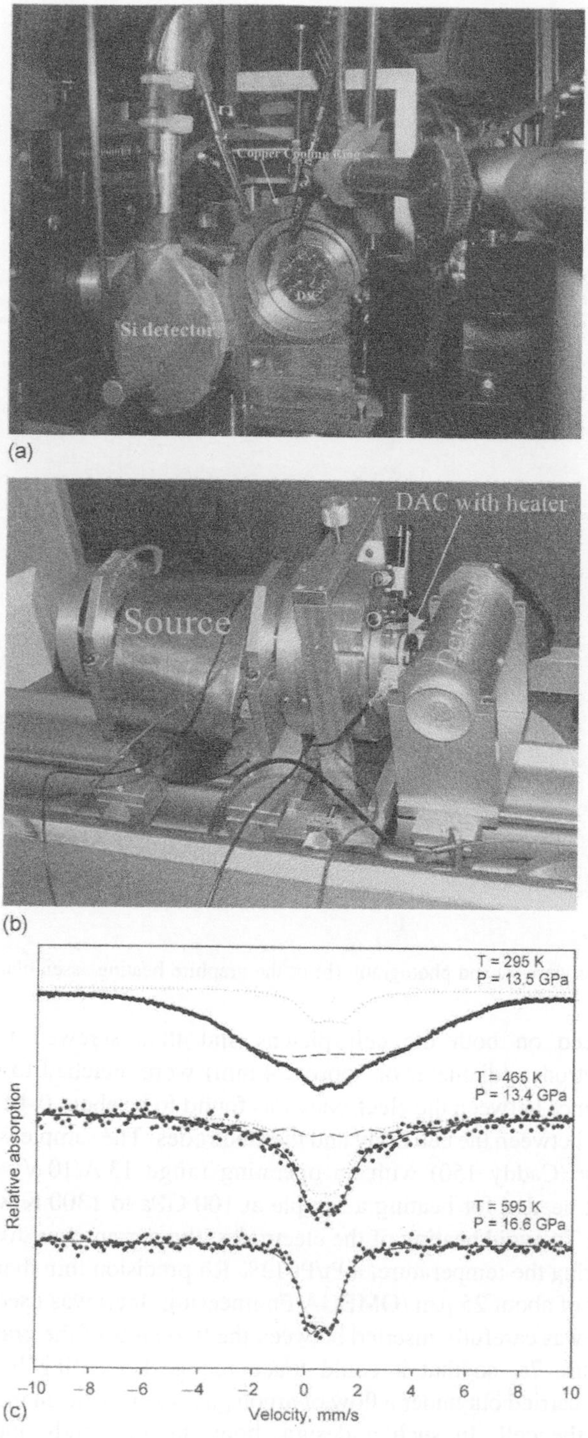
Copper Cooling Ring
DAC
Si detector
(a)
DAC with heater
Source
Detector
(b)
T = 295 K
P = 13.5 GPa
T = 465 K
P = 13.4 GPa
T = 595 K
P = 16.6 GPa
Relative absorption
−10 −8 −6 −4 −2 0 2 4 6 8 10
Velocity, mm/s
(c)

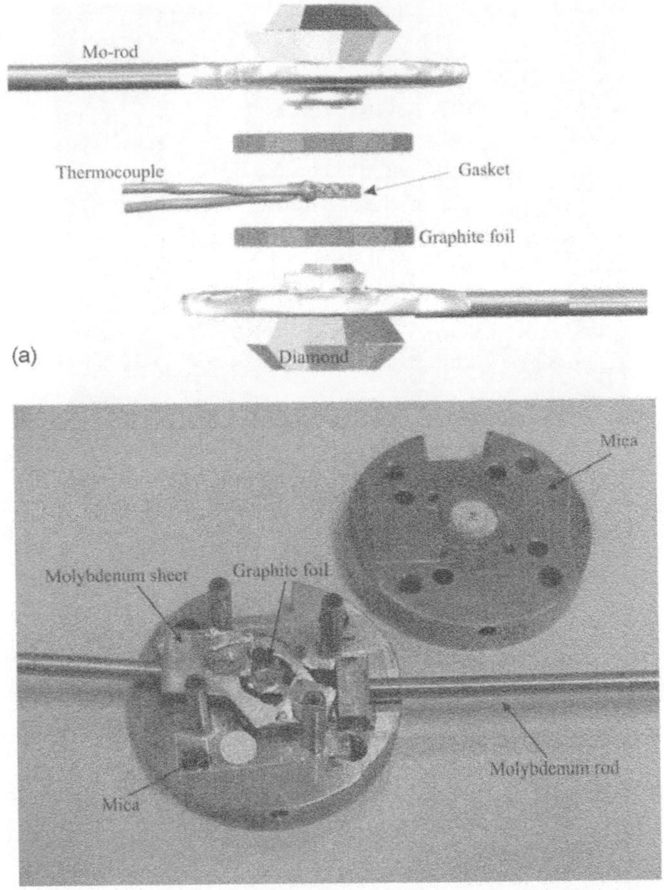

(a)

(b)

Figure 7. Schematic view (a) and photograph (b) of the graphite heating assemblage.

(20–50 μm) placed on both the cell platens and then screwed to the cell. Two molybdenum electrodes (diameter of approx. 4 mm) were attached to the molybdenum sheets. The resistance between the electrodes was found to be about 0.4 Ω, and it was very high (Megaohms) between the cell body and the electrodes. The sample was heated using a DC power supply (Caddy 150) with an operating range 13 A/10 V–150 A/26 V. The maximum current needed for heating a sample at 100 GPa to 1300 K was around 120 A (~150 W power). To avoid heating of the electrodes, the current was divided into various loops. For measuring the temperature, a Pt/Pt 13% Rh precision thin thermocouple with a junction diameter of about 25 μm (OMEGA Engineering, Inc.) was used. The junction of the thermocouple was carefully inserted between the two parts of the graphite heater foils, as shown in Figure 7a, so that it could touch the gasket sitting between them. All experiments were carried out under a flow of argon gas, so creating an inert atmosphere, as well as cooling the cell. In such a design, both the diamonds and the gasket are homogeneously heated up to a temperature of 1500 K, while the maximum temperature measured near the outer surface of the diamonds by inserting a K-type thermocouple was found to be around 800 K.

Although a graphite heating assemblage provides stable, reproducible, long-time (several hours) heating (temperature stability is 10° at 1000°C), it has a few drawbacks. Preparation of the heater is time consuming and the same heater cannot be used twice. Experiments can be conducted only when pressure increases (during heating graphite sheets become thinner and the load of the body of the cell is necessary to ensure they remain intact. Significant temperature gradients in diamonds often result in their stress and destruction. But the most important problem is variation of pressure during heating due to uneven thermal expansion of different parts of the cell (diamonds, diamond-seats, etc.). For example, heating to 1000 K in a DAC with the graphite heating assemblage (Dubrovinsky et al., 1999) can double the pressure (Fig. 8). Note that, attempts to reduce unwanted changes in pressure due to differential expansion of different parts of the DAC lead to an increase in a size of the DAC assemblages. One reason that some of the externally heated DACs, especially the hydrothermal DAC, are large and springs can be placed well away from the heated sample to minimize the effect. Internal chambers and tubes for air cooling the parts of the cell to reduce differential expansion also take up space. However, stabilizing the load on various parts of the DAC assemblage does not completely prevent from changes in pressure upon heating.

3.3. Whole-cell heater for DAC

One of the major goals in the development of a high-pressure–high-temperature whole-cell heated DAC for reliable, routine use to 1300 K and 100 GPa was its capability for conducting (quasi)isobaric experiments (Dubrovinskaia and Dubrovinsky, 2003). The high-pressure–high-temperature system comprises three subunits, namely, the anvil assembly, mechanical loading mechanism, and heaters (Fig. 9).

As a prototype for the anvil assembly so-called TAU and Merrill–Bassett type cells were used (Merrill and Bassett, 1974; Sterer et al., 1990; Dubrovinskaia and Dubrovinsky, 2003).

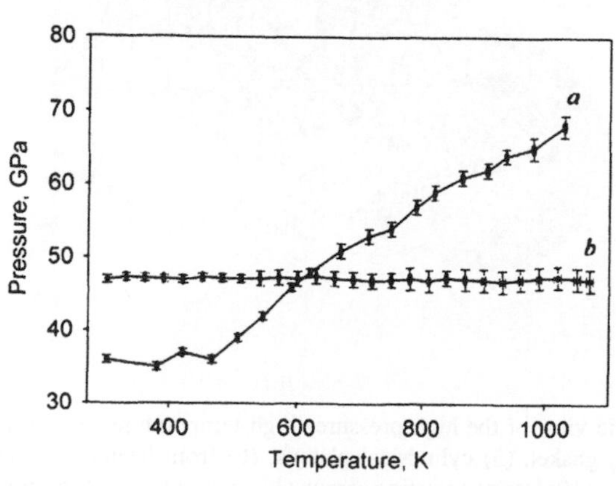

Figure 8. Pressure variation, on heating, in the graphite heater assemblage (a) and in the whole-cell external heating system (b). In both experiments a mixture of Pt and NaCl was used and pressure was determined from the Pt thermal equation of state.

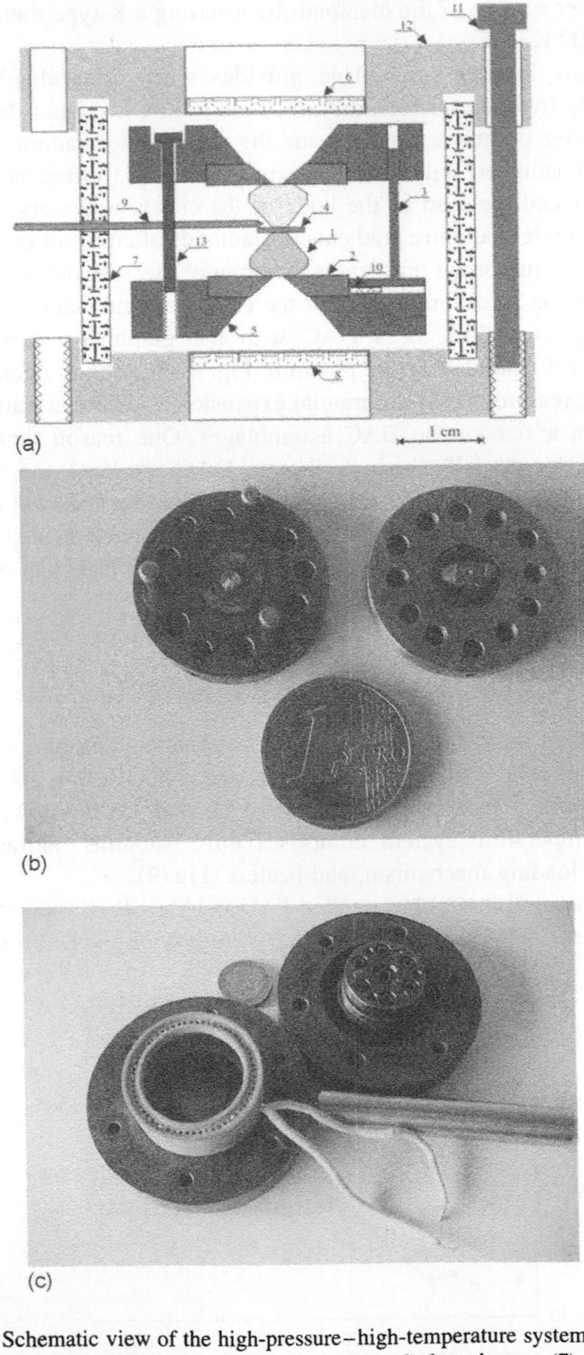

Figure 9. (a) Schematic view of the high-pressure–high-temperature system: (1) diamond, (2) backing plate, (3) guide pin, (4) gasket, (5) cylindrical platens, (6) front heater, (7) cylindrical heater, (8) rear heater, (9) thermocouple, (10) lateral adjusting screw, (11) one of the six M5 screws with hexagonal heads for applying a load during heating, (12) cylindrical plate, (13) one of the six M3 Allen screws used for room-temperature operations. (b) The diamond anvil assembly. (c) The mechanical loading mechanism and heaters.

The anvil assembly (Fig. 9) consists of two cylindrical platens, backing plates, diamond anvils, and a gasket. For room-temperature operations six Allen M3 tightening screws are used. On one of the otherwise identical platens there are three guide pins (Fig. 9). The lateral positions of the backing plates are adjustable and locked with three fine adjusting screws with M2 threads. Different alloys were tested as construction materials – Inconel 718, Udimet 700, Republica, and Nimonic. The best performance was demonstrated by cells made of Nimonic – up to 850°C in air, and up to 1100°C in an inert atmosphere.

Two massive cylindrical plates are used to apply mechanical loading to the anvil assembly (Fig. 9). One of the plates has six holes, while the other has six holes with M5 threads for tension screws. Screws made with hexagonal heads are easily accessible and can be adjusted even during heating. Both plates are made of Nimonic. The plates also have cylindrical holes and a trench for heaters (Fig. 9).

The system includes three heaters – two mounted directly on the outside of the front and back platens, and one placed around the cell (Fig. 9). The front- and rear heaters are flat and have a hole (about 10 mm in diameter in the rear heater, and 20 mm in the front heater) for optical access to the sample. Heaters made of pyrophilite or zirconia ceramics were found to be satisfactory to 1200°C. As heating elements, either a nichrome wire (working range up to 800°C in air), a molybdenum wire (tested to 1300°C in an inert atmosphere), or a platinum wire (working equally well in air and in an Ar atmosphere to highest temperature reached at high-pressure experiments – 1100°C) can be used.

Total resistance of the heating wires varies from 50 to 100 Ohm. For heating, a power supply consisting of a Eurotherm controller (2404) and a thyristor plate (TE 10A) was used. The thyristor operates in phase angle firing mode and is appropriate for 230 V heating elements. The maximum output current is 16 A. The maximum output power is 3600 W.

A thermocouple (S type) is placed near the center of the high pressure cell, next to the gasket-sample interface (Fig. 9). It was found that on heating to ~1000°C the temperature gradient in the center of the high-pressure assemblage was negligible. In the 5 mm diameter area around the diamonds, the temperature varies no more than 2–3°C.

During heating to 300–350°C between 5 and 100 GPa, the pressure remains constant to within the accuracy of measurements, while at higher temperatures fine mechanical adjustment of the load is necessary to maintain constant pressure (which is controlled using the equation of state of an internal standard, for example, Pt (Fig. 8)), so opening the way for conducting almost isobaric high-temperature studies. As an example, Fig. 10 shows the results of measurements for the thermal expansion of the ϵ-$Fe_{0.95}Ni_{0.05}$ alloy at 37(1), 64(1) and 92(2) GPa at temperatures up to 1200 K.

4. Conclusions

Finally, a comparison of some advantages and disadvantages of external electrical and laser heating techniques for DACs is presented (Internal electrical heating has basically the same advantages and disadvantages as laser heating (Table 1)). It is clear from the table

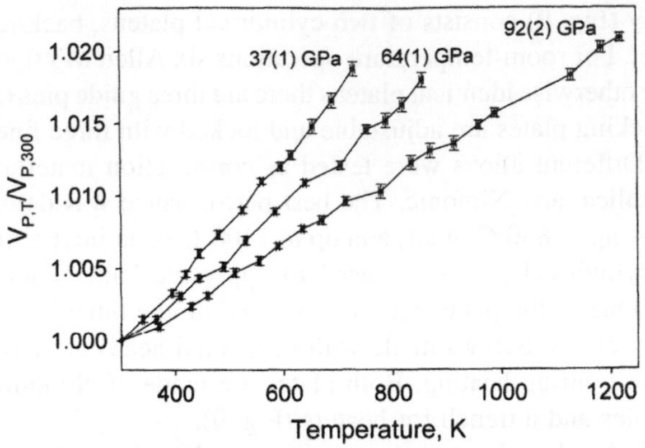

Figure 10. The thermal expansion of the $Fe_{0.95}Ni_{0.05}$ alloy at different pressures. The powdered sample, together with platinum, was loaded into the cell and compressed above 10 GPa, when initially the α- (bcc) phase transforms to ε-$Fe_{0.95}Ni_{0.05}$ (hcp-structure). Thermal expansion of ε-$Fe_{0.95}Ni_{0.05}$ was then measured at 37(1), 64(1) and 92(2) GPa at temperatures up to 1200 K. Corresponding values of the thermal expansion coefficients (in K^{-1}) are: 4.4(1) × 10^{-5}, 3.2(1) × 10^{-5}, 2.3(1) × 10^{-5}.

Table 1. Comparison of external electrical and laser heating techniques for DACs.

	Electrical (external) heating	Laser heating
Pressure range (GPa)	Up to 300	Up to 250
Temperature range (K)	From ambient to 1500	From about 1200 to over 5000
Heated area (μm)	Whole pressure chamber	20–40
Homogeneity of heating	Small temperature gradients (<50 K)	Significant temperature gradients (100–500 K)
Stability of heating	Stable heating over tens of hours	Temperature fluctuates over a timescale of minutes
Effect on stresses within the sample	Relaxes stresses at $T > 900–1000$ K	Adds thermal stresses

that both methods, electrical and laser heating, have certain unique characteristics and in the future, combination of the two experimental techniques could open new possibilities for investigation of materials at extreme conditions.

Acknowledgements

The authors thank F. Bromiley for her help in preparation of the manuscript.

References

Adams, D.M., Christy, A.G., 1992. High Pressure–High Temp. 24, 1–11.

Bassett, W.A., Shen, A.H., Bucknum, M., Chou, I.-M., 1993. A new diamond anvil cell for hydrothermal studies to 2.5 GPa and from − 190 to 1200°C. Rev. Sci. Instrum. 64, 2340–2345.

Boehler, R., 1986. The phase diagram of iron to 430 kbar. Geophys. Res. Lett. 13, 1153–1156.
Boehler, R., Nicol, M., Zha, C.S., Jonson, M.L., 1986. Resistance heating of Fe and W in diamond-anvil cells. Physica B 139–140, 916–918.
Dubrovinskaia, N.A., Dubrovinsky, L.S., 2003. Whole-cell heater for the diamond anvil cell. Rev. Sci. Instrum. 74, 3433–3477.
Dubrovinskaia, N.A., Dubrovinsky, L.S., Saxena, S.K., Sundman, B., 1997. Thermal expansion of chromium (Cr) to melting temperature. Calphad 21(4), 497–508.
Dubrovinsky, L.S., Saxena, S.K., 1998. Iron at extreme conditions: in situ X-ray study of $P-V-T$ relations in internally heated diamond anvil cell and modeling of interatomic potential of iron. Petrology 6, 535–545.
Dubrovinsky, L.S., Saxena, S.K., Lazor, P., 1998. High-pressure and high-temperature in situ X-ray diffraction study of iron and corundum to 68 GPa using internally heated diamond anvil cell. Phys. Chem. Miner. 25, 434–441.
Dubrovinsky, L.S., Saxena, S.K., Tutti, F., Le Bihan, T., 1999. In situ X-ray study of thermal expansion of iron at multimegabar pressure. High Pressure–High Temp. 31, 553–559.
Dubrovinsky, L.S., Saxena, S.K., Tutti, F., Le Bihan, T., 2000. X-ray study of thermal expansion and phase transition of iron at multimegabar pressure. Phys. Rev. Lett. 84, 1720–1723.
Dubrovinsky, L., Annersten, H., Dubrovinskaia, N., Westman, F., Harryson, H., Fabrichnaya, O., Carlson, S., 2001. Chemical interaction of iron and corundum as a source of heterogeneity at the core–mantle boundary. Nature 412, 527–529.
Dubrovinsky, L., Dubrovinskaia, N., Langenhorst, F., Dobson, D., Rubie, D., Geβmann, C., Abrikosov, I., Johansson, B., 2003. Iron–silica interaction at extreme conditions and the nature of the electrically conducting layer at the base of Earth's mantle. Nature 422, 58–61.
Eremets, M., 1996. High Pressure Experimental Methods. Oxford University Press, New York.
Fei, Y., 1996. In: Dyar, M.D., McCommon, C., Schaefer, M.W. (Eds), Mineral Spectroscopy: A tribute to Roger Burns. Geochemical Society, pp. 243, Special Publication.
Fei, Y., Mao, H.K., 1994. Science 266, 1678.
Harlow, G.E., (Ed.), 1998. The Nature of Diamonds. Cambridge University Press, New York.
Hazen, R.M., Finger, L.W., 1982. Comparative Crystal Chemistry. Wiley, New York.
Hemley, R.J., (Ed.), 1998. Ultrahigh-Pressure Mineralogy. Rev. Mineral., Vol. 37, New York.
Kikegawa, T., 1987. In: Manghnani, M.H., Syono, Y. (Eds), High-Pressure Research in Mineral Physics. TERRAPUB, Tokyo, pp. 61–68.
Liu, L., Bassett, W.A., 1975. The melting of iron up to 200 kbar. J. Geophys. Res. B 80, 3777–3782.
Mao, H.K., Bell, P.M., Hadidiacos, C., 1987. Experimental phase relations of iron to 360 kbar, 1400°C, determined in an internally heated diamond-anvil apparatus. In: Manghnani, M.H., Syono, Y. (Eds), High-Pressure Research in Mineral Physics. Terra Scientific Publishing Co., Tokyo, pp. 135–138.
Merrill, L., Bassett, W.A., 1974. Miniature diamond anvil pressure cell for single crystal X-ray diffraction studies. Rev. Sci. Instrum. 45, 290–297.
Moore, M.J., Sorensen, D.B., DeVries, R.C., 1970. A simple heating device for diamond anvil high pressure cells. Rev. Sci. Instrum. 41, 1665–1666.
Rekhi, S., Dubrovinsky, L.S., Saxena, S.K., 1999. Study of temperature-induced ruby fluorescence shifts up to a pressure 15 GPa in an externally heated diamond anvil cell. High Pressure–High Temp. 31, 299–305.
Schiferl, D., 1987. Temperature compensated high-temperature/high-pressure Merrill–Bassett diamond anvil cell. Rev. Sci. Instrum. 58, 1316–1317.
Sterer, E., Pasternak, M.P., Taylor, R.D., 1990. A multipurpose miniature diamond anvil cell. Rev. Sci. Instrum. 61, 1117–1119.
Williams, Q., Knittle, E., 1991. The high-pressure melting curve of iron: technical discussion. J. Geophys. Res. B 96, 2171–2184.
Zha, C.S., Bassett, W.A., 2003. Internal resistive heating in diamond anvil cell for in situ X-ray diffraction and Raman scattering. Rev. Sci. Instrum. 74, 1255–1262.

Advances in High-Pressure Technology for Geophysical Applications
Jiuhua Chen, Yanbin Wang, T.S. Duffy, Guoyin Shen, L.F. Dobrzhinetskaya, editors

Chapter 26

A new gasket material for higher resolution NMR in diamond anvil cells

Takuo Okuchi, Ho-kwang Mao and Russell J. Hemley

Abstract

We propose a new gasket material which gives higher resolution in high pressure NMR spectroscopy carried out in diamond anvil cells. The magnetic susceptibility of the gasket is similar to that of the diamond and the sample, therefore significantly reducing the disturbance of the static sample field. NMR linewidths are reduced to 0.8 ppm for methanol samples at 0.9 GPa, which enables resolution of the methyl and hydroxyl protons in the observed spectra. The spin–spin relaxation times of these resonances were separately determined.

1. Introduction

Nuclear magnetic resonance (NMR) is a radio frequency (rf) spectroscopy that gives unique information about local structure and dynamics surrounding specific nuclear spins. NMR spectrum gives the local structure and the spin relaxation time gives the local dynamics. Both are mutually exclusive and important for exploring the microscopic nature of materials. NMR, therefore, plays a complementary role to many diffraction, scattering, and other spectroscopic methods in material sciences, including those of minerals, silicate melts, and fluids. Though its use has been relatively limited, NMR has also provided unique information at very high pressures using diamond anvil cell (DAC) techniques (Lee et al., 1989; Bertani et al., 1990; Halvorson et al., 1991; Marzke et al., 1994; Yarger et al., 1995; Pravica and Silvera, 1998a). Technical challenges of combining NMR with the DAC techniques have been answered in a variety of approaches (Lee et al., 1987; 1992; Bertani et al., 1992; Pravica and Silvera, 1998b; Okuchi, 2004; Okuchi et al., 2004). These issues include (i) intrinsic low sensitivity of NMR, (ii) small sample volume of conventional DACs, (iii) shielding effects of the metallic gasket, (iv) prohibition of using ferromagnetic components, and (v) the limited working space in the magnet. However, one of the tightest constraints defined by the earliest work has remained: disturbance of the static field by the DAC components, such as the diamonds and especially the gasket.

NMR spectra and spin relaxation times are both affected by this disturbance. A NMR linewidth at half maximum in the frequency domain spectrum is given by

$$\nu_{1/2} = 1/\pi T_2^*. \quad (1)$$

T_2^* is a time that includes two contributions to broaden the line, the spin–spin relaxation with relaxation time T_2, and static field inhomogeneity ΔH_0

$$1/T_2^* = 1/T_2 + \gamma\,\Delta H_0/2, \tag{2}$$

where γ is the gyromagnetic ratio (Farrar and Becker, 1971). The peak amplitude of each NMR line in a frequency spectrum is directly proportional to the T_2^*. The spin relaxation time of each NMR line is separately determined from relaxation of the peak amplitude as a function of time, if there are multiple lines in the spectrum. Thus, a large ΔH_0 gives a small T_2^*, degrading both the NMR spectrum and the intrinsic relaxation time.

The field within a small DAC sample volume can be quite uniform, but it is easily disturbed by the magnetic properties of the surrounding components. Magnetic susceptibilities of diamond for the anvils, nonmagnetic (not ferromagnetic) metals for the gasket, and representative samples are shown in Table 1. In contrast to diamond, which has a slightly negative susceptibility, the gasket may have variable susceptibility from negative to highly positive values.

Two nonmagnetic metals were selected as gasket material in early work: high strength beryllium copper (UNS C17200 or alloy 25; hereafter HSBC) and rhenium (Lee et al., 1987). T_2^* for HSBC was twice that of rhenium, as measured at 82.9 MHz for methanol–ethanol samples in sample chambers of 0.5 mm diameter in these gaskets. Later experiments at 340 MHz showed that T_2^* was 900 μs with HSBC, and 100 μs with rhenium for liquid H_2, in sample chambers of similar size (Lee et al., 1989). It was demonstrated that rhenium, and thus presumably all transition metals with open d- or f-shells, has too high of a susceptibility to be a suitable gasket for NMR.

The field disturbance by the gasket is minimized if its susceptibility is matched to the diamonds and to the sample, both of which have negative susceptibility in this study. Therefore, even the HSBC gasket, which gives a larger T_2^* than rhenium, is less suitable for NMR because it has a positive susceptibility. It should be noted that elemental beryllium and copper have negative susceptibilities. HSBC may contain up to 0.6 wt.% of Fe and Co, however, which tend to form ferromagnetic centers; those give apparently positive susceptibilities. We, therefore, focused attention on types of beryllium copper alloy having enough strength to act as a gasket but with lower Fe and Co concentrations. High conductivity beryllium copper (UNS C17510 or alloy 3; hereafter HCBC) was found to be a commercially available material that satisfied these criteria.

In this chapter, we demonstrate that the HCBC gasket can give large T_2^* of 2.0 ms, which was calculated from observed linewidth of 0.8 ppm in a liquid methanol spectrum measured at 200 MHz Larmor frequency. Two NMR lines of methyl and hydroxyl protons were separately observed, and spin–spin relaxation times of these resonances were separately determined. These technical improvements are important for extending the applicability of NMR at high pressure.

2. Experimental

The specially designed DAC made of 6-4 titanium alloy and NMR system used in the experiments are described in detail elsewhere (Okuchi, 2004). Preparation and installation

Table 1. Volume susceptibilities of DAC components.

Material	Composition (wt. %)	Volume susceptibility ($\times 10^{-6}$ SI unit)	Measured at (T)	Refernces
Diamond	C	-21.7		Lide (2003)
Beryllium	Be	-23.2		Lide (2003)
Copper	Cu	-9.7	7	Doty et al. (1998)
Molybdenum	Mo	120	7	Doty et al. (1998)
Tungsten	W	80	7	Doty et al. (1998)
Rhenium	Re	95		Lide (2003)
HSBC	Be 1.8–2.0, Cu balance, Co + Ni + Fe 0.60 max	42	0.3	NGK Insulators, LTD. http://www.ngk.co.jp/MTL/english/products/bery/index.html
HCBC	Be 0.2–0.6, Cu balance, Ni 1.4–2.2	-5	7	Doty et al. (1998), NGK Insulators, LTD http://www.ngk.co.jp/MTL/english/products/bery/index.html
Water	H_2O	-9.1		Doty et al. (1998), Lide (2003)
Methanol	CH_3OH	-6.7		Doty et al. (1998), Lide (2003)
Ethanol	C_2H_5OH	-7.3		Doty et al. (1998), Lide (2003)
Toluene	C_7H_8	-7.8		Doty et al. (1998), Lide (2003)
Hydrogen	H_2	-2.4		Lide (2003)

techniques of the rf circuit in the DAC are also described. We operated this NMR system with a 200 MHz Oxford wide bore magnet. Diamond anvils having 1.0 mm culets were mounted with minimal epoxy, the only source of spurious proton signals. A HCBC foil that was 254 μm in thickness was supplied from NGK Metals Corp. (Sweetwater, TN) for sample evaluation. HCBC gaskets were slightly preindented at 2 GPa before drilling 0.6 mm sample chambers. Anhydrous methanol (#41838, Alfa Aesar) was loaded into the sample chamber with a syringe and pressurized quickly in a dry glove box. The pressure was measured by ruby fluorescence. Due to the length of time of the measurements, the pressure was measured before and after data collection. Its reproducibility was <0.1 GPa.

NMR measurements were made at room temperature (21°C). A new rf coil design with higher sensitivity for liquid samples (Okuchi et al., 2004) was used to obtain spin echoes for the linewidth (T_2^*) measurements and to obtain spin echo trains for relaxation time (T_2) measurements. Spurious signals from epoxy were excluded because they do not produce echoes. A standard echo sequence with waiting time τ(90°$-\tau-$180°$-\tau-$echo) was applied to obtain the spin echoes. The Carr–Purcell–Meiboom–Gill pulse sequence (90°$-\tau-$180°$y-2\tau-$180°$y - 2\tau-\cdots$) was applied to obtain the spin echo trains. The power of the transmitter was 50 W at maximum. It was reduced when longer pulses were acceptable, because trimmer capacitors could be broken during hours of signal accumulations. The 90° pulse width was determined by applying Carr–Purcell pulse sequences (90°$x-\tau-$180°$x-2\tau-$180°$x-2\tau-\cdots$) with varying pulse lengths (Fukushima and Roeder, 1981). Each whole echo was separately Fourier transformed, which gives the pure absorption spectrum without phase correction (Bax et al., 1979). No window function was applied to these echoes before the Fourier transform.

3. Results and discussion

Figure 1 shows an observed ^{1}H NMR spectrum of methanol at 0.9 GPa. A total of 2923 transients with a 10 kHz filter bandwidth were averaged, giving an acquisition time of 97 min. The 90° pulse length was 40 μs, which was long because of the reduced transmitter power. The methyl proton line was arbitrarily set to zero frequency.

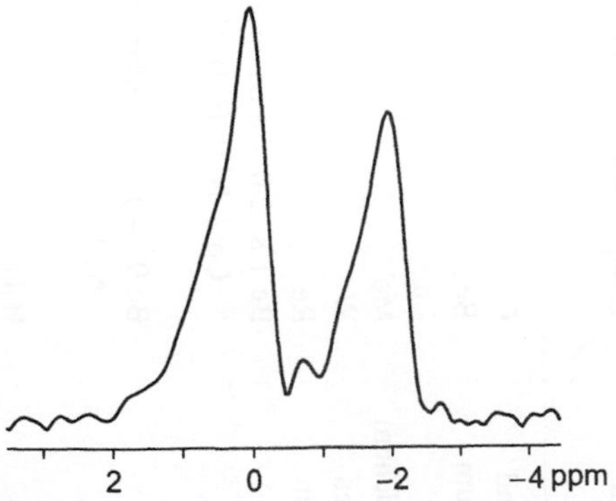

Figure 1. ^{1}H NMR spectrum of methanol obtained at 200 MHz at 0.9 GPa pressure. The left line is methyl protons and the right line is hydroxyl protons.

To increase the number of transients, the repetition rate was only set at twice the spin-lattice relaxation time of the methyl protons, so that the amplitude of this line was slightly reduced. The hydroxyl line is observed at the right. The frequency difference between methyl and hydroxyl lines was 2.0 ppm. It increased by 0.4 ppm compared with ambient pressure, indicating that hydroxyl protons were more deshielded as hydrogen bonding became stronger with increasing pressure (Czeslik and Jonas, 1999).

The observed linewidth of 0.8 ppm in Figure 1 gives $T_2^* = 2.0$ ms at 200 MHz. This width is definitely smaller than those of previous works using HSBC as the gasket; Yarger et al., 1995 reported 2.25 ppm width at 400 MHz for toluene sample which has diamagnetic susceptibility close to methanol (Table 1). Lee et al. (1989) reported $T_2^* = 0.9$ ms at 340 MHz, equivalent to 1.0 ppm width, for liquid hydrogen sample which also has diamagnetic susceptibility but is smaller than methanol. Thus, the effectiveness of HCBC as the gasket depends on the actual susceptibility of the sample. This suggests that fine susceptibility tuning of the gasket will be more useful for each specific sample.

Figure 2 shows the relaxation of the same spectrum as time after the first 90° pulse increases. Two thousand transients with a 40 kHz filter bandwidth were averaged to obtain the spin echo train which continued for more than 500 ms, taking 3 h to acquire. The 90° pulse length was 9.6 μs. The amplitude of each line shows the magnetization along the axis perpendicular to the static field, which exponentially decreases to zero by spin–spin relaxation through molecular motion. The linewidth is broader than in Figure 1 because the sampling time of each echo was smaller. However, the methyl and hydroxyl lines are still separately observed throughout the relaxation occurred, which assures exact determination of intrinsic relaxation time for each line. The repetition rate was five times the spin–lattice relaxation time of methyl protons at the relevant condition.

Figure 3 shows the exact amplitude of the two lines in Figure 2. The hydroxyl protons ($T_2^{OH} = 200$ ms) relax faster than the methyl protons ($T_2^{CH_3} = 330$ ms). Both of these T_2 times are much larger than the T_2^* of 2.0 ms, so that the spin–spin relaxation was not responsible for the observed linewidth. Since faster relaxation results from slower motion of methanol molecules, the hydroxyl protons are more restricted by hydrogen bonding than the methyl protons, which appear to freely rotate even at these pressures. Measurement of these spin relaxation times as a function of pressure will give new

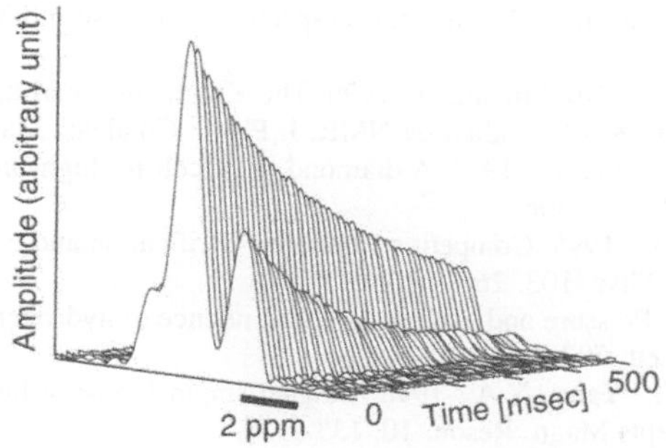

Figure 2. Relaxation of the ¹H NMR spectrum obtained with the Carr–Purcell–Meiboom–Gill sequence with the waiting time $\tau = 5$ ms. The smaller hydroxyl line relaxed faster than the larger methyl line.

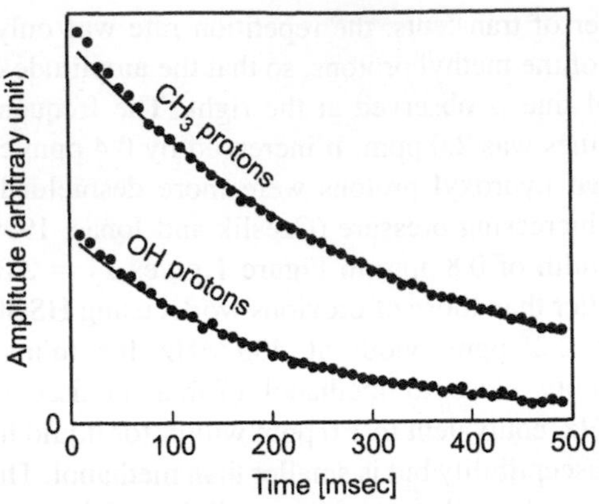

Figure 3. The exact amplitude of methyl and hydroxyl proton lines in Figure 2 as a function of time. The data were fit with a simple exponential function, $A \exp(t/T_2)$, where A is a constant, t is the time from the first 90° pulse and T is the spin–spin relaxation time.

insights into the microscopic dynamics of liquid methanol, which transforms to a glassy state at several gigapascals (Brugmans and Vos, 1995). We will report extended results at higher pressures in the near future.

Acknowledgements

The authors thank F. D. Doty at Doty Scientific, Inc. for comments on magnetic susceptibility of copper based alloys. S. Tozer at Florida State University provided critical comments on the manuscript. T.O. was supported by JSPS Postdoctoral Fellowships for Research Abroad. This work was supported by NSF and DOE (CDAC).

References

Bax, A., Mehlkopf, A.F., Smidt, J., 1979. Absorption spectra from phase-modulated spin echoes. J. Magn. Reson. 35, 373–377.

Bertani, R., Mali, M., Roos, J., Brinkmann, D., 1990. The Knight shift and diffusion in solid lithium and sodium at pressures up to 8 GPa studied by NMR. J. Phys.: Condens. Matter 2, 7911–7923.

Bertani, R., Mali, M., Brinkmann, D., 1992. A diamond anvil cell for high-pressure NMR investigations. Rev. Sci. Instrum. 63, 3303–3306.

Brugmans, M.J.P., Vos, W.L., 1995. Competition between vitrification and crystallization of methanol at high pressure. J. Chem. Phys. 103, 2661–2669.

Czeslik, C., Jonas, J., 1999. Pressure and temperature dependence of hydrogen-bond strength in methanol clusters. Chem. Phys. Lett. 302, 633–638.

Doty, F.D., Entzminger, G., Yang, Y.A., 1998. Magnetism in high-resolution NMR probe design. I: general methods. Concepts Magn. Reson. 10, 133–156.

Farrar, T.C., Becker, E.D., 1971. Pulse and Fourier Transform NMR: Introduction to Theory and Methods. Academic Press, New York.

Fukushima, E., Roeder, S.B.W., 1981. Experimental Pulse NMR: A Nuts and Bolts Approach. Addison-Wesley, Reading, MA.

Halvorson, K.E., Raffaelle, D.P., Wolf, G.H., Marzke, R.F., 1991. Proton NMR chemical shifts in organic liquids measured at high pressure using the diamond anvil cell. In: Hochheimer, H.D., Etters, R.D. (Eds), Frontiers of High-Pressure Research. Plenum Press, New York, pp. 217–226.

Lee, S.-H., Luszczynski, K., Norberg, R.E., Conradi, M.S., 1987. NMR in a diamond anvil cell. Rev. Sci. Instrum. 58, 415–417.

Lee, S.-H., Conradi, M.S., Norberg, R.E., 1989. Molecular motion in solid H_2 at high pressures. Phys. Rev. B. 40, 12492–12498.

Lee, S.-H., Conradi, M.S., Norberg, R.E., 1992. Improved NMR resonator for diamond anvil cells. Rev. Sci. Instrum. 63, 3674–3676.

Lide, D.R., 2003. Handbook of Chemistry and Physics, 84th edn. CRC, Boca Raton, FL, pp. 3-700; 4-141.

Marzke, R.F., Raffaelle, D.P., Halvorson, K.E., Wolf, G.H., 1994. A ^{1}H NMR study of grycerol at high pressure. J. Non-Cryst. Solids 172–174, 401–407.

Okuchi, T., 2004. A new type of nonmagnetic diamond anvil cell for nuclear magnetic resonance spectroscopy. Phys. Earth Planet. In. 143–144, 611–616.

Okuchi, T., Hemley, R.J., Mao, H.K., 2005. Radio frequency probe with improved sensitivity for diamond anvil cell nuclear magnetic resonance. Rev. Sci. Instrum. 76, 026111.

Pravica, M.G., Silvera, I.F., 1998a. NMR study of ortho–para conversion at high pressure in hydrogen. Phys. Rev. Lett. 81, 4180–4183.

Pravica, M.G., Silvera, I.F., 1998b. Nuclear magnetic resonance in a diamond anvil cell at very high pressure. Rev. Sci. Instrum. 69, 479–484.

Yarger, J.L., Nieman, R.A., Wolf, G.H., Marzke, R.F., 1995. High-pressure ^{1}H and ^{13}C nuclear magnetic resonance in a diamond anvil cell. J. Magn. Reson., A 114, 255–257.

Author index

Subject index

Printed and bound by CPI Group (UK) Ltd, Croydon, CR0 4YY
03/10/2024
01040330-0013